VIII International Congress of Engineering Physics 2016

AA006466

Journal of Physics: Conference Series Volume 792

Merida, Mexico
7 – 11 November 2016

Editors:

Ernesto Rodrigo Vazquez Ceron
Ines Riech Mendez
Oscar Olvera Neria
Milenis Acosta Diaz
Roberto Tito Hernandez Lopez
Cesar Renan Acosta
Jaime Granados Samaniego
Jose Mendez Gamboa
Anatolio Martinez Jimenez Luis
Enrique Norena Franco

ISBN: 978-1-5108-3753-9
ISSN: 1742-6588

Printed from e-media with permission by:

Curran Associates, Inc.
57 Morehouse Lane
Red Hook, NY 12571

Some format issues inherent in the e-media version may also appear in this print version.

Copyright© (2016) by the Institute of Physics
All rights reserved. The material featured in this book is subject to
IOP copyright protection, unless otherwise indicated.

Printed by Curran Associates, Inc. (2017)

For permission requests, please contact the Institute of Physics
at the address below.

Institute of Physics
Dirac House, Temple Back
Bristol BS1 6BE UK

Phone: 44 1 17 929 7481
Fax: 44 1 17 920 0979

techtracking@iop.org

Additional copies of this publication are available from:

Curran Associates, Inc.
57 Morehouse Lane
Red Hook, NY 12571 USA
Phone: 845-758-0400
Fax: 845-758-2633
Email: curran@proceedings.com
Web: www.proceedings.com

Table of contents

Volume 792

VIII International Congress of Engineering Physics7–11 November 2016, Mérida, Yucatán, Mexico

Accepted papers received: 14 December 2016Published online: 8 March 2017

Preface

Papers

IOP Conf. Series: Journal of Physics: Conf. Series **792** (2017) 011001 doi:10.1088/1742-6596/792/1/011001

Introduction

The Physicist Engineering is a multidisciplinary profession in which an undergraduate student integrates and applies their knowledge of Physics, Mathematics, and Computing to adapt and innovate all kind of technologies, as well as to development and apply theoretical knowledge to different fields of science and engineering.

A Physicist Engineer, supported by a solid preparation in science and engineering is available to study, analyze, identify, and solve problems in different fields of science and technology for industries, hospitals, research laboratories, institutes and governmental organizations focused on metrology, optics, energy, materials, instrumentation of sophisticated equipment. The Physicist Engineering bachelor degree is a national and international reference to encouraging all those students to apply knowledge through active participation in the operation, planning, and management of projects.

The technological and scientific impact of a Physicist Engineering undergraduate and graduate students, as well as the tutorial of the research professor that are interested in Physics and Physics Engineering, may be found in a biennial event like the International Conference on Physicist Engineering, through the diffusion of both experimental and theoretical research and teaching.

In order to promote and encourage strength academic goals among universities and research institutions, the International Conference on Physics Engineering Committee, integrated by a faculty meeting of the Universidad Autónoma Metropolitana, Campus Azcapotzalco, has decided to grant the VIII International Congress of Physics Engineering to the Universidad Autónoma de Yucatán, which is going to take place on Mérida, November 7th to 11th, 2016.

This conference considers different topics as:

Renewables Energies Engineering
Materials Technology
Nanotechnology
Medical Physics
Education in Physics Engineering
Nuclear Engineering
High Precision Instrumentation
Atmosphere Physics
Optics Engineering
Physics History
Acoustics
Related Topics

The conference integrates lectures on top trending topics with pre-congress workshops, which are given by recognized scientists with an outstanding academic record. The lectures and workshops allow to share experiences and create research networks. The Conference also encourages professional mobility among all universities and research institutes.

Content from this work may be used under the terms of the Creative Commons Attribution 3.0 licence. Any further distribution of this work must maintain attribution to the author(s) and the title of the work, journal citation and DOI.
Published under licence by IOP Publishing Ltd

IOP Conf. Series: Journal of Physics: Conf. Series **792** (2017) 011001 doi:10.1088/1742-6596/792/1/011001

CIIF2016 Organizing and Editorial Committee

Dr. Ernesto Rodrigo Vázquez Cerón
Dra. Inés Riech Méndez
Dr. Oscar Olvera Neria
Dra. Milenis Acosta Díaz
Dr. Roberto Tito Hernández López
Dr. Cesar Renán Acosta
Prof. Jaime Granados Samaniego
Dr. José Méndez Gamboa
Dr. Anatolio Martínez Jiménez
Dr. Luis Enrique Noreña Franco

Peer review statement

All papers published in this volume of *Journal of Physics: Conference Series* have been peer reviewed through processes administered by the proceedings Editors. Reviews were conducted by expert referees to the professional and scientific standards expected of a proceedings journal published by IOP Publishing.

Content from this work may be used under the terms of the Creative Commons Attribution 3.0 licence. Any further distribution of this work must maintain attribution to the author(s) and the title of the work, journal citation and DOI.
Published under licence by IOP Publishing Ltd

IOP Conf. Series: Journal of Physics: Conf. Series **792** (2017) 012001 doi:10.1088/1742-6596/792/1/012001

Optimal allocation and effectiveness of the heat exchangers for a power plant with two like-Carnot cycles

G Aragón-González and A León-Galicia
Programa de Desarrollo Profesional en Automatización, Universidad Autónoma Metropolitana-Azcapotzalco, México, D. F. 02200, México.

E-mail: gag@correo.azc.uam.mx

Abstract. A stationary power plant with two Carnot-like cycles is optimized. Each cycle has the following irreversibilities: finite rate heat transfer between the working fluid and the external heat sources, internal dissipation of the working fluid, and heat leak between reservoirs; is extended to two or more of this combined model. Using the Bellman' Principle, we find the optimal recurrence relations for the allocation of the heat exchangers for this power plant. The optimal allocation or effectiveness of the heat exganchers of power plant is determined by two design rules: internal thermal conductance fixed; or areas fixed. The optimal obtained are invariant to the power and efficiency and to the heat transfer law.

1. Introduction

Recently in [1], a methodology of optimization was applied to an irreversible Carnot-like stationary power plant, where the characteristic parameters were: the allocation or effectiveness of the heat exchangers of this plant. The above standard irreversible Carnot-like cycle has been studied at length for many objective functions, different transfer heat laws and several characteristic parameters (see [2]-[24] for more details).

On the other hand, the maximum power and efficiency have been obtained in [3]-[6]. In general, these optimizations were performed with respect to only one characteristic parameter: internal isentropic temperature ratio (x) including the time too [18]. However, Bejan and Lewins [15, 21] have considered the optimization with respect to other parameters (see also the reviews of [18-19] for more details): "the allocation and effectiveness of the heat exchangers of the hot and cold sides".

In [24-27] have been presented a n-stage combined Carnot cycle optimizing the specific power and efficiency for the characteristic parameters: isentropic temperature ratios x_j; $j = 1, 2$, and effectiveness. In this last case the law of heat conduction used.

In this paper we extend the results of allocation or effectiveness for one cycle to a combined cycle with two like-Carnot cycle using the Bellman' Principle [24] which has been successfully applied in [15 and 26]. We found optimal relations, for two constraints (design rules): constrained internal thermal conductance or fixed total area of the heat exchangers from hot and cold side, and the power and efficiency maximum for a stationary power plant with two Carnot-like cycles.

Content from this work may be used under the terms of the Creative Commons Attribution 3.0 licence. Any further distribution of this work must maintain attribution to the author(s) and the title of the work, journal citation and DOI.
Published under licence by IOP Publishing Ltd

2. Power plant with two Carnot-like cycles

The stationary power plant with two Carnot-like cycles is shown in Figure 1. Each cycle satisfies the conditions expressed in [2, 4]: leak heat $\dot{Q}$ and finite heat transfer rates $\dot{Q}_i$ and internal dissipations of the working fluid expressed by constants I_i $(i = 1,2)$ such that $I_i = \frac{\Delta S_{i+1}}{\Delta S_i} \geq 1; i = 1,2$ [4] which make the Claussius inequality to become equality (for details see [21]). Each cycle of the power plant consists of two isothermal and two adiabatic processes. Denoting, for each cycle, the temperatures of the working fluid in the hot and cold isothermal processes as T_i and T_{i+1} $(i=1,2,3)$, respectively, and the end temperatures as T_H and T_L.

Following to [1, 26], the thermal efficiency of the power plant is given by (this functional form for only a cycle has appeared in [1, 2, and 18]):

$$\eta = \frac{P}{f(x_1, x_2)P + \dot{Q}} \tag{1}$$

where $x_i = \frac{T_{i+1}}{T_i}$ correspond to the isentropic temperature ratio for each cycle i $(i = 1,2)$, P is the power of the plant, $\dot{Q}$ is the leak heat and the function

$$f(x_1, x_2) = \frac{1}{1 - I_1 I_2 x_1 x_2} \tag{2}$$

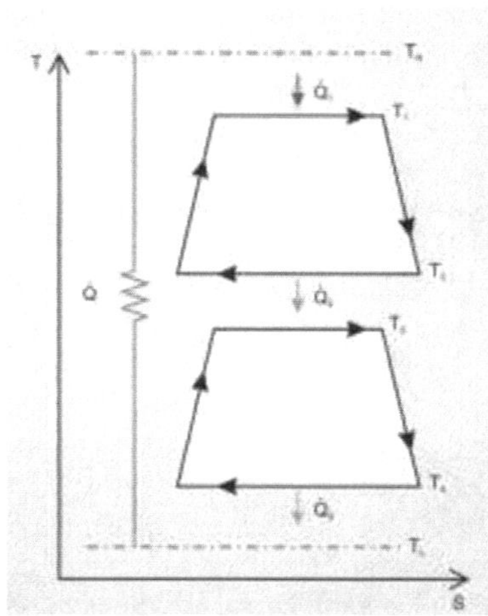

Figure 1: A power plant stationary of two Carnot-like cycles, with linear leak heat and finite heat transfer rates, and internal dissipations of the working fluid

Indeed, according to the second law of Thermodynamics, for each $i = 1,2$, we have [26]: $\dot{Q}_2 = I_1 x_1 \dot{Q}_1$; $\dot{Q}_3 = I_2 x_2 \dot{Q}_2$. Thus, $\dot{Q}_3 = I_2 x_2 \dot{Q}_2 = I_1 x_1 \dot{Q}_1$

So that, the efficiency is given by,

$$\eta = \frac{P}{\dot{Q}_H} = \frac{\dot{Q}_H - \dot{Q}_L}{\dot{Q}_H} = \frac{\dot{Q}_1 - \dot{Q}_3}{\dot{Q}_H} = \left(1 - \frac{\dot{Q}_3}{\dot{Q}_1}\right)\frac{\dot{Q}_1}{\dot{Q}_H} = \frac{(1 - I_1 I_2 x_1 x_2)\dot{Q}_1}{\dot{Q}_H} \quad (3)$$

From the equation (5),

$$\dot{Q}_1 = \frac{P}{1 - I_1 I_2 x_1 x_2} \quad (4)$$

And from the equation (4), (2) is obtained.

Now, let $z \neq x_i \ (i = 1, 2)$ the Corollary given at [1] can extends to:

"The power P achieves a maximum value in z_{mp} if and only if the efficiency η achieves a maximum value in the value: $z_{mp} = z_{me}$"

And the proof is similar to the presented in [1]. It is easy see that in the calculations of the proof of the Corollary none transfer heat law has been used. Thus, in the optimization of power plant with two Carnot-like cycles, with respect to z, it is enough to find the maximum power by,

$$\left.\frac{\partial P}{\partial z}\right|_{z_{mp} = z_{me}} = 0 \text{ and } \left.\frac{\partial^2 P}{\partial z^2}\right|_{z_{mp} = z_{me}} < 0 \quad (5)$$

The optimization performed, with respect to z, will be: a property independent from the heat transfer law. Henceforth, $z = \phi$ or ψ, alternatively, depending of which design rule is applied.

3. Optimal relations for the allocation and effectiveness of the heat exchangers of the power plant

In this section, x_1, x_2 will be fixed and we will assume that the law of heat transfer can be any law, included also the heat leak. Next, we will treat the following two design rules, alternatively: *internal thermal conductance fixed*; or *areas fixed for heat exchangers*; which will be applied alternatively.

The first design rule is that the internal conductance of the Carnot-like cycle is constrained to: $\sum_{i=1}^{3} \alpha_i = \Gamma$;where Γ is a constant, which is applied to the allocation of the heat exchangers from hot and cold side with the same overall heat transfer coefficient U by unit of area A in both ends of the cycle i, and $\alpha_i, \alpha_{i+1} \ (i = 1, 2)$ are the thermal conductance correspondent to the finite heat transfers of the hot/cold sides for this cycle. Thus, $\sum_{i=1}^{3} UA_i = \Gamma$ where A_i, A_{i+1} are heat transfer areas on hot/cold sides of the cycle $i \ (i = 1, 2)$. Alternatively, the second design rule is that the total area is constrained by: $\sum_{i=1}^{3} A_i = A$; where A_i, A_{i+1} are heat transfer areas on hot and cold side for the cycle $i \ (i = 1, 2)$. Now, the total area (A) is fixed, but when distributed it has distinct overall heat transfer coefficients and hence different effectiveness on each one of the hot and cold sides. As $\alpha_i = U_i A_i$ (see [1, 12]), then: $A = \sum_{i=1}^{3} A_i = \sum_{i=1}^{3} \frac{\alpha_i}{U_i}$; where U_i, U_{i+1} are the overall heat transfer coefficients on hot/cold sides of the cycle $i \ (i = 1, 2)$. From the Corollary of the section 2, it is enough to found *"The maximum power of each Carnot-like cycle $i \ (i = 1, 2)$ for these two design rules alternatively.* Indeed, from [3] the dimensionless power output for the cycle i is given by:

$$p_i = \frac{P_i}{\alpha_i T_{i-1}} = \frac{1 - I_1 I_2 x_1 x_2}{\frac{1}{\alpha_i} + \frac{I_i}{\alpha_{i+1}}} \quad (6)$$

where $\mu_i = \frac{T_{i+1}}{T_{i-1}}; i = 1,2$ The numerator of (10) is obtained of apply the conduction heat law and have optimized with respect to x_i ; $i = 1,2$.

Next, applying the Bellman' Principle [21]:"*to state that every part of an optimum path is optimal"*.

3.1 Constrained internal thermal conductance

The three thermal conductance can be written as: $\alpha_i = UA_i$; $i = 1,2,3$; where U is overall heat transfer coefficient and A_i; $i = 1,2,3$ are the available areas for heat transfer. Thus, for the first optimization we can take for this first design rule: $\alpha_1 + \alpha_2 = \Gamma_1$;where $\Gamma_1 = \Gamma - \alpha_3$ is supposed to be a constant. Equivalently, $\frac{\alpha_1 + \alpha_2}{\Gamma_1} = 1$. Fixing the temperature T_j $(j = 3....n+1)$ and applying only for the first cycle: $\frac{a_1\alpha_1 + \alpha_2}{\Gamma_1} = 1$; $a_1 = 1$. In parameterizing, $\phi = \frac{\alpha_1}{\Gamma_1}$; $1 - a_1\phi = \frac{\alpha_2}{\Gamma_1}$. According to the equation (6), in optimizing $\left(\frac{1}{\frac{a_1}{\phi} + \frac{I_1}{1-a_1\phi}}\right)$ with respect to ϕ ; we obtain: $\phi_1 = \frac{1}{1+\sqrt{I_1}}$. Solving, $\frac{1-a_1\phi_1}{\phi_1} = \sqrt{I_1} = \frac{\alpha_2}{\alpha_1}$; so $\alpha_2 = \sqrt{I_1}\alpha_1$. For the second optimization, the constraint is now: $\frac{a_2\alpha_1 + \alpha_3}{\Gamma_2} = 1$; with $a_2 = 1 + \sqrt{I_1}$. In parameterizing, $\phi = \frac{\alpha_1}{\Gamma_2}$; $1 - a_2\phi = \frac{\alpha_3}{\Gamma_2}$. Similarly, in optimizing $\left(\frac{1}{\frac{a_2}{\phi} + \frac{I_2}{1-a_2\phi}}\right)$ with respect to ϕ ; we obtain: $\phi_2 = \frac{1}{a_2 + \sqrt{I_2}}$. Solving, $\frac{1-a_2\phi_2}{\phi_2} = \sqrt{I_2} = \frac{\alpha_3}{\alpha_1}$; so, $\alpha_3 = \sqrt{I_2}\alpha_1$. In summary, the optimal conductances, are:

$$\alpha_2 = \sqrt{I_1}\alpha_1; \alpha_3 = \sqrt{I_2}\alpha_1 \quad (7)$$

3.2 Constrained areas of heat exchangers

Applying the design second design rule $A_1 + A_2 = A_1$; where $A_1 = A - A_3$. How, $A_1 = \frac{\alpha_1}{U_1}$; $A_2 = \frac{\alpha_2}{U_2}$; then $A = a_1\alpha_1 + u_1\alpha_2$; where $u_1 = \frac{U_1}{U_2}$; $a_1 = 1$. In parameterizing, $\psi = \frac{\alpha_1}{A}$; $1 - a_1\psi = \frac{\alpha_2}{A}$. From according to the equation (10): $\frac{1}{\frac{a_1}{\psi} + \frac{I_1}{(1-a_1\psi)u_1}}$, the first optimization with respect to ψ , gives: $\psi_1 = \frac{\sqrt{u_1}}{1 + a_1\sqrt{u_1}}$. Then, $\frac{1-\psi_1}{\psi_1} = \frac{1+\sqrt{u_1}(a_1 - 1)}{\sqrt{u_1}}$. Solving, $\frac{\alpha_1}{\sqrt{a_1u_1}} = \alpha_2$; so, $A_2 = \frac{A_1}{\sqrt{a_1u_1}}$; and the results of [1] are recovered. For the second optimization, the constraint is now, $a_2A_1 + \frac{A_3}{u_2} = A$; where $a_2 = \frac{1+\sqrt{a_1u_1}}{\sqrt{a_1u_1}}$; $u_2 = \frac{U_1}{U_3}$. In parameterizing, $\psi = \frac{A_1}{A}$; $1 - a_2\psi = \frac{A_3}{A}$. From (6): $\frac{1}{\frac{a_2}{\psi} + \frac{I_2}{(1-a_2\psi)u_2}}$. The second optimization with respect to ψ , gives, $\psi_2 = \frac{\sqrt{u_2}}{1 + a_2\sqrt{u_2}}$ and $\frac{1-\psi_2}{\psi_2} = \frac{1+\sqrt{u_2}(a_2 - 1)}{1 + a_2\sqrt{u_2}}$.

Thus, $A_3 = \frac{1+\sqrt{u_2}(a_2-1)}{\sqrt{u_2}}A_1$. So that, the optimal areas, are:

$$A_2 = \frac{A_1}{\sqrt{a_1 u_1}};\ A_3 = \frac{1+\sqrt{u_2}(a_2-1)}{\sqrt{u_2}}A_1 \tag{8}$$

5. Conclusions

We have found and determined the optimal allocation and effectiveness of heat exchangers of a combined cycle with two Carnot-like cycles. Moreover, these relations can be satisfied for other operation regimes, e.g. algebraic combination of power and/or efficiency that have thermodynamic meaning and satisfy imposed power conditions (equation (5)). Nevertheless, the optimal isentropic temperatures ratios depend on the heat transfer law and the operation regime of the engine as was discussed in [1].

Finally, in [15] the maximum power and its corresponding efficiency to maximum power for two plants were calculated without internal irreversibilities, as:

$$P_1 + P_2 = \frac{1}{9}\Gamma_3 \eta_{mp}^2;\ \eta_{mp} = 1-\sqrt{\mu}$$

The above equation can be extended to the model presented here. However, the latter requires a comprehensive study of the implications for the power plant considered. We will study such implications as future work.

6. References

[1] Aragón G, Canales A, Galicia A and Rivera J M 2012 On a methodology of optimisation for an irreversible Carnot-like power plant, *J. Energy Inst.* **85** (4) 201-208.

[2] Aragón G, Canales A, Galicia A and Rivera J M 2009 The fundamental optimal relations of the allocation, cost and effectiveness of the heat exchangers of a Carnot-like power plant *J. Phys. A: Math. Theor.* **42** 425205

[3] Aragón G, Canales A, Galicia A and Morales J R 2008 Maximum Power Ecological Function and Efficiency of an Irreversible Carnot Cycle. A Cost and Effectiveness Optimization. Braz *J. of Phys.* **38** (4) 543-550

[4] Chen J 1994 The maximum power output and maximum efficiency of an irreversible Carnot heat engine *J. Phys. D: Appl. Phys.* **27** 1144-1149

[5] Yan Z and Chen L 1995 The fundamental optimal relation and the bounds of power output and efficiency for and irreversible Carnot engine *J. Phys. A: Math. Gen* **28** 6167-6175

[6] Aragón G, Canales A, Galicia A and Musharrafie M 2003 A criterion to maximize the irreversible efficiency in heat engines *J. Phys. D: Appl. Phys.* **36**, 280-87

[7] Wu C, Kiang L R, Lopardo J V and Karpouzian N G 1993 Finite-time thermodynamics and endoreversible heat engines *Int. J. Mech. Eng.* **21** 337-346

[8] Cegonho N and Domingos T 2005 New and old regimes for the endoreversible heat engine. *J. Non-Equilib. Thermodyn.* **30** 113-127.

[9] Chen L, Zhou J, Sun F and Wu C A 2004 Optimal configuration and performance for a generalized Carnot cycle assuming the heat-transfer generalized law *Open Sys. & Information Dyn.* **10** 351-375

[10] Chen L, Zhou J, Sun F and Wu C 2008 Generalized irreversible heat-engine experiencing a complex heat-transfer law *Appl. Energ.* **85** 52-60

[11] Gordon J M 1991 On optimizing maximum power heat engines *J. Appl. Phys.* **69** 1-7

[11] Bejan A 1982 *Entropy Generation through Heat and Fluid Flow* ed Wiley (New York)

[12] Bejan A 1988 *Advanced Engineering Thermodynamics* ed Wiley (New York)

[13] Bejan A 1995 Theory of heat transfer-irreversible power plants II. The optimal allocation of heat exchange equipment *Int. J. Heat Mass Transfer.* **38** (3) 433-44

[14] Bejan A 1996 *Entropy Generation Minimization ed* CRC Press (Boca Raton F L)

[15] Curzon F L and Ahlborn B 1975 Efficiency of a Carnot engine at maximum power output. *Am. J. Phys.* **43** 22-24

[16] Rubin M H 1979 Optimal configurations of a class of irreversible heat engines. *Phys. Rev.* **15** 2094-2102

[17] Hoffman K H, Burzler J M and Shuberth S 1997 Endoreversible Thermodynamics. *J. Non-Equilib. Thermodyn.* **22** (4) 311-355

[18] Durmayaz A, Sogut O S, Sahin B and Yavuz H 2004 Optimization of thermal systems based on finite time thermodynamics and thermoeconomics *Progr. Energ. and Combus. Sci.* **30** 175-217

[19] Aragón G, Canales-Palma A, León A and Morales J R 2006 Optimization of an irreversible Carnot engine in finite time and finite size. *Rev. Mex. Fis.* **52** (4) 309-314

[20] Lewins J D 2000 The endo-reversible thermal engine: cost and effectiveness optimization *Int. J. Mech. Eng. Edu.* **28** (1) 41-46

[21] De Vos A 1985 Efficiency of some heat-engines at maximum-power conditions. *Am. J. Phys.* **53** (6) 570-573

[22] Calvo A, Medina A, Roco J M, White J A and Velasco S2001 Unified optimization criterion for energy converters *Phys. Rev. E.* **63** (4) 0371021-5

[23] Aragón G, Canales A and León A 2000 Maximum irreversible work and efficiency in power cycles *J. Phys. D: Appl. Phys.* **33** 1403-10

[24] Lewins J. D 1999 Optimizing cascades of endo-reversible heat engines *Int. J. Mech. Eng. Edu.* **27** (2) 91-101

[26] Chen J A 1998 universal model of an irreversible combined Carnot cycle system and its general performance characteristics *J. Phys. A: Math. Theor.* **3** 3383-3394.

[27] Bandyopadhyay S, Bera N C and Bhattacharyya S 2001 Thermoeconomic optimization of combined cycle power plants *Energ. Convers. Manag.* **42** 359-371

Maximum power for a power plant with two Carnot-like cycles

G Aragón-González and A León-Galicia
Programa de Desarrollo Profesional en Automatización, Universidad Autónoma Metropolitana-Azcapotzalco, México, D. F. 02200, México.

E-mail: gag@correo.azc.uam.mx

Abstract. A stationary power plant with two Carnot-like cycles is optimized. Each cycle has the following irreversibilities: finite rate heat transfer between the working fluid and the external heat sources, internal dissipation of the working fluid, and heat leak between reservoirs. The optimal allocation or effectiveness of the heat exchangers for this power plant is determined by applying, two alternating design rules: fixed internal thermal conductance or fixed areas. The optimal relations obtained are substituted in the power and the maximum power, according to the isentropic ratio of each one of the Carnot-like cycles of the power plant, is calculated. Additionally, the efficiency to maximum power is presented.

1. Introduction

Recently in [1], a methodology of optimization was applied to an irreversible Carnot-like stationary power plant, where the characteristic parameter was the allocation or effectiveness of the heat exchangers of this plant. This standard irreversible Carnot-like cycle has been studied at length for many objective functions, different transfer heat laws and several characteristic parameters (see [2]-[7] for further details).

[8-11] have presented a n-stage combined Carnot cycle optimizing the specific power and efficiency for the isentropic temperature ratios x_j; $j = 1, \ldots\ldots n$, and effectiveness. In this last case, the law of heat conduction was used.

In this paper, we extend the results of allocation or effectiveness from one cycle to a combined cycle with two Carnot-like cycles using the Bellman's Principle [9], which has been successfully applied in [8 and 9]. We found optimal relations for two constraints (i.e. design rules): constrained internal thermal conductance or fixed total area of the heat exchangers from the hot and cold sides. Finally, the maximum power and the efficiency of maximum power for the stationary power plant with two Carnot-like cycles are calculated.

2. Power plant with two Carnot-like cycles

The stationary power plant with two Carnot-like cycles is shown in Figure 1. Each cycle satisfies the conditions expressed in [2, 4]: leak heat $\dot{Q}$ and finite heat transfer rates $\dot{Q}_i$, and internal dissipations of the working fluid expressed by constants I_i $(i = 1, 2)$, such that $I_i = \frac{\Delta S_{i+1}}{\Delta S_i} \geq 1; i = 1, 2$ [4] in which the Claussius inequality becomes equality (see [3]). Each cycle of the power plant consists of two isothermal and two adiabatic processes, denoting,for each cycle the temperatures of the working fluid during the hot and cold isothermal processes as T_i and T_{i+1} ($i=1,2,3$), respectively, and the end temperatures as T_H and T_L.

Following [1 and 5], the thermal efficiency of the power plant is given by:

$$\eta = \frac{P}{f(x_1, x_2)P + \dot{Q}}$$

Content from this work may be used under the terms of the Creative Commons Attribution 3.0 licence. Any further distribution of this work must maintain attribution to the author(s) and the title of the work, journal citation and DOI.
Published under licence by IOP Publishing Ltd

IOP Conf. Series: Journal of Physics: Conf. Series **792** (2017) 012002 doi:10.1088/1742-6596/792/1/012002

(1)

where $x_i = \frac{T_{2i}}{T_{2i-1}}$ correspond to the isentropic temperature ratio for each cycle i $(i = 1, 2)$, P is the power of the plant, $\dot{Q}$ is the leak heat and the function $f(x_1, x_2) = \frac{1}{1 - I_1 I_2 x_1 x_2}$. Note that this functional form for one cycle has only appeared in [1, 2, and 5]

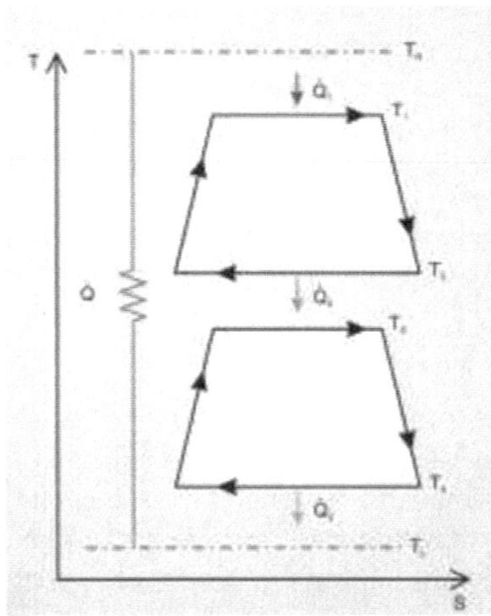

Figure 1: A power plant stationary of two Carnot-like cycles with linear leak heat and finite heat transfer rates, and internal dissipations of the working fluid

According to the second law of Thermodynamics, for each $i = 1, 2$, we have [26]: $\dot{Q}_2 = I_1 x_1 \dot{Q}_1$; $\dot{Q}_3 = I_2 x_2 \dot{Q}_2$; thus, $\dot{Q}_3 = I_2 x_2 \dot{Q}_2 = I_1 x_1 \dot{Q}_1$ so that, the efficiency is given by:

$$\eta = \frac{P}{\dot{Q}_H} = \frac{\dot{Q}_H - \dot{Q}_L}{\dot{Q}_H} = \frac{\dot{Q}_1 - \dot{Q}_3}{\dot{Q}_H} = \left(1 - \frac{\dot{Q}_3}{\dot{Q}_1}\right)\frac{\dot{Q}_1}{\dot{Q}_H} = \frac{(1 - I_1 I_2 x_1 x_2)\dot{Q}_1}{\dot{Q}_H} \tag{2}$$

From equation (5),

$$\dot{Q}_1 = \frac{P}{1 - I_1 I_2 x_1 x_2} \tag{3}$$

and equation (3), (1) is obtained.

Now, let $z \neq x_i$ $(i = 1, 2)$, the Corollary given at [1], extend to:

"The power P achieves a maximum value in z_{mp} if and only if the efficiency η achieves a maximum value in the value: $z_{mp} = z_{me}$"

The proof is similar to that presented in [1]. It is easy to see that in the computations of the proof for the Corollary, no transfer heat law has been used. Thus, in the optimization of the power plant with two Carnot-like cycles, with respect to z, it is enough to find the maximum power by,

$$\frac{\partial P}{\partial z}\Big|_{z_{mp}=z_{me}} = 0 \text{ and } \frac{\partial^2 P}{\partial z^2}\Big|_{z_{mp}=z_{me}} < 0 \tag{4}$$

The optimization performed, with respect to z, is a property independent from the heat transfer law. As a result, $z = \phi$ or ψ, depending on which design rule is applied.

3. Optimal relations for the allocation and effectiveness of the power plant's heat exchangers

In this section, x_1, x_2 will be fixed and we will assume that the law of heat transfer can be any law, including heat leak. Next, we will discuss the following two design rules: *fixed internal thermal conductance* or *fixed areas for the heat exchangers*, which will be alternately applied.

The first design rule is that the internal conductance of the Carnot-like cycle is constrained to: $\sum_{i=1}^{3} \alpha_i = \Gamma$, where Γ is a constant applied to the allocation of the heat exchangers from the hot and cold sides; with the same overall heat transfer coefficient U by unit of area A in both ends of the cycle $i\,(i = 1,2)$ and $\alpha_i\,(i = 1,2,3)$ are the thermal conductance correspondent to the finite heat transfers of the hot/cold sides for this cycle, respectively. Thus, $\sum_{i=1}^{3} UA_i = \Gamma$ where A_i, A_{i+1} are heat transfer areas on hot/cold sides of the cycle $i\,(i = 1,2)$. The second design rule is that the total area is constrained by: $\sum_{i=1}^{3} A_i = A$; where $A_i\,(i = 1,2,3)$ are the heat transfer areas on the hot and cold sides for the cycle $i\,(i = 1,2)$, respectively. Now, the total area A is fixed but, when distributed, it has different overall heat transfer coefficients and hence different effectiveness on each one of the hot and cold sides. How $\alpha_i = U_i A_i$ (see [1, 12]), then $A = \sum_{i=1}^{3} A_i = \sum_{i=1}^{3} \frac{\alpha_i}{U_i}$, where $U_i\,(i = 1,2,3)$ are the overall heat transfer coefficients on the hot and cold sides of the cycle $i\,(i = 1,2)$, respectively.

From the Corollary of the section 2, it is enough to find *"the maximum power of each Carnot-like cycle* $i\,(i = 1,2)$ *for these two design rules applied alternatel"*. Indeed, from [3] the dimensionless power output for the cycle, i is given by:

$$p_i = \frac{P_i}{\alpha_i T_{2i-1}} = \frac{1 - \sqrt{I_i \mu_i}}{\frac{1}{\alpha_i} + \frac{I_i}{\alpha_{i+1}}}, \tag{5}$$

where μ_i is the end temperature ratio of the cycle $i = 1,2$. The numerator of (5) is obtained from applying the conduction heat law and optimizing it with respect to x_i; $i = 1,2$.

Next, if we apply the Bellman' Principle [9]: "*to state that every part of an optimum path is optimal*", we determine the optimal allocation or effectiveness of the heat exchangers.

3.1 Constrained internal thermal conductance

The three thermal conductances can be written as $\alpha_i = UA_i;\ i = 1,2,3$, where U is overall heat transfer coefficient and $A_i;\ i = 1,2,3$ are the available areas for heat transfer. Thus, for the first optimization we can assume for this first design rule: $\alpha_1 + \alpha_2 = \Gamma_1$, where $\Gamma_1 = \Gamma - \alpha_3$ is supposed to be a constant. Equivalently, $\frac{\alpha_1 + \alpha_2}{\Gamma_1} = 1$. Fixing the temperature $T_j\ (j = 3,4)$ and applying only for the first cycle, $\frac{a_1\alpha_1 + \alpha_2}{\Gamma_1} = 1;\ a_1 = 1$. In parameterizing, $\phi = \frac{\alpha_1}{\Gamma_1};\ 1 - a_1\phi = \frac{\alpha_2}{\Gamma_1}$. According to the equation (6), in optimizing $\left(\frac{1}{\frac{a_1}{\phi} + \frac{I_1}{1-a_1\phi}}\right)$ according to ϕ, we obtain $\phi_1 = \frac{1}{1+\sqrt{I_1}}$. Solving, $\frac{1 - a_1\phi_1}{\phi_1} = \sqrt{I_1} = \frac{\alpha_2}{\alpha_1}$; so $\alpha_2 = \sqrt{I_1}\alpha_1$. For the second optimization, the constraint is now $\frac{a_2\alpha_1 + \alpha_3}{\Gamma_2} = 1$, with $a_2 = 1 + \sqrt{I_1}$. In parameterizing, $\phi = \frac{\alpha_1}{\Gamma_2}; 1 - a_2\phi = \frac{\alpha_3}{\Gamma_2}$. Similarly, in optimizing $\left(\frac{1}{\frac{a_2}{\phi} + \frac{I_2}{1-a_2\phi}}\right)$ according to ϕ, we obtain $\phi_2 = \frac{1}{a_2 + \sqrt{I_2}}$. Solving, $\frac{1 - a_2\phi_2}{\phi_2} = \sqrt{I_2} = \frac{\alpha_3}{\alpha_1}$; so that $\alpha_3 = \sqrt{I_2}\alpha_1$.

In summary, the optimal conductance is:

$$\alpha_2 = \sqrt{I_1}\alpha_1; \alpha_3 = \sqrt{I_2}\alpha_1 \tag{6}$$

and $a_3 = \sum_{j=0}^{2} \sqrt{I_j}, I_0 = 1$.

3.2 Constrained areas of heat exchangers

For simplicity, we suppose $I_2 = I_1 = 1$. Applying the second design rule $A_1 + A_2 = \mathrm{A}_1$ where $\mathrm{A}_1 = A - A_3$, $A_1 = \frac{\alpha_1}{U_1};\ A_2 = \frac{\alpha_2}{U_2}$, then $\mathrm{A} = a_1\alpha_1 + u_1\alpha_2$, where $u_1 = \frac{U_1}{U_2};\ a_1 = 1$. In parameterizing, $\psi = \frac{\alpha_1}{A}; 1 - a_1\psi = \frac{\alpha_2}{A}$. From equation (5) $\frac{1}{\frac{a_1}{\psi} + \frac{1}{(1-a_1\psi)u_1}}$, the first optimization with respect to ψ, gives $\psi_1 = \frac{\sqrt{u_1}}{1 + a_1\sqrt{u_1}}$. Then, $\frac{1 - a_1\psi_1}{\psi_1} = \frac{1 + \sqrt{u_1}(a_1 - 1)}{\sqrt{u_1}}$. Solving, $\frac{\alpha_1}{\sqrt{a_1 u_1}} = \alpha_2$; so that $A_2 = \frac{A_1}{\sqrt{a_1 u_1}}$, and the results of [1] are recovered. For the second optimization, the constraint is now $a_2 A_1 + \frac{A_3}{u_2} = A$, where $a_2 = \frac{1+\sqrt{a_1 u_1}}{\sqrt{a_1 u_1}}$; $u_2 = \frac{U_1}{U_3}$. In parameterizing, $\psi = \frac{A_1}{A}; 1 - a_2\psi = \frac{A_3}{A}$. Now, from (5) $\frac{1}{\frac{a_2}{\psi} + \frac{1}{(1-a_2\psi)u_2}}$. The second optimization with respect to ψ, gives, $\psi_2 = \frac{\sqrt{u_2}}{1 + a_2\sqrt{u_2}}$ and

$\frac{1-a_2\psi_2}{\psi_2}=\frac{1+\sqrt{u_2}(a_2-1)}{1+a_2\sqrt{u_2}}$. Thus, $A_3=\frac{1+\sqrt{u_2}(a_2-1)}{\sqrt{u_2}}A_1$. So that, the optimal areas, are:

$$A_2=\frac{A_1}{\sqrt{a_1u_1}};\ A_3=\frac{1+\sqrt{u_2}(a_2-1)}{\sqrt{u_2}}A_1 \tag{7}$$

and $a_3=1+\frac{\sqrt{I_2}+\sqrt{u_2}(a_2-1)}{\sqrt{I_2}+a_2\sqrt{u_2}}$.

4. Maximum power and efficiency to maximum power

Let P_1, P_2 be the power of cycles 1 and 2, respectively, of Figure 1. For the first design rule, the maximum power of the power plant will be given by:

$$(P_1+P_2)_{max}=\frac{\sqrt{I_2}UA\left(1-\sqrt{I\mu}\right)^2}{a_3\left(\sqrt{I_2}+I_1a_2\right)}\quad; \tag{8}$$

and for the second design rule, the maximum power will be given by:

$$(P_1+P_2)_{max}=\frac{T_Ha_3a_2(a_2-1)AU_1\left(1-\sqrt{I\mu}\right)^2}{\left(I_1u_1a_2(a_2-1)+a_3\right)\left(a_2(a_2-1)+a_3\right)}\quad, \tag{9}$$

where $I=I_1I_2$; $\mu=\frac{T_L}{T_H}$ using the notation of the subsections 3.1 and 3.2, respectively.

We can show (8) as follow:

$$\begin{aligned}\frac{P_1+P_2}{\alpha_1T_H}&=\left(1-\sqrt{I_1\mu_1}\right)^2+\frac{\alpha_3}{\alpha_1a_2}\frac{T_2}{T_H}\left(1-\sqrt{I_2\mu_2}\right)^2\\&=\left(1-\sqrt{I_1\mu_1}\right)^2+\frac{\sqrt{I_2}}{I_1a_2}I_1\frac{T_2}{T_H}\left(1-\sqrt{I_2\mu_2}\right)^2\\&=\left(1-\sqrt{I_1\mu_1}\right)^2+\frac{\alpha_3}{I_1a_2\alpha_1}\left(\sqrt{I_1\mu_1}-\sqrt{I_1I_2\mu}\right)^2\end{aligned}$$

where we apply equation (6). In optimizing according to $\sqrt{I_1\mu_1}$:

$$\begin{aligned}(P_1+P_2)_{max}&=\alpha_3\alpha_1T_H\frac{\left(1-\sqrt{I_1I_2\mu}\right)^2}{I_1a_2\alpha_1+\alpha_3}\\&=T_H\frac{\left(1-\sqrt{I_1I_2\mu}\right)^2}{\frac{1}{\alpha_1}+\frac{I_1a_2}{\alpha_3}}=\frac{I_2UA\left(1-\sqrt{I_1I_2\mu_1\mu_2}\right)^2}{a_3\left(I_1a_2+\sqrt{I_2}\right)}\end{aligned}$$

because of $UA=\alpha_1+\alpha_2+\alpha_3=\alpha_1+\sqrt{I_1}\alpha_1+\sqrt{I_2}\alpha_1=\alpha_1\left(1+\sqrt{I_1}+\sqrt{I_2}\right)$, so, $\alpha_1=\frac{UA}{a_3}$.

Now from [10], we can obtain the efficiency to maximum power:

$$\eta_{mp}=\frac{\left(1-\sqrt{I\mu}\right)}{1+\frac{Q_L}{\Lambda T_H}}\quad, \tag{10}$$

where $\Lambda = \sum_{i=1}^{3} \frac{1}{\alpha_i}$. Equation (10) is obtained similarly.

5. Conclusions

We have found and determined the optimal allocation and effectiveness of heat exchangers of a combined cycle with two Carnot-like cycles. Moreover, these relations can be satisfied for other operation regimes, e.g. algebraic combination of power and/or efficiency that have thermodynamic meaning and satisfy imposed power conditions (equation (5)). Nevertheless, the optimal isentropic temperatures ratios depend on the heat transfer law and the operation regime of the engine as discussed in [1]. Additionally, the maximum power and its corresponding efficiency to maximum were calculated. The equations (6-10) can be extended to a power plant with *n*-Carnot-like cycles of the model presented here. However, the latter requires a comprehensive study of the implications for the power plant considered. We will study such implications in our future work.

6. References

[1] Aragón G, Canales A, Galicia A and Rivera J M 2012 On a methodology of optimisation for an irreversible Carnot-like power plant, *J. Energy Inst.* **85** (4) 201-208.
[2] Aragón G, Canales A, Galicia A and Rivera J M 2009 The fundamental optimal relations of the allocation, cost and effectiveness of the heat exchangers of a Carnot-like power plant *J. Phys. A: Math. Theor.* **42** 425205
[3] Aragón G, Canales A, Galicia A and Morales J R 2008 Maximum Power Ecological Function and Efficiency of an Irreversible Carnot Cycle. A Cost and Effectiveness Optimization. Braz *J. of Phys.* **38** (4) 543-550
[4] Aragón G, Canales A, Galicia A and Musharrafie M 2003 A criterion to maximize the irreversible efficiency in heat engines *J. Phys. D: Appl. Phys.* **36**, 280-87
[5] Hoffman K H, Burzler J M and Shuberth S 1997 Endoreversible Thermodynamics. *J. Non-Equilib. Thermodyn.* **22** (4) 311-355
[6] Aragón G, Canales-Palma A, León A and Morales J R 2006 Optimization of an irreversible Carnot engine in finite time and finite size. *Rev. Mex. Fis.* **52** (4) 309-314
[7] Aragón G, Canales A and León A 2000 Maximum irreversible work and efficiency in power cycles *J. Phys. D: Appl. Phys.* **33** 1403-10
[8] Bejan A 1999 Entropy Generation Minimization. CRC Press, Boca Raton
[9] Lewins J. D 1999 Optimizing cascades of endo-reversible heat engines *Int. J. Mech. Eng. Edu.* **27** (2) 91-101
[10] Chen J A 1998 universal model of an irreversible combined Carnot cycle system and its general performance characteristics. *J. Phys. A: Math. Theor.* **3** 3383-3394
[11] Bandyopadhyay S, Bera, N C and Bhattacharyya S 2001 Thermoeconomic optimization of combined cycle power plants. *Energ. Convers. Manag.* **42** 359-371

IOP Conf. Series: Journal of Physics: Conf. Series **792** (2017) 012003 doi:10.1088/1742-6596/792/1/012003

Stages in water loss in fruit and vegetables

Antonio de Ita A and Georgina Flores

Área de Ciencia de Materiales, Universidad Autónoma Metropolitana, Av. San Pablo 180, Azcapotzalco, Ciudad de México

E-mail: add@correo,azc.uam.mx

Abstract. We analyzes the different stages in the dehydration of fruits and vegetables. It was found to vary from a lettuce: as a sharp drop at low temperature, (arround 150 °C) to five stages in total, with a loss almost continuous to about 300 ° C, such as grapefruit, papaya and fig.The first section in your paper.

1. Introduction

In previous studies [1,2] we have determined the loss of water in fruits and vegetables and the results obtained are generally consistent with the literature, where there is much variety in certain specific values of the water loss. This is because most literature is found on the internet, is not very scientific, or made without a safe and repeatable method, except in the case reported by the Department of Health of Great Britain [3]. Which explains in detail how the various parameters were determined, the only problem is that the list of materials is very short compared to existing, for example, in Mexico.

As is known [1.2] fruits in "general" are formed by an average of 90% water, with an 8% carbohydrates or sugars, fiber 1.5% and the rest in protein, vitamins, minerals, flavors and pigments. While vegetables contain average just less water, 75% plus 10% carbohydrates or sugars, a little more fiber than fruit, 4%, also a little more proteins 3%, also contains besides lipids, calcium, carotenes, vitamins, flavors and pigments.

It is clear that these percentages vary a little in vegetables. Thus in the case of water, potatoes contain 77 % while envy has 95 %. Carbohydrates are present with 19 % in potatoes while spinach has only 1.3 %. In the case of fiber, beans have 6.5 % but zucchini only 1.2 %. Parsley has 4.4 % of proteins while lettuce contains only 0.9%. And finally in the case of lipids parsley has 0.4 % but lettuce has only 0.1 %. Fruits have a similar behavior; there are fruits with more and less water and so on with the different components.

2. Methodology

The water loss in fruits and vegetables were made in a Shimadzu Thermal Analysis, DTG-60 model. We obtained TGA, Thermo Gravimetric Analysis, and also DTA, Differential Thermal Analysis, although these curves are not shown here. The latter technique can determine transition temperatures. The temperature was varied from ambient to 300°C. The heating rate was 10°C per minute under nitrogen flow rate of 10 cm^3 per minute. The mass used was around 15 mg, only pulp and always shelled without outer skin or bone were used, except in some cases for comparison purposes.

Content from this work may be used under the terms of the Creative Commons Attribution 3.0 licence. Any further distribution of this work must maintain attribution to the author(s) and the title of the work, journal citation and DOI.
Published under licence by IOP Publishing Ltd

3. Results

In previous studies we have shown curves of weight loss have a single step as lettuce or poblano chile, as shown in figure 1. From the room temperature to 300°C a loss of almost 95% occurs in the case of lettuce and 92% in the case of poblano chile.

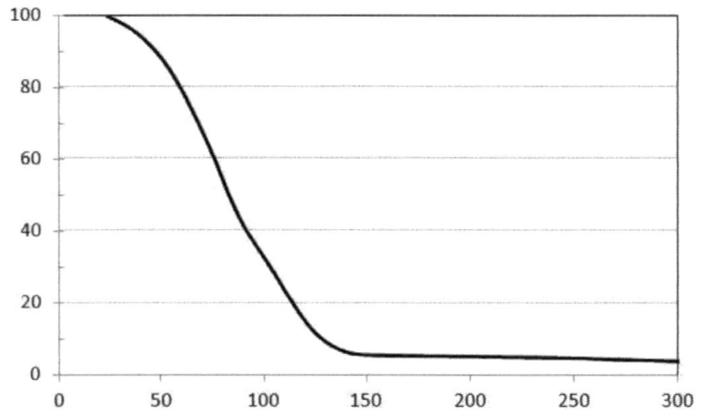

Figure 1. Graph of weight loss versus temperature of lettuce

But there are curves such as strawberry or melon where two stages very clearly, the first a period make a loss of over 82% and in the second stage falls only a 10% in both stages the fall is observed in the second part the loss is slow but continuous. In melon the first drop is more drastic but lighter in the second only 5% at 85°C. The strawberry result is shown in Figure 2.

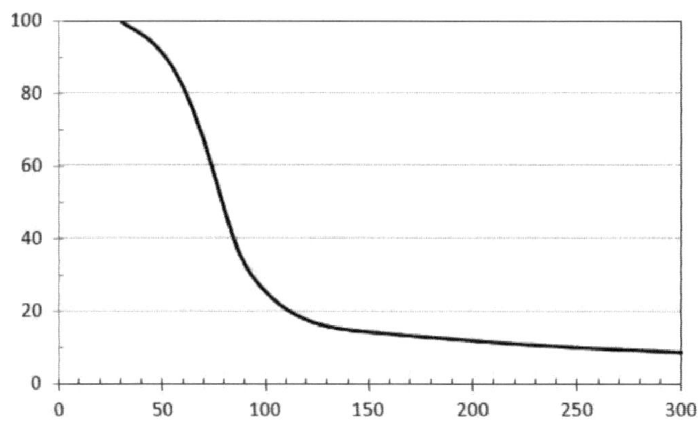

Figure 2. Graph of weight loss versus temperature of the strawberry

Three stages can be observed in pear and guava, as shown in figure 3 for the pear. The first loss is 84 % weight loss for about 140 °C, the second ends for 200 °C with a total loss of 88 % and finally the third stage though was not fully complete but at 300 °C there was a total loss of almost 95 %.

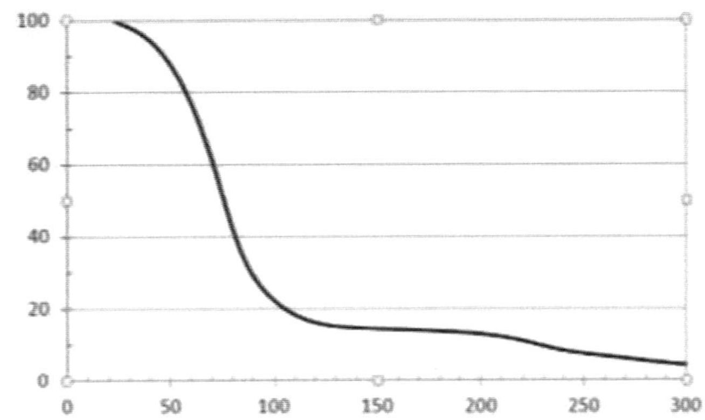

Figure 3. Graph of weight loss versus temperature of the pear

As one can see the values are very approximate therefore we develop a procedure to recognize in more detail what happens in the curve of weight loss. In this work, lines are drawn in each part of the curve and where two lines intersect a critical point, characterized by a temperature and weight loss was determined.

The idea is to determine whether there is a correlation in the different curves and identify more precisely the different stages. Thus in figure 4 is shown the grapefruit weight loss curve. Using straight lines a drop of 18 % is found at 58 °C, a second fall is related at 134 °C with a 85 %, total loss, the third stage is observed at 196 °C with a 90 %, total loss, at 229 °C a fourth stage with 93 % loss, finally the fifth loss is located where the experiment ends, at 300 °C with a 95 % total loss.

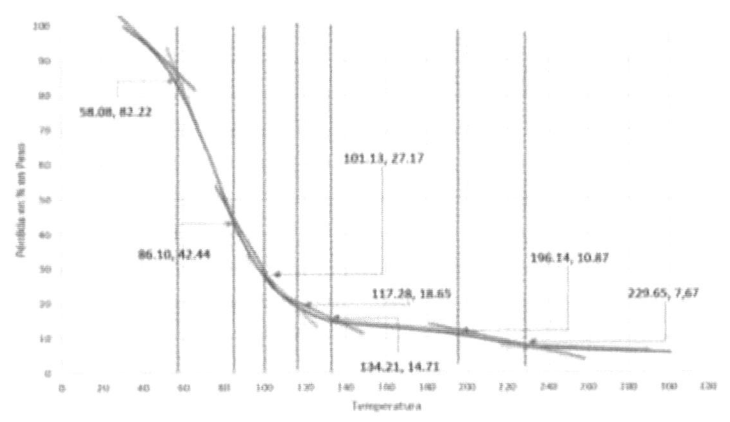

Figure 4. Graph of weight loss versus temperature for grapefruit

In figure 5 is another fruit, in this case, papaya is analyzed, the first stage is at 56 °C with a 13 % loss, the second is at 142 °C with 82% loss. The third stage is located at 200 °C with 85 % loss, the fourth at 224 °C and 86 % loss, the fifth finally, again located where the experiment ended at 300 °C and 90 % is the total loss.

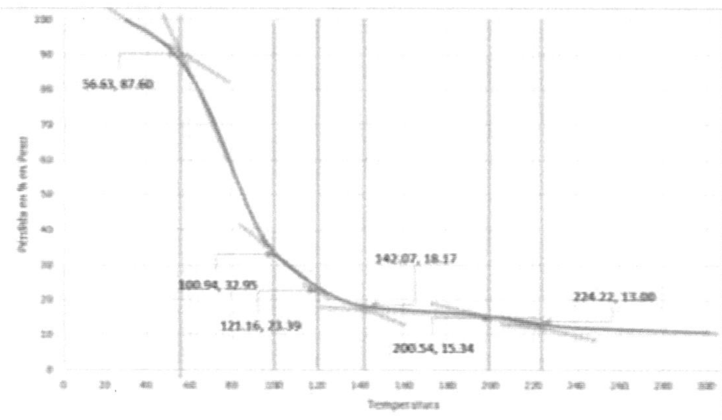

Figure 5. Graph of weight loss versus temperature for papaya

Fig fruit is presented in figure 6. Here we can also found in five weight loss stages. The first is at 77 °C with a 4 % loss, the second is at 146 °C with a 62 % total loss. The third ends at 206°C and represents a 4%, partial loss. The fourth stage is at 241 °C and a 6 % partial loss. The fifth and last is located where the experiment ends, at 300 °C and a 74 % total loss. In the last three cases it does not seem to finish the last stage and is variable for each of the tested materials.

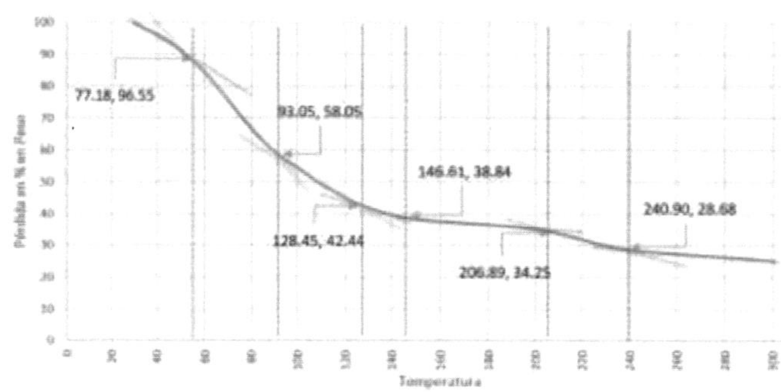

Figure 5. Graph of weight loss versus temperature for fig frut

4. Discussion

More stages especially in fruits were detected in the weight loss curves. On the one hand, it is recognized that because the major component present in fruits is water, the weight loss should be in less stages, but in fact is not. For the latter reasoning is that makes very reasonable that in the second stage is where is present the largest contribution to the weight loss and occurs between 100°C and 140°C in the last three examples shown here. In the grapefruit and papaya the second loss is 82% and only 62 % for the fig frut. For the final stage near to 300 °C the contributions are very small: 10 % for grapefruit, 8 % for papaya and 13 % for the fig frut.

Another interesting point to remember is that the curves of water loss can be classified by the lost amount and are: as banana or low, about 70 %. The medium or the average of fruits is about 90 % , the typical case is the poblano chile and finally the grater o height, with almost a total loss of 98 %, such as lettuce. This data is independent of the number of steps shown by the curves.

According to its curve evolution, or the number of stages in which the weigh is lost, we belied that the two first stages can be associated to water loss..

To analyze whether the second component, carbohydrates, have an important contribution and there is a possibility that the other minor components such as fiber, protein, vitamins, minerals, flavors and pigments have a chance to contribute somehow weigh losses is to investigate. It is possible because their molecules can decompose with temperature, but as we know their presence is minimal.

5. Conclusions

We proposed a new graphical procedure, which can be determined up to 5 stages in the weight loss. The first two stages, below 150°C can be associated with water loss.

6. References

[1] De Ita A, Flores G and Franco F Fruits and vegetables dehydration. 2015 VII International Congress of Engineering Physics, *IOP Publishing, Journal of Physics: Conference Series.* **582** 1-5 012065, doi:10.1088/1742-6596/582/1 /012065

[2] Antonio de Ita, Flores G and Franco F 2015 Determinación de la Pérdida de Agua de Frutas y Verduras por medio de Termogravimetría VI Congreso Internacional de Docencia e Investigación en Química 23 al 25 de Septiembre, Universidad Autónoma Metropolitana Azcapotzalco

[3] Department Health 2013 Nutrient analysis of fruit and vegetables (U.K)

[4] Harlan J R 1987 Les plantes cultivées et l'homme ed Presses Universitaires de France (Paris-France)

IOP Conf. Series: Journal of Physics: Conf. Series **792** (2017) 012004 doi:10.1088/1742-6596/792/1/012004

Photoluminescence and structure study in mixture of ZnO and carbon nanoparticles during mechanical activation

E Velázquez-Lozada[1], T Torchynska[2] and G M Camacho-González[3]
[1] SEPI – ESIME – Instituto Politécnico Nacional, Ciudad de México. 07738, México
[2] ESFM – Instituto Politécnico Nacional, Ciudad de México. 07738, México
[3] ESIME – Instituto Politécnico Nacional, Ciudad de México. 07738, México

E-mail: evlozada5@yahoo.com.mx

Abstract. The photoluminescence and X ray diffraction of the mixture of ZnO + 1.0% C nanocrystals (NCs) have been studied before and after intensive mechanical processing (MP) with the aim to identify the nature of emission bands and radiative defects. The study reflects the diversity of physical processes occurring during MP: the destruction of primary ZnO NC aggregates, crushing individual ZnO NCs from the size of 250 nm down to 14 nm, the interaction of carbon atoms with oxygen in a treatment chamber and with the surface of ZnO nanoparticles etc. The new PL band peaked at 2.84-2.95 eV has been revealed in PL spectra after 9 min of MP. The origin of this emission in ZnO has not been conclusively established. The new PL band peaked at 2.84-2.95 eV has been studded and its nature is discussed.

1. Introduction

ZnO nanocrystals (NCs) with the native defects or different type dopants are the one of most popular new systems (1). ZnO NCs with a direct band gap (3.37 eV) and a high exciton binding energy (60 meV) at 300K promise numerous applications in optical and electronic devices (1, 2), such as: catalytic materials (3), solar cells (4), field emission cathodes (5, 6), luminescent materials (7) etc. Important electronic properties have been revealed in the nanocrystalline composites ZnO + xC recently, which permit improving the field emission properties of cathodes (5), engineering room-temperature ferromagnetism (8), modify luminescence properties (9), and to create the selective solar absorbed coatings (10). The creation of efficient electronic devices requires to study the emission nature and the mechanisms of defect generation in ZnO NCs. In present paper the transformation of photoluminescence (PL) and its temperature dependences have been investigated for the mixture of ZnO + 1.0% C NCs with the goal to study the nature of PL bands, the PL quenching mechanism, the radiative defect nature and the process of defect creation at mechanical processing (MP).

2. Experimental details

ZnO NCs (99.5 % purity, Reasol; with the particle size of dZnO ~ 250 nm) and carbon nanoparticles (EPRUI Nanoparticles & Microspheres Co. Ltd.) were mixed. High resolution TEM images (performed earlier) have shown that the carbon nanoparticles are characterized by a spherical-like shape with an average diameter of $d_C \sim 50$ nm. Mechanical processing (MP) of ZnO + xC mixtures with x = 1.0 % wt. was carried out in a Planetary Ball-Mill (PM 400/2, Retsch Inc.). The grinding chamber of 50 ml with the tungsten carbide balls (3 of 20 mm and 10 of 10 mm) were used. The weight ratio of balls to mixture powder equal to 28:1 has been used. MP was carried out with the rotation speed 400 rev/min for processing times (tMP) of: 1, 3, 9, 30, 90 and 390 min. The early X ray diffraction study has shown that the average ZnO NC sizes, estimated using Sherrer formula (11), decreases from 250 nm in an initial state down to 14 nm for tMP = 390 min.

Content from this work may be used under the terms of the Creative Commons Attribution 3.0 licence. Any further distribution of this work must maintain attribution to the author(s) and the title of the work, journal citation and DOI.
Published under licence by IOP Publishing Ltd

PL spectra were measured at 300K and the excitation by a He-Cd laser with a wavelength of 325 nm and a beam power of 46 mW using a PL setup on a base of spectrometer SPEX500 described in (12,13) in the temperature range of 10-300K.

3. Experimental results and discussion

PL spectra of the ZnO + 1% C mixture at 300 K measured for the different MP times are shown in figure 1.

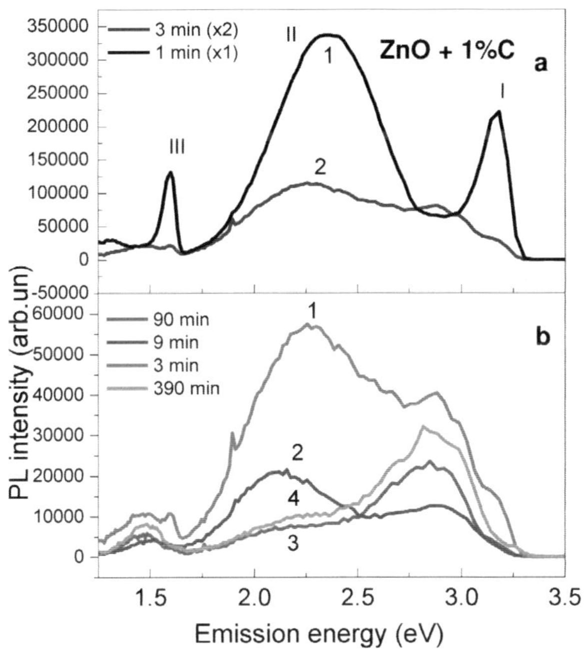

Figure 1. PL spectra of ZnO+1% C NCs at different moments of MP: (a) 1-1min, 2-3min, (b) 1-3min, 2-9min, 3-90min, 4-390min

It is clear that PL spectra are complex and can be represented as a set of PL bands. The PL spectrum measured after 1 min of MP can be decomposed on three PL bands with the peaks at: 3.14 eV (I), 2.42-2.50 eV (II) and 1.57 eV (III) (Fig.1a, curve 1). The UV and visible PL bands in ZnO NCs are attributed to the near-band-edge (NBE) (I) and defect related (II) emissions (14-18).

NBE emission in ZnO NCs, as a rule, is related to the free or defect bound excitons with their phonon replicas, and/or the donor-acceptor pairs (14,15). It was shown earlier in (16,17) that free exciton emission or its replicas dominate in ZnO NC spectra at room temperature owing to the high free exciton binding energy (60 meV) at 300K and the fast decay of the PL intensities of bound exciton and donor-accepter pair emissions with increasing temperature in the range of 10-300K. Thus the high PL intensity and a small half width of 3.14 eV PL band in PL spectra of the ZnO + 1% C mixture at 300K permit assigning NBE emission to the LO phonon replicas of free exciton emission in ZnO NCs. The nature of IR PL band peaked at 1.57 eV (III) is not clear. Note that in our experiments (presented below) this PL band intensity varies by the same way as the PL intensity of 3.14 eV PL band. Thus we can attribute the IR PL band at 1.57 eV (III) to the second order diffraction peak of the 3.14 eV PL band (15-18).

The PL intensity mentioned above PL bands decreases dramatically at MP and the new PL bands with peaks at 2.84 -2.95 eV, 2.10-2.20 eV and 1.42-1.47 eV have appeared (Fig. 1b). The variation kinetics of the integrated PL intensities for all PL bands at MP are presented in figure 2. As it is clear there are two stages in this kinetics depending on the duration of MP process.

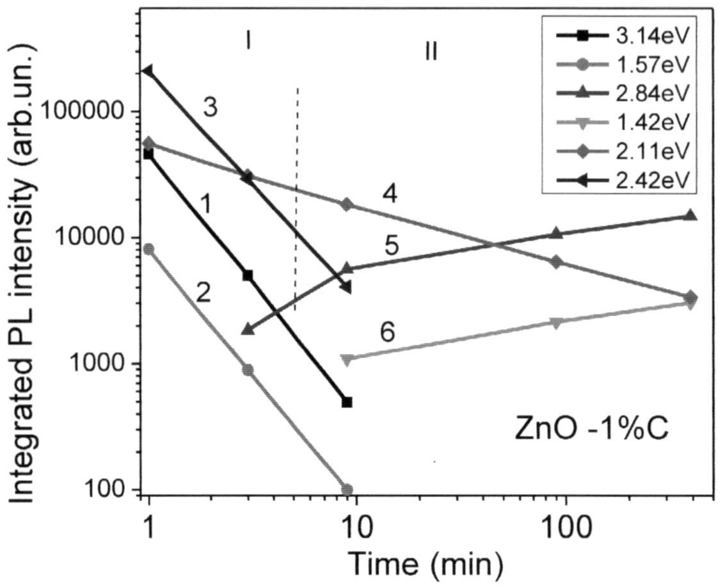

Figure 2. The variation of integrated PL intensities versus MP times for the PL bands:1 -3.14 eV, 2-1.57 eV, 3-2.42 eV, 4-2.11 eV, 5-2.84 eV, 6-1.42 eV

First stage.The PL intensity decreases upon the first MP stage (1-9 min) with nearly the same rates for the NBE and defect related PL bands. This effect can be assigned to the generation of nonradiative recombination centers (NR) at MP. The PL intensity is proportional to the quantum efficiency (η) of visible recombination (19). The increase of NR center concentrations stimulates the rise of NR recombination rate together with falling down the PL intensity at MP. Linear dependences of the PL intensity decay for the stage I, presented in logarithmic scales (Fig. 2), testify on the hyperbolic PL intensity variation versus time. Note that NR centers can be assigned to the dangling bonds on the surface of ZnO NCs created at crushing the individual ZnO nanoparticles at MP. Simultaneously, in the first MP stage the peak of defect related PL band (II) shifts to 2.11-2.25 eV and the new PL bands peaked at 2.84-2.95eV and 1.42-1.47 eV appear (Figures 1b).

Second stage. The new PL bands with the peaks at 2.11-2.25 eV, 2.84-2.95 eV and 1.42-1.47 eV dominate in the PL spectra of all samples after 9 min of MP (Figures 1b).The defects, responsible for the 2.11-2.25 eV PL band, can exist in the original ZnO NCs or their concentration can increase at MP as well. The IR PL band peaked at 1.42-1.47 eV is attributed to the second order diffraction peak of 2.84-2.95 eV PL band. The high energy PL band is characterized by the peak at 2.84-2.95 eV in the different samples of ZnO+1%C NC mixture.

In order to gain more insight on the light-emitting mechanism of the ZnO+1%C NC mixture and to recognize the different types of optical transitions, we have performed the temperature dependent PL measurements in the range of 20-300K (Fig.3).

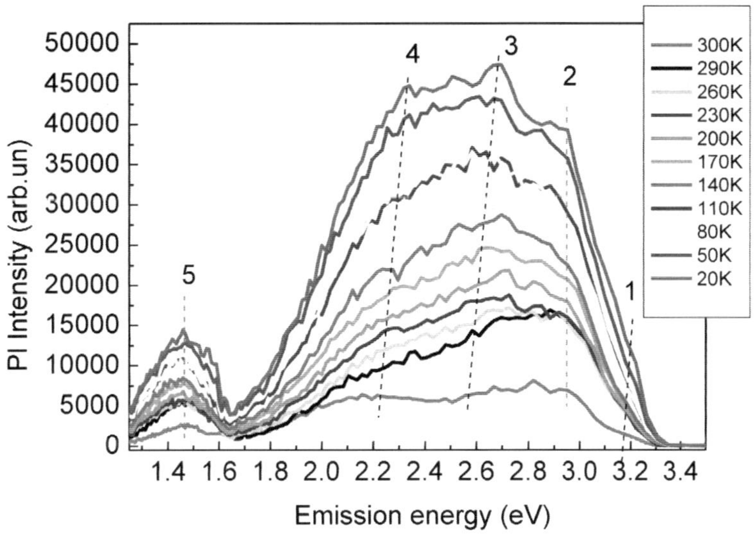

Figure 3. PL spectra of studied ZnO+1% C NCs measured at different temperatures after MP for 390 min

The deconvolution procedure has been applied to all PL spectra measured at different temperatures from the range 20-300K to gain the detailed analysis of their temperature behavior. The five PL bands have been used at 300K for the deconvolution process: 3.14, 2.92, 2.58, 2.23 and 1.46 eV (Fig.4a). The change of PL peak positions for the mentioned PL bands at the temperature variation is presented in figure 4b. As one can see from Fig.4b, the PL band (2) peaked at 2.92 eV does not change its PL peak position with temperature. In contrary the other PL bands (1-3.14eV, 3-2.58eV and 4-2.23eV at 300K) monotonically shift to higher energy at the temperature decreasing. The last behavior is typical for the PL bands in the ZnO NCs and related to shrinking the energy band gap at higher temperatures (17). Note that the analysis of the PL peak shift versus temperature (Fig.4b) has been done using the well-known Varshni formula $E(T) = E_o - aT^2/(T+b)$ (20), where E_o is the band gap at the absolute temperature T = 0 K; "a" and "b" are the Varshni thermal coefficients. The fitting parameters "a" and "b" were estimated in studied samples as a= 0.975 meV/K and b = 750K. These values are very close to fitting parameters estimated earlier for the band gap shrinkage in the bulk ZnO (α equal 0.981meV/K and β equal to 650K) (21). Thus we can conclude that the PL bands peaked at 3.14, 2.58 and 2.23 at 300K related to emission in ZnO NCs.The nature of the 2.92eV PL band needs to be discussed additionally.

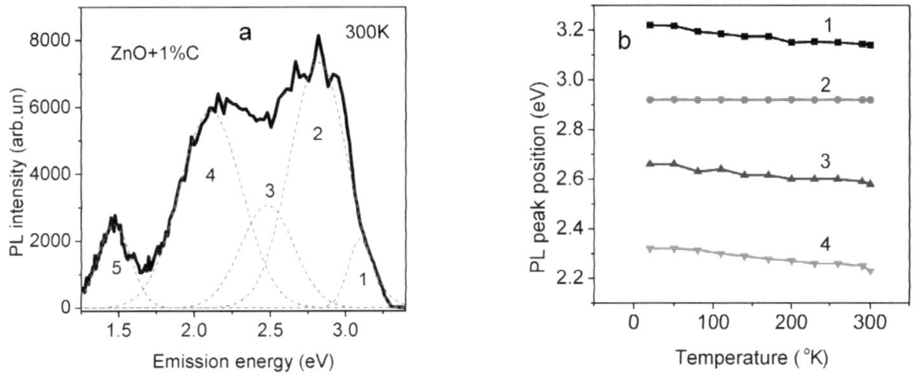

Figure 4. Deconvolution procedure applied to the PL spectrum measured at 300K of the ZnO+1% C NCs after MP for 390min (a), and (b) The variation of PL peak positions versus temperature for the PL bands 1,2,3 and 4

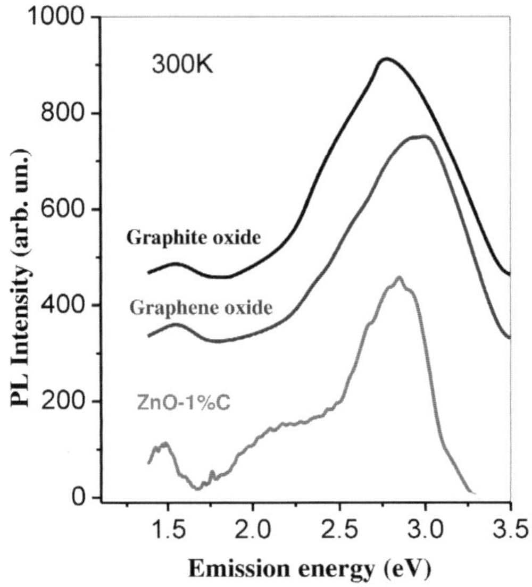

Figure 5. PL spectra of graphene and graphite oxides [25], as well as the studied ZnO+1%C NCs after MP for 390min

A variety of emission bands in UV and visible spectral ranges have been detected in ZnO NCs earlier (14,18). The origin of these emissions has not been conclusively established. The blue PL band with the peak at 2.80-2.90 eV in ZnO is attributed earlier to Cu related defects in (21,22), to Zn interstitials (23) or to the donor–acceptor pairs included the shallow donor and oxygen vacancy (24). In our case we can attribute the 2.92 eV PL band to emission of the graphene or graphite oxides covered the surface of ZnO NCs (25-27). Note that the PL spectra of carbon quantum dots were detected earlier in the spectral range 2.75-2.95eV as well (28).

The PL spectra of graphene and graphite oxides [25], as well as of the studied ZnO+1%C NC mixture after MP are presented in figure 5 for the comparison. The corresponding PL peak positions are: 2.79 eV in the graphite oxide, 2.95eV in the graphene oxide and 2.92 eV in the studied mixture. Thus the new high energy PL band in our samples is characterized by the intermedium position between the positions of luminescence bands typical for the graphite (graphene) oxides. Note that the graphite (graphene) oxides can be created at the oxidation of carbon nanoparticles at prolonged MP. To make the final conclusion concerning the nature of 2.84-2.95 eV PL bands the additional deep study is necessary.

4. Conclusion

The transformations of PL spectra and XRD diagrams in the ZnO NCs + 1%C mixture at mechanical processing have been investigated. Two stages of PL spectrum transformation have been revealed and discussed. It is shown that the first stage is connected with the quenching of PL intensities of all PL bands owing to the nonradiative center appearing at crushing the individual ZnO NCs and decreasing their sizes from 250 nm down to 14nm. The second stage of PL spectrum transformation at MP is connected with appearing the new 2.84-2.95 eV and 2.11-2.25 eV PL bands. The former one is attributed to emission in the graphite (graphene) oxides.

5. References

[1] Morkoç H and Özgur Ü 2009 Zinc Oxide: Fundamentals, Materials and Device Technology ed WILEY-VCH (Weinheim)
[2] Özgür Ü, Alivov Ya I, Liu, Teke A and Reshchikov M A 2005 *J Appl Phys* **98** 041301
[3] Gurwitz R, Cohen R and I Shalis h C 2014 *J Appl Phys* **115** 033701
[4] Fan Z and Lu J 2005 G *J Nanosci Nanotechnol.* **5** 1561
[5] Pan N, H Xue H, M Yu, Cui X, X Wang, Hou J G and Huang J 2010 *Nanotechnology* **21** 2257
[6] Won Il Park, Jin Suk Kim, Gyu-Chul Yi, M. H. Bae and H.-J. Lee , *Appl. Phys. Lett.* **85**, 5052 (2004).
[7] M A Reshchikov, H. Morkoc, B Nemeth, J Nause, J Xie, and B. Hertog 2007 Osinsky *Physica B* **401–402** 358
[8] H S Hsu, Y Tung, Y J Chen, M G Chen, J S Lee and S J Sun 2011 P*hys status solidi* (RRL) **5** 44
[9] Y Hu and H-J Chen 2007 *J Appl Phys* **101** 124902
[10] Katumba G, Olumekor L, Forbes A, Makiwa G and Mwakikunga B 2008 *Sol Energy Mater Sol Cells.* **92** 1285
[11] Klug H P and Alexander L E 1974 *X-ray Diffraction Procedures for Polycrystalline and Amorphous Materials* ed Wiley and Sons (NewYork)
[12] Torchynska T V, Dybiec M and Ostapenko M 2005 *Phys. Rev. B.* **72** 195341
[13] Korsunska N, Khomenkova L, Sheinkman M. K, Stara T, Yuhimchuk V, Torchynska T V and Vivas A 2005 *J. Lumines.* **115** 117
[14] Djurišic A B, Ng A M C and Chen X Y 2014 *Progress in Quantum Electronics* **34** 191
[15] Torchynska T V and B. El Filali 2014 *J. of Lumines.* **149** 54
[16] Diaz A I, B El Filali, Torchynska T V and Casas J L 2013 *Physica E*, **51** 24
[17] Velázquez E, Torchynska T V, Casas J L, Perez B and L. Castañeda 2014 *Physica B*. **453** 111
[18] Diaz A I, B El Falali, Torchynska T V and Casas B *2013 J. Phys. Chem. Solids.* **74** 431
[19] Torchynska T V 2011 *Physica E.* **44** 56
[20] Varshni Y P 1967 *Physica* **34** 149
[21] J. Lv, Ch. Liu, W. Gong, Zh. Zi, X. Chen, K. Huang, T. Wang, G. He, Sh. Shi, X. Song and Zh. Sun 2012 *Opt. Mater.* **34** 1917
[22] Reshchikova M A, Morkoc H, Nemeth B, J. Nause J, Xie J, Hertog B and Osinsky A 2007 *Physica B: Condens. Matter.* **401–402** 358
[23] Patra M K, Manzoor K, Manoth M, Vadera S P and Kumar N 2008 *J. Lumines.* **128** (2) 267
[24] D.H. Zhang, Z.Y. Xue and Q.P. Wang 2012 *J. Phys. D: Appl. Phys.* **35** (21) 2837
[25] Singh G, Choudhary A, Haranath D, Amish G. Joshi, Singh N, Singh S and Pasricha R 2012 *Carbon.* **50** (2) 3852
[26] Chih-Tao Chien, Shao-Sian Li, Wei-Jung Lai, Yun-Chieh Yeh and Hsin-An Chen 2012 *Angew. Chem. Int. Ed.* **51** 6662
[27] Fuchi Liu, Yong Cao, Mingdong Yi, Linghai Xie, Wei Huang, Nujiang Tang, Wei Zhong and Youwei Du 2013 *Crystals.* **3** 28
[28] L. Bao, Z.L. Zhang, Z.Q. Tian, L. Zhang. Ch. Liu and B. Qi 2011 *Advanced Materials.* **28** 5801
[29] Strite S and Morkoc H 1992 *J. Vac. Sci. Technol.* B **10** 1237

IOP Conf. Series: Journal of Physics: Conf. Series **792** (2017) 012005 doi:10.1088/1742-6596/792/1/012005

Numerical study of the *3D*-shape of a drop immersed in a fluid under an elongational flow with vorticity

A S Sanjuan[1], M A H Reyes [2], A A Minzoni[3] and E Geffroy[1]
[1]Instituto de Investigaciones en Materiales, Universidad Nacional Autónoma de México, Mexico City.
[2]Departamento de Termofluidos, Facultad de Ingeniería. Universidad Nacional Autónoma de México, Mexico City.
[3]Instituto de Investigaciones en Matemáticas Aplicadas y en Sistemas., Universidad Nacional Autónoma de México, Mexico City.

E-mail: alfrsj07@gmail.com

Abstract. This work focuses on a three-dimensional analysis of the deformation of a drop — immersed in a Newtonian fluid— generated by a 2D elongational flow with vorticity. The study of steady-state deformations of the cross-section of the drop shows a prevalent non-circular shape. The axisymmetric idealization of the ellipsoid is not observed nor the linear dependency between capillary number and deformation of the drop, as Taylor and Cox theory predicted. Our numerical results are consistent with experiments and other numerical simulations. However, in the latter cases, measurements of the cross section of the drop are few while a limited class of flows is applied. In this work, deformations induced by general two-dimensional flows upon the 3D drop shape are presented with special emphasis about the length scale along the third axis —perpendicular to the plane of the applied flow field.

1. Introduction

Flow-induced drop deformation is an important topic in Fluid Mechanics [1,2] due to applications in industry as emulsion processing, micro-fluidics technologies [3-5], and others. Since the work of Taylor [1,2] the study of drop deformation has focused in the effects induced by the flow applied on the continuum phase. As Rallison explains [3], studies of drop deformations have focused principally when it is subjected to shear flows or extensional flows. The analysis of these two-dimensional flows permitted the theoretical analysis of small-drop deformation [1,3,5,6], assuming an spheroidal shape with the cross-section remaining axisymmetric. However, Guido et all [7] found that cross-sections in simple shear flow are non circular. Numerical studies of drop deformation showed this conditions [5,8], but these studies did not analyze the drop cross-section.

The study of drop deformation in two-dimensional flows is still an open question because an ample family of 2D-flows is not easily carried out in the laboratory [9-11]. As we will see in the next Section, a family of 2D-flows ranging from simple shear to pure extensional flow can be obtained as a mixture of rotation and elongation parts [1,12]. With this family of flows, it is possible to study the drop deformations induced from simple shear to extensional flow, while observing the contribution of added vorticity in the flow. The numerical experiments here presented can supplement the experimental data to understand better the dynamics of drop deformations.

This work focuses in flows with kinematics close to those of simple shear flow but, with less vorticity —more elongational effects. To calibrate the numerical experiments, the experimental data of Rosas et al., [10] are used.

Content from this work may be used under the terms of the Creative Commons Attribution 3.0 licence. Any further distribution of this work must maintain attribution to the author(s) and the title of the work, journal citation and DOI.
Published under licence by IOP Publishing Ltd

IOP Conf. Series: Journal of Physics: Conf. Series **792** (2017) 012005 doi:10.1088/1742-6596/792/1/012005

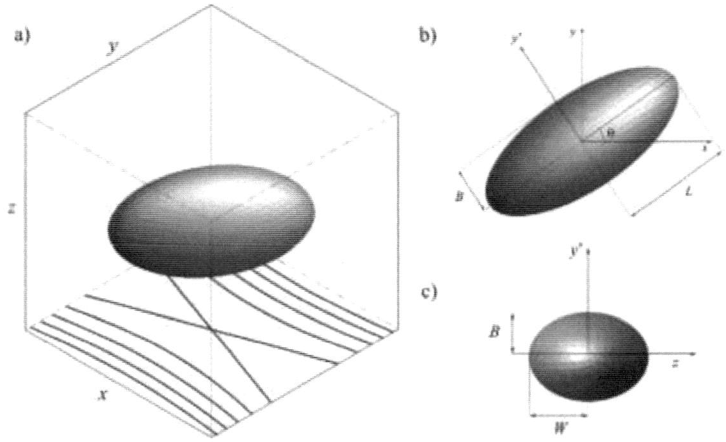

Figure 1. Scheme of a two-phase fluid. Drop in stationary shape. $\lambda_\mu = 0.012$, $\alpha = 0.13$, $Ca = 0.4$.
a) 3D-view view of the drop within a 2D-flow applied.
b) Conventional perspective on the *xy* plane.
c) Cross section of the drop.

2. Drop deformation in 2-D linear flow

The analysis of drop deformation assumes two Newtonian fluids, figure 1. The first (inner) phase is the drop with viscosity μ_0 and surface tension σ. The continuum phase has a viscosity μ_1. The viscosity ratio is $\lambda_\mu = \mu_0/\mu_1$. The two fluids have the same density and both are immiscible. There is not surfactant and there are not Marangoni stresses present. The creeping-flow conditions are assumed. In the creeping-flow regime the fluid motion is governed by the Stokes equations [12]

$$\mu\nabla^2\boldsymbol{u} = \nabla p,$$
$$\nabla\cdot\boldsymbol{u} = 0. \tag{1}$$

The velocity $\boldsymbol{u}$ is continuous at the drop interface *S*. The tractions exerted on the two sides of the interface between the two fluids have two different values, with a corresponding discontinuity:

$$\Delta f = f_1 - f_0 = (\Pi_1 - \Pi_0)\cdot\hat{\boldsymbol{n}}, \tag{2}$$

where $\hat{\boldsymbol{n}}$ is the normal vector pointing out of *S*. In this work, a constant value surface tension [13] is assumed; i.e., $\boldsymbol{\nabla}\cdot\hat{\boldsymbol{n}}$ is equal to twice the mean curvature κ_m at that point on the interface,

$$\Delta f = \sigma\hat{\boldsymbol{n}}\nabla\cdot\hat{\boldsymbol{n}} = 2\sigma\kappa_m\hat{\boldsymbol{n}} \tag{3}$$

The drop is subject to a two dimensional linear incident flow

$$\boldsymbol{u}(x,y) = \frac{G}{(1+\alpha)}(\alpha y, x), \tag{4}$$

where G is the intensity of the rate of deformation tensor [15] and the flow parameter is α [9-11, 14].

$$G = \left(\left|II_{2D}\right|\right)^{1/2}. \tag{5}$$

Equation (4) assumes that the α parameter goes from zero to one. The case $\alpha = 0$ corresponds to simple shear flow with the velocity gradient in the *y* direction, the value of G is equal to shear rate $\dot{\gamma}$. For $\alpha = 1$ we have an extensional flow in two-dimension with the principal axes of deformation at $x = y$ —the compressional axes being $x = -y$; the intensity of the flow is the strain rate $\dot{\varepsilon}$.

The dynamic of drop deformation is characterized by three dimensionless numbers. The viscosity ratio λ_μ, the capillary number *Ca* equation (6) and The flow parameter α equation (7). The capillary number characterizes the ratio between viscous stresses —imposed by the flow— and capillary forces that resist the deformation and drive the drop towards the equilibrium shape, were r_0 is the non-deformed radius of the drop.

$$Ca = \frac{r_0\mu_1 G}{\sigma}, \tag{6}$$

Finally, the flow parameter α characterizes the magnitude of the rotational component relative to the extensional component of the external flow. Here, results are presented for a range of values of the flow-parameter close to simple shear flow, i.e., $\alpha = 0.03$, $\alpha = 0.05$ and $\alpha = 0.13$. The applied flow covers a range of the Capillary number from *0.05* to *0.40*. For values less than the minimum, the drop shape does not present different values of its principal conditions; for values of *0.4* or less, the deformation is not significant to cause drop breakup. The continuum phase is more viscous than the drop viscosity: $\lambda_\mu = 0.012$; the value is chosen so that numerical predictions match the values of the experiment results of Rosas [10].

The measure of drop deformation used corresponds to Taylor Deformation:

$$D_T = \frac{L - B}{L + B}, \tag{7}$$

were L and B are the axes of the drop shown in figure 1. Additionally, we present the time evolution of the lengths of all principal axes of the drop *vs.* the initial radius; see figure 2.

3. Numerical Method

The numerical method used is the 3D collocation boundary element method [13]. Using the Lorentz reciprocal theorem for Stokes regime [12,13], particular solutions can be expressed in terms of known solutions as point sources on the interface where $\boldsymbol{u}_0$ is the flow on the surface of the drop and $\boldsymbol{f}_0$ is the force on the surface of the drop.

$$\int_S (\mu_1 \boldsymbol{u}_1 \boldsymbol{f}_0 - \mu_0 \boldsymbol{u}_0 \boldsymbol{f}_1) dS = 0, \tag{8}$$

were $\boldsymbol{f} = \boldsymbol{\sigma} \cdot \hat{\boldsymbol{n}}$, considering on the surface of the drop point sources whose velocity and stresses are the stokeslet and the stresslet respectively [5,12,13]. Then, the numerical scheme is represented as

$$\boldsymbol{u}^\infty(x_0) - \frac{1}{8\pi\mu_1}\int_S \Delta \boldsymbol{f}(x)\cdot \boldsymbol{G}(x, x_0)\, dS(x) + \frac{1-\lambda_\mu}{8\pi}\int_S \boldsymbol{u}_1(x_0)\cdot \boldsymbol{T}(x, x_0)\cdot \hat{\boldsymbol{n}}(x)\, dS(x) = \begin{cases} \boldsymbol{u}_1(x_0) & \text{a}), \\ \dfrac{1+\lambda_\mu}{8\pi}\boldsymbol{u}_1(x_0) & \text{b}), \text{ or} \\ \lambda_\mu \boldsymbol{u}_2(x_0) & \text{c}). \end{cases} \tag{9}$$

In this equation there are three cases. The case, (b) Solves the velocity field when x_0 is on the surface of the drop. Then with this information the case (a) calculates the velocity field when x_0 is outside of the drop —exterior flow. Finally, (c) calculates the velocity field when x_0 is inside the drop.

The flow imposed on the continuum phase is determined by $\boldsymbol{u}^\infty(x_0)$. The second term on the left side of equation (9) is known as the Single Layer Potential and evaluates stresses across the interface. The third term is known as the Double Layer Potential and is used to evaluate the velocity field at the interface. The boundary element method has the advantage that it reduces the 3D computation problem into a 2D-evaluation, i.e., the method requires to solve the velocity field on the surface of the drop, and (b) with this information is possible to have the velocity field outside or inside the drop.

3.1 Numerical implementation of the computation of stresses and velocity fields

The numerical scheme solves the Stokes flow equation at an instant of time, equation (9 b). The evolution of the drop shape is carried out using the velocities obtained at the collocation points. By means of an interpolation subroutine the velocity at all nodes is calculated and then the new position of the drop is evaluated. For the evaluation of the Single Layer Potential, an approximation of the local curvature of the drop is used, using equation 3 with curved triangles. A quadratic interpolation throughout these triangles is performed to obtain the mean curvature of the elements of the drop. In this manner, the mesh of the interface and the algebraic system of equations of the boundary elements is smaller than those of other more conventional numerical schemes, such as e.g., finite differences.

3.2 Numerical Accuracy

The numerical simulation presented in this paper uses a surface of *2048* elements. To obtain the *time evolution* of the drop deformation, time is advanced using a fourth-order Adams-Bashforth-Moulton method [16]. The total time required to reach the stationary shape of the drop is $T = Gt$, reaching the final stationary deformation in all numerical experiments. These numerical results were compared with the data reported by Rosas [10] under the same flow regime.

4. Results

The numerical experiments explore effects of the applied flow on the drop shape. Figure 2 shows the evolution of the drop shape with the application of steady flow conditions. The drop evolves from the initial spherical shape to a final ellipsoidal form. The simulation is stopped when the drop deformation no longer changes; see figures 2 and 3. The deformation values attained numerically are those obtained experimentally from Rosas [10] within $\pm 5\%$ for $0.05 \leq Ca \leq 0.25$ and $\pm 1\%$ for $0.3 \leq Ca \leq 0.4$.

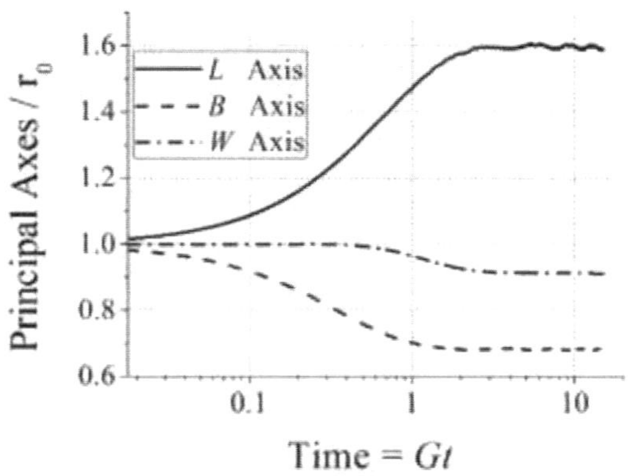

Figure 2. Evolution of the three principal axes of the drop in a flow with $\alpha = 0.13$, $Ca = 0.4$ and Total time $Gt = 15$.

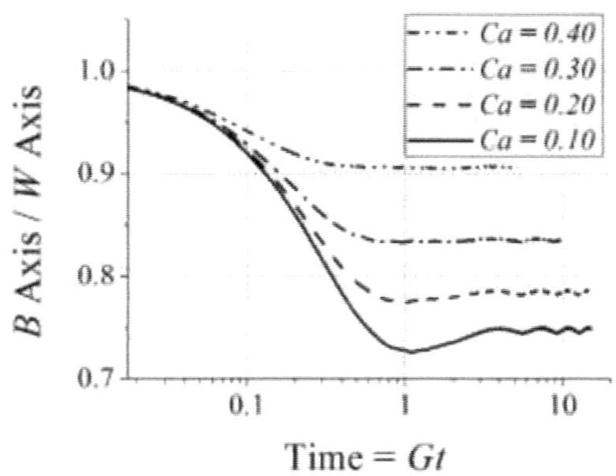

Figure 3. Ratio of the *B* wrt the *W* axes in numerical simulations for a flow with $\alpha = 0.13$, and Total time $Gt = 15$; 3 different values of *Ca*.

Figure 2 shows that the stationary state is reached as early as *10%* of the total simulation time; the drop remains in the stationary shape for the rest of the simulation. Although small oscillations in the stationary values persist, the amplitude of these oscillations is less than $\pm 1.5\%$ of the average long-time value.

As well, data for the *W*-axis indicates that the cross section of the drop is not circular, for *B* values are significantly smaller. To observe the non-symmetric form of the drop, the ratio of *B* vs. *W* is presented in figure 3. On one hand, it can be observed that, with $Ca = 0.4$, in the initial *10%* of the evolutions *B*/*W* decreases and undershoots its stationary value. On the other hand, the rates of change for *B* and *L* are very similar, and are mainly determined by the magnitude of the vorticity. However, the increment of *W* wrt *B* clearly occurs at a faster pace; actually, based upon figure 3, the rate appears to be about four times faster for $Ca = 0.4$. The time evolution of the *B* and *W*-axis are different, as seen by the rate of change for the cross section axes, i.e., the *B*-axis changes at a rate similar to that of as the *L*-axis, while the rate of rotation of the *W*-axis begins to change immediately at the onset of the flow. However, the *W*-axis lags in time. Up to the time $Gt = 1$, the *W*-axis does not change. Then the *W*-axis changes about the time when the other two axes have reached their stationary value.

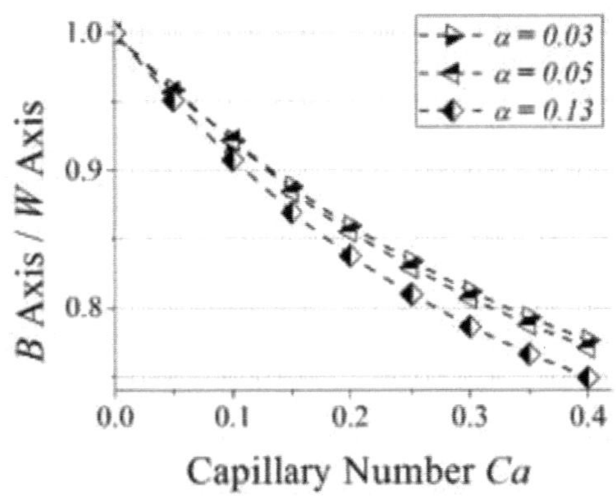

Figure 4. Stationary ratio of B wrt W axes in numerical simulations for a flow with $\alpha = 0.03$, $\alpha = 0.05$ and $\alpha = 0.13$ for different values of Ca.

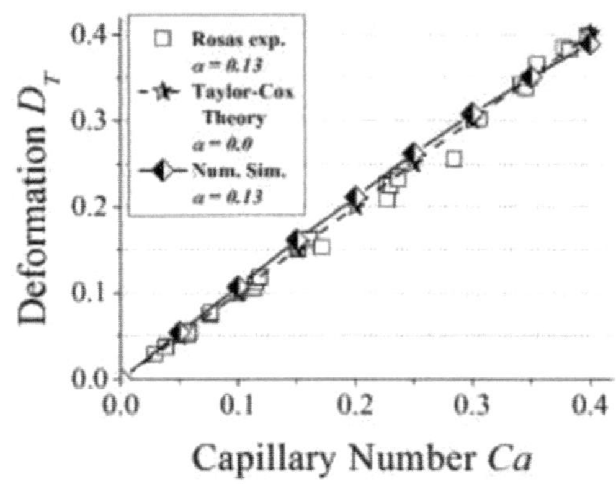

Figure 5. Stationary Taylor Deformation values for numerical simulations and experimental data for $\alpha = 0.13$ and analytical results of Taylor-Cox at $\alpha = 0.0$ for different values of Ca.

In figure 4 the ratio between axes of the cross section of the drop are given for three different flow strengths. As the capillary number increases, the eccentricity of the ellipsoid of the cross section increases as well. This ratio also depends on the flow-type parameter α becoming more eccentric for flows close to simple shear flow. For simple shear flows, these simulations produce values for the ratios of the cross section that are similar to the numerical results reported in References [7,8], having as well the same general behavior, even though the viscosity ratio reported is larger. Qualitatively it is possible to see the similarity [8]. Figure 5 presents a comparison of experimental results [10] and the simulated values for $\alpha = 0.13$, and the analytical results of Taylor-Cox for $\alpha = 0.0$ for the stationary Taylor Deformation value in the *xy*-plane; see figure 1(b).

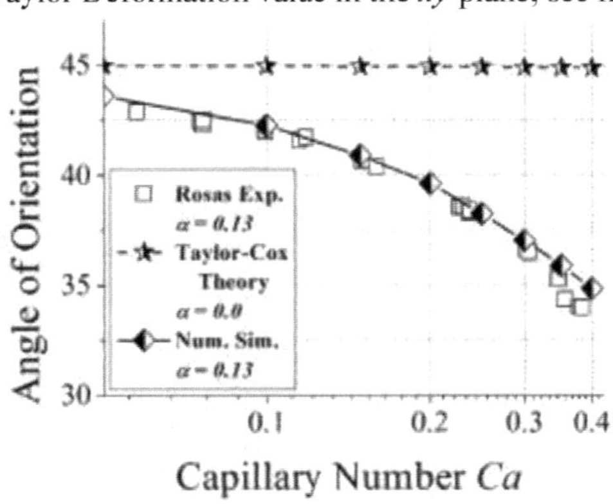

Figure 6. Orientation of the drop for numerical simulations and experimental data for $\alpha = 0.13$ and analytical results of Taylor-Cox $\alpha = 0.0$ for different values of Ca.

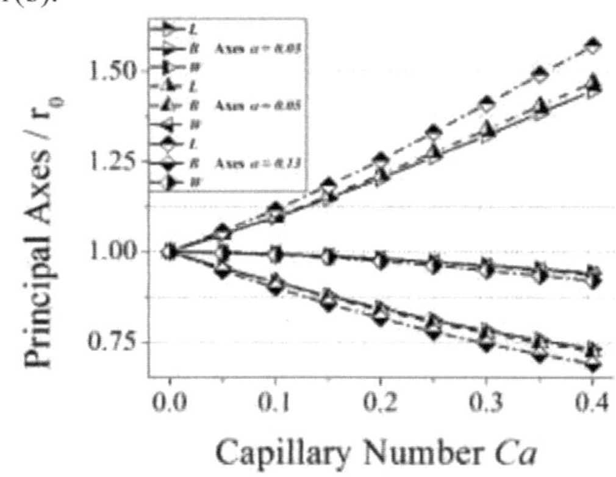

Figure 7. Stationary values of the principal axes of the drop for a flow with $\alpha = 0.03$, $\alpha = 0.05$ and $\alpha = 0.13$ for different values of Ca.

The stationary deformation of the drop depends upon the capillary number as well as the flow type. When $Ca = 0.4$, and for flows with $\alpha = 0.03$, the deformation achieved is less than 85% of the values reached for flows of $\alpha = 0.13$. Cox [5] and Taylor [1] theory for simple shear flows establishes a linear relationship between deformation *vs.* the capillary number. However, the dependence observed with simulations of stronger flows is no longer linear, the later being in agreement with the experiments of Rosas [10, 11].

Figure 6 shows the rotation of the principal axis of the drop; see figure 1 (b). In figure 6, the orientation is comparable with the experimental data for similar capillary numbers. The theory based in simple shear flow does not match with the experimental data. Finally, the measurements of the principal axes of the drop are shown in figure 7. These values were obtained in the stationary shape of the drop.

5. Conclusions

The simulations here presented for drop forms induced in 2D-flows do not appear to match the axisymmetric case established by G. I. Taylor and Cox, as neither did Kennedy et al., and Guido et al., for the simple shear flows. This is likely the case for capillary number less than *0.4*.

For $Ca > 0.4$, the deformation in the *xy*-plane increases, but, neither as numerical simulation show, there is a correlation with the Taylor-Cox model. On the other hand, the simulated behavior appears consistent with the experimental data of Rosas.

The difference between the simple shear flow, $\alpha = 0.0$, and flows with stronger degree of elongation induced larger drop deformations while the angle of orientation (principal axis) rotates away from *45* degrees.

The analysis of the *L*, *B* and *W*-axes of the ellipsoidal drop, in steady state, clearly show deviations from the axisymmetric shape as the flow type parameter α increases. However, the change of *W* remains unchanged for the values $\alpha = 0.03 - 0.13$, as a result of the vorticity of the applied 2D-flow.

6. Acknowledgments

The authors acknowledge funding provided by Project No. IG100714 of PAPIIT-UNAM Program. ASS gratefully acknowledges the financial support provided by CONACyT and PCeIM-UNAM during his PhD studies.

7. References

[1] Taylor G I 1932 *Proc. Roy. Soc.* A **138** 41
[2] Taylor G I 193. *Proc. Roy. Soc,* A **146** 501
[3] Rallison J M 1984 *Ann. Rev. Fluid Mech.* **16** 45
[4] Prosperetti A and Tryggvasson G 2009 *Computational Methods for Multiphase Flow* (Cambridge, New York, Melbourne: Cambridge University Press)
[5] Cox R G 1969 *J. Fluid Mech.* **37** 601
[6] Hinch E J and Acrivos A 1980 *J. Fluid Mech.* **98** 305
[7] Guido S and Villone M 1999 *J. Rheol.* **42**(2) 395
[8] Kennedy M R, Pozrikidis C and Skalak R 1994 *Computers Fluids.* **23**(2) 251
[9] Bentley B J and Leal L G 1986 *J. Fluid Mech.* **167** 241
[10] Rosas I Y, Reyes M A H, Minzoni A A and Geffroy E 2015 *Exp. Therm. Fluid Sci.* **60** 54
[11] Escalante C A, Reyes M A H, Rosas I Y and Geffroy E 2015 *J. Phys.: Conf. Ser.* **582** 012014
[12] Leal L G 2007 *Advanced Transport Phenomena* (Cambridge, New York: Cambridge University Press)
[13] Pozrikidis C 1997 *Introduction to Theoretical and Computational Fluid dynamics* (New York, Oxford: Oxford University Press)
[14] Reyes M A H, Minzoni A A and Geffroy E 2011 *J. Eng. Math.* **71**(2) 185
[15] Macosko C W 1994 *Rheology Principles, Measurements, and Applications* ed Wiley-VCH (Chichester New York) p21-23 68-74
[16] Lambert J D 1973 *Computational Methods in Ordinary Differential Equations* ed Wiley & Sons (Chichester New York)

Preparation of nanostructured Bi-modified TiO_2 thin films by crossed-beam laser ablation plasmas

L Escobar-Alarcon[1,*], D A Solís-Casados[2], F González-Zavala[2], S Romero[1], M Fernandez[3] and E Haro-Poniatowski[4]

[1] Departamento de Física, Instituto Nacional de Investigaciones Nucleares, Apdo. Postal 18-1027, México DF 11801, México.
[2] Centro Conjunto de Investigación en Química Sustentable UAEM-UNAM, Carretera Toluca-Atlacomulco Km 14.5, Toluca, Estado de México, 50200, México.
[3] Departamento de Aceleradores, Instituto Nacional de Investigaciones Nucleares, Apdo. Postal 18-1027, México DF 11801, México.
[4] Departamento de Física, Universidad Autónoma Metropolitana Iztapalapa, Apdo. Postal 55-534, México, D.F., México.

E-mail: luis.escobar@inin.gob.mx

Abstract. The preparation and characterization of titanium dioxide thin films modified with different amounts of bismuth using a two laser ablation plasmas configuration is reported. The plasmas were produced ablating simultaneously two different targets, one of bismuth and other of titanium dioxide, using a Nd:YAG laser with emission in the fundamental line. The elemental composition, together with the vibrational and optical properties of the deposited films were investigated as a function of the parameters of the bismuth plasma. The composition of the thin films was determined from measurements of X-ray photoelectron spectroscopy (XPS) as well as by Rutherford backscattering spectroscopy (RBS). The structural modification of the deposited material, due to the incorporation of Bi, was characterized by Raman spectroscopy. The optical properties were determined from UV-Vis spectroscopy measurements. It is found that bismuth incorporation has an important effect on the optical properties of TiO_2 narrowing the band gap from 3.2 to 2.5 eV.

1. Introduction

The treatment of water polluted with recalcitrant organic molecules, has become an issue of current interest in which different areas of knowledge have converged in the search for possible solutions. In particular, the development of new nanostructured materials, powders and thin films, to be used as photocatalysts in the degradation of organic contaminants has shown outstanding results [1].

It is worth mentioning that most of these photocatalysts are only active with ultraviolet light, which limits them to use only a fraction, about 5%, of the solar radiation. This has increased research in the development of active photocatalysts with wavelengths in the visible region of the electromagnetic spectrum, taking advantage up to 45% of the total solar radiation reaching the Earth's surface.

In recent years, titanium dioxide (TiO_2) thin films have been used as photocatalysts, for the degradation of different organic pollutants present in wastewaters, owing to its high stability in aqueous solution and its high oxidizing power [2]. However, one of the major disadvantages of TiO_2 is its relatively high band gap value (E_g = 3.2 eV for the anatase phase) which impedes its activation with wavelengths greater than 387 nm, this is a drawback if sunlight is used to activate its photocatalytic response. The modification of TiO_2 with transition metals, is a technique frequently used to reduce its

Content from this work may be used under the terms of the Creative Commons Attribution 3.0 licence. Any further distribution of this work must maintain attribution to the author(s) and the title of the work, journal citation and DOI.
Published under licence by IOP Publishing Ltd

band gap [3], allowing to use visible light for its activation and consequently take advantage of sunlight for this purpose.

In order to prepare TiO_2 thin films modified with metals, various deposition techniques, resulting in materials with promising photocatalytic properties, have been used. However, there is still much work to improve the characteristics of the deposits; specifically, the control of the composition is essential since the properties of the obtained materials depend strongly on the concentration of the modifying element.

This work reports on the use of a variant of the laser ablation technique, the so-called Crossed Beam Pulsed Laser Deposition (CBPLD) which has been successfully used to prepare thin films of TiAlN [5], allowing the incorporation of Al in the TiN lattice in a controlled way. It should be noted that the features of the employed configuration could result in nanomaterials with novel or improved properties.

2. Experimental Procedure

The thin films were deposited on glass substrates using the CBPLD technique [4], in which two different laser ablation plasmas are produced ablating simultaneously a TiO_2 and a Bi high purity (99.99%) targets. A Nd: YAG laser with emission in the fundamental line (1064 nm, 5 ns pulse duration) was used to generate the plasmas.

In order to obtain films with different bismuth content, the ion mean kinetic energy of the Bi plasma was varied from 56 to 195 eV, keeping constant the TiO_2 plasma parameters. The target to substrate distance and the deposition time were set at 5 cm and 90 min respectively. The mean Bi-ion kinetic energy was calculated using the time of flight technique (TOF) from measurements performed with a Langmuir planar probe (6 mm diameter). The films were deposited at room temperature under the same experimental conditions as the diagnosed plasma.

The physical properties of the thin films were studied as follows: composition was determined by Rutherford backscattering spectroscopy (RBS) and X-ray photoelectron spectroscopy (XPS). RBS measurements were carried out in a Tandem Van de Graff accelerator using a beam of ^7Li (4.0 MeV) and a detection angle of 165°, while XPS measurements were performed with a Jeol JS 9200 spectrometer, with an X-ray source (K□ of aluminum) acquiring spectra in the low and high resolution regimes. The microstructure of the films was studied by Raman spectroscopy, the spectra were obtained using a micro-Raman LabRam 800 system, equipped with a confocal microscope, Olympus BX40, using the second harmonic of a Nd: YAG laser (532 nm) focusing the laser beam with a 50X objective onto the sample surface. A cooled CCD camera was used to record the spectra, typically an average of 50 accumulations of 10 sec was done in order to improve the signal to noise ratio. All spectra were calibrated using the 521 cm^{-1} line of monocrystalline silicon. Optical properties were determined from UV-Vis measurements carried out on a Perkin Elmer lambda 35 spectrophotometer.

3. Results and Discussion

3.1 Plasmas characterization

The ion mean kinetic energy of the Bi ions was calculated using the time of flight technique (TOF) from measurements made with a Langmuir planar probe placed at the substrate position. Figure 1, show the TOF curves corresponding to plasmas produced with different laser fluences. It is seen that as the laser fluence decreases, the maximum of the curves shifts to higher times and its intensity decreases, indicating that the Bi^+ kinetic energy and the plasma density decrease.

The ion mean kinetic energy and plasma density calculations were performed assuming that the ions arriving to the Langmuir probe were predominantly Bi^+. The results indicate that Bi ions with kinetic energies from 56.5 to 195.2 eV are present. The plasma density was determined from the maximum values of current collected by the probe; it was found that it varies from 3.4 x 10^{11} to 4.3 x 10^{13} cm^{-3}. Figure 2 shows that these two parameters are related and follow an exponential tendency under the experimental conditions for the present experiments. Due to the fact that the plasma density and the kinetic energy of Bi^+ are correlated, each sample was deposited under the conditions defined by each pair of values, as shown in table 1.

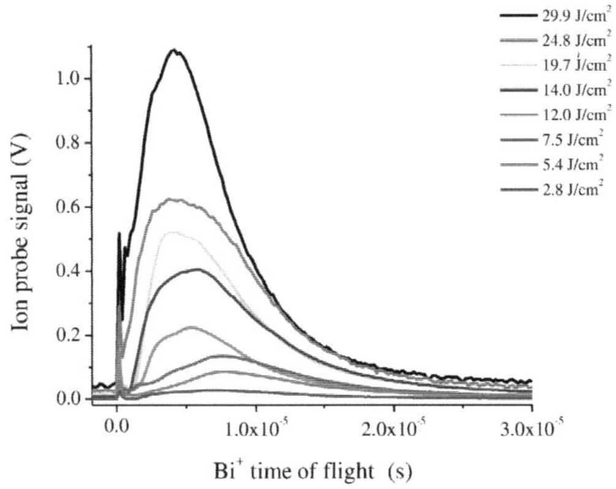

Figure 1. Bi^+ time of flight curves at different laser fluences

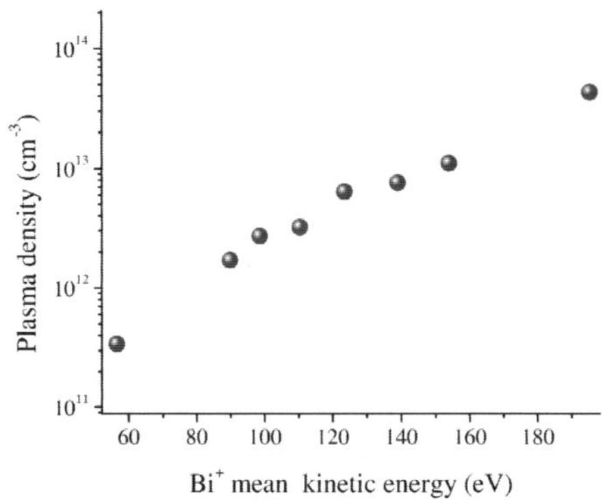

Figure 2. Plasma density as a function of the Bi^+ mean kinetic energy

Table 1. Plasma parameters employed for thin film deposition

Laser fluence (J/cm^2)	Bi^+ Kinetic energy (eV)	Plasma density (cm^{-3})
2.8	56.5	3.4×10^{11}
5.4	89.8	1.7×10^{12}
7.5	98.6	2.7×10^{12}
12.0	110.3	3.2×10^{12}
14.0	123.5	6.4×10^{12}
19.7	139.0	7.6×10^{12}
24.8	153.9	1.1×10^{13}
29.9	195.2	4.3×10^{13}

3.2 Composition

RBS results showed that the atomic concentration of bismuth increases from 12 to 28 at. % in a monotonous way followed by a decrease from 28 to 18 at. % as shown in figure 3. In order to confirm the observed tendency, some samples were measured by XPS and the results revealed a good agreement with the RBS data. This figure shows two regions in which Bi content is incorporated in a different way. For kinetic energies lower than 140 eV, Bi is incorporated in an approximately linear way, whereas kinetic energies greater than 140 eV produce a decrease in the Bi content. According to data of table 1, the higher is the kinetic energy, the higher the plasma density, implying more species of bismuth arriving to the substrate that can be incorporated into the film. This is observed for kinetic energies lower than 140 eV, for greater energies a small amount of bismuth is incorporated, possibly because of re-sputtering processes or due to a decrease of the Bi sticking coefficient with energy.

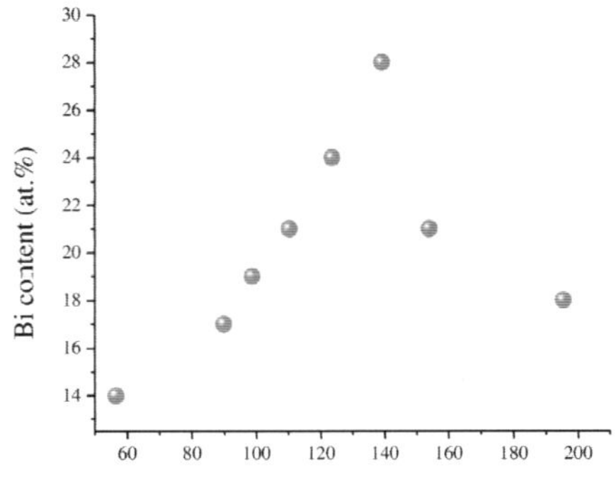

Figure 3. Bismuth content as a function of the Bi^+ mean kinetic energy

Figure 4 shows the Bi/Ti atomic ratio as a function of the Bi^+ kinetic energy. Is clearly observed that at low kinetic energies the Ti atoms predominate, indicating that titanium dioxide is present as the main phase. As the Bi^+ kinetic energy increases the Bi atoms begin to increase equalling the number of Ti atoms at an energy close to 140 eV, the bismuth oxide becoming the dominating phase as is shown in figure 4.

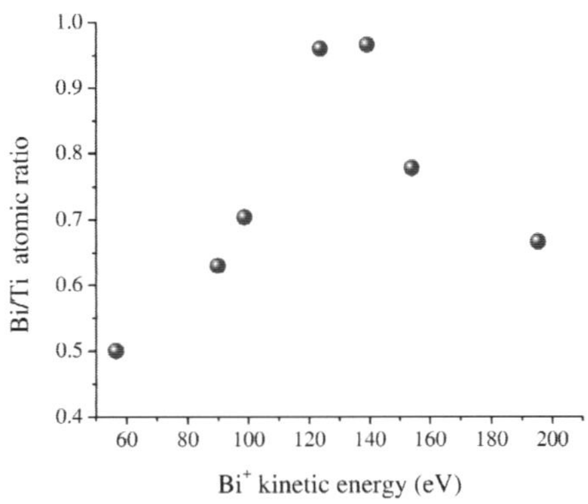

Figure 4. Bi/Ti atomic ratio as a function of the Bi^+ kinetic energy

3.3 Microstructure

The Raman Spectroscopy results are shown in figure 5a. The spectra of films with Bi contents up to 19 at.% are characterized by peaks at 144, 395, 515 and 638 cm^{-1} assigned to the anatase phase of TiO_2 [6]. It is observed that as the Bi content increases, the peaks become less intense and broader indicative of a decrease of the crystallinity degree attributed to distortions caused by the Bi incorporation into the TiO_2 lattice. In fact, for Bi contents of 21 and 24 at.% the Raman spectra do not show any features suggesting amorphization of the material induced by the disorder introduced by the presence of Bi atoms. At the highest Bi content, 28 at.%, the obtained films are formed by bismuth oxide (Bi_2O_3) coexisting the □ and □ crystalline phases, as is shown in figure 5b [7]. When the films are deposited using higher Bi^+ kinetic energies and plasma densities, the Bi contents diminish to values of 21 and 18 at.% and the Raman spectra show that, additionally to the peaks of Bi_2O_3, Raman signals at 440 and 610 cm^{-1} associated with the rutile phase of TiO_2 appear, suggesting that higher ionic bombardment favors the anatase-rutile phase transition.

IOP Conf. Series: Journal of Physics: Conf. Series **792** (2017) 012006 doi:10.1088/1742-6596/792/1/012006

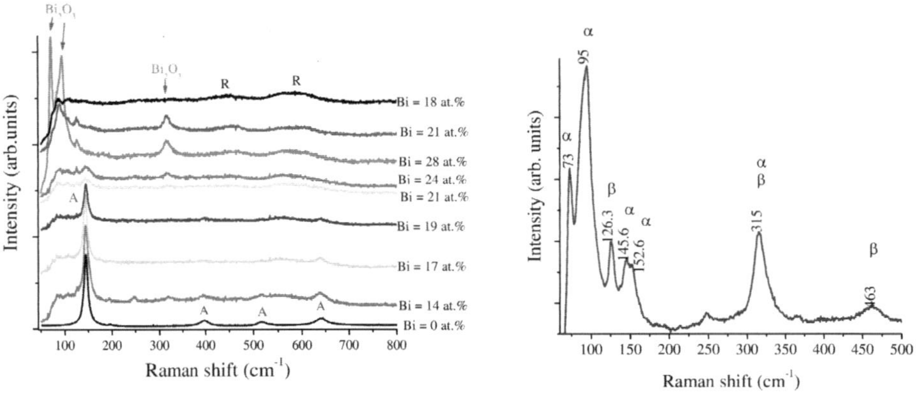

a) b)

Figure 5. a) Raman spectra corresponding to thin films with different content of bismuth. b) Raman spectrum of the film with a Bi content of 28 at.%

3.3 Optical properties

UV-Vis transmittance spectra of deposited films are presented in figure 6. An important effect due to the Bi incorporation is observed: with increased content of Bi, the absorption edge shifts to higher wavelengths, indicative of a decrease in the gap band. This was calculated using the Tauc method [7]; the obtained results are presented in table 2. It is observed that the incorporation of Bi effectively narrows the band gap value from 3.2 to 2.5 eV. This is an interesting result because it suggests that these materials could be activated using visible light, such as the sunlight, in photocatalytic processes to degrade recalcitrant organic compounds. Further studies regarding this effect are underway.

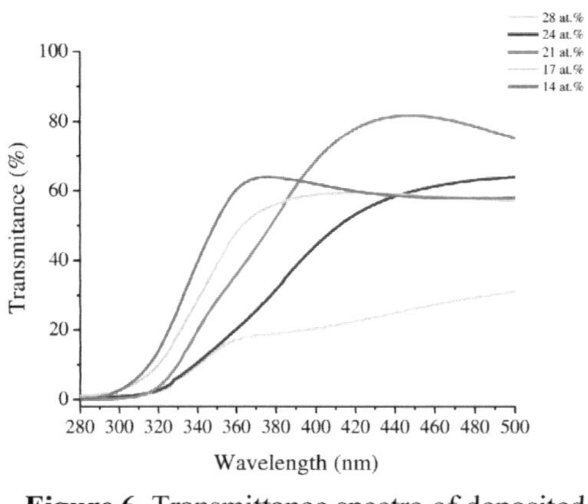

Figure 6. Transmittance spectra of deposited films with different content of bismuth

4. Conclusions

Bismuth was incorporated into the TiO_2 lattice, from 12 to 28 at. % with important effects on the microstructural and optical properties of the resulting material. Raman investigations reveal a phase transition from the anatase phase of TiO_2 to a mixture of the and phases of Bi_2O_3. The incorporation of bismuth has an important effect on the optical properties of TiO_2; narrowing the band gap from 3.2 to 2.5 eV. This is an important result because these materials can be activated with wavelengths in the visible region of the electromagnetic spectrum. This can be useful to carry out reactions of degradation of organic compounds in aqueous solutions using solar light.

Table 2. Band gap energy as a function of the Bi content

Bi content (at.%)	Band gap energy (eV)
0	3.2
14	2.8
17	2.9
19	2.7
21	3.1
24	2.7
28	2.4
21	2.6
18	2.5

5. Acknowledgements
This research project was partially supported by the CONACYT project CB-240998.

6. References
[1] Obare S O and Meyer G J 2004 *J. Environ. Sci. Heal.* A **39** 2549
[2] Fujishima A, Zhang X and Tryk D A 2008 *Surf. Sci. Rep.* **63** 515
[3] Rehman S, Ullah R, Butt A M and Gohar N D 2009 *J. Hazard. Mater.* **170** 560
[4] Escobar-Alarcón L, Camps E, Romero S, Muhl S, Camps I and Haro-Poniatowski E 2010 *Appl. Phys.* A **101** 771
[5] Escobar-Alarcón L, Solís-Casados D A, Romero S, Fernández M, Pérez-Alvarez J and Haro-Poniatowski E 2013 *Appl. Surf. Sci.* **283** 808
[6] Haro-Poniatowski E, Rodríguez Talavera R, de la Cruz Heredia M, Cano-Corona O and Arroyo-Murillo R 1994 *J. Mater. Res.* **9** 2102
[7] Tauc J, Grigorovici R and Vancu A 1966 *Phys. Stat. Sol.* **15** 627

Thermoluminescent response of C-modified Al_2O_3 thin films deposited by parallel laser ablation plasmas

J Garcés[1,2], L Escobar-Alarcón[1], P R Gonzalez-Martinez[1], D A Solís-Casados[3], S Romero[1], F Gonzalez-ZAvala[1] and E Haro-Poniatowski[4]

[1] Departamento de Física, Instituto Nacional de Investigaciones Nucleares, Apdo. Postal 18-1027, México DF 11801, México.

[2] Facultad de Medicina, Universidad Autónoma del Estado de México, México.

[3] Centro Conjunto de Investigación en Química Sustentable UAEM-UNAM, Toluca, Estado de México, 50200, México.

[4] Departamento de Física, Universidad Autónoma Metropolitana-Iztapalapa, Apdo. Postal 18-1027, México DF 11801, México.

E-mail: luis.escobar@inin.gob.mx

Abstract. Aluminium oxide thin films modified with different amounts of carbon were prepared using a parallel laser ablation plasmas configuration. The effect of the amount of carbon incorporated in the films on their compositional, morphological, structural, and thermoluminescent properties was studied. The results showed that films with different C content, from 11 to 33 at. %, were obtained. The structural characterization revealed the growth of an amorphous material. Surface morphology of the obtained thin films showed smooth surfaces. The films were exposed to UV and gamma radiation (Co-60) in order to study their thermoluminescence response. The results tend to indicate that carbon incorporation into the alumina favours the increase of a high temperature TL peak.

1. Introduction

Thermoluminescence (TL) properties of thin films have a wide spectrum of potential applications in dosimetry for both ionizing and nonionizing radiation. The interest to study the TL response of thin films has been motivated because of their importance in the measurements of absorbed doses from weakly penetrating radiation as well as in the study of dose distributions in interfaces [1]. Particularly, Al_2O_3 is an important TL material studied for its application as a radiation dosimeter due to its high thermal and chemical stability and low effective atomic number [2]. It has been reported elsewhere [3] that the improvement of the TL response of Al_2O_3 can be achieved by the intentional introduction of defects such as oxygen vacancies and impurity centers formed by carbon. In particular, doping of α-Al_2O_3 with carbon improves its TL response resulting in a material with good dosimetric properties with potential applications in various areas of dosimetry, such as personnel, neutron, space, medical, environmental, and emergency dosimetry [4]. Although C doped Al_2O_3 is being used and investigated extensively, the role of carbon in rendering this phosphor remarkably sensitive towards radiation is not well understood [5]. Therefore, a lot of work is still necessary to investigate this issue.

Content from this work may be used under the terms of the Creative Commons Attribution 3.0 licence. Any further distribution of this work must maintain attribution to the author(s) and the title of the work, journal citation and DOI.
Published under licence by IOP Publishing Ltd

Among the thin film deposition techniques, laser ablation has been employed to prepare thin films of carbon nitride [6], as well as aluminum oxide with thermoluminescent response when subjected to UV, gamma and beta radiation [7, 8, 1]. In particular, the versatility of the laser ablation technique allows the possibility of implementing different alternative configurations in order to improve some characteristics of the deposits as has been reported elsewhere [9, 10]. In the present work, the preparation of aluminium oxide thin films modified with different amounts of carbon, with TL properties, using a parallel laser ablation plasmas deposition configuration is reported.

2. Experimental Procedure

Thin films were deposited by the combination of two laser ablation plasmas produced sequentially. The plasma plumes were generated by laser ablation using the third harmonic of a Nd:YAG laser with emission at 355 nm and 10 ns pulse duration at a frequency of 10 Hz. The laser beam was focused onto the targets with a 50-cm focal length spherical lens. The targets were disks of high purity (99.99%) of aluminum oxide (Al_2O_3) and carbon (50 mm diameter and 2 mm thick). The laser fluence delivered on both targets was kept constant at 8.2 J/cm^2. By varying the ratio of the ablated areas of carbon and alumina (AA_C/AA_A), thin films modified with different amounts of carbon were deposited. The target to substrate distance was set at 5 cm and the deposition time was 60 min. Thin films were deposited at room temperature onto glass and silicon substrates (1x1 inch). Determination of the elements presents in the deposits as well as their bonding features were done by X-ray Photoelectron Spectroscopy (XPS) using a Jeol JPS 9200 XPS. The structural properties of the synthesized films were studied by Raman spectroscopy using an HR LabRam 800 system equipped with an Olympus BX40 confocal microscope and a Nd:YAG laser (532 nm). The surface morphology of the films was analyzed using a Phillips XL30 scanning electron microscope. The TL glow curves were obtained using a Harshaw 4000 TL reader using a cycle composed by a preheat at 50 °C for 5 s followed by an acquisition from 50 to 300 °C at a heating rate of 10 °C s^{-1}. The TL measurements were performed under a N_2 atmosphere. The readout cycle was repeated after each evaluation in order to guarantee that no remaining TL signal was still present on the sample. For the UV irradiation a low pressure Hg UV lamp with main emission at 254 nm and an energy density of approximately 1350 μW/cm^2 was used. Gamma irradiations were performed using a Co-60 source (E = 1.17 and 1.33 MeV) at an absorbed dose of 10 Gy. All the irradiations and TL measurements were performed at room temperature and the irradiated samples were kept in the dark in order to avoid any influence of the environmental light.

3. Results and Discussion

3.1 Compositional characterization

The XPS results showed that the elements present in the film were aluminium, oxygen and carbon. Figure 1a shows the carbon content as a function of the ratio of the carbon to alumina ablated areas (AA_C/AA_A). It is observed that the carbon content in the film follows a monotonic behaviour, increasing its content as the AA_C/AA_A increases. In this way it was possible to vary the carbon content from 11 to 33 at.%.

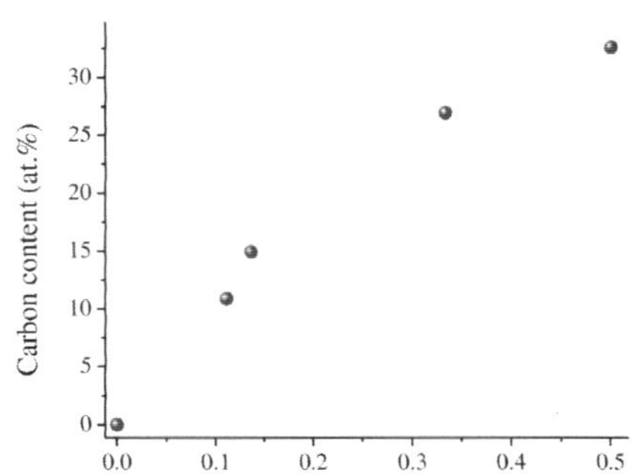

Figure 1. Carbon content as a function of the ratio of the carbon to alumina ablated areas

Figure 2 shows the high-resolution XPS spectrum of the O1s region of the samples prepared with different carbon content. The peak at 531.5 eV observed in the spectrum of the sample without carbon is assigned to O1s in Al_2O_3. As the carbon is incorporated into the film the position of this peak is shifted to 532 eV revealing that oxygen is bonded to carbon atoms forming O=C bonds.

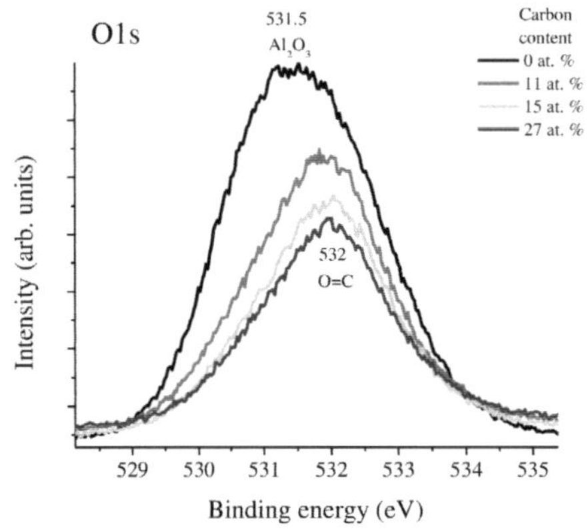

Figure 2. High-resolution XPS spectrum of the O1s region

3.2 Raman characterization

In figure 3 the Raman spectrum corresponding to an aluminium oxide thin film with a carbon content of 33.0 at.% is presented. The peaks at 300, 521 and 900 cm^{-1} correspond to the silicon substrate. The broad band from 1000 to 1800 cm^{-1} is due to carbon. The absence of peaks characteristics of alumina indicates that the material is amorphous.

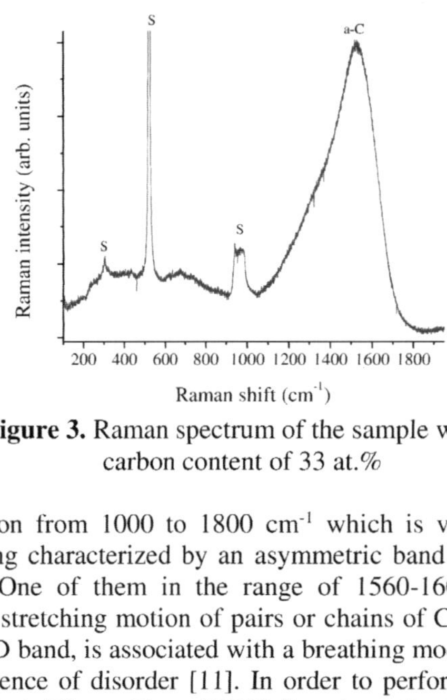

Figure 3. Raman spectrum of the sample with a carbon content of 33 at.%

Figure 4 shows the region from 1000 to 1800 cm^{-1} which is very similar to those previously reported for DLC films being characterized by an asymmetric band composed of two sub-bands in different proportions [11]. One of them in the range of 1560-1600 cm^{-1}, the so-called G band associated with the in-plane stretching motion of pairs or chains of C sp2 bonded atoms. The second band around 1350 cm^{-1}, the D band, is associated with a breathing mode of six fold aromatic rings and only becomes active in presence of disorder [11]. In order to perform the data analysis, the Raman spectra were fitted using a Breit-Wigner-Fano (BWF) line shape for the G peak and a Lorentzian line shape for the D peak. The interpretation of the Raman results was performed following the model proposed by Ferrari [12], which give information about the content of sp3 bonded carbon (diamond-like degree). The obtained results were: G position = 1560 cm^{-1} and $I_D/I_G = 1.1$ which according to the Ferrari´s model correspond to a C-sp3 content of approximately 30 %.

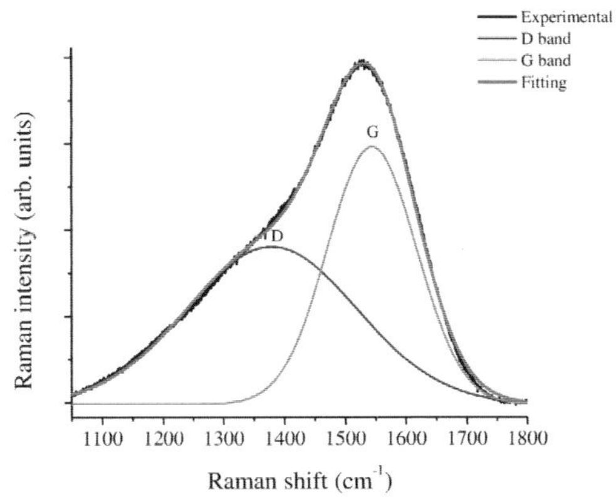

Figure 4. The 1000 to 1800 cm^{-1} Raman spectrum region showing the fitting performed

3.3 Surface morphology

The SEM images of films without carbon and with the maximum content of carbon are shown in figures 5a and 5b respectively. It is clearly seen that films with very smooth surfaces are obtained suggesting no effects on the surface morphology due to the carbon incorporation.

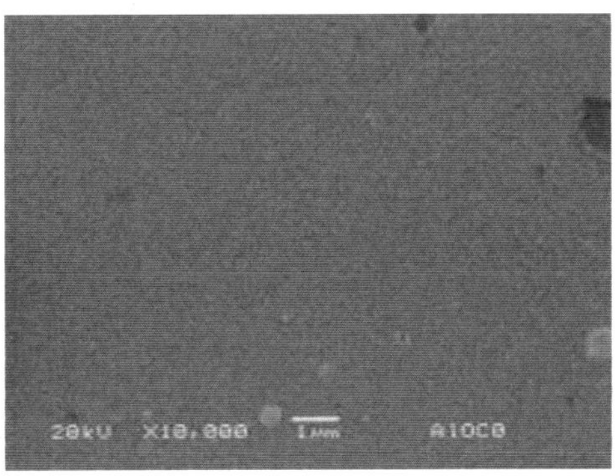

a)

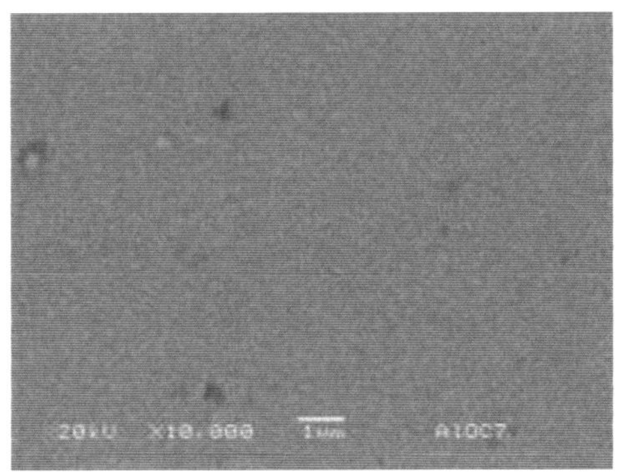

b)

Figure 5. SEM images of films, a) without carbon and b) with a carbon content of 33 at.%

3.4 TL response

Two series of experiments were performed, one to investigate the thermoluminescence response of the C-modified aluminium oxide thin films subjected to UV radiation and the other to investigate the TL response due to gamma radiation. All the irradiations and TL measurements were performed at room temperature and the irradiated samples were kept in the dark in order to avoid any influence of the environmental light. It is worth noting that TL measurements of non-irradiated samples revealed that the films have an intrinsic TL response attributed to defects induced during the deposition process.

Figure 6a shows the characteristic TL glow curves of UV irradiated films with different carbon content. The curves exhibit very broad peaks centered approximately at 156 °C for the sample without carbon and 183 °C for the C-modified alumina films. The differences observed in the peak positions can be attributed to the presence of carbon and the shift is indicative of the presence of impurities or to a slightly different composition of the materials used in each case. The typical TL response (considered as the integrated signal of the TL band) of the UV irradiated films as a function of the irradiation time is shown in Figure 6b. From this figure two main points can be emphasised: a) clearly, the film without carbon exhibits the higher TL response and for carbon contents greater than 15 at.% the TL response remains almost constant; b) it is evident that the prepared material present different behaviours of the TL intensity with dose (irradiation time) depending on the carbon content, the sample without carbon and with 11 at.% show and increase for irradiation times up to 30 min approaching saturation at higher doses. This is a sublinear-like dose dependence effect in which the TL signal starts increasing at a high rate for low doses and then increase gradually slower until it reaches a final value. The films with carbon contents higher than 15 at.% show a monotonic behavior increasing the TL signal as the irradiation time increases.

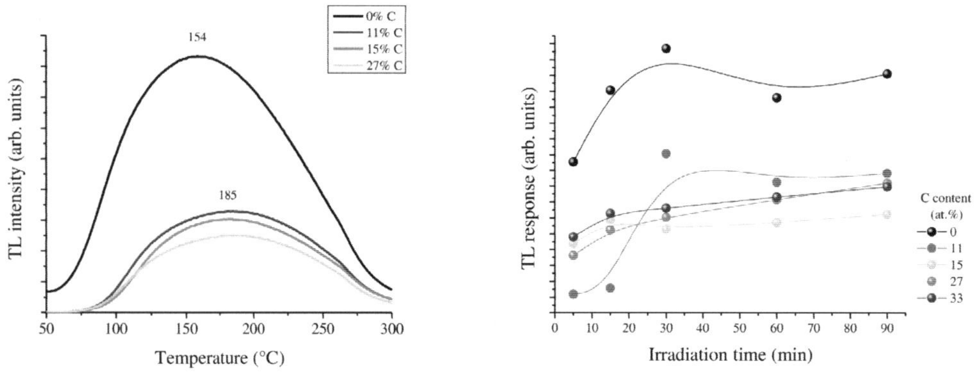

a) b)

Figure 6. a) Typical TL glow curves of UV irradiated films, b) TL response as function of the UV irradiation time

In Figure 7a the glow curves of the films irradiated at a Co-60 dose of 10 Gy are shown. Two clear bands peaking at 170 and 280 °C are observed. These glow curves have very different shapes compared with the curves obtained under UV irradiation, suggesting that different kind of traps are activated depending on the kind of radiation used. It is worth mentioning that deep traps are more useful for dosimetry purposes due to their higher stability.

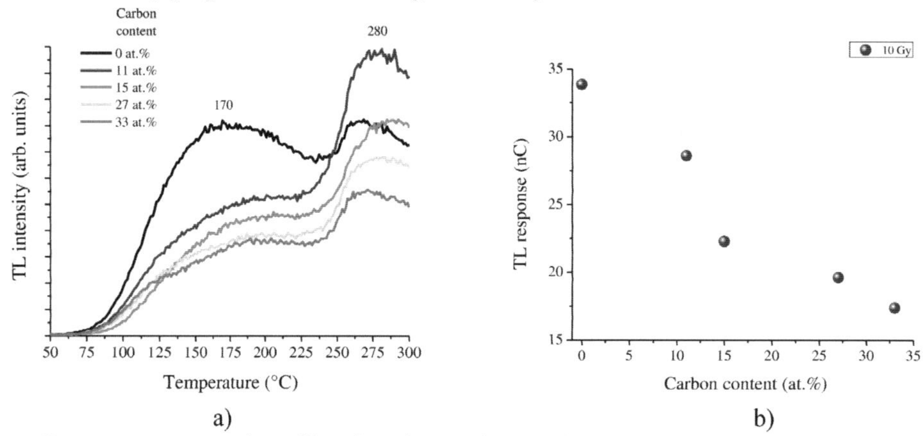

a) b)

Figure 7. a) glow curves of the films irradiated with Co-60 gammas, b) TL response as a function of the carbon content of samples irradiated at 10 Gy of gamma radiation

Additionally, it is clearly seen that carbon content favors the increase of the high temperature TL peak. In this case the material shows a linear relationship between carbon content and TL response for the dose studied, as it is observed in Figure 7b. It is worth noting that for samples containing carbon the TL response decreases as the C content increases.

4. Conclusions

The results presented in this work show that it is possible to obtain C-modified Al_2O_3 thin films which exhibit TL response to UV and gamma radiation. The deposition configuration employed in which the combination of two laser ablation sequentially produced plasmas allows to control the incorporation of carbon in a simple setup. In spite of the fact that unmodified alumina films exhibit a higher TL response, the obtained results seem to indicate that carbon incorporation into the alumina favours the increase of a TL peak of high temperature indicative of the presence of deep traps more useful for dosimetric purposes due to their better stability. A more detailed characterization of the thin films as well as the effects on the thin film properties depending on the deposition conditions in order to optimize the TL response is under investigation.

5. Acknowledgements
This research project was partially supported by CONACYT under the contract CB-240998

6. References
[1] Villarreal J E, Escobar L, González P R, Camps E and Barboza-Flores 2002 *M; Rad. Meas.* **35** 355
[2] Rieke J K and Daniels F J 1957 *J. Phys. Chem.* **61** 629
[3] Akselrod M S, Kortov V S, Kravetsky D J and Gotlib V I 1990 *Radiat. Prot. Dosim.* **32** 15
[4] Yukihara E G and McKeever S W S 2011 *Optically Stimulated Luminescence: Fundamentals and Applications* ed John Wiley & Sons (New York)
[5] Yang X, Li H, Bi Q, Cheng Y, Tang Q and Xu J 2008 *J. Appl. Phys.* **104** 123112
[6] Escobar-Alarcón L, Villarreal Barajas J E, Camps E, Muhl S, Haro-Poniatowski E, Romero S and Salinas B 2001 *Mater. Sci. and Eng.* B **90** 79
[7] Escobar-Alarcón L, Villagrán E, Camps E, Romero S, Villarreal-Barajas J E and González PR 2003 *Thin Solid Films* **433** 126
[8] Arrieta A, Escobar-Alarcón L, Camps E, González P R 2007 *Radiat. Eff. Defect. S.* **162** 745
[9] Escobar-Alarcón L, Solis-Casados D A, Romero S, Fernandez M, Perez-Alvarez J, and Haro-Poniatowski E, 2013 Appl. Surf. Sci. **283** 808
[10] Escobar-Alarcón L, Solís-Casados D A, Romero S, Morales-Mendez J G and Haro-Poniatowski E; 2014 *Appl. Phys.* A. **117** 31
[11] Ferrari A C, Rodil S E and Robertson J 2003 *Phys. Rev.* B **67** 155306
[12] Ferrari A C and Robertson J 2000 *Phys. Rev.* B **61** 14095

Transmittance spectra in one-dimensional dielectric photonic crystals with defects

H Vinck-Posada[1] **and F A Segovia-Chaves**[1, 2]
[1] Departamento de Física, Universidad Nacional de Colombia Sede Bogotá, Bogotá D.C, Colombia
[2] Facultad de Ciencias Exactas y Naturales, Universidad Surcolombiana, Neiva, Colombia

E-mail: francis.segoviac@gmail.com

Abstract. We have calculated the dependence of transmittance spectra with wavelength for TE wave at normal incidence in one-dimensional dielectric photonic crystals (1DPC) using the transfer matrix method. We demonstrate that when consider two defects layers spatial breaking periodicity in the 1DPC, the presence of defect modes located within the photonic band gap can be found. Additionally we have calculated the effect on the transmittance spectrum by varying the thickness of the defects in the 1DPC.

1. Introduction

In 1987 two works were published independently, which marked the birth of what today is known as photonic crystals (PC) [1, 2]. Such crystals are characterized by having a periodic modulation in space of its dielectric constant and are essentially act as light semiconductors, in them the light always find some direction from which it can propagate. The light waves that are allowed to propagate in the crystal are known as modes, modes groups form the bands. An element of interest for the design and subsequent use of the PC is to determine the regions of existence of forbidden photonic bands (PBG), in which no mode with energy found in this region will propagate [3]. Among the numerical methods allowing to obtain information from the band structure, the plane wave expansion method, the finite difference method in the time domain and of the dispersion matrix method [4, 5]. The existence of PBG in PC gives rise to different optical phenomena as the inhibition of spontaneous emission, waveguides with low losses and Fabry Perot resonators.

The introduction of defects that broken the spatial periodicity of the PC, may be point, linear or planar, and produce electromagnetic modes locating or the radiation guidance within the PBG. The defects are introduced in the PC by changing the thickness of the layers, inserting another dielectric structure, or removing a layer of PC. It is in this field where PCs are attractive, planar microcavities using with such systems with high quality factors [6]. Photonic crystals can be formally described with Maxwell's electromagnetic theory, which makes them systems of scalable features, the electromagnetic properties of a microscopic crystal whose work is in the visible spectral range are maintained if the crystal is scaled to macroscopic size in the range of microwaves. The purpose of this work is to investigate the transmittance spectrum using the method of the transfer matrix when two defects are introduced in the one-dimensional dielectric photonic crystals (1DPC).

Content from this work may be used under the terms of the Creative Commons Attribution 3.0 licence. Any further distribution of this work must maintain attribution to the author(s) and the title of the work, journal citation and DOI.
Published under licence by IOP Publishing Ltd

2. Transfer matrix method

Consider first a light beam incident with an angle θ from the left of a 1DPC, z direction, as shown in figure 1. In 1DPC, N is the number of periods of the bilayers, H and L are layers of refractive index, n_H high and n_L low, respectively. The medium surrounding the dielectric 1DPC heterostructure is air.

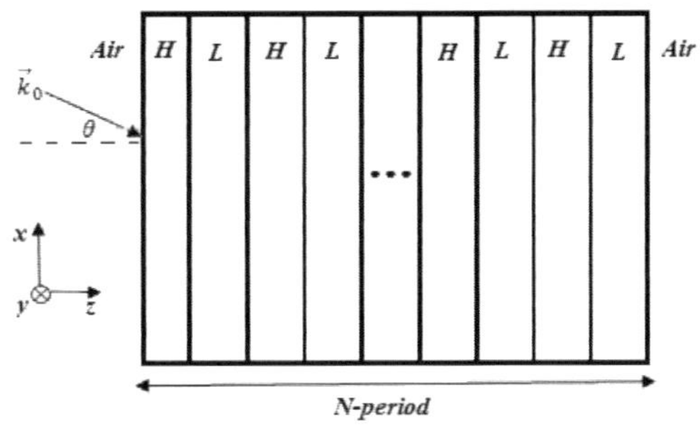

Figure 1. Schematic diagram a one-dimensional photonic crystal

The electric field for TE mode satisfies the Helmholtz equation:

$$\partial_z^2 E(z) + \frac{w^2}{c^2}\epsilon(z)E(z) = 0, \tag{1}$$

where $\epsilon(z)$ is the dielectric constant related to the refractive index $\epsilon(z) = n^2(z)$. The solution of equation (1) for the *j-th* layer is given by

$$\vec{E}_j(x,z) = (A_j e^{ik_{j,z}z} + B_j e^{-ik_{j,z}z})e^{-ik_x x}\hat{e}_y, \tag{2}$$

where the TE mode is defined as the component of electric field parallel to interface 1DPC, $k_{j,z} = \sqrt{(w/c)^2\epsilon_j - k_x^2}$ y ϵ_j, are the z component of the wave vector and the dielectric constant in the *j-th* layer, respectively. The transverse component of the wave vector is $k_x = k_0 sin\theta$, with k_0 the wave vector in the incident medium.

The A_j and B_j parameters are calculated using the continuity conditions of the tangential components of electromagnetic fields at each interface of 1DPC. However, requires solving a substantial number of algebraic equations, therefore it is necessary to use the transfer matrix method (TMM) [7]. According to the TMM, each layer may be represented by a matrix.

$$M_j = D_j P_j D_j^{-1}\,; j = H, L. \tag{3}$$

In equation (3), propagation matrix is defined as

$$P_j = \begin{pmatrix} e^{i\varphi_j} & 0 \\ 0 & e^{-i\varphi_j} \end{pmatrix}, \tag{4}$$

where the φ_j phase is

$$\varphi_j = k_{j,z} l_j = \frac{2\pi d_j}{\lambda} n_j cos\theta_j. \tag{5}$$

In equation (5), d_j, n_j and θ_j are the thickness, the refractive index and the angle at the *j-th* layer, respectively. The dynamical matrix is given by

$$D_j^{TE} = \begin{pmatrix} 1 & 1 \\ n_j cos\theta_j & -n_j cos\theta_j \end{pmatrix}, \tag{6}$$

for *TE* waves, and

For the structure, $Air/(HL)^N/Air$, the total transfer matrix is

$$M = \begin{pmatrix} M_{11} & M_{12} \\ M_{21} & M_{22} \end{pmatrix} = D_0^{-1}[M_H M_L]^N D_0, \quad (7)$$

where D_0 is the dynamical matrix for air.

The reflectance R and transmittance T in the case of normal incidence on the structure is given by

$$R = \left|\frac{M_{21}}{M_{11}}\right|^2 \text{ and } T = \left|\frac{1}{M_{11}}\right|^2, \quad (8)$$

where M_{11} and M_{21} represent the matrix elements of the transfer matrix, equation (7).

3. Results and discussion

In the following calculations, we consider a 1DPC structure with a PBG in the visible region. Here, H is taken as TiO_2 with a refractive index $n_{TiO_2} = 2.2$ y SiO_2 with $n_{SiO_2} = 1.45$ for layer L, the period is $N = 6$, [8]. In addition, both layers are a quarter wavelength, this is,

$$n_{TiO_2} d_{TiO_2} = n_{SiO_2} d_{SiO_2} = \frac{\lambda_0}{4}, \quad (9)$$

where λ_0 is the designed wavelength which is assumed to be $500\ nm$. Our study is restricted to the case of normal incidence to TE waves, where the left band edge λ_L and right band edge λ_R can be calculated by the theory of Bragg reflector, using the following equations [9],

$$\lambda_L = \frac{\pi(n_{TiO_2} d_{TiO_2} + n_{SiO_2} d_{SiO_2})}{cos^{-1}(-\rho)}, \quad \lambda_R = \frac{\pi(n_{TiO_2} d_{TiO_2} + n_{SiO_2} d_{SiO_2})}{cos^{-1}(\rho)}, \quad (10)$$

where $\rho = (n_{TiO_2} - n_{SiO_2})/(n_{TiO_2} + n_{SiO_2})$ is the coefficient of Fresnel. Based on equation (10), together with parameters of the material, we find that for a 1DPC, $Air/(HL)^N/Air$, the edges of the band are $\lambda_L = 441.79\ nm$ and $\lambda_R = 575.87\ nm$, as shown on figure 2.

If we introduce a defect located in the center of the 1DPC, the total transfer matrix for the structure $Air/(HL)^N/D/(HL)^N/Air$, is

$$M = D_0^{-1}[M_H M_L]^N M_D [M_H M_L]^N D_0, \quad (11)$$

in our calculations we consider that the defect D is the layer of low refractive index L, SiO_2. The transmittance of this structure is shown on figure 3, a transmission peak is observed with a transmittance of 87.2% located in λ_0 within the *PBG* called defect mode, [8].

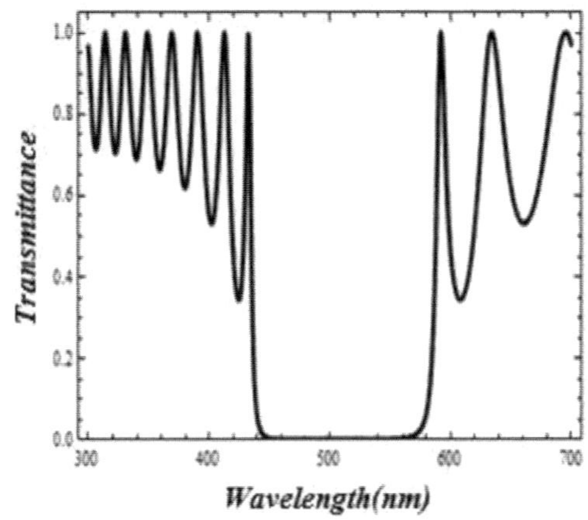

Figure 2. Transmittance spectrum for 1DPC, $Air/(HL)^6/Air$

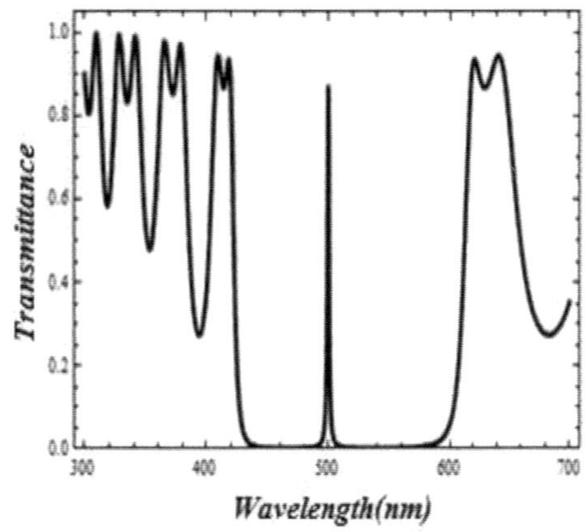

Figure 3. Transmittance spectrum for 1DPC with a defect of SiO_2, $Air/(HL)^6/D/(HL)^6/Air$

In the following calculations, we will investigate the introduction of two defects of SiO_2 to 1DPC in the transmittance spectrum, as shown in figure 4.

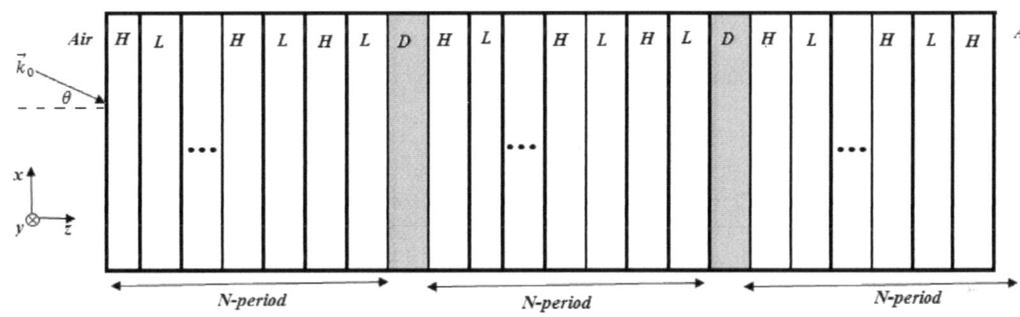

Figure 4. Diagram of a 1DPC with two defects, $Air/(HL)^N/D/(HL)^N/D/(HL)^N/Air$

It can be seen from figure 5, two transmission peaks located around λ_0 within the PBG with a transmittance of 87.4% and 85.1%, at wavelengths of 503 nm and 507.5 nm, respectively. The TE mode with a wavelength value mentioned above, will propagate through the 1DPC structure and will not be fully reflected in the PBG. The appearance of these peaks is used to function as narrowband transmission filters multilayers which are also known as Fabry Perot multilayers resonators.

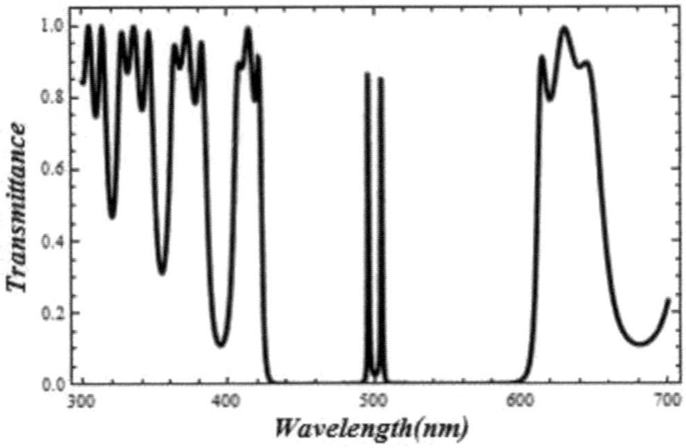

Figure 5. Transmittance spectrum for 1DPC, $Air/(HL)^6/D/(HL)^6/D/(HL)^6/Air$

Let us final examine the effect of thickness on the defect modes in 1DPC, $(HL)^N/mD/(HL)^N/mD/(HL)^N$, where m representing the repeated number of defects with the condition $m < 6$. For an odd number of m, $m = 3$ see figure 6 and $m = 5$ see figure 7, the two defect modes are fixed around the λ_0 design wavelength. However, for m even, pronounced transmission peaks that are located within the PBG have transmittance values of 97.8% at wavelengths of 449.3 nm and 562.4 nm, as shown in figure 8. By increasing m, the separation of defect modes decreases, figure 9 shows the presence within the PBG of four defect modes with a transmittance of 95% for wavelengths of 444.5 nm, 457.4 nm, 551.9 nm and 574.4 nm.

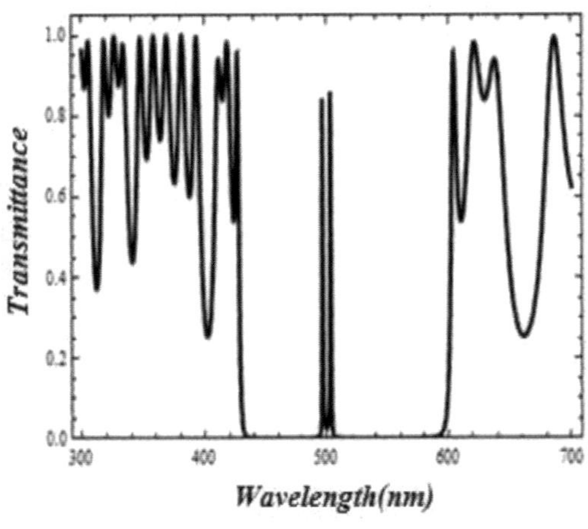

Figure 6. Transmittance spectrum for 1DPC, $Air/(HL)^6/mL/(HL)^6/mL/(HL)^6/Air$, with *m=3*

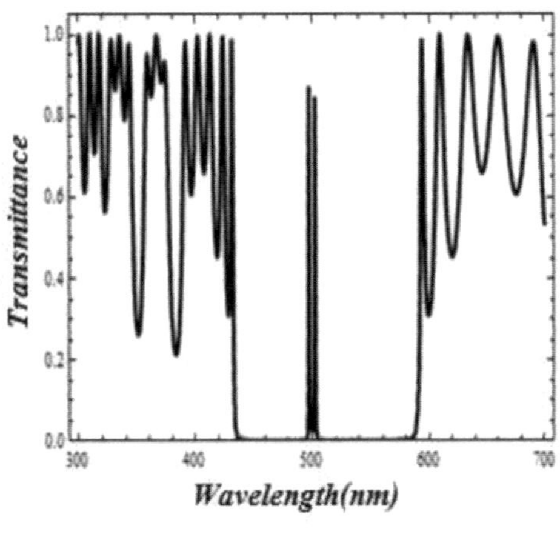

Figure 7. Transmittance spectrum for 1DPC, $Air/(HL)^6/mL/(HL)^6/mL/(HL)^6/Air$, with *m=5*

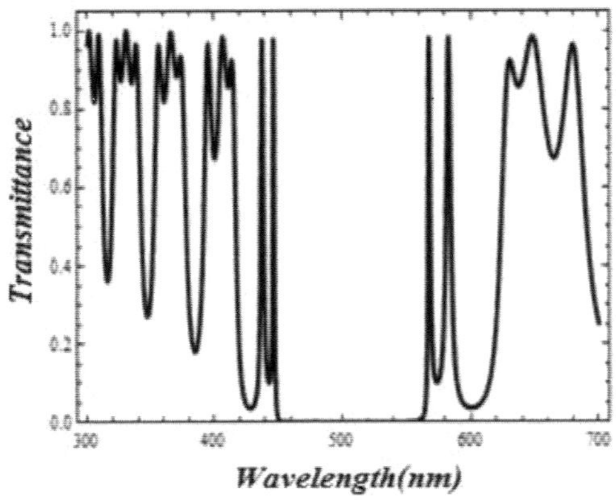

Figure 8. Transmittance spectrum for 1DPC, $Air/(HL)^6/mL/(HL)^6/mL/(HL)^6/Air$, with $m = 2$

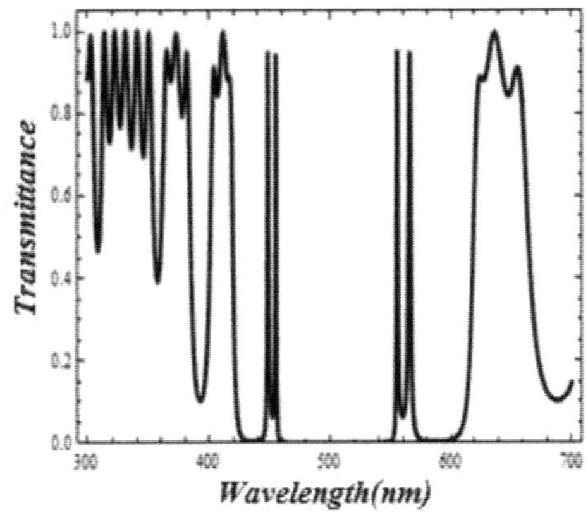

Figure 9. Transmittance spectrum for 1DPC, $Air/(HL)^6/mL/(HL)^6/mL/(HL)^6/Air$, with $m = 4$

4. Conclusions

The unusual property of photonic crystals to allow propagation of light for certain ranges of wavelengths, is demonstrated using the transfer matrix method when calculating the transmission spectrum for normal incidence TE waves in 1DPC. Introducing defects in 1DPC, causes the appearance of transmittance peaks which are located in the PBG, called defect modes. Unlike a 1DPC structure with a defect, where only a defect mode appears located at the design wavelength λ_0. Introducing two defects in the 1DPC causes two transmittance peaks around λ_0. The transmittance spectrum depends on the number of times that the defects are repeated. For odd values of the thickness, the defect modes maintain their position around λ_0. While for even of values of thicknesses, one can find the presence of additional defect modes in the PBG, whose separation decreases as the thickness of such modes increases.

5. Acknowledgments

This work has been supported by Colciencias within the project with code 110156933525, contract number 026-2013 and HERMES code 17432.

6. References

[1] Yablanovitch E 1987 Inhibited spontaneous emission in solid state physics and electronics *Phys. Rev. Lett.* **58** 2059
[2] John S 1987 Localization of light *Phys. Rev. Lett.* **58** 2486
[3] Yee K 1966 Numerical solution of initial boundary value problems involving Maxwell's equations in isotropic media *IEEE Trans. Antennas Propag.* ***14*** 302
[4] Shi S, Chen C and Prather D 2004 Plane-wave expansion method for calculating band structure of photonic crystal slabs with perfectly matched layers *J. Opt. Soc. Am. A.* **21** 1769
[5] Brown E, Yablonovitch E and Parker C 1993 Radiation properties of a planar antenna on a photonic-crystal substrate *J. Opt. Soc. Am. B* **10** 404
[6] Kavokin A, Baumberg J, Malpuech G and Laussy F 2011 *Microcavities* (Oxford University press)
[7] Yeh P 1991 *Optical Waves in Layered Media* ed Jhon Wiley and Sons (New York)
[8] Chang Y, Jhu Y and Wu C 2012 Temperature dependence of defect mode in a defective photonic crystal. *Optics Communications*. **285** 1501
[9] Orfandis S 2008 *Electromagnetic Waves and Antennas* (Rutger University)

Thermal and Exergetic Analysis of a Solar Still

S Chávez, H Terres, A Lizardi, R López, A Lara
Universidad Autónoma Metropolitana, Unidad Azcapotzalco, Departamento de Energía, Área de Termofluidos, San Pablo 180, Col. Reynosa Tamaulipas,

E-mail: scs@correo.azc.uam.mx

Abstract. In this work an energy and exergy analysis of solar distillation process is presented. The analysis is based on the experimental observation of the simple basin type solar stills for 4 different initial volumes (5.5, 6.5, 7.5 and 8.5 L). Energy and exergy balance equations have been written for all components of the solar still including glass cover, brine and absorber plate. The thermodynamic models for the energy and exergy analysis are presented on the critical heat transfer correlations in literatures for the simple basin type solar still. The results show that maximum values are reached in the energy efficiency of 45.6, 41.5, 35.7 and 31.8%, however exergetic efficiency for maximum values are 7.5, 7.2, 7 and 5.4%, corresponding to volumes 5.5, 6.5, 7.5 and 8.5 L respectively.

1. Introduction

The availability of drinking water is reducing day by day; where as the requirement of drinking water is increasing rapidly. To overcome this problem there is a need for some sustainable source for the water distillation (purification). Solar still is a useful device that can be used for the distilling of brackish water for drinking purposes. The energy and exergy performance of solar stills has been investigated by many researchers. Mehdizadeh [1], used energy and exergy calculations to determine the optimum operating conditions for a solar still that uses the benefits of both reverse osmosis and nano-filtration technologies. They showed promising results for the use of the integrated reverse osmosis and nano-filtration process. Dehghan et al [2], showed a study for thermodynamic modeling of a novel portable solar still through the first and second laws analysis, where the daily average energy and exergy efficiencies of the solar still are 19.8% and 0.95%, respectively, also they found that the rate of exergy destructions in solar still components is proportional to the incident solar intensity. The largest exergy destruction in concentrated in the thermoelectric module, with a value of 63.4% of the total exergy destruction, while the glass cover has the smallest value share with 0.8%. On the other hand, Basel [3], designed and build a simple transportable hemispherical solar still and evaluated its performance experimentally under outdoors of Dhahram climatic conditions. In his experiments found that the solar still from 2.8 to 5.71 L/m^2 day and with an average efficiency of 33%.

Torchia et al. [4], they made a theoretical exergy analysis on stationary and transient state of a solar still, focused on the study of the destruction of exergy in the distellers' components, such as, plate collector, brine and glass cover. They solve energy balances to find the temperatures of each component and turn these temperatures are used to calculate the energy and exergy flows. The results show that the exergetic efficiency it is: 12.9%, 6% and 5%, for the collector, brine and solar still respectively. Ranjan [5], show thermodynamic models for the energy and exergy analysis in a simple basin type solar still. The exergetic efficiencies are estimated to be between 19% and 26% for a triple effect system, 17–20% for a double effect system, and less than 5% for a single effect system. Productivity increases significantly by the use of integrated solar stills with better efficiency.

The overall energy and exergy efficiency of the integrated systems rises up to 62% and 8.5%, respectively, using single effect solar stills. Energy efficiency and productivity of the conventional

Content from this work may be used under the terms of the Creative Commons Attribution 3.0 licence. Any further distribution of this work must maintain attribution to the author(s) and the title of the work, journal citation and DOI.
Published under licence by IOP Publishing Ltd

solar stills is found to below in the range of 20-46% and less than 6 L/m^2/day, respectively, for most cases, even under optimized operating conditions. In this work, a study for first and second law applied to the distillation process using a solar still is shown. The results obtained can be used to evaluate the process and indicate what is the real energy quantity used in the distillation process of brine.

2. Experimental procedure

The solar still employed is a closed container with a transparent glass cover which has a tilt angle of 40°, which allows the passage of the solar radiation. Inside the zone of the solar still it's found an absorber plate painted in matte black, is where the fluid is deposited to distill (brine), with an area of 0.36 m^2, at the same time there are 4 gutters connected to the absorber plate to collect the condensate. In the solar still 4 experimental tests were performed, using an initial volume of brine different, for each test data from temperature and the solar radiation incident on the device were obtained. The instrumentation used for obtaining temperatures, was a Compact FieldPoint equipment, type K thermocouples and a Pyranometer Eppley model 8-48, they are obtained recording data every 10 minutes during an interval of time from 10:00 am to 18:00 pm, since in this period of time there is more insolation. In figure 1 the solar still is shown.

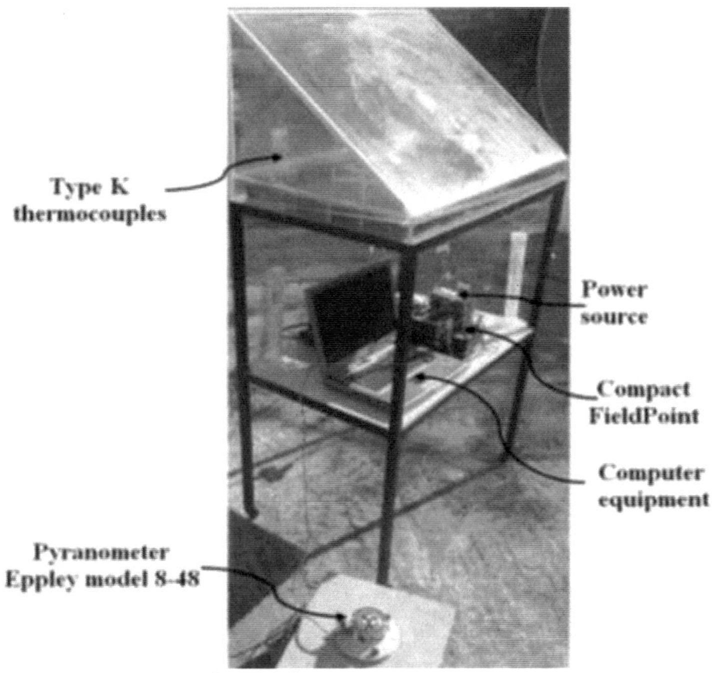

Figure 1. Solar Still and instrumentation equipment

3. Energy balance equations

The energy balance is analysed from different parts of the system such as: (a) Glass Cover, (b) Fluid and (c) Absorber plate, for which a set of equations that depend only on the temperature are obtained.

The total energy received by the glass cover is equal to the incident solar radiation and evaporative, convective and radiative heat transfer from the brine minus the sum of the energy loss between the glass cover and the sky by radiative and convective heat transfers and the energy accumulation within the glass cover. The equation for the glass cover is given by:

$$Q_{t,g} = \left(\alpha_g I(t) A_g + Q_{r,w-g} + Q_{ev,w-g} + Q_{c,w-g}\right) - \left(Q_{c,g-\infty} + Q_{r,g-\infty} + \left(m_g C_{P,g}\right)\left(dT_g / dt\right)\right) \quad (1)$$

The total energy received by the brine in the solar still is equal to the incident solar radiation and convective heat transfer between absorber plate and water minus the sum of the energy from the water to the glass cover by radiation, convection and evaporation and the energy accumulation within the brine. The equation for the brine is given by:

$$Q_{t,w} = \left(\alpha_w I(t) A_w + Q_{c,p-w}\right) - \left(Q_{r,w-g} + Q_{ev,w-g} + Q_{c,w-g} + \left(m_w C_{P,w}\right)\left(dT_w / dt\right)\right) \tag{2}$$

The total energy received by the absorber plate is equal to the incident solar radiation minus the sum of the energy by convective heat transfer between absorber plate and brine, energy losses to the environment and the energy accumulation within the absorber plate. The equation for this case is given by:

$$Q_{t,p} = \left(\alpha_w I(t) A_p\right) - \left(Q_{c,p-w} + Q_{loss,p-\infty} + \left(m_p C_{P,p}\right)\left(dT_p / dt\right)\right) \tag{3}$$

The solar still overall thermal efficiency is considered to be the ratio of evaporative heat transfer to the solar irradiance on the absorber plate and it is given by the following equation.

$$\eta_I = \frac{h_{ev,w-g} A_w \left(T_w - T_g\right)}{I(t) A_p} = \frac{Q_{ev,w-g}}{I(t) A_p} \tag{4}$$

4. Exergy balance equations

The exergy balance equations for the different parts of the system are given here.

The total exergy of the glass cover is received by the sun and brine minus the exergy from the glass cover to environment, exergy destruction of the glass cover and the accumulated exergy within the glass cover. The equation for the glass cover is given by:

$$Ex_{t,g} = \alpha_g Ex_{sun} + Ex_{w-g} - Ex_{g-\infty} - dEx_v / dt \tag{5}$$

The total exergy of the brine is received by the sun and absorber plate minus the evaporation exergy from the brine to the glass cover, exergy destruction of the brine and the accumulated exergy within the brine. The equation for the brine is given by:

$$Ex_{t-w} = \alpha_w \tau_w Ex_{sun} + Ex_{p-w} - Ex_{w-g} - dEx_a / dt \tag{6}$$

The total exergy of the absorber plate is received by the sun minus the exergy destruction of the absorber plate, exergy transfer from absorber plate to the brine, exergy from the absorber plate to the environment and the accumulated exergy within the absorber plate. The equation for the absorber plate is given by:

$$Ex_{t-p} = \alpha_p \tau_w \tau_p Ex_{sol} - Ex_{p-w} - dEx_p / dt \tag{7}$$

Hence, the expression for the overall exergy efficiency of the solar still can be defined as the ratio of the rate of useful exergy output of solar still to the rate of total input exergy. The rate of useful exergy output of the solar still is the total rate of useful exergy associated with the useful evaporation heat transfer rate.

$$\eta_{II} = \frac{h_{e,w-g} A_w \left(T_w - T_g\right)}{A_w I(t)} \times \frac{\left(1 - \dfrac{T_\infty + 273}{T_w + 273}\right)}{\left[1 + \dfrac{1}{3}\left(\dfrac{T_a + 273}{6000}\right)^4 - \dfrac{4}{3}\left(\dfrac{T_a + 273}{6000}\right)\right]} \tag{8}$$

4. Results

With the experimental data that they have for each of the tests and making use of the equations that have previously been mentioned, it is possible to obtain the curve of energy efficiency and exergetic efficiency. The results for efficiency corresponding to each test are shown in figure 2 to 5.

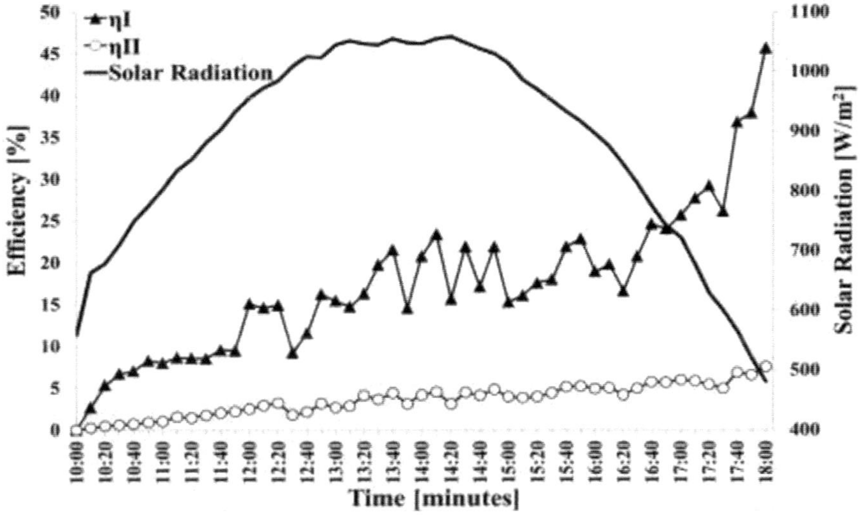

Figure 2. Variation of energetic and exergetic efficiency and solar radiation for 5.5 L of brine

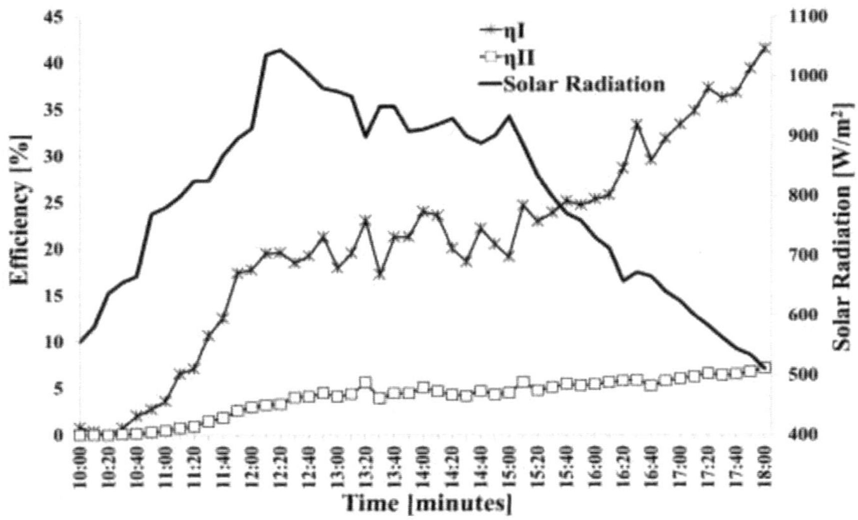

Figure 3. Variation of energetic and exergetic efficiency and solar radiation for 6.5 L of brine

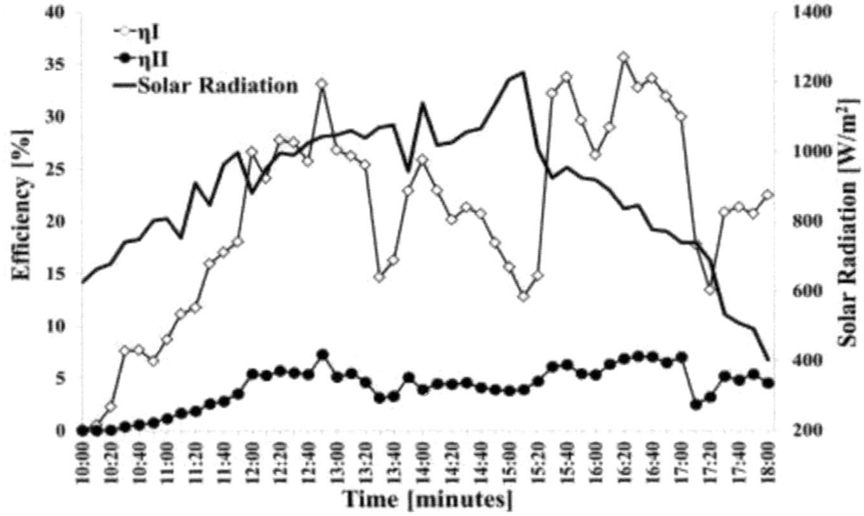

Figure 4. Variation of energetic and exergetic efficiency and solar radiation for 7.5 L of brine

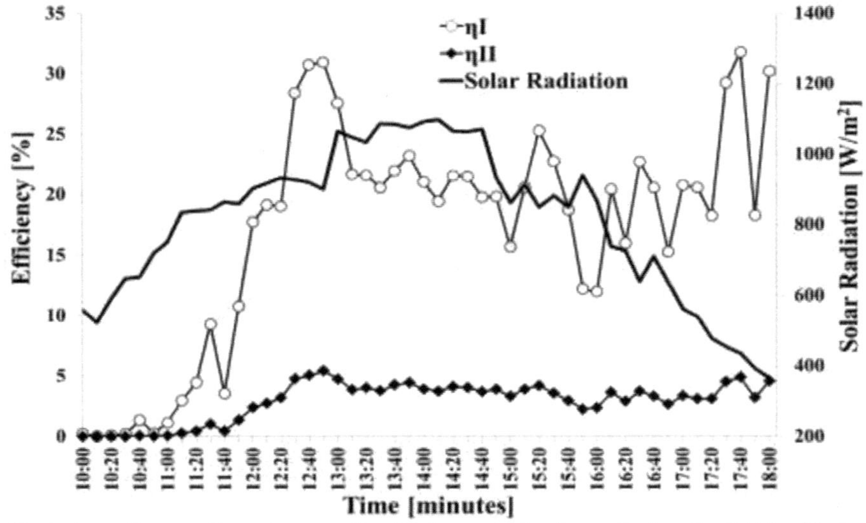

Figure 5. Variation of energetic and exergetic efficiency and solar radiation for 8.5 L of brine

5. Discussion

The solar still show a maximum energy efficiency of 45.6% for 5.5 L of brine; however, for 8.5 L of brine only an efficiency of 31.8 % was reached. In figures 2 to 5, is showed an increase in efficiency, reaching the highest values at the end of the test, this is because that the energy is stored in the device as time passed. It was also found that the average energy efficiency of the solar still decreased by 13.8% when the saline water depth increased by 39%.

The variations that are showing efficiency is due to weather conditions, since the solar radiation is changing because that is not constant and varies its intensity due to cloudiness, dust, pollution, etc., due to these changes the heat transfer by evaporation is affected and therefore the thermal efficiency of the device. Therefore the device on average only used the 38.7% of the energy proceeding from the sun. According to the analysis and the results obtained, the exergy efficiency is much smaller compared with the energy results, reaching maximum values in the exergetic efficiency of 7.5, 7.2, 7 and 5.4%, corresponding to volumes 5.5, 6.5, 7.5 and 8.5 L respectively. However, it is observed that there is a difference of 0.2 %, for exergy efficiency when the device is operated with 5.5, 6.5, and 7.5 L, so for these cases, the depth of brine to be distilled is not relevant compared with the test performed with 8.5 L of brine. It is seen that the exergy efficiency is much lower than the energy efficiency.

According to the study carried, the device efficiency is influenced by the brine depth showing that in volumes over 7.5 L, a bigger energy supply is required, thus the process becomes slower and in consequence the energetic and exergetic efficiency decreases. As a result, wide areas with small thickness are indicatives of a better evaporation and therefore a better efficiency.

6. Conclusion
The values of exergetic efficiency (second law of thermodynamics) in solar stills are quite low compared to energy efficiency (first law of thermodynamics). The efficiency of the solar still is increased if the temperature value is higher in the fluid, so the heat fluxes that take advantage are higher and, therefore, the efficiencies increase. The energy efficiency and exergetic, have different behaviors and this depends on the climatic conditions of operation, i.e., if the energy and exergy analysis are compared, the latter is better since it gives a real insight into the working of the device to carry out the distillation process. So the exergy analysis for the solar still represents the quality of the energy that is contained in this. That indicates how much of the energy coming from the Sun is taken advantage in the system.

7. Appendices
Nomenclature

A	Area, m^2
C_p	Specific heat, J/kg K
Ex	Exergy, W/m^2
h	Heat Transfer coefficient, W/m^2 K
$I(t)$	Solar Radiation, W/m^2
Q	Heat Transfer, W
T	Temperature, °C

Greek letters

α	Absorptivity
τ	Transmissivity
η	Efficiency

Subscript

c	Convection
ev	Evaporation
g	Glass
p	Absorber plate
r	Radiation
sun	Sun
t	Total
w	Water
∞	Environment

8. References
[1] Mehdizadeh H 2006 Membrane desalination plants from an energy–exergy viewpoint *Desalination.* **191** 200–209.
[2] Ali A, Abbas A and Nader R 2015 Thermal modeling and exergetic analysis of a thermoelectric assisted solar still *Solar Energy.* **115** 277–288.
[3] Basel I, 2009 Design and performance of a transportable hemispherical solar still *Renewable Energy.* **34** 145–150
[4] Torchia N, Porta G, and Cervantes G, 2008 Exergy analysis of a passive solar still *Renewable Energy.* **33** 608–616
[5] Ranjan K and Kaushik S 2013 Energy, exergy and thermo-economic analysis of solar distillation systems: A review *Renewable and Sustainable Energy Reviews.* **27** 709–723

IOP Conf. Series: Journal of Physics: Conf. Series **792** (2017) 012010 doi:10.1088/1742-6596/792/1/012010

Irreversibility analysis in the process of solar distillation

S Chávez, H Terres, A Lizardi, R López, A Lara
Universidad Autónoma Metropolitana, Unidad Azcapotzalco, Departamento de Energía, Área de Termofluidos, San Pablo 180, Col. Reynosa Tamaulipas,

E-mail: scs@correo.azc.uam.mx

Abstract. In this work an irreversibility analysis for the thermal process of solar distillation of three different substances is presented, for which it employs a solar still of a slope where three experimental tests with 5.5 L of brine, river water and $MgCl_2$ were performed. Temperature data principally in the glass cover, absorber plate, fluid, environment and the incident solar radiation on the device were obtained. With measurements of temperature, solar radiation and exergetic balance, irreversibilities are found on the device. The results show that the highest values of irreversibilities are concentrated in the absorber plate with an average of 321 W, 342 W and 276 W, followed by the cover glass with an average of 75.8 W, 80.4 W and 86.7 W and finally the fluid with 15.3 W, 15.9 W and 16 W, for 5.5 L of brine, river water and $MgCl_2$.

1. Introduction

In the industrial processes, the distillation has an infinite number of applications, from petrochemical industry to the pharmacist, however to realize the stills separation electrical are used, which consume a large amount of conventional energy or condensing the steam water that comes from the boilers, and to turn these consume hydrocarbons from the oil, this process is very expensive and it pollutes the atmosphere so the use of the renewable energy sources is an alternative to the distillation process by means of the application of the solar energy. This way the solar distillation has turned into a wide expansion, thanks to the fact that by means of the solar stills, it is possible to obtain drinking water.

The design of a solar device implies the evaluation of all the parameter that have on them, the irreversibility is important to evaluate the performance of the solar stills. In the search of the solar still topic literature several authors have worked on the exergy and irreversibility as applied to solar devices. Some results are shown in what follows.

Torchia et al. [1], they conducted a study of exergy in the stationary and transient state of a solar still, focused on the study of the destruction of exergy in the distillers components, such as, plate collector, brine and glass cover. They solve energy balances to find the temperatures of each component and turn these temperatures are used to calculate the energy and exergy flows. The results show that in the steady state the irreversibilities produced in the collector are 615 W/m^2 for a 935 W/m^2 solar energy input, whereas irreversibilities rates in the brine and in the glass cover can be neglected. On the other hand they found that the exergetic efficiency is: 12.9%, 6% and 5%, for the collector, brine and solar still respectively. Dehghan et al [2], showed a study for thermodynamic modeling of a novel portable solar still through the first and second laws analysis, where the daily average energy and exergy efficiencies of the solar still are 19.8% and 0.95%, respectively, also they found that the rate of exergy destructions in solar still components is proportional to the incident solar intensity. The largest exergy destruction in concentred in the thermoelectric module, with a value of 63.4% of the total exergy destruction, while the glass cover has the smallest value share with 0.8%. On the other hand, Ranjan [3], shows thermodynamic models for the energy and exergy analysis in a simple basin type solar still based on the fundamental heat transfer for a triple effect system, double effect system and a single effect system for the solar still. The results show that the energy efficiency

Content from this work may be used under the terms of the Creative Commons Attribution 3.0 licence. Any further distribution of this work must maintain attribution to the author(s) and the title of the work, journal citation and DOI.
Published under licence by IOP Publishing Ltd

and productivity of the solar still are 20-46% and 6 L/m^2 day, respectively for all cases, for the exergetic efficiencies are estimated to be between 19% and 26% for a triple effect system, for a double effect system are between 17–20%, and less than 5% for a single effect system. He also found that the productivity increases significantly by the use of integrated solar stills with better efficiency.

The overall energy and exergy efficiency of the integrated systems rises up to 62% and 8.5%, respectively, using single effect solar stills. In addition, it has also been made works on economic and thermo-economic analysis of solar stills, where the cost of desalination through solar stills is reported in the range of US $0.014 to 0.237/L, therefore it decreases further with the increase in efficiency. Kaushik et al [4] develop a thermodynamic model to estimate the overall exergy efficiency of the single effect horizontal basin type ideal passive solar stills. The solar still was evaluated in June with water depth of 0.04 m and found that the results of energy efficiency were from 8 to 87.2% and the daily energy and exergy efficiency of the solar still is 20.7 and 1.31%, respectively. It is also confirmed that the overall exergy efficiency increases with the increase of water temperature and decreases with the increase of ambient temperature. Basel [5], designed and built a simple transportable hemispherical solar still and evaluated its performance experimentally under outdoors of Dhahram climatic conditions. In his experiments found that the solar still from 2.8 to 5.71 L/m^2 day and with an average efficiency of 33%. Kumar and Tiwari [6], presented an expression for instantaneous exergy efficiency of a passive solar still, considering the effect of the design, climatic parameters, effective absorptivity of basin liner, glass cover tilt and wind velocity. The results showed that with a decrease in absorptivity with time is from 0.9–0.6, the energetic and exergetic efficiencies decrease by 21.8% and 36.7% respectively, also with the effect of the glass cover tilt the efficiencies decreased by 0.75% and 0.47% per degree increased in tilt and these efficiencies increased rapidly up to a wind velocity of 2 m/s.

In this work the irreversibility analysis in a solar still are shown, these are obtained when the operation of a solar still with 5.5 L of brine, river water and $MgCl_2$ is evaluated.

2. Experimental procedure

The solar distillation process is: the solar still is partially filled with some fluid (brine, river water and $MgCl_2$) in an absorber plate which is a black surface used to absorb incoming radiation after it passes through the glass cover and the fluid, the absorber plate increases its temperature and transfers heat to the fluid, after this, water evaporates at the free surface. A natural convection flow of humid air circulating inside the enclosure takes place due to the temperature difference between the free surface of the heated fluid and the upper cool cover, and this inclined glass cover serves as a condensing plate where the distillate water runs by gravity along its internal face to a small collector gutter. The solar still employed has a tilt angle of 40°, with an area of 0.36 m^2 for the absorber plate. The solar still considered as reference in this work is shown in figure 1.

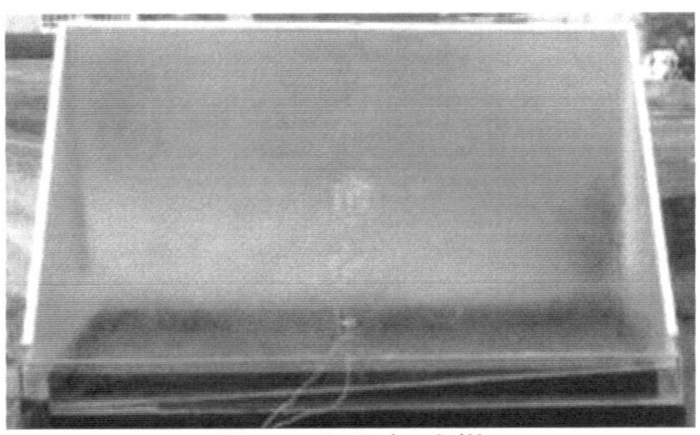

Figure 1. Solar Still

The instrumentation used for obtaining temperatures when the solar still is evaluated with 5.5 L of brine, river water and $MgCl_2$, is with an equipment Compact FieldPoint, type K thermocouples and a Pyranometer Eppley model 8-48, they are obtained recording data every 15 minutes during an interval of time from 10:00 am to 18:00 pm, since in this period of time there are more insolation.

3. Irreversibility for the solar still

One of the important uses of the second law of thermodynamics in engineering is to determine the best theoretical performance of systems. By comparing actual performance with the best theoretical performance, insights often can be gained into the potential for improvement. A process is called irreversible if the system and all parts of its surroundings cannot be exactly restored to their respective initial states after the process has occurred.

The irreversibility balance is analysed from different parts of the system such as: (a) Glass Cover, (b) Fluid and (c) Absorber plate, for which a set of equations that depend only on the temperature are obtained and can be calculated according to reference [1, 2, and 3]. The exergy accumulation in the glass cover, fluid and absorber plate is assumed to be negligible.

The irreversibility balance in the glass cover is evaluated considering the exergy flow through glass cover as shown in figure 2 and equation (1).

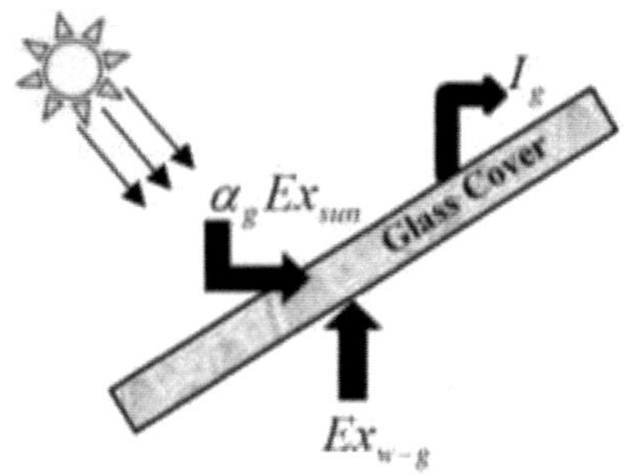

Figure 2. Exergy flow of the glass cover

$$I_g = Ex_{g-\infty} + Exd_g = \alpha_g Ex_{sun} + Ex_{w-g} \quad (1)$$

Where $\alpha_g Ex_{sun}$, is the amount of the solar exergy absorbed by glass cover, Ex_{w-g}, is the sum of exergy associated with the heat exchanges between the water surface and the glass cover, and $Ex_{g-\infty}$, represents the exergy losses from the glass cover to the environment by the heat exchange processes of radiation and convection.

The irreversibility balance in the fluid (brine, river water and $MgCl_2$), is evaluated considering the exergy flow through the fluid as shown in figure 3 and equation (2).

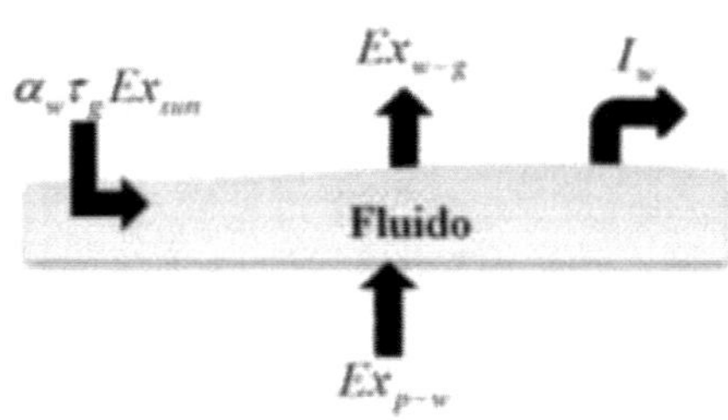

Figure 3. Exergy flow of the fluid

$$I_w = Exd_w = \alpha_w \tau_g Ex_{sun} + Ex_{p-w} - Ex_{w-g} \quad (2)$$

Where, $\alpha_w \tau_g Ex_{sun}$, is the part of the incident solar exergy absorbed by water and useful exergy from the absorber plate Ex_{p-w}, a fraction of it is used as the exergy involved during heat transfers between the water surface and glass cover inside the solar still Ex_{w-g} and remaining is destroyed Exd_w.

The irreversibility balance in the absorber plate is evaluated considering the exergy flow through the absorber plate as shown in figure 4 and equation (3).

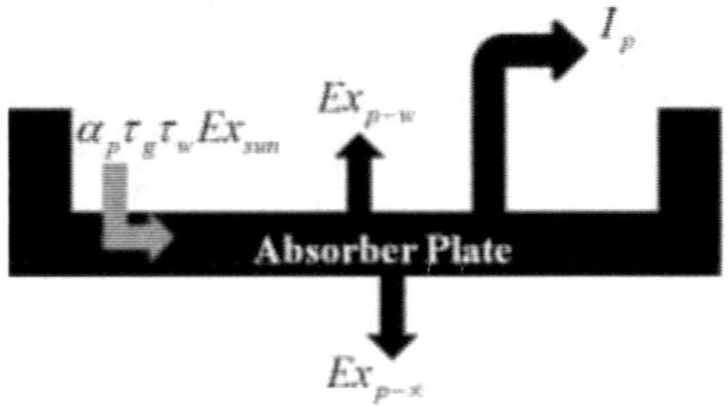

Figure 4. Exergy flow of the absorber plate

$$I_p = Ex_{p-\infty} + Exd_p = \alpha_p \tau_g \tau_w Ex_{sun} - Ex_{p-w} \tag{3}$$

The absorber plate of the solar still absorbs the solar exergy quantity $\alpha_p \tau_g \tau_w Ex_{sun}$. This amount is shared between the brackish water heated by Ex_{p-w}, and the exergy losses from the absorber plate to the environment $Ex_{p-\infty}$.

4. Results

The temperatures obtained by the experimentation are shown in figure 5 to 7.

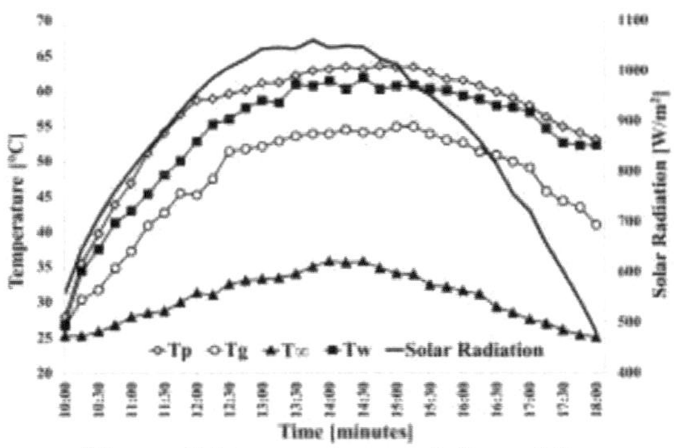

Figure 5. Temperature variation of the solar still for 5.5 L of brine

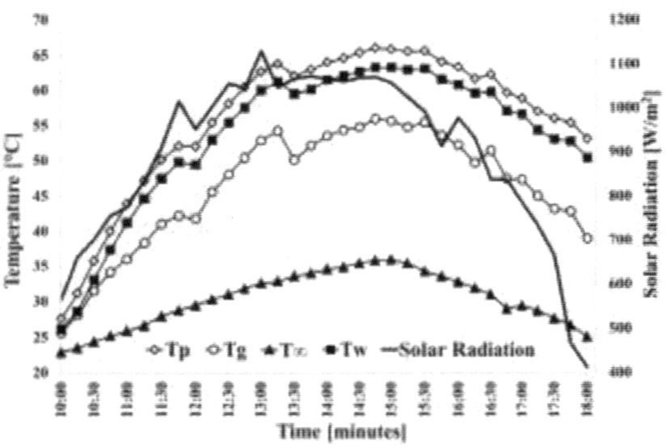

Figure 6. Temperature variation of the solar still for 5.5 L of river water

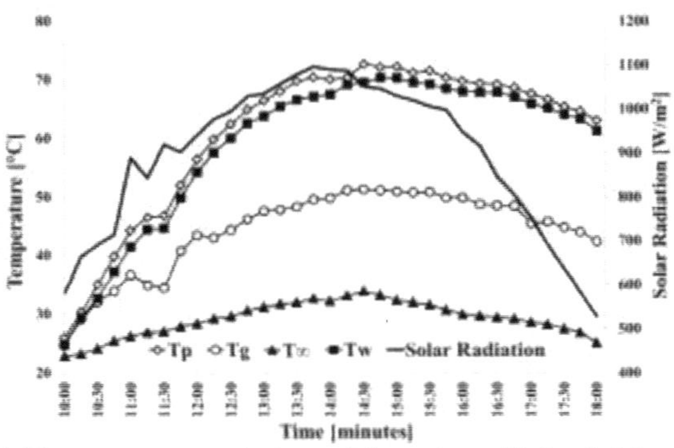

Figure 7. Temperature variation of the solar still for 5.5 L of $MgCl_2$

Equations (1) to (3) are applied to the data obtained experimentally and the irreversibility is determined for each of the tests. In figures 8 to 10 are shown.

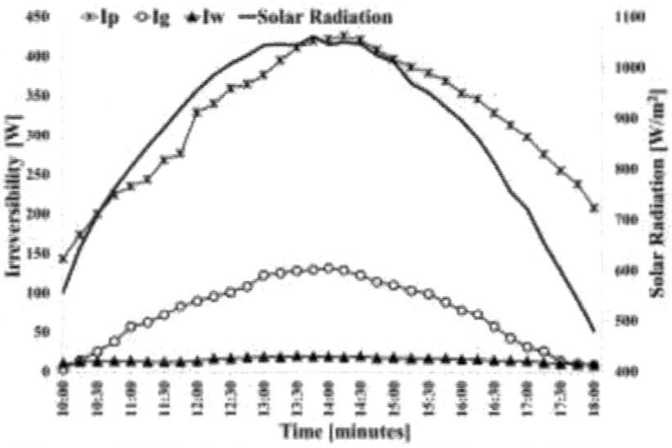

Figure 8. Variation of irreversibility of different parts of the solar still and solar radiation for 5.5 L of brine

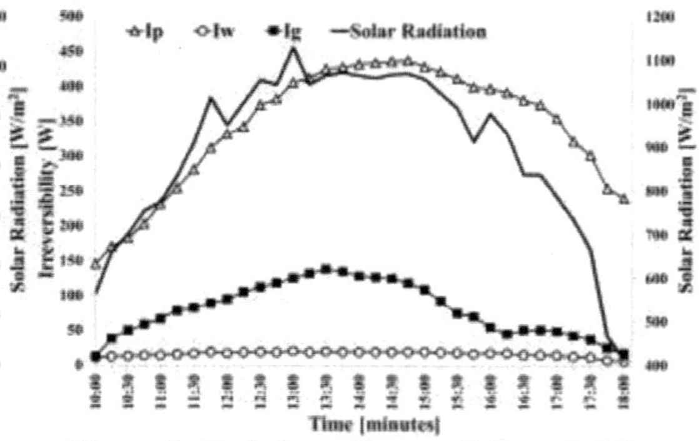

Figure 9. Variation of irreversibility of different parts of the solar still and solar radiation for 5.5 L of river water

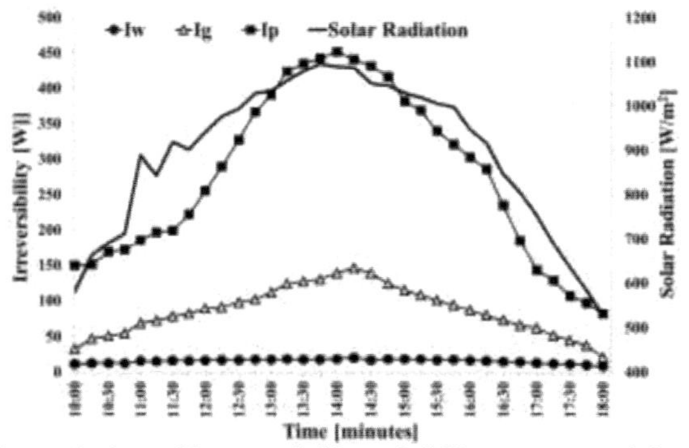

Figure 10. Variation of irreversibility of different parts of the solar still and solar radiation for 5.5 L of $MgCl_2$

5. Discussion

For the process of solar distillation with different substances, it is possible to obtain results of temperature and radiation. In figures 5 to 7, the distribution of temperature and radiation with respect to time is shown, where the behavior in the temperatures measured are: $T_x < T_g < T_w < T_p$, reaching maximum values in the fluid of 61.8 °C, 63.3 °C and 70.4 °C to 5.5 L of brine, river water and $MgCl_2$, respectively. The test with $MgCl_2$, acquires the highest temperature values, and this is due to the capacity of each one of the substances has to store the heat. According to these results, it is observed that it is not necessary that the solution to distill reaches its boiling point to achieve condensation. In the solar radiation values were obtained in the range of 400-1129 W/m^2, undergoing changes with time and therefore temperatures are affected.

Irreversibility variation for different parts of the solar still, such as: glass cover, fluid and absorber plate is shown in figures 8 to 10. As it can be seen, the irreversibility more important is found in the plate absorber, later he follows the glass cover and finally the fluid, such behavior is the same for the 3 tests performed. On the plate absorber, is reached the highest values in a range of 350 to 450 W in a time of 12:30 to 15:30 pm, in all three cases, however for the glass cover only reach values maximum of 100 to 146 W, for the fluid is concentrated irreversibility with smaller values with only 20.6 W, this being the maximum for all tests.

6. Conclusion

In this work, a study of irreversibility applied to process of solar still was done. In general for the solar distillation process using 3 different substances, the average irreversibility for absorber plate, glass cover, and fluid is 314.3 W, 80.96 W and 15.73 W respectively, and the total irreversibility of the whole system is 410.99 W. Therefore, 76.47% of the total irreversibility is related to the absorber plate which shows that the selection of an appropriate absorber plate has a great impact on the solar still performance. The reason behind the higher irreversibility of the absorber plate could be the temperature difference between the absorber plate and the sun and also the exergy dissipation from the absorber plate to the surrounding. Therefore, the selection of an appropriate absorber plate plays an important role in construction of solar stills. In addition, this work allows viewing how the concepts of the thermodynamics can be applied in the heating water process when solar cookers are used. Finally, the results obtained in this work can be useful in the taking of decisions for new designs of solar cookers.

7. Appendices

Nomenclature

Ex	Exergy, W
I	Irreversibility, W

Greek letters

α	Absorptivity
τ	Transmissivity

Subscript

g	Glass
p	Absorber plate
sun	Sun
w	Water
∞	Environment

8. References

[1] Torchia N, Porta G, and Cervantes G 2008 Exergy analysis of a passive solar still *Renewable Energy.* **33** 608–616

[2] Dehghan A, Afshari A, Rahbar N 2015 Thermal modeling and exergetic analysis of a thermoelectric assisted solar still *Solar Energy.* **115** 277–288

[3] Ranjan K and Kaushik S 2013 Energy, exergy and thermo-economic analysis of solar distillation systems: A review, *Renewable and Sustainable Energy Reviews* **27** 709–723

[4] Kaushik S, Ranjan K and Panwar N 2013 Optimum exergy efficiency of single-effect ideal passive solar stills *Energy Efficiency.* **6** 595–606

[5] Basel I 2009 Design and performance of a transportable hemispherical solar still *Renewable Energy.* **34** 145–150

[6] Kumar S and Tiwari G 2011 Analytical expression for instantaneous exergy efficiency of a shallow basin passive solar still *International Journal of Thermal Sciences.* **50** 2543–2549

Estimation of the instable composition areas and its dependence on the thickness of GaInAsSb layers grown on different substrates

Castillo E G[1] Castillo A K Michournyi V A De Anda F and Yu Gorbatchev A
Universidad Autónoma de San Luis Potosí, Instituto de Investigación en Comunicación Óptica, Av. Karacorum 1470, Col. Lomas 4ª Sec., CP 78210, San Luis Potosí, México.

E-mail: baldivia_17@hotmail.com

Abstract. In this work we inform about the results of estimations of the changes of the spinodal decomposition areas and its dependence on the thickness of epitaxial layers. The calculations have been performed using the CALPHAD method and the SGTE data taking into account the elastic energy generated by the lattice mismatch between forming solid solutions and substrate. We have shown that in thin layers the elastic energy may serve both as a stabilizing factor for compositions inside the immiscibility region but lattice matched, or close, to the substrate and, on the contrary, as a reason of instability for the compositions laying in the periphery of the miscible regions.

1. Introduction

In different III-V solid solutions systems, growth within decomposition spinodal areas can produce layers with composition modulated regions [1-3] of the order of several tens of nanometers [4,5]. Such layers can serve to form low-dimensional heterostructures in the fabrication of advanced optoelectronic devices [1]. However, the instability of the solid phase inside the spinodal decomposition area can impose restrictions on the growth of alloys with certain compositions. These restrictions limit the wavelengths in which photovoltaic devices, lasers or photodetectors could be fabricated. This is clearly seen on the example of the III-V solid solutions containing Sb [2,3]. At different times, different authors have calculated spinodal decomposition areas [4-6]. However, the results of experimental studies show some disagreement with the calculated data. In addition, it was shown experimentally that instability of layers grown within the miscibility gap depend on their thickness [7].

The object of this work is to estimate the variation of the miscibility gaps when the thickness of the grown layers change, taking into account their mismatch with the substrate. The calculations have been performed for the GaInAsSb quaternary system, as an example. GaAs, InAs, InP and InSb were chosen as substrates for the formation of the layers.

[1] To whom any correspondence should be addressed.

Content from this work may be used under the terms of the Creative Commons Attribution 3.0 licence. Any further distribution of this work must maintain attribution to the author(s) and the title of the work, journal citation and DOI.
Published under licence by IOP Publishing Ltd

2. Calculation method

The criterion for thermodynamic stability for the quaternary III-V ($A_xB_{1-x}C_yD_{1-y}$) alloys is chosen as [8]:

$$\left(\frac{\partial^2 G}{\partial x^2}\right)\left(\frac{\partial^2 G}{\partial y^2}\right)-\left(\frac{\partial^2 G}{\partial x \partial y}\right)^2 \geq 0 \tag{1}$$

where G is the Gibbs free energy per mole of the solution, x and y are the molar fractions of the binary constituents.

The total Gibbs energy G^{total} of the film is given as a sum of the strain-free energy G^S of the solid phase and the elastic energy G^{el} due to the lattice mismatch between the solid solutions and the substrate was calculated with:

$$G^{total} = G^S + G^{el} \tag{2}$$

In the case of regular solutions the Gibbs energy for $Ga_xIn_{1-x}As_ySb_{1-y}$ is written [9]

$$\begin{aligned} G^S = {} & G^o_{GaAs}xy + G^o_{GaSb}x(1-y) + G^o_{InSb}(1-x)(1-y) + G^o_{InAs}(1-x)y \\ & + RT[xlnx + ylny + (1-x)ln(1-x) + (1-y)ln(1-y)] \\ & + x(1-x)yL^s_{GaAs-InAs} + x(1-x)(1-y)L^s_{GaSb-InSb} + xy(1-y)L^s_{GaAs-GaSb} \\ & + (1-x)y(1-y)L^s_{InAs-InSb} \end{aligned} \tag{3}$$

where G^0_{ij} are the Gibbs energies of the binary solid compounds, L^s_{m-n} denote the interaction parameters of the pseudobinary systems in the solid phase solutions. To calculate the strain-free energy of the solid phase we used the CALPHAD method [10]. The Gibbs energy of each binary solid compound is calculated with an expression including the sum of the enthalpies H^0_i of the stable state of the constituting pure elements at 298.15 K and an empirical function of temperature $GHSER_i$ [11]

$$G^0_{ij} = H^0_i + H^0_j + GHSER_{ij} \tag{4}$$

The values of the parameters used in the calculations are listed in Table 1.

In the epitaxial layers the elastic energy G^{el} is generated by the lattice mismatch with the substrate. To calculate G^{el} we used the procedure of Ohtani [12] that takes into account its dependence on the thickness, h, when it exceeds the critical thickness h_c

$$G^{el} = \frac{\mu}{2}\frac{1+v}{1-v}N_A a^3 f^2 \quad (h < h_c) \tag{6}$$

$$G^{el} = \frac{N_A a^3}{4h}\Delta\left(1 + ln\frac{h}{h_c}\right) \quad (h \geq h_c) \tag{7}$$

where a is the lattice parameter, μ is the shear modulus, N_A is the Avogadro's number. The critical thickness h_c is calculated with the equation

$$h_c = \frac{\Delta}{2\mu f^2}\frac{1-v}{1+v} \tag{8}$$

using the value of 0.8 J/m^2 for the energy barrier Δ [13] as corresponds to the case of growth on substrates with orientation (001). The lattice misfit parameter f is given by

$$f = \frac{a - a_s}{a_s} \tag{9}$$

where a_s is lattice parameter of the substrate.

Table 1. The thermodynamic parameters used in the calculations.

Parameters	Values (J/mol)	Ref.
$H^0_{Ga\_Ort}$	5572	14
$H^0_{In\_Tet\_A6}$	6610	14
$H^0_{As\_Rho\_A7}$	5117.032	14
$H^0_{Sb\_Rho\_A7}$	5870.152	14
$GHSER_{GaAs}$	$-104352 + 265.43256T - 48.681258TlnT - 1.1158 \times 10^{-3}T^2 + 127670 \times T$	12
$GHSER_{InAs}$	$-73057.2 + 230.91896T - 45.187942TlnT - 7.73 \times 10^{-3}T^2 + 69438 \times T^{-1}$ -	12
$GHSER_{GaSb}$	$-59774.701 + 267.809609T - 51.1966138TlnT + 5.14355 \times 10^{-3}T^2 - 1770$ $\times 10^{-6}T^3 + 1.645 \times 10^{23}T^{-9}$	15
$GHSER_{InSb}$	$-47974.978 + 281.217555T - 55.3821132TlnT + 8.29484 \times 10^{-3}T^2 - 1110$ $\times 10^{-6}T^3 + 3.53116 \times 10^{22}T^{-9}$	15
$L^s_{GaAs-InAs}$	$13360 - 5T$	12
$L^s_{GaSb-InSb}$	$9093 - 2.8698T$	15
$L^s_{GaAs-GaSb}$	$24824 - 7.74301T + 4774(X_{GaSb} - X_{GaAs})$	16
$L^s_{InAs-InSb}$	$13196.3 - 3.2915T + (20336.8 - 21.8105T)(X_{InAs} - X_{InSb})$	17

3. Calculated results and discussions

Firstly, we have studied the influence of the layers thickness on the modification of the spinodal decomposition area for $Ga_xIn_{1-x}As_ySb_{1-y}$ compounds grown on GaSb substrates (Fig 1: b – 3 nm, c – 15 nm, d – 90 nm). For comparison Fig. 1a shows the spinodal decomposition area calculated without taking in account the elastic energy G^{el}. For layers with a thickness of 3 nm (Fig. 1b), corresponding approximately to 1 ML, almost all compositions are stable, except a little region close to pure GaAs. In Figs. 1c and 1d it is possible to see that a band of stable compositions appears along the line of the solid solutions $(GaSb)_x(InAs_{0.91}Sb_{0.09})_{1-x}$ perfectly matched with the substrate, this bands narrows as the layer become thicker. Our calculation predicts that this band disappears at a thickness of around 1 um. In fact, several authors have already predicted the stabilizing effect of the elastic energy before [18] when

$$\left(\frac{\partial^2 G^{el}}{\partial x^2}\right)\left(\frac{\partial^2 G^{el}}{\partial y^2}\right) - \left(\frac{\partial^2 G^{el}}{\partial x \partial y}\right)^2 \geq \left(\frac{\partial^2 G^S}{\partial x^2}\right)\left(\frac{\partial^2 G^S}{\partial y^2}\right) - \left(\frac{\partial^2 G^S}{\partial x \partial y}\right)^2 \tag{10}$$

This stabilizing effect can be explained considering a competition between elastic and strain-free energies of solids. The calculations show that the positive contribution of the elastic energy increases in the less tensioned layers and decreases when the layers become thicker.

Fig. 1 also shows some experimental points [19] that represent the composition of epitaxial layers of high crystalline quality grown by SS MBE on GaSb substrates, point 1 corresponds to a 12 nm and points 2-4 to 14.5 nm thick layers. The compositions of these layers lay inside or close to the spinodal decomposition area when the elastic energy is not considered in the calculations (Fig.1a). However, they lay in the stable regions when the elastic energy is taken into account for a layer thickness of 15 nm (Fig.1 c).

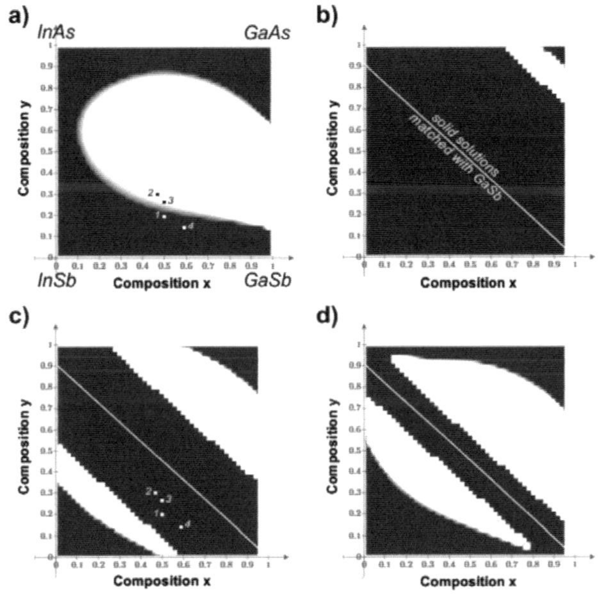

Figure 1. Areas of the instable, in white, GaInAsSb compositions for the layers grown on GaSb substrate at 570 °C (a) – strain-free, b) – L= 3 nm, c) – 15 nm, d) – 90 nm); 1-4 – experimental data of Shterengas et al.

The stabilizing effect of the elastic energy can be observed also for thin GaInAsSb layers grown on other substrates. Fig.2 shows the spinodal decomposition areas in the case of layers grown at 570 °C on GaAs substrates. Fig. 2a, 2b and 2c correspond to layers of thickness 3 nm, 15 nm and 90 nm, respectively. The stabilizing effect of the elastic energy can be seen on layers with thicknesses of 3 nm and 15 nm close top right point corresponding to binary GaAs.

Also, we would like to mention another effect caused by the competition between elastic and strain-free energies for compositions on the periphery of the spinodal decomposition area. We have found that, at small thicknesses, new immiscibility regions, originated by the elastic energy, appear in stable compounds and that the immiscibility area expands in unstable compounds. Moreover, the instability is displayed only in some range of the thicknesses of the layers. So, to illustrate this effect we considered disclosing the spinodal decomposition areas in GaInAs ternary solutions in the case of growing on various substrates at 570 °C. The results are shown in fig. 3a for GaAs substrate, fig. 3b for InAs, fig. 3c for InP and fig. 3d for InSb substrates. This system is interesting because it is considered as a stable. However, the calculation demonstrated, in the case of growing on GaAs and InAs substrates, that an unstable area appears in the side close to the opposite corner to the one corresponding to the substrate and moves backwards as the layer's thickness increases. The layers with thicknesses larger than 55 nm are stable in all compositions. When there is an important lattice mismatch between the layers and the substrate the immiscibility regions shrinks as the layer's thickness decrease. As an example we choose for consideration the system GaInAs on InSb substrate.

IOP Conf. Series: Journal of Physics: Conf. Series **792** (2017) 012011 doi:10.1088/1742-6596/792/1/012011

It can be seen that all compositions are stable, except only the interval $0.3 < x < 0.6$ and for thicknesses up to 2 ML.

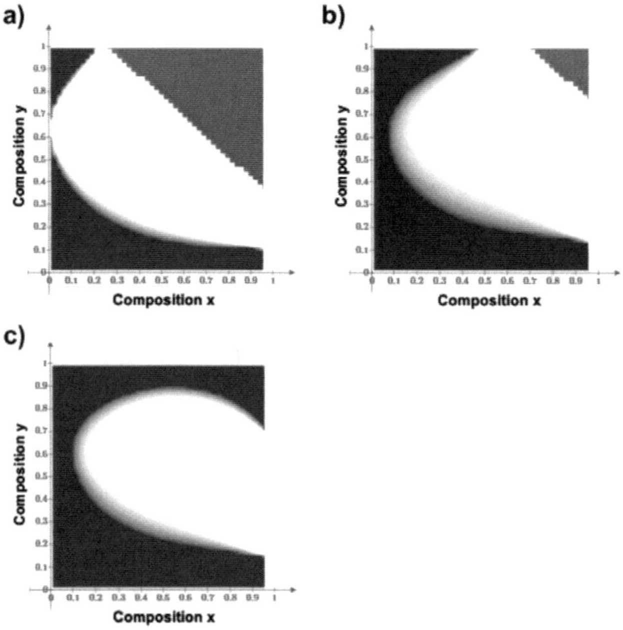

Figure 2. Areas of the instable GaInAsSb compositions, in white, for the layers grown at 570 °C on GaAs substrate (a) – L= 3 nm, b) – 15 nm, c) – 90 nm)

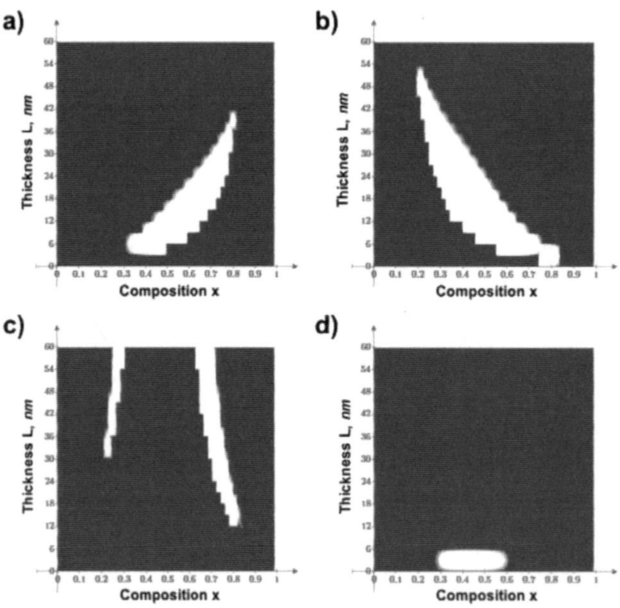

Figure 3. Areas of the instable GaInAs compositions, in white, for the layers grown at 570 °C grown on different substrates: a) – GaAs, b) – InAs, c) – InP, d) – InSb.

Also, at thicknesses of several monolayers, instable narrow areas appear in both sides of isoperiodical composition to InP substrate (Fig. 3c). This behavior of spinodal decomposition areas resembles the one we have observed for GaInAsSb grown on GaSb considering the stabilizing effect of the elastic energy. However, the difference is that, if we consider only the strain-free energy, GaInAs solid solutions do not contain instable areas.

4. Conclusions

Using the CALPHAD method and the SGTE data, we have estimated the changes of spinodal decomposition areas, induced by variations of the thicknesses, of GaInAsSb layers grown on InAs, GaAs, GaSb, InSb and InP substrates taking in account the elastic energy associated with the lattice mismatch. We have shown that the elastic energy may induce two opposite effects, from one side it can serve as a stabilizing factor and causes the apparition of a band of stable compositions which lay inside the spinodal decomposition area of the unstrained alloys. From the other side, the competition of elastic and strain-free energies may cause the instability of some compositions which are in the periphery of spinodal decomposition areas and even induce the apparition of immiscibility regions in normally stable alloys. The influence of elastic energy becomes stronger as the thickness of the layers decrease. We have to note that only the thermodynamic type of instability was studied in our work without considering any kinetic effects during the growth processes and the contribution of dislocations to the energy changes. Nevertheless the results are confirmed experimentally with data from the growth of actual epitaxial layers.

5. Acknowledgements

This work was partially supported by CONACYT and PROMEP.

References

[1] Shchukin V A and Bimberg D 1999 *Reviews of Modern Physics* **71** 1125
[2] Andreev I A Kunitsyna E V Mikha lova M P and Yakovlev Yu P 1999 *Semiconductors* **33** 216.
[3] Semenov A N Solov'ev V A Meltser B Ya Terent'ev Ya V Prokopova L G and Ivanov S V 2005 *J. Crystal Growth* **278** 203
[4] Stringfellow G B 1983 *J. Appl. Phys.*, **54** 404
[5] Lazzari J L Tournie E Pitard F and Joullie A 1991 *Mater. Sci. Eng B* **9** 125.
[6] Sorokin V S Sorokin S V Semenov A N Meltser B Ya and Ivanov S V 2000 *J. Cryst. Growth* **216** 87.
[7] Semenov A N Terent'ev Ya V Meltser B Ya Solov'ev V A Popova T V Nashchekin A V Andreev I A Kunitsina E V Usikova A A Yakovleva Yu P and Ivanov S I 2010 *Semiconductors* **44** 672.
[8] Cahn John W 1962 *Acta Metall* **10** 179.
[9] Dutta P S and Miller T 2000 *J. of Electron. Mater* **29** 956.
[10] Lukas H L Weiss J and Henig E Th 1982 *CALPHAD* **6** 229
[11] Ansara I Chatillon C Lukas H L Nishizawa T Ohtani H Ishida K Hillert M Sundman B Argent B B Watson A Chart T G and Anderson T 1994 *CALPHAD* **18** 177
[12] Ohtani H Kobayashi K and Ishita K 2001 *J. of Phase Equilibria* **22** 276
[13] Yokogawa T Sato H and Ogura M 1988 *Appl. Phys. Lett.* **52** 1678
[14] Dinsdale A T 1991 *CALPHAD* **15** 317
[15] Jianrongv Yang and Watson Andy *CALPHAD* **18** 165
[16] Li J B Zhang W Li Ch and Du Zh 1998 *J. of Phase Equilibria* **19** 466
[17] Li J B Zhang W Li Ch and Du Zh 1998 *J. of Phase Equilibria* **19** 473
[18] Ipatova I P Malyshkin V G Maslov A Yu and Shchukin V A 1993 *Fizika I Technika Poluprov* **27** 285
[19] Shterengas L Belenky G L Kim J G and Martinelli R U 2004 *Semicond. Sci. Technol.* **19** 655

IOP Conf. Series: Journal of Physics: Conf. Series **792** (2017) 012012 doi:10.1088/1742-6596/792/1/012012

Exergy applied to the heat conduction analysis in glass covers of a solar cooker box-type with internal and external reflectors

Terres H[1] Lizardi A Chávez S López R and M Vaca
[1]Universidad Autónoma Metropolitana, Unidad Azcapotzalco, Departamento de Energía, Area de Termofluidos, San Pablo 180, Col. Reynosa Tamaulipas, Azcapotzalco, 02200, México, D. F.

E-mail: tph@correo.azc.uam.mx

Abstract. In this work, an exergy evaluation to determine the energy availability across to glass covers, place where the solar radiation enters toward a solar cooker box-type is done. Considering the heating process of water, the energy not used is quantified by means of exergy. The results allow identifying the glasses in the cover as the zone where the solar cooker could be improved. The conduction heat transfer losses for the glasses is most big than 75%. Because the values for the conduction heat losses are around 90%, which are very important, this allows to identify the cover glass as the area where improvements could be made in this type of solar cookers.

1. Introduction

When solar cookers are constructed, is very important to consider the heat losses in the same ones. The conduction, convection and radiation mechanisms happen when a solar cooker is operating during the heating process of water. According to previous works by Terres et al. [1, 2, 3], the conduction is the most important because of their losses are bigger than convection and radiation in the heating process in a solar cooker box-type. Pandey et al. [4] realized an experimental study of solar cookers based on the exergy analysis. They showed that for a solar cooker box-type when the volume of water used in tests increases, the exergy efficiency increases too. However, for a solar cooker paraboloid-type, the exergy efficiency is higher than the solar cooker box-type for the same increment in the volume of water. Panwar [5] present a study for an animal feed solar cooker. He shows a thermal model and an experimental validation for the same one. He considered an experiment for 9 months. Agree to his model, it is capable of predicting the water temperature close to experimental values. Also in his work energy and exergy assessment for the cooker was also carried out. Nemati and Javanmardi [6] considered the exergy theory to study a domestic solar cylindrical-parabolic cooker, they determined the 1st and 2nd law efficiencies for the cooker. They establish that the 1st law efficiency is very low and the 2nd law efficiency is even lower. Taking in account these results, the authors considered necessary the performance optimization in domestic solar cylindrical-parabolic cookers. Kumar et al. [7] proposed parameters based in exergy to study solar cookers. Their results are shown by graphs between exergy output power and temperature difference. They determined the exergy power with variation of the temperature difference. The slope of the straight line obtained through curve fitting represents the heat loss coefficient of the cooker. In this work, the exergy is applied to determine the useful energy across the glasses inside of the cover of a solar cooker box-type. The results are important for improve designs and studies in solar cooker cookers box-type.

Content from this work may be used under the terms of the Creative Commons Attribution 3.0 licence. Any further distribution of this work must maintain attribution to the author(s) and the title of the work, journal citation and DOI.
Published under licence by IOP Publishing Ltd

2. Experimental procedure and experimental results

For the experiment, the solar cooker box-type used consists of the following elements: 1. a cover with two flat glasses with a clearance between them. 2. Reflectors placed on the cover outer of the cooker 3. Internal reflectors made in commercial aluminium paper placed to different tilt angles, 4. Thermal insulator placed in the lateral part of the same one, and 5. Recipient contains the product to cook. The solar cooker is locked tightly; this allows reaching considerable temperatures in the water, which is used as test fluid. 2 liters of water were used. The general dimensions of the solar cookers 0.7 x 0.7 x 0.35 m. In addition, a radiometer Eppley model 8-49 was used for the solar radiation and thermocouples type-k were used for the temperature measures. An acquisition system data integrated by a compact Field Point of National Instruments and Lab View software were instrumented for the test. The interval considered was 4 hours (10:00 to 14:00) and the day was April 10, 2016. The test was done in Mexico City. In figures 1 and 2, the solar cooker evaluated is shown.

Figure 1. Solar cooker box-type with exterior and external

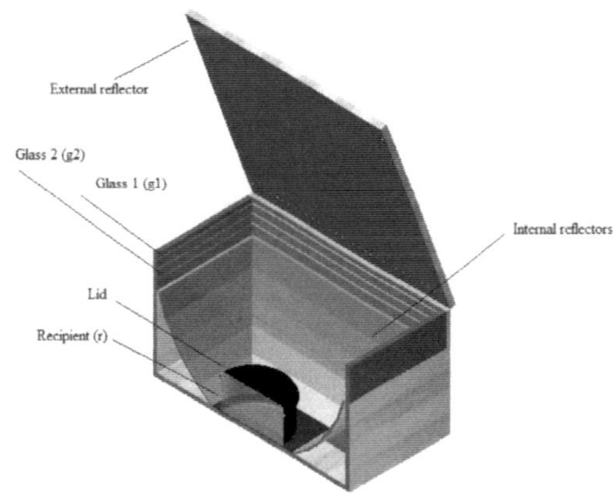

Figure 2. Solar cooker box-type with exterior and external (cut presentation)

Thermocouples k-type were placed in the cooker to measure: 1. Temperatures in glass covers (T_{g1_upper}, $T_{g1\ bottom}$, $T_{g2\ upper}$, $T_{g2\ bottom}$), 2. lid of the recipient (T_{lid}), 3. body of the recipient (T_r), 4. Water (T_w), 5. internal environment (T_{int_env}), 6. external environment (T_{ext_env}). In figures 3, the values for the temperature distribution are shown.

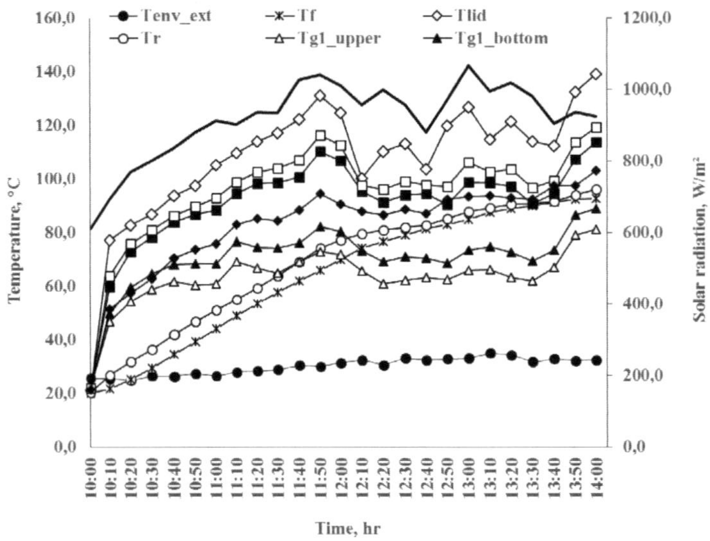

Figure 3. Temperatures for the solar cooker

3. Conduction heat transfer and destroyed exergy in the cover glasses

To evaluate the heat transfer and the destroyed exergy across the cover glasses in the solar cooker, the conduction heat transfer and the exergy for each glass is determined. For this, next equations are applied [8]

Conduction heat transfer in glasses

$$\dot{Q}_{g1} = \frac{kA_{g1}}{e}\left[T_{\text{g1\_bottom}} - T_{\text{g1\_upper}}\right] \quad [\text{W}] \tag{1}$$

$$\dot{Q}_{g2} = \frac{kA_{g2}}{e}\left[T_{\text{g2\_bottom}} - T_{\text{g2\_upper}}\right] \quad [\text{W}] \tag{2}$$

Destroyed Exergy

$$\dot{X}_{g1} = \dot{Q}_{g1}\left[1 - \frac{T_0}{T_{\text{g1\_bottom}}}\right] - \dot{Q}_{g1}\left[1 - \frac{T_0}{T_{\text{g1\_upper}}}\right] \quad [\text{W}] \tag{3}$$

$$\dot{X}_{g2} = \dot{Q}_{g2}\left[1 - \frac{T_0}{T_{\text{g2\_bottom}}}\right] - \dot{Q}_{g2}\left[1 - \frac{T_0}{T_{\text{g2\_upper}}}\right] \quad [\text{W}] \tag{4}$$

Where, k is thermal conductivity, W/m K; T_o is temperature in dead state, K in this case, this temperature correspond to the environment temperature. In figures, 4 and 5 the conduction heat transfer and the destroyed exergy are shown.

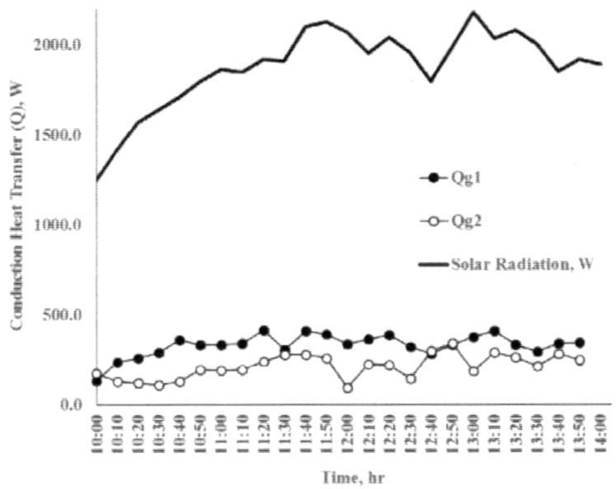

Figure 4. Conduction heat transfer for glasses *g1* and *g2* in the cover of the solar cooker

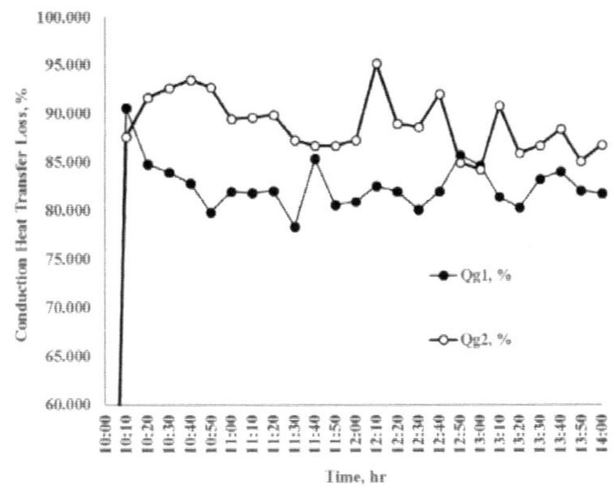

Figure 5. Conduction heat transfer losses for glasses *g1* and *g2* in the cover of the solar cooker

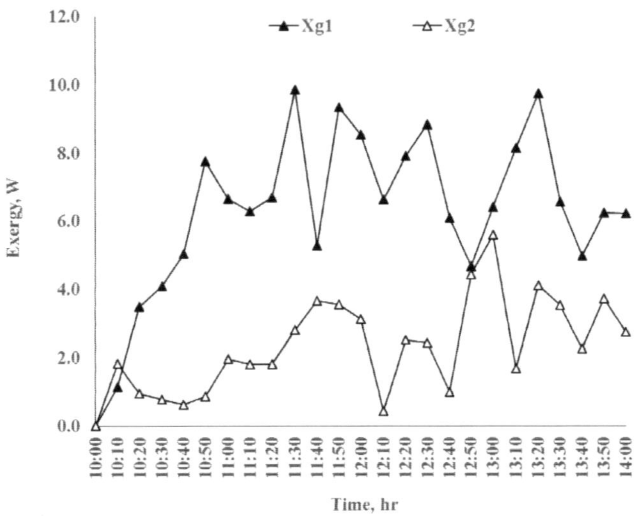

Figure 6. Exergy for the glasses *g1* and *g2* in the cover of the solar cooker

5. Discussion

As one can see in figure 3, temperatures for the glasses in the cover are high; 81.3°C, 89.0 °C, 113.7 °C and 119.2 °C that correspond to T_{g1_upper}, T_{g1_bottom}, T_{g2_upper} and T_{g2_bottom} respectively. This is the reason why the cover is identified as the zone where the heat losses is the most important. However, for the reached temperatures in the cover glass covers, the temperature for the water was 92.7 °C near to

saturation temperature in Mexico City. This temperature could be enough for domestic proposes, but is necessary to establish how much of the energy is not used in the heating process.
Considering figure 4, the comparative between conduction heat transfer in the glasses and the solar radiation in Watts (Solar Radiation/Ag) shows the glass 1 is where the conduction heat loss is biggest. This is explained due to glass 1 is exposed directly to the external environment and here the temperature is affected for the convection heat transfer.
As one can see in figure 5, the conduction heat transfer losses for the glasses is most big than 75%. For the glass 1, the conduction heat losses values are around to 90 %, which is very important and shows this is the zone where improvements for the solar cooker could be made. Here, it should be noted that the temperature inside the solar cooker is increasing during the heating process. This situation is important to take into account, since otherwise; this might be misinterpreted under the second law's concept and establish that the inlet energy in the cooker must be equal to what comes out of it. Thus, the heat energy outlet is the energy inlet plus the energy cumulated and released during the heating process.
Figure 6 indicates the quantity of useful energy, which is represented by the exergy. Their values are consistent with the conduction heat transfer losses in the glasses.
The results allow identifying the glasses as the zone where the solar cooker could be improved. In a first approximation, the thickness could be a parameter to study; because it has an important impact in the thermic resistance of the conduction heat transfer.

6. Conclusions

An evaluation of exergy considering the conduction heat transfer in glass covers of a solar cooker box-type was done. Results show that a great quantity of the entering energy to solar cooker during the heating process is not used. With the results is possible to identify how the cover glasses is the zone where the solar cooker could be improved. Almost 90% of the energy is not used during the heating process. The results can be useful in design or improvements of solar cooker box-type.

References

[1] Terres H Gordon M Lizardi A Lara A Morales J and López R 2011, Estudio de pérdidas en una estufa solar tipo caja con reflectores internos *35 Semana Nacional de Energía Solar* 1-8
[2] Terres H Morales J Lizardi A López R and Portillo R Estudio numérico de eficiencias en estufas solares tipo caja con reflectores internos *Revista internacional de métodos numéricos para cálculo y diseño de ingeniería* **29 (2)** 122-128
[3] Terres H Chávez S López R Lizardi A Lara A and Morales J 2015 Irreversibility and second law analysis in a solar cooker box-type *Proceedings of the ASME 2015 9th International Conference on Energy Sustainability* 1-6
[4] Pandey A Tyagi V Park S and Tyagi S 2012 Comparative experimental study of solar cookers using exergy analysis *J Therm Anal Calorim* **109** 425–431
[5] Panwar N 2013 Thermal modeling energy and exergy analysis of animal feed solar cooker *J. Renewable Sustainable Energy* **5** 1-17
[6] Nematia H and Javanmardi M 2012 Exergy optimization of domestic solar cylindrical-parabolic cooker *J. Renewable Sustainable Energy* **4** 1-11
[7] Kumar N and Vishwanath G 2012 An exergy based unified test protocol for solar cookers of different geometries *Renewable Energy* **44** 457-462
[8] Cengel Y and Boles M 2012 *Thermodynamics. An Engineering Approch* 7th ed McGraw-Hill Co

IOP Conf. Series: Journal of Physics: Conf. Series **792** (2017) 012013 doi:10.1088/1742-6596/792/1/012013

Evaluation of the cooking power in three different solar cookers box-type

H Terres[1] A Lizardi S Chávez R López and M Vaca
[1]Universidad Autónoma Metropolitana, Unidad Azcapotzalco, Departamento de Energía, Area de Termofluidos, San Pablo 180, Col. Reynosa Tamaulipas, Azcapotzalco, 02200, México, D. F.

E-mail: tph@correo.azc.uam.mx

Abstract. In the present work calculations for the cooking power in three different solar cookers are shown. The designs considered for the solar cookers are square, rectangular and octagonal. Agree to the results, a solar cooker with larger area for the solar radiation inlet has the biggest cooking power. The cooking powers obtained for the solar cookers are 4.04 W (0.49 m^2), 2.06 W (0.15 m^2) and 0.88 W (0.19 m^2) which correspond to square, rectangular and octagonal designs respectively. For the evaluation, the standard ASAE S580 JAN03 was considered to evaluate the cooking power in the solar cookers. Following the method established in this standard was possible to calculate the cooking power and evaluate the solar cookers at the same time. This activity except for what has been done in the standard, have not been done in other works.

1. Introduction

In the evaluation of solar cookers, different parameters are considered to establish their yield and operation. Among those parameters, the geometry and materials are evaluated frequently. However, in many cases, the cooking power has not been used as the major indicator. Among authors that have considered the cooking power, highlights the followings.

Funk [1] worked on standard presented at the Third World Conference on Solar Cooking (Coimbatore, India, January, 1997). His evaluation was aimed to determine the practical uses for solar cookers designs and possible differences when it used in variable conditions as locations and geometry or the solar cookers.

El-Sebaii and Ibrahim [2] present results for the experimental tests on a solar cooker box-type considering the cooking power standard procedure. Their results allow to establish the cooking power related to the temperature and solar radiation. This work was done for a one case of solar box cooker.

Sethi et al. [3] did a study about qualitative and quantitative variables that intervene in the operation of a solar cooker with parallelepiped cooking vessel design. Their results are identified with figures called figures of merit F_1 and F_2. Also, they developed a mathematical model to compute the effect of the solar radiation in the solar cooker.

Harmin et al. [4] evaluated a design for a optogeometrical arrangement in a solar cooker box-type. Their device was built using local materials in Alegian Sahara. Agree to their results, its effectiveness allow

Content from this work may be used under the terms of the Creative Commons Attribution 3.0 licence. Any further distribution of this work must maintain attribution to the author(s) and the title of the work, journal citation and DOI.
Published under licence by IOP Publishing Ltd

IOP Conf. Series: Journal of Physics: Conf. Series **792** (2017) 012013 doi:10.1088/1742-6596/792/1/012013

to cook meals for a family of four persons. In their experimental tests the cooking power was considered as the principal parameter for the results.
Purohit and Purohit [5] studied a paraboloid concentrator type solar cooker. In this study they consider parameters as efficiency factor, heat loss factor, overall heat loss coefficient and cooking power for their experimental tests and results. They highlighted the importance of the cooking power due to it indicates the real behavior and yield in a solar cooker.
Esteves et al. [6] presented a study for a solar furnace considering qualitative and quantitative parameters for the same one. The solar furnace operates in Nacunan Mendoza, Argentina. The solar absorber, reflectors, gate, interior chamber and cooking power were evaluated and their results allow to establish if the use time could have impact on them.
Servin et al. [7] determined the cooking power in a solar cooker called Jorhejpatarantskua. According to their results, the cooking power obtained is very important due to the value of 136 W is maximum value reached for a solar box cooker in similar designs.
Gonzalez and Servin [8] presents a mathematical model, which is adjusted with data obtained from an experimental design. This model is useful to estimate some parameters and allows comparing the performance of solar cookers considering testing protocols and international standards such as: standardized cooking power and thermal efficiency.
In this work, a comparative of the cooking power for three different solar cookers are shown. This work in difference of other previous works, takes in account three solar cooker in operation at the same time, which have not been done for other authors.

2. Experimental procedure

In the experimental tests three solar cookers box-type were used. These devices are integrated by the following elements: 1. A cover with two flat glasses (which are identified as T_{g1} (external glass) and T_{g2} (internal glass)) with a clearance between them. 2. Reflectors placed on the cover outer of the cooker 3. Internal reflectors made in commercial aluminium paper placed to different tilt angles, 4. Thermal insulator placed in the lateral part of the same one, and 5. Recipient contains the product to cook. The solar cookers are locked tightly; this allows reaching considerable temperatures in the water.
In figure 1, the solar cookers are shown in figure 1. In figures 2, 3 and 4 the solar cookers: Square, Rectangular and Octagonal are presented respectively. The solar cooker with octagon design does not have external reflectors as the other cookers evaluated. It was selected to calculate the effect of the reflectors in the heating process.
In the tests, 1.5 liters of water were used as the fluid to be heating. The general dimensions for the solar cookers are 0.7 x 0.7 x 0.40 m (square), 0.5 x 0.3 x 0.35 m (rectangular) and 0.5-diameter m (octagonal) with 0.4-high m.
A radiometer Eppley model 8-49 was used for measure the solar radiation in horizontal plane. Thermocouples type-k were used for the temperature measures. An acquisition system data integrated by a compact Field Point of National Instruments and Lab View software were instrumented for the tests.
The tests were done on April 1, 2016 and the place Mexico City. The interval considered was 4 hours and 10 minutes (10:00 to 14:10).

Figure 1. Solar cookers box-type: with exterior and interior reflectors.

Figure 2. Solar cooker: Square design (cut presentation)

Figure 3. Solar cooker: Rectangular design (cut presentation)

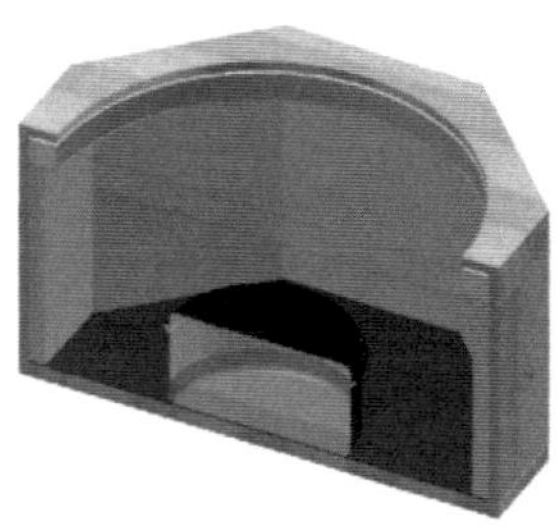

Figure 4. Solar cooker: Octagonal design (cut presentation)

3. Experimental results

To measure temperatures, thermocouples in specific points on glass covers were placed. The temperatures measured are: 1. glasses of the cover ($T_{g1\ upper}$, $T_{g2\ bottom}$), 2. lid of the recipient (T_{lid}), 3. body of the recipient (T_r), 4. Water (T_w), 5. internal environment (T_{int_env}), 6. external environment ($T_{ext\ env}$). In figures 5, 6 and 7 the temperature distributions values for the solar cookers: square, rectangular and octagonal are shown respectively. In figure 8, a comparative of the temperature water for the solar cookers is shown.

IOP Conf. Series: Journal of Physics: Conf. Series **792** (2017) 012013 doi:10.1088/1742-6596/792/1/012013

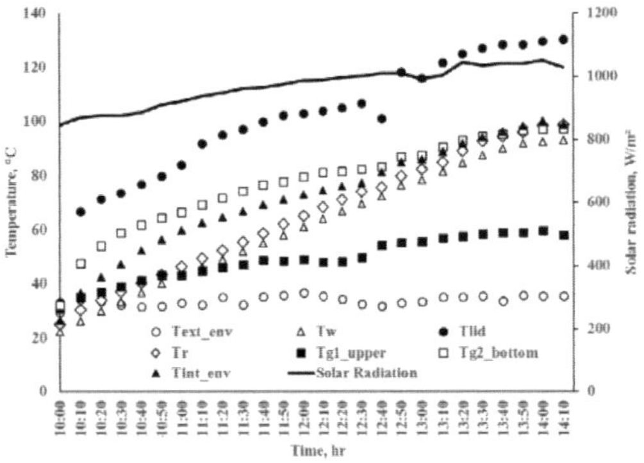

Figure 5. Temperatures for Solar cooker: Square design

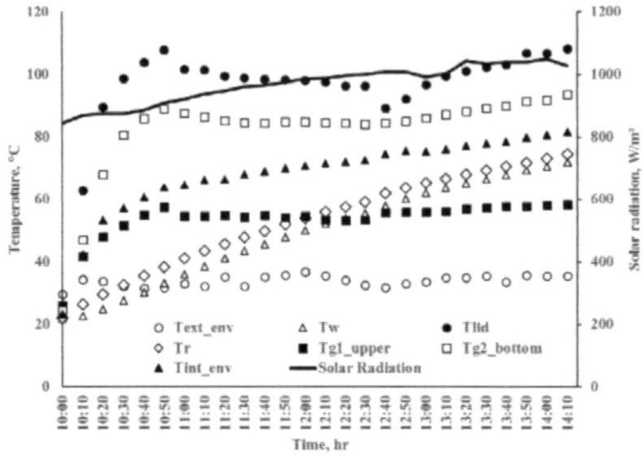

Figure 6. Temperatures for Solar cooker: Rectangular design

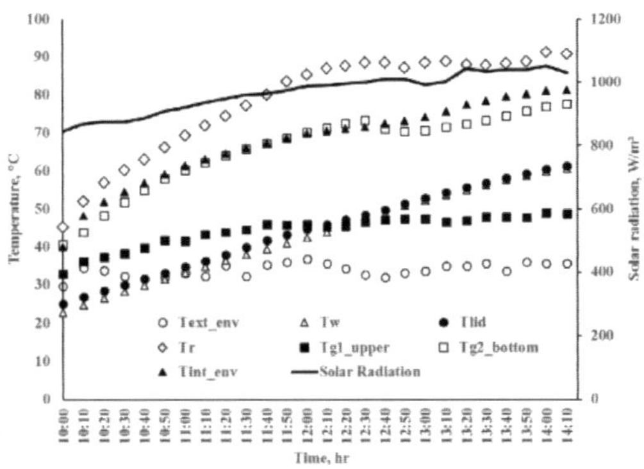

Figure 7. Temperatures for Solar cooker: Octagonal design

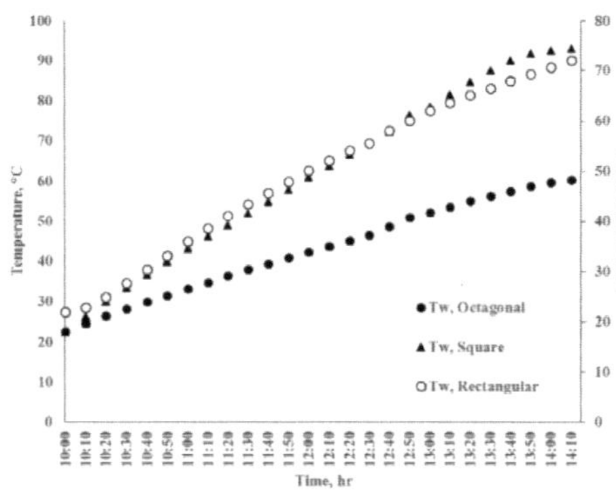

Figure 8. Temperatures comparative in the water for the solar cookers

4. Determination of cooking power

For the cooking power calculation, the standard ASAE S580 JAN03 was considered. According to this standard some conditions must be controlled: 1. Wind velocity. During tests, the velocity should be less than 1.0 m/s. This happened in the test. 2. Ambient temperature. Tests should be conducted when ambient temperatures are between 20 and 35 °C. As one can see, the test reports this interval. 3. Insolation. Available solar energy over the solar cooker must be measured using a radiation pyranometer. The device to measure solar radiation was an Eppley pyranometer. Some equations are considered

$$P_i = mC(P_2 - P_1) \tag{1}$$

Where, P_i = cooking power (W), T_2 = final water temperature, T_1 = initial water temperature, m = water mass (kg), C = heat capacity (4186 J/[kg·K]). The time, 600 seconds represent a ten-minute interval.

$$P_s = P_i(700/I_i) \tag{2}$$

Where, P_s = standardized cooking power (W), P_i = interval cooking power (W), calculated in (1), I_i = interval average solar insolation (W/m^2).

$$T_d = T_w - T_a \tag{3}$$

Where, T_d = temperature difference (°C), T_w = water temperature (°C), T_a = ambient air temperature (°C). After to apply equations (1), (2) and (3) to the experimental data, P_s is plotted against T_d for each time interval. A linear regression of the plotted points is used to find the relationship between cooking power and temperature difference in terms of intercept a (W) and slope b (W/°C) or $P_s = (a + b)\ T_d$. In figures 8, 9 and 10 the plotted for the solar cookers square, rectangular and octagonal are shown respectively.

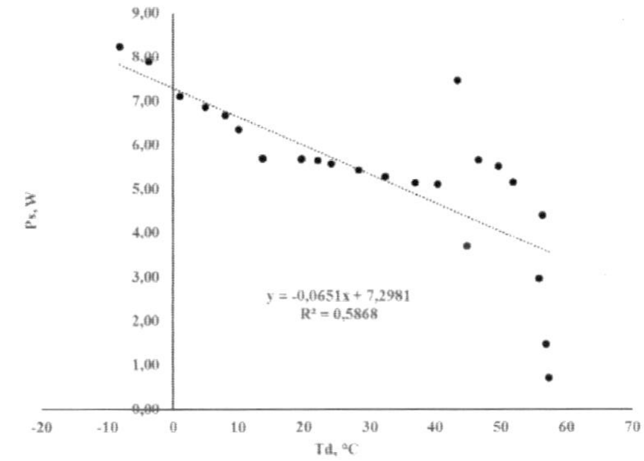

Figure 8. Td vs Ps and regression for Solar cooker Square

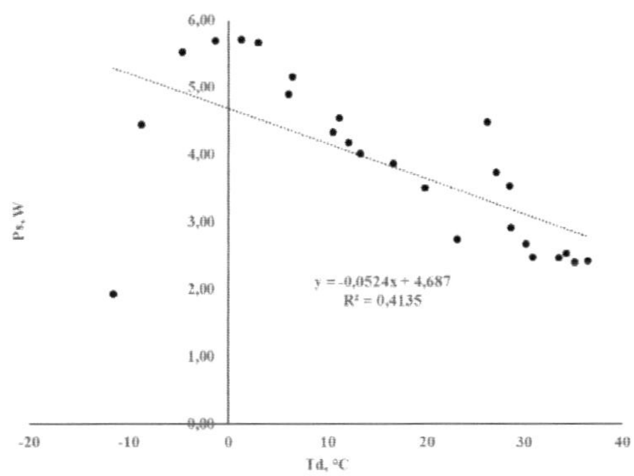

Figure 9. Td vs Ps and regression for Solar cooker Rectangular

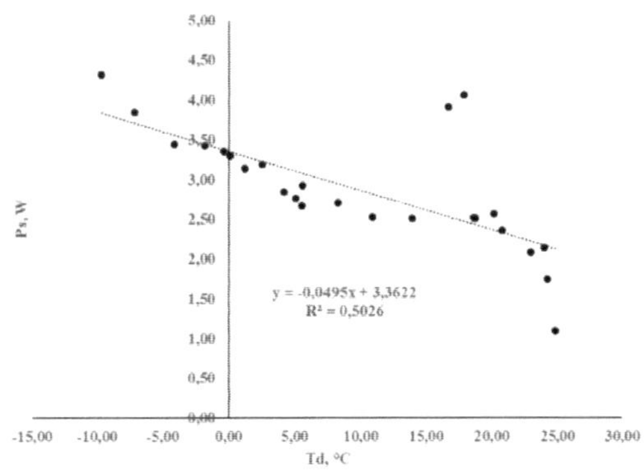

Figure 10. Td vs Ps and regression for Solar cooker Octagonal

5. Discussion

Agree to the results, the maximum reached temperature for the water correspond to the solar cooker with square design, its value is 92.9 °C, while the minimum temperature for the water is for the solar cooker with octagonal design with 60.4 °C. These results can be associated to the external reflector, which is present in the square design, which augment the solar radiation toward to the cooker. In addition, the reflexion area inside of the square design is bigger than the octagonal cooker. The solar radiation that impacts over the solar cookers is 1050 W/m² which is very important for April, due to May is the month with biggest radiation in Mexico City. Considering the equations by regression for the solar cookers showed in figures 8, 9 and 10 and evaluating for a temperature of 50 °C (what is required by the standard), the cooking powers are 4.04 W (0.49 m^2), 2.06 W (0.15 m^2) and 0.88 W(0.19 m^2) for the solar cookers: square, rectangular and octagonal respectively. These results must be evaluated taking in account the area for each case. As one can see, when the area increases, the cooking power increases. However, the cooker with octagonal design has a cooking power bigger than the cooker with rectangular form. In this case, it is explained because the reflection area inside of the octagonal cooker is bigger than the rectangular cooker. By other hand, it is very important to point out, that the values reported by other authors for the cooking power in solar cookers is near to 50 W, but the areas considered for those cases is bigger than 2.5 m^2. According to the obtained results, the solar cooker with better cooking power is the solar cooker with square design.

6. Conclusions

Calculations for the cooking power in three different solar cookers were determined. According to results, the solar cooker with larger area has the biggest cooking power. In this case, a solar cooker with a square design showed the biggest cooking power. The method used in this work can be useful to determine different solar cookers working at the same time. Searching improve the results obtained in this work, new configurations for solar cookers will be considered. It will allow to establish a complete study for several cases of solar cookers.

References

[1] Funk P 2000 Evaluating the international standard procedure for testing solar cookers and reporting performance *Solar Energy* **68(1)** 1–7

[2] El-Sebaii A and Ibrahim A 2005 Experimental testing of a box-type solar cooker using the standard procedure of cooking power *Renewable Energy* **30** 1861–1871

[3] Sethi V Pal D and Sumathy K 2014 Performance evaluation and solar radiation capture of optimally inclined box type solar cooker with parallepiped cooking vessel design *Energy Convers. Manage.* **81** 231-241

[4] Harmim A Merzouk M Boukar M 2013 Design and experimental testing of an innovative building-integrated box type solar cooker *Solar Energy* **98** 422-433

[5] Purohit I and Purohit P 2009 Instrumentation error analysis of a paraboloid concentrator type solar cooker *Energy Sustainable Dev.* **13** 255-264

[6] Esteves A Buenanueva F Orduna D and Cuitiño G 2008 Estudio del comportamiento de hornos solares *Avances en Energías Renovables y Medio Ambiente* **12** 71-78

[7] Servín H González M López Sosa L and Pérez D 2012 Estimación de la potencia de cocción de la estufa solar Jorhejpatarantskua *36 Semana Nacional de Energía Solar* 472-475

[8] González M and Servín H 2014 Thermal model to estimate the standardized cooking power of solar cookers *Revista Electronica Nova Scientia* 49-65

[9] ASAE S580 JAN03 2013 Testing and Reporting Solar Cooker Performance *American Society of Agricultural Engineers* 824-826

Effect on hydrogen adsorption due to a lonely or a pair of carbon vacancies on the graphene layer.

Arellano J S[1]
[1]Departamento de Ciencias Básicas, Universidad Autónoma Metropolitana Azcapotzalco, Av. San Pablo 180 Col. Reynosa Tamaulipas, 02200 México City.

E-mail: jsap@correo.azc.uam.mx

Abstract. The influence on the hydrogen molecule adsorption on a pristine and a defective graphene layer is compared. The different lengths for the C-C bonds on the graphene layer with one vacancy are visualized and compared respect to pristine graphene. The energy of formation of one vacancy is calculated and a comparison of the binding energy for the hydrogen molecule is presented when the molecule is adsorbed on pristine graphene or on the defective graphene layer. The adsorption is studied for a single vacancy and at least for two different pairs of carbon vacancies. The qualitative general result, and contrary to the expected effect of the carbon vacancies on the hydrogen adsorption is that the rearrangement of the carbon atoms on the defective graphene layer allows only a relatively small increase in the magnitude of the binding energy for the hydrogen molecule.

1. Introduction

The hydrogen adsorption by graphitic materials has been studied from the theoretical point of view at least since the 2000 year. Now we look for the effect of some carbon vacancies at the graphene layer, on the possible new results for the hydrogen molecule adsorption compared with those obtained previously for pristine graphene. In the next section a resume of the pristine graphene calculation is given. Then the one carbon vacancy graphene layer is designed and relaxed. The hydrogen molecule is added above the defective graphene layer and the entire system formed by the hydrogen molecule and the mono vacancy graphene layer is relaxed and the results are studied and discussed. There are discussions for at least two initial and different configurations, both with a pair of vacancies. The principal difference is that the initial distances between the vacancies are different for those configurations. These can be compared with the given figures. They allow appreciating that due to the differences in the sizes of the cavities it could be find some differences for the final hydrogen adsorption energies for the studied configurations. These results can be compared with those obtained for a lonely or a pair of vacancies on the BN layer [1]. For that case the studied systems were the entire BN layer, B vacancy, N vacancy and B and N vacancies (a pair of vacancies). As it was expected the hydrogen adsorption energy was in some cases greater for the BN systems mentioned above, respect to the defective graphene layer.

2. Pristine graphene layer.

Quantum espresso computer code was used for all the computer calculations given here. The pristine graphene layer has been designed with the equilibrium cell parameter a = 4.6117 a.u., that was obtained before [2]. Again c.fhi.UPF and h.cpi.UPF, LDA pseudopotentials were used for all the calculations. The pristine graphene supercell is formed with 32 atoms, with C-C distance 2.6626 a.u. (1.4090 Å). The total static energy, (E_{tot}) for the pristine supercell is -365.58814088 Ry, with Fermi energy of -1.4977 eV. The cell parameter is the distance between any pair of second neighbor carbon

Content from this work may be used under the terms of the Creative Commons Attribution 3.0 licence. Any further distribution of this work must maintain attribution to the author(s) and the title of the work, journal citation and DOI.
Published under licence by IOP Publishing Ltd

atoms. There is a distance of 25 a.u. (13.2294 Å), between adjacent graphene layers. This distance is close to that used in a similar study [3] and it is big enough to assume there is no interaction between the images of the graphene layers.

3. One vacancy graphene layer.

To obtain the relaxed system with a vacancy on the graphene layer, schematically illustrated in figure 1, it was simply removed in the pristine configuration one carbon atom that in the final and relaxed configuration appears as an empty space between carbon number 14 and 15 along the horizontal axis of the supercell. By symmetry the vacancy site is equidistant to carbon atoms with number 8, 15 and 21 on figure 1. A repetition of the supercell shows that there are at least two entire hexagonal cavities in any direction around the cavity of the vacancy. One interesting thing to point out is about the calculation of the energy of formation (E_f) of a single vacancy for the graphene layer. To calculate that energy, it is necessary to take into account three energies: a) the total static energy E_{tot}, obtained from the final coordinates for the relaxed entire graphene layer system, b) the non-relaxed initial energy for the defective layer (E_{iwr},) and c) the relaxed final energy (E_{fr},) for the vacancy graphene layer. The calculated values for those energies are E_{iwr} = -353.54841261 Ry and E_{fr} = -353.56568555 Ry. Following the examples of defects in solids given by Alessandra Satta [4], to calculate the energy of formation of a single vacancy in silicon, here the energies 0.6150 and 0.5978 Ry were obtained for the unrelaxed and relaxed energies of formation of a single vacancy for the graphene supercell. Those energies are 8.368 and 8.133 eV, respectively, and are in good agreement with the formation energy obtained in other work, [5], for single and double vacancy on graphene. Compared with the corresponding energies in silicon, 3.27 and 3.26 eV, it can be seen that the energies are close to 2.5 times larger for carbon in graphene than for the silicon structure. For silicon and for the graphene vacancy, the relaxed energies are more negative than for the non-relaxed systems. Figure 2 shows the distortions of the relaxed graphene layer due to the carbon vacancy. The energy for pristine graphene layer is 12.039728 Ry more negative than for the graphene layer with one vacancy. This energy difference is equal to 163.8088 eV. Something similar occurs for silicon energy because this is around 8.1626018 Ry (111.05784 eV) more negative than for the system with a single vacancy.

There are some differences for C-C bond lengths and angles between the carbon atoms when one compares the pristine graphene layer and the final relaxed vacancy system. To clarify this, take into account the correspondence between the positions of each one of the carbon atoms on figure 1, and their positions in the final configuration for the relaxed defective layer, shown on figure 2. Numbers from 1 to 31 are assigned consecutively from left to right and from above to below the array of carbon atoms. Because the symmetry above and below the line joining the more left carbon atom (number 13) and the righter carbon atom (number 19) along the horizontal axis of the supercell, it is expected equal results for lengths of C-C bonds and for angles between 3 adjacent carbon atoms if they are in symmetrical positions respect to the horizontal axis of the supercell. This is confirmed with measured distances given in table 1, for different pairs of neighboring carbon atoms. The distance d(angstroms), between the carbon atoms Ci and Cj, is denoted by d (Ci, Cj).

Here we comment about the lengths of the C-C bonds for the relaxed vacancy graphene layer. There are nine different distances between pairs of closer carbon atoms; the smaller (bigger) is 1.3802 Å (1.4328 Å). There are for example, six (two) pairs of bonds with the smaller (biggest) bond length. Above the horizontal axis of the supercell the smaller bond length is for the pairs of atoms (3, 8), (7, 8)

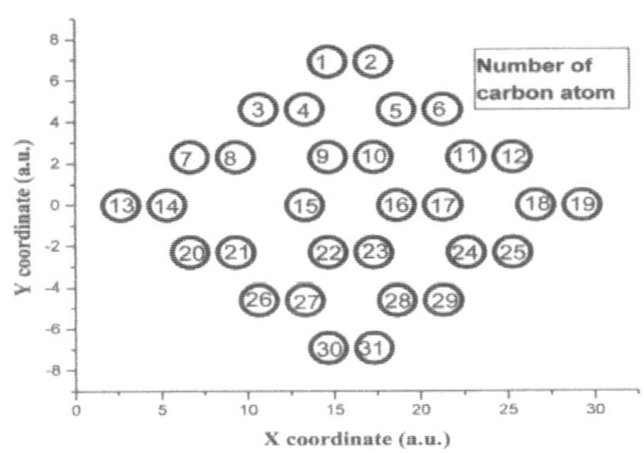

Figure 1. Positions of the carbon atoms in the graphene cell in the relaxed final configuration with one vacancy. Numbers from 1 to 31 identify each one of the carbon atoms in the system. The bond between adjacent carbon atoms can be seen in a better way in figure 2.

and (9, 15). It was used Xcrysden [6] to build figure 2. Displaying the final coordinates of the atoms, it could be seen that the biggest bonds were between carbon atoms pairs (9, 10) and (22, 23). Those are respectively the two horizontal bonds of the central hexagonal cavity on that figure. The angles around the cavity are (in the clockwise direction, beginning at carbon site 14), 116.717^0, 123.197^0, 123.110^0, 123.197^0, 116.717^0, 123.196^0, 123.108^0, 123.196^0, 116.717^0, 123.197^0, 123.111^0, and 123.197^0. The first angle, 116.717^0, has as vertex the site for carbon number 14 and is defined by the bonds formed by carbon number 20 and 14 and between carbon number 14 and 7. The angle with vertex at site 7 is the next one, 123.197^0, and so on. In the pristine graphene layer all the angles are 120^0. The maximum angular deviation respect to that angle is 3.197^0 or -3.283^0. The relatively small differences of lengths and angles respect to the pristine graphene layer will have some perhaps little change on hydrogen adsorption that is discussed in the next section.

4. Hydrogen adsorption on a single vacancy graphene layer.

The hydrogen adsorption energy is obtained using equation (1)

$$E_{ad}(H_2) = E(H_2 \text{ on gl-1v}) - E(\text{gl-1v}) - E(H_2) \quad (1)$$

E (H_2 on gl-1v) is the total energy for the relaxed system formed by one H_2 molecule and the graphene layer (gl) with one vacancy (1v): (gl-1v), E (H_2) is the total energy of only the hydrogen atoms in the final configuration and in a similar way, E (gl-1v) is the total energy of only the carbon atoms in their final positions corresponding to the final stage of the relaxation of the entire system. With the calculated values, E (H_2 on gl-1v) = -355.83591945 Ry, E (H_2) = -2.26356214 Ry and E (gl-1v) = -353.56555313 Ry, the hydrogen adsorption energy is E_{ad} (H_2) = -0.00680418 Ry (-0.09257 eV). These data are given on table 2. This adsorption energy is a small energy, is less than one tenth of one electronvolt. The hydrogen molecule was adsorbed on pristine graphene with an energy around -0.070 eV [2]. The negative sign for the energy means the hydrogen molecule is adsorbed to the defective graphene layer. Panels a, b and c of figure 3 are different views for the final configuration of the hydrogen molecule above the vacancy graphene layer. The initial position of the molecule was parallel to the layer. As can be seen in the final configuration the right hydrogen atom is closer to the layer than the left one. The H-H distance is 1.4704 a.u., a little bigger than for the hydrogen molecule in equilibrium. The distances of the hydrogen atoms to the layer are around 4.49 and 5.79 a.u.

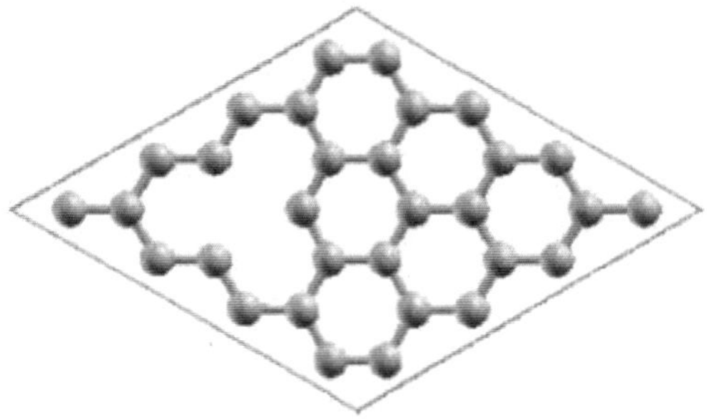

Figure 2. Relaxed final configuration for the one vacancy graphene layer. There are nine different values for C-C lengths. The smaller bond length appears for six carbon bonds, between the pairs of atoms (3, 8), (7, 8), (9, 15), (26, 21), (20,21) and (22,15). The value for that length is 1.3802 Å. In the pristine graphene layer the C-C length is 1.4090 Å.

Table 1. Distances d, (angstroms), between pairs of carbon atoms. Distances goes from the smaller, 1.3802 to the biggest, 1.4328 Å. The differences for these distances respect to the theoretical equilibrium distance d(C-C) for the pristine graphene layer, 2.6626 a.u. (1.4090 Å), are -2.044 and 1.689 % respectively.

d(angstroms)	Distance between pairs of carbon atoms, d (Ci, Cj)		
1.3802	d(7, 8) = d(20, 21)	d(3, 8) = d(21, 26)	d(9, 15) = d(15, 22)
1.3940	d(13, 14)	d(1, 4) = d(27, 30)	
1.4038	d(18,19)	d(2, 5) = d(28, 31)	
1.4056	d(11, 12) = d(24, 25)	d(10, 16) = d(16,23)	d(6, 11) = d(24, 29)
1.4071	d(3, 4) = d(26, 27)	d(7, 14)= d(14, 20)	d(4, 9) = d(22, 27)
1.4104	d(1, 2) = d(30, 31)		
1.4128	d(16,17)	d(11, 17) = d(17,24)	
1.4180	d(5, 6) = d(28, 29)	d(5, 10) = d(23, 28)	d(12, 18) = d(18,25)
1.4328	d(9, 10) = d(22, 23)		

Table 2. The hydrogen adsorption energy on the single vacancy graphene layer, Ead (H_2), is the sum of the energy for the hydrogen molecule on the graphene layer with one vacancy, E (H_2 on gl-1v) minus the sum of the energies of the graphene layer with one vacancy, E (gl-1v), and the energy of the hydrogen molecule, E(H_2). Energies in Ry units.

Ead (H_2)	E (gl-1v)	E (H_2)	E (H_2 on gl-1v)
-0.0068042	-353.566	-2.2636	-355.8359195

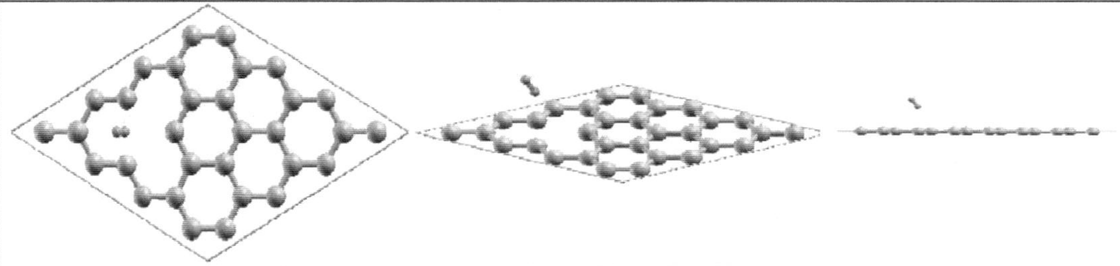

Figure 3. Final configuration for the hydrogen molecule above one vacancy graphene layer.

5. Hydrogen adsorption on two vacancies of the graphene layer.

Next we see if a second carbon vacancy can increase the adsorption energy for the hydrogen molecule. There are several possible elections for the second carbon vacancy. The selected pair of vacancies discussed here is only one of the possibilities. The initial and relaxed graphene layer with double vacancy is those of figure 4a and 4b. The total energy becomes 0.533 Ry more negative for the relaxed configuration respect to the initial one, without relaxation. Greater distortion of the graphene layer is obtained for this relaxed configuration with two carbon vacancies. When the hydrogen molecule is added and placed above the graphene layer with that pair of vacancies, the initial and final relaxed configurations are those of figure 4c and 4d. Besides the differences of this system with the lonely vacancy configuration and contrary to the expected, the adsorption energy for the hydrogen molecule is now -0.0917 eV, very close as the obtained with only one carbon vacancy. When the second vacancy is close to the first one, and both on the principal axes of the graphene supercell a high symmetry is obtained for the relaxed configuration. Relaxing the system with one hydrogen molecule above this new double vacancy, the hydrogen molecule is adsorbed and still remains parallel to the graphene layer with an interatomic distance d(H-H) of 1.47 a.u. The distance between the hydrogen molecule and the double vacancy layer is reduced to 4.5 a.u. This can be compared with the corresponding distance 5.07 a.u., obtained for the pristine graphene layer [2]. Then, the double vacancy allows an approach much closer between the hydrogen molecule and the defective layer compared with the entire graphene layer without vacancies. Table 3 contains the different energy contributions to obtain the hydrogen energy adsorption, $Ead(H_2)$. This is the energy of the hydrogen molecule on the graphene layer with two vacancies, given by E (H_2 on gl-2v) minus E(gl-2v), the energy of the graphene layer with two vacancies, and minus $E(H_2)$, the energy of the hydrogen molecule. The hydrogen energy adsorption is close to - 0.0076 Ry. This means that there is a relatively small increase (10.5%) in the hydrogen adsorption energy obtained from the value -0.0068 to -0.0076 Ry, for one and two vacancies. Some other configurations, not shown here, but that have been studied, show in general that the hydrogen adsorption energies are very close to those given before even the final orientation for the hydrogen molecule could be very different between one and other. For one of those configurations the hydrogen adsorption energy was -0.0075 Ry (-0.103 eV). Contrary to the expected and the bigger size of the cavity created by the pair of carbon vacancies, still the adsorption energy is too low, and almost the same as with only one vacancy.

Table 3. The hydrogen adsorption energy on the double vacancy graphene layer, Ead (H_2), is the sum of the energy for the hydrogen molecule on the graphene layer with two vacancies, E (H_2 on gl-2v), the negative of the energy of the graphene layer with two vacancies, E(gl-2v) and the negative of the energy of the hydrogen molecule, $E(H_2)$. Energies in Ry units.

Ead (H_2)	E (gl-1v)	E (H_2)	E (H_2 on gl-1v)
-0.00757685	-342.11511875	-2.26356367	-344.38625927

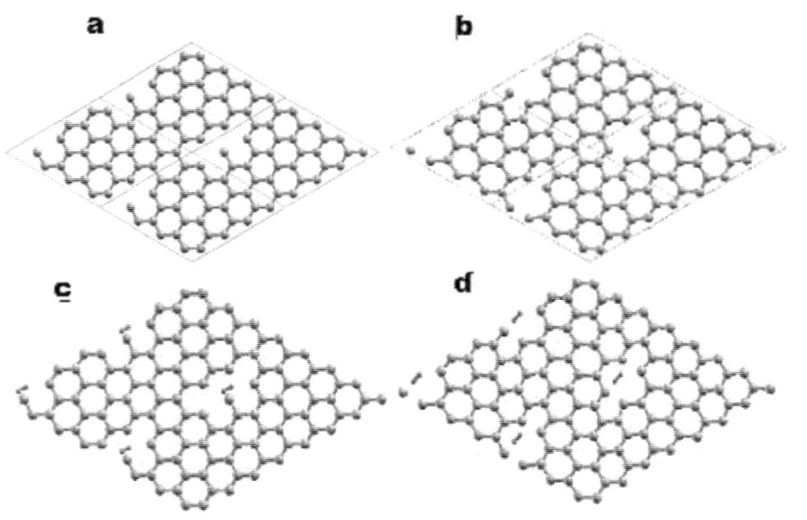

Figure 4. Initial double carbon vacancy configuration (a) and relaxed graphene layer with the pair of vacancies (b). Initial double carbon vacancy with one hydrogen molecule above the unrelaxed graphene layer (c). Final configuration for the hydrogen molecule and the pair of carbon vacancies on the graphene layer (d). The hydrogen atoms are 4.63 and 4.98 a.u. above the double carbon vacancy. The hydrogen molecule length is 1.455 a.u.

5. Conclusions.

A general trend about the adsorption of hydrogen by defective graphene layer and BN layer [1] is that there is only a small increase of around 10% in the adsorption energy.

References.

[1] Arellano J S 2015 Double vacancy on BN layer: A natural trap for Hydrogen Molecule*.J. Phys: Conf. Ser.* **582** 1742-6596

[2] Arellano J S Molina L M Rubio A and Alonso J A 2000 *J. Chem. Phys.* **112** 8114

[3] Granja A Alonso J Cabria I and López M. 2015 *RSC Adv.* **5** 47945-47953

[4] Satta A 2005 Hands-on tutorial on the Quantum-Espresso Package *Examples of defects in solids from Espresso*

[5] Borisova D Antonov V and Proykova A 2013 Hydrogen Sulfide Adsorption on a Defective Graphene *Int. J. Quantum Chem.* **113** 786–791

[6] Kokalj A 1999 XCrySDen –a new program for displaying crystalline structures and electron densities *J. Mol. Graphics Modell.* **17** 176-179.

Density of states of adsorbed sulphur atoms on pristine and defective graphene layers.

Arellano J S[1]
[1]Departamento de Ciencias Básicas, Universidad Autónoma Metropolitana Azcapotzalco, Av. San Pablo 180 Col. Reynosa Tamaulipas, 02200 México City.

E-mail: jsap@correo.azc.uam.mx

Abstract. The density of states for adsorbed sulphur atom on a graphene layer system is discussed for pristine graphene layer and for mono and divacancies on the graphene layer. To our knowledge this is the first time that an entire adsorption of the sulphur atom is reported at the plane of the carbon atoms, when there is a pair of closer vacancies at the graphene layer.

Introduction

The adsorption of sulphur atoms on a graphene layer and on mono and divacancies on the graphene layer is discussed showing how the changes on the electronic density of states are. The most important result is the fact that due to the presence of sulphur atoms, there are many available states beyond the Fermi level. This change promises new possibilities about adsorption or capture of some other pollutants atoms or molecules. A brief resume for pristine graphene layer and of one and two carbon vacancies on the graphene layer is given. Then the adsorption of one or two sulphur atoms on those systems is given with a comparison and discussion of their electronic densities of states. One work about this class of study explores simultaneous doping of the graphene layer with S and N atoms, for example [1]. Also several factors could be studied about the performance of a sulfur graphene composite cathode material [2]; this was applied in a lithium-sulfur battery. Also the adsorption of molecules such as NO, NO_2 or O_2 were studied theoretically over S-doped graphene [3]. It is expected that this kind of novel systems could be used as gas sensors for those or other polluting gases. Even more close to this work it was found an article about hydrogen sulfide adsorption on defective graphene [4]. The adsorption is studied as a function of the carbon vacancy concentration. There is covalent binding between the S atom of the hydrogen sulfide molecule and the three closer carbon atoms around the graphene vacancy. In a very recently accepted paper [5], there is a DFT study for H_2O_2 adsorption and activation on a graphene surface (formed by 73 C atoms), when this is doped with different atoms such as N, B, S, Pd, Pt, Au, Ag and Cu. The Gaussian calculation includes 21 H atoms around an equal amount of the more external C atoms of the graphene layer.

1. Pristine graphene layer.

The computer free code Quantum espresso is used to do the computer calculations. A 4x4 graphene layer (32 carbon atoms) with cell parameter a = 4.6117 a.u. [6] was used to build the initial pristine graphene layer. The pseudopotentials used for the calculations were c.fhi.UPF and S.pz-van_ak.UPF.

For the optimized cell parameter, the total energy, (E_{tot}) was -365.58814088 Ry, with Fermi energy -1.4977 eV. It was enough to use 25 a.u. for the distance between graphene layers. The calculated DOS (electronic density of states), figure 2, shows the known result: DOS at the Fermi level is zero.

Content from this work may be used under the terms of the Creative Commons Attribution 3.0 licence. Any further distribution of this work must maintain attribution to the author(s) and the title of the work, journal citation and DOI.
Published under licence by IOP Publishing Ltd

IOP Conf. Series: Journal of Physics: Conf. Series **792** (2017) 012015 doi:10.1088/1742-6596/792/1/012015

2. Sulphur atom above one carbon atom on pristine graphene layer.

To begin the study of the interaction of sulphur atoms with the graphene layer, only one sulphur atom was placed above one of the carbon atoms of the pristine graphene layer. The relaxation of the system show that S atom was adsorbed 2.1938 Å above the C atom of the graphene layer with adsorption energy of -1.73 eV. Figure 1 shows the studied system. The corresponding DOS is shown on figure 2. Due to the S-C bond, there are now a few available electronic states above the Fermi state.

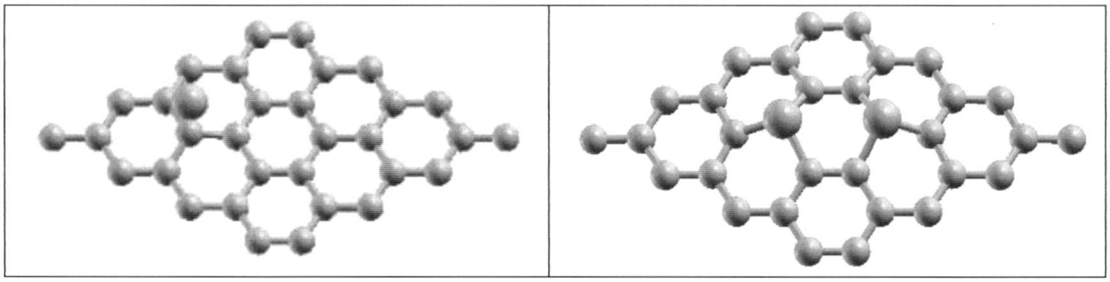

Figure 1. Relaxed system formed by one sulphur atom adsorbed above one carbon atom of the graphene layer, (left panel). Double carbon vacancy with a pair of adsorbed sulphur atoms, (right panel). Sulphur atoms have bigger size than carbon atoms in this and other figures.

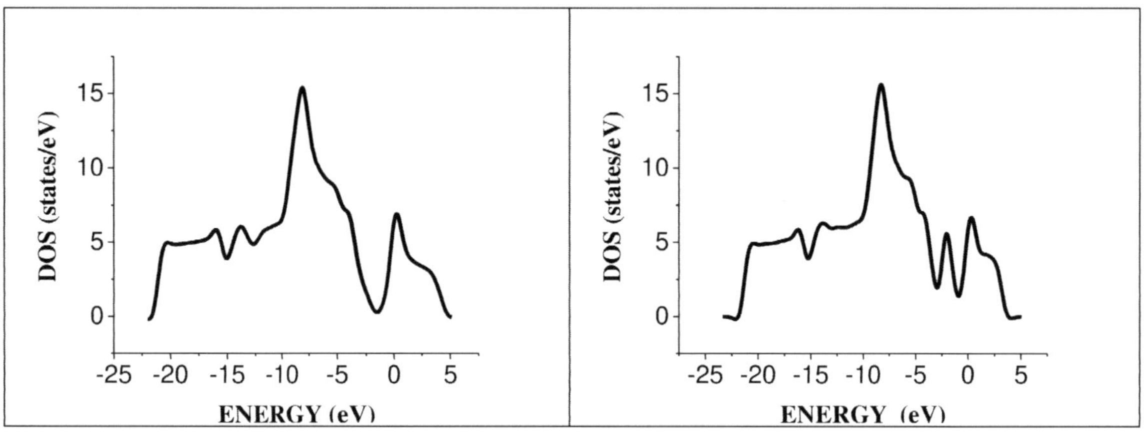

Figure 2. a) Density of electronic states for pristine graphene, Fermi energy = -1.533 eV; b) Density of electronic states for a sulphur atom adsorbed above one of the carbon atoms of pristine graphene, Fermi energy = -1.4818 eV.

3. Adsorption of one S atom above the center of one hexagonal cavity of the graphene layer.

Compared with the relaxation of a single S atom above a C atom, it was hoped that one sulphur atom could be adsorbed to a closer distance due to the space at the center of the hexagonal cavity. But contrary to the expected, the S atom remains 2.833 Å above the center of the cavity and the adsorption energy for S atom is only -0.625 eV, much less than -1.73 eV for the adsorption above a C atom. The first panel on figure 3 shows the density of states of a sulphur atom adsorbed above one of the hexagonal cavities of pristine graphene layer. For this case there are even more available states above the Fermi level, compared to the sulphur adsorption above one carbon atom.

4. Sulphur atom above one carbon vacancy graphene layer.

The design of one or two carbon vacancies on the graphene layer, follows very nearly the same as was done before [7], for the BN layer. Here are given only some details. After extracting one or two carbon atoms from the pristine graphene layer, one has to completely relax the defective carbon structure and obtain the relaxed total energy. One compares those energies with that obtained for the pristine graphene layer. For example, for the unrelaxed and relaxed energies of formation of a single vacancy for the graphene supercell, we obtained the energies 0.6150 and 0.5978 Ry [8]. DOS for one carbon vacancy graphene layer is shown on figure 3, panel b. There is a relative maximum value for DOS at the Fermi level, at -1.9274 eV. Again, there are electronic available states beyond the Fermi level. Besides here it is not shown a figure for the final configuration when only one sulphur atom is relaxed above the one carbon vacancy graphene layer, the final configuration is very nearly the same as figure 3 of reference [4]. In that case the hydrogen atoms of the hydrogen sulfide molecule are repelled by the defective graphene layer and form one hydrogen molecule above the sulphur atom and the carbon atoms of the vacancy graphene layer. The S atom makes protrude bonds out of the graphene plane, with the three closer carbon atoms of the vacancy. The hydrogen molecule remains above the sulphur atom.

5. Adsorption of S atom at a double carbon vacancy site.

Relaxing one sulphur atom above the center of a double carbon vacancy, created when two closer carbon atoms are omitted in the pristine graphene layer gives an interesting configuration. There are two pentagonal and two hexagonal cavities around the adsorbed sulphur atom that is almost entirely adsorbed on the same plane of the carbon atoms (see figure 5). Besides there is a distortion of the carbon atoms around the S atom, the changes are in such a way that in the final configuration the S atom makes bonds with four carbons atoms, the nearest to sulphur atom. The electron density of states goes beyond the Fermi energy level, localized at -1.4719 eV. This is shown on figure 4. To obtain that result, first it was necessary to do a scf (self-consistent-calculation), followed by an nscf (non-self-consistent-calculation), the dos (density of states) calculation and finally the calculation of the pdos (projected density of states). The same sequence of calculations was done for the different studied configurations. The final z coordinate of the sulphur atom was 0.165 Å and some of the carbon atoms around the S atom have z coordinates around 0.11 Å; that is why it can be said that the S atom is adsorbed nearly at the same plane of the carbon atoms of the defective graphene layer. It was thought that due to the bonding of the sulphur atom with four carbon atoms the DOS could be different as that obtained in figure 4. It was believed that necessarily at the Fermi level the DOS could have values far from zero. This was a non-expected result.

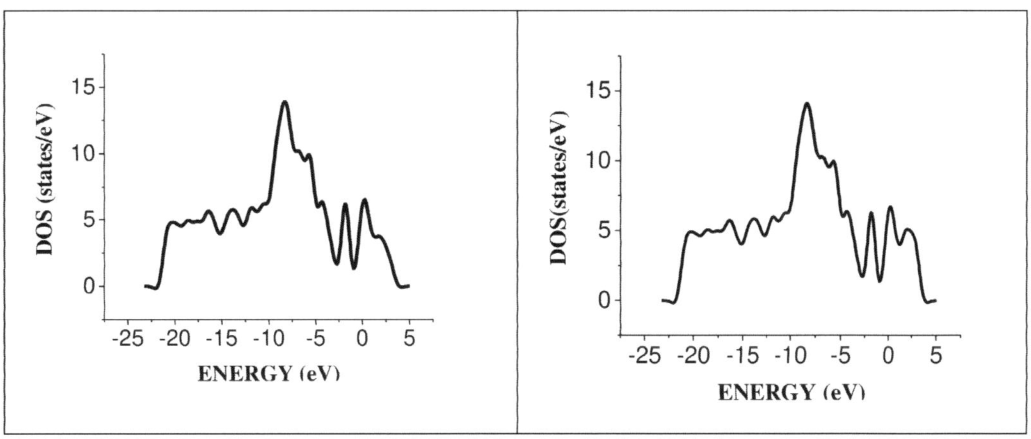

Figure 3. a) Density of electronic states of a sulphur atom adsorbed above one of the hexagonal cavities of pristine graphene layer, Fermi energy = -2.3157 eV; b) Density of electronic states created by a lonely carbon vacancy on the graphene layer, Fermi energy = -1.9274 eV.

IOP Conf. Series: Journal of Physics: Conf. Series **792** (2017) 012015 doi:10.1088/1742-6596/792/1/012015

7. Adsorption of two S atoms at a double carbon vacancy site.

Two S atoms were relaxed above a pair of carbon vacancies that correspond to two opposite carbon atoms at one of the hexagonal cavities in the pristine graphene layer. Besides the distance between the carbon vacancies, there was not enough space to allow the complete adsorption of both S atoms at the graphene layer. The final configuration, (second panel of figure 1), is a very symmetrical one. The S atoms stay a little above a distorted graphene layer.

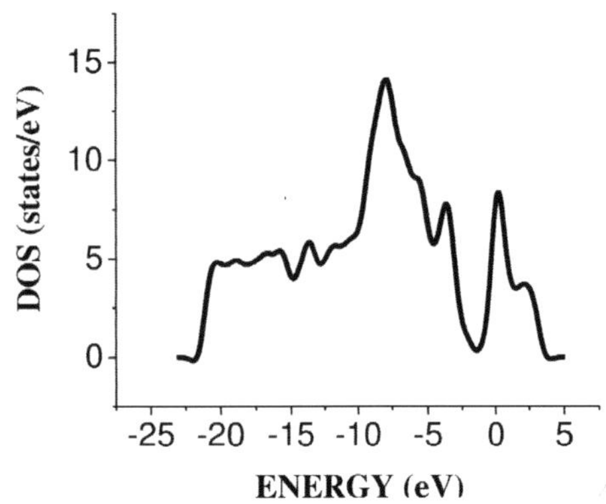

Figure 4. Density of states for one sulphur atom adsorbed at the center of a double carbon vacancy, Fermi energy = - 1.4719 eV. The corresponding system is shown on figure 5.

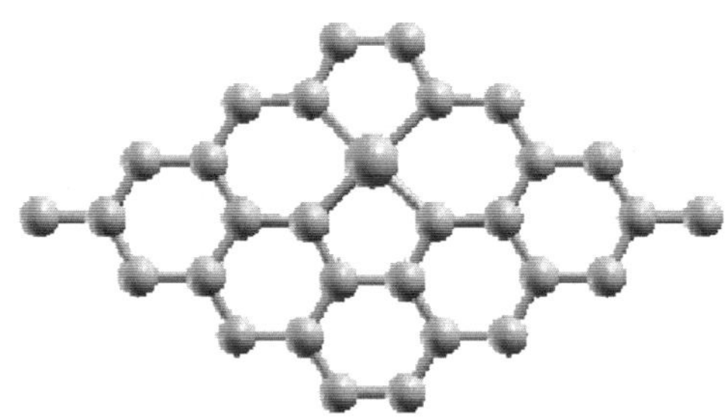

Figure 5. Adsorption of a sulphur atom at the center of the carbon double vacancy. There are two opposite pentagonal and two hexagonal cavities around the S atom.

8. Conclusions.

Results were obtained with Quantum espresso computer code, with ultrasoft pseudopotential S.pz-van_ak.UPF for sulphur atom and LDA (local density approximation) pseudopotential and c.fhi, for carbon atoms. This pseudopotential was used before to study the hydrogen molecule adsorption both, on a graphene layer [6] and on the (5, 5) and (6, 6) carbon nanotubes [9]. Cutoff energies as high as 80 Ry were used in the scf and nscf calculations. Other configurations for the adsorption of the sulphur atom on the pristine and defective graphene layer were studied. For an entire pristine graphene layer, it is impossible that some sulphur atom could be at the plane of the carbon atoms of the graphene layer. There is not enough space in the hexagonal cavity to find one sulphur atom at its centre or near that point. In some configurations the sulphur atom remains some distance above the centre of the hexagonal cavity or above of one of the carbon atoms. Some of the explored configurations considered simultaneous adsorption on both sides of the graphene layer, above and below the same point or different points of the layer.

Taking into account the results given here and other configurations in which S-C bonding breaks the bidimensionality of graphene layer, can be used not only to modify but even control or regulate electronic and other kind of properties of these and similar systems.

Taking into account similar results [1-5], those obtained with colleagues [6, 9], those reported previously [7], and the reported in this work, allows to propose that the defective graphene layer, and in particular the double vacancy, as shown here can be used to capture, retain and storage not only sulphur atoms. Because for this kind of application it is not necessary a high quality of graphitic material, the cost of this one to capture and storage different kind of molecules or atoms as sulphur atoms for example, it could be a non-expensive one in contrast with other applications.

References

[1] Liang J Jiao Y Jaroniec M and Zhang S 2012 *Angew. Chem. Int.* **51** 11496–11500

[2] Wang H. Yang Y Liang Y Robinson J T Li Y Jackson A Cui Y and Dai H 2011 Graphene-wrapped sulfur particles as a rechargeable lithium–sulfur battery cathode material with high capacity and cycling stability *Nano letters* **11(7)** 2644-2647

[3] Dai J Yuan J and Giannozzi P 2009 Gas adsorption on graphene doped with B, N, Al, and S: A theoretical study *Appl. Phys. Lett.* **95** 232105

[4] Borisova D Antonov V and Proykova A 2013 Hydrogen Sulfide Adsorption on a Defective Graphene *Int. J. Quantum Chem.* **113** 786–791

[5] Düzenli D 2016 A Comparative Density Functional Study of Hydrogen Peroxide Adsorption and Activation on the Graphene Surface Doped with N, B, S, Pd, Pt Au, Ag and Cu Atoms *J. Phys. Chem. C* (*Preprint* acs.jpcc.6b06131)

[6] Arellano J S Molina L M Rubio A and Alonso J A 2000 *J. Chem. Phys.* **112** 8114

[7] Arellano J S 2015 *J. Phys.: Conf. Ser.* **582** 1

[8] Satta A 2005 Hands-on tutorial on the Quantum-Espresso Package *Examples of defects in solids from Espresso*

[9] Arellano J S Molina L M Rubio A López M J and Alonso J A 2002 Interaction of molecular and atomic hydrogen with (5,5) and (6,6) single-wall carbon nanotubes *J. Chem. Phys.* **117** 2281

Construction of a low-cost detector to identify dissolved metals in aqueous media by fluorescence spectroscopy: design and perspectives.

González M[1] Montaño M[2] and Hoyo C[3]

[1]Desarrollo Científico y Tecnológico para la Sociedad, [2]Departamento de Física, [3]Departamento de farmacología. CENTRO DE INVESTIGACION Y DE ESTUDIOS AVANZADOS DEL IPN. Av. Instituto Politécnico Nacional No. 2508, Sn. Pedro Zacatenco. C.P. 07360, Cd. de México.

E-mail: mgonzalezg@cinvestav.mx

Abstract. We have constructed a low cost fluorescence detector model to determine the presence of some heavy metals in an aqueous medium. In particular, we focus on metals which cause public health problems in our country. We did the first tests with standard samples of Hg (II). The innovative features of this instrument are its small dimensions (9 dm^3) and the low cost of materials used in its construction.

1. Introduction

The determination of contaminants of the heavy metal kinds in aqueous media is supported through American and European procedures and rules [1, 2, 3]. These rules guide the traceability and repeatability of measurements of these contaminant. In them, sampling and conservation are included, which implies an increase in the costs of such analysis.

There exist some specific laboratory instruments to detect metal contaminants dissolved in aqueous media. All of them are made for a specific size and high cost for acquisition data and its operation [4] to smaller size but also very expensive [5]. In the case of metal contaminants in aqueous media, spectrometry equipment is based on atomic absorption or plasma-mass technologies. All these use the spectral detection method, i.e. the emission spectrum of an aqueous sample previously excited by a high frequency source (plasma) or non-ionizing radiation (UV light), is detected and interpreted. Taking this last exciting method, we have developed a low cost instrument that can be used "in situ" requiring only that the sample is stabilized without conservation or transportation.

The initial plan to construct this instrument for water analyses was divided in two steps: First, an analytic method to prepare the samples. Second, the electronic prototype development itself. In the first step, we made use of the chemical properties of the 8-hydroxyquinoline to build the chelates in metals previously dissolved in aqueous media to obtain fluorescence when irradiating [4, 5]. Our prototype was built to have control mechanisms to excite the sample by a UV source and to detect the fluorescence light into the 390-410 nm range [6].

2. Design and construction

We our design on the concept of having a device able to detect the fluorescence emission coming from a physic-chemical reaction produced by a high power UV source (figure 1). This emission is sent to a

Content from this work may be used under the terms of the Creative Commons Attribution 3.0 licence. Any further distribution of this work must maintain attribution to the author(s) and the title of the work, journal citation and DOI.
Published under licence by IOP Publishing Ltd

IOP Conf. Series: Journal of Physics: Conf. Series **792** (2017) 012016 doi:10.1088/1742-6596/792/1/012016

ATMega 2560 microprocessor coupled to a graphic interface which show qualitatively and quantitatively the contaminants in the analyzed sample [7, 8, 9].

For an analog electronic level, we selected three IN, one OUT and five control signals, all shown in the control interface. To detect the non-ionizing radiation, we implemented the ferricyanide silica ($SiFe(CN)_2$) photo-detector with three sensitivity channels working from 0.1 to 0.6 mV range (figure 2).

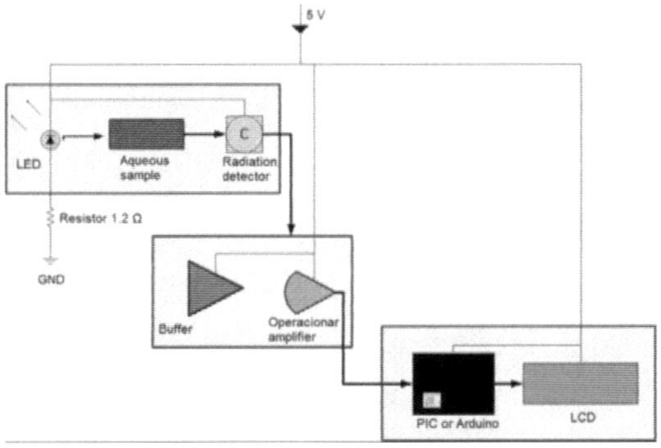

Figure 1. Conceptual design of the device.

Figure 2. Three-channel radiation detector

For the UV source we acquired a high power LED (1W) in the 390-400 nm range mounted in a detection cell to avoid energy losses. This cell was designed to have all the components in a complete dark environment with the possibility, if required, to replace easily the emission source and the photo-detector (figure 3).

Figure 3. Detector cell and the emission source integrated

Once detected, the signal is amplified by a complement electronic circuit to avoid voltage drops. Then, the amplified signal enters the interpretation circuit zone where the corresponding algorithms were programed to show the presence of the contaminant source (figure 4).

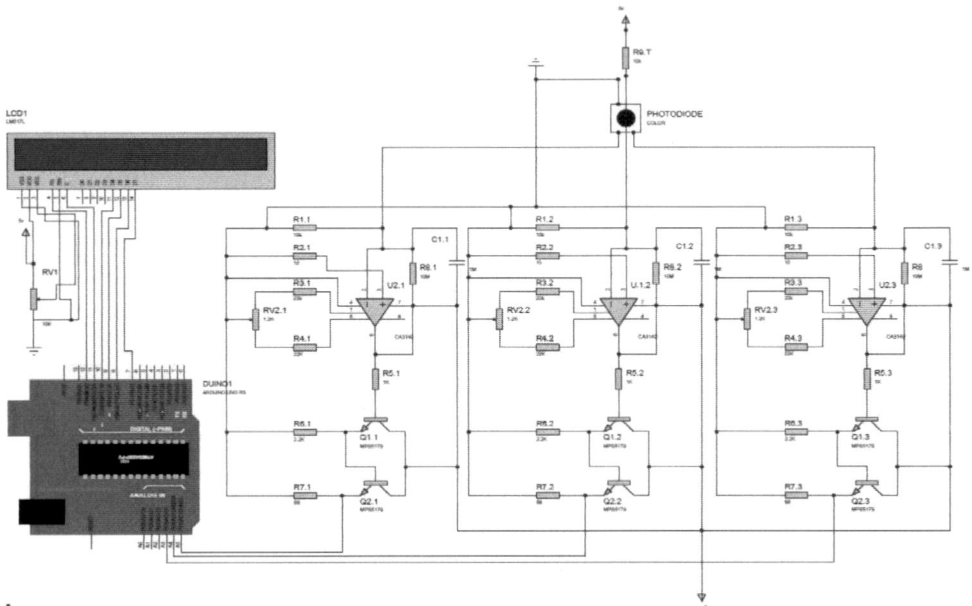

Figure 4. Electronic diagram of the microprocessor, photodetector and interface

To improve the detection, we again use the three signals from the photodetector for the graphical interface to identify the contribution of each of them to establish a statistical weight among them. The intention with this analysis is to obtain a cleaner final result of the real presence of the contaminant. All this is connected to a touch screen that functions as an interface between the control electronics, cell detection and UV emission source.

3. Results.

For the three-channel sensor, we measured the spectral response of ultrapure water to test the sensibility limits of our prototype. For this, we tested the response of the photodetector with different LEDs of different wavelengths. We try to check if the photodetector is able to give us a consistent signal to avoid saturation or a change in the linearity of its response and, if appropriate, seek alternatives to address this problem (figure 5).

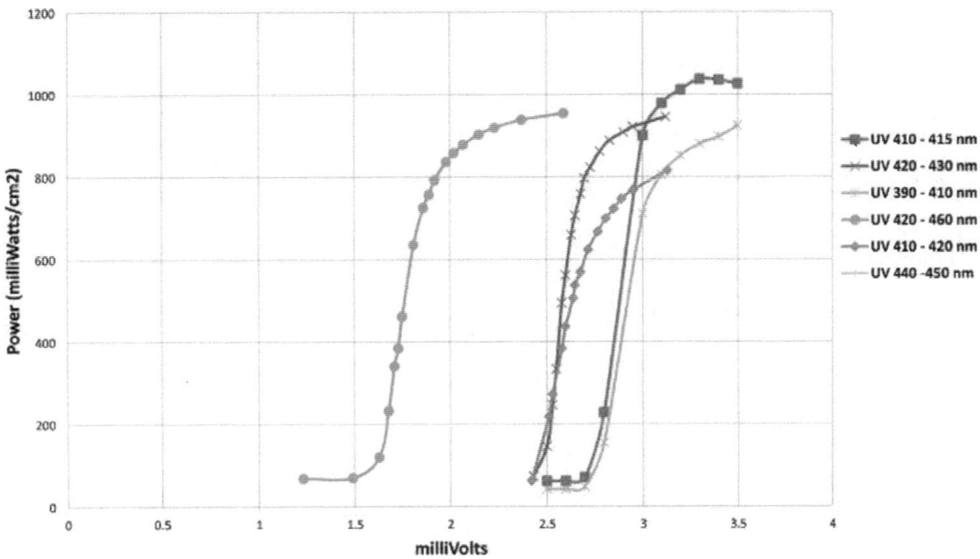

Figure 5. Photodetector response at different wavelengths for each high power LED

To obtain the calibration curve, we measured Hg standard samples of different concentrations to establish the detection methodology. Then, we test this methodology using a mixture of 8-hydroxyquinoline with pH of 8.5 and Tris-hydroxyaminometane [9]. We use high-power LED in the range of 390-410 nm with a concentration range for 0.02 to 0.2 mg/l obtaining few significant deviations compare to different sample dilutions; we developed 9 tests of this kind (figure 6).

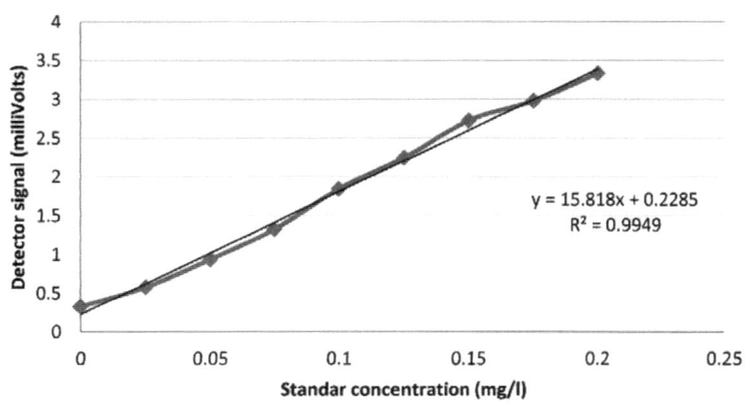

Figure 6. Calibration curve for Hg.

The limit of detection was obtained by calculating three times the standard deviation of ten measurements reagent blank between the slope "m" obtained by the calibration curve; in case of LD = 0.0021 mg / l, and the uncertainties presented values range from 0.13% for the reagent blank to 0.25% for the standard of 0.2 mg / l of Hg.

We say that there are variations in measurements due to the presence of chemical reagents. In this sense, we performed a series of tests at very high concentrations of two species of the same analyte; mercuric chloride and mercuric nitrate, in order to try to differentiate a soluble a partially insoluble substances in a basic medium. Notice that the behavior detection is not entirely linear and does not lose continuity (figure 7).

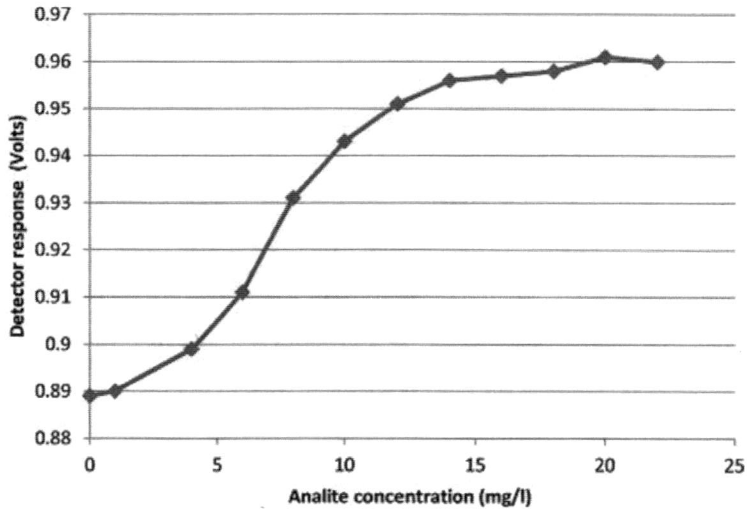

Figure 7: Detector response to high concentrations for mixture for two analytes.

With this procedure of applying fluorescence phenomena to identify contaminants in aqueous media, chemically previously treated, that our prototype can be applied to detect other metallic contaminants dissolved in water [10, 11].

4. Discussion

Fluorescence of metal complexes dissolved in aqueous media is a property that can be used to determine its presence not only qualitative, but quantitative. In this sense, the metal complexes of Cd, Zn, Cu, Pb and As, formed with chelating solutions also can be detected with our instrument. The instrument response to these complexes have still to be characterized for, first, to know if the detection is entirely possible and, second, to find the detection limits of the instrument. It is clear that these metals in water, which represent a public health problem in our country, are not found in pure form. They appear always in association with other components as oxides, sulfates, nitrates or acids; and these chemical combinations contribute as a detection problem.

5. Conclusions

By achieving fluorescence emissions detection for Hg for ranges between 0.02-0.25 mg / l we verified the usefulness of our instrument. The range of fluorescence of the sensor used, powered by the operational amplifier ensures applicability for detecting metal contaminants dissolved in water with a low price and fully portable device. The portability of the instrument, which is one of the added value, allows us to appreciate its application directly "in situ" to take immediate actions to contribute on the problem of water contamination. This instrument is not intended to replace the studies conducted by certified laboratories and centers; rather it is a viable option for presumptive analysis in areas or places causing health problems due to water pollution.

References

[1] NMX-AA-051-SCFI-2001 Análisis de Agua – Determinación de metales por absorción atómica en aguas naturales, potables, residuales y residuales tratadas – Método de prueba SEMARNAT
[2] 2005 *Water Analysis Handbook* 3 ed HACH
[3] Clesceri L Greenberg A and Eaton A 2008 Methods for the examination of water and wastewater 20th edition *Oxford Academic Press* **1475**
[4] Lakowicz J 2006 Principles of Fluorescence Spectroscopy *Springer-Verlag* **954**
[5] IOE-CP Spectrometers 2015 *http://www.spectro.com/products/optical-emission-spectroscopy-metal-analyzer/trace-metal-analysis-spectrolab* [last access: February 8 2016]
[6] Fluorescence Method Development 2015 *http://www.dionex.com/en-us/webdocs/113591-Handbk-FLD-3000-Method-Develop-AN7* [last access: February 17 2016]
[7] Zhang H Qiang-Li and Jiang Y 2007 8-Methoxyquinoline based turn-on metal fluoroionophores *Tetrahedron Letters* **48** 3959 – 3962
[8] Valuer B and Brochon J. C 2001 New Trends in Fluorescence Spectroscopy. Applications to Chemical and Life Sciences *Springer- Verlag* **496**
[9] Soraka K Vithanage R and Phillips D 1997 Fluorescence Properties of Metal Complexes of 8-Hydroxyquinoline-5-sulfonic Acid and Cromatographic Applications *Anal. Chem.* **59** 629–636
[10] García-Colín L Dagdug L and Picquard M 2010 Introducción a la Física Biológica *El Colegio Nacional* **1295**
[11] Yan Z and Yan-Bin J 2015 Combined Toxic Effects of Heavy Metals and Antibiotics on a Pseudomonas fluorescens Strain ZY2 Isolated from Swine Wastewater Int.J.Mol.Sci. **16** 2839-2850

In a 21-2n deformed stainless steel influence of recovery temperature

De Ita A Ugalde P and Flores D
Área de Ciencia de Materiales, Departamento de Materiales UAM - Azcapotzalco, Av. San Pablo # 180, CP 02200, México, C. de M.

Email: add@correo.azc.uam.mx

Abstract. We present the influence high heat treatment temperature of a nitrogen austenitic stainless steel, deform by cold compression, in 10 different percentages. The steel contains high chromium (19.25 %), nickel (1.5 %) and nitrogen (0.2 %). The typical applications for this alloy are automobile parts and special valves for his excellent mechanical properties and corrosion resistance. Produced by hot rolling, they were subjected homogenized treatment at 975 °C for 45 minutes. Subsequently, deformed, by cold compression. We get ten different deformations, from 3 % to 22 %. These samples then to a heat treatment at 750 °C for one, 2 and 4 hours respectively. To observe the microstructure all samples were metallographic study and measured also their Rockwell C hardness. The initial sample has an austenitic matrix with a small amount of precipitates with a 42 RC average hardness. The homogenized sample had a 39 RC hardness. The deformed samples increased their hardness with a maximum of 49 RC. The samples with the treatment, showed a lower hardness with longer time with high dispersion. The decreased of hardness is due to the elimination of residual stresses and precipitates increasing size.

1. Introduction

In the car, industry different and special metals are use in certain parts. The valves of combustion motors have to be heat, corrosion and wear resistance. In the search of better performance the austenitic stainless steel 21 - 2 N has been used for this application. However, not all the mechanical o thermal characteristics of these alloys are well known [1]. We are want to find the time and temperature dependence in which the alloy can work with the best properties. In previous works we were study other heat temperature treatment, a lower one 650 °C [2-4].

2. Experimental

Our austenitic stainless steel has a nominal composition: 0.50 % C, 8.0 % Mn, 0.25 % Si, 0.05 % P, 0.06 % S, and the main components: 20.50 % Cr, 1.50 % Ni, and 0.3% N. The steel was originally obtained by hot rolled in cylindrical bar form. Then with a solution treatment at 975 °C for 45 minutes and finally air - cooled. To test the recrystallization process we compressed samples from 3 % to 22 % in a 100 ton machine in 10 different deformation steeps. Then three samples from each deformation grade were heat - treating at 750 °C for one, 2, 3 and 4 hours respectively.

To observe the microstructure all samples were metallographic study, to reveal the internal structure we measured as a main characteristics their Rockwell C hardness. We had as much as 21 cylindrical samples, 13.5 mm diameter and 9 mm height.

Content from this work may be used under the terms of the Creative Commons Attribution 3.0 licence. Any further distribution of this work must maintain attribution to the author(s) and the title of the work, journal citation and DOI.
Published under licence by IOP Publishing Ltd

3. Results

The initial sample has an austenitic matrix with a small amount of precipitates with a 42 RC average hardness as shown in figure 1.

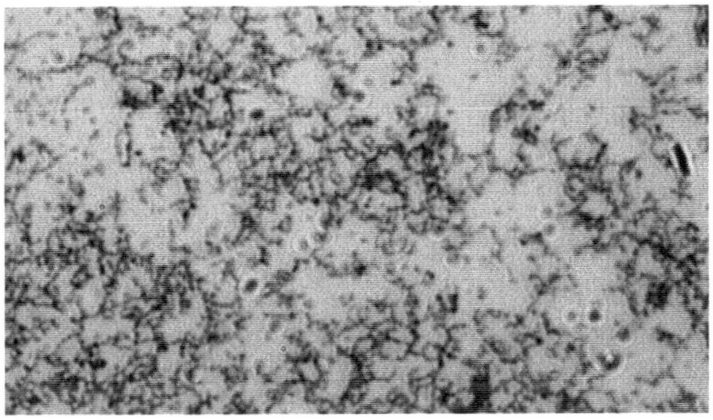

Figure 1. Hot rolling structure, 100 X, 42 Rc.

In figure 1, we can see mainly austenitic phase, a small part of martensitic phase and carbides in and out of the grains. After the solution treatment at 975 °C for 45 minutes and finally air - cooled, the hardness is reduces just a little to 39 RC. The changes in the structure were only small also. However, the martensitic phase is not present.

The next step was the deformation in compression. We used a very hard materials as base or support for the compression because the hardness of the original material. We used a 100 ton Satec machine. We compressed samples from 3 % to 22 % in 10 steeps. The hardness increased with the deformation and the result is present in the figure 2. The best fit was a polynomic curve. It is clear the carbide increased precipitation and small reorientations of the grains, as shown in figure 3.

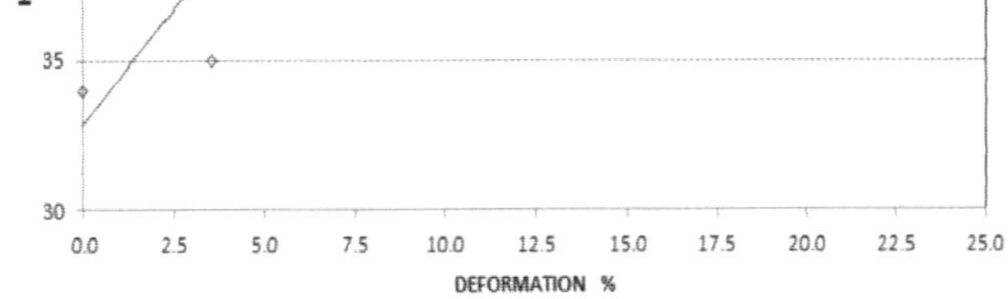

Figure 2. Hardness vs. deformation in compression.

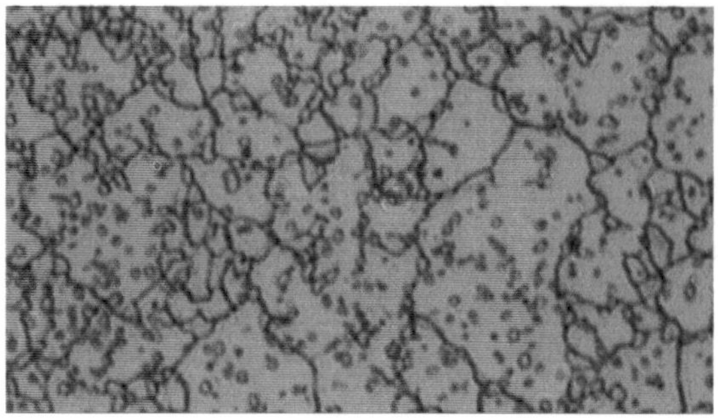

Figure 3. Compressed structure, 22 %, 100 X, 49 Rc.

Then two samples from each deformation grade were heat - treating at 750 °C for one, 2 and 3 and 4 hours respectively.

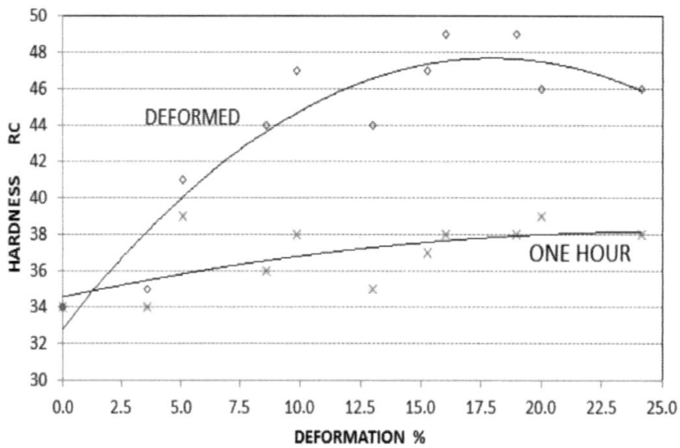

Figure 4. Hardness vs. deformation, as deformed and by one hour heat - treating.

In figure 4, we have the one - hour result and the as deformed hardness. We used a polynomial for the bet fir of the curve. In figure 5 are the 2, 3 and 4 hour heat treatment. The best fit was a second grade polynomial. The r values were between 0.6 and the best was 0.9 for the 2 hours. Great data dispersion was observed

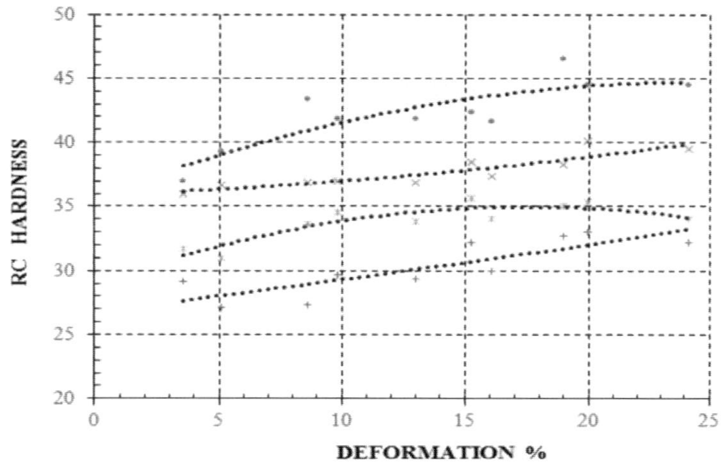

Figure 5. Hardness vs. deformation, for one, 2, 3 and 4 hour heat - treating at 750 °C.

4. Discussion

As in general, the hardness lows with the heat treatment time. More or less 5 points were measured between the intervals of 2 hours heat treatment. The origin of the dispersion data was to be related with the bars fabrication process. It was hot rolling and despite of the homogenized treatment it was impossible to obtain a very homogeneous grain size of the bars. It was greater corm at the outside than the inside of the bar. The means values had a big variation.

5. Conclusion

The austenitic stainless steel 21 - 2 N, normal automotive valve material steel shows the hardness increase with the deformation in compression and a decreasing behaviour by heat - treating. The first effect is more pronounce in the first process than the second one. That means a 10 HRC values for hour of heat treatment at almost every percent of deformation. The best fit was a polynomial approach.

6. References

[1] Dulis E 1990 Age-Hardening Austenitic Stainless *The metallurgical evolution of stainless steels* American Society for Metals and the Metals Society USA

[2] Olmos A De Ita A and Ugalde P 2015 Influence of recovery time in a 21-2N deformed stainless steel (Proc. XXIV IMRC, Cancún) S5B-P058

[3] Valdespino N 2010 Aging 21-2N (Universidad Autónoma Metropolitana Azcapotzalco, México) Lic. Tesis.

[4] Gaeta M 2000 Efecto de la pre-deformación y envejecimiento en el acero 21-2N (Universidad Autónoma Metropolitana Azcapotzalco, México) Lic. Tesis.

Instabilities in the Poiseuille flow of a magnetorheological carbonyl-iron suspension.

Méndez A Pérez L Rivera I and Paniagua A
Escuela Superior de Física y Matemáticas, Instituto Politécnico Nacional, Edif. 9
Unidad Profesional Adolfo López Mateos, Col. San Pedro Zacatenco, C. P. 07730,
Ciudad de México, MÉXICO.

E-mail: arturo@esfm.ipn.mx, leopt@esfm.ipn.mx

Abstract. The flow of a magnetorheological suspension in the Poiseuille flow is studied in this work. A capillary rheometer was constructed and designated in order to adapt a magnetic field perpendicular to the flow direction covering the entire capillary zone; this restricts the maximum length of the glass capillary. A Newtonian suspension of carbonyl iron in glycerol at 20% in volume was studied at several intensities of magnetic field. The results showed a close-Bingham behavior at lower intensities of magnetic field. But a deviation is seen at higher intensities accompanied with a clearly shear thinning behavior with flow instabilities. These instabilities were evidenced by oscillations in shear rate and increased with the magnetic field and the shear stress.

1. Introduction

Magnetorheological fluids (MRF) are composed of magnetic particles dispersed in to a non-magnetic continuum phase like silicon oils. These suspensions are characterized by an abrupt increase of viscosity as well as the presence of yield stress under the presence of an external magnetic field, as result of a structure formation of the magnetic particles in the magnetic field direction. Its physical properties decrease when the magnetic field is removed due to the breaking structure. The magnetorheological suspensions have a Newtonian behavior but under the presence of the magnetic field, a close-Bingham fluid is the typical rheological behavior seen [1-5]. However, a variety of models are used to adjust the shear thinning observed on them [6-8]. In addition, the formation of the structured chain gives a high anisotropic behavior, and it is expected that phenomena like slip are observed in this kind of MRF.

Shorey et al. [1] designed and tested a parallel-plate magnetorheometer and observed poor data reproducibility when a smooth aluminum surface and sand paper were used. They used magnetic particles embedded in the surface by using a thin layer of epoxy in order to avoid slip in the boundary. de Vicente et al. [9] investigated the wall slip phenomena at low magnetic fields in stainless steel parallel plates by using different rheometer gaps and found that the wall slip thickness increased as a function of the Mason number. Volkova et al. [10] showed that a Carbonyl iron in silicon oil suspension gave the same yield stress but the flow curve differed at low shear rates ($1\text{-}10 s^{-1}$) when a stainless-steel cone-plate is replaced by a glass one. Those authors suggest that the observed behavior is due to the aggregates slipping on the walls but the structure is remained. On the other hand, Pappas and Klingenberg [11] simulated the flow of a magnetorheological suspension in the Poiseuille plane considering interparticle magnetic forces approximated at the point-dipole limit as well as magnetizable walls. They showed that slip boundary condition fails to capture the Bingham-like behavior, and thick lamellar microstructures oriented in the flow direction were developed. In contrast, the no-slip boundary condition reproduces the Bingham-like behavior and thin lamellar aggregates were observed in the microstructure. These differences may be caused by the boundary nature (magnetic and non-magnetic walls). However, more

Content from this work may be used under the terms of the Creative Commons Attribution 3.0 licence. Any further distribution of this work must maintain attribution to the author(s) and the title of the work, journal citation and DOI.
Published under licence by IOP Publishing Ltd

research should be done in this direction, because it is very likely that phenomena like those appear in melts (instabilities like: spurt and oscillations among others) could be experimentally seen in MRF.

Since the variety of designs and applications of this kind of suspension like prosthesis, clutches, etc, it is important to study their behavior in ducts and capillary geometry. However, the majority of researches are performed in rotational rheometers (simple shear flow) as pointed out by Pappas and Klingenberg [11], due to the relative facility of incorporating a uniform magnetic field to the flow. There are few researches about magnetorheological suspensions in capillary flow [6, 7, 11, 12, 13]. Nevertheless, only the Pappas and Klinglenberg [11] simulation deals with the pressure-driven flow (Poiseuille flow). So, the goal of this work is to analyze a 20%vol MR suspension of carbonyl iron in glycerin at several intensities of magnetic field, by using a capillary magnetorheometer constructed with a pressure-driven flow.

2. Experimental work

2.1. Capillary rheometer

The theory of the Poiseuille flow is well known and can be found elsewhere [14]. Figure 1 show the experimental setup used. The constituent parts of the rheometer constructed were similar to that employed by one of us [15] but modifications were made in order to adjust the magnetic field.

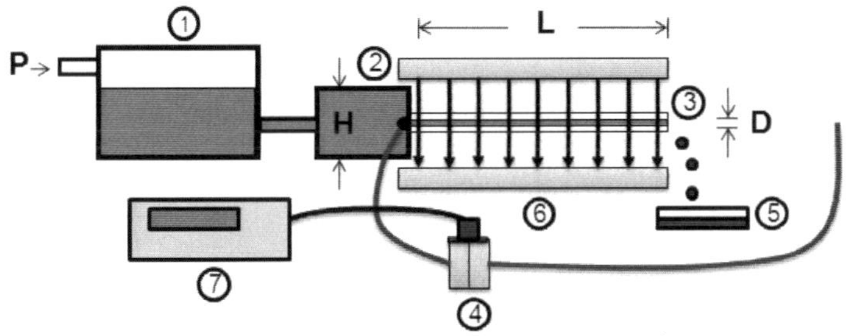

Figure 1. Experimental arrangement. (1) Reservoir at constant pressure, (2) fluid container, (3) glass capillary, (4) pressure transducer, (5) collector container, (6) electromagnet, (7) pressure displayed.

The constructed rheometer consists of a pressurized constant reservoir with the fluid under study. The fluid moves from the container of high H through the glass capillary and it is collected by the catch and measure technique in order to obtain the volumetric flow (Q). The geometrical parameters of the capillary are L=0.035m of length and D=0.00125m of diameter. In this case, the length to diameter ratio was L/D=28, and the contraction ratio was H/D>26. The pressure (ΔP) between the ends capillary is measured by a differential pressure transducer. The wall shear stress (τ_w) and the apparent shear rate ($\dot{\gamma}_a$) were calculated by using the well known formulas [16]:

$$\tau_w = \frac{D\Delta P}{4L} \tag{1}$$

$$\dot{\gamma}_a = \frac{32Q}{\pi D^3} \tag{2}$$

The capillary rheometer calibration was performed using Newtonian oil. The experiments were carried out at 22°C and the viscosity results were compared with those obtained using a Thermo Electron rolling ball viscometer, the difference not exceed 0.5%

2.2. Magnetic field

An electromagnet was adapted on the capillary region in order to apply a magnetic field perpendicular to the flow direction. The electromagnet is composed of soft iron core of square shape and 0.04m of side. Considering these dimensions the maximum length of the capillary is restricted to 0.04m. Two coils with 250 turns, resistance of 0.6 Ohms and 2.2mH inductance each. In addition, a DC power supply was employed of 24V and 0-10A constructed in our electronic laboratory. The electromagnet was located at 0.0225m from the capillary center in such a way that it covers the entire capillary region. Figure 2 shows the magnetic field lines with the help of iron filings, in actual scale. It can be seen that the field is homogeneous throughout the capillary, ie in the area where carried out flow measurements. The magnetic flux density was obtained by direct measurement in the capillary area using a Leybold brand model 51662 teslameter as a function of the applied current. Electric currents of 1, 2, 3, 4 and 5 ampers were used and the corresponding values of magnetic field strength B were 12.4, 24.8, 36.9, 48.7 and 60.95 mT.

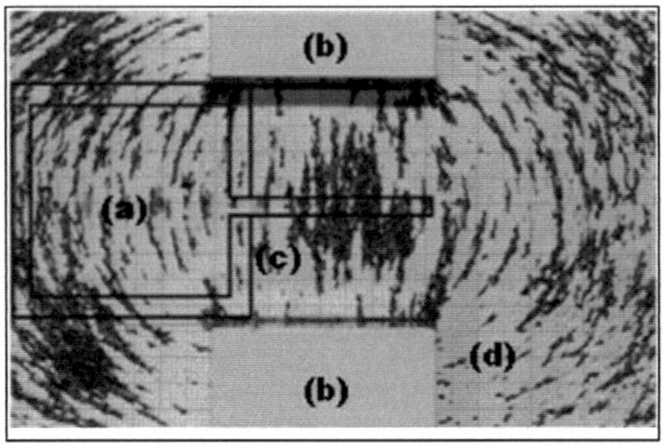

Figure 2. Magnetic field lines in actual measurement area, B=12.4 mT. (a) Reservoir at constant pressure, (b) electromagnet, (c) glass capillary, (d) iron filings tangents to the magnetic field lines.

2.3. MR experiments

MR suspension used in this work consists of spherical Carbonil Iron particles with 3.56 g/cm^3 density. SEM observations yielded a diameter that does not exceed 2 microns. Particles were dispersed at 20% volume in glycerol with 1.241 g/cm^3 density. The suspension was vigorous shaken and sedimentation effects were seen after 8 hrs. This time was enough to perform capillary rheometry.

Experiments with the MRF were carried out at 24°C, pressure injection was kept constant. Then, the capillary ends pressure and the volumetric flow rate were obtained without magnetic field. After that, the magnetic field was applied during 10 minutes, while capillary ends pressure and flow rate were measured. Subsequently, the magnetic field was removed and pressure and the flow rate were determined once more. These values were consistent with those measured before to apply the magnetic field. This protocol was repeated by increasing the injection pressure. To produce different magnetic fields, currents of 1, 2, 3 and 4 amperes were used. The currents values were chosen because a two-phase separation of the MRF was seen in the container (number 2 in figure 1) due to the magnetic field gradient when a current of 5 A is used. In fact, for times longer than 20 minutes, a decrease in flow followed by blocking was observed. Figure 3 shows the shear rate decrease until blocking and the corresponding shear stress increase. For times shorter than 20 minutes, shear stress is kept constant and shear rate is oscillating. At later times, shear rate dramatically decreases.

IOP Conf. Series: Journal of Physics: Conf. Series **792** (2017) 012018 doi:10.1088/1742-6596/792/1/012018

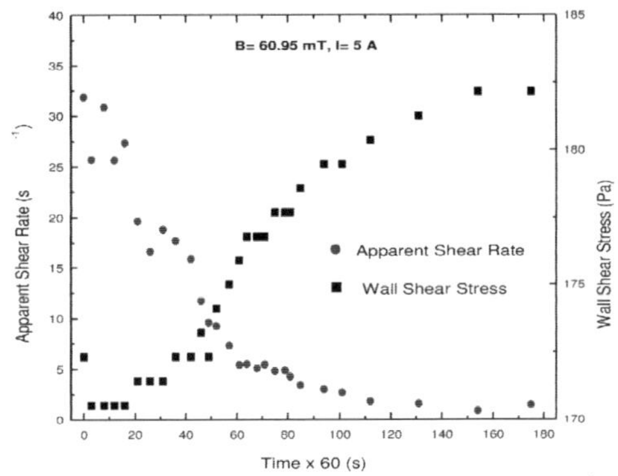

Figure 3. Shear rate and wall shear stress transients.

3. Results and Discussion

Figure 4 shows the flow curve at several current intensities. Corresponding fitting, where possible, is included. Flow curve at B=0 mT, shows Newtonian expected behavior, with μ=0.87 Pa·s and the interjection with the shear stress axis is within the experimental error (τ_y=-1.76±2.18 Pa). The flow curve at B=14.26 mT, shows Bingham like behavior, with τ_y=20.70±0.56 Pa the yield stress value. However, at B=28.57 mT the obtained flow curve observes a substantial deviation of the Bingham model. In this case, the behavior is close to the Hershel-Bulkley model, since the flow index decrease (n=0.72) showing a shear thinning behavior with yield stress (τ_y=77.38±4.71 Pa). At B=42.85 and 57.14 mT, the yield stress values were τ_y=210.3 and 415.4 Pa, respectively. Moreover, flow instabilities evidenced by the presence of shear rate oscillations were seen from these values. In this case, experimental data corresponds to the mean value on the oscillations and in such cases a fitting was considered inappropriate as the Rabinowitsch correction. To our knowledge the presence of flow instabilities evidenced by oscillations on shear rate on MRF have not been reported in literature. These oscillations are comparable to those seen in melts due to wall slip [17].

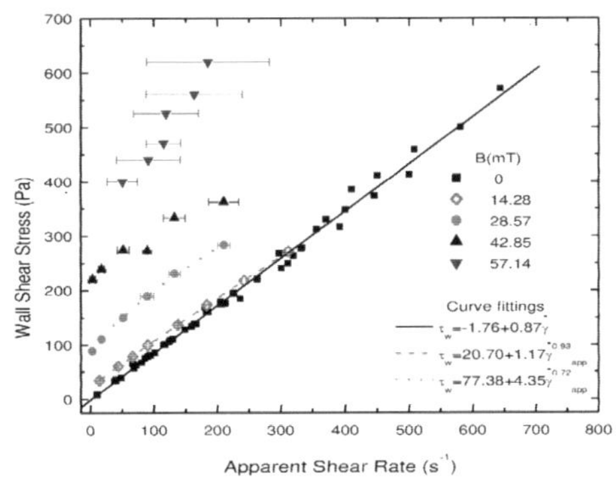

Figure 4. Flow curves at different magnetic fields. Lines represent curves fittings. Error bars correspond to the amplitude of the oscillations in shear rate values.

Furthermore, the shear rate oscillations amplitude are increased when shear stress increases at fixed magnetic field, as well as oscillations are enlarged when magnetic field strength is increased to the same shear rate value. For example, oscillations amplitude goes from 2 s^{-1} at B=14.28 mT and until 96 s^{-1} at B=57.14 mT were observed and these are shown in figures 5 and 6, respectively. Pappas and Klingenberg [11] in their simulation, attribute the fluctuation velocity to breaking and reformation of structures of the magnetized particles. In our case, we believe that a decrease in shear rate is the result of the capillary blockage due to the structures formation of the particles on the magnetic field direction. On the other hand, the increase on shear rate results from the transient breaking structured chain on the capillary due to the pressure increment that outcomes from the blockage on the capillary. This hypothesis might be supported by the presence of pressure oscillations (like spurt in melts). We suspected slight pressure oscillations, but we were not able to quantify them, due to these being in the experimental error. It is important to mention that at this time we are not able to distinguish if the breaking chain is full or partial (due to an adhesion failure at the wall), because the chain can slip due to the roughness of the wall without beginning to break [10]. In addition, some kind of slip could be involved in these results, but we have doubts about real wall slip, since Pappas and Klingenberg [11] have shown that wall slip boundary does not reproduces the yield stress behavior observed in this kind of suspensions. In our experiments, a non-magnetic wall was employed (glass capillary). At this point more research should be done in order to elucidate the slip nature.

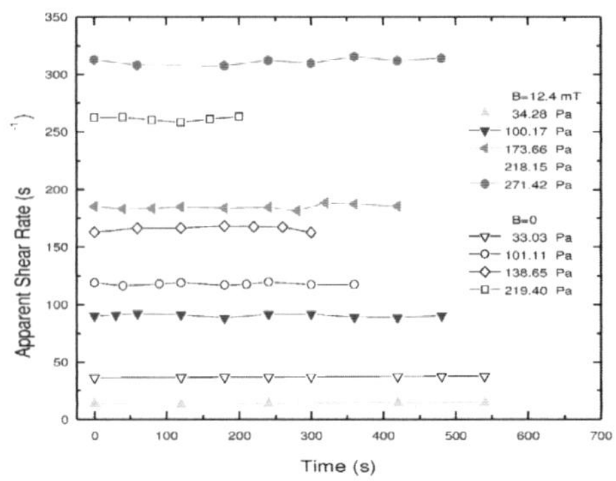

Figure 5. Apparent shear rate values with lowest magnetic field applied (full symbols), and without magnetic field (open symbols).

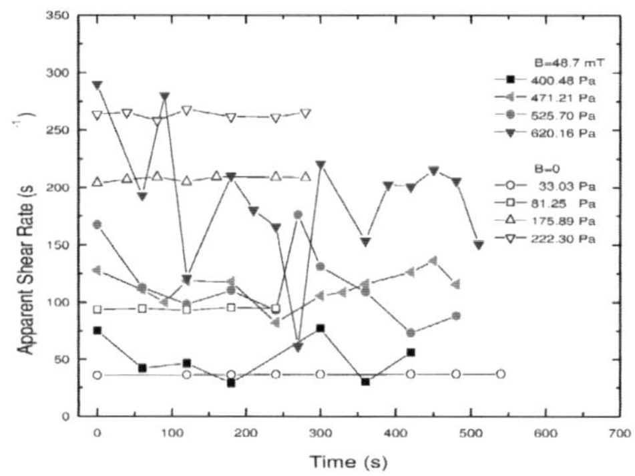

Figure 6. Apparent shear rate values with highest magnetic field applied (full symbols), and without magnetic field (open symbols). Amplitude of the oscillations is evident.

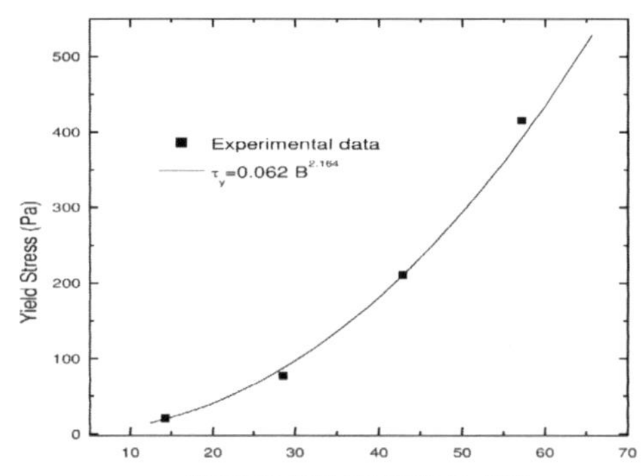

Figure 7. Yield stress vs Magnetic field induction. Solid line corresponds to a power-law fit.

Figure 7 plots yield stress versus magnetic field induction; a non-linear behavior can be seen and is shown by the included fitting in the same figure. The power index of B was 2.164 and is due to the local saturation of the magnetized particles [18], and this value is in accord with low applied magnetic field [18, 19] and a micromechanics and statistical approach of the MRF model [20].

4. Conclusions:

A pressure driven capillary was adapted to study the MR behavior and a protocol was established to determine the rheological properties of the 20%vol of CI in Glycerol. At low intensities of magnetic field a close Bingham behavior was seen but deviates from it, when the magnetic field is increased and a shear thinning Hershel-Bulkley behavior is obtained. Also a close-quadratic relation between magnetic field induction and yield stress was obtained.

Flow instabilities, not reported before to MR fluids, were observed evidenced by the presence of shear rate oscillations, which increased with shear stress and magnetic field induction. This result may help to better understanding about mechanisms that occur when this phenomenon is presented in other complex fluids. Finally, a direct application of the results obtained in this work could be that the presence of flow instabilities could improve the operation of dampers and suspension systems, although more research must be done.

5. References

[1] Shorey A B Kordonski W I Gorodkin S R Jacobs S D Gans R F Kwong K M and Farny C H 1999 *Rev. Sci. Inst.* **70** 4200-06

[2] López-López M T Kuzhir P Lacis S Bosis G González-Caballero F and Duran J D G 2006 *J. Phys.: Condens. Matter* **18** S2803-13

[3] Ngatu G T Wereley N M Karli J O and Bell R C 2008 *Smart Mater. Struct.* **17** 045022

[4] Kuzhir P López-López M T Vertelov G Pradille C and Bosis G 2008 *Rheol. Acta* **47** 179-87

[5] Galindo-Gonzalez C Ponton A Bee A Chavalet J Talbot D Perzynski R and Dubois E 2016 *J. Rheol. Acta* **55** 67-81

[6] Laun H M Kormann C and Willenbacher N 1996 *Rheol. Acta* **35** 417-32

[7] Jha S and Jain V K 2009 *Int. J. Adv. Manuf. Technol.* **42** (7-8) 656-68

[8] Lim S T Cho M S Jang I B Choi H J 2004 *Journal of Magnetism and Magnetic Materials* **282** 170-3

[9] De Vicente J López-López M T Durán J D G and González-Caballero F 2004 *Rheol. Acta* **44** 94-103

[10] Volkova O Bossis G Guyot M Bashtovoi V and Reks A 2000 *J. Rheol* **44**(1):91-104

[11] Pappas Y and Klingenberg D J 2006 *Rheol. Acta* **45** 621-29

[12] Wang X and Gordaninejad F 2006 *Rheol. Acta* **45** 899-908

[13] Gabriel C and Laun H M 2009 *Rheol. Acta* **48** 755-78

[14] Walters K 1975 *Rheometry* (John Wiley & Sons NY)

[15] Méndez-Sánchez A F Pérez-González J de Vargas L Castrejón-Pita J R Castrejón-Pita A A and Huelsz G 2003 *J. Rheol.* **47**(6) 1445-66

[16] Macosko C W 1994 *Rheology principles, measurements and applications* (Wiley-VCH, Inc. NY)

[17] Pérez-Trejo L 2005 *Triboelectrificación de polietilenos fundidos en extrusión continua* Ph. D. Thesis (ESIQIE Instituto Politécnico Nacional, México D. F., México)

[18] Liu Y D and Choi H J 2014 *Magnetorheology Advances and Applications* (Norman Wereley RSC Smart Material No 6.The Royal Society of Chemestry)

[19] Yang L Duan F and Eriksson A 2008 *Smart Mater. Struct.* **17** 015047

[20] Peng X and Li H 2007 *Smart Mater. Struct.* **16** 2477-85

Microstructural Evaluation of Salt Nitrocarburizing H13 Steel as a Function of Time

Vázquez A[1] Ramírez A L[1] and Juárez J A[1]

[1] Departamento de Materiales Metálicos y Cerámicos, Instituto de Investigaciones en Materiales, UNAM, Ciudad Universitaria, Coyoacán, 04510, México, Cd. Mx.

E-mail: alberto.vazram@gmail.com

Abstract. A set of H13 steels were nitrocarburized with the aim of improving their performance as matrices and pins. The nitrocraburizing process was performed in an ETSA liquid salt bath furnace at 570°C for 45, 60, 90, 120, 150 and 180 minutes in order to evaluate the effect of time during the nitrocarburizing treatment on the quality of the nitrocarburizing layer. Microstructural evolution of the compound and diffusion layers were studied by scanning electron microscopy and by X-ray diffraction analysis. From microstructural characterization, it was observed mainly a continuous nitrocraburizing ε-$Fe_{2-3}(C,N)$ layer with nitrocraburizing times between 45 to 60 minutes showing better wear performance as compared with specimen with nitrocarburizing times between 90 to 180 minutes.

1. Introduction

Nitrocarburizing is a surface heat treatment employed to improve hardness and resistance to wear and fatigue [1] in which the elements nitrogen and carbon are diffused into the surface of steels, commonly in the temperature range of 520 to 580°C [2]. The nitrocarburized surface of steels present two different structures known as the compound layer and diffusion zone from surface to core respectively. The compound layer consists of ε-$Fe_{2-3}(C,N)$ and γ'-$Fe_4(C,N)$. The wear and corrosion resistance is attributed to the presence of the ε-$Fe_{2-3}(C,N)$ carbonitride [3], while the diffusion zone improves fatigue strength when compared to an untreated material. In this zone, N and C atoms are dissolved interstitially in the ferritic lattice and form nitride precipitates [4]. Regarding the tribological performance, the optimum compound layer will be that which consists of ε-$Fe_{2-3}(C,N)$, although other phases can be present, mainly γ'-$Fe_4(C,N)$; from the nitrocarburizing methods [5], liquid nitrocarburizing is regarded as effective and low cost method with many advantages such as low treatment temperature, short treatment time, high degree of shape, dimensional stability and reproducibility.

This work present results of microstructural characterization, microhardness and wear resistance of liquid nitrocarburized H13 tool steel as a function of treatment time.

2. Experimental procedure

The chemical composition of the steel H13 under study is presented in Table 1. Previous to the nitrocarburizing treatment, the tool steel was austenized at 1015 °C for 180 minutes, quenched in high vacuum with nitrogen followed by two tempering temperatures at 540°C and 560°C for 180 minutes each one. The liquid nitrocarburizing process was performed in an ETSA liquid salt bath furnace (with 3% cyanide, 36.7% cyanates and 15.3% carbonates) at start temperature of 350°C, when the salt bath temperature reached 580°C, specimens were withdraw at 45, 60, 90, 120, 150 and 180 minutes of treatment time. Hardness Vickers measurements were performed in a microhardness tester Shimadzu

Content from this work may be used under the terms of the Creative Commons Attribution 3.0 licence. Any further distribution of this work must maintain attribution to the author(s) and the title of the work, journal citation and DOI.
Published under licence by IOP Publishing Ltd

at a load of 0.1 mN for 15 sec. After grinding and polishing, the specimens were etched with 5% nital. Microstructural characterization was performed by using a Leica 440 scanning electron microscope operated a 20 kV and 3 A. The phase composition of layer was determined using a Siemens 5000 diffractometer at a scanning rate of 1 deg/min in characteristic Kα Co-radiation and the wear test was performed on pin-on-disk tribometer according with ASTM standard G-99 under a normal load of 2 N of load with 0.1 m/s of speed and 1×10^4 cycles with track radius of 10 and 8 mm. The pin used was of Cr6-steel ball with diameter of ¼ in, in dry condition at atmospherically pressure and room temperature.

Table 1. Chemical composition of steel H13 (in wt. %).

C	Si	Mn	V	Cr	Mo	Fe
0.40	0.97	0.41	0.98	5.60	1.38	Bal.

3. Results and Discussion

Figure 1 shows the evolution of the compound layer after liquid nitrocarburizing of tool steel H13 as a function of time, for instance, it is clearly seen that there is an increase in thickness as the nitrocarburizing treatment time increased. This compound layer shows mainly the presence of the ε-$Fe_{2-3}(C,N)$ at 45 and 60 minutes of treatment time, and as the time increase, it was observed the presence of γ'-$Fe_4(C,N)$ associated with porosity, as is observed in figures 1e and 1f.

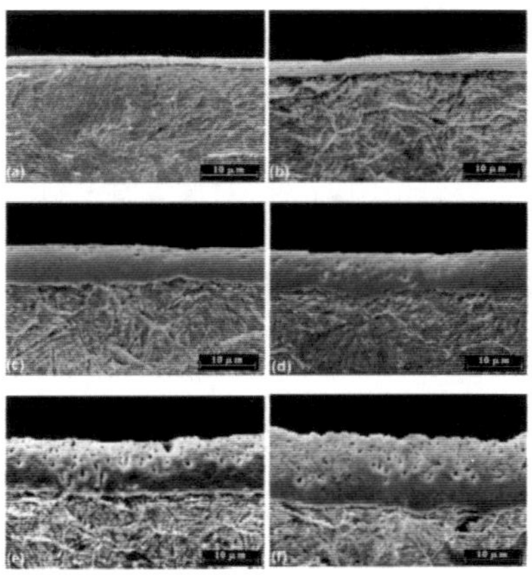

Figure 1. SEM micrographs of the compound layer after nitrocarburizing treatment: (a) 45, (b) 60, (c) 90, (d) 120, (e) 150 and (f) 180 minutes of treatment time.

As reported in [6], the ε-$Fe_{2-3}(C,N)$ carbonitride shows very good wear and corrosion resistance, compared to an untreated steel, therefore it is expected a better wear performance for nitrocarburizing specimen when the amount of γ′-$Fe_4(C,N)$ is lower and there is no presence of porosity as is observed mainly in specimens with treatment time with 150 and 180 minutes.

Regarding thickness increment of the compound layer as a function of nitrocarburizing time has been attributed to the diffusion mechanism that controls growth rate of this layer [7, 8], quantifying ~2 μm in thickness for the lowest nitrocarburizing time and ~14 μm for the longest as is shown in figure 2.

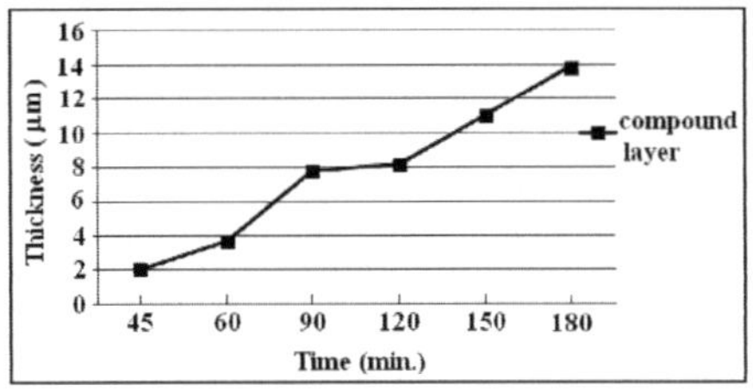

Figure 2. Thickness of compound layer as a function of time.

The microstructure of the diffusion zone is shown in figure 3 and its thickness is plot in figure 4, where a thickness of ~43 µm was measured for the shortest nitrocarburizing time and ~126 µm for the longest. As is shown in figure 4, the diffusion zone depth followed a lineal path.

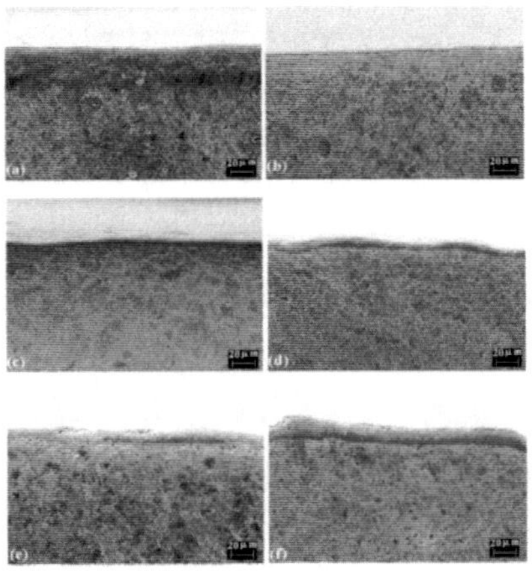

Figure 3. SEM micrographs of the compound layer and diffusion zone after nitrocarburizing: (a) 45, (b) 60, (c) 90, (d) 120, (e) 150 and (f) 180 minutes.

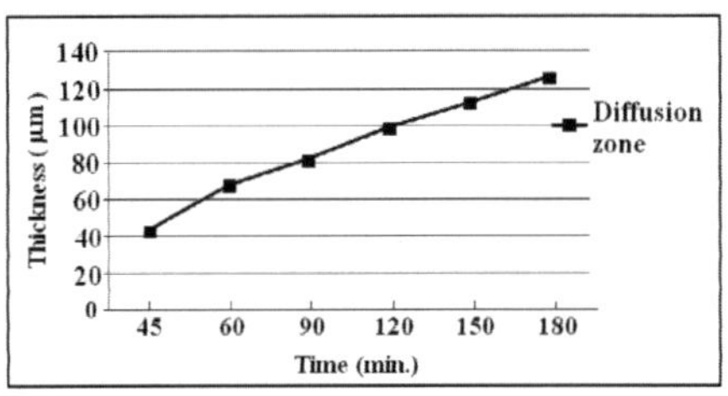

Figure 4. Thickness of diffusion zone as a function of time.

Examination of the microstructure of the diffusion zone reveals in the uppermost region, beneath the compound layer, the presence of precipitates at grain boundaries which were identified by SEM-microanalysis (see figure 5) as (Cr,V,Mo) (C, N), being chromium the main alloying element in these precipitates, because chromium has relatively a strong interaction with nitrogen during the nitrocarburization process.

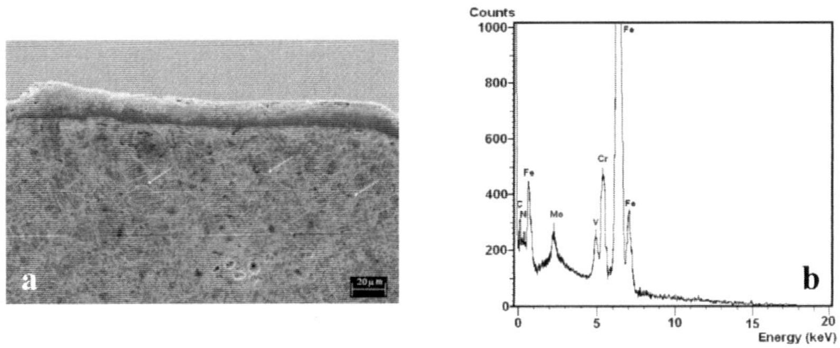

Figure 5. a) SEM micrograph showing by arrows the presence of precipitates at grain boundaries, b) SEM microanalysis of precipitates.

During the nitrocarburizing process, more nitrogen atoms than carbon atoms are introduced in the steel and the phase compositions found are more similar to the ones found in the Fe-N system than in the Fe-C system [9]. After ferritic nitrocarburizing the microstructure consists of three zones, the compound layer at the surface, the diffusion zone just below it and further the unaffected base material. The compound layer is typically 2–30 μm thick and consists of iron carbonitrides ε-Fe_{2-3}(C,N) and γ'-Fe_4(C,N). The ε-phase has a hexagonal crystal structure whereas γ'-phase is cubic [10]. The diffusion zone can reach 100 to 500 μm in thickness and consists of carbon and nitrogen in interstitial solid solution in the base material combined with small carbide and nitride precipitates.

The growth rate of the compound layer is governed by which phases it contains. In accordance with aisothermal section of the Fe-C-N phase diagram at 580 °C, the γ'-phase is a stoichiometric compound with a narrow solubility range for carbon and nitrogen. This makes the growth rate of γ'-phase relatively slow while the ε-phase has a higher solubility range for both nitrogen and carbon and a high nitrogen mobility, which lets it grow quicker. For specimens treated for 90 to 180 minutes, it was observed that the outer part of the compound layer contains pores, that means that long process time increases the amount of porosity, this phenomenon occurs particularly at grain boundaries and other discontinuities [11].

The results of phase analysis by X-ray diffraction for the H13 tool steel nitrocarburized from 45 to 180 minutes are shown in figure 6 where an important feature observed is the difference in the quantity of ε-Fe_{2-3}(C,N) and γ'-Fe_4(C,N) phases. For instance when the steel is nitrocarburized with a holding time between 45 to 60 minutes, the main detected phase corresponded to the ε-Fe_{2-3}(C,N) and as the holding time increases, it is observed that the diffracted peaks for the γ'-Fe_4(C,N) phase increased.

It has been reported in references [12, 13] that ε-Fe_{2-3}(C,N) phase is hard and brittle, while the γ'-Fe_4(C,N) phase is hard and tough and that the usefulness of the formed compound layer depends on mainly two factors: 1) the homogeneity which can produce higher ductility of the structure of the compound layer and 2) its thickness. Therefore, the optimum properties will be obtained when compound layer is mono phased (ε-Fe_{2-3}(C,N) and γ'-Fe_4(C,N)) and has an optimum thickness, which is necessary to meet specific property requirements such as wear resistance.

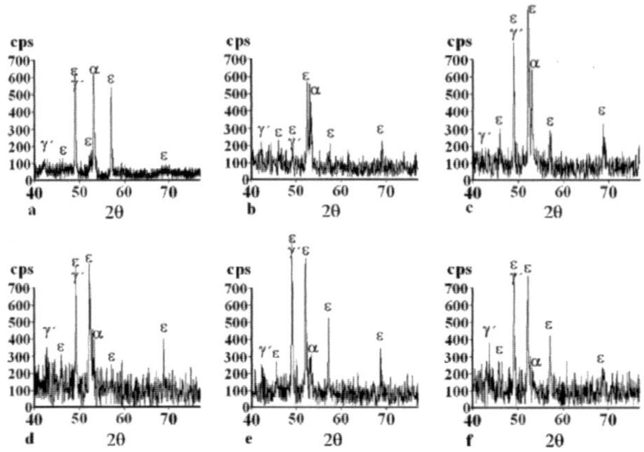

Figure 6. X-ray diffraction patterns. (a) 45, (b) 60, (c) 90, (d) 120, (e) 150 and (f) 190 minutes.

Microhardness distribution of compound layer, diffusion zone and core are shown in figure 7. As can be observed, the highest microhardness value of compound layer corresponded to specimens kept for 45 to 60 minutes in the nitrocarburized liquid bath. Diminution of microhardness as the nitrocarburized treatment time increased is attributed to the presence of both the γ'-$Fe_4(C,N)$ phase and porosity. The microhardness values of the compound layer for each nitrocarburized time has almost a constant hardness value throughout most of its cross section, however in the outer region of specimens kept for 90 to 180 minutes microhardness values lowered presumable by the presence porosity.

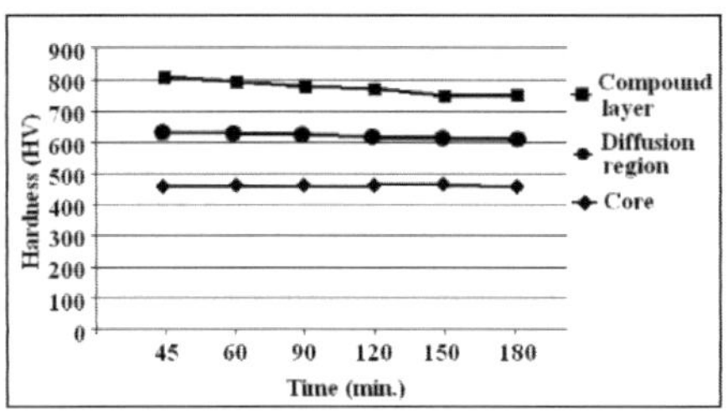

Figure 7. Hardness HV of compound layer and diffusion zone as a function of time.

These microhardness values can be related directly to the steel grade and its microhadness increment as the alloy content increases, for instance, the compound layer has been reported [14] to have a microhardness value of around 700 HV in low alloy steels and about 800 HV for steels with 1.5 wt% chromium content. Regarding the microhardness values of the diffusion zone follow the concentration gradients of nitrogen and carbon.

Disks of H13 tool steel with 60 minutes of nitrocarburzing time were tested in dry, sliding wear against a stationary ruby ball (pin) up to 1×10^4 cycles as is shown in figure 8 where is shown the development of friction throughout the course of ~250 m wear tests. An almost constant friction behavior was observed up to 5000 cycles afterwards is seen an increasing or decreasing friction behavior between the ruby pin and the nitrocarburized sample.

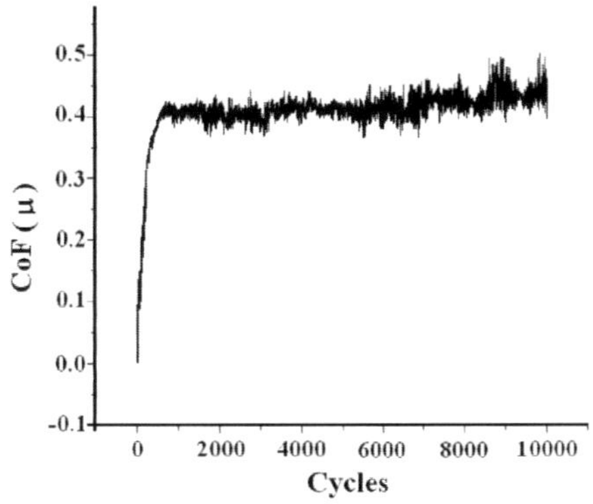

Figure 8. Friction coefficient of H13 tool steel nitrocarburized for 60 minutes.

Wear resistance with sample subjected to nitrocarburizing has been associated with the presence of the compound layer in the structure, although its thinner thickness, with the predominant ε-Fe_{2-3}(C,N) phase with high nitrogen content, preferably affecting the tribological properties [9] and other steel properties [10, 11].

4. Conclusions

Liquid nitrocarburizing H13 tool steel from 45 to 60 minutes showed to be more effective than applying longer times, in terms of forming an almost uniform 2 to 4 μm compound layer formed mainly of the ε-Fe_{2-3}(C,N) phase, which resulted in microhadness values close to 810 HV and a coefficient friction between 0.4 to 0.5. The increment on treatment time provokes an increment of the γ'-Fe_4(C,N) phase which is associated with the presence of porosity.

References

[1] Alsaran A Karakan M Çelik A Bulbul F and Efeoglu I 2003 *J. Mater. Sci. Lett.* **22** 1759
[2] Du H and Ågren J 1996 *Metall. Mater. Trans. A* **27A** 1073
[3] Bell T, Sun Y and Suhadi A 2000 *Vacuum* **59** 14
[4] Ameen H A and Hassan K S 2010 *American Journal of Scientific and Industrial Research* **1** 320
[5] Wahl G 1995 *Heat Treat. Met.* **22** 65
[6] Visuttipitukul P Paa-rai C and Kuwahara H 2006 *Acta Metallurgica Slovaca* **12** 264
[7] Bell T 1999 Gaseous and Plasma Nitrocarburizing of Steels *ASM Handbook* (*Heat Treating* vol 4) ed ASM International (Ohio: ASM International) p 425-436
[8] Du H 1993 *J. Phase Equilib.* **14** 682
[9] Somers M A J 2000 *Heat Treat. Met.* **4** 92
[10] Wells A 1985 *J. Mater. Sci.* **20** 2439
[11] Baranowska J and Wysiecki M 2000 *Surf. Coat. Technol.* **125** 30
[12] Friehling P B Poulsen F W and Somers M A J (2001) *Z. Metallkd.* **92** 589
[13] Fattah M and Mahboubi F 2010 *Mater. Des.* **31** 3915
[14] Karakan M Alsaran A and Çelik A 2004 *Mater. Des.* **25** 349
[15] Ebersbach U and Naumann J 2000 *HTM* **55** 345

IOP Conf. Series: Journal of Physics: Conf. Series **792** (2017) 012020 doi:10.1088/1742-6596/792/1/012020

Experimental evaluation of a heat pump for the water-supply heating of a public swimming pool

R López, M Vaca, H Terres, A Lizardi, J Morales and S Chávez
Universidad Autónoma Metropolitana-Azcapotzalco
Av. San Pablo No. 180, Col. Reynosa-Tamaulipas, Del. Azcapotzalco, México, 02200, D.F. México,
E-mail: rlc@correo.azc.uam.mx

Abstract. In this work the analysis of the thermodynamic behavior of heat pumps (HP) which supply the energy needed in the public pool at the Aquatic Center of Azcapotzalco was performed. There are 18 installed HP´s but only those needed to provide the energy required are alternately activated. The evaluation was conducted during May and June of 2015. We selected one of the HP to implement temperature and pressure gauges at the inlet and outlet of the compressor. The measurements were made every day at three times, 6:30, 13:00 and 18:00 hours. In a period of 24 hours, 1 000 L evaporated, there was no variation registered overnight, since the pool was covered with plastic to avoid loss of the fluid. The heat pump provided 150 kW to maintain the water temperature at the right level of operation, namely 28 °C. The coefficients of performance (COP) of the HP were 6.39 at 6:30, 7.42 at 13:00 and 7:32 at 18:00 hrs., values which are very close to the one provided by the manufacturer.

1 Introduction

The supply of hot water in Mexico is part of services requiring considerable amounts of energy both in households and buildings intended for commercial and recreational use. This supply requires the consumption of fossil fuels (natural and LP gas or diesel) and is used in private and public pools (sports, hotels and houses); the need of this hot water is essential both for the pool itself and the bath services. In other countries electric power is used to achieve this goal [1], with electric heaters and other elements such as heat pumps, which are significantly more efficient than traditional systems [2]. Other sources of energy have also been used such as solar energy in combination with other transformation processes in order to improve the thermal efficiency, an example of these devices are heat pumps [2-6]. Currently, mechanical vapor compression heat pumps are extensively applied in many manufacturing industries, the efficiency has improved considerably, primarily in four areas: compressor performance, new multistage cycles, new refrigerants, and combined ejector compression [4]. Many of the commercial heat pump systems comprise a single-stage vapor compression cycle, only one evaporator is used for cooling and dehumidifying, and recovering the heat from a heat source [7]. A compound system consists of two or more compression stages connected in series. It may have one high-stage and one low-stage compressor or several compressors connected in series, a multistage one has a smaller compression ratio and higher compression efficiency for each stage of compression, greater refrigeration effect, lower discharge temperature at the high-stage compressor, and greater flexibility [8]. A compound multi-stage HP system is a favorable option to improve system COP (coefficients of performance). The most recent refrigeration compressor is known as 'Revolving Vane (RV) compressor', the principal feature of this innovative design involves the use of a rotating cylinder that moves together with the compressing mechanism to cut down on energy loss, consequently, frictional and leakage losses are effectively reduced [9]. The use of refrigerant mixtures in heat pumps and air conditioning systems constantly poses new challenges to engineers, the occurrence of glide during the phase change present problems to heat transfer engineers, among which incomplete condensation is considered the most relevant one [10]. Recently, the performance of the R410A refrigerant in an air conditioning unit was evaluated, resulting in about 12% higher efficiency

Content from this work may be used under the terms of the Creative Commons Attribution 3.0 licence. Any further distribution of this work must maintain attribution to the author(s) and the title of the work, journal citation and DOI.
Published under licence by IOP Publishing Ltd

than that of R22 unit, R410A refrigerant absorbs and releases heat more efficiently which means systems using it will require less electricity and therefore will operate more efficiently [11]. An ejector-compression heat pump employs low-grade thermal energy to provide space cooling and heating, and depends on the geometrical, aero-dynamical, and mechanical design of the ejector; a theoretical study has shown that COP of an ejector-compression heat pump can yield improved performance of up to 21% over the vapor compression standard cycle [12].
With reference to the commitment made in Europe in terms of energy consumption which states that by the 2020 year, 20% of the current should be saved [13], the energy consumed in all places was calculated and the pools and ice skating tracks were found to occupy the first consumption places [14]. There are some studies on the energy analyses of indoor swimming pools, where the open-cycle absorption heat pump system was compared to the mechanical heat pump, the absorption system operated by chemical dehumidification on the exhausted air of the swimming pool was analyzed; although considerable energy saving effect was demonstrated, the promotion of this technology was restricted by the product maturity and reliability [15]. The objective of the present work was to experimentally evaluate the thermodynamic behavior of a heat pump by mechanical compression which operates to provide hot water to a public pool.

2 System description.

The public pool is located at Aquatic Center of Azcapotzalco (19° 30' 01.2" N, 99° 10' 59.4" W and 2240 meters above sea level). This is a swimming school with a pool of Olympic dimensions, which offers daily service to 450 users, on a schedule from 6:00 to 21:00 hours, in shifts of 50 min-classes and 10 min-recess, from Monday to Saturday. The design of the pool was regulated in compliance with the Mexican regulations (namely, Norma Oficial Mexicana NOM-245-SSA1-2010) [16]. The dimensions of the pool and its operating conditions are recorded in table 1, its photograph is presented in fig. 1.

Table 1. Pool characteristics

Length	50 m
Width	25 m
Number of lanes	10
Lane width	2.5 m
Minimum depth	1.55 m
Maximum depth	1.90 m
Water temperature	25-28 °C
Environment temperature	26-27 °C
Relative humidity	50-60 %

The pool water should be between 25 and 28 °C according to the existing standards, [17]; the ASHRAE Handbook [18], recommends that the temperature of the space of the pool should remain close to 28 °C, and the fluctuation range of relative humidity should be 50-60 %, this is in order to reduce the evaporation of water from the pool. The evaluation was conducted between the months of

May and June, 2015, the average recorded temperature of the environment was between 24 and 25 °C and relative humidity at this period was 50 %.

3 Heat pump energy

For the supply of hot water, the Azcapotzalco Aquatic Center has 18 C5HP3-model Heat Siphon-brand heat pumps. They are air-water type, i.e., they take energy from the environment and together with the energy of the air compressor, it is transferred to water. Only the necessary number of HPs to provide energy to the water are activated, in this assessment we used up to 8 units alternately. To keep the water flowing through the HP a centrifugal pump which distributes it among the heat pumps that are required is used; water is turned back to the pool at operation temperature (30 °C), R 410A is the refrigerant and the theoretical coefficient of performance (COP) is 7.6 (manufacturer data). The distribution of HPs in the machine room can be observed in fig. 2. The amount of energy required was calculated as follow.

Figure 1. Azcapotzalco Aquatic Center pool.

Figure 2. Distribution of heat pumps

Total heat which must be supplied to the pool shall be equal to the observed loss of heat, which is:

$$Q = Q_s + Q_l + Q_f + Q_{conv} \quad (1)$$

where Q_s is the loss of heat by conduction in solid surface that surrounds the pool; Q_l is the loss of heat by evaporation of the water; Q_f is the energy of the water replenishment; Q_{conv} are the losses by convection on the free surface of the water.

Heat losses by conduction of water from the pool to the walls and the floor forming it, are usually small, they are calculated as,

$$Q_s = U_p A_p (T_w - T_{wg}) \quad (2)$$

where U_p is the overall pool wall heat transfer coefficient; A_p is the overall pool wall and bottom surface area; T_w and T_g are the pool water temperature and the ground temperature, respectively. The heat loss by evaporation is given by:

$$Q_l = W\,Y \quad (3)$$

where W is the rate of evaporation at water surface, and Y is the latent heat of water evaporation at the water surface. The most widely used correlation for water evaporation rate is reported in the ASHRAE Application Handbook as follows [17],

$$W = (0.089 + 0.782 v_a)(p_w - p_a)\frac{A_p}{Y} \quad (4)$$

where v_a is the air velocity above the water surface; p_w is the saturated vapor pressure at the water surface temperature; p_a is the vapor pressure at the indoor air dew point; A_p is the total area of the pool water surface.

The rate of energy for daily renovated feed water heating depends on the pool volume, feed water temperature and the designed pool water temperature. This can be calculated by taking the 24-h average, i.e.

$$Q_f = C_w\, q_d\, \rho(T_w - T_f)/24 \quad (5)$$

where C_w is the specific heat of water; q_d is the daily feed water flow rate (is generally 3 % of the total pool volume, it was used in this study); T_f is the supplementary feed water temperature. The heat flow rate by convection can be calculated with the Newton's formula, i.e.

$$Q_{conv} = h_{conv} A_p (T_w - T_a) \quad (6)$$

where h_{conv} is the convective heat transfer coefficient, and T_a is the indoor air temperature.

The heating capacity of the condenser can be calculated by,

$$Q_{cond} = \dot{m}_r (h_o - h_i) \quad (7)$$

where $\dot{m}_r$ is the mass flow rate of refrigerant; h_o and h_i are the refrigerant specific enthalpy at the outlet and inlet of the condenser. This heat is equal to the heat transfer to water is determined by,

$$Q_w = \dot{m}_w C_w (T_{wi} - T_{wo}) \quad (8)$$

where $\dot{m}_w$ is the mass flow rate of hot water; T_{wi} and T_{wo} are the inlet and outlet hot water temperatures. The energy consumption Q_{hp} of the heat pump unit is the total energy consumed by the compressor, the fans, the water pumps and the controller during the period, i.e.

$$Q_{hp} = \Sigma(P_c + P_f + P_p) \quad (9)$$

where P_c, P_f, P_p are is the power consumptions of the compressor, the fan and the water pumps respectively. The coefficient of performance (COP) of the overall heating system is defined as

$$COP_{sys} = Q_{cond}/Q_{hp} \quad (10)$$

4 Experimental measurements.

Some modifications were carried out to the studied heat pump with the aim of obtaining its experimental evaluation; the points at which the measuring devices were installed are presented in fig. 3. The temperature was measured at the entrance of the compressor (T1), condenser (T2), expansion (T3) and the evaporator valves (T4); pressure at the inlet (p1) and outlet (p2) of the compressor; capacitor the water supply to the pool entrances to the condenser at temperature (T5) and exhausts at temperature (T6); in the evaporator the flow mass of used air coming out at temperature (T7) was measured; and the temperature of the air in the environment (T8). The consumption of electrical energy used by the compressor, fan and water recirculation pump was also evaluated. The mean values of the temperatures in the heat pump for an average day of work for the month of June 2015 at three times are listed in table 2, the first line corresponds to 6:30 a.m., he second to 13:00 hrs., and third to 18:00 hrs.

IOP Conf. Series: Journal of Physics: Conf. Series **792** (2017) 012020 doi:10.1088/1742-6596/792/1/012020

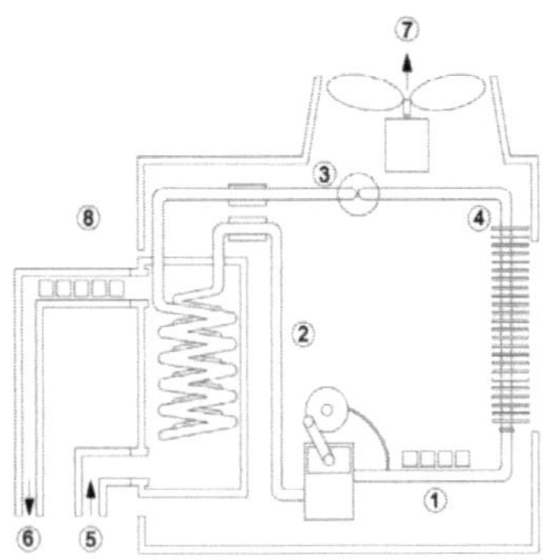

Figure 3. Diagram of the heat pump and the places where the measurements were made.

Table 2. Temperatures measured at points considered in °C, at 6:30, 13:00 and 18:00 hrs. T (°C) and p (MPa).

$T1$ ($p1$)	$T2$ ($p2$)	$T3$	$T4$	$T5$	$T6$	$T7$	$T8$
8.8 (0.85)	88.1 (4.5)	36.9	3.9	23.4	29.2	3.8	13
9.5 (0.85)	87.3 (4.5)	35.8	3.4	29.1	32	3.2	21
9.1 (0.85)	87.4 (4.5)	36.1	3.1	28.7	31.8	3.0	23

5 Results

The amount of water evaporated in the 15 hours of service by the pool was 990 L, in the first hours 85 liters per hour evaporated and at the end only 60 liters every hour, fig. 4. This amount of water is replaced from the municipal service, a portion is dragged by air entering the interior of building as moisture coming out at the other end and the rest stays inside the premises. To the extent that the temperature difference between the pool water and the interior air is less, the amount of evaporated water will diminish. During the night the free surface of the water of the pool is covered with a plastic canvas to prevent both heat losses from water and the evaporation of part of it. Also from the graph presented in fig. 4 we can obtain the speed at which the water evaporates, it is simply the volume of the water evaporated over time.

The amount of heat that is required to supply to the water is approximately 150 kW per day, its distribution can be observed in the graph of fig. 5. In the first four hours a supply of 13 kW of heat (with 8 working HPs) is needed, this is due to the temperature difference between the pool water and the environment which is high; then this demand diminishes to a value of approximately 8 kW (with 5 working HPs), finally it increases slightly towards the end of the day to a value of 9.6 kW (with 6 working HPs).

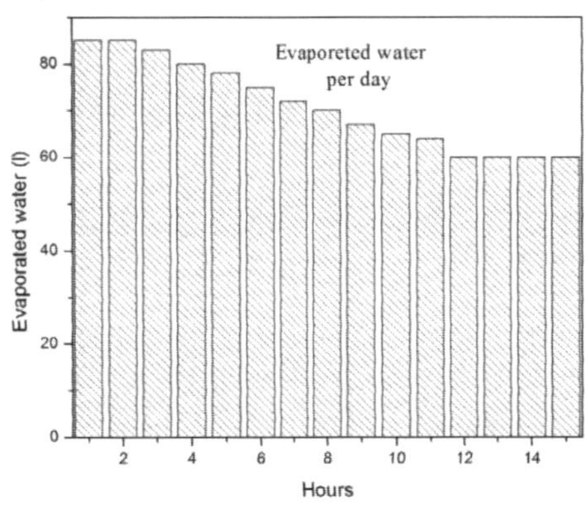

Fig. 4. Evaporated water per day.

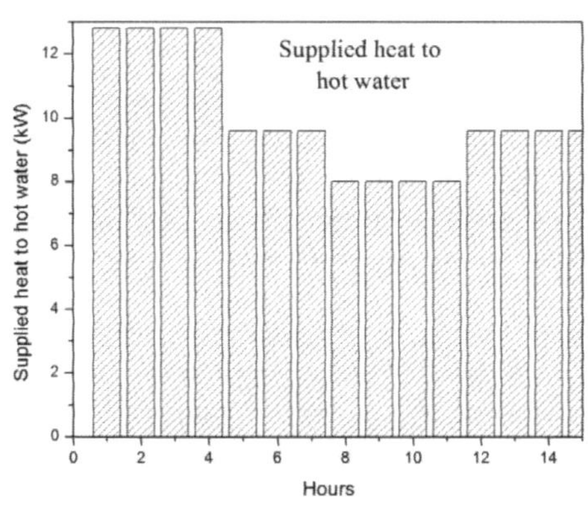

Fig. 5. Supplied heat to hot water per day.

The electrical energy consumed during this period of time by the HP compressor, the evaporator fan and the water recirculation pump motor is presented in graph of fig. 6. The total consumption was 25 kW in the 15 hours service-period of the pool. The compressor consumed 50% of the energy, 30% was consumed by the pump motor and the remaining 20% corresponded to the drive of the fan motor.
Using data from the intermediate line of table 2, as well as those of the pressure at the entrance, namely 0.85 MPa, at the exit of the compressor, 4.5 MPa and using the thermodynamic diagram of R-410A refrigerant (Fig 7), enthalpy values were obtained to determine with the help of the equations provided above. The COP of heat pump at 6:30 was 6.39; at 13:00 was 7.42 and finally 7.32 at 18:00 hrs. These values are within a very close range of the one provided by the manufacturer of the HP which is 7.6.

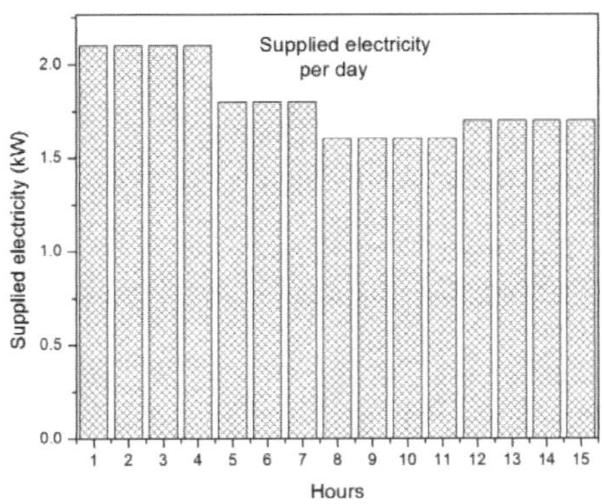

Fig. 6. Electrical power supplied to the compressor, fan and water recirculation pump engines.

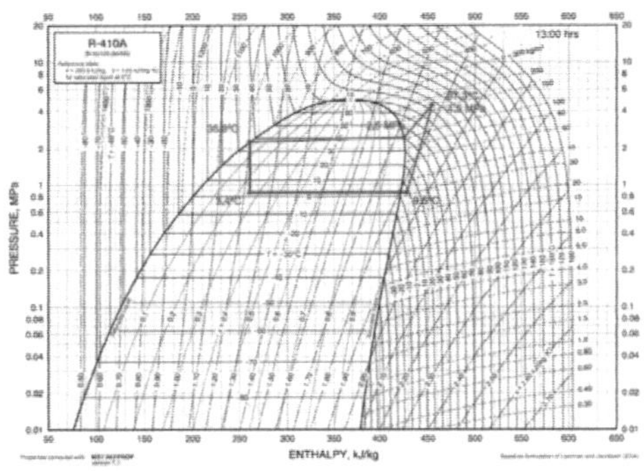

Fig. 7. Diagram of heat pump thermodynamic cycle (DuPont® Suva® 410A).

6 Conclusions.

In this work the analysis of the thermodynamic behavior of heat pumps that supply the energy needed in the public pool at the Aquatic Center of Azcapotzalco was performed. The values used corresponded to the months of May and June of the year 2015. We selected one of the HP to implement temperature and pressure gauges at the inlet and outlet of the compressor. The measurements were made every day at three times, 6:30, 13:00 and 18:00 hours. During 24 hours, almost 1 000 L evaporated, there was no variation overnight since the pool is completely covered with a plastic canvas to avoid loss of the fluid. The HP provided 150 kW to maintain the water temperature at the right level of operation, i. e., 28 °C. The consumption of required electrical energy turned out to be 25 kW. The COP of the HP was 6.39 at 6:30, 7.42 at 13:00 and 7.32 at 18:00 hs., very close values to the one provided by the manufacturer.

7 References

[1] Kokkinides L and Sachs H M 2002. *Toward market transformation: commercial heat pump water heaters for the ''New York energy Smarts Region''* ACEEE and NYSERDA.

[2] Kim M, Kim M S and Chung J D 2004 *Transient thermal behavior of a water heater system driven by a heat pump.* International Journal of Refrigeration 27, pp 415–21.

[3] Cervantes J G and Torres-Reyes E 2002.Experiments on a solar-assisted heat pump and an exergy analysis of the system. Applied Thermal Engineering 22, pp 1289–1297.

[4] Chua K J, Chou S K and Yang W M 2010 *Advances in heat pump systems*: review. Appl. Energy 87, pp 3611–3624.

[5] Chow T T, Bai Y, Fong K F and Lin Z 2012 *Analysis of a solar assisted heat pump system for indoor swimming pool water and space heating*. Appl. Energy 100, pp 309–317.

[6] Huan-Liang T 2015. *Modeling and validation of refrigerant-based PVT-assisted heat pump water heating (PVTA–HPWH) system* Appl. Sol. Energy 122, pp 36–47.

[7] Ma Y and Zhao X 2008 *Experimental study of a heat pump system with flash-tank coupled with scroll compressor.* Sol. Energy Appl. Build. 40, pp 697–701.

[8] Chen L, Li J, Sun F, Wu C 2008. *Performance optimization for a two-stage thermoelectric heat-pump with internal and external irreversibilities.* Appl Energy. 85, pp 641–649.

[9] Teh L and Ooi T, 2009 *Experimental study of the revolving vane (RV) compressor*. Appl Therm Eng. 29, pp 3235–3245.

[10] Pan G and Li Z, 2006 *Investigation on incomplete condensation of non-azeotropic working fluids in high temperature heat pumps* Energy Convers. Manage. 47, pp 1884–1893.

[11] Han D H, Lee K J and Kim Y H 2003 *Experiments on the characteristics of evaporation of R41 Refrigerating and Air-Conditioning Engineers, Inc., Atlanta, GA.0A in brazed plate heat exchangers with different geometric configurations.* Appl Therm Eng. 23, pp 1209–1225.

[12] Yari M, Sirousazar M 2007 *Performance analysis of the ejector-vapour compression refrigeration cycle.* Proc Inst Mech Eng, Part A: J Power Energy. 221, pp 1089–1098.

[13] European Commission, The EU climate and energy package. http://ec.europa.eu/clima/policies/brief/eu/index_en.htm.

[14] http://www.orsa.se/wk_custom/documents/%7B8a31df1e-7390-4221-8e5c-5a91111706dd%7D_energi_i_idrottsanlaggningar.pdf

[15] Johansson L L and Westerlund L 2001 *Energy savings in indoor swimming pools: comparison between different heat recovery systems* Appl. Energy. 70, pp 281–303.

[16] FINA, 2015. Facilities Rules, part IX.

[17] ASHRAE Handbook, 2003. ASHRAE Handbook–HVAC Applications, American Society of Heating, Refrigerating and Air-Conditioning Engineers, Inc., Atlanta, GA.

Surface photovoltage spectroscopy characterization of AlGaAs/GaAs laser structures

Y González, A Abelenda and M Sánchez[1]

Facultad de Física, Universidad de La Habana. San Lázaro y L, Vedado, 10400. La Habana, Cuba.

E-mail: maruchy@fisica.uh.cu

Abstract. The characterization of AlGaAs/GaAs laser structures using surface photovoltage spectroscopy (SPS) is presented. The experiments were carried out at room temperature in the simplified variant of metal–insulator–semiconductor (MIS) configuration by means of a lab-made automated experimental setup. With this technique was possible to identify all the layers forming the heterostructure and important parameters have been extracted. The aluminium composition of the layers, obtained in this way agrees very well with the intentional growth parameters. Also from the surface photovoltage spectra was possible to obtain the lasing wavelength in each structure, which is in a good agreement with the values obtained by measuring the electroluminescence spectra. Results underline the power of SPS in characterization of actual laser devices in a contactless nondestructive way.

1. Introduction

The Surface Photovoltage Spectroscopy (SPS) is a nondestructive, contactless and versatile technique for characterizing semiconductors structures and devices, which can give important information even at room temperature [1]. SPS is based on the analysis of the induced change in the surface photovoltage (SPV) due to the external illumination. This technique allows determine parameters like: the bandgap energy, minority carrier diffusion length and life time [2], and also can provide a detailed picture of the surface and interfaces band diagrams. A detailed explanation of the basics of this technique can be found in the review paper of L. Kronik and Y. Shapira referenced in [3].

Compared to other spectroscopic techniques such as, Photoreflectance and Photoluminescence, SPS has the advantage that is less expensive and simplest since not requires pump laser or detector.

[1] maruchy@fisica.uh.cu.

Content from this work may be used under the terms of the Creative Commons Attribution 3.0 licence. Any further distribution of this work must maintain attribution to the author(s) and the title of the work, journal citation and DOI.
Published under licence by IOP Publishing Ltd

IOP Conf. Series: Journal of Physics: Conf. Series **792** (2017) 012021 doi:10.1088/1742-6596/792/1/012021

Moreover, since SPS is intrinsically insensitive to reflection and scattering process, is very useful for the study of micro and nanocrystalline materials.

SPV measurements are usually performed in two configurations: Kelvin probe and the metal–insulator–semiconductor (MIS) operation mode. In the last decade, the SPS has been developed considerably in terms of methods for data analysis and experimental arrangements respectively [4, 5]. However, despite its advantages, this technique has not been fully exploited and its use has been relatively limited.

In this paper, AlGaAs/GaAs separate confinement laser structures with different aluminium content in the active region are studied. The experiments were carried out at room temperature by means of a lab-made automated experimental setup. Measurements were performed in a modification of conventional MIS operation mode. In the SPV spectra was possible to identify all the layers forming the laser heterostructure. A proposal of a practical criterion for spectra processing and extraction of the parameters of interest is presented and the extracted parameters are in a very good agreement with the intended growth parameters. The lasing wavelength in each structure obtained from SPV spectra agree very well with the values measured from electroluminescence spectra.

2. Experimental

The surface photovoltage signal is defined as the difference between the surface potential under illumination and in the dark:

$$SPV = V_S\left(light\right) - V_S\left(dark\right)$$

The experimental set up used to measure the SPV spectra comprises a monochromator, a lock-in amplifier, a chopper, and a sample holder. In Figure 1 it's schematically represented.

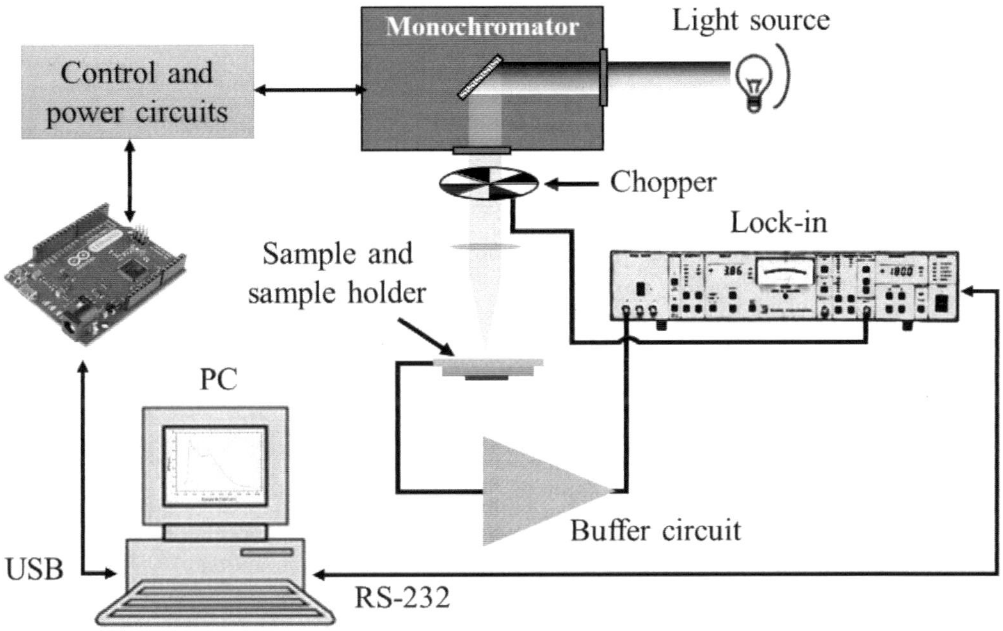

Figure 1. Sketch of the experimental setup for Surface Photovoltage Spectroscopy measurements.

The system is fully automated and allows measurements in a wide spectral range. The monochromator is controlled from the PC through an Arduino Leonardo, a power circuit and a stepper motor. The whole experimental setup is interfaced with a PC through a RS232 serial port and the control software.

The light coming from the lamp passes through a monochromator, and is focused onto the sample after being modulated by an optical chopper with a variable frequency. Thus, the photon flux impinging on the sample causes a periodic generation of excess carriers and their subsequent redistribution change the surface potential inducing the SPV signal. This signal goes through to a high impedance preamplifier and is registered by a lock-in amplifier. The SPV is measured as a function of incident photon energy providing the SPV spectra.

For measurements, the sample is positioned between a transparent indium–tin–oxide (ITO) electrode and a copper plate with a spring which acts as a sample holder and the bottom ground electrode. Figure 2 shown a simple diagram of the used sample holder.

Unlike the conventional configuration of MIS where it is left a thin air gap between the sample and the ITO electrode to form a capacitor, in our setup the ITO is directly in contact with the sample. This allows the direct measurement of the SPV signal, whereas in conventional MIS method the voltage measured is affected by the capacitance and must be corrected in some way to obtain the real SPV signal.

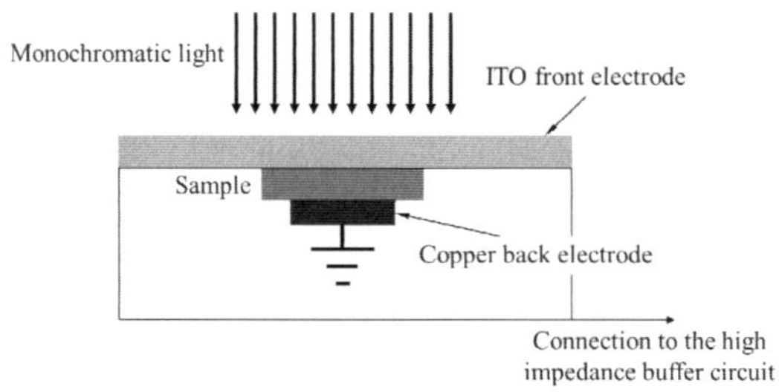

Figure 2. Schematic diagram of the sample holder used in SPV measurements.

The ITO plate acts as both the top electrode and a window for light to pass through it. The copper bottom contact is covered with conductive silver paste to improve the electric contact. Experiments were performed at room temperature and in air ambient.

3. Results and discussion

The AlGaAs/GaAs laser diodes studied here are separate confinement heterostructures in the metal stripe configuration, and consist of layers of different compositions, thicknesses and doping level respectively. In Figure 3 it is shown the schematic of the laser structures with the corresponding energy band diagram.

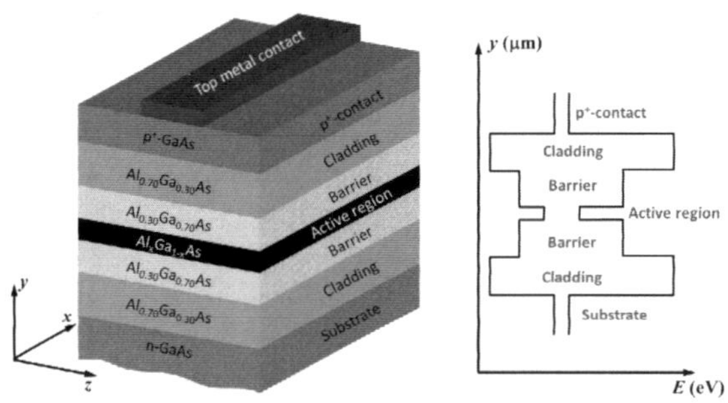

Figure 3. Left, schematic of AlGaAs/GaAs laser structures. Right, approximately energy band diagram.

In Figure 4 the features corresponding to: the GaAs substrate (1.42 eV), the $Al_{0.18}Ga_{0.82}As$ active region (~ 1.63 eV) and the $Al_{0.3}Ga_{0.70}As$ barriers (~ 1.73 eV) are identified.

Figure 5 shows the SPV spectra for two laser structures with different aluminium content (x) in the active region. As can be seen the contribution of the different layers forming the heterostructure can be clearly distinguished. The band gap of each layer can be identified as the slope changes in the SPV spectrum. This can be understood by the abrupt change of the absorption coefficient in the vicinity of the bandgap energy in semiconductors, causing a significant change in the signal SPV. In this case, only the signature of the active region and barriers was seen. Peaks corresponding to the active region in each structure can be clearly differentiated and appear at 1.46 eV and at 1.66 eV respectively. The band gap is higher in sample 192 indicating the higher Al content in this structure, whereas, in both spectra the peak corresponding to the $Al_{0.3}Ga_{0.70}As$ barriers are at the same energy, around 1.73 eV.

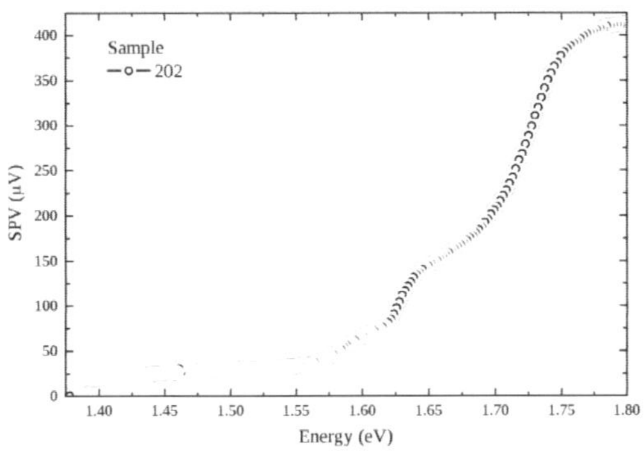

Figure 4. SPV spectra measured for sample 202. Features corresponding to the GaAs substrate, the $Al_{0.18}Ga_{0.82}$ As active region and the $Al_{0.3}Ga_{0.70}As$ barriers are visibly identified.

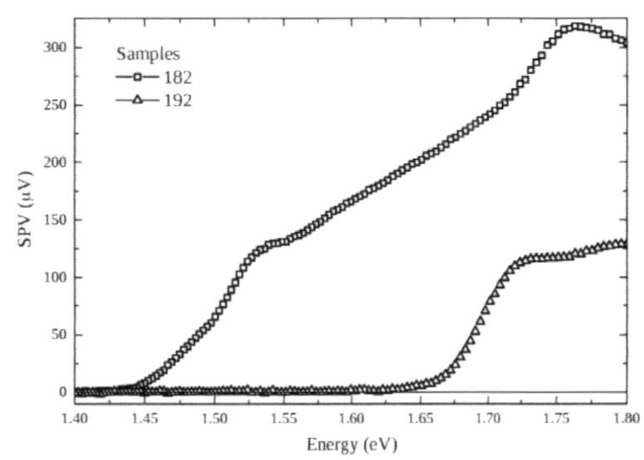

Figure 5. SPV spectra of structures with different aluminium content in the active region. Energy peaks of the active region are clearly different in the two structures.

The band gap energies obtained in this way was used for estimation of the Al mole fraction in the structure active region. The intended growth parameters are: x = 0.05 and 0.2 for structures 182 and 192 respectively, which according to the dependency of the band gap with the mole fraction Al for of $Al_xGa_{1-x}As$ for x< 0.45[6]:

$$E_g(x) = (1.42280 + 1.31804 \cdot x)\ \text{eV}$$

results in transition energies of 1.48 eV and 1.68 eV in each case, which is in very good agreement with the values extracted from SPV spectra.

With the values of the transition energies of the active region obtained in this way the emission wavelength of the structure was estimated and compared with that obtained from the electroluminescence (EL) emission spectra of laser diodes coming from these structures. In Table 1 these results are presented.

Table 1. Comparison of the lasing wavelength obtained from SPV and electroluminescent emission spectra.

	Lasing wavelength (nm)	
Sample	**SPV**	**Electroluminescent spectra**
182	844	849
192	743	750

The electroluminescence spectra were measured by forward biasing the laser diodes by pulsing the current source at 1Hz with a 0.1% duty cycle square wave. The emission resulting from radiative recombination is collected by a lens and focused into the monochromator. The detection was made with a Silicon photodiode and a lock-in amplifier registers the signal.

As shown in Table 1, results obtained from both methods are in very good agreement, demonstrating than SPV spectroscopy can be used to obtain performance parameters of actual laser devices in a simple manner.

Recently some authors have reported that, the numerical derivative of SPV spectrum can provide the transition energies in a heterostructure in a more accurately way [4, 7]. In fact, assuming that SPV signal is proportional to the open-circuit photovoltage we can write [4]:

$$SPV \ \mu \ V_{oc} \ \mu \ \frac{\alpha}{h\nu}$$

$$\frac{\mathrm{d}\left[E \times SPV(E)\right]}{\mathrm{d}E} = \frac{\mathrm{d}\alpha}{\mathrm{d}E}$$

From Equation (4) it can be seen that a plot of the numerical derivative (d[E x SPV(E)]/dE) versus the incident photon energy E, provides the transition points in absorption for the different layers. Figure 6 show a comparison between the SPV spectra and the numerical derivative (d[E x SPV(E)]/dE) versus Energy for structures 185 and 175 respectively.

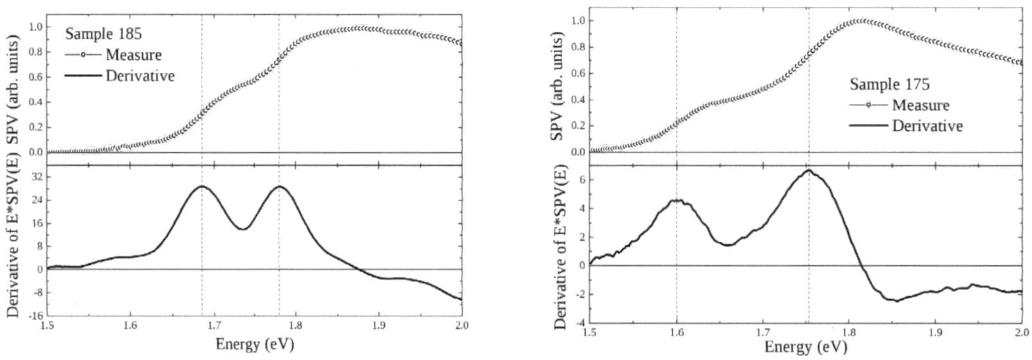

Figure 6. Comparison of the SPV spectrum and numerical derivative of SPV versus *E* for two structures: (a) sample 185 and (b) sample 175. The transition energies are indicated by dashing lines.

As can be seen, in the derivative spectra is possible appreciate much better the points where the slope changes occur. In this case, the transitions energies points correspond to the maxima in the derivative spectra and results in 1.68 and 1.60 eV respectively.

Is worth mentioning that the SPS has been applied before to the study of both, vertical cavity surface emitting laser (VCSEL) [8, 9] and edge emitting laser diodes [7, 10].

In case of edge emitting lasers, Leibitovich et al [10] used a commercial Kelvin probe unit for SPV measurements on InGaAs QW structures. The authors extracted the structure parameters by a

Lorentzian fitting of the SPV spectra. Muralidharan [7] did SPV measurements in MIS geometry to characterize AlGaAs laser structures and extracts the transition energy points from the numerical derivative of the SPV spectra. In this case, the measurement was done by successively etching of the individual layers and subsequent measurement of SPV. Their results were compared with photoluminescence (PL) measurements.

In our work, the SPV measurements were carried out in a simplified MIS configuration using a lab-made automated experimental setup. Our structures are fully metalized allowing us to do EL measurements to compare results. The results show that is possible to obtain the composition of the layers in the laser structure, without the necessity of complicated procedures such as etching away the individual layers.

4. Conclusions

In this work, characterization of AlGaAs/GaAs laser structures using surface photovoltage spectroscopy (SPS) is presented. Measurements were carried out in a simplified MIS configuration using a lab-made automated experimental setup. The ITO electrode is directly in contact with the sample allowing the direct measurement of the SPV signal. With this technique was possible to get information regarding each layer forming the $Al_xGa_{1-x}As/GaAs$ heterostructure and the results are in agreement with the intended growth parameters. Also from the SPV spectra was obtained the lasing wavelength in each structure, which is in a good agreement with the values obtained from the electroluminescent spectra of laser diodes coming from the structures.

Our results show that it is possible evaluates the composition of AlGaAs layers in laser structures without the necessity of a chemical etching process. These results emphasize the potentialities of the SPS as a simple tool for semiconductor devices characterization, compared to other more expensive spectroscopic techniques.

References

[1] Cavalcoli D and Cavallin A 2010 *Phys. Status Solidi* **C7** 1293
[2] Schroder-Dieter K 2001 *Meas. Sci. Technol.* **12** R16
[3] Kronik L and Shapira Y 2001 *Surf. Interface Anal.* **31** 954
[4] Datta Shouvik, Ghosh Sandip and Arora B M 2014 *Review of scientific instruments* **85** 043909
[5] Donchev V, Kirilov K, Ivanov T and Germanova K 2006 *Materials Science and Engineering B* **129** 186
[6] Sánchez M, Díaz P, González J C, Marín E and Prutskij T 1995 *Rev. Mex. Fís.* **41** 739
[7] Muralidharan R, Ramesh V, Mishra Puspashree and Srinivasan T 2014 *Semicond. Sci. Technol.* **29** 035002
[8] Liang J. S, Huang Y. S , Tien C. W, Chang Y. M, Chen C W, Li N Y, Li P W and Pollak F H 2001 *Appl. Phys. Lett.* **79** 3228
[9] Huang Y S, Malikova L, Pollak F, Debray J P, Hoffman R, Amtout A and Stall R A
[10] Ashkenasy N, Leibovitch M, Shapira Y, Pollak F, Burnham G and Tand X Wang 1998 *J. Appl. Phys.* **83** 1146

IOP Conf. Series: Journal of Physics: Conf. Series **792** (2017) 012022 doi:10.1088/1742-6596/792/1/012022

Experimental and numerical analysis of convective flow in a square cavity with internal protuberances

A Lizardi,[1] H Terres, R López, M Vaca, S Chávez., A Lara, J R Morales
Universidad Autónoma Metropolitana-Azcapotzalco, Departamento de Energía, Área de Termofluidos. Av. San Pablo 180, Col. Reynosa Tamaulipas, Del. Azcapotzalco, Ciudad de México.

E-mail: arlr@azc.uam.mx

Abstract. A comparison between experimental and numerical results for natural convection flow generated inside a square cavity filled with water, which has internal protuberances, is presented. The cavity is formed by vertical-isothermal and horizontal-adiabatic walls. The built prototype is integrated by: a clear glass square cavity, where internal protuberances are located; a cavity with water which maintains one of its walls at low temperature; an electrical resistance which preserves another wall at high temperature; and an array of mirrors. The experimental flow development was studied with a Particle Image Velocimetry (PIV) device. Furthermore, ANSYS software was used and mass, momentum and energy equations were numerically solved. Results were compared using a system with rectangular, semi-circular, and triangular protuberances versus a system without protuberances; it was found that: a) the maximum positive value of the vertical velocity decreased 7.12, 3.33 and 3.03%, respectively, for the experimental case, and 4.52, 2.26 and 1.27%, respectively, for the numerical case; b) the maximum positive value of the horizontal velocity decreased 18.37, 11.89 and 4.59%, respectively, for the experimental case, and 6.46, 4.75 and 2.47%, respectively, for the numerical case; c) the average Nusselt number decreased 10.52, 7.95 and 6.06%, respectively, for the experimental case, and 12.01, 9.06 and 3.02%, respectively, for the numerical case.

1. Introduction

In natural convection fluid movement results from buoyancy forces imposed on the fluid when its density in the vicinity of the heat transfer surface decreases or increases as a result of warming or cooling processes. Natural convection is a process of energy transport by the combined action of heat conduction, energy storage, and movement of matter. The complexity of the majority of the cases involving the transfer of heat by natural convection makes it impossible to have an exact analysis of the equations of conservation, having to resort to numerical simulations or experiments. The objective of this study is to compare the experimental and numerical results of the flow field in a square cavity filled with water, in four cases: without internal protuberances, and with internal rectangular, semi-circular and triangular protuberances, located symmetrically on the lower part of the system, as illustrated in Figure 1. The experimental analysis is accomplished by building a model that consists of: a square cavity of transparent glass that houses the internal protuberances under study; a cavity with water that maintains one of its walls at low temperature; an electric heater that keeps its wall at high temperature; and an arrangement of mirrors to aid light reflection. The experimental development of

[1] To whom any correspondence should be addressed.

Content from this work may be used under the terms of the Creative Commons Attribution 3.0 licence. Any further distribution of this work must maintain attribution to the author(s) and the title of the work, journal citation and DOI.
Published under licence by IOP Publishing Ltd

the flow is done with a Velocimetry by Images of Particles device (PIV). Regarding, numerical solution, the equations of conservation of mass, momentum and energy are solved for permanent, two-dimensional, flow, together with the boundary conditions, using the finite element numerical method, by means of the ANSYS software.

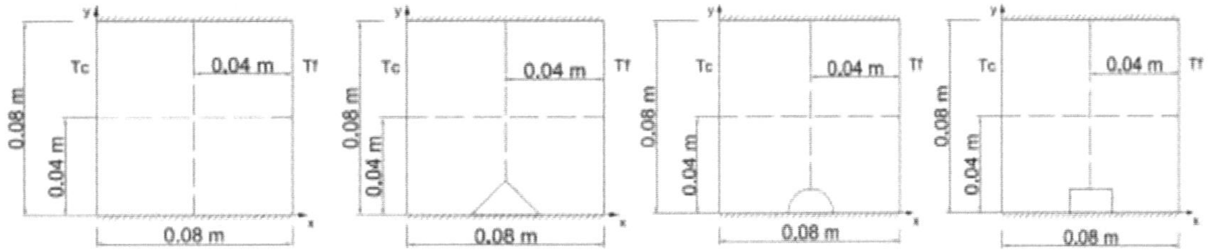

Figure 1. Physical representation of the problem and reference axes (dashed lines) for the analysis.

From the experimental point of view, the phenomenon of natural convection in cavities has been studied by various researchers. Some studies relating to this work are: Corvaro and Paroncini [1], experimentally analysed the natural convection in a square cavity full of air, with a rectangular fountain located at the bottom. The structure of the flow was analyzed for different numbers of Rayleigh and different positions of power source. The experimental apparatus used was a 2D-PIV, which determined the fields of streamlines and velocity maps. Corvaro et al. [2], experimentally analyzed the natural convection in a square cavity filled with air at atmospheric pressure using a PIV and holographic interferometry. Two (hot and cold) plates on the vertical sides of the box were placed. They investigated the relationship between the dynamic and temperature fields to describe how the flow and transfer of heat in the interior of the cavity were influenced by the temperature of the hot plate and the position of the cold plate. Nardini and Paroncini. [3], analyzed the effects on natural convection caused by the variation of the dimensions and positions of heat sources on the side walls of a square cavity filled with air. Temperature distributions were obtained, heat transfer coefficients were experimentally measured, using the holographic interferometry technique and compared to numerical results. On the other hand, reports relating to this topic, but from the numerical point of view are: Kaviany [4], numerically analyzed temperature fields and streamlines in a square cavity filled with air, with a semi-circular protuberance. The vertical walls were considered at constant temperature. The top and bottom surfaces were treated as adiabatic. Results of the number of Nusselt and maximum values of the stream function for a range of $10^1<Ra<10^4$ Rayleigh number and radii of the protuberance of 0, 0.2, and 0.4, are presented. Qi-Hong [5], numerically analyzed natural convection in a square cavity filled with air with two or three pairs of sources of heat in the vertical side walls. The analysis focuses on the effects of size and arrangement of sources and sinks on flow and heat transfer characteristics. Wang and Pepper [6], analyzed laminar natural convection in vertical channels with obstructions inside the channels. The results illustrate the temperature fields and streamlines for different parameters (number of Rayleigh, geometric relationship and location of obstructions) for smooth and blocked channels.

2. Experimental apparatus

In order to determine the properties of the flow in a square cavity inside with protuberances on the inside, we designed and built a test bench consisting of the following elements (Figure 2 - 3):

- A model, built in 0.006 m-thick glass, 0.4 m long, 0.22 m in height and 0.05 m in width. The model contains in itself: a system of mirrors which lights up the inside of the cavity and prevents the generation of shadows; a water tank of 0.14x0.1x0.05 m with an aluminum wall that provides the low temperature of the system; a 300 W electrical resistance mounted on an aluminum plate that provides the high temperature of the system; and a cavity of 0.08x0.1x0.05 m, which is the area of study. The protuberances were built in black-painted PVC to avoid light reflections. The rectangular protuberance is 0.02 m base by 0.01 m in height, semi-circular protuberance has 0.010 m radius, and triangular protuberance is 0.028 m base by 0.014 m height.

- An electrical resistance of 300 W in the form of rectangular plate of 0.05x0.10 m.
- A voltage regulator of 120 V input, 0-140 V output and 10 A.
- A 380801 model EXTECH power Analyzer, with 0-300 W ±0.1 W, 0-200 V ±0.1 V and 0-2 A ±0.001A ranges.
- Eight type K thermocouples with a temperature range from -200°C to +1372°C with a sensitivity of 41 μV/°C. Four thermocouples are placed inside the cavity, on the low-temperature aluminum plate with four others on the high-temperature aluminum plate with high temperature.
- A computer with the 8.1 Lab-View software installed. In this case, the program that allows the acquisition of data from eight thermocouples was previously prepared.
- A Dantec Dynamics Velocimetry by Images of Particles (VIP) apparatus.

Figure 2. Equipment calibration.

Figure 3. Velocimetry equipment and model.

3. Experimental procedure

Prior to experimentation, electrical resistance was calibrated with the power analyzer, finding that when the voltage regulator was operated at 20%, the electrical power was 50 W. This power was managed in all the experiments. The temperatures of the thermocouples were measured in steady state, from the hot and cold plates, being these of 22.3 and 19.3 °C, respectively. To determine the Nusselt number the temperature in the vicinity of the hot wall was determined by placing eight thermocouples along its vertical distance. The time that the equipment took achieve steady state was 10 minutes. Once these parameters were determined, the methodology described in Lizardi et al. [7] was used.

4. Mathematical formulations

The physical model consists of an enclosure of height and width H, the left vertical wall is at a high temperature Tc and the right temperature, at a low temperature Tf. The upper and lower walls are thermally insulated. Four settings were handled: cavity without protuberances on the inside and with rectangular, semi-circular and triangular protuberance in the background, as illustrated in Figure 1. The cavity was full of an incompressible Newtonian fluid, with constant physical properties, except for density in the term of buoyancy (Boussinesq approximation). Under these conditions, the density is considered as a linear function of the temperature, $\rho = \rho_o\left[1-\beta\left(T-T_o\right)\right]$, Incropera and DeWitt [8]. The mathematical approach to the problem of flow promoted by natural convection is defined by the equations of continuity (1), quantity of movement in the x and y axis (2) and energy (3). The approximation of Boussinesq that considers the physical properties of the fluid to be independent of temperature is used in this study, Bejan [9]

$$\frac{\partial u}{\partial x}+\frac{\partial v}{\partial y}=0 \tag{1}$$

$$\rho_o\left(u\frac{\partial u}{\partial x}+v\frac{\partial u}{\partial y}\right)=-\frac{\partial P}{\partial x}+\mu\left(\frac{\partial^2 u}{\partial x^2}+\frac{\partial^2 u}{\partial y^2}\right), \quad \rho_o\left(u\frac{\partial v}{\partial x}+v\frac{\partial v}{\partial y}\right)=-\frac{\partial P}{\partial y}+\rho_o g\beta(T-T_o)+\mu\left(\frac{\partial^2 v}{\partial x^2}+\frac{\partial^2 v}{\partial y^2}\right) \tag{2}$$

$$\rho_o C\left(u\frac{\partial T}{\partial x}+v\frac{\partial T}{\partial y}\right)=k\left(\frac{\partial^2 T}{\partial x^2}+\frac{\partial^2 T}{\partial y^2}\right) \quad (3)$$

Appropriate boundary conditions for the system, Figure 1, are

$x=0$, $u=v=0$, $T=Tc$; $x=H$, $u=v=0$, $T=Tf$; $y=0$, $u=v=0$; $y=H$, $u=v=0$ (4)

5. Numerical treatment

To solve the equations of motion governing inside of system (1), (2) and (3), together with the boundary conditions (4), the method of the ANSYS finite element software was used. The following methodology was used, Moaveni and Saeed [10]:

a) The analysis, in this case FLOTRAN type is selected in the software.

b) The geometry to be analyzed is drawn, and elements in the boundaries are generated. In this case the cavity was dimensioned with a height and width of 0.08 m. The generation of nodes and elements was as follows: in the system without protuberance 3364 nodes with 6498 elements; in the rectangular protuberance 3428 nodes with 6604 elements; in the semi-circular protuberance 3449 nodes with 6664 elements; and in the triangular protuberance 3372 nodes with 6512 elements.

c) The value of the properties of the fluid, in this case water, which remain constant: dynamic viscosity, specific heat and thermal conductivity, volumetric expansion coefficient was fed. The density was handled as variable and pressure was modeled with the algorithm of relaxation (TDMA).

d) Boundary conditions were inserted and the software, was programmed for a maximum of 500 iterations. The foregoing was sufficient to reach the convergence of the problem.

It is worth mentioning that when the values of the thermophysical properties of water and the conditions of the cavities were fed, we obtained numbers of Prandtl of 6.87 and Rayleigh 2.705×10^6.

6. Results and discussion

The experimental and numerical results of the fields of velocity, in m/s, are respectively presented in Figures 4-11. It can be observed that in the vicinity of the walls with high and low temperature fluid moves vertically towards the top and bottom boundary, respectively. In upper and lower adiabatic walls movement of the fluid is to the right and left, respectively. The combination of these flows generates a rotary motion in a clockwise direction. In Figure 12 the numerical and experimental results of vertical velocity, in m/s, for different positions in the x-axis and a location on the shaft "y" of 0.04 m, can be observed. For the non-protuberance system, the component of velocity starts out at zero. Following it can be observed that the flow of the hydrodynamic boundary layer is directed vertically toward the upper border. It can be seen that the magnitude of the vertical velocity increases up to a maximum of 6.60×10^{-4} and 7.07×10^{-4} m/s, respectively, and from there it begins to decrease to a value close to zero. From this point there are small fluctuations of the component of velocity along the central part of the system. Finally, the area of the other hydrodynamic boundary layer where now the flow is directed toward the lower border is observed. In this part the vertical velocity increases up to a maximum negative of -6.83×10^{-4} and -6.99×10^{-4} m/s, respectively, and from there again its magnitude decreases until it reaches zero. The same behavior is observed for the systems with rectangular, semi-circular, and triangular protuberance but with different values, in this case the maximum positive magnitudes are 6.13×10^{-4} and 6.75×10^{-4} m/s, 6.38×10^{-4} and 6.91×10^{-4} m/s, 6.40×10^{-4} and 6.98×10^{-4} m/s, respectively, and the negative are -6.18×10^{-4} and -6.73×10^{-4} m/s, -5.92×10^{-4} and -7.32×10^{-4} m/s, -6.70×10^{-4} and -6.81×10^{-4} m/s, respectively. In Figure 13 the numerical and experimental results of horizontal velocity, in m/s, for different positions in the shaft "y" and for a location on the x-axis of 0.04 m are presented. For the non-protuberance system, the component of velocity starts out at zero. Then, the zone of the hydrodynamic boundary layer is subsequently observed where flow is directed horizontally towards the left border. It can be seen that the magnitude of the horizontal velocity increases up to a maximum negative of -6.40×10^{-4} and -7.95×10^{-4} m/s, respectively, and from there it decreases down to a value close to zero; from that point there are small fluctuations of the component of velocity along the central part of the system. The area of the other hydrodynamic boundary layer is observed; now the flow is directed towards the right border. There, the horizontal velocity increases up to a positive maximum of $4.79x10^{-4}$ and $5.26x10^{-4}$ m/s, respectively, and from there again it decreases down to zero. The same behavior is observed for the systems with rectangular, semi-circular, and triangular protuberance but with different values, in this case the maximum negative magnitudes are -$4.91x10^{-4}$ and -$5.28x10^{-4}$ m/s, -$5.73x10^{-4}$ and -$6.78x10^{-4}$ m/s, -$6.01x10^{-4}$ and -$7.52x10^{-4}$ m/s, respectively, and the positive are $3.91x10^{-4}$ and $4.92x10^{-4}$ m/s, $4.22x10^{-4}$ and $5.01x10^{-4}$ m/s, $4.57x10^{-4}$ and $5.13x10^{-4}$ m/s, respectively.

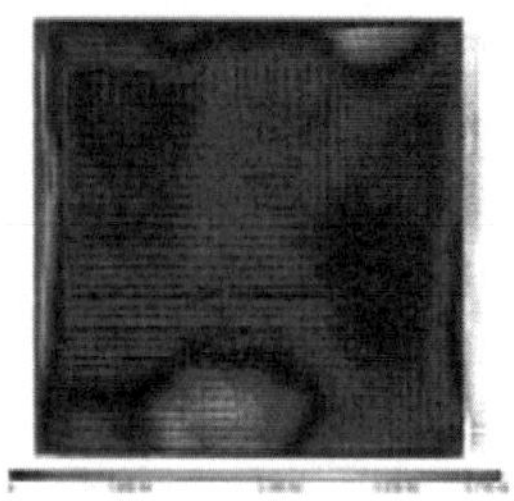

Figure 4. Experimental velocity field, ∇, without protuberance.

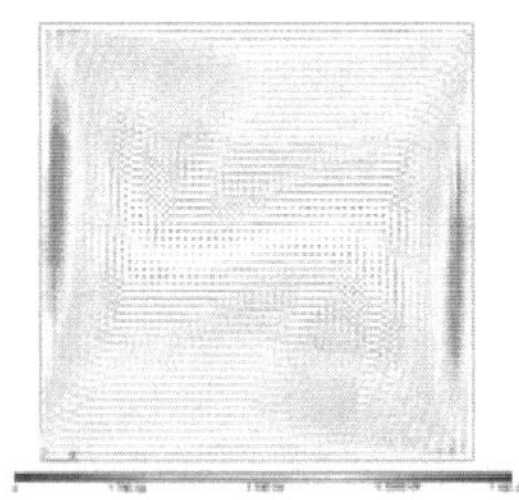

Figure 5. Numerical velocity field, ∇, without protuberance.

Figure 6. Experimental velocity field, ∇, with rectangular protub.

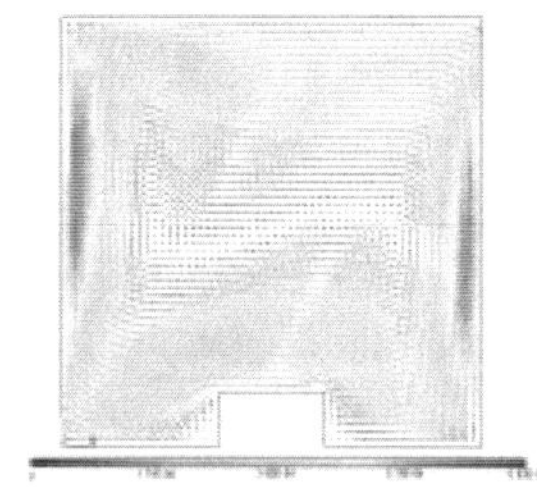

Figure 7. Numerical velocity field, ∇, with rectangular protub.

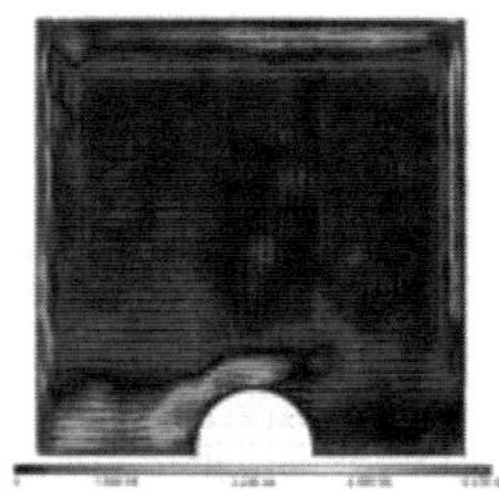

Figure 8. Experimental velocity field, ∇, with semi-circular protub.

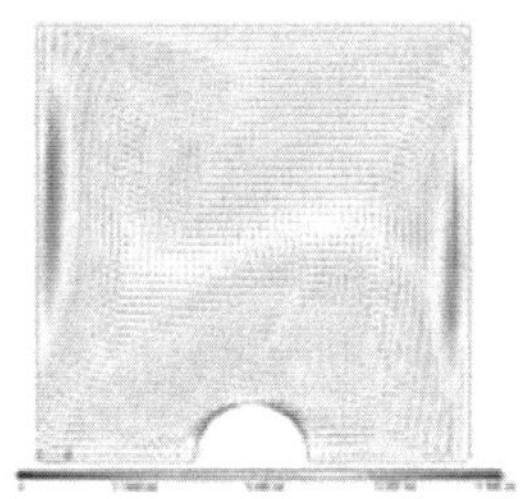

Figure 9. Numerical velocity field, ∇, with semi-circular protub.

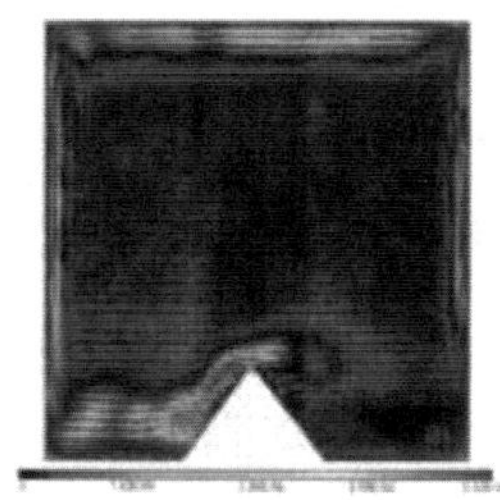

Figure 10. Experimental velocity field, ∇, with triangular protuberance.

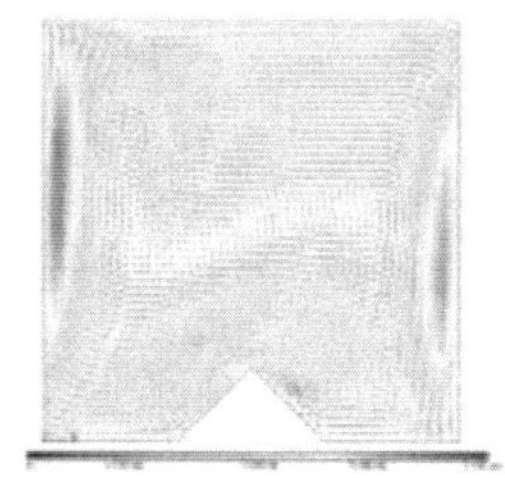

Figure 11. Numerical velocity field, ∇, with triangular protuberance.

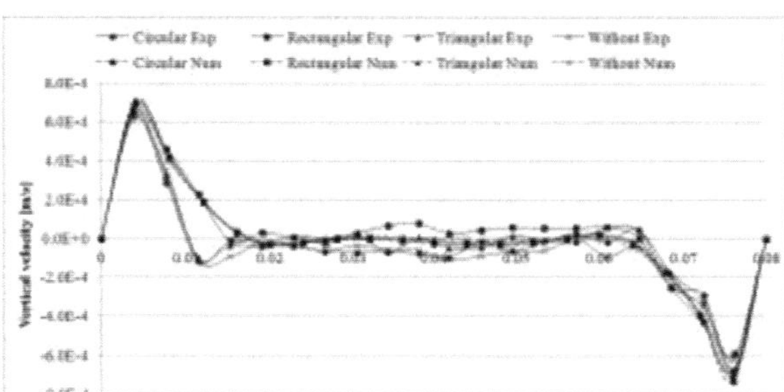

Figure 12. Distribution of horizontal velocity (m/s) for the x=0.04 m position and four protuberances.

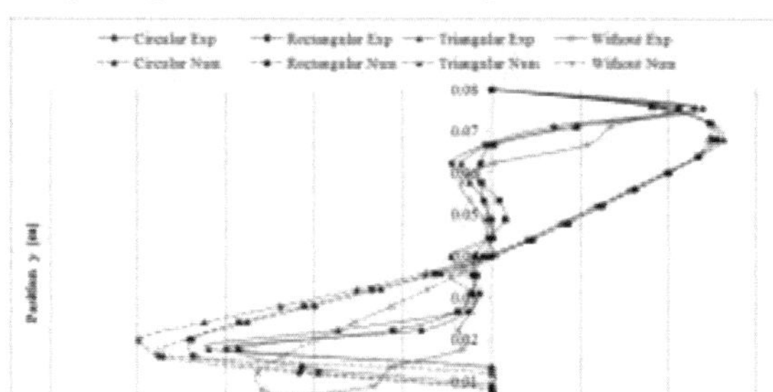

Figure 13 Distribution of vertical velocity (m/s) for the y=0.04 m position and protuberances.

On the other hand, for the calculation of the number of Rayleigh, local Nusselt number and average Nusselt number the equations (7) are used. The average Nusselt numbers are shown in Table 1.

$$Ra = \frac{g\,\beta\,(Tc - Tf)\,H^3\,Pr}{\nu^2} \qquad Nu_{(Y)} = -\left.\frac{\partial\theta}{\partial X}\right|_{X=0} \qquad Nu = \int_0^{\varepsilon} \left(\frac{\partial\theta}{\partial X}\right)_{X=0} dY \tag{7}$$

Table 1. Average Nusselt's numbers

	Without protub.	Triangular protub.	Semi-circular protub.	Rectangular protub.
Experimental	13.21	12.41	12.16	11.82
Numerical	13.90	13.48	12.64	12.23

The experimental and numerical results showed that both the vertical velocity component and horizontal are diminished by placing the protuberances, finding the highest value in the cavity without protuberance and lowest in the rectangular protuberance. The presence of protuberances generated that in thermal convection occurred a decrease in the average number of Nusselt both for numeric and experimental cases.

7. Conclusions

When comparing the results between systems with rectangular, semi-circular and triangular protuberance, respect to that with no protuberance, it was found that: a) the maximum positive value of the vertical velocity was reduced: 7.12, 3.33, and 3.03%, respectively, for the experimental case and 4.52, 2.26, and 1.27 %, respectively, for the numerical case, b) the maximum positive value of the horizontal velocity was reduced: 18.37, 11.89 and 4.59%, respectively, for the experimental case and 6.46, 4.75 and 2.47%, respectively, for the numerical case, c) the average values of Nusselt number decreased: 10.52, 7.95 and 6.06%, respectively, for the experimental case and 12.01, 9.06 and 3.02%, respectively, for the numerical case. The results can be used to design different equipment heat transfer by natural convection, for example, solar collectors, electric transformers, etc.

8. References

[1] Corvaro F, Paroncini M 2009 An experimental study of natural convection in a differentially heated cavity through a 2D-PIV system *International Journal of Heat and Mass Transfer* Vol. 52 pp 355–365

[2] Corvaro F, Paroncini M, Sotte M 2011 Experimental PIV and interferometric analysis of natural convection in a square enclosure with partially active hot and cold walls *International Journal of Thermal Sciences* 50 pp 1629-1638

[3] Nardini G, Paroncini M 2012 Heat transfer experiment on natural convection in a square cavity with discrete sources *Heat Mass Transfer* 48 pp 1855–1865 DOI 10.1007/s00231-012-1026-6 Springer-Verlag

[4] Kaviany M 1984 Effect of a Protuberance on Thermal Convection in a Square Cavity *Journal of Heat Transfer* Vol. 106 pp 830-834

[5] Qi-Hong Deng 2008 Fluid flow and heat transfer characteristics of natural convection in square cavities due to discrete source–sink pairs *International Journal of Heat and Mass Transfer* 51 pp 5949–5957

[6] Wang X and Pepper D W 2009 Numerical simulation for natural convection in vertical channels *International Journal of Heat and Mass Transfer* 52 pp 4095–4102

[7] Lizardi A, Terres H, López R, Hernández A, Morales J R, Lara A., Zavala C N 2012 Análisis experimental del flujo en cavidades con protuberancias en su interior VII Congreso Bolivariano de Ingeniería Mecánica Cusco Perú

[8] Incropera F P, DeWitt D P 1999 *Fundamentos de Transferencia de Calor* México Prentice Hall

[9] Bejan A 1995 *Convection Heat Transfer* New York Wiley Chap. 4

[10] Moaveni, Saeed 2003 *Finite Element Analysis Theory and Application with ANSYS* 2nd Ed Pearson Education USA

Measurement of the transmittance of edible films of aloe vera (barbadensis miller) and cassava starch using optical fibers trifurcated

L Díaz[1], J Reales[2] and C Torres[1]
[1]Grupo de Óptica e Informática (LOI), Universidad Popular del Cesar, Valledupar, Cesar, Colombia.
[2]Grupo BIAT, Universidad Popular del Cesar, Valledupar, Cesar, Colombia.

E-mail: leonardodiaz@unicesar.edu.co

Abstract. In Colombia, especially in the Atlantic Coast it is produced and marketed the costeño cheese, an indigenous product of the gastronomy of this region, but the prolonged exposure of this product to the environment leads to microbial contamination and non-enzymatic rancidity. For this reason the transmittance of an edible coating based in aloe vera gel and cassava starch to preserve costeño cheese was evaluated using trifurcated optical fibers. The results become a tool for the selection of treatments in making edible films and their subsequent use in coatings for various types of food products.

Introduction

The consumer demand for innocuous food with use of materials environmentally friendly packaging have prompted industries and researchers to develop new strategies for processing, handling and packaging. Therefore, the edible films and based proteins, lipids and polysaccharides coatings are being studied and applied to food products as a potential packaging material that can act as a barrier against water vapor, gases, volatile compounds and serve as carriers of antimicrobial substances, antioxidants, flavorings and vitamins, which can increase the quality, safety, functionality and product lifetime [1].

In Colombia there are a variety of dairy products such as cheese costeño, considered a consumer commodity in the Colombian Caribbean region and a great source of protein. It is in fresh the main form of consumption, which is why their shelf life is short because of its high moisture content, so they look for new alternatives for their conservation, without losing quality and provide consumers a safe food. An alternative for preserving of the cheese is the application of an aloe vera gel based edible coating containing antioxidants and antimicrobial properties.

This paper shows transmittance measurements of edible films based gel of Aloe Vera (barbadensis miller) and cassava starch for use in the coating of costeño cheese.

Content from this work may be used under the terms of the Creative Commons Attribution 3.0 licence. Any further distribution of this work must maintain attribution to the author(s) and the title of the work, journal citation and DOI.
Published under licence by IOP Publishing Ltd

Methodology

The measurement of absorption-based fiber-optic sensors is achieved by comparing the light power output from the absorbing sample with the initial light power input into the sample; the stability of light power input to the absorbing sample is critical to the accuracy of the sensor [2]. The gel samples used in the study were extracted from leaves of aloe vera (barbadensis miller) from an intensive cultivation in the Island farm, which is located in the municipality of Valledupar Cesar and stabilized at the Popular University of Cesar in the pilot plant of fruit and vegetables. The optoelectronic mounting used the absorption and reflection properties of the materials under study [3].
When Light passing through a material can be absorbed by interaction with the molecules and atoms of the host material or with impurities and defects. At a given wavelength λ; the intensity of the light passing through the material is attenuated as an exponential function of its path length x .

This is expressed by the Beer–Lambert law:

$$I_\lambda(x)=I_\lambda(0)\exp(-\alpha_\lambda[M]x), \quad (1)$$

Where $I_\lambda(x)$ is the light intensity at a distance x , $I_\lambda(0)$ the incident light intensity at $x=0$, α_λ the absorption coefficient and $[M]$ the concentration of the absorbing material. Absorption-based fiber-optic sensors work on this principle [2].

1.1. *Preparation of stabilized Aloe Vera gel*

To determine the optimal concentration of each of the components included in the edible film, was made a design completely randomized, where he was studied the concentration of aloe vera gel (Aloe barbadensis miller) and cassava starch. The coating composition consists of 1% glycerol as a plasticizer, distilled water and the concentration of aloe vera gel and cassava starch according design of experiments. The design was carried out according to Table 1.

Tabla 1. Design of Experiments

Treatments	**Aloe Vera Gel (%)**	**Cassava Starch (%)**
T_1	5	3
T_2	15	3
T_3	25	3

1.2. *Preparing film*

Cassava starch is mixed with half distilled water with constant stirring in a heating plate and stirring of mark IKA C-MAG HS7 Model: ETS-D5, submitting at heating at a temperature of 70 °C until reach the gelatinization. On the other hand the aloe vera gel is mixed with glycerol and the remainder of distilled water and added slowly until the whole mixture reached a temperature of 75 °C for 3 minutes to avoid loss of the biological properties of the gel. The thickness of the films was determined using an micrometer analogous brand StarrettCo with measures 25 mm and 1", with rotating spindle and precision of ± 0,004 mm or ± 0,00015", with an average of 12 random measurements of the surface of each film sample according to [4]. Before testing the films, a visual inspection was performed to ensure that the films have a continuous and homogeneous surface, ie, had no cracks after the drying process, or submit insoluble particles or open pores, then movies they were stored in desiccators at

25°C and relative humidity of 58% measured with a thermo-hygrometer brand Fisher Scientific (NaBr saturated solution), until use.

1.3. *Measuring the transmittance of the films*

The transmittance of the thin films of about 80 µm thick were determined preliminarily by absorption using trifurcated optical fibers, three lasers with wavelengths 632.8 nm, 532 nm, 405 nm, lower powers at 10 mW and an optical power meter (Newport) model 1928-C. See Figure 1. The procedure was performed under the same lighting conditions, temperature and measuring distances between light source (laser) the samples (films) and detector, taking as variable of interest to measure the intensity variations generated by analyzing the samples after it a light beam impinges in them.

a) b) c)

Figure 1. Measuring the transmittance of the films. a) λ= 632.8 nm b) λ= 532 nm c) λ= 405 nm

Results

The transmittance measurements for each treatment used in the manufacture of edible films are presented in Figures 2, 3 and 4. In Fig. 2 T1 and T3 are those that present higher transmittance compared to T2.

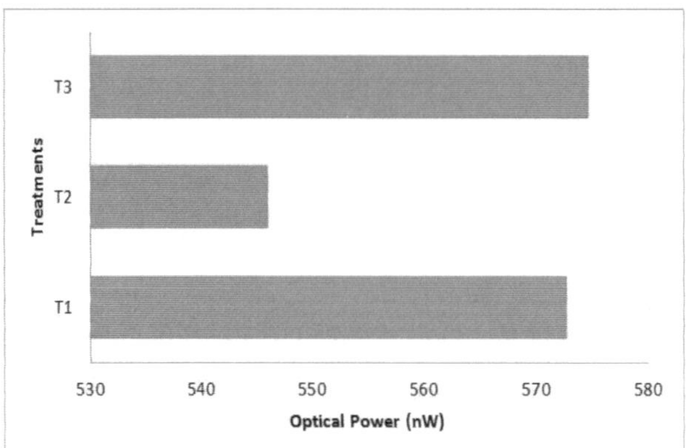

Figure 2. Transmittance at 632.8 nm.

Upon interaction between the beam and the film, part of the incident radiation was absorbed for each wavelength used, so that the emergent beam intensity is different from that of the incident beam. According to the ANOVA performed for test data subjected to the wavelengths used, there are no statistically significant differences. However, when examining the data transmittance at 532 nm was found that there is statistically significant difference between treatments. This is the attributed to the affinity and particular characteristics to this lambda with the compounds present in each treatment, thus presenting an optical absorption by the films. The treatment that presents higher transmittance was the T3.

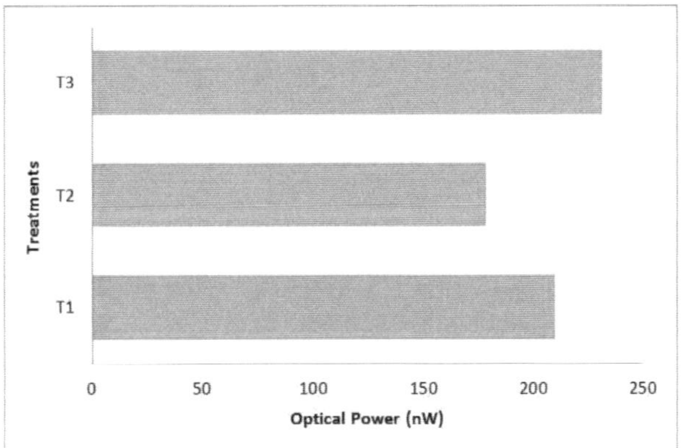

Figure 3. Transmittance at 532 nm.

To determine which treatment is with more transmittance, must take into account that a transparent object is one that transmits light with relatively little absorption [5]. In the Figure 4, it is observed that the treatment T2 has little transmittance in comparison with T3.

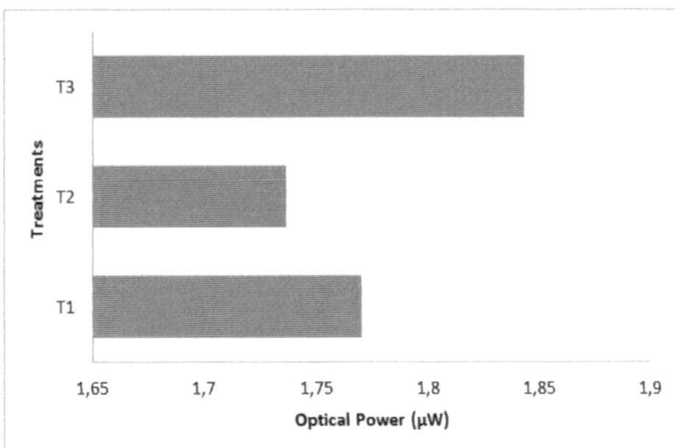

Figure 4. Transmittance at 405 nm.

According to the above, the treatment T3 has a greater transparency when comparing the results obtained for the wavelengths used, is better appreciate the variations of optical power in the graphics, which impacts advantages in terms of marketing in the sense that if this formulation is used to coat the foods like fruits and cheese, the film formed is practically imperceptible and the concentration of this gel allows better use of their properties for food preservation [6].

Conclusions

Were performed of transmittance measurements a several treatments of edible films aloe vera and cassava starch for coating cheese costeño using trifurcated optical fibers. Are notable the variations of optical power obtained according to the analyzed samples. These variations were the decision criterion in determining the best treatment to be used as coating of costeño cheese, in our case, treatment T3 meets the best features. This research becomes a tool for the selection of treatments in making edible films and their subsequent use in coatings for various types of food products.

Acknowledgments
The authors thank at the University Popular of Cesar for their support in this research.

References

[1] Galus S and Kadzińska J 2015. Food applications of emulsion-based edible films and coatings *Trends Food Sci. Technol.* **45**(2), 273-283

[2] XuaY, Barrie-Jonesa N, Fothergilla J C and Hanningb C D 2001 Error analysis of two-wavelength absorption-based fibre-optic sensors *Opt. Lasers Eng.* ***36*** 607–615

[3] Díaz L, Morales Y, Mattos L and Torres C O 2013 Detection of wavelengths in the visible range using fiber optic sensors *Proceedings SPIE.* **8785** 87852L

[4] Khoshgozaran-Abrasa S, Hossein M., Hamidya Z and Bagheripoor-Fallahb N 2012. Physicochemical and color properties of chitosan based-films as a function of Aloe vera gel incorporation *Carbohydr. Polym.* **87**, 2058– 2062

[5] Alvarado-González J S, Chanona-Pérez J J, Welti-Chanes J S, Calderón-Domínguez G, Arzate-Vázquez I, Pacheco-Alcalá S U, Garibay-Febles V and Gutiérrez-López G F 2012 Optical, microstructural, functional and nanomechanical properties of Aloe vera gel/gellan gum edible films. *Rev. Mex. Ing. Quím.* **11**(2), 193-210

[6] Cerqueira M, Sousa M, Macedo I, Rodriguez R, Souza B, Teixeira J and Vicente A 2010 Use of galactomannan edible coating application and storage temperature for prolonging shelf-life of "Regional" cheese. *J. Food Eng.* **97** 87-94

IOP Conf. Series: Journal of Physics: Conf. Series **792** (2017) 012024 doi:10.1088/1742-6596/792/1/012024

Simulation of the flow field of water in an Olympic swimming pool.

R López, M Vaca, H Terres, A Lizardi, S Chávez and E Meza.

Universidad Autónoma Metropolitana -Azcapotzalco
Av. San Pablo No. 180, Col. Reynosa-Tamaulipas Del. Azcapotzalco, México, 02200, D.F. México,

E-mail: rlc@correo.azc.uam.mx

Abstract. This paper presents the modelling of hot-water flow at the inlet of a swimming pool in the absence of swimmers, in order to find the option that allowed the best distribution of the fluid with less temperature loss. Two different input velocities, 3 and 4 m/s and two injection-nozzles inclinations, 60° and 30°, were used in the simulation. COMSOL Multiphysics software, with appropriate boundary conditions that were obtained by direct measurements in the pool, was used. The best condition was observed for an input velocity of 4 m/s and an injection-nozzle inclination of 30°, since it presented the best distribution of the fluid at the temperature of 29.86 °C in most of the considered site.

1 Introduction

The pattern of the flow in swimming pools depends on the number and the position of the inputs and outputs of water, the geometry of the pool and feed flow rate. Currently, there are no universal regulations that govern the design of a pool, the International Swimming Federation (FINA, Fédération Internationale de Natation), has only imposed geometries of 25 m x 15 m x 1.6 m for regional competitions and 50 m x 21 m x 1.8 m in national and international competitions [1], without specifying any function related to the number of inputs and outputs of water. When a competition is going to take place organizers demand that water does not have any movement and that temperature is uniform across the pool. Pools are used for different purposes: learning and recreational swimming practice, training and competition, or for therapeutic purposes, all of which have different needs regarding water temperature. According to the American Society of Heating, Refrigeration and Air conditioning Engineers (ASHRAE), the average temperature for comfortable swimming is 27 °C, although it may vary by 5 °C [2].

Some studies have been conducted to obtain a reliable methodology for the prediction of energy losses from the pool and the property where it is located. Li et al., [3] investigated the relationship existing between the evaporation of water and the movement of air, thus determining the evaporation mass flow rate to achieve the approppriate sizing of ventilation systems for enclosures intended for swimming. Models of evaporation of water on the surface of the pool have been developed through computational fluid dynamics in two and three dimensions.

Mančić et al. [4] developed a mathematical model of a swimming pool, created with the TRNSYS software to determine the demand for energy within the enclosure of the pool, including losses by evaporation of the fluid, which were analyzed for different temperatures of the water in the pool and the surrounding air. The simulation showed that the pool water heating accounted for about 22% of the heat demand, while heating and ventilation consumed around 60% of energy within the enclosure, evaporation loss was in the range of 46% to 54% of the total losses from the pool.

Also, Cloteaux et al. [5] using experimental and computational fluid dynamics found the effect of the design of the pool on the hydraulic behavior, considering the pool as a chemical reactor with its

Content from this work may be used under the terms of the Creative Commons Attribution 3.0 licence. Any further distribution of this work must maintain attribution to the author(s) and the title of the work, journal citation and DOI.
Published under licence by IOP Publishing Ltd

hydraulic characteristics. They were able to develop a mathematical model based on the principle of an agitated reactor, which could be used as a first approximation to describe the hydraulic behavior of regular swimming pools, being a suitable example for the study of physical and chemical phenomena with long characteristic times, including the prediction of the concentrations of chemical species, as well as the optimization of the chlorination and the daily renewal of water. The present paper examines the way in which the flow field is affected when varying the angle of the injection of fluid into 30° and 60° (120° and 150° from horizontal plane) in the supply of hot water, in order to find the best temperature distribution.

2 Theoretical analyses

The hypotheses about the behavior of density with changes of pressure and/or temperature are relevant in the study of flows. Multiphysics interface of *non-isothermal flow* [6] is used to simulate the coupling between modules of *heat transfer in fluids* [7] and *k-ε turbulent flow* [6]. *Non-isothermal flow multi-physics* interface contains the formulation of the equations of continuity and momentum,

$$\frac{\partial \mathbf{u}}{\partial t} + \nabla \cdot (\rho \mathbf{u}) = 0$$

$$\rho \frac{\partial \mathbf{u}}{\partial t} + \rho \mathbf{u} \cdot \nabla \mathbf{u} = -\nabla p + \nabla \cdot \left(\mu(\nabla \mathbf{u} + (\nabla \mathbf{u})^T) - \frac{2}{3}\mu(\nabla \cdot \mathbf{u}) \right) + \mathbf{F} \quad (1)$$

where ρ is the density of the working fluid, u is the velocity vector, p is the pressure, μ is the dynamic viscosity of the fluid, and F is the vector of body strength. The basic law that governs all heat transfer is the first law of thermodynamics, commonly known as the conservation of energy principle. For a fluid, the resulting equation is:

$$\rho C_p \left(\frac{\partial T}{\partial t} + (u \cdot \nabla)T \right) = -(\nabla \cdot \mathrm{q}) + \tau \colon \mathrm{S} - \frac{T}{\rho} \frac{\partial \rho}{\partial T}\bigg|_p \left(\frac{\partial p}{\partial t} + (u \cdot \nabla)p \right) + Q \quad (2)$$

where, C_p describes the amount of heat energy required to produce a change in temperature per mass unit, T is the absolute temperature, q is the flow of heat by conduction, τ is the tensor of viscous efforts and S is the tensor of deformation velocity. *Heat transfer in fluids* interface uses the following version of the heat equation to model heat transfer in fluids, either by conduction, convection or radiation:

$$\rho C_p \frac{\partial T}{\partial t} + \rho C_p \mathbf{u} \cdot \nabla T = -(k \nabla T) + Q \quad (3)$$

The thermal conductivity of the fluid k is a scalar or tensor and the heat source (or sink) is represented by Q. In steady state, where temperature does not change with time and velocity is set to zero, the equation that governs the conductive heat transfer is:

$$\rho C_p \frac{\partial T}{\partial t} + \nabla \cdot (-k \nabla T) = Q \quad (4)$$

This expression describes the local evolution of temperature with respect to time $(\partial T / \partial t)$ motivated by a heat conduction due to a difference in temperature between the point of study and its immediate surroundings $\nabla(-k\nabla T)$ and local heat Q generation. Turbulence is a property of the flow field and is characterized by a wide range of scales of flow, with a non-stationary and three-dimensional nature. The Navier-Stokes equations completely describe fluid flow. $k - \varepsilon$ *turbulent flow* interface, is one of turbulence models most commonly used for industrial applications. This module introduces two equations of transport and two dependent variables: the turbulent kinetic energy, k and turbulence dissipation rate, ε. The turbulent viscosity is modeled as

$$\mu_T = \rho C_\mu \frac{k^2}{\varepsilon} \tag{5}$$

where μ_T is the Eddy viscosity, also known as viscosity of turbulence and C_μ is a constant of the model.

3 Geometrical descriptions

The geometry of the public pool at the Aquatic Center of Azcapotzalco is Olympic-type, and its actual dimensions are: 50 m-length, 25 m-width and 1.3 m-depth. This pool operates in a closed circuit, i.e., the water that comes out of it is fed again, after adding the amount of lost energy. The water is collected 10 cm below the free surface through two overflow channels (spillways), located on the North and South outer walls. Fourteen circular- section entries are aimed to supply hot fluid to pool, seven of which are housed in each of the longer sides. An experimental period for the pool water recirculation was not considered, since a steady-state study was interpreted.

The numerical simulation to the pool was done cautiously modifying the model, this allowed us to work with only a portion of it, gradually assuming that in some of the adjacent entities, the geometry is symmetrical.

In this model, the geometry was developed from a work cell, representing approximately half of the longitudinal section of the complete pool. Therefore, the domain was established by a main rectangle and two secondary rectangles, one for each water outlet (weir), as well as four tiny cylinders, representing the hot water supply line. Such a procedure has made it possible to reduce the modeling space and therefore, achieve a refinement in the discretization mesh with the minimum amount of operable items. The model dimensions are: 12.5 m- length, 6.25 m-width, and 1.3 m-depth. Weirs are 5 cm height by 5.25 m long, while the nominal diameter of the pipe is only 0.05 m. The three-dimensional model is shown in Figure 1.

4 Boundary conditions

Boundary conditions used in the simulation were obtained from measurements made in the Azcapotzalco Aquatic Center pool, they are: temperatures of water in different parts of the swimming pool and the velocity of flow in the supply line, just at the entrance of the swimming pool, the values are 3 and 4 m/s. The simulation was done with COMSOL Multiphysics software using k-ε turbulence model. The objective was to obtain the temperature distribution in the bulk of the fluid considering the velocity of injection to the pool. Temperatures are: 30 °C at the entrance, the operation one of 28 °C and the third condition is on the free surface of the fluid in the pool, 29 °C. The obtained mesh had 45,662 elements, Fig. 2.

IOP Conf. Series: Journal of Physics: Conf. Series **792** (2017) 012024 doi:10.1088/1742-6596/792/1/012024

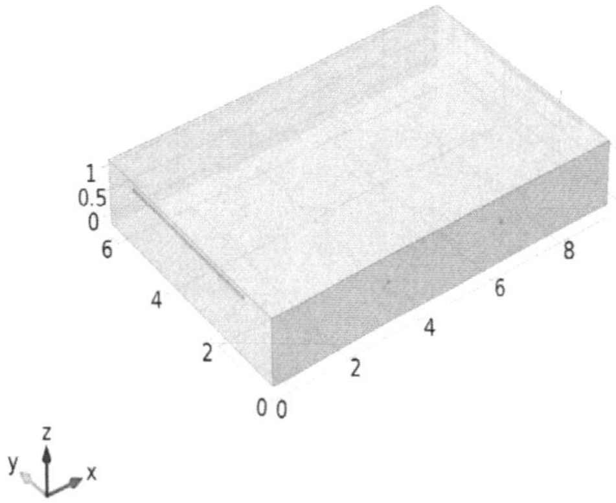

Figure 1. Selected model-cell.

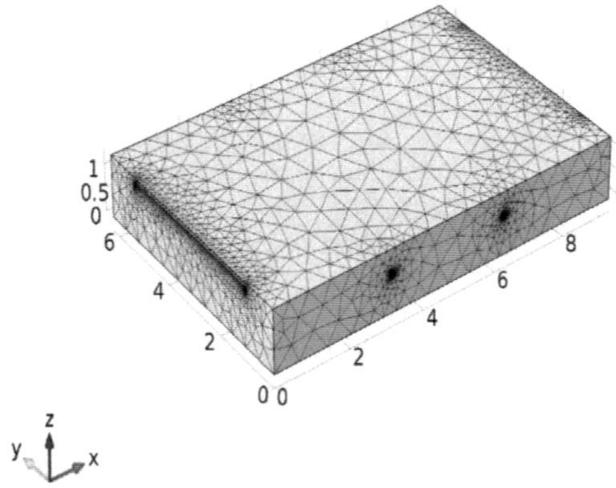

Figure 2. Hexametrical meshing of the geometrical model.

5 Results analysis and discussion

The results obtained when modelling the fluid entry with four injection nozzles at 60° are presented in Figure 3. The top-view distribution of the flow at a velocity of injection of 3 m/s in the x-y plane is observed as a partially radial behavior, caused by the tilt angle of the nozzles. Since most of the flow is directed towards the surroundings, regions of low velocity originate in the middle part of the geometry, with a section plane right to the center of the injection nozzle; this is 0.5 m below the free surface of the fluid. Central vortex is located at low velocities, with values between 0.02 m/s and 0.1 m/s; this could almost be considered as a region of stagnation, since the fluid bumped into the middle part of the walls (the vectors are preserved with a rectilinear motion), it did not adhere completely on the edges of the pool. With respect to the distribution of the temperature in the same x-y plane, isotherm curves for the same velocity of 3 m/s are presented in Fig. 4. The temperature in the central region has a vast field in which the value remains close to 29.70 °C and in the periphery it is 29.73 °C; only in the inlet region the observed value is 29.80 °C but it is a very small region.

The results for an injection velocity of 4 m/s are presented in Fig. 5. They show a parabolic behavior on the periphery of the cell, where the greatest amount of flow is grouped. The main vortex is observed in the middle part of the model, as well as small regions with geometric similarity at the corners of the swimming pool, which are considered areas of stagnation. Lowest veloçities range between 0.03 m/s and 0.1 m/s, and in the input section it reaches 0.7 m/s. A more uniform distribution of temperature is observed throughout the region, with 29.80 °C in the center and 29.82 °C towards the edges of the pool; in very few places it reaches 29.83 °C. In the inlet section the fluid presents a 29.85 °C maximum temperature, the geometry of the corresponding isotherms is elliptical, in the central region, however they are slightly distorted towards the corners, Fig. 6.

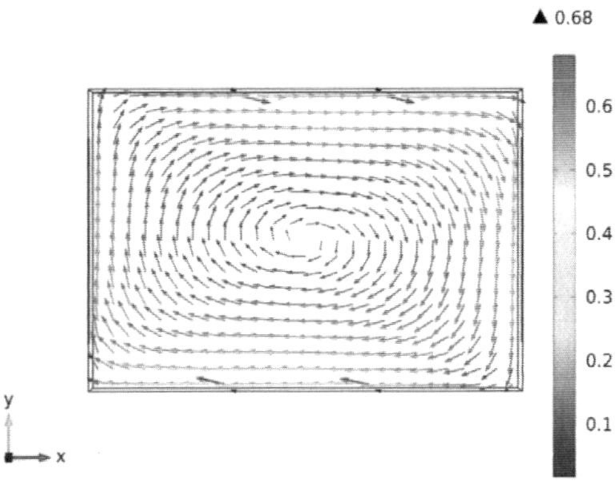

Figure 3. Flow model at an entrance velocity of 3 m/s, 60° nozzle inclination.

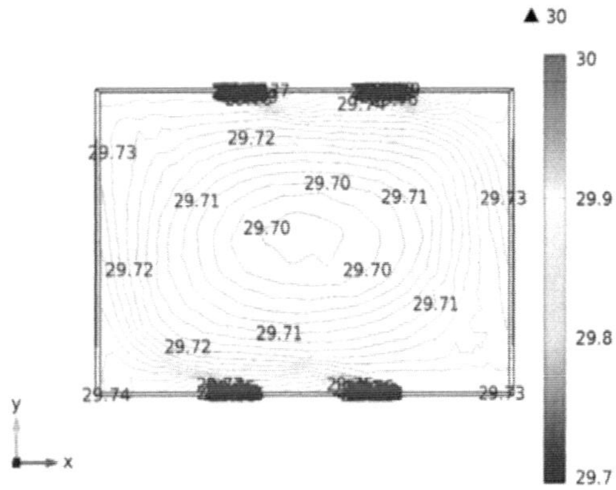

Figure 4. Temperature distribution for a velocity of 3 m/s, 60° nozzle inclination.

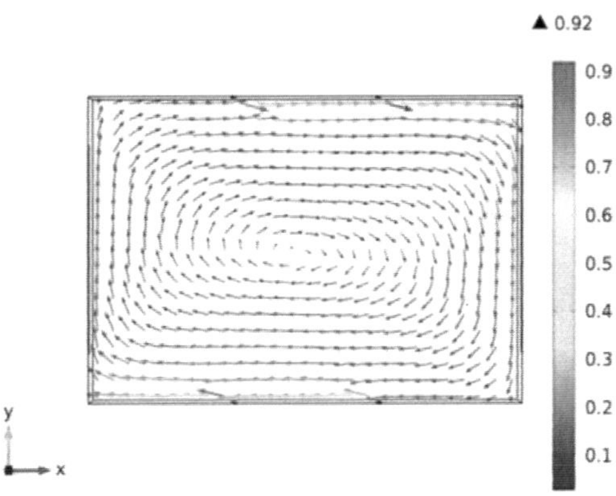

Figure 5. Flow model at an entrance velocity of 4 m/s, 60° nozzle inclination.

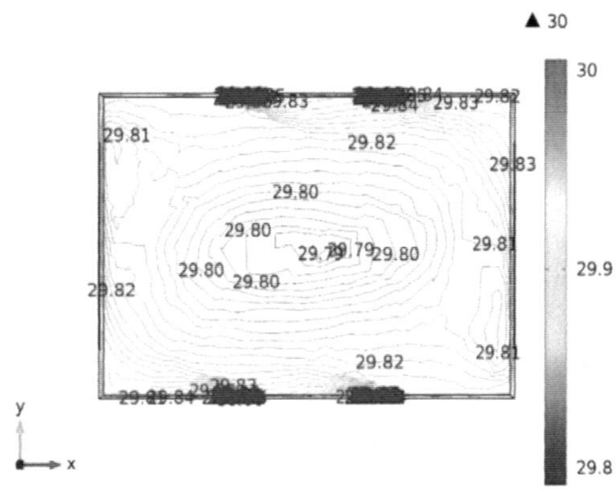

Figure 6. Temperature distribution for a velocity of 4 m/s, 60° nozzle inclination.

When the angle of the nozzle at the inlet was modified to 30°, the fluid tended to move very close to the walls of the pool and towards the center, its value decreased considerably. For a velocity of 3 m/s the distribution of velocities had a significant change with respect to the values obtained previously, then the flow presented a lower velocity, reaching values close to 0.1 m/s in a wide section of the enclosure and towards the center this value was close to 0.01 m/s, near the corners the fluid it remained almost static with a value of 0.001 m/s, Fig. 7. With regard to the temperature distribution a slight increase with respect to the original status when the fluid entered at 60°, was observed, the value in a broad cross section of the enclosure was 29.74 °C and towards the corners, 29.77 °C, Fig. 8.0.

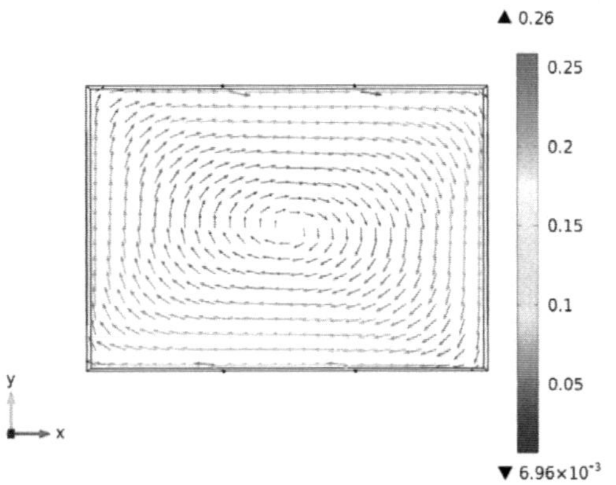

Figure 7. Flow model at an entrance velocity of 3 m/s, 30° nozzle inclination.

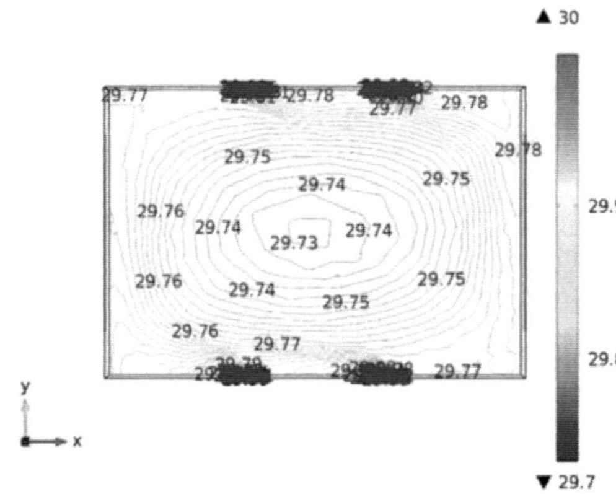

Figure 8. Temperature distribution for a velocity of 3 m/s, 30° nozzle inclination.

For an entry velocity of 4 m/s, the flow pattern remained elliptical and presented a slight modification increasing its value on the walls of the pool to 0.12 m/s; but towards the center it decreased again, down to 0.08 m/s in a considerable section of the enclosure, Fig. 9. The distribution of temperature reached the highest values between 29.86 and 29.88 °C for the quasi-totality of the fluid. From the results obtained for the four conditions imposed, two of velocity variation and two angles of entry of the fluid, it can be concluded that the best option for the injection of the hot water to the pool would be a velocity of 4 m/s and nozzles inclination of 30°, since this produces a distribution of the fluid with less temperature loss.

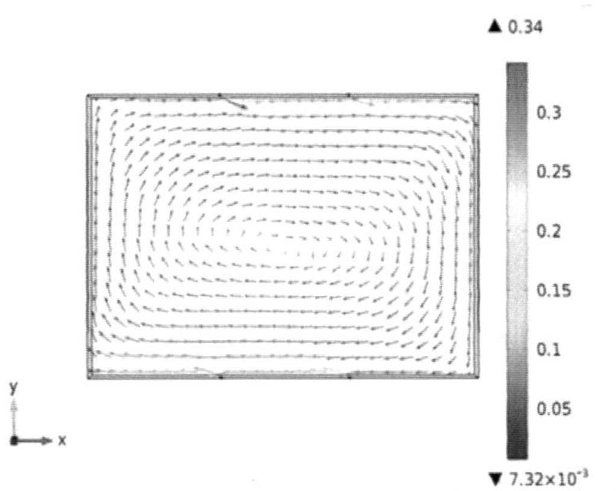

Figure 9. Direction vectors, with an initial velocity of 4 m/s.

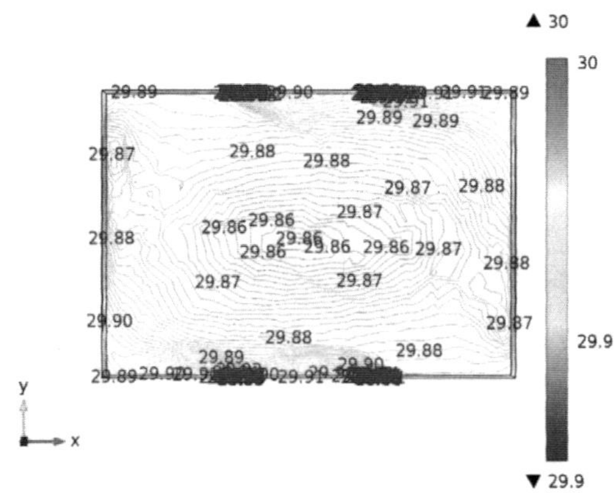

Figure 10. Temperatures distribution for an initial velocity of 4 m/s.

6 Conclusions

This paper presents the modelling of the flow of hot water at the entry of a pool in the absence of swimmers, seeking the option that allows the best distribution of the fluid with less temperature loss. Two different input velocities were simulated: 3 and 4 m/s and two nozzle- inclinations: 60° and 30°. COMSOL Multiphysics software, with appropriate boundary conditions that were obtained by direct measurements in the Aquatic Center of Azcapotzalco, was used. The best condition was an entry velocity of 4 m/s and an injection nozzle inclination of 30 °, since the best distribution of the fluid at 29.86 ° C in most of the considered site was obtained.

7 References

[1] FINA. (2009). http://www.fina.org/
[2] ANSI, "ANSI/NSPI-1 (2003) Standards for Public Swimming Pools," 2003. www.webtore.ansi.org
[3] Li Z and Heiselberg K 2005 CFD Simulations for Water Evaporation and Airflow Movement in Swimming Baths. Instituttet for Bygningsteknik: Aalborg Universitet.
[4] Mančić V, Živković S, Milosavljević M and Todorović N 2014 Mathematical Modelling and Simulation of th Thermal Performance of a Solar Heated Indoor Swimming Pool. *Thermal Science*, **18(3)**, 999-1010.
[5] Cloteaux A, Gérardin F and Midoux N (2013). Influence of Swimming Pool Design on Hydraulic Behavior: A Numerical and Experimental Study *Eng.* **5**(5), 511-524.
[6] COMSOL Multiphysics. (2014). CFD Module User's Guide. Versión 5.0. COMSOL AB.
[7] COMSOL Multiphysics. (2014). Heat Transfer Module User's Guide. Versión 5.0. COMSOL AB.
[8] Wilcox D C (1998). Turbulence Modeling for CFD. DCW Industries.

Design, construction, characterization, and use of a detector to measure time of flight of cosmic rays

Araujo A C[1,2], **Felix J** [1,3]

[1] Laboratorio de partículas elementales, Departamento de Física, División de Ciencias e Ingenierías, Campus León, Universidad de Guanajuato, Loma del Bosque 103, Lomas del Campestre, 37150, León, Guanajuato, México

[2] araujoma2012@licifug.ugto.mx

[3] felix@fisica.ugto.mx

Abstract. In the study of cosmic rays, measurements of time of flight and momentum have been used to identify incident particles from its physical properties, like mass. In this document we present the design, construction, characterization, and operation of a detector to measure time of flight of cosmic rays. The device is comprised of three small plates of plastic scintillator arranged in vertical straight line, coupled to one photomultiplier tube. The analogical output has been connected to a data acquisition system to obtain the number of digital pulses per millisecond. We present details of design, construction, operation, and preliminary results.

1. Introduction

Cosmic rays were discovered in 1912 by Victor Hess [1], at the highest level of the Earth's atmosphere. They consist of stable charged particles and nuclei with lifetimes of order 10^6 years or longer [2]. When cosmic rays collide with atmospheric atoms they produce secondary particles, mainly pions. Muons are emitted in pion decay, they penetrate below ground, and interact weakly with matter [1].

There are several techniques to detect cosmic ray radiation:

2. Scintillation

This method is used to detect radiation in general. It applies photomultipliers coupled to scintillators, which produce a light flash when interacting with radiation. The flashes are detected by the photomultiplier tubes, so the output pulses are electrical pulses proportional to the amount of energy deposited by the incident particles [3].

Content from this work may be used under the terms of the Creative Commons Attribution 3.0 licence. Any further distribution of this work must maintain attribution to the author(s) and the title of the work, journal citation and DOI.
Published under licence by IOP Publishing Ltd

Plastic scintillators belong to the organic group of scintillators, with maximum scintillation emission around 425 nm, they are sensitive to X-rays, gamma rays, and charged particles [4].

3. Time of flight detectors

Scintillation technique is applied in time of flight detectors to measure the velocity of particles from the difference of time obtained with plastic scintillators counters [3]. Velocity can be combined with measures of momentum to determine the mass of particles, and identify them [5].

4. Motivation

The aim of designing, constructing, using, and characterizing a detector to measure the time of flight, is to identify the type of particles that interact with it. Furthermore, a better understanding of the physical functionality of plastic scintillators and photomultiplier tubes can be acquired developing this device.

5. Design

The main components of the detector are three plates of plastic scintillators produced by Sumitomo Chemical, arranged in a vertical straight line, separated by one meter using two extensor aluminum plates as is shown in figure 1.

The plastic scintillators are connected to the photomultiplier tube Hamamatsu (H10493-012) showed in figure 2, which has a photosensitive area of 25 mm diameter [6], optical fibers of 3.15 m long were used for the two scintillators at the ends, and a guide of light for the one in the middle part.

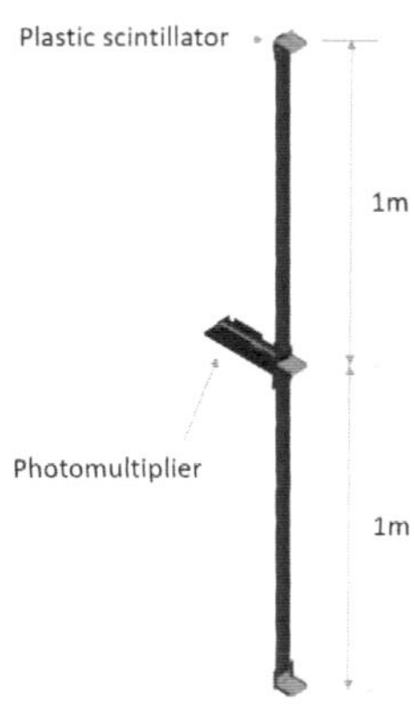

Figure 1. Design of the detector.

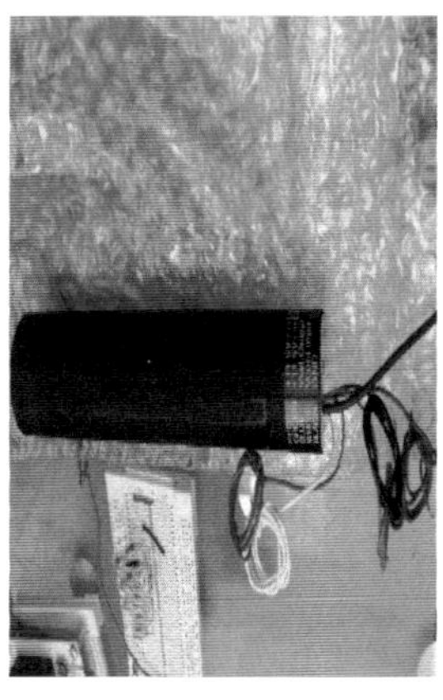

Figure 2. Photomultiplier tube Hamamatsu H10493-012.

6. Plastic scintillators

The dimensions of the plastic scintillators are 5 cm length, 0.6 cm width, and 5 cm height. They are covered with a box made of aluminum plates to block out external light, and protect them from damage, this is showed in figure 3.

The inside walls of aluminum boxes are polished to mirror level to reflect photons inside, and ensure that most of them are guided toward the photomultiplier tube. For each box, one rectangle face has a perforation that is used to connect the optical fibers and the guide of light.

6.1.1. Supports. Three pieces with a set square form were designed as a support to place the plastic scintillators over them, and fixed them to the extensor aluminum plates. Figure 4 shows the supports for the upper and lower plastic scintillators, the central one has the same shape as the upper.

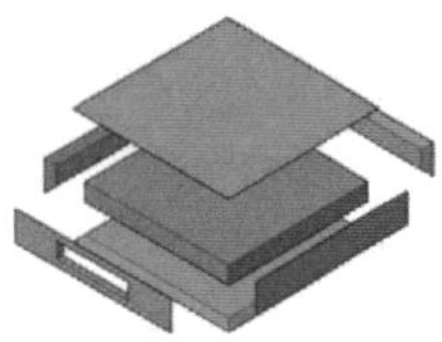

Figure 3. Design of the boxes for the plastic scintillators.

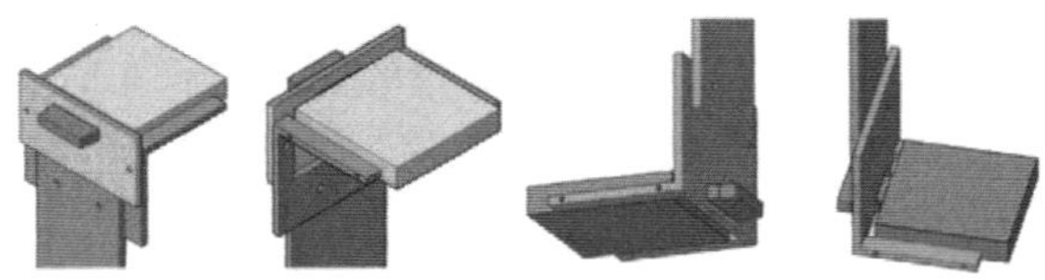

Figure 4. Support for plastic scintillators.

7. Middle part of the detector

The base for the photomultiplier tube is in the center of the detector. As it can be seen in figure 5, one aluminum tube is used to put inside the photomultiplier tube, it is placed over three aluminum base plates, and two setsquares are used to hold it. There are another two setsquare pieces used that keep fixed the larger plates to the middle part. Two aluminum squares have the function of covering both sides of the tube, but one of them has three perforations to place the two optical fibers and the guide of light.

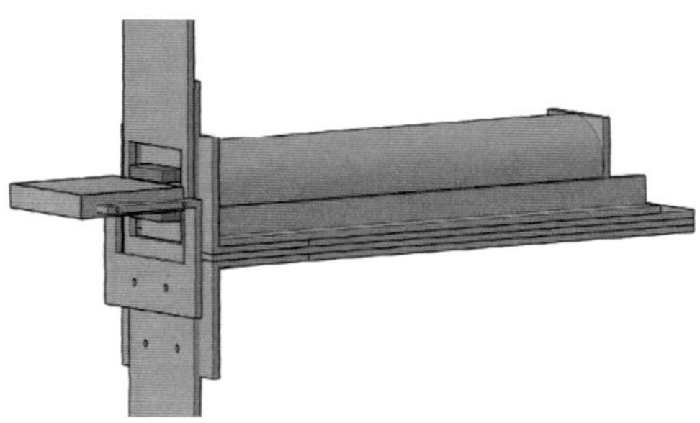

Figure 5. Middle part of the detector.

8. Construction

The three plates of plastic scintillators were cut using a laser cutter machine. The guide of light was cut using a hacksaw and the edges were polished with sandpaper. All the aluminum pieces were cut using a bandsaw, a plate 1.6 mm thick was used to draw the flat parts of the design, an angle for the supports, and a tube.

The aluminum plates of the boxes for the plastic scintillators were polished until dimensions were appropriate, this is shown in figure 6. The next step was making 45° cuts on the edges of each plate using a Belt Disc Sander. Three rectangular plates were drilled, two with a perforation of 16.00 mm diameter, 3.20 mm height, and the last one with a perforation of 25.00 mm of diameter, 3.00 mm height.

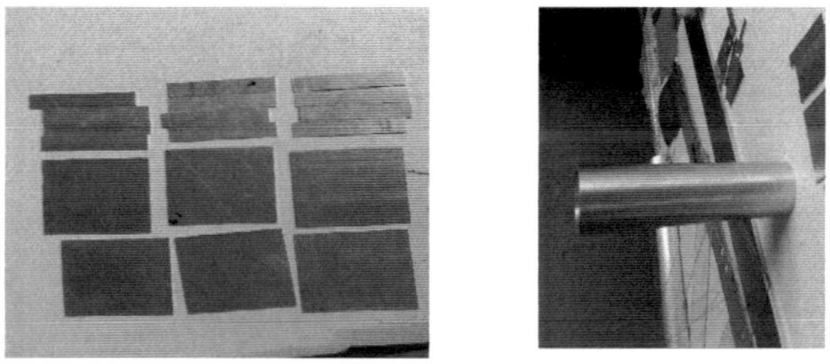

a) **b)**

Figure 6. Aluminum pieces. a) Plates made to cover the plastic scintillators. b) Tube for the photomultiplier.

In order to polish one face of each plate to mirror finish, sandpapers of different grit sizes were used in the following order: 320, 600, 1000, and 2000. Finally, the plates were polished with a microfiber towel and metal polish cream. The final result is shown in figure 7.

The boxes for the aluminum plates were assembled as it follows: the rectangle plates were stuck together in pairs making a right angle using epoxy adhesive, and fixed with an angle press tool. After one hour, the two angles were glued into a frame.

Then, a square plate was stuck below the frame and the plastic scintillator was put inside. This process is shown in figure 8. Finally, another square plate was glued on the top completing the box. The epoxy adhesive on the surface of the aluminum plates was polished with sandpaper, the finished boxes are shown in figure 9.

Figure 7. Aluminum plates with 45° cuts on edges and mirror finishing.

IOP Conf. Series: Journal of Physics: Conf. Series **792** (2017) 012025 doi:10.1088/1742-6596/792/1/012025

Figure 8. Assembling of the boxes for the plastic scintillators. a) Angle press tool used to stick two rectangle plates. b) Frame of the box. c) Box with plastic scintillator inside.

Figure 9. Finished boxes with the plastic scintillators inside. From left to right, the first box is for the middle and the latest two are for the up and down part of the detector.

The edges of the boxes were covered with aluminum tape, the final dimensions were 54.0 cm length, 1.0 cm width, and 54.0 cm height. The same process for the boxes of plastic scintillator was applied to the aluminum plates of the guide of light.

Screws of 1/4 in shank hole size were used to set the extensor aluminum plates to the middle part of the detector. For the aluminum base plates and supports of the plastic scintillators the shank size of the screws was 1/8 in.

Square perforations 4.2 cm length and 4.2 cm height were made in the extensor aluminum plates; they were needed for the input window of the photomultiplier tube.

The photomultiplier tube was put inside the aluminum tube and covered with the aluminum plate, in the output cable was added a SMA connector. Finally, the optical fibers were connected to the plastic scintillators and the photomultiplier tube, so it was the guide of light. All the pieces were assembled. The middle part of the detector is shown in figure 10.

Figure 10. Middle part of the detection system.

The detector was fixed to a cabinet using aluminum tape, which was also used to cover the parts of the detector to block out external light. The complete detector is presented in figure 11.

Figure 11. Complete system of detection.

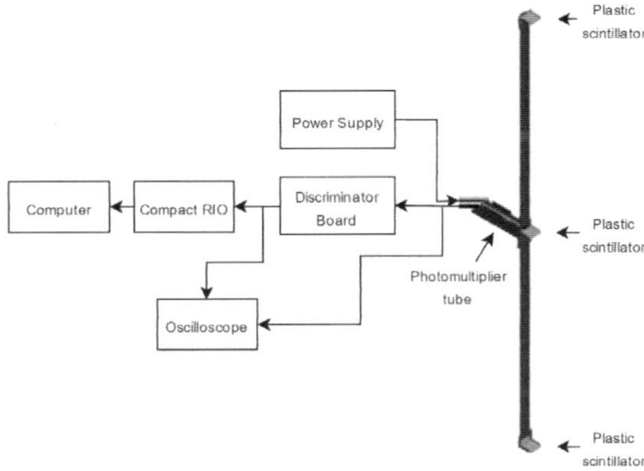

Figure 12. Block diagram for the connection of the detection system.

9. Tests

The block diagram for the connection of the detector is shown in figure 12. The materials used for tests were the following: three power suppliers TENMA 72-8335A, an oscilloscope Tektronix TDS 1001C-EDU, a discriminator board V3 of four channels, BNC coaxial cables, CompactRIO, a computer, and a multimeter.

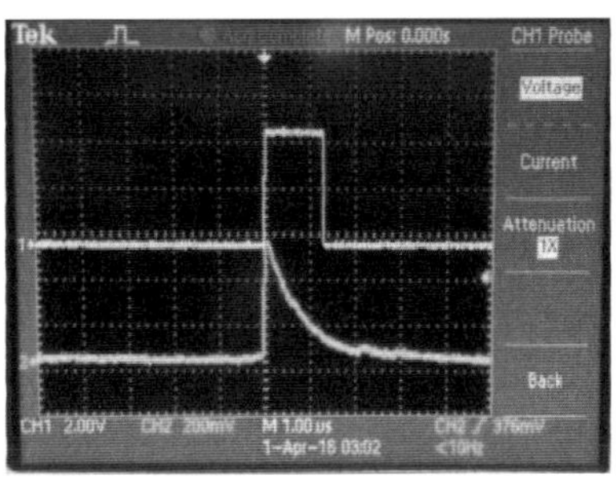

Figure 13. Oscilloscope display for the detector signals. Digital output is channel 1 at the top. Analogical output is channel 2 at the bottom.

The sensitivity adjustment method used for the photomultiplier was the voltage programming suggested in the data sheet. The analogical signal of the detector was connected to the discriminator board, whose voltage supply was -5 V, and + 5 V. The digital signal obtained was connected to the data acquisition system, it means the CompactRIO of National Instruments that was programmed to obtain the number of counts per millisecond.

The operation voltages applied were: Low Voltage Input (Red) +14.01 V, Low Voltage Input (Green) -14.04 V, Vcont Input (White) +1.135 V, and Trigger +266.9 mV.

10. Results

Data recordings were done using the CompactRIO. Analog and digital outputs are shown in figure 13, where the oscilloscope configuration parameters were: channel 1 (digital) 2.00 V, channel 2 (analogic) 200 mV, and time 1.00 μs. Data of a 30 minutes recording is shown in figures 14, and 15.

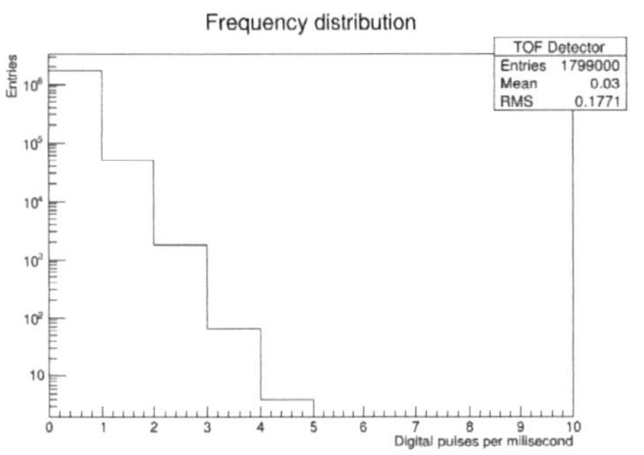

Figure 14. Frequency distribution of digital pulses per millisecond in a recording of 30 minutes.

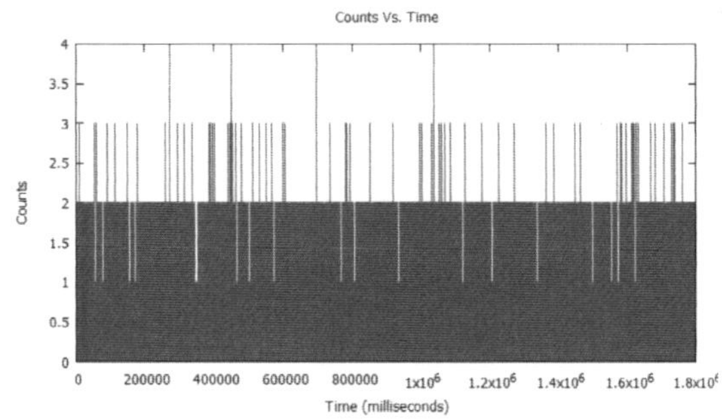

Figure 15. Counts per millisecond vs time in a recording of 30 minutes.

11. Conclusion

In the first tests, analogic signals with amplitudes close to 400 mV in a time scale of 1 μs were observed. Modifications in the design are needed to reduce the noise caused by the leak of electromagnetic radiation to the plastic scintillators and the photomultiplier tube. New tests should be done to probe if there were improvements in the operation of the detector.

Furthermore, a faster acquisition system is required to measure the difference in time of the three digital pulses produced by plastic scintillators, because this difference is in the nanosecond range.

Acknowledgements

To the project 223179, CB-2013-01, Fondo SEP CONACYT. And, to the working group of the Laboratorio de Partículas Elementales of the Universidad de Guanajuato.

References

[1] Cosmic rays: particles from outer space | CERN [Internet]. 2016 [cited 28 June 2016]. Available from: http://home.cern/about/physics/cosmic-rays-particles-outer-space.
[2] Cosmic rays | Particle Data Group [Internet]. 2016 [cited 28 June 2016]. Available from: http://pdg.lbl.gov/2014/reviews/rpp2014-rev-cosmic-rays.pdf
[3] Photomultiplier Tubes. Basics and Applications | Hamamatsu Photonics [Internet]. 2016 [cited 28 June 2016]. Available from: https://www.hamamatsu.com/resources/pdf/etd/PMT_handbook_v3aE.pdf
[4] Organic Scintillation materials | Saint-Gobain Crystals [Internet]. 2016 [cited 28 June 2016]. Available from: http://www.crystals.saint-gobain.com/uploadedFiles/SG-Crystals/Documents/SGC%20Organics%20Brochure.pdf
[5] How a detector works | CERN [Internet]. Home.cern. 2016 [cited 28 June 2016]. Available from: http://home.cern/about/how-detector-works
[6] Photosensor module H10493-012 | Hamamatsu Photonics [Internet]. 2016 [cited 28 June 2016]. Available from: http://www.hamamatsu.com/us/en/product/category/3100/3003/3045/H10493-012

Mechanical and tribological behavior of red clay ceramic tiles coated with fly ash powders by thermal spraying technique.

G Peña-Rodríguez[1,*], H Dulce-Moreno[1], J Daza-Ramírez[2], S. Orozco-Hernández[2] and F Vargas-Galvis[3]

[1] Faculty of Basic Sciences. Department of Physics. GITEC group. Francisco de Paula Santander University. Cúcuta - Colombia. Av. Gran Col. # 12E-96

[2] Industrial Engineering. Francisco de Paula Santander University. Cucuta - Colombia.

[3] Pyrometallurgical and Materials Research Group (GIPIMME), University of Antioquia, Medellin Colombia.

E- mail: gabrielpr@ufps.edu.co, ggabrielp@yahoo.com

Abstract. The mechanical and tribological performance of red clay ceramic tiles uncoated and coated by oxy-fuel thermal spraying process from fly ash powders was evaluated. The ceramic tile substrates were manufactured by uniaxial pressing at 26.17 bar pressure, and sintered at 1100 °C. The coating thickness was determined based on the number of projection-cycles oxyacetylene flame over substrate. Coal fly ash coatings were deposited, with average thickness of 56.18±12.18 µm, 180.42±20.32 µm, and 258.26±25.88µm. The mechanical resistance to bending and wear by abrasion deep, were studied using ISO 10545-4 standards and ISO 10545-6 respectively; adhesion was measured using Elcometer equipment Type III according to ASTM D-4541-02 and the average roughness (Ra) was found according to ASTM standard D7127-13, using the profilometer Mitutoyo SJ 201. The surface morphology presented the heterogeneous molten or semi molten splats with average size of 35.262±3.48 micrometers with good adhesion, justifying increased mechanical resistance to bending by 5%, as well as wear by abrasion deep. These results contribute to the development of ceramic products with added value, to be used in various technological applications.

Keywords: Ceramic tiles, fly ash coatings, thermal spray, Mechanical resistance to flexion, deep abrasion wear, adherence, rugosity.

1. Introduction

In thermal power plants for energy production, fly ash are generated, which have applications in the construction industry and applications such as glass-ceramic materials[1-5]. Fly ash is composed mainly of mineral matter (70-80%) as glassy particles with a small proportion of crystalline phase and unburned coal. The most important minerals found in fly ash are mullite, magnetite, hematite and Silicas [5-9]. Fly ash contains amorphous spheres called cenospheres or plerospheres. Its low cost, light weight and low thermal conductivity, make them suitable for applications such as insulating blocks and as coatings for thermal protection of substrates [9-13].

Content from this work may be used under the terms of the Creative Commons Attribution 3.0 licence. Any further distribution of this work must maintain attribution to the author(s) and the title of the work, journal citation and DOI.
Published under licence by IOP Publishing Ltd

Surface treatments are considered important to improve the durability of a piece, either increasing its thermal resistance, corrosion and wear, providing material durability improving their physical appearance and their application [14]. Oxyacetylene thermal spraying (PTO) is a technique used in the recovery of worn parts, and consists of projecting small molten particles, semi-molten, hot or even cold (cold spray) which are successively deposited to a surface. The aim is to provide a surface treatment to parts which are going to be subjected to extremes of friction, wear, heat and / or mechanical stress. Choosing the coating material, the process for coating and its thickness [14], is one of the most difficult decisions for both; the manufacturer and responsible for maintenance. In addition, the PTO can cover parts of various dimensions thanks to its versatility [14-22].

Studies related to the coating of fly ash on ceramic substrates paste by the technique of PTO, were not found in the literature review of the last decade, therefore, it broke the results of studies made previously as reference studies for the determination of some working conditions, as well as previous laboratory tests.

2. Materials and Methods

The substrates were made from a paste of atomized red clay supplied by the company Ceramics Italia SA, (Cucuta, Colombia), with approximate dimensions of length L = 10 cm, width A = 5 cm, and thickness H = 0.8 cm, which they were manufactured by the uniaxial pressing, using hydraulic press, with average pressure bars 26,17 (≈30,6 Kgf / cm^2), and humidity between 5 and 9% by weight; heat treatment (baking) was conducted using brand muffle furnace Ney D-130® with a firing curve at a maximum temperature of 1100 ° C, with a speed of 5 ° C / min.

For the preparation of coatings fly ash supplied by the central Termotasajero SA (Norte de Santander-Colombia) were used, which were sieved through sieve No. 270 (53 µm) and retained mesh sieve No 325 (45 µm) according to ASTM E-11-09.

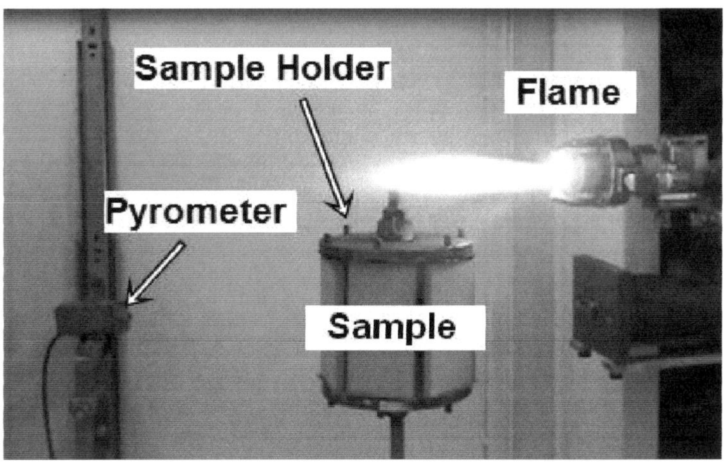

Figure 1. Sample holder, pyrometer and Terodyn 2000 TM torch.
Source: University of Antioquia group GIPIMME.

The development of the coatings was carried out using the ARESTE I chamber containing the torch Eutectic Castodin Terodyn 2000 ™ and sample holder hexagonal seats six (6) samples (see Fig. 1). For the screening, it was not necessary to apply anchor layer (both coat) between the substrate and projected powders, because the substrates have optimal roughness that allow adhesion of the molten particles. The projection parameters were determined by simulation using the Jets et Poudres software and pilot tests. Preheating three passes, a distance (sample-torch) of 8 cm, rotation speed of 116.0 rpm holder, vertical speed torch 0.72 cm / s, pressure and nitrogen flow of 40 psi were used and 16 L/min, respectively, while the pressure and flow of oxygen was 50 psi and 94 L/min, finally 12 psi and 22 L/min were used for the pressure and flow of acetylene, and a constant flow of powder 6 g / min.

These parameters were similar to those reported in reference [23, 24], for alumina and silica coatings. Three projections 5, 9 and 14 passes oxyacetylene flame on the substrates, values found by pilot tests and following reported by Araque et al were performed, for alumina powders on the same type of substrates [23]. The samples obtained in each projection were coded as 5P, 9P and 14P respectively.
The test for resistance to abrasion was performed according to ISO 10545-6, using the equipment Gabrielli®-CAP-3. The modulus of rupture and flexural strength was carried out according to ISO 10545-4, using equipment brand Gabrielli, Flexi-1000 LX-650 model. Adhesion test for the Pull-off type was applied, using the Elcometer Type III equipment according to ASTM D-4541-02. Microindentation hardness Vickers was performed according to standard ASTM C1327-08, which employs as diamond penetrator indentor quadrangular pyramid shaped. The average roughness (Ra) was found according to ASTM D7127-13 standard, using the profilometer Mitutoyo SJ201, for which 10 measurements and randomly oriented in different directions, with a sweep of 12.5mm was performed. The test water absorption was realized according to the UNE-EN ISO 10545-3.

3. Results and discussion

Table 1 shows the Modulus of rupture by flexural, volume footprint deep abrasion, roughness and adhesion to the substrate and the three coatings 5P, 14P and 9P. It is observed that the modulus of rupture has similar behavior for the three coatings, and increases by approximately 6% compared to the modulus of rupture of the substrate. Furthermore these values comply with the technical characteristics required for wall coverings industrial level [25, 26], and are very similar to those reported by Araque et al [23], for alumina powders. It was found that as the number of passes increases the thickness increased from 56.18 ± 12.18 µm, 180.42 ± 20.32 µm and 258.26 ± 25.88 µm, for 5P samples, 9P and 14P respectively, this is consistent with deep abrasion wear, because it decreases with the increasing in the number of passes, protecting the substrate up to 23.54%, since greater film thickness is better anchoring it provides to the substrate. It is also observed that the projected amount of fly ash influences the microhardness, being smaller as the number of passes or amount of fly ash increases projected. By comparing the results of the microhardness with those reported in the literature, we found that are very similar to those found in reference [24] where Vickers microhardness values are reported between 509.85 kgf/mm^2 and 713.80 kgf/mm^2. On the other hand, Adhesion results are very close and range between 3.6 and 5.5 MPa, presenting cohesive failure type, showing detachment substrate and not the coating, and such values are in the range of those reported by Araque et al. [23], who reported that this parameter should be greater than 2 MPa and can be increased by heat treatment at high temperatures.

Table 1. Modulus of rupture by flexural, the average volume of footprint by deep abrasion, adhesion. Microhardness and roughness to the substrate and the three coatings 5P, 14P and 9P.

SAMPLE	SURFACE ROUGHNESS Ra (µm)	VICKERS MICROHARDNESS (Kgf/mm^2)	ADHERENCE (MPa)	DEEP ABRASION (mm^3)	MODULUS OF RUPTURE (N/mm^2)	WATER ABSORPTION %
Substrate	4.73±1.27	140.00±19.89	-	357.55±9.40	12.20±0.42	12,54
5P	15.84±0.96	616.00±95.29	4.80±0.1	128.90±2.69	12.95±0.78	12,69
9P	17.60±0.98	602.00±59.79	5.50±0.2	89.10±2.55	12.90±0.89	14,34
14P	20.44±2.41	571.00±45.3	3.60±0,3	84.15±8.69	13.05±0.92	13,77

Figure 2, shows the image of Scanning Electron Microscopy (SEM) of the surface and at Figure 3, the picture shows the SEM of a cross section for the sample 9P, to 20kV and 500X presented in these a heterogeneous morphology and molten semimolten starting powders of particles and porosities due to stacking of particles as the number of passes increases. It can be stated that as fly ash powders are deposited, its roughness is increased, protecting the substrate by deep abrasion wear, improved mechanical and tribological properties. It was found that the coating projection 2 with 9 passes, is the one with the best mechanical and tribological behavior.

Figure 2. SEM micrograph surface, 9P projection.

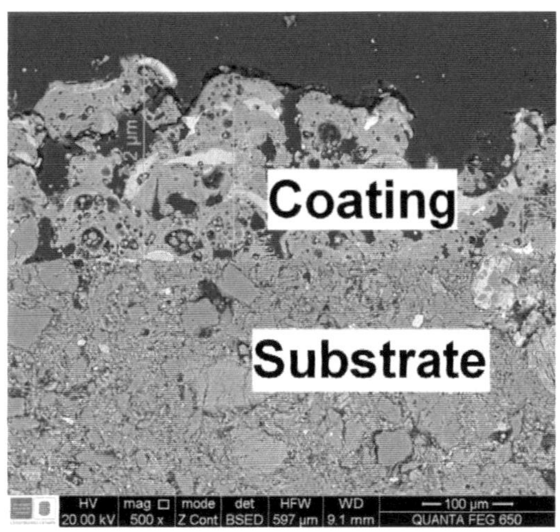

Figure 3. SEM micrograph cross, 9P projection.

4. Conclusions

PTO technique was applied, to produce powder coatings fly ash on substrates of atomized paste prepared by uniaxial pressing, in conditions of constant thermal spraying, selected by pilot tests. The ceramic substrate does not require any preparation because this has the characteristic of being porous, rough and chemically compatible with fly ash powders. Therefore the technique PTO on red clay substrates is viable because they allow direct anchor, getting good percentages of glassy material being the porous substrate mechanically and chemically compatible; also good results were achieved in improving the mechanical and tribological properties. Due to the homogeneity on the surface of samples no preparation was not performed. It only was made a verification that the test surface was parallel to the surface of the base or support. For the samples used, the water absorption % showed an increment of 10%, indicating that according to the UNE-EN ISO 10545-3, for tile dry pressed ceramic, are categorized within the group BIII and ISO 10545-4 according to the standard module breakage or bending strength of 12 N / mm2 (equivalent to 122.40 kgf / cm2), the suggested application for covered tiles is wall covering. These results give way to future trials, which could be exploited surface porosity, it could be thinking of adding oxides to these coatings and thus provide safety features, hydrophobicity and antibacterial qualities to the samples. Finally, it is concluded that, with Projection 2 (9 passes) coating the best mechanical and tribological properties of the substrate-coating system was obtained. It is the coating having the most irregular surface topography.

Acknowledgments
Authors are thankful to FINU-UFPS for financial support by Contract No. 026-2014, and No. 015-2015 and CODI-University of Antioquia for the financial support given to GIPIMME group through both: the PR 14-1-04 project and sustainability strategy-2014

References
[1] Chávez A, Vargas G, Almanza J M 2008 Caracterización y preparación de ceniza volante para su uso como recubrimientos cerámicos sobre substratos metálicos. *Bol. Soc. Esp. Ceram.* V., 47, 6, 333-8
[2] Cheng T W and Chen Y S 2004 Characterisation of glass ceramics made from incinerator fly ash Departamento de Materiales y Recursos Minerales Ingeniería de la Universidad Nacional de Tecnología de Taipei, Taipei, Taiwán, República de China *Ceram. Int.* **30** 343–9
[3] Ulrichs C, Schmidt U, Mucha-Pelzer T, Goswami A and Mewis I 2009 Hard Coal Fly Ash and Silica-Effect of Fine Particulate Matter Deposits on Brassica chinensis *Am. J. Agric. Bio. Sci.* **4** (1): 24-31, India
[4] Shanthakumar S, Singh D N and Phadke R C 2008 Characterization of Fly Ash from Various Locations of Electrostatic Precipitator. The 12th *International Conference of International Association for Computer Methods and Advances in Geomechanics* (IACMAG), October Goa, India
[5] Arango-Martínez N, Pérez-Bueno J, Gochi-Ponce1 Y and Sánchez-Rubiol M 2011 Zeolitas a partir de cenizas volantes generadas en las centrales carboeléctricas para el tratamiento de agua contaminada con Mn y Pb. Ciencia Ed (IMIQ) Vol. **26** núm. 1.
[6] Chávez-Valdez A, Arizmendi-Morquecho A, Vargas G, Almanza J M and Alvarez-Quintana J 2011 Ultra-low thermal conductivity thermal barrier coatings from recycled fly-ash cenospheres *Acta Mater.* **59** 2556–62.
[7] Khanra S, Mallick D, Dutta S N and Chaudhuri S K 1998 Studies on the phase mineralogy and leaching characteristics of coal fly ash, *Water Air Soil Pollut* **107**:251–75
[8] Peña-Rodríguez G and Ortega-Triana L R Morphological and structural characterization of fly ash powders 2014 *I+D* Vol. **14**, No. 2, pp. 14-19
[9] Vucini D, Miljanovic I, Rosic A and Lazic P 2003 Effect of Na2O/SiO2 mole ratio on the crystal type of zeolite synthesized from coal fly ash *J.Serb.Chem.Soc.* 68(6)471–78
[10] Gomes S and François M 2000 Characterization of mullite in silicoaluminous fly ash by XRD, TEM and 29Si MAS NMR, *Cement Concr. Res.* 30, 175–81

[11] Valencia S L 2001 Caracterización física química y mineralógica de las cenizas volantes Ciencia e Ingeniería Neogranadina, núm. **10**, julio, pp. 47-62, Universidad Militar Nueva Granada, Colombia

[12] Norton G, Markuszewskl R, and Shanks H 1986 Morphological and Chemical Characterization of Iron-Rich Fly Ash Fractions *Envlron. Sci. Technol.* 1986, 20, 409-413 Ames Laboratory, Iowa State University, Ames, Iowa

[13] Ruiz-Román J M, Santos C, Cambronero L, Corpas F, Alfonso M and Morano A J 2000 Aprovechamiento de las cenizas volantes, clase F, de centrales térmicas para la fabricación de materiales cerámicos. *Bol. Soc. Esp. Ceram.* V., 39 [3] pp. 229-31

[14] Lira-Olivares J, González-Rodulfo S, De Abreu J 2011 Recubrimientos metálicos y cerámicos solucionan el problema económico y ambiental de países desarrollados y en desarrollo *Rev. Latinam. Metal. Mat.* **31** (2): 106-121.

[15] Morales J, Olaya-Flórez J and Rojas-Molano H 2012 An approach to thermal spray technology AVANCES *Inv. Ing.* Vol. **9** - No. 2 pp. 60-71

[16] Chata S, Sidhu H, and Sidhu B 2012 Characterisation and corrosión-Erosion Behaviour of Carbide based Thermal Spray Coatings *J. Miner. Mater. Charact. Eng.* Vol. **11**, No. 6, pp. 569-86

[17] Vargas F 2010 Élaboration de couches céramiques épaisses á structures micrométriques et nanométriques par projections thermiques pour des applications tribologiques Tesis de Doctorado. Faculté des sicences et techniques Université de Limoges

[18] González A, Henao J, Díaz A, López E and Vargas F 2013 Influence of thermal spray parameters on the microestructure of Zirconia-Aluminia and Zirconia-Ceria coatings used as a thermal barrier *Rev. Latinam.Metal.Mat, 33*(2), 272-81

[19] Cano, C 2008 "Recubrimientos cerámicos con aplicación en barreras térmicas y ambientales" Tesis doctoral Universidad Autónoma de Madrid, Madrid, España

[20] Observatorio Tecnológico del Metal 2008 Proyección Térmica para Matrices. AIMME-Instituto Tecnológico Metalmecánico. Parque Tecnológico

[21] Restrepo E, Vargas F, López M, Cardona R and Duarte G 2012 Elaboración de recubrimientos cerámicos mediante proyección térmica por combustión a partir de residuos sólidos industriales *Rev. Mat.* Vol. **17**, n.4, pp. 1176-85

[22] Pawlowski L 2008 The Science and Engineering of Thermal Spray Coatings. Second Edition Ecole Nationale Supérieure de Chimie de Lille, Villeneuve d'Ascq, France

[23] Araque-Pabón M, Peña-Rodríguez G and Vargas-Galvis F 2015 Desempeño mecánico y tribológico de baldosas cerámicas de arcilla roja recubiertas por proyección térmica a partir de alúmina *Tecno Lógicas*, ISSN 0123-7799, Vol. 18, No. 35 pp 125-35

[24] Vallejo-Pérez M, Gómez-García I C, Zuluaga-Castrillon D, Hernández-Ruiz J F, Vargas-Galvis F and López-Gómez M E 2013 Estudio de la estructura de recubrimientos de silicio depositados mediante proyección térmica por llama oxiacetilénica *Rev. Col. Mat.* No. **5**, pp. 145-151

[25] http://www.ceramicaitalia.com/inf/images/stories/WebPageFiles/FichaTecnicaRV.pdf, consultado abril 01 de 2016

[26] https://alfa.com.co/uploads/fichas/99099cf35dbaf3e6c02f2ee692fa3bedb7e8c238.pdf, consultado abril 01 de 2016

IOP Conf. Series: Journal of Physics: Conf. Series **792** (2017) 012027 doi:10.1088/1742-6596/792/1/012027

Abiotic degradation of plastic films

Y G Ángeles-López, A M Gutiérrez-Mayen, M Velasco-Pérez*, M Beltrán-Villavicencio, A Vázquez-Morillas and M Cano-Blanco

Universidad Autónoma Metropolitana Unidad Azcapotzalco, Avenida San Pablo Xalpa 180, Azcapotzalco, Reynosa Tamaulipas, 02200 Ciudad de México, D.F., México

E-mail*: mvp@correo.azc.uam.mx

Abstract. Degradable plastics have been promoted as an option to mitigate the environmental impacts of plastic waste. However, there is no certainty about its degradability under different environmental conditions. The effect of accelerated weathering (AW), natural weathering (NW) and thermal oxidation (TO) on different plastics (high density polyethylene, HDPE; oxodegradable high density polyethylene, HDPE-oxo; compostable plastic, Ecovio ®; metalized polypropylene, PP; and oxodegradable metalized polypropylene, PP-oxo) was studied. Plastics films were exposed to AW per 110 hours; to NW per 90 days; and to TO per 30 days. Plastic films exposed to AW and NW showed a general loss on mechanical properties. The highest reduction in elongation at break on AW occurred to HDPE-oxo (from 400.4% to 20.9%) and was higher than 90% for HDPE, HDPE-oxo, Ecovio ® and PP-oxo in NW. No substantial evidence of degradation was found on plastics exposed to TO. Oxo-plastics showed higher degradation rates than their conventional counterparts, and the compostable plastic was resistant to degradation in the studied abiotic conditions. This study shows that degradation of plastics in real life conditions will vary depending in both, their composition and the environment.

1. Introduction

Plastics are polymers constituted mainly by atoms of carbon and hydrogen. The versatility of their applications has significantly increased production and consumption of plastics. In 2013 production of plastics worldwide reached 299 million of ton; a 46% increase from 2002 [1]. Plastic carrier bags occupy an important place on the plastic market. An European citizen uses an average of 198 plastic bags per year [2]. In Mexico, 41% of the plastic industry produces bags and flexible films, employing 49 000 people in the country and manufacturing 600 000 tons per year of these commodities [3].

Plastics can be classified, according to their origin, in synthetic or biobased. Synthetic plastics are manufactured from petroleum or natural gas, both non-renewable resources; for instance, polyethylene

Content from this work may be used under the terms of the Creative Commons Attribution 3.0 licence. Any further distribution of this work must maintain attribution to the author(s) and the title of the work, journal citation and DOI.
Published under licence by IOP Publishing Ltd

(PE), polypropylene (PP), polyvinyl chloride (PVC) and polyethylene terephthalate (PET). Biobased polymers can be natural or synthetic. Natural biobased polymers are synthetized by a living organism (starch, cellulose and bacterial polyhydroxyalkanoates) and synthetic biobased polymers are chemically transformed naturally occurring monomers. One example is the production of polylactic acid (PLA): starch is hydrolyzed to lactic acid, then lactic acid is transformed to polylactic acid, PLA. Some synthetic plastics have their bio-based equivalent, some examples are bio-PE and bio-PVC.

Accelerated consumption, inadequate disposal and difficulties in achieving recycling targets of plastic bags made chemical stability of these polymers an issue when they are discarded. Oxo- and bio-degradable plastic bags have been promoted to reduce environmental impacts of this commodity. Oxodegradable bags are manufactured from conventional synthetic polymers, such as polyethylene, and an additive that promotes oxidation. The pro-oxidant additive catalyzes the natural degrading process of plastics, and it is activated trough ultraviolet radiation, temperature, humidity, mechanical tension or a combination of them. In the other hand, biodegradable plastics are capable of being degraded by living organisms. Biodegradability is independent of the origin of the plastic; bio-PE and PE from petroleum will degrade at the same rate. One biodegradable plastic commercially available in Mexico is Ecovio ®, a biobased plastic film produced by BASF® that is designed biodegrade through composting.

Plastic industry has evolved to produce materials compatible with environmental aspirations of consumers, such as oxo and biodegradable films. However, it is not clear how these environmental credentials will affect the functionality of plastics during their useful life. This study investigates the effect of three oxidation processes (accelerated weathering, AW; natural weathering, NW; and thermal oxidation, TO) on five types of plastic films (high density polyethylene, HDPE; high density polyethylene with an oxidant additive, HDPE-oxo; compostable plastic film, Ecovio ®; metalized polypropylene, PP; and metalized polypropylene with an oxidant additive, PP-oxo). Its aim is to assess if their degradability through oxidation or biological processes will affect its functionality.

2. Methodology

This study was carried out in the laboratory of Sustainable Technologies at the Autonomous Metropolitan University, Campus Azcapotzalco. The main stages of methodology were: preparation of plastic films, exposure of plastic films to oxidation processes, characterization of plastic films and determination of the kinetics of oxidation processes.

3. Preparation of plastic films

HDPE, HDPE-oxo and Ecovio ® were obtained from commercial plastic film rolls, whereas PP and PP-oxo were obtained from post-consumer potato chip bags. Post-consumer films were cleaned with deionized water and paper towels to remove residuals. Finally, all types of plastics were cut into probes (1 x 15 cm).

4. Exposure of plastic films to the oxidation processes

Oxidation processes were studied for different periods; AW for 110 hours, NW for 90 days and TD for 30 days. Table 1 presents the sampling times for each oxidation process. Experiments were run with 20 replicates for each sampling time, type of oxidation process and type of plastic film.

Table 1. Sampling times for each oxidation process.

Oxidation process	Symbol	Units of time	Sampling times				
Accelerated weathering	AW	Hours	50	70	80	100	110
Natural weathering	NW	Days	30	45	60	75	90
Thermal oxidation	TO	Days	3	6	12	18	30

4.1.1. Accelerated weathering. An accelerated weathering test chamber was custom made by the Laboratory of Development in Professional Automation at the Autonomous Metropolitan University [4]. The chamber has a volume of 0.145 m^3, and eight ultraviolet lamps with wavelength between 350 nm and 400 nm. The internal walls of the chamber are covered with stainless steel sheets AISI 304 with mirror finish. The chamber has a heating system capable of regulating temperature from 55 °C to 70 °C, and relative humidity can be set up to 50%. For this experiment temperature and humidity were set at 70 °C and > 50 %, respectively. In the chamber, samples are placed in a carousel to ensure all plastic film probes are exposed to similar conditions (radiation, temperature and humidity).

2.2.2 Natural weathering. Plastic films probes were placed on a metal rack. They were set parallel to the floor and held to the rack with adhesive tape. The rack was installed outdoors, facing south, on the roof of a four story university building.

2.2.3 Thermal oxidation. Plastic film probes were placed on a metal rack in a laboratory room with controlled temperature and humidity. Room conditions were set with a Norit humidifier at 30 °C and 80% relative humidity.

5. Characterization of plastic films

Elongation at break, weight and carbonyl index were used as indicators of degradation. Percentage of elongation at break and weight were measured before the oxidation process and for each sampling time, while carbonyl index was calculated only before and after the oxidation process. The average and standard deviation for 20 samples is reported for each sampling time.

2.3.1 Percentage of elongation at break. Elongation at break reflects the capability of plastics to resist changes of shape without formation of cracks. It is a direct indication of the loss of mechanical properties due to plastic degradation. Elongation at break was measured with a universal testing machine (UTM), model LP Plus from Lloyd instruments [5]. Plastic film probes were placed in the grips of the UTM, gauge length was set at 50 mm and velocity of the mobile grip was set at 50 mm/min.

2.3.2 Weight. Loss in weight is also produced through degradation processes. It was measured with an analytical balance, model DV114C from Ohaus, with a precision of ± 0.1 mg.

2.3.3 Carbonyl index. It is a measure of the presence of carbonyl groups (C=O), a product of oxidation of photodegradable polymers such as oxo-polyethylene and oxo-polypropylene. Carbonyl index was determined with an infrared spectrometer FT-IR model Nicolet iS10 from Thermo Scientific, dividing the height of the peak at a wavelength of 1751 cm^{-1} (C-O and C=O bonds) by the height of the peak at a wavelength of 1895 cm^{-1} (C-C bonds).

6. Kinetics of oxidation processes
Kinetics for each oxidation process was determined with the integrated rate equations, percentage of elongation at break was the parameter used to follow the reaction rate.

7. Results

8. Accelerated weathering
All types of plastic films showed a decrease in percentage of elongation at break (figure 1a). The highest loss occurred for HDPE-oxo (95% loss, from 400.4% to 20.9%) the smallest loss occurred for PP (30% loss, from 133.4% to 93.3%). The decrease in elongation at break was higher for the polymers with the oxidant additive; i.e. higher for HDPE-oxo than for HDPE and higher for PP-oxo than PP.

Weight of plastic films is shown in figure 2b, the only plastic film that presented a loss in weight was HDPE. Increase of weight on HDPE-oxo, Ecovio, PP and PP-oxo can be explained by adsorption of water.

There was an increase on the carbonyl index for HDPE, HDPE-oxo and PP-oxo (figure 1c), this reflects an increase on the carbonyl groups due to oxidation. Also, Ecovio ® and PP showed only marginal changes on the carbonyl index. Since Ecovio® degradation process begins with hydrolysis, instead oxidation, no changes were expected on the carbonyl index for Ecovio®.

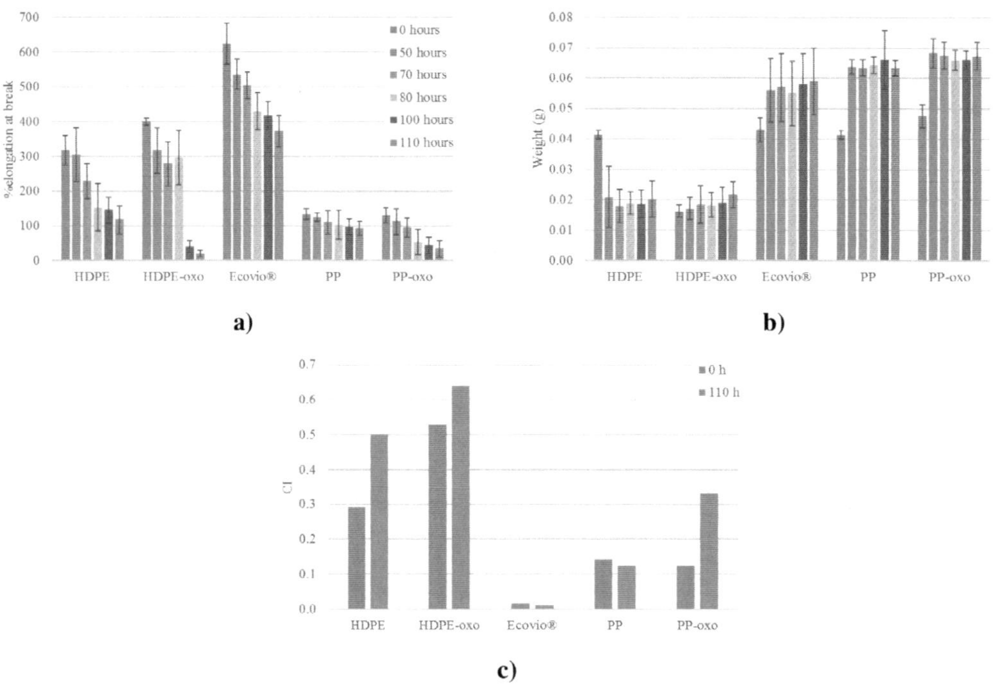

Figure 1. Accelerated weathering, a) % elongation at break, b) loss of weight, and c) carbonyl index (CI) for HDPE, HDPE-oxo, Ecovio ®, PP and PP-oxo.

9. Natural weathering

During the 90 days that natural weathering of plastic films was carried out, average temperature varied from 15 °C to 22 °C and average relative humidity from 30 % to 80 %. Percentage elongation at break decreased in all types of plastics (figure 2a). The loss in elongation at break was higher than 90% for HDPE, HDPE-oxo, Ecovio ® and PP-oxo., and was 30.4% for PP. In this oxidation process, loss of elongation at break was also higher for the polymers with the oxidant additive.

Weight of plastic films in natural weathering is presented in figure 2b; HDPE, HDPE-oxo and PP-oxo showed a decrease in weight, while Ecovio ® and PP showed a slight increase.

There was an increase on the carbonyl index on all types of plastic films, increase was higher in PP (200%, from 0.125 to 0.375) and Ecovio ® (150%, from 0.016 to 0.040).

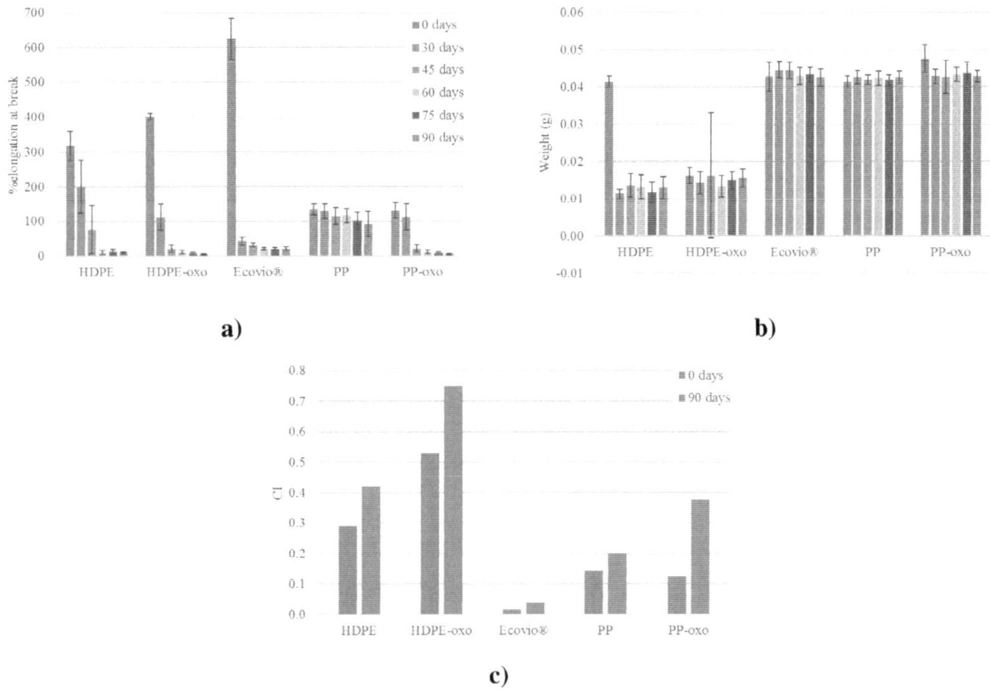

Figure 2. Natural weathering, a) % elongation at break, b) loss of weight, and c) carbonyl index (CI) for HDPE, HDPE-oxo, Ecovio ®, PP and PP-oxo.

10. Thermal oxidation

Average temperature varied from 19 °C to 23 °C and relative humidity from 90% to 100% during the 30 days that this experiment lasted. Percentage of elongation at break did not present any significant change (figure 3a). HDPE was the only type of plastic film that presented a decrease on weight (figure 3b). Finally, HDPE, PP and PP-oxo showed a slightly increase on the carbonyl index.

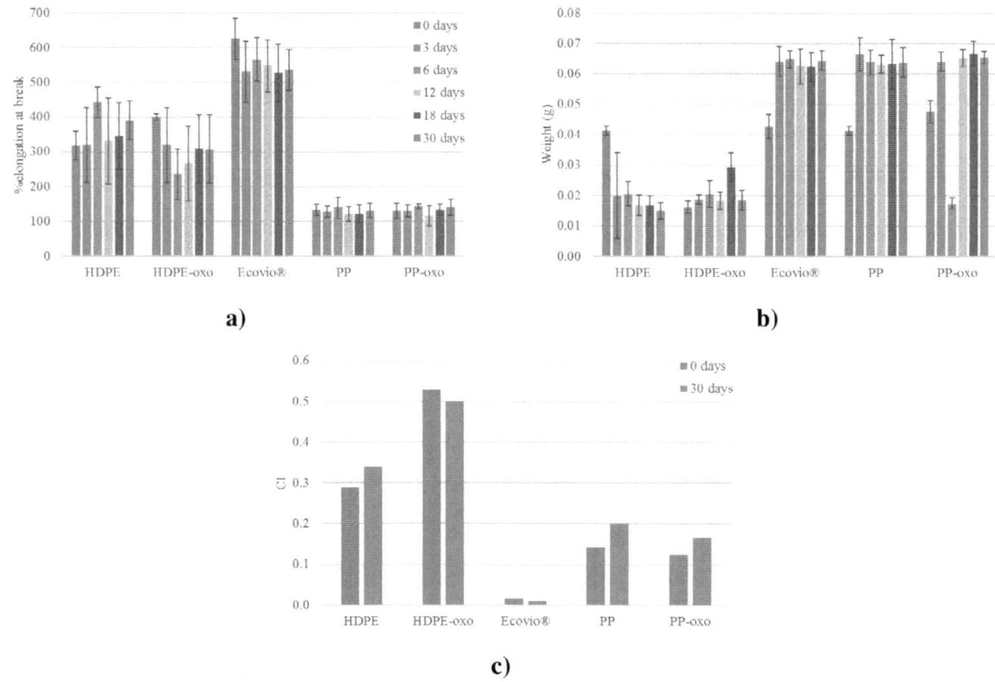

Figure 3. Thermal oxidation, a) % elongation at break, b) weight and c) carbonyl index (CI) for HDPE, HDPE-oxo, Ecovio ®, PP and PP-oxo.

11. Kinetics of degradation processes

Table 2 presents the reaction order and the kinetic constant by type of plastic for the oxidation process of accelerated and natural weathering. Kinetics parameters were not calculated for thermal oxidation because there were no significant changes in elongation at break for this oxidation process. In AW -elongation at break followed a cero order reaction for all plastic films studied. Kinetic constants were higher for HDPE-oxo than for HDPE, and for PP-oxo than for PP. In NW a cero order reaction was followed by HDPE and PP and a first order reaction was followed by HDPE-oxo and PP-oxo.

Table 2. Kinetics of degradation for accelerated and natural weathering.

Oxidation process	**Plastic film type**	**Reaction order**	**Kinetic constant**	**Units**	**Correlation coefficient**
Accelerated weathering	HDPE	0	1.980	% Elongation h^{-1}	0.8395
	HDPE-oxo	0	3.438	% Elongation h^{-1}	0.7573
	Ecovio®	0	2.28	% Elongation h^{-1}	0.9563
	PP	0	0.39	% Elongation h^{-1}	0.9367
	PP-oxo	0	0.93	% Elongation h^{-1}	0.8723
Natural weathering	HDPE	0	0.154	% Elongation h^{-1}	0.8822
	HDPE-oxo	1	2.11E-03	h^{-1}	0.9509
	Ecovio®	2	2.23E-05	% Elongation^{-1} h^{-1}	0.9178
	PP	0	1.90E-02	% Elongation h^{-1}	0.9153
	PP-oxo	1	1.67E-03	h^{-1}	0.9069

12. Conclusions

During accelerated and natural weathering processes occurred a loss of mechanical properties of all types of plastic film studied (HDPE, HDPE-oxo, Ecovio®, PP and PP-oxo), evidencing their degradation. No substantial evidence of degradation was found on thermal oxidation. The loss of mechanical properties on accelerated and natural weathering was higher on HDPE-oxo than on HDPE and higher in PP-oxo than on PP. This was expected since oxo additives catalyse oxidation of synthetic plastics.

Ecovio®, a bio-degradable and bio-based polymer, presented loss of mechanical properties. However, this plastic film is designed to degrade through hydrolysis. High relative humidity and temperature may have triggered degradation of Ecovio®; loss of mechanical properties was higher in natural weathering (15 °C – 22 °C and 30 % - 80% relative humidity) than in accelerated weathering (70 °C and > 50% relative humidity).

Kinetics was found of cero order for HDPE, HDPE-oxo, Ecovio®, PP and PP-oxo for AW and only for HDPE and PP for NW. The value of the cero order kinetic constant of HDPE and PP was higher for AW than for NW, as expected. In AW reaction rates were higher for oxo equivalents of conventional polymers (HDPE-oxo > HDPE and PP-oxo > PP). Kinetic reaction of oxo polymers was found of first order for NW, hence reaction rate depends directly on elongation at break. Kinetic constant is on the same magnitude order for both, HDPE-oxo and PP-oxo.

Although synthetic, synthetic with oxidant additives and bio-based bio-degradable polymers showed evidence of degradation on accelerated and natural weathering conditions. The results show that the extent and rate of degradation are dependent on the type of plastics and the environment conditions. Thus, the benefits of using degradable plastics must be assessed for each specific situation where they will be used

References

[1] Plastics Europe Association of Plastics Manufacturers, Plastics – the Facts 2014/2015, Brissels, Belgium, 2015. http://www.plasticseurope.org/home.aspx.

[2] C. Barbiere, EU to halve plastic bag use by 2019, EurActiv. (2015). http://www.euractiv.com/section/sustainable-dev/news/eu-to-halve-plastic-bag-use-by-2019/.

[3] INEGI Instituto Nacional de Estadística y Geografía, Censos económicos 2014, Sensos Económicos Sect. Ind. Manuf. (2014). http://www.inegi.org.mx/est/contenidos/Proyectos/ce/ce2014/doc/tabulados.html (accessed March 30, 2016).

[4] G. Aragón González, A. León Galicia, V.M. A., Manual del Usuario de la cámara de intemperismo acelerado con radiación UV, Univ. Autónoma Metrop. Unidad Azcapotzalco. División d (2011).

[5] American Society of the International Association for Testing and Materials, ASTM D882 - 10, Standard Test Method for Tensile Properties of Thin Plastic Sheeting, United States of America, n.d.

IOP Conf. Series: Journal of Physics: Conf. Series **792** (2017) 012028 doi:10.1088/1742-6596/792/1/012028

Fast conical surface evaluation via randomized algorithm in the null-screen test

D Aguirre-Aguirre[1,2] , R Díaz-Uribe[1] and B Villalobos-Mendoza[2,3]
[1]Centro de Ciencias Aplicadas y Desarrollo Tecnológico, Universidad Nacional Autónoma de México, Cd. Universitaria, Apdo. Postal 70-186, C.P. 04510, Cd Mx, México
[2]Polo Universitario de Tecnología Avanzada, Universidad Nacional Autónoma de México, Vía de la Innovación No. 410, Autopista Monterrey-Aeropuerto Km. 10, PIIT C.P. 66629 Apodaca, N. L.
[3]Centro de Investigación y de Estudios Avanzados, Unidad Monterrey, Apodaca N.L. 66629, México

E-mail: daniel.aguirre@ccadet.unam.mx, rufino.diaz@ccadet.unam.mx

Abstract. This work shows a method to recover the shape of the surface via randomized algorithms when the null-screen test is used, instead of the integration process that is commonly performed. This, because the majority of the errors are added during the reconstruction of the surface (or the integration process). This kind of large surfaces are widely used in the aerospace sector and industry in general, and a big problem exists when these surfaces have to be tested. The null-screen method is a low-cost test, and a complete surface analysis can be done by using this method. In this paper, we show the simulations done for the analysis of fast conic surfaces, where it was proved that the quality and shape of a surface under study can be recovered with a percentage error < 2.

1. Introduction

In recent years, monitoring systems and real-time inspection have been certainly very attractive tools for the manufacturing industry and industry in general. When these inspection systems are not destructive and have a fast image capture and processing time, besides small dimensions and low weight, their interest increases greatly. Nowadays, the demand for new and more precise non-destructive industrial control techniques that allow automatic or almost automatic evaluation of the object to be inspected has been growing. This requires accurate and fast methods able to perform in situ measurements?i.e. in the production line, in a workshop or in a place where the piece is measured sometimes outside of a laboratory environment.

Content from this work may be used under the terms of the Creative Commons Attribution 3.0 licence. Any further distribution of this work must maintain attribution to the author(s) and the title of the work, journal citation and DOI.
Published under licence by IOP Publishing Ltd

For this purpose, we decided to use the technique of null-screens [1, 2]. This technique allows us to evaluate specular or partially reflective surfaces without contact, it is very robust and easy to handle, and besides that this method is capable of analyzing very large surfaces with a wide range of shapes [3, 4].

2. Null-screen method

The basic concept of this test is to design null-screens starting from the ideal shape of the surface under study. These null-screens are printed with geometric shapes which can be stripes, dots, drops, among others (Figure 1). This drawing is located on the screen in such way that, when it is reflected by the surface under test, if the surface is ideal or has a high quality, the observed image that also depends on the used geometry, can be stripes or dots perfectly aligned without deformation [1] (Figure 2a).

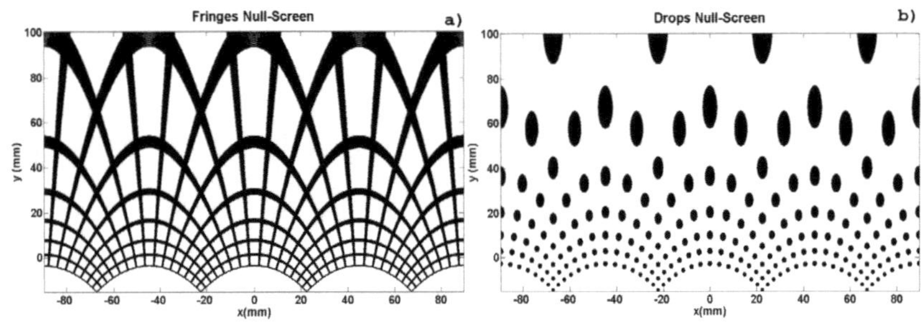

Figure 1. Null-screens a) fringes and b) drops.

However, if the surface is not ideal, or is not the surface for which the screen was designed, the observed image could have stripes or points that do not follow a perfect grid (Figure 2b). From the differences between the perfect grid (theoretical) and the distorted grid (experimental) can be obtained relevant information about the shape of the surface that is being analyzed.

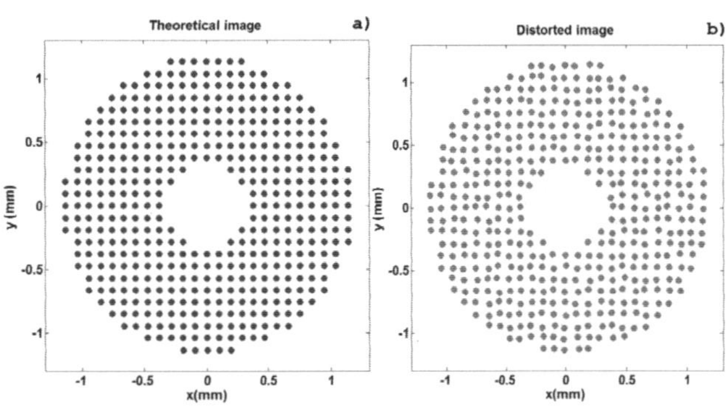

Figure 2. Centroids on the CCD, from the surface a) theoretical and b) distorted.

IOP Conf. Series: Journal of Physics: Conf. Series **792** (2017) 012028 doi:10.1088/1742-6596/792/1/012028

3. Mathematical basis

To determine the coordinates where each drop will be located on the null-screen, which will generate a perfect grid on the CCD, we make a reversed ray-tracing (figure 3). These rays leave the CCD $P_1(x_1, y_1, z_1)$ passing through the pinhole $P(0, 0, b)$ and intersect the surface at the point $P_2(x_2, y_2, z_2)$.

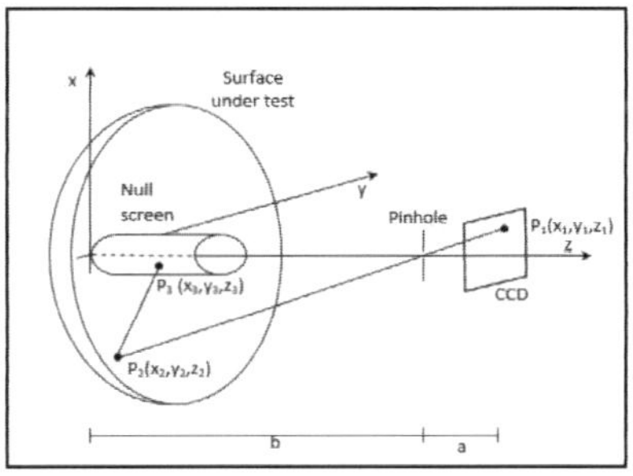

Figure 3. Setup for calculating a null-screen.

From equation of the sagitta which describes this type of surface given by

$$Q(z_2 - z_0)^2 \pm 2r(z_2 - z_0) + (y_2 - y_0)^2 + (x_2 - x_0)^2 = 0,$$

where r is the paraxial radius of curvature, x_0, y_0, z_0 are the decentering of the surface and $Q = (k\text{-}1)$, where k is defined by the conic constant.

To obtain the coordinates where the reflected ray intersects the cylinder, we start from the equation of the cylinder

$$(x_3 - x_0')^2 - (y_3 - y_0')^2 = C^2.$$

The parametric equations of the reflected ray are defined by

$$x_3 = x_2 + R_x\lambda$$
$$y_3 = y_2 + R_y\lambda$$
$$z_3 = z_2 + R_z\lambda.$$

where

$$R_x = -x_1 + 2(x_2 - x_0)\left[\frac{(x_2 - x_0)x_1 + (y_2 - y_0)y_1 + a(Q(z_2 - z_0) \pm r)}{(x_2 - x_0)^2 + (y_2 - y_0)^2 + (Q(z_2 - z_0) \pm r)^2}\right]$$

$$R_y = -y_1 + 2(y_2 - y_0)\left[\frac{(x_2 - x_0)x_1 + (y_2 - y_0)y_1 + a(Q(z_2 - z_0) \pm r)}{(x_2 - x_0)^2 + (y_2 - y_0)^2 + (Q(z_2 - z_0) \pm r)^2}\right]$$

$$R_z = -a + 2(z_2 - z_0)\left[\frac{(x_2 - x_0)x_1 + (y_2 - y_0)y_1 + a(Q(z_2 - z_0) \pm r)}{(x_2 - x_0)^2 + (y_2 - y_0)^2 + (Q(z_2 - z_0) \pm r)^2}\right].$$

Substituting the equation into , matching up λ terms we have

$$O\lambda^2 + P\lambda + T = 0,$$

where

$$O = R_x^2 + R_y^2$$
$$P = 2x_2R_x - 2x_0R_x + 2y_2R_y - 2y_0R_y$$
$$T = x_2^2 + x_0^2 - 2x_2x_0 + y_2^2 + y_0^2 - 2y_2y_0 + C^2.$$

Solving with the quadratic formula for λ

$$\lambda = \frac{-P \pm \sqrt{P^2 - 4OT}}{2O}.$$

With this, we obtain the coordinates x_3, y_3, and z_3, which correspond to the positions where the rays reflected by the surface intersect the display.
It should be noted that the diameter of the null-screen in the case of convex surfaces is greater than the diameter of the surface.

4. Surface analysis

In this section, the method used for the analysis of synthetic patterns is described.

4.1 Algorithm used for the analysis

It was decided to use a probabilistic algorithm for the surface analysis. The algorithm has three basics steps: two cycles to found the nearest solution to the problem, and a third cycle was only for the acquisition of the entrance data. The three steps of the algorithm are described next:

Step 1. *Parameter acquisition*
In this step, the theoretical surface data are acquired (i.e., radius of curvature, conic constant, diameter, *a* and *b* parameters, the size of the CCD, cylinder diameter, and experimental centroids positions), all this could be written in a .txt or .jpg file.

This data are necessary to calculate the theoretical centroids position of every spot in the null-screen. This task is done with the aim of making a comparison between the positions of every spot in the null-screen, with respect to the calculated position regarding the experimental centroids image while the surface form is varying.

Step 2. *"Approximation" cycle*
In this cycle, the conic constant, radius of curvature, surface, and screen decentering (x_0, y_0, z_0) and (x_0', y_0', z_0'), nearest to the real solution are found. This solution is found by modifying a theoretical surface by changing in a random manner the coefficients mentioned before. The found coefficients are introduced in the "limits reduction" cycle which is described next.

Step 3. *"Limits reduction" cycle*
In this cycle, the coefficients found in the approximation cycle are optimized. This process of optimization is done in our case by using the next equation

$$Coef = V - (V * n) + (V * n) 2rand,$$

where V is the value found in the approximation cycle, n is the percent of V where the search was done, and rand is a random value. With this process, the search of the solution is reduced to a $\pm n\%$ of V. At the end of this cycle, the values found for every coefficient are taken as the values that describe the surface under test.

4.2 Decentered surface ($x_0 \neq 0$, $y_0 \neq 0$, $z_0 \neq 0$)

In this subsection, we are going to talk about the decentered surface problem. This is a typical problem in the laboratory when a surface is being tested. The synthetic surface and the decentered coefficients used in this analysis are shown in Table 1.

Table 1. Design parameters of the Synthetic surface.

Surf. Design		Surf. Decentering	
r	8.75 mm	x_0	0.41 mm
k	-3.65	y_0	-0.63 mm
D	35.0 mm	z_0	-0.95 mm

A comparison between the centroids positions without decentering (blue) and the change in the centroids positions when a decentering is introduced in x, y and z (red) are shown in Figure 4a, where the centered and decentered centroids positions can be seen.

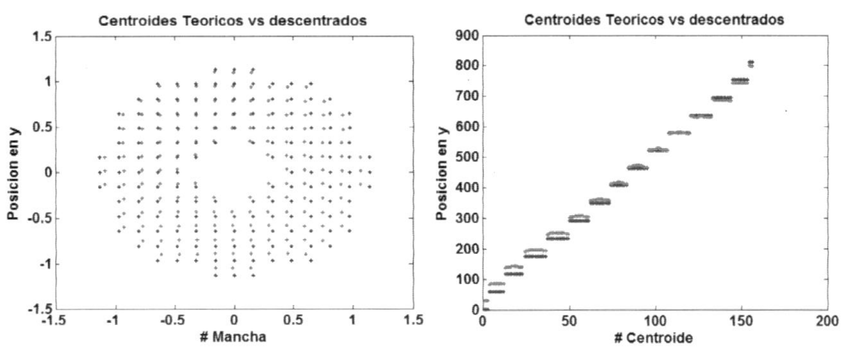

Figure 4. Surface centroids on the CCD, a) comparison between theoretical and decentered centroids position; and b) plot of the y position of each centroid.

In Figure 4b, the y positions of every centroid with respect to its assigned value are shown. This is done to prove that the correspondence between centroids was well done. This correspondence is done as follows: the coordinates in pixels of one centroid is taken from the real image, then the nearest centroid to this position is searched in the decentered image. To prove that this correspondence was well done, the *y* positions of each centroid was plotted.

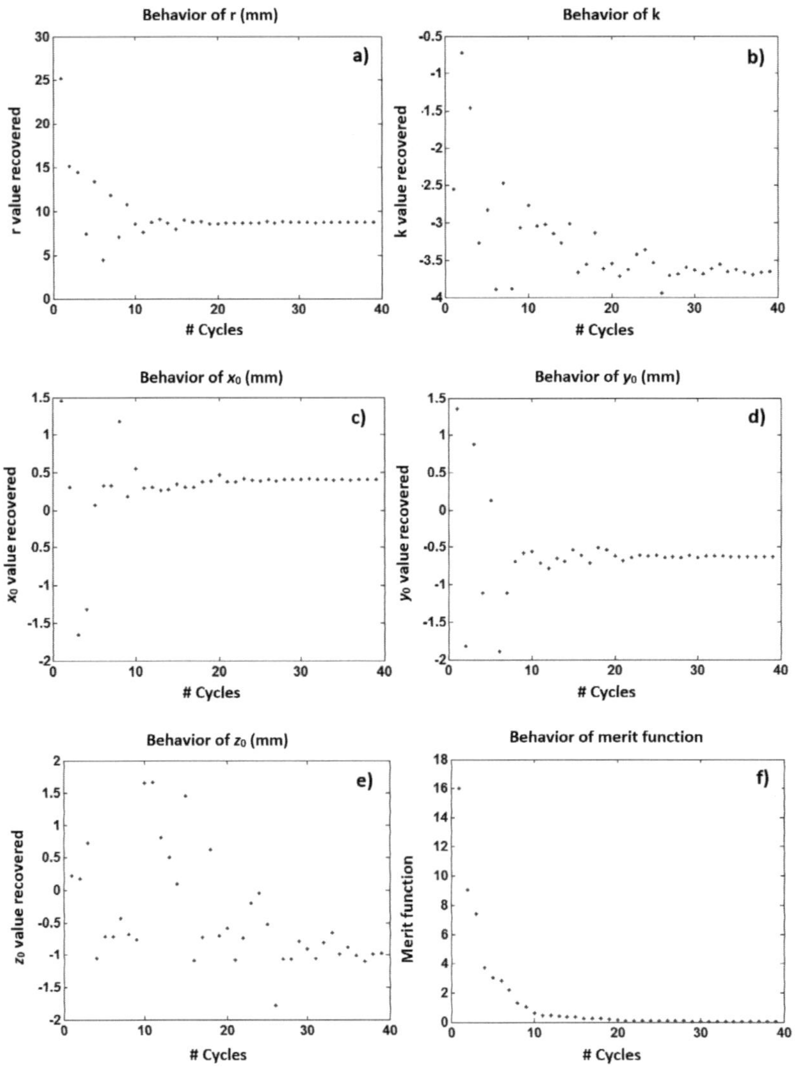

Figure 5. The behavior of, a) radius of curvature, b) conic constant, decentering in c) x, d) y, e) z and f) merit function when the program converged.

The recovered values by our modified algorithm are shown in Table 2. It can be seen that the percentage errors of the decentered coefficient in z and the conic constant was 1.89% and 0.28%, respectively.

Table 2. Comparison between the design and the recovered values.

	Design value	Recovered value	Percentage error
r	8.75 mm	8.7491 mm	0.01
k	-3.65	-3.6605	0.28
x_0	0.41 mm	0.4060 mm	0.97
y_0	-0.63 mm	-0.6274 mm	0.41
z_0	-0.95 mm	-0.9680 mm	1.89

5. Conclusions

In this paper, an algorithm for obtaining the shape of a conical concave or convex synthetic surface was presented. We described the null-screen design procedure and the randomized algorithm used for the surface analysis, where instead of using an integration process as is usually performed, the shape of the surface is found in a direct and in a random way.

We found that the surface decentering coefficients were recovered with a maximum percentage error less than 2%, while for the case of the coefficients which describe the surface shape (r and k), the maximum percentage of error was less than 0.3%. With our method, we eliminated the errors that are introduced during the integration process, therefore, our percentage error is reduced in a great manner. On the other hand, this algorithm can be modified for the analysis of surfaces of any size and shape due the hardiness of our method, where we can modify the parameters of the surface that we want to analyze, i.e. from surfaces of the human eye to the new generation of telescopes surfaces.

6. References

[1] Díaz-Uribe R and Campos-García M 2000 *Appl. Opt.* **39** 2670-2677
[2] Díaz-Uribe R 2000 *Appl. Opt.* **39** 2790-2804
[3] Díaz-Uribe R, Campos-García M and Granados-Agustín F S 2002 *Proc. SPIE* **4818** 63
[4] Moreno-Oliva V, Campos-García M, Román-Hernández E and Santiago-Alvarado A 2014 *Opt. Eng.* **53** 11

IOP Conf. Series: Journal of Physics: Conf. Series **792** (2017) 012029 doi:10.1088/1742-6596/792/1/012029

Performance simulation of a radial flow type impeller of centrifugal pumps using CFD

R López, M Vaca, H Terres, A Lizardi, S Chávez and M García.
Universidad Autónoma Metropolitana -Azcapotzalco
Av. San Pablo No. 180, Col. Reynosa-Tamaulipas Del. Azcapotzalco, México, 02200, D.F. México,

E-mail: arlr@correo.azc.uam.mx

Abstract. The numerical simulation of a centrifugal impeller that had previously been designed and manufactured is presented in this paper. The following operating conditions were determined: 0.50 m^3/min volumetric flow at a load of 25 m, velocity of rotation of 1750 rpm, and specific velocity of 0. 27. The ANSYS CFX 14.5 software with the k-ε turbulence model was used for simulation with appropriate boundary conditions. The distributions of velocities in the flow field in addition to the distribution of pressures on the entire impeller were obtained. The simulation showed no negative values for the pressure at the entrance of the impeller. The curve of hydrodynamic behaviour of the impeller, which contains the point of operation in which the pump will work was also developed.

1. Introduction

A wide variety of centrifugal pump types has been designed and constructed for uses in industries, agriculture, and domestic applications, but their design and performance prediction process is still a difficult task. This is due to the number of free geometric parameters, because their effect cannot be directly evaluated. Experimental model testing is one of the solutions for prediction of performance but it is tedious, time consuming, and costly. The theoretical method gives a value, but it is unable to determine the cause of the poor performance. Recently computational fluid dynamics (CFD) analysis has been used in the design and construction stages of this turbomachinery [1-3]. The significant costs and time of the trial-and-error process incurred when constructing and testing physical prototypes, reduces the profit margins of the pump manufacturers. In contrast, numerical simulation can provide accurate information on the fluid behavior inside the machine, and thus help the engineer to obtain a performance evaluation of a particular design. Previous numerical studies were focused on the study of the off-design performance of centrifugal pump impellers; the complex internal flows through the different components of pumps can be studied at different operating conditions, which help in the performance improvement [4]. The impeller curve profile and stator angle too have been numerically

Content from this work may be used under the terms of the Creative Commons Attribution 3.0 licence. Any further distribution of this work must maintain attribution to the author(s) and the title of the work, journal citation and DOI.
Published under licence by IOP Publishing Ltd

analyzed for, with the conclusion that the nature of the head and the power versus capacity curves obtained was similar to that of standard mixed flow pumps and the efficiency was predicted within ± 5% range at the working point [5]. The objective of this work is to simulate and analyze through CFD the impeller performance of a centrifugal pump with outlet diameter of 0.15 m, 0.05 m inlet diameter and 6 blades, this impeller was designed and constructed for a specific velocity of 0.27.

2 Mathematical model

For three-dimensional incompressible flow in the impeller, a continuity equation is:

$$\nabla \cdot \vec{V} = 0 \tag{1}$$

where $\vec{V}$ is the velocity vector, and the moment equation is:

$$\rho[\vec{V} \cdot (\nabla\vec{V})] = -\nabla p + \mu_{ef}(\nabla^2\vec{V}) \tag{2}$$

where ρ is the density of the fluid, μ_{ef} is the viscosity of the fluid, and p is the pressure. These equations were solved using ANSYS CFX 14.5 software and the $k - \varepsilon$ turbulence model.

3 Impeller geometry

The geometric characteristics of impeller used in the simulation are presented in table 1. This impeller was designed and cast in carbon steel, the model of the impeller can be observed in Fig. 1. The design parameters were 0.50 m^3/min volumetric flow, 25 m head, velocity of rotation of 1750 rpm, and specific velocity of 0. 27. A sectioning of the real impeller showing the position of the blades is presented in Fig. 2.

Fig. 1. Radial-flow impeller model.

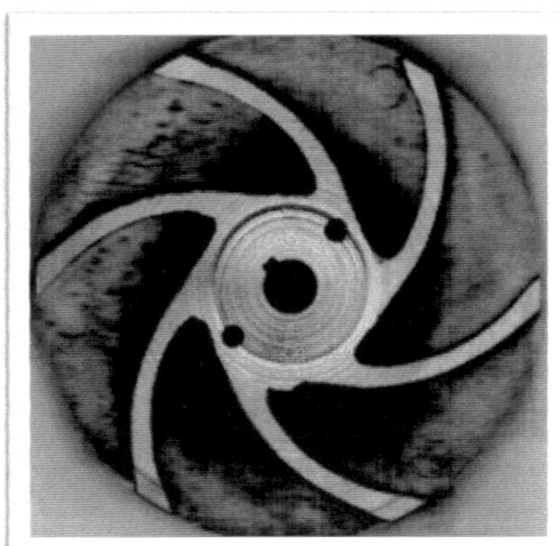

Fig. 2. Sectioning of the real impeller showing the position of the blades.

IOP Conf. Series: Journal of Physics: Conf. Series **792** (2017) 012029 doi:10.1088/1742-6596/792/1/012029

The impeller configured in the software is presented in Fig. 3. The displayed channels are those corresponding to the clearance between the impeller vanes, ochre-colored section is the space through which fluid is flowing. The section indicated with red color corresponds to the eye of the impeller where the fluid enters and the blue area is the outlet section. The total number of nodes generated for the control volume including all sections is 70 000, the mesh is shown in Fig. 4. The boundary conditions used in the solution can be observed in Fig. 5.

Table 1 Physical dimensions of the impeller.

Inlet diameter	(D_1)	0.052 m
Outlet diameter	(D_2)	0.15 m
Blades number		6
Inlet angle	(β_1)	46°
Outlet angle	(β_2)	24°
Inlet thickness	(b_1)	0.012 m
Outlet thickness	(b_2)	0.007 m

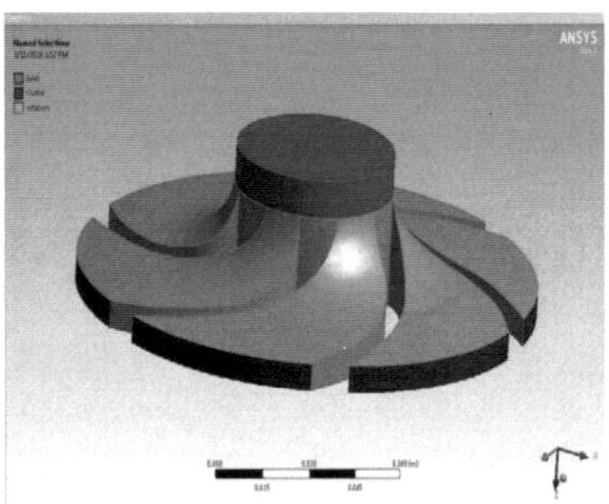

Fig. 3. Configuration of the impeller, generated by the software.

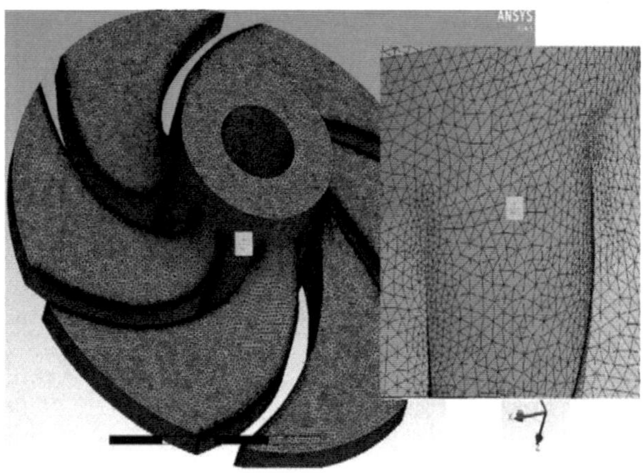

Fig. 4. Meshing used in the impeller.

Table Domain Physics for Copy of CFX

Domain - Default Domain	
Type	Fluid
Location	B210
Materials	
Water	
Fluid Definition	Material Library
Morphology	Continuous Fluid
Settings	
Buoyancy Model	Non Buoyant
Domain Motion	Rotating
Angular Velocity	-3.5500e+03 [rev min^-1]
Axis Definition	Coordinate Axis
Rotation Axis	Coord 0.3
Reference Pressure	0.0000e+00 [atm]
Heat Transfer Model	Isothermal
Fluid Temperature	2.5000e+01 [C]
Turbulence Model	k epsilon
Turbulent Wall Functions	Scalable

Table Boundary Physics for Copy of CFX

Domain	**Boundaries**	
Default Domain	**Boundary - Inlet**	
	Type	INLET
	Location	Inlet
	Settings	
	Flow Direction	Normal to Boundary Condition
	Flow Regime	Subsonic
	Mass And Momentum	Stationary Frame Total Pressure
	Relative Pressure	0.0000e+00 [Pa]
	Turbulence	Medium Intensity and Eddy Viscosity Ratio
	Boundary - Outlet	
	Type	OUTLET
	Location	Outlet
	Settings	
	Flow Regime	Subsonic
	Mass And Momentum	Mass Flow Rate
	Mass Flow Rate	8.3333e+00 [kg s^-1]
	Boundary - Fixed	
	Type	WALL
	Location	Fixed
	Settings	
	Mass And Momentum	No Slip Wall
	Wall Velocity	Counter Rotating Wall
	Wall Roughness	Smooth Wall
	Boundary - Rotation	
	Type	WALL
	Location	Botton, Top, Blades
	Settings	
	Mass And Momentum	No Slip Wall
	Wall Roughness	Smooth Wall

Fig. 5. The boundary conditions.

4 Results

The distribution of the fluid velocity inside the impeller is shown in Fig. 6. The fluid that is in contact with the blade surface on the left side shows the highest values produced as a result of vortices formation, which are best appreciated when the flow lines are plotted on the opposite side, with some areas of stagnation with very low velocity, Fig. 7. The velocity of the fluid at the impeller inlet is practically constant, but once the fluid contacts the blade and moves away from the impeller eye it acquires a certain velocity and the aforementioned vortices arise, this behavior is reported in [3-6].
Pressure distribution shown in Fig. 8, the value in the impeller eye is between 0.00 and 19.00 kPa, it increases as the fluid moves away from the eye up to a maximum value in the periphery of 380.00 kPa. Note that in this case values of vacuum are not reached in the eye of the impeller. This distribution is better appreciated when performing a meridional sectioning to the impeller, as the vertical axis is representative of the radius, the influence of the increase of the radius is clear; such behavior is reported in same previous references.

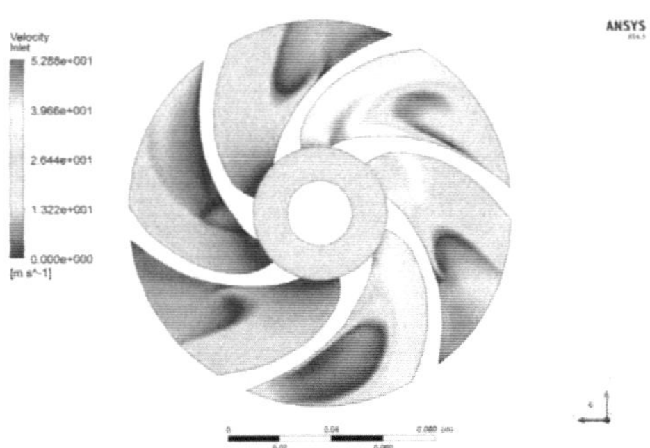

Fig. 6. Field of velocities

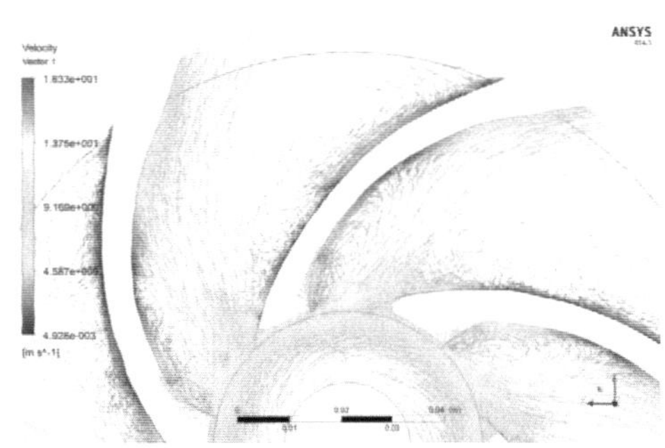

Fig. 7. Field of velocity vectors in the impeller.

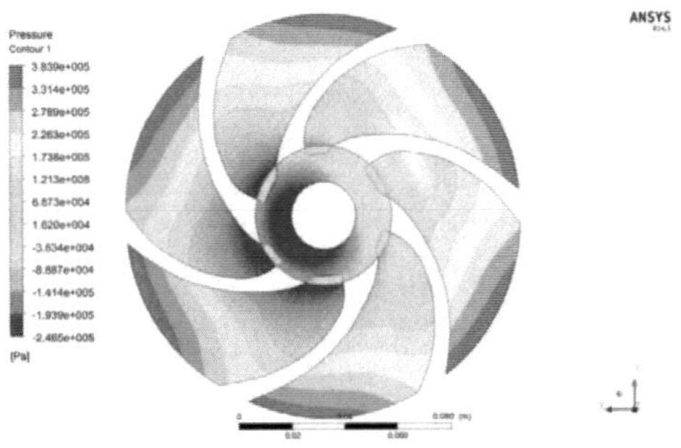

Fig. 8. Pressure distribution in the impeller

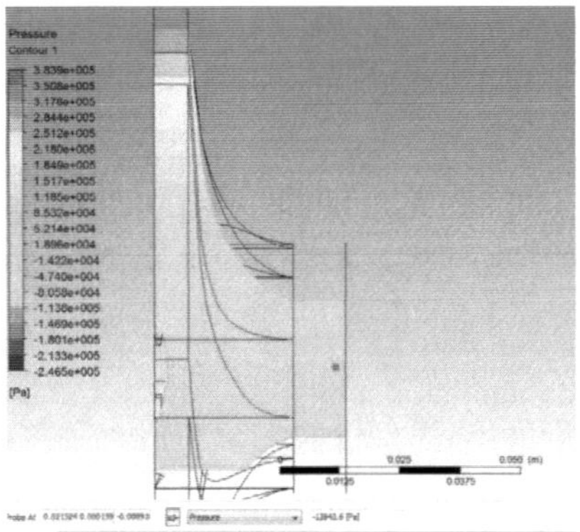

Fig. 9. Pressure distribution in the impeller, meridional sectioning.

With the calculated values the impeller head vs. flow curve is obtained. This curve is used for inspecting the hydrodynamic behavior of the impeller, the results are presented in Fig. 10. The maximum load is 38.0 m, this value occurs when the volumetric flow passing through the impeller is null. The program is made to vary the volumetric flow from zero to a value a little beyond the point of design, in this case the design flow is 0.500 m^3/ min, with 25 m head, the maximum value selected is 1.2 m^3/min. Since the curve is theoretical, slope is substantially constant.

5 Conclusions

The numerical simulation of a centrifugal impeller that had previously been designed and manufactured was performed, the volumetric flow rate is 0.50 m^3/min and a head of 25 m, the velocity of rotation is 1750 rpm and the specific velocity is 0. 27. For simulation, the ANSYS CFX 14.5 software with the k-ε turbulence model and appropriate boundary conditions was used. The distributions of velocities in the flow hood as well as the distribution of pressures all over the impeller were obtained. The simulation showed no negative values for the pressure at the impeller inlet. With the results obtained the hydrodynamic behavior curve of the impeller, which contains the operation point in which the pump will work, was calculated.

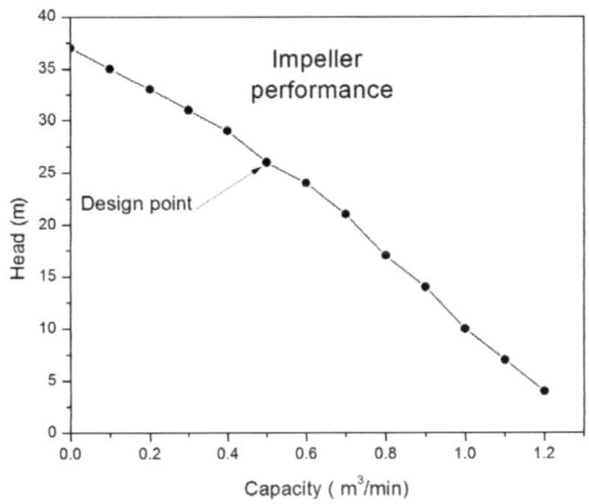

Fig. 10. Curve of the hydrodynamic behavior of the impeller.

6 References

[1] Kaewnai S, Chamaoot M and Wongwises S 2009. Predicting performance of radial flow type impeller of centrifugal pump using CFD *J. Mech. Sci. Techno.* **23**, pp 1620-1627.

[2] Jafarzadeh B, Hajari A, Alishahi M M and Akbari M H 2011 The flow simulation of a low-specific-velocity high-velocity centrifugal pump *Appl. Math. Mode.* **35** pp 242-249.

[3] Sung K, Kyoung-Yong L, Joon-Hyung K, Jin-Hyuk K, Uk-Hee J and Young-Seok C 2015. High performance hydraulic design techniques of mixed-flow pump impeller and diffuser. *J. Mech. Sci. Techno.* **29**, pp 227-240.

[4] Mentzos M, Filios A, Margaris P and Papanikas D 2005. CFD predictions of flow through a centrifugal pump impeller. Proceedings of International Conf. Experiments/Process/System Modelling/ Simulation/Optimization. Athens, pp. 1-8.

[5] Patel K and Ramakrishnan N, 2006. CFD analysis of mixed flow pump. International ANSYS Conference Proceedings.

[6] Anagnostopoulos J, 2006. CFD Analysis and Design Effects in a Radial Pump Impeller WSEAS Transactions on Fluid Mechanics, **1,** pp 763-770

Micrometric displacement sensor based on the strain of a fiber Bragg grating with heterodyne detection of intensity in a Mach-Zehnder interferometer.

H Gonzalez [1, 2], F J Alvarez [1, 2] and G E Sandoval. [2]

[1]Programa de Maestría y Doctorado en Ingeniería, Universidad Nacional Autónoma de México, Ciudad Universitaria, Av. Universidad 3000, Primer piso Edificio T "Bernardo Quintana Arrioja", Coyoacán, CP 04510, Ciudad de México.

[2]Centro de Ciencias Aplicadas y Desarrollo Tecnológico, Universidad Nacional Autónoma de México, Ciudad Universitaria, Apdo. Postal 70-186, Coyoacán, CP 04510, Ciudad de México.

hely.gonzalez@ccadet.unam.mx

Abstract. This paper describes the implementation and characterization of a displacement sensor according to the strain of a fiber Bragg grating. The system is able to obtain measurements on micrometer scales through heterodyne detection of intensity in a Mach-Zehnder interferometry in fiber optic. For the displacement characterization, the fiber Bragg grating was attached from one end meanwhile at the other end a tension was induced by a high-precision rotation mount. Experimental results are presented for which varying the angular displacement we obtain a linear response in the amplitude of the intensity, determining a sensitivity of 0.77 pW/μm with a displacement size of 48.80 μm in a dynamic range from 0 μm to 488 μm.

1. Introduction

The fiber optic, besides the data transmission through long distances, is able to provide additional information about the environment where it is located. This characteristic is possible by measurements in the properties of the confined light as intensity, wavelength, phase and polarization [1]. Most fiber optic sensors based their operation on interferometric principles, being the Mach-Zehnder interferometer the setting most used owing to they have extremely flexible geometries and high sensitivities that allow the possibility of sensing a wide variety of parameters [2, 3]. The Mach-Zehnder interferometer is an optical instrument capable of high-resolution phase analysis [4]. The basic elements of a fiber optic Mach-Zehnder interferometer consist in a spectral light source to feed the device, a fiber beam splitter to produce two different optical paths; the sensing and reference path. A transducer element is placed on each interferometer arm where one will be under the environmental

Content from this work may be used under the terms of the Creative Commons Attribution 3.0 licence. Any further distribution of this work must maintain attribution to the author(s) and the title of the work, journal citation and DOI.
Published under licence by IOP Publishing Ltd

effects whereas the other arm will be isolated to be used as a reference. Finally homodyne or heterodyne demodulator is used to estimate the change in optical paths which is a function of the physical variable to be measured [5]. The transducer most used to detect strain is the fiber Bragg grating (FBG) and based on this parameter the displacement is indirectly measured. The FBG is a periodic variation of the refractive index into the core of an optical fiber. This periodic structure will reflect a specific optical wavelength dependent on the periodicity. To characterize the FBG as displacement sensor, as a strain is applied to the FBG periodicity change, vary the periodicity vary the reflected wavelength known as Bragg wavelength [6]. The strain applied to the FBG is expressed in microstrains (με) which is a dimensionless unit that roughly describes the elongation of 1 μm in an optical fiber with a length of 1 m [5].

The electromagnetic waves at the output of the interferometer are detected with a photodetector and the generated photocurrent is given by the square of the total electric field, this is described by Equation 1.

$$I(t) = |A_R \cos(\omega_R t + \phi_R) + A_S \cos(\omega_S t + \phi_S)|^2 \tag{1}$$

Where the subscripts *R* and *S* indicate that each parameter describes the reference and sensing wave respectively. A is the amplitude of each wave, ω is the angular frequency, while ϕ is the relative phase for the reference and sensing wave. When a strain is applied each interferometer arm experiences an increase in relative phase, which is expressed in Equation 2.

$$\phi = \frac{2\pi n_{eff} L}{\lambda_B} \tag{2}$$

The relative phase of an optical beam depends on the optical pathlength [1]. n_{eff} is the effective refractive index of the optical fiber, L indicates the length of the optical path and λ_B is the Bragg wavelength of the FBG. We dropped the losses effect by fiber couplers. Therefore, the interferometric output signal in steady state is given by

$$I(\phi) = A\cos(\Delta\phi) \tag{3}$$

Where

$$\Delta\phi = \phi_R - \phi_S \tag{4}$$

The amplitude A depends on A_R, A_S and the responsivity of the photodetector [7]. As a strain is applied the phase shifts change and this is clearly observable in the intensity detected. Substituting Equation 2 in Equation 4 and the resulting of it in Equation 3, we obtain

$$I(\phi) = A\left[\cos\left(\frac{2\pi n_{eff}(\lambda_{BS}L_R - \lambda_{BR}L_S)}{\lambda_{BR}\lambda_{BS}}\right)\right] \tag{5}$$

Where λ_{BR} and λ_{BS} are the Bragg wavelength in nm of the reference FBG and the sensing FBG, respectively. Meanwhile L_R and L_S are the pathlength of each interferometer arm.

2. Materials and methods

A FBG of OEMARK brand was used in each arm of the Mach-Zehnder interferometer. By means of an interrogation system O/E Land OEFSS-200a we characterize each FBG to know their properties. The reference FBG has a Bragg wavelength of 1549.50 nm and the Bragg wavelength of the sensing FBG was 1550.08 nm. A strain was applied in the sensing FBG in order to appreciate the Bragg wavelength shift. The fiber optic interferometer, as shown in Figure 1, was built with a single mode fiber SMF-28 and a super-luminescent diode (SLD) was used as a source. The SLD has a spectral range from 1535 nm to 1565 nm and a central wavelength of 1547 nm. The optical power used was of 8 mW. To compose the interferometers arms we used a fiber splitter (C1), later the sensing FBG was coupled. In a second splitter (C2) the reference FBG was coupled. The reflected wavelength in each FBG contains a part of the spectrum and it was combined by a 2x1 coupler (C3). The intensity at the exit of the coupler C3 was measured by a Thorlabs S155C photodetector (PD). Through a Thorlabs PM100USB interface we send the reading data to PC where by an application developed by Thorlabs in LabVIEW, and improved for us, the signal conditioning and processing of information in a database is achieved.

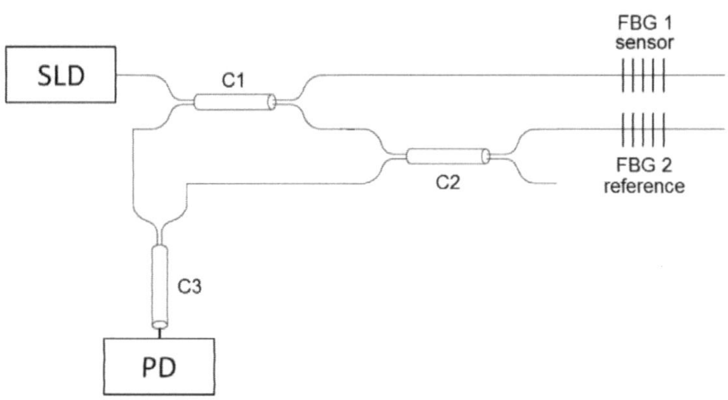

Figure 1. Experimental setup.

To displacement characterization based on the FBG strain, we attached with scotch tape the ends of a FBG section with a length of 27 cm. One end was fixed on an optical post assembly meanwhile the other end was attached on a high-precision rotation mount in order to strain the FBG section. The minimum angular displacement unit of the high-precision rotation mount was equivalent to a strain of 90.37 με and in turn this was equivalent to a linear displacement of 24.40 μm.

3. Experimental results

The reflectivity spectral of the sensing and reference FBG were obtained by means of an interrogation scheme. Figure 2 shows the reflectivity spectral of each FBG.

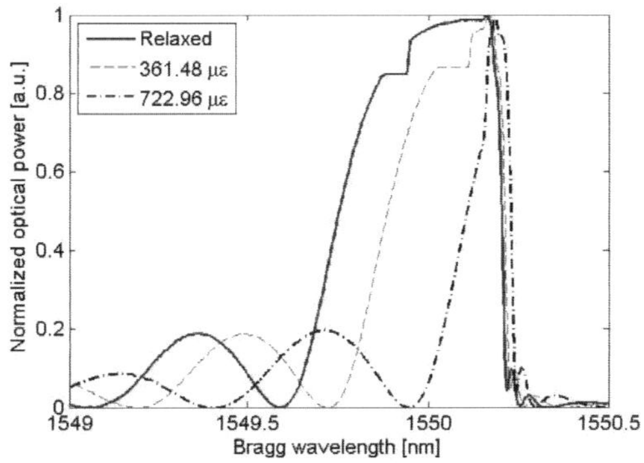

Figure 2. Reflectivity spectral of the sensor FBG and reference FBG.

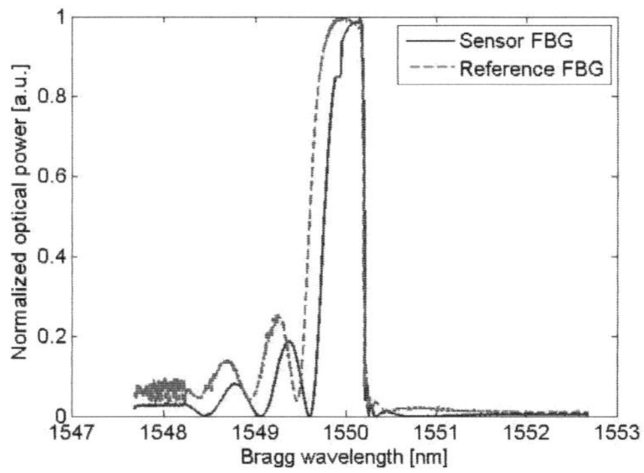

Figure 3. Displacement of Bragg wavelength in function of the strain applied in the sensor FBG.

The reflectivity spectrum was normalized owing to splices in the optical fiber the optical power amplitude varies in each FBG. The displacement of Bragg wavelength is measurable in function of the applied strain. In Figure 3 we appreciate the reflectivity spectral and the Bragg wavelength when the FBG is relaxed but whenever a measurable strain is applied the reflectivity spectral is reduced and a shift occurs in the Bragg wavelength.

To know the relation between the applied strain and the shift Bragg wavelength we realized a series of measurements with the sensor FBG. In a strain range from 0 με to 903.70 με with a strain size of 90.37 με, shown in Figure 4, an almost linear behaviour was obtained that fits in Equation 6.

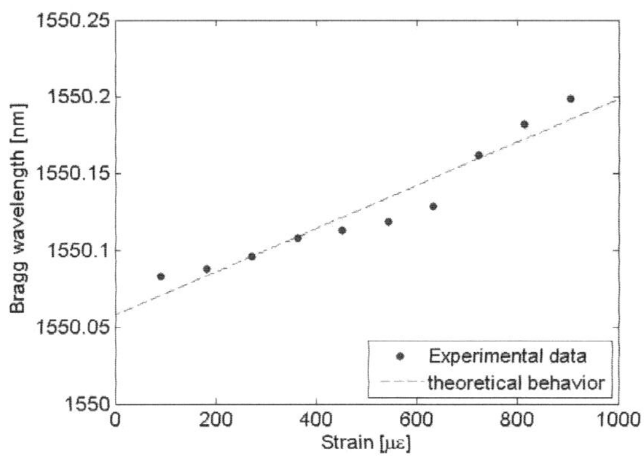

Figure 4. Bragg wavelength in function of applied strain.

$$\lambda_B(\mu\varepsilon) = 0.00014 \left[\frac{\mathrm{nm}}{\mu\varepsilon}\right] x[\mu\varepsilon] + 1550.06\ [\mathrm{nm}] \quad (6)$$

Knowing the relation between strain and shift wavelength we demonstrated that the minimum appreciable displacement was a strain of 180.74 με equivalent to a lineal displacement of 48.80 μm. We realize the characterization of the sensor through a series of measurements, as shown in Figure 5, which consisted in step size of 180.74 με for a total of 10 steps and returning to the relax state of the sensing FBG. For a linear fitting we did an average of every intensity data step in order to obtain a mathematical expression of the sensor behaviour.

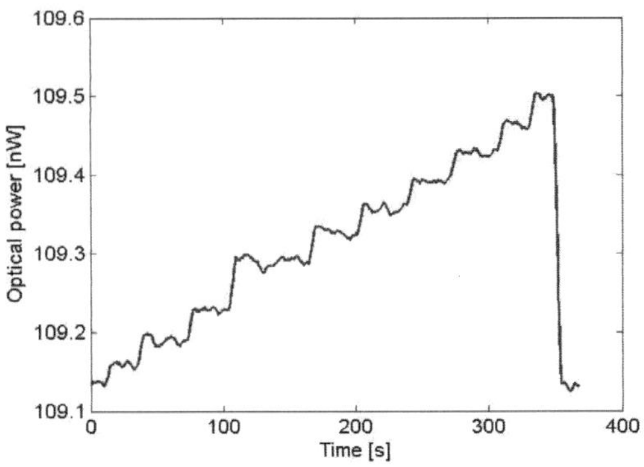

Figure 5. Measurement series from 180.74 με to 1807.40 με.

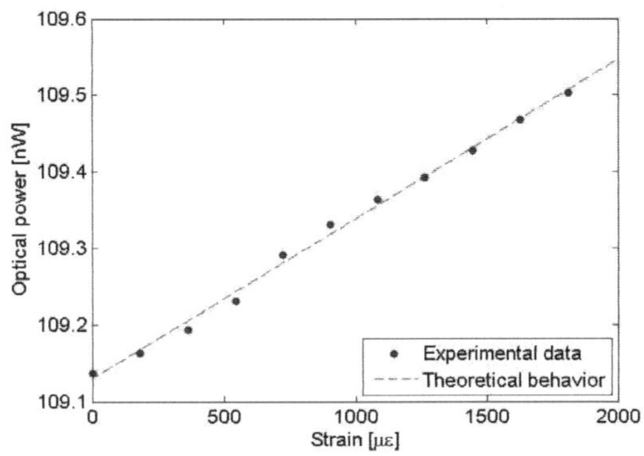

Figure 6. Linear fitting of the displacement sensor based on strain.

With a dynamic range from 0 με to 1807.40 με and strain size of 180.74 με, we obtain another linear behaviour described by Equation 7.

$$I(\mu\varepsilon) = 0.00021 \left[\frac{\mathrm{nW}}{\mu\varepsilon}\right] x[\mu\varepsilon] + 109.13\ [\mathrm{nW}] \tag{7}$$

An interpretation of Equation 7 is that for a strain in the sensing FBG of 1 με we obtain a change in the intensity of 0.21 pW at the exit of the Mach-Zehnder interferometer. This change was imperceptible but to appreciate a considerable change, as shown in Figure 5, we applied a strain of 180.74 με and in turn this was equivalent to an intensity change of 37.95 pW. For a description of the intensity in function of the lineal displacement, we obtained a characteristic expression described by Equation 8.

$$I(\mu m) = 0.00077 \left[\frac{\mathrm{nW}}{\mu\mathrm{m}}\right] x[\mu\mathrm{m}] + 109.13\ [\mathrm{nW}] \tag{8}$$

In the same way a linear behaviour was appreciated with a sensibility of 0.77 pW/μm, where the minimum lineal displacement was 48.80 μm in a dynamic range from 0 μm to 488 μm.

4. Conclusions

The interferometric sensor for strain described bellows provides a simple method that allows us the possibility to measure other parameters in the same system. Experimental results show that this technique has an acceptable sensibility. Mathematically we obtained a sensibility of almost 0.21

pW/με but the resolution that the system detect is as 180 times the obtained sensibility. In a displacement characterization we obtain an indirect sensibility 0.77 pW/μm. The system resolution depends on the photodetector responsivity but if we want improve our resolution we need to considerate all the variables that are introducing noise in the system. Furthermore the fiber optic interferometric system has a lot of fiber splices and couplers who generate losses that affects it performance.

Acknowledgment
This work was supported by DGAPA-UNAM through financing of project PAPIIT IT-101515. Besides we thank to Consejo Nacional de Ciencia y Tecnologia (CONACyT) for its financial support in form of a scholarship for the Master and PhD Engineering Program at Universidad Nacional Autonoma de Mexico.

References

[1] Bellil H and Abushagur M 2000 *Heterodyne detection for Fiber Bragg Grating sensors, Optics & Laser Technology* **32** 5 pp. 297–300 .
[2] Ghatak A and Thyagarajan K 2000 *Introduction To Fiber Optics* 2nd edition Cambridge University Press
[3] Maity A B 2013 *Optoelectronics and Fiber Optics Sensors*, PHI Learning
[4] JacksonD A Dandridge A and Sheem S K Measurement of small phase shifts using a single-mode optical-fiber interferometer 1980 Opt Lett **5**(4) pp. 139–141
[5] Yin S, Ruffin P B and Yu F T S 2008 *Fiber Optic Sensors* 2nd edition CRC Press
[6] Cusano A, Cutolo A and Albert J 2011 *Fiber Bragg Grating Sensors: Recent Advancements, Industrial Applications And Market Exploitation* Bentham Books
[7] Wu Q et. Al 2010 *High resolution temperature insensitive interrogation technique for FBG sensors, Optics & Laser Technology* **42** pp. 653–656 (2009).

IOP Conf. Series: Journal of Physics: Conf. Series **792** (2017) 012031 doi:10.1088/1742-6596/792/1/012031

Decaying steady states of the one-dimensional nonlinear Schrödinger equation with a constant potential

Torres-Vega G

Physics Department, Cinvestav, Apdo. postal 14-740, 07000 Mexico city, Mexico

gabino@fis.cinvestav.mx

Abstract. We introduce a set of non-linear functions of real, exponential type that can be useful for dealing with some physical problems. We use these functions to find steady states of the one-dimensional non-linear Schrödinger equation for a condensate between an infinite barrier and a step potential in the case where the height of the step is larger than the chemical potential of the condensate.

1. Introduction

The present work is concerned with the solutions to the nonlinear Schrodinger equation in one dimension[1,2] of the real, exponential type. The nonlinear Schrödinger equation models wave propagation phenomena in Bose-Einstein condensates as well as many other systems.[3-7] A large amount of early work was performed on the steady states of the nonlinear Schrödinger equation for a free condensate in several dimensions (see Ref. [8] and references there in), such as a spherically symmetric, positive function that decays for large distances and is related to the existence, uniqueness and stability of the ground state.

A set of real, periodic, steady states of the one-dimensional nonlinear Schrödinger equation is given in terms of Abel's and elliptic functions[9-19] and has been used to find the stationary states of a Bose-Einstein condensate in a box[17,18] and in a double square well[20] when the chemical potential of the Bose condensate is larger than the height of the potential barrier encountered by the condensate. These solutions include bright, gray and dark soliton solutions as special cases. A linear superposition of Jacobi's elliptic functions with constant terms added to their arguments is one useful result of this approach.[11]

However, non-periodic, decaying solutions can also be useful. These functions would be nonlinear versions of the functions $e^{\pm x}$ and are also similar to the elliptic functions. These functions can be useful for situations such as when the condensate encounters step potentials and potential barriers.

In section 2, we introduce non-linear exponential-type functions and determine some of their properties. Later, in section 3, we use these functions to find the steady states for a Bose-Einstein condensate between an infinite barrier and a step potential. In the final section, we offer some concluding remarks.

2. Non-linear exponential-type functions

Let us introduce the following nonlinear functions of exponential type:

Content from this work may be used under the terms of the Creative Commons Attribution 3.0 licence. Any further distribution of this work must maintain attribution to the author(s) and the title of the work, journal citation and DOI.
Published under licence by IOP Publishing Ltd

$$pn(u)=e^x \ , \qquad mn(u)=e^{-x} \ , \qquad fn(u)=a e^x+b e^{-x} \ , \tag{1}$$

$$gn(u)=a e^x - b e^{-x} \ , \qquad rn(x)=\sqrt{1+m(a e^x - b e^{-x})^2} \ , \qquad nf(u)=\frac{1}{fn(u)} \ , \tag{2}$$

$$ng(u)=\frac{1}{gn(u)} \ , \quad nr(u)=\frac{1}{rn(u)} \ , \tag{3}$$

with u and x related as

$$u=\int_0^x \frac{dt}{\sqrt{1+m(a e^x - b e^{-x})^2}} \ , \tag{4}$$

where $m>0$ to avoid branching points.

We note that $rn(u)\neq rn(-u)$ and then that $mn(u)$ is not the mirror image of $pn(u)$, i.e., $mn(u)\neq pn(-u)$ unless $a=b$. A plot of these functions is found in Fig. 1 for a set of values of the parameters a, b and m. The values of a and b are related to the mirror symmetry between the functions $pn(u)$ and $mn(u)$, with $a=b$ as the more symmetric case (which would be the case of Jacobi's elliptic functions with complex arguments). The value of m causes these functions to decay or increase more rapidly than the regular exponential functions. The domain of these functions is finite unless $m=0$; in fact, increasing the magnitude of x beyond, for instance, $\ln(10^4/2a\sqrt{m})$ does not significantly increase the magnitude of u. One can extend the domain of these functions by setting the value of the function to zero or infinity for larger u V , making them non-periodic functions on the real axes. We also note that some of these functions are actually bounded.

We can easily verify the following properties that are similar to those for the elliptic functions. The squares of these functions are related as

$$4ab=fn^2(u)-gn^2(u) \ , \tag{5}$$

$$rn^2(u)-1=m\,gn^2(u)=m[rn^2(u)-4ab] \ , \tag{6}$$

$$fn(u)\,gn(u)=a^2 pn^2(u)-b^2 mn^2(u) \ , \tag{7}$$

$$fn^2(u)+gn^2(u)=2[b^2 mn^2(u)+a^2 pn^2(u)] \ , \tag{8}$$

The derivatives of these functions are

$$pn'(u)=pn(u)\,rn(u) \ , \qquad mn'(u)=mn(u)\,rn(u) \ , \qquad fn'(u)=gn(u)\,rn(u) \ , \tag{9}$$

$$gn'(u)=fn(u)\,rn(u) \ , \qquad rn'(u)=m\,fn(u)\,gn(u) \ , \qquad nf'(u)=-gn(u)\,nf^2(u)\,rn(u) \ , \tag{10}$$

$$ng'(u)=-fn(u)\,ng^2(u)\,rn(u) \ , \qquad nr'(u)=-m\,fn(u)\,gn(u)\,nr^2(u) \ , \tag{11}$$

As we can see from these derivatives, the rate of increase or decrease of the functions is not constant along u and would be constant only for the case of $m=0$.

IOP Conf. Series: Journal of Physics: Conf. Series **792** (2017) 012031 doi:10.1088/1742-6596/792/1/012031

We also find that the derivatives of the inverse functions are given by

$$\frac{d\,pn^{-1}(y)}{dy}=\frac{1}{\sqrt{y^2+m\left(a\,y^2-b\right)^2}}\ ,\quad \frac{d\,mn^{-1}(y)}{dy}=\frac{-1}{\sqrt{y^2+m\left(a-b\,y^2\right)^2}}\ , \tag{12}$$

$$\frac{d\,gn^{-1}(y)}{dy}=\frac{1}{\sqrt{\left(y^2+4ab\right)\left(1+m\,y^2\right)}}\ ,\quad \frac{d\,nf^{-1}(y)}{dy}=\frac{-1}{\sqrt{\left(1-4\,ab\,y^2\right)\left(n_2\,y^2+m\right)}}\ , \tag{13}$$

$$\frac{d\,ng^{-1}(y)}{dy}=\frac{-1}{\sqrt{\left(1+4\,ab\,y^2\right)\left(y^2+m\right)}}\ ,\quad \frac{d\,nr^{-1}(y)}{dy}=\frac{-1}{\sqrt{\left(1-y^2\right)\left(1-n_2\,y^2\right)}}\ , \tag{14}$$

where

$$n_1=1-8\,mab\ ,\quad n_2=1-4\,mab\ ,\quad n_3=1-2\,mab\ ,\quad n_4=1+4\,mab\ . \tag{15}$$

As expected, from these derivatives, we can see that these functions also invert the same integral functions as Jacobi [9,12].

The second derivatives are

$$pn''(u)-pn(u)\left[n_3+2m\,a^2\,pn^2(u)\right]=0\ ,\qquad mn''(u)-mn(u)\left[n_3+2\,m\,b^2\,mn^2(u)\right]=0\ , \tag{16}$$

$$fn''(u)-fn(u)\left[n_1+2\,m\,fn^2(u)\right]=0\ ,\qquad gn''(u)-gn(u)\left[n_4+2\,m\,gn^2(u)\right]=0\ , \tag{17}$$

$$rn''(u)+2\,rn(u)\left[n_3-rn^2(u)\right]=0\ ,\qquad fn''(u)-nf(u)\left[n_1-8\,ab\,n_2\,nf^2(u)\right]=0\ , \tag{18}$$

$$ng''(u)-ng(u)\left[n_4+8\,ab\,ng^2(u)\right]=0\ ,\qquad nr''(u)-2\,nr(u)\left[n_2\,nr^2(u)-n_3\right]=0\ , \tag{19}$$

Then, the functions that we have just introduced are solutions of the nonlinear second order differential equations with the one-dimensional Gross-Pitaevskii equation form for a constant potential and real functions.

Additionally, the corresponding energy, or the Liapunov functions are given by

$$\left[pn'(u)\right]^2-pn^2(u)\left[n_3+m\,a^2\,pn^2(u)\right]=m\,b^2\ , \tag{20}$$

$$\left[mn'(u)\right]^2-mn^2(u)\left[n_3+m\,b^2\,mn^2(u)\right]=m\,a^2\ , \tag{21}$$

$$\left[fn'(u)\right]^2-fn^2(u)\left[n_1+m\,fn^2(u)\right]=-4\,ab\,n_2\ , \tag{22}$$

$$\left[gn'(u)\right]^2-gn^2(u)\left[n_4+m\,gn^2(u)\right]=4\,ab\ , \tag{23}$$

$$\left[rn'(u)\right]^2+2\,rn^2(u)\left[n_3-rn^2(u)\right]=n_2\ , \tag{24}$$

$$\left[nf'(u)\right]^2-nf^2(u)\left[n_1-4\,ab\,n_2\,nf^2(u)\right]=m\ , \tag{25}$$

$$\left[ng'(u)\right]^2-ng^2(u)\left[n_4+4\,ab\,ng^2(u)\right]=m\ , \tag{26}$$

$$\left[nr'(u)\right]^2-nr^2(u)\left[n_2\,nr^2(u)-2\,n_3\right]=1\ , \tag{27}$$

IOP Conf. Series: Journal of Physics: Conf. Series **792** (2017) 012031 doi:10.1088/1742-6596/792/1/012031

where we have made use of the relationships between the squares of the functions. We note that the functions $nf(u)$ and $ng(u)$ have the same energy, whereas the functions $pn(u)$ and $mn(u)$ would have the same energy if $b=a$.

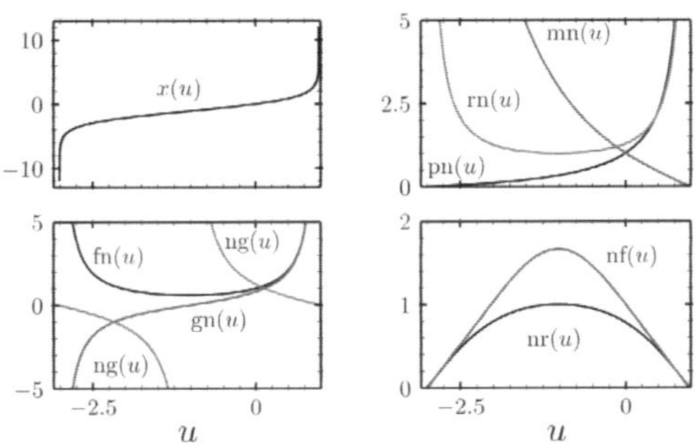

Fig. 1. Nonlinear exponential-type functions for $m=1$, $a=0.1$ and $b=0.9$.

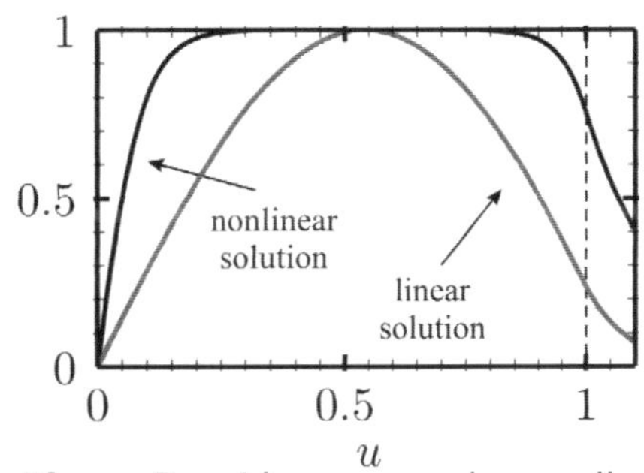

Fig. 2. Linear and nonlinear steady states (unnormalized), for the wall-step potential. The dashed vertical line indicates where the start of the step potential. Parameter values are $a=-0.3$, $m=0.9$, and $k_l=5.4$.

When $a=b=1$, the nonlinear functions are reduced to Jacobi's elliptic functions with a complex argument,

$$u=\int_0^x \frac{dt}{\sqrt{1+4m\sinh^2(t)}}=-iF(ix;4m) \, , \tag{28}$$

where F is elliptic integral of the first kind.

In the solutions of physical problems, these functions can be chosen when functions that are small in one region of the real axes and large in other region are required or when functions that are non-zero at the ends of some interval of the real axes are required. They can also be useful when functions that decay at both ends of an interval are necessary.

This is the minimum set of properties of the functions that are necessary to give exponential-type solutions of the one-dimensional nonlinear Schrödinger equation with a constant potential. In the next section, we apply these functions to this case.

3. Wall-step potential and the nonlinear Schrödinger equation

Similarly to the treatment of the linear Schrödinger equation, we can use the functions $pn(u)$ and $mn(u)$ (now together with $rn(u)$) to solve the nonlinear Schrödinger equation when the condensate encounters or leaves a step potential, whereas the functions $fn(u)$ and $gn(u)$ (and $rn(u)$) can be used when the condensate is crossing a potential barrier.

Now, let us use these functions to find the steady state for the one-dimensional Gross-Pitaevskii equation with a wall and a step potential, i.e., with the potential function given by

$$V(u)=\begin{cases}\infty, & x<0,\\ 0, & 0\le x\le 1,\\ V_0, & 1<x.\end{cases} \tag{29}$$

Here u is a dimensionless length variable. The height of the step is larger than the chemical potential of the condensate, $\mu<V_0$.

We write the time independent Gross-Pitaevskii equation as

$$\frac{d^2\psi(u)}{du^2}+\frac{2ML^2}{\hbar^2}\left(\mu-V_0-\frac{NU_0}{A^2}|\psi(u)|^2\right)\psi(u)=0, \tag{30}$$

where $\psi(u)$ is the unnormalized steady state for the Bose-Einstein condensate (BEC), M is the mass of a single atom, N is the number of atoms in the condensate, $U_0=4\pi\hbar^2 a/M$ characterizes the atom-atom interaction, a is the scattering length, L is a scaling length corresponding to the width of the potential well, A^2 is the integral of the square of the wave function magnitude , μ is the chemical potential and V_0 is a constant external potential.

For $0\le u\le 1$ (we call this region I), we can use the usual Jacobi elliptic function, $\psi_I(u)=sn(k_I u)$, with k_I and m_I constants to be determined later. This function satisfy the second order differential equation

$$\frac{d^2}{du^2}sn(k_I u)-k_I^2\left[2m_I\, sn(k_I u)^3-(1+m_I)sn(k_I u)\right]=0. \tag{31}$$

Then, by comparison between Eq. (30) and Eq. (31), we find that

$$\mu=V_0+\frac{\hbar^2 k_I^2}{2ML^2}+\frac{NU_0}{2A^2}\quad,\quad m_I=\frac{ML^2NU_0}{\hbar^2 k_I^2 A^2}\quad. \tag{32}$$

Since it only depends on the characteristics of the condensate, the last term in the expression for μ is the self-energy of the condensate when the condensate is found in a region of constant potential. This result is in agreement with conjecture 1 of Ref. [21] that the self-energy $NU_0/2A^2$ is the difference between the chemical potential and the linear energy.

The relationship between u and x , in region I, is given by

$$u=\int_0^x\frac{dt}{\sqrt{1-m_I\sin^2(k_I t)}}\quad, \tag{33}$$

with $0\le u\le 1$.

When $u>1$ (we call this region II), we use the decaying function $mn(u)$ introduced in this work, with the origin at $u=1$, and $\psi_{II}(u)=T\,mn(k_{II}(u-1))=T\,e^{-k_{II}y}$, where u and y are related by

$$u-1=\int_0^y \frac{dt}{\sqrt{1+m_{II}\sinh^2(k_{II}t)}} \quad , \tag{34}$$

Here, y is zero for the start of the step at $u=1$. Since we want to perform a comparison with the linear solution, we have set $b=a=1/2$.

The comparison between Eq. (30) and the differential equation for $\psi_{II}(u)$ (see Eq. (16))

$$\frac{d^2}{du^2}\psi_{II}(u)-k_{II}^2\psi_{II}(u)\left[1-\frac{m_{II}}{2}+\frac{m_{II}}{2T^2}\psi_{II}{}^2(u)\right]=0, \tag{35}$$

indicates that

$$\mu=V_0-\frac{\hbar^2k_{II}^2}{2ML^2}+\frac{T^2NU_0}{A^2} \quad , \qquad m_{II}=\frac{4ML^2T^2NU_0}{\hbar^2k_{II}^2A^2} \quad . \tag{36}$$

Eqs. (32}) and (36) should be consistent. This leads to

$$k_{II}^2=\frac{2ML^2NU_0}{\hbar^2A^2}\left(T^2-\frac{1}{2}\right)-k_I^2 \quad . \tag{37}$$

The energy relations are given by

$$\left(\frac{d}{du}\psi_I(u)\right)^2+k_I^2\psi_I(u)\left[1+m_I-m_I\psi_I^2(u)\right]=k_I^2 \quad , \tag{38}$$

and

$$\left(\frac{d}{du}\psi_{II}(u)\right)^2-k_{II}^2\psi_{II}(u)\left[1-\frac{m_{II}}{2}+\frac{m_{II}}{4T^2}\psi_{II}^2(u)\right]=T^2k_{II}^2\frac{m_{II}}{4} \quad . \tag{39}$$

Thus, requiring the energy to be the same in both regions leads to $k_{II}^2=4k_I^2/m_{II}T^2$.

We now address the boundary conditions. With this choice of functions, $\psi(u)$ already satisfies the boundary condition at $u=0$, i.e., $\psi_I(u=0)=0$, and at $u=\infty$, $\psi_{II}(u=\infty)=0$. We therefore only should take care of the boundary conditions at $u=1$. Since the wave function should be continuous at $u=1$, we obtain $sn(k_I)=T\,mn(0)$, i.e., $\sin(k_Ix_1)=T$, where x_1 is the value of x that corresponds to $u=1$, i.e.,

$$1=\int_0^{x_1}\frac{dt}{\sqrt{1-m_I\sin^2(k_I t)}}. \tag{40}$$

Now, equating the derivatives of $\psi_I(u)$ and $\psi_{II}(u)$ at $u=1$ leads to $k_I cn(k_I)dn(k_I)=-k_{II}T\,mn(0)rn(0)$, i.e., $k_I\cos(k_I x_1)\sqrt{1-m_I\sin^2(k_I x_1)}=-k_{II}T$. The combination of the two conditions at $u=1$ gives the relationship between the parameters k_I and k_{II} that should be satisfied, $k_{II}=-k_I\cot(k_I x_1)\sqrt{1-m_I\sin^2(k_I x_1)}$.

The normalization of the wave function is given by

$$A^2=\int_0^\infty du|\psi(u)|^2=\frac{1}{k_I m_I}(k_I-E(k_I;m_I))+\frac{T^2}{k_{II}}\int_0^\infty dw\,mn^2(w) , \tag{41}$$

where $E(u;m)$ is the elliptic integral of the second type. Then, the normalized wave function is obtained by dividing $\psi_I(u)$ and $\psi_{II}(u)$ by A .

Thus, a procedure for numerically finding the solutions starts with the setting of values for a,b,m_I and k_I , followed by the calculation of the value of x_1 that corresponds to $u=1$ using Eq. (40). We then calculate k_{II} using $k_{II}=-k_I\cot(k_I x_1)\sqrt{1-m_I\sin^2(k_I x_1)}$, $T=\sin(k_I x_1)$ and $k_{II}^2=4k_I^2/m_{II}T^2$. This gives all the parameters necessary for defining the functions. Figure 2 shows a typical steady state together with the corresponding linear eigenfunction for comparison (calculated in such a way that their maxima are located at the same point). We found that the nonlinear wave function is less concentrated around its maximum than the linear state.

4. Remarks

The linear Schrödinger equation for a free particle has several sets of solutions, namely, trigonometric functions, real and complex exponential functions and their linear combinations. The functions introduced in this work will enable us to nonlinearly treat the same range of solutions as that for the linear equation. We are interested in these expression because they allow us to obtain physically appealing solutions in the presence of a decay of the wave function.

The nonlinear combination of other physical appealing functions will be explored in another paper.

References

[1] Pitaevskii L 1961 *Sov. Phys. JETP* **13** 451
[2] Gross E 1961 *Nuovo Cimento* **20** 454
[3] Dalfovo F, Giorgini S, Pitaevskii L and Stringari S 1999 *Rev. Mod. Phys.* **71** 463
[4] Kivshar Y S and Yanf X 1994 *Phys. Rev. E* **49** 1657
[5] Hasimoto H 1972 *J. Fluid Mech.* **51** 477
[6] Scott M M, Kalinikos B A and Patton C E 2000 *Phys. Rev. Lett.* **84** 4697
[7] Deutsch I H, Garrison J C, Ciao R Y and Wright E W 1993 *Frontiers in Nonlinear Optics: the Serge Akhmanov Memorial Volume* (Bristol: Institute of Physics Publishing)

[8] Sulem C and Sulem P L 1999 *The Nonlinear Schrödinger Equation. Self-focusing and Wave Collapse (Applied Mathematical Sciences* vol 139*)* (New York: Springer)
[9] Cayley A 1961 *An elementary treatise on Elliptic functions. Second edition* (New York: Dover Publications, Inc.)
[10] Abramowitz M and Stegun I A 1964 *Handbook of Mathematical Functions* (New York: Dover Publications, Inc.)
[11] Khare A and Sukhatme U 2002 *Phys. Rev. Lett.* **88** 244101
[12] Markushevich A 1992 *Introduction to the Classical Theory of Abelian Functions (Translations of Mathematical Monographs* vol. 96*)* (Providence, Rhode Island: American Mathematical Society)
[13] Belokolos E D and Enolskii V Z 2001 *J. Math. Sc.* **106** 3395
[14] Belokolos E D and Enolskii V Z 2002 *J. Math. Sc.* **108** 295
[15] Khare A and Saxena A 2014 *J. Math. Phys.* **55** 032701
[16] Zhang Y and Li J B 2015 *J. Math. Phys.* **56** 084101
[17] Khare A and Saxena A 2015 *J. Math. Phys.* **56** 113510
[18] Carr L D, Charles W C and Reinhardt W P 2000 *Phys. Rev. A* **62** 063610
[19] Carr L D, Charles W C and Reinhardt W P 2000 *Phys. Rev. A* **62** 063611
[20] Mahmud KW, Kutz JN and Reinhardt W P 2002 *Phys. Rev. A* **66** 063607
[21] D'Agosta R, Malomed B A and Presilla C 2000 *Phys. Lett. A* **275** 424

Ultrasonic and conventional synthesis of NaA zeolite from rice husk ash

Farías T[1],de Ménorval L C [2], Picazo O[1] and Jordán R[3]

[1] Department of Chemistry and Technology of Materials (QUITMAT), Institute of Materials Science and Technology (IMRE), University of Havana, 10400 Havana, Cuba.

[2] Institut Charles Gerhardt Montpellier, UMR-5253 CNRS-UM, Equipe Agrégats, Interface, et Matériaux pour l'Energie (AIME), Université de Montpellier, C.C. 1502 Place Eugène Bataillon, 34095 Montpellier Cedex 5, France.

[3] Geoscience Department, State University of Sonora, Hermosillo, Sonora, Mexico.

E-mail: tania@imre.uh.cu

Abstract. In the present work, a simple synthetic route for the production of single phase NaA zeolite is demonstrated. Rice husk ash (RHA) as alternative silica source was employed for the synthesis by conventional hydrothermal and non-conventional ultrasound methods. The zeolite was also synthesized using commercially available silicate for comparison. The effect of the reaction time (2, 4 and 6 h) at a fixed temperature of 70 °C was investigated. The elemental, structural and morphological characterization of the ashes and the synthesized zeolites was performed by X-ray diffraction (XRD), scanning electron microscopy (SEM), energy dispersive spectroscopy (EDS) and thermogravimetric analysis (TG/DSC).

1. Introduction

Zeolites are crystalline microporous aluminosilicates which are constructed from tetrahedral AlO_4 and SiO_4 units. The porous structure of zeolites yields valuable physicochemical properties for cation exchange, molecular sieving, catalysis and adsorption whereby this are good candidates for a large number of applications . The search for new materials with improved properties for different applications has motivated the research in the synthesis of zeolites and related materials since some decades and today continues to be a topic very addressed in the scientific literature . Conventionally zeolites are synthesized under hydrothermal conditions , and the reaction gel medium contains the framework atoms, solvents, structure-directing agents (SDAs) and mineralizers.

Ultrasound is a new method for the synthesis of a wide range of nano- and micro-particles. It is very simple, fast and does not need any complicated facilities. This technique has also been introduced in the field of zeolite synthesis. In a recent review Askari et al. summarized the effects of ultrasound on the synthesis of various zeolites. The most important mechanism enhancing crystallization is ultrasonic cavitation, a phenomenon that can be defined as the growth and explosive collapse of microscopic bubbles. The use of ultrasonic treatment in synthesis reactions offers several practical

Content from this work may be used under the terms of the Creative Commons Attribution 3.0 licence. Any further distribution of this work must maintain attribution to the author(s) and the title of the work, journal citation and DOI.
Published under licence by IOP Publishing Ltd

advantages among which the lower-temperature and the shorter time required for the synthesis process are the most significant .

Rice husk (RH) is an agricultural waste material abundantly available in rice-producing countries. It is known that rice husk ash (RHA) usually contains over 80% of silica, reason why it has been investigated as an alternative source of silica for different applications. For example RHA has been widely used as additive for construction material and as an adsorbent for organic dyes and inorganic metals . RHA can be also an economically raw material for the production of zeolites. Several researchers have been report the synthesis of zeolites ZSM-5 , ZSM-11 , Beta , EMT , Y , mordenite , NaA and NaX starting from RHA. However, only few papers combined the use of RHA as silica source and the ultrasound energy for the synthesis of zeolites .

The present work aims to evaluate the applicability of ultrasound method to the synthesis of Na-LTA zeolite using a low cost, abundant and renewable source of amorphous silica extracted from a waste of the Cuban agricultural industry, the rice husk. Furthermore, the characterization of the obtained materials is carried out in order to evaluate its potential in later applications.

2. Experimental

2.1. Rice husk ashes

Rice husk of national production was kindly provided by the *Research Institute of Grain* of Cuba. It was washed with distilled water to remove dust and mud and calcined at 500 and 700 °C for 2 hours in a muffle to obtain rice husk ashes (RHA-500 and RHA-700, respectively). The chemical compositions of the RHAs obtained, expressed in weigh %, are shown in Table 1.

Table 1. Chemical composition of rice husk ashes (wt. %)

	SiO_2	C	Na_2O	K_2O	CaO	MgO	MnO	Al_2O_3	P_2O_5
RHA-500	81.4	6.4	1.9	3.6	1.0	1.5	0.9	1.2	2.0
RHA-700	92.5	--	0.5	2.1	1.0	1.0	1.0	0.9	1.1

To optimize the extraction of silicon from the ash a known amount of RHA-700 was mixed with a 2 M NaOH solution at a solid/liquid ratio of 1 g / 10 ml and was left under vigorous stirring for different times at 55 °C. Then, the solution was separated from the mixture by centrifugation and the solid residues were washed with distilled water and dry until constant weight.

2.2. Synthesis of zeolites

For the zeolite synthesis the following composition of the initial gel was selected: $3.5Na_2O$: Al_2O_3: $2SiO_2$: $145H_2O$. Such composition is typically employed for LTA zeolite synthesis , although other reports can be found in the literature with different contents of water and Na_2O.

Two different series (SI and SII) were synthesized starting from RHA and commercial sodium silicate, respectively. The sodium silicate solution for SI was prepared mixing 1 g of RHA-700 with 10 ml of 2 M NaOH solution followed by vigorous stirring for 4 h at 55 °C. Then the solution was left to cool and solid residue was separated by centrifugation. The sodium silicate solution for SII was prepared dissolving 3.14 g of $Na_2SiO_3x5H_2O$ in 8.67 ml of H_2O. A clear aluminate solution was prepared by mixing 1.75 g of sodium aluminate with 4 ml of 2 M NaOH solution and 5 ml of H_2O followed by stirring at room temperature until complete dissolution. The silicate solution was added slowly into the aluminate solution while stirring vigorously and then the obtained gels were heating at 70 °C for different times (2, 4 and 6 h) in a conventional ultrasonic bath and in an oven under static conditions. After the crystallization time, the solids were separated by centrifugation, washed repeatedly with distilled water, dried at 70 °C and stored for further characterization. For the ultrasonic synthesis a commercial ultrasonic cleaner (VWR) with a frequency of 35 kHz and temperature control was employed. The samples were labelled as SI-HT-t and SI-US-t for samples of SI series and SII-HT-t and SII-US-t for samples of SII series obtained by hydrothermal and ultrasound methods, respectively, where t correspond to the crystallization time.

2.3. Characterization techniques

The characterization of the RHAs and the synthesized zeolites was performed by X-ray diffraction (XRD), scanning electron microscopy (SEM), energy dispersive spectroscopy (EDS) and thermogravimetric analysis (TG/DSC). For qualitative XRD analysis a Philips Xpert diffractometer was employed, using Cu Kα radiation (λ = 1.541838 Å). The 2θ range from 1° to 60° was swept in a continuous way at 2 °min^{-1}. The morphologies of the samples and chemical analysis were investigated by SEM and EDS on a JEOL 5600 microscope. TG/DSC analysis was carried out with the aid of a NETZSCH STA 409 PC/PG thermal analyzer. Samples were heated form 25 to 800 °C at a heating rate of 10 °C/min under dry air at a purging flow rate of 50 mL/min. The sensitivity of the thermobalance was ± 1 µg. A solid sample of about 15 mg was used in each test.

3. Results and Discussion

4. Rice husk ashes

RH calcined at both temperatures (500 and 700 °C) produces amorphous silica as is evident from a broad hump at the 2θ of around 15–30° in the X-ray diffraction patterns (Figure 1). This result is very suitable for further synthesis because the silica is rendered active in its amorphous form to produce zeolites. Similar results have been reported by other authors for calcinations temperatures smaller than 700 °C for short times . However, for higher temperatures or larger times crystalline phases of SiO_2 such as cristobalite, tridymite and α-quartz are obtained .

Figure 2 shows the results of the thermogravimetric analysis of RH and RHA ashes. RH undergoes a total mass loss of about 85 % that occurs in three steps, at 78, 292 and 470 °C. The first loss of mass is related to the loss of water and the two others with the decomposition of the organic matter (mainly cellulose, hemicellulose and lignin). The total loss of mass for RHA-500 and RHA-700 is of 5 and 1 %, respectively. In RHA-700 this loss of mass corresponds to the loss of water while in RHA-500 is due to the decomposition of organic matter. Taking into account the content of SiO_2 and that the unburned organic matter could interfere in the subsequent synthesis of the zeolite, we selected as silica source the ash obtained at 700 °C.

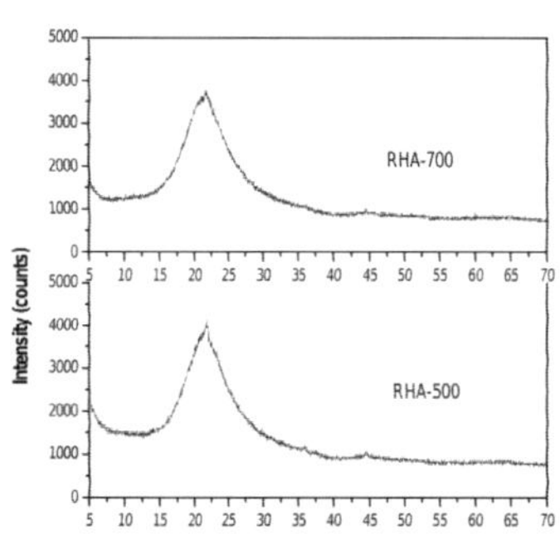

Figure 1. XRD patterns of RHAs obtained

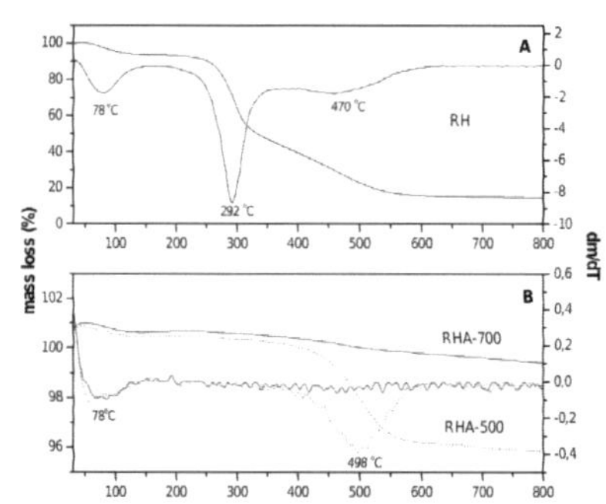

Figure 2. TG/DTG curves of rice husk

by rice husk calcination at 500 and 700 °C. (A) and rice husk ashes (B).

Figure 3 shows the percent of silicon extracted from RHA-700 in function of the time using a 2M NaOH solution at 55 °C. The amount of silicon extracted increased with the increasing of time. However, no significant increase was observed at longer extraction time than 4 h. It was found that in these conditions is possible to extract about 90 % of the content of silicon present in the ash, so that we select 4 hours as optimum extraction time.

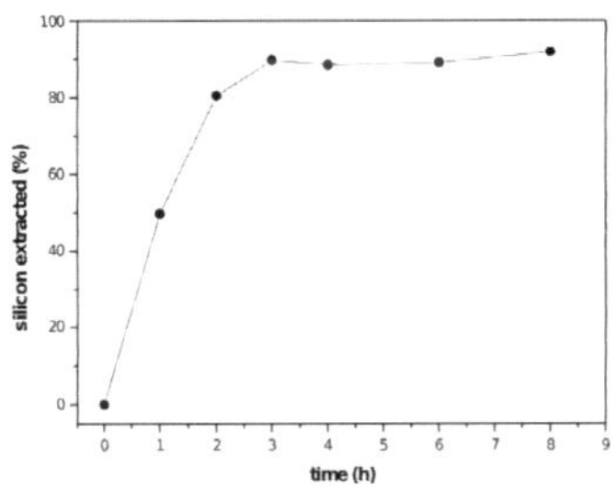

Figure 3. Percent of silicon extracted from RHA-700 in function of time employing a 2M NaOH solution at 55 °C.

5. Characterization of the synthesized zeolites

The XRD patterns of the synthesized zeolites are shown in Figures 4 and 5. The diffractograms of all the samples matched the reference peaks of NaA zeolite (PDF: 01-089-8015) and no others zeolitic phases appear, which ascertain the successful synthesis of high purity NaA zeolitic phase. It can be noted that RHA-700 is a suitable silica source for the synthesis of NaA. For both methods it was possible to synthesize the desired zeolite to a moderate temperature, short crystallization time and without seeding and aging.

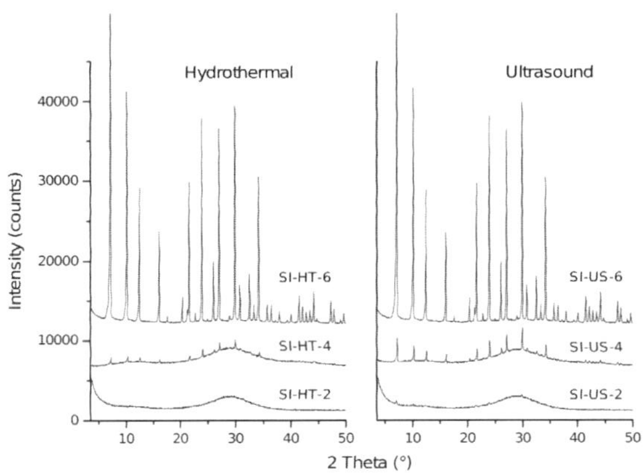

Figure 4. XRD patterns of NaA zeolites synthesized from RHA-700 as silica source by hydrothermal and ultrasound methods for different times.

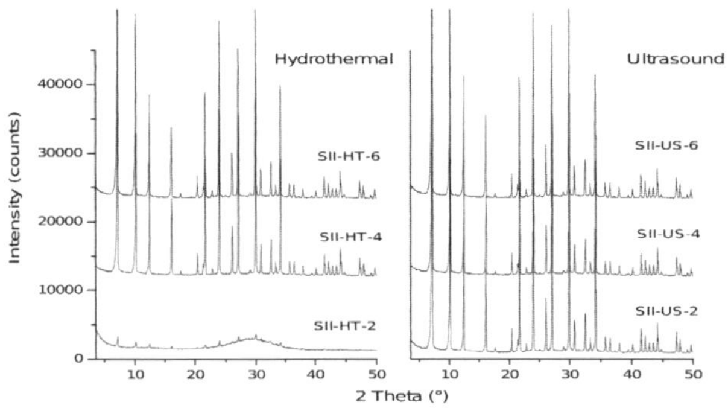

Figure 5. XRD patterns of NaA zeolites synthesized from sodium silicate by hydrothermal and ultrasound methods for different times.

The comparison of the diffraction patterns allows affirming that for both methods a larger crystallization time is needed for SI series (samples obtained starting from the ash), which could be related with the presence of impurities in the extracted silica. Also, for both series the crystallinity of zeolite A started to develop at earlier synthesis times, indicating an increase in the nucleation rates in the presence of ultrasound. These statements are better illustrated in Figure 6, where the degree of crystallinity in function of the crystallization time is presented. The crystallinity was calculated as the sum of the intensities of the representative diffraction planes of NaA, normalized by assuming that this sum for SII-HT-6 corresponds to 100 % of crystallinity. Park *et al.* found similar results when they studied the conventional and ultrasound synthesis of zeolite 4A from kaolin at 60 and 70 °C. Andaç *et al.* reported that nucleation and crystallization rates, as well as the yield of zeolite A increased as a result of the application of ultrasound.

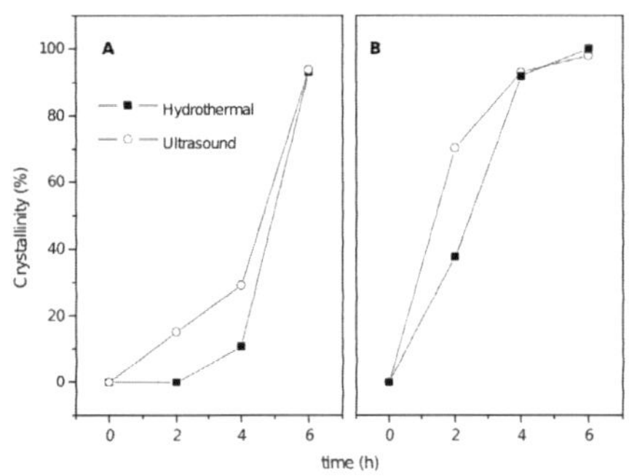

Figure 6. Degree of crystallinity *vs.* crystallization time for zeolites synthesized starting from (A) RHA-700 and (B) sodium silicate as silica source by hydrothermal and ultrasound methods.

The SEM images of the zeolites are shown in Figure 7. A uniform particles morphology -cubic of smoothed edge- can be observed, which are characteristics of zeolite A. Contrary to other works not significant differences in the particle size were observed for hydrothermal and ultrasound methods. It is interesting to note that as early as 2 hours for the ultrasonic synthesis of the SII series a large number of crystals appears.

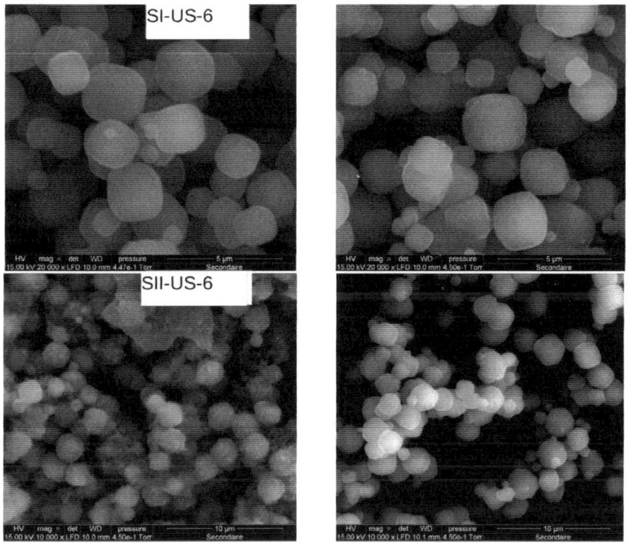

Figure 7. SEM pictures of the NaA zeolites synthesized starting from RHA-700 and sodium silicate as silica source by hydrothermal and ultrasound methods.

The chemical composition of the samples (determined by EDS) are presented in Table 2. The calculated Si/Al ratios are around 1. Among the metals found as traces in the ash only the potassium is involved in the synthesis of the zeolite, appearing in very small amounts in the final chemical composition.

Table 2. Chemical composition of the synthesized zeolites determined by EDS. The values are expressed in wt. %.

Sample	Si	Al	O	Na	K	Si/Al
SI-HT-2	12.65	12.55	62.83	11.64	0.29	1.007
SI-HT-4	12.44	12.47	62.89	11.88	0.31	0.998
SI-HT-6	11.67	12.04	64.57	11.50	0.17	0.969
SI-US-2	13.16	13.27	62.12	11.09	0.35	0.992
SI-US-4	12.24	12.39	62.98	12.12	0.25	0.988
SI-US-6	11.80	12.29	64.26	11.47	0.16	0.960
SII-HT-2	12.24	12.39	62.05	13.26	0.01	0.988
SII-HT-4	12.08	12.40	64.03	11.46	0.00	0.974
SII-HT-6	11.58	11.94	65.05	11.41	0.01	0.970
SII-US-2	12.26	12.51	62.73	12.48	0.01	0.980
SII-US-4	10.84	11.34	66.79	10.98	0.03	0.956
SII-US-6	11.18	11.63	65.75	11.41	0.02	0.961

The DTG curves of the samples of SI and SII series are given in Figure 8 and 9, respectively. It can be observed that the curves for the amorphous samples were quite different than those of the crystalline and partly crystalline samples. In all curves there are a minimum at about 95 °C, which corresponds to the removal of moisture. The intensity of this temperature minimum was higher for the amorphous sample and decreases with the increases of crystallinity. For samples of higher crystallinity the DTG curves exhibited others two minimums at around 170 and 410°C, which have been related with the loss of water molecules coordinated to the Na ions and inside the sodalite cage cavities, respectively .

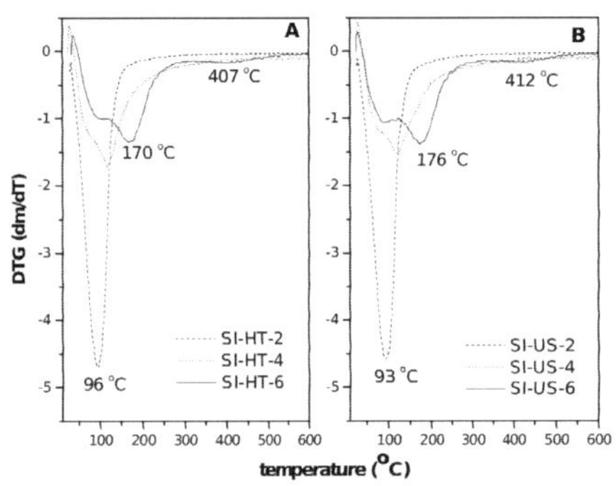

Figure 8. DTG curves of zeolites obtained from RHA-700 (SI series) by hydrothermal (A) and ultrasound (B) methods.

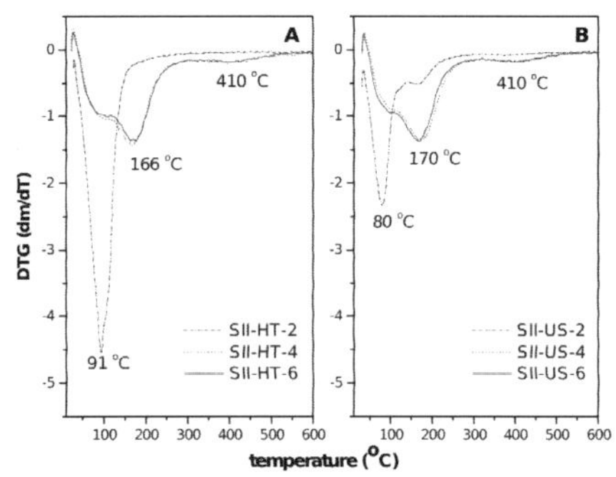

Figure 9. DTG curves of zeolites obtained from sodium silicate (SII series) by hydrothermal (A) and ultrasound (B) methods.

6. Conclusions

It was verified that it is feasible to use the silicon extracted from rice husk ash as starting material for the synthesis of NaA zeolite with good purity, suitable morphology and with a Si/Al ratio close to 1. The results shows that the ultrasound synthesis favors the kinetics of the reaction, yielding more crystalline samples in less time.

References

[1] R. M. Barrer 1978 in *Natural Zeolites: Ocurrence, Properties and Uses*, L. B. Sand and F. A. Mumpton, Eds., Oxford-New York: Pergamon Press.
[2] F. A. Mumpton 1999 *Proc. Natl. Acad. Sci. USA* **96** 3463-3470.
[3] M. Zaarour, B. Dong, I. Naydenova, R. Retoux, and S. Mintova 2014 *Micropor. Mesopor. Mater.* **189** 11-21.
[4] S. S. Bukhari, J. Behin, H. Kazemian, and S. Rohani 2015 *Fuel* **140** 250-266.
[5] J. Li, A. Corma, and J. Yu 2015 *Chemical Society Reviews* **44** 7112-7127.
[6] R. M. Barrer, 1982 *Hydrothermal Chemistry of Zeolites*. London: Academic Press
[7] S. Askari, S. Miar Alipour, R. Halladj, M. Hossein, and D. Abadi Farahani 2013 *J. Porous Mater.* **20** 285-302.
[8] P. Chindaprasirt, C. Jaturapitakkul, and U. Rattanasak 2009 *Fuel* **88** 158-162.
[9] K. Ganesan, K. Rajagopal, and K. Thangavel 2008 *Constr. Build Mater.* **22** 1675-1683.
[10] M. Nehdi, J. Duquette, and A. E. Damatty 2003 *Cement Concrete Res* **33** 1203-1210.
[11] U. R. Lakshmi, V. C. Srivastava, I. D. Mall, and D. H. Lataye 2009 *J Environ Manage* **90** 710-720.
[12] Q. G. Feng, Q. Y. Lin, F. Z. Gong, S. C. Sugita, and M. Shoya 2004 *J. Colloid Interf. Sci.* **278** 1-8.
[13] V. C. Srivastava, I. D. Mall, and I. M. Mishra 2008 *Colloids Surf. A* **312** 172-184.
[14] Z. Ramli and H. Bahruji 2003 *Malaysian Journal of Chemistry* **5** 48-55.
[15] M. K. Naskar, D. Kundu, and M. Chatterjee 2012 *J. Am. Ceram. Soc.* **95** 925-930.
[16] K. P. Dey, S. Ghosh, and M. K. Naskar 2012 *Materials Letters* **87** 87-89.
[17] D. Prasetyoko, Z. Ramli, S. Endud, and H. Hamdan 2006 *Waste Management* **26** 1173-1179.
[18] E.-P. Ng, H. Awala, K.-H. Tan, F. Adam, R. Retoux, and S. Mintova 2015 *Microporous and Mesoporous Materials 204 (2015) 204–209* **204** 204-209.
[19] Z. Ramli, E. Listiorini, and H. Hamdan 1996 *Jurnal Teknologi* **25** 27-35.
[20] P. K. Bajpai, M. S. Rao, and K. V. G. K. Gokhale 1981 *Ind. Eng. Chem. Prod. Res. Dev.* **20** 721-726.
[21] Z. Ghasemi and H. Younesi 2011 *J Nanomater* 1-8.
[22] T. Wajima, O. Kiguchi, K. Sugawara, and T. Sugawara 2009 *J. Chem. Eng. Jpn* **42** s61-s66.
[23] Z. Ghasemi and H. Younesi 2012 *Waste Biomass Valor* **3** 61-74.
[24] R. W. Thompson and K. C. Franklin, 2001 *Verified Synthesis of Zeolitic Materials*. Amsterdam: Elsevier
[25] H. Hamdan, M. N. M. Muhid, S. Endud, E. Listiorini, and Z. Ramli 1997 *Journal of Non-Crystalline Solids* **211** 126-131.
[26] E. F. Ledesma, C. R. Acosta, M. L. Garrido, M. C. González, M. M. Bernal, X. R. Pileta, *et al.* 2012 *J. Mater. Environ. Sci.* **3** 760-765.
[27] J. Park, B. C. Kim, S. S. Park, and H. C. Park 2001 *J. Mater. Sci. Letter* **20** 531-533.
[28] Ö. Andaç, M. Tatlıer, A. Sirkecioğlu, I. Ece, and A. Erdem-Şenatalar 2005 *Micropor. Mesopor. Mater.* **79** 225-233.
[29] A. Dyer and M. J. Wilson 1973 *Thermochim Acta* **5** 91.

IOP Conf. Series: Journal of Physics: Conf. Series **792** (2017) 012033 doi:10.1088/1742-6596/792/1/012033

National Pyranometers comparison of solar thermal labs in Mexico

H A Castillo-Matadamas [1], J C Molina-Vazquez [1], G Moreno-Quintanar [2], A Fuentes-Toledo [2], N Ortega-Avila [2], J M Rodríguez-González [3], J A Barrón-Mancilla[4], J J Navarrete-Gonzalez [4].

[1] Centro Nacional de Metrología, Dirección de óptica y radiometría, El Marques Querétaro, México, C.P. 76246, [2] Instituto de energías renovables-UNAM, [3]Universidad Autónoma del Estado de México, [d] Instituto de Investigación y Desarrollo de Energías Renovables y Eficiencia Energética, [4] Laboratorio Mexicano de pruebas solares.

E-mail: hcastill@cenam.mx

Abstract. The results of the first national comparison of pyranometers used in testing laboratories of solar water heating are reported. In the comparison carried out at the facilities of Centro Nacional de Metrología (CENAM-México) participated three testing laboratories, a university and CENAM with seven secondary standards and first class pyranometers. The measurement results for all instruments were adequate, considering that the deviations found in all cases for global irradiance measurements greater than 500 W / m^2 were in a band of +/- 2.5%, even though pyranometers have different dates of calibration.

1. Introduction

In order to support the implementation of quality infrastructure in the México market of solar water heating systems (SWH)- which is understood as the interrelation between standardization, conformity assessment and metrology - Mexican test laboratories of solar water heating systems were invited to participate in a pyranometers comparison exercise in the facilities of Centro Nacional de Metrología (CENAM) as part of the PTB-German Mexican project: "Fortalecimiento de la Infraestructura de la Calidad para Energías Renovables y Eficiencia Energética".
CENAM was chosen to host the comparison due to their development in measurement capabilities for instruments related with renewable energy, including calibration of pyranometers in laboratory with traceability of measurements to the international system of units via a cryogenic radiometer. Metrological support developed in the country is critical to the national net of secondary laboratories that test photovoltaic devices and solar water heaters. The metrological support ensures the measurement values and specifications that consumers receive on efficiency labels of products.

Content from this work may be used under the terms of the Creative Commons Attribution 3.0 licence. Any further distribution of this work must maintain attribution to the author(s) and the title of the work, journal citation and DOI.
Published under licence by IOP Publishing Ltd

IOP Conf. Series: Journal of Physics: Conf. Series **792** (2017) 012033 doi:10.1088/1742-6596/792/1/012033

The pyranometers comparison exercise was held from August 31st to September 2nd 2015 on the facilities of CENAM located at 20° 32.246' North latitude, 100° 15.320' west longitude. The exercise was thinking as a first pyranometers comparison between laboratories, regardless of the time of the last calibration of their instruments. Short measurement periods were considered to evaluate the agreement between the measurements of the pyranometers that laboratories use for routinary testing of SWH.

Since the CENAM does not have facilities specifically designed for outside solar radiation measurements, the exercise place was the roof of the building D of the optical and radiometry section. In this location a steel platform levelled and prepared to accommodate up to nine pyranometers with a distance between them of 0.43 m was mounted. The height of the platform with respect to other buildings not allow measurements for large angles of incidence with respect to normal, however in the measurements, were not considered the response of instruments for angles greater than 60 degrees from the normal and even some measurements were made near the noon, for periods up to 90 minutes.

2. Participants in the comparison

In the comparison participated testing laboratories, a university and CENAM. The participants were:

- Laboratorio de Pruebas de Equipos de Calentamiento Solar from Instituto de Energías Renovables, IER-UNAM, who participated with two pyranometers, identified as IER1 e IER2.
- Instituto de Investigación y Desarrollo de Energías Renovables y Eficiencia Energética (IIDEREE), who participated with one pyranometer, identified as IIDEREE.
- Laboratorio Mexicano de Pruebas Solares, (MEXOLAB), who participated with one pyranometer, identified as MEXOLAB.
- Centro Nacional de Metrología, (CENAM), who participated with two pyranometers, identified as CENAM1 y CENAM2, calibrated inside of laboratory in the previous months of the comparison [1].
- Universidad Autónoma del Estado de México (UAEMéx), who participated with one pyranometer, identified as UAEM.

The objective of this first comparison was to obtain the deviation of the measurements of each participant pyranometer with reference to the arithmetic mean of the group without a reference instrument. The participants defined prior to exercise, the best sensitivity coefficient of their instruments in the units of [$\mu V / (W / m^2)$] and also its measurement uncertainty. With this coefficient the global solar irradiance values were calculated and also the deviation of each instrument compared to the group average.

According to [2] the CMP6 instruments are first class pyranometers and CMP11 and CMP21 are secondary standards with technical specification compatible with high accuracy measurements.

Table 1. Identification of participant's instruments in the comparison.

ID	Measurement number	Manufacturer	Model	Serial	Sensitivity coefficient [μV m²/W]	Uncertainty of coefficient [%], k=2	Response time* [s]
CENAM 1	4	KIPP & ZONEN	CMP 21	140336	9.43	2.0	< 15
CENAM 2	5	KIPP & ZONEN	CMP 6	140666	18.87	2.0	< 30
IER 1	5	KIPP & ZONEN	CMP 6	140542	15.58	2.6	< 30
IER 2	5	KIPP & ZONEN	CMP 6	140541	17.06	2.6	< 30
UAEM	5	KIPP & ZONEN	CMP 11	126917	9.50	2.6	< 15
IIDEREE**	2	KIPP & ZONEN	CMP 6	123269	12.39	2.6	< 30
MEXO LAB	1	KIPP & ZONEN	CMP 21	110594	8.67	1.5	< 15

* These data were obtained from the manufacturer.
** The laboratory changes the sensitivity value of their pyranometer during comparison.

3. Measurement conditions

The air temperature during the pyranometers measurement period ranged between (22-28) °C and barometric pressure measured every minute in the pressure laboratory of CENAM had maximum and minimum variations between values (81 396 to 80 990) Pa. During measurements there was cloudy sky, clear sky for short periods and even light rain.

Levelling of the instruments with respect to the horizontal was adjusted slightly every morning and also water cleaning condensation of domes was performed.

The measurement exercise considered only simultaneous comparisons of instruments for short measurement times because the interest was to know the results immediately to conduct a preliminary analysis of the measurements, and if necessary to find the causes of widely scattered data to stablish calibration needs.

Figure 1. Experimental setup for a pyranometer calibration

4. Measurement results

The benchmarking exercise was conducted over three days, with four measurement periods between 60 and 90 minutes each. In addition, during the early hours of September 1st a measurement was performed to confirm the behaviour of the instruments in the absence of solar radiation (night measurement) and the mean value of all the instruments was in the band of (-1 to -0.6) W/m^2.

Because the laboratory MEXOLAB had no connection cable for their pyranometer, he participated in a single measurement, corresponding to the morning of September 1st, using the pyranometer cable of CENAM1. The pyranometer IIDEREE participated in two measurements of September 1st. And pyranometers IER1, IER2, UAEM and CENAM2 participated in all measures of comparison.

The graphs of global irradiance measurements and relative deviations of each instrument to the average of the group are shown in Figures 2 to 5.

The measurement of August 31st is shown in left of Figure 2, which was de first test. It is a common irradiance curve of a cloudy day, since the irradiance value has sudden changes. In Figure 2, the right graph is showing the error percentage taking as the reference value the average value of all pyranometers. The maximum error values are ±10%, which correspond to abruptly changes of irradiance. In the stable periods the errors are from -4% to 2.5%. The following graphs will show the results for two other test days.

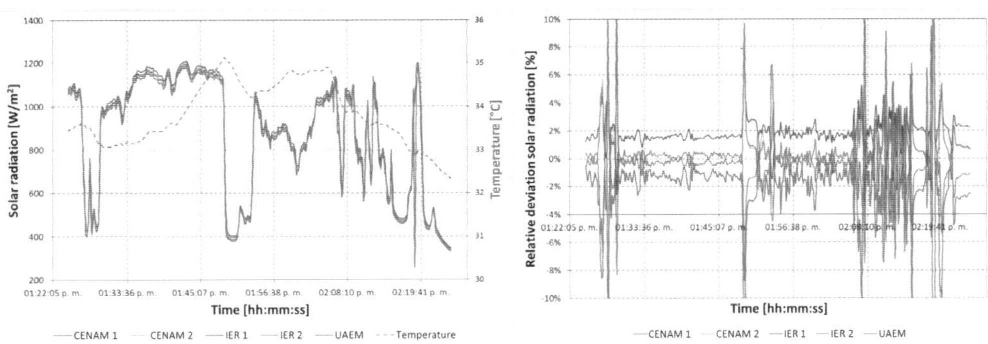

Figure 2. August 31st measurement and results.

Figures 5 shows three different stable periods obtained during two different test days. This Figure shows the irradiance of morning September 1st, the values are between (640 to 740) W/m2, the

evening measurements are around 1200 W/m2. On September 2nd was obtained irradiation between (400 to 500) W/m2. The comparison was performed in a variety of irradiance values.
In case b of the Figure 6, between 10:16 y 10:19 a difference in the pyranometer path is appreciated. Two pyranometer experiments a greater increase of irradiance than the others (UAEM and Mexolab), a peak can see in the graph, the other irradiance paths has peaks but are smoother, this is due to the differences in the response time.

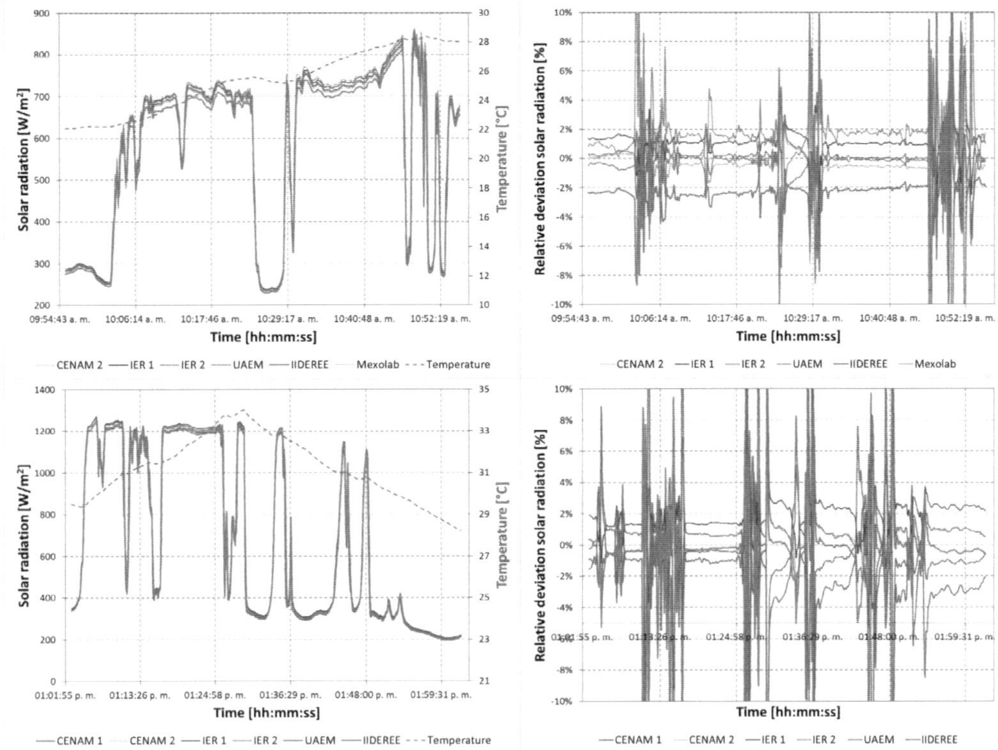

Figure 3. September 1st measurement and results.

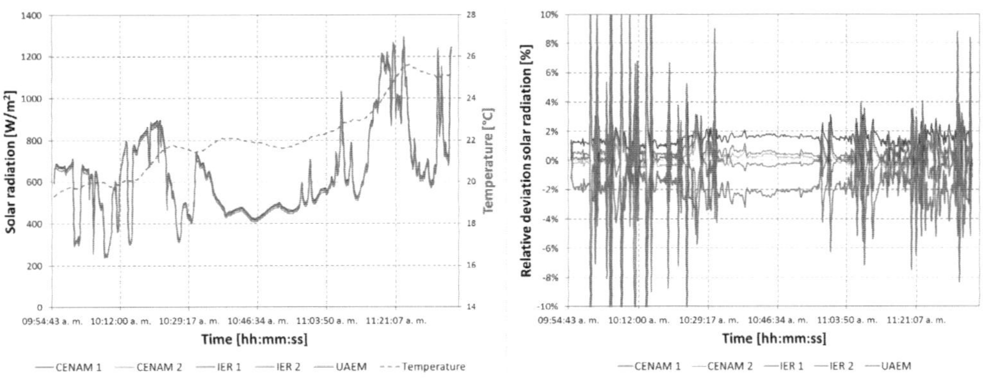

Figure 4. September 2nd measurement and results.

The high dynamic of global irradiance measurement during exercise allowed only a few windows of time with stable solar irradiance, however this type of weather offered the opportunity to appreciate low irradiance values during measurements.

The results for all participants during periods of high irradiance and half irradiance above 500 W/m2 show lower deviation +/- 2.5% in all cases and there was unnecessary to discard or analyze any outlier data in the exercise.

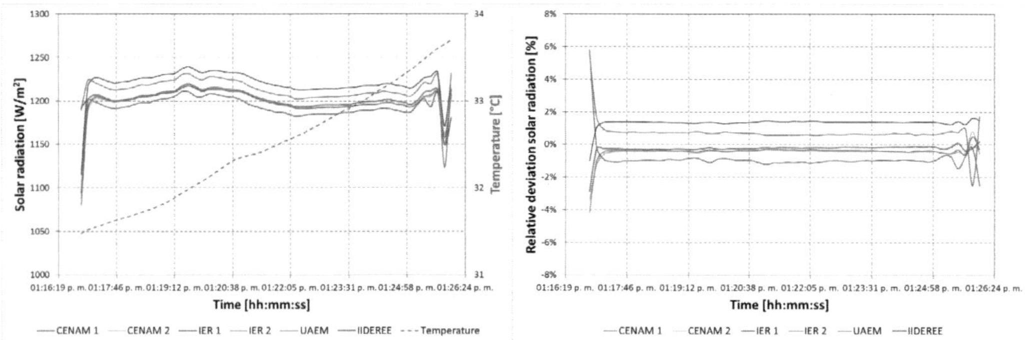

(a) 1200 W/m^2 stable period (September 1st. 2015).

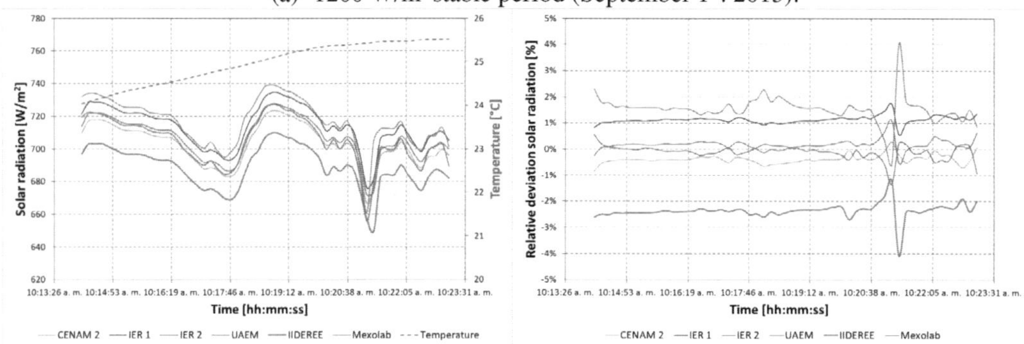

(b) 700 W/m^2 stable period (September 1st. 2015).

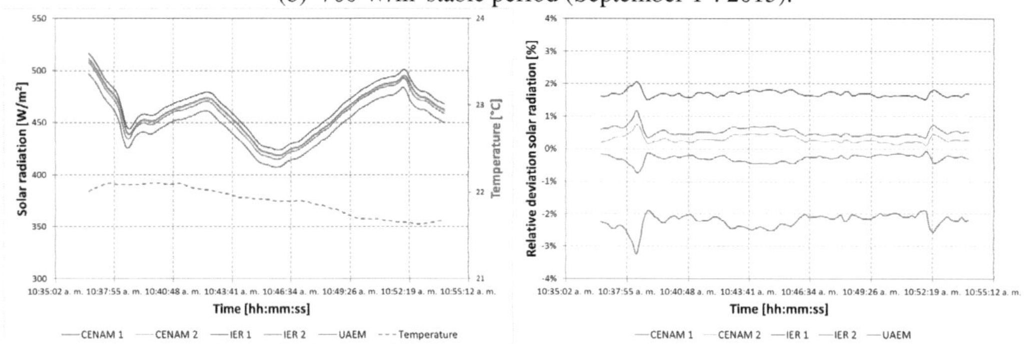

(c) 450 W/m^2 stable period (September 2nd. 2015).

Figure 5. Measurement and results during three different stable periods.

To global irradiance values below 500 W/m^2 results are generally more dispersed with high variability of the fast CMP11 and CMP21 pyranometers. Scattering of measurements increases with decreasing irradiance values which is evident in the final part of the afternoon September 1st measurement. This phenomenon is related to the fact that the sensitivity coefficient is not linear with respect to intensity especially at low irradiance values [3], [4]. It is therefore important to know the specific sensitivity coefficient of these instruments at the ends of the scale (200 W/m^2 and 1200 W/m^2) if is desired to reduce measurement uncertainty below 3% in its operating range.

5. Conclusions

A first pyranometers comparison exercise between test labs of SWH systems was perform from 31 August to 2 September 2015 at CENAM. The results show high dynamic in global solar irradiance values due to the clouding weather during measurements.

The comparison was made on the roof of a CENAM's building using a levelled steel platform as support for seven pyranometers, four pyranometers from test laboratories of SWH systems, one from a university meteorological institute and two pyranometers from CENAM.

The deviation of the results obtained for medium and high irradiance levels of above 500 W/m^2 is +/- 2.5% at each instrument under test respect to the average of the group when the stability in the band of 100 W/m^2 . The dispersion values for measurements increases below 500 W/m^2, due to effects of nonlinearity in the sensitivity coefficient of the pyranometers regarding the irradiance.

The deviation obtained from measurements of the two CENAM's pyranometers for medium and high irradiance is not greater than the uncertainty reported in the calibration laboratory, however for low irradiance, the measurements deviation respect to the average of the group is greater than 3% because it was considered a constant sensitivity coefficient at low irradiance, so it is necessary to study the CENAM1 pyranometer in subsequent calibrations at low intensity. The CMP21 and CMP11 instruments and participants in comparison, have fast measurement changes in irradiance while CMP6 instruments tend to integrate measurements in short time so they remain quasi-stable for rapid changes in irradiance values due to cloudy weather.

Overall this first comparison shows good results for instruments with highly variable calibration dates and confirms the need for traceability of pyranometers measurements to reduce the measurement uncertainty in the testing results for solar water heaters systems carried by the test labs in Mexico.

Acknowledgments

We thanks to the project: "Fortalecimiento de la Infraestructura de la Calidad para Energías Renovables y Eficiencia Energética" supported by the PTB (National Metrology Institute of Germany) and especially to Dr. Harald Drück of SWT Stutgart for their valuable advice for conducting the comparison. Also the authors acknowledge the financial support received from CONACYT through the Mexican Center for Innovation in Solar Energy (CeMIE-Sol), in the framework of the call 2013-02, of the CONACYT – SENER – SUSTAINABLE ENERGY found within the project strategic P13 "Laboratorios de pruebas para baja y media temperatura, laboratorio para el diseño e integración de sistemas termosolares asistido por computadora".

6. References

[1] Castillo-Matadamas H A, Molina-Vazquez J C and Quintero-Torres R, *Unexplored Indoors method for pyranometers calibration traceable to SI* 2015 J. Phys.: Conf. Ser. 582 012 051

[2] National standard NMX-ES-J-9060-NORMEX-ANCE-2015 base on international standard ISO 9060 (1990).

[3] The coefficient of sensitivity in low and high irradiance changes as a function of pyranometer model and have been measured with differences in coefficient of up to 1.5% in the calibrations performed by the CENAM.

[4] Pyranometers specifications. Brochure KippZonen Pyranometers, 2015. (http://www.kippzonen.com/Download-Center/Download-search?s=true&productid=13)

IOP Conf. Series: Journal of Physics: Conf. Series **792** (2017) 012034 doi:10.1088/1742-6596/792/1/012034

A method for correcting the structural instability of time-dependent atmospheric trajectory models under perturbations of the mass balance

M A Núñez and R Mendoza
Departamento de Física, Universidad Autónoma Metropolitana Iztapalapa, A. P. 55-534, C. P. 09340, D. F.

E-mail: manp@xanum.uam mx, oicor_192000@hotmail.com

Abstract. Lagrangian trajectory models of atmospheric fluid parcels have been used over a wide range of spatial scales to study the transport and dispersion of pollutants, radioactive materials, ash clouds produced by volcanic eruptions or modeling global carbon cycle. It was pointed out trajectory calculations have three error sources: error in the gridded data from measurements error or from approximations in Eulerian numerical models, error from the spatial and temporal resolution of data, and truncation error from numerical integration of the velocity field. In a recent work we showed that trajectory models can be structurally unstable under perturbations of the mass balance, e.g., the flow can go from hyperbolic to elliptic and vice versa. In this work we propose a mass-consistent approach to correct this structural instability. The orthogonal projection character of this approach guarantees the correction even for large perturbations of the mass-balance. This is illustrated by means of numerical examples. Mass-consistent models have been considered as diagnostic models of the wind field but the results of this work show that such a models can be used for four-dimensional data assimilation of nonstationary flows.

1. Introduction

Trajectory models of an atmospheric fluid particle are used for long-range atmospheric transport and dispersion calculations [1-3]. These models are obtained by interpolation and integration of velocity data given by meteorological networks and global numerical models. However, the sole interpolation does not guarantee a correct mass balance and in a previous work [4] it was shown that a small perturbation of the mass balance can change substantially the local structure of a flow. In regions with a complex topography an interpolated horizontal velocity field $\mathbf{v}_h=u\mathbf{i}+v\mathbf{j}$ is used to estimate the vertical velocity w by means of the equation $\nabla \cdot \mathbf{v} = 0$ with $\mathbf{v}=\mathbf{v}_h+w\mathbf{k}$ Although this procedure yields a mass-balanced flow $\mathbf{v}$, the horizontal flow may have an incorrect structure because of a mass imbalance. In this work we show that the structure of a time-dependent horizontal field $\mathbf{v}_h$ can be corrected by means of the so-called variational mass consistent models (VMCM's) [5,6].

Content from this work may be used under the terms of the Creative Commons Attribution 3.0 licence. Any further distribution of this work must maintain attribution to the author(s) and the title of the work, journal citation and DOI.
Published under licence by IOP Publishing Ltd

Among the diagnostic models of the wind field VMCM's have been used because of its simplicity and capacity to accept several measurements in a region with complex terrain [5,6], but these models have been ignored by a wide community devoted to the estimation of trajectory models [1-3]. In this work we show that VMCM's can be used to compute trajectory models from time-dependent data provided by operational networks and global numerical models.

2. Variational mass-consistent models as a projection method

An accepted approximation of the continuity equation that considers the vertical density variation is $\nabla \cdot (\rho_0 \hat{\mathbf{v}})=0$ where ρ_0 is a reference density. In terms of the field $\mathbf{v} = \rho_0 \hat{\mathbf{v}}$ we get the continuity equation $\nabla \cdot \mathbf{v}=0$. In a previous work we show that a perturbation $\mathbf{u}^{\delta}$ with divergence of order $\nabla \cdot \mathbf{u}^{\delta}$ $=10^{-5}$ can yield a divergent field $\mathbf{v}^{\delta}=\mathbf{v}+\mathbf{u}^{\delta}$ with a substantially different local structure, e.g., if $\mathbf{v}$ is elliptic then $\mathbf{v}^{\delta}$ can be hyperbolic. This poses the problem of recovering $\mathbf{v}$ from $\mathbf{v}^{\delta}$. A way is given by the following variational formulation. Consider the inner product between two vector fields $\langle \mathbf{a}, \mathbf{b} \rangle = \int_R \mathbf{a}(\mathbf{r}) \cdot \mathbf{b}(\mathbf{r})\, dR$ in a region *R*. The norm is $\|\mathbf{a} - \mathbf{b}\| = \langle \mathbf{a}, \mathbf{b} \rangle^{1/2}$ and the distance between $\mathbf{a}$ and $\mathbf{b}$ is $\mathrm{d}(\mathbf{a}, \mathbf{b}) \equiv \|\mathbf{a} - \mathbf{b}\|$. A variational mass-consistent model of the wind field is an (adjusted) velocity field $\mathbf{v}^{a}$ that minimizes the functional $\mathrm{J}(\mathbf{w}) = \|\mathbf{w} - \mathbf{v}^{\delta}\|^2$ subject to the condition $\nabla \cdot \mathbf{w} = 0$ [5,6]; that is, $\mathbf{v}^{a}$ is the nondivergent field that is closer to the field $\mathbf{v}^{\delta}$. The solution of this variational problem is

$$\mathbf{v}^a = \mathbf{v}^{\delta} + \nabla \lambda \qquad (1)$$

where λ is determined by the equation

$$\nabla^2 \lambda = \nabla \cdot \mathbf{v}^{\delta}\,. \qquad (2)$$

To get a unique solution λ we impose, by simplicity, the Dirichlet boundary conditions (DBC's)

$$\lambda = 0 \ \text{ on } \ \Gamma \qquad (3)$$

where Γ denotes the boundary of region *R*. A discussion of the improved boundary conditions in region with a complex topography is given in Ref. [7]. The power of VMCM's for trajectory calculations can be explained in terms of the projection properties of this variational approach. Let $\mathbf{L}^2(\mathrm{R})$ be the space of vector fields **w** with components in $L^2(\mathrm{R})$ (the set of square-integrable functions in *R*). This space has the orthogonal decomposition $\mathbf{L}^2(\mathrm{R})$=**W**+**A** where **W** is a set of nondivergent fields **w** and **A** is the set of gradient fields $\nabla\Phi$ where Φ satisfies the DBC (3). The sets **W** and **A** are orthogonal, $\langle \mathbf{w}, \nabla\Phi \rangle = 0$ holds for **w**, $\nabla\Phi$, in **W** and **A**, respectively, and each vector field $\mathbf{u}$ in $\mathbf{L}^2(\mathrm{R})$ is the sum of a *unique* pair $\mathbf{w}^a$,$\nabla\Phi$: $\mathbf{u} = \mathbf{w}^a - \nabla\Phi$ [8]. The orthogonality and uniqueness of $\mathbf{w}_{\mathbf{u}}$ implies it is the orthogonal projection of **u** on the set **W** and the projection theorem for Hilbert spaces asserts that $\mathbf{w}^a$ is the unique element in **W** that minimizes the distance between **u** and the set **W**:

$d(\mathbf{u},\mathbf{w}^a) < d(\mathbf{u},\mathbf{w})$ for **w** in **W**. In the case of a divergent field $\mathbf{v}^\delta = \mathbf{v} + \mathbf{u}^\delta$ obtained by perturbation of a nondivergent one $\mathbf{v}$, we have the decomposition

$$\mathbf{v}^\delta = \mathbf{v} + \mathbf{u}^\delta = \mathbf{v}^a - \nabla\lambda$$

with $\mathbf{v}^a$ in **W** and λ is solution of the problem (2), (3). Thus the VMCM $\mathbf{v}^a$ (1) is the orthogonal projection of $\mathbf{v}^\delta$ on the set of nondivergent fields **W**. It should be emphasized that $\mathbf{v}^\delta$ is unique and independently of the size of a divergent perturbation $\mathbf{u}^\delta$. The examples reported below show that $\mathbf{v}^a$ is almost equal to $\mathbf{v}$ even for large perturbations $\mathbf{u}^\delta$, except in a vicinity of the boundary Γ because of the DBC (2) [7].

In this work we consider velocity fields in a rectangular region $R = [0, x_M] \times [0, y_M]$ and the problem (2), (3), is replaced by the approximated problem

$$\nabla^2\lambda_{mn} = F_{mn}$$

with the Fourier series

$$\lambda_{mn} = \sum_{i=1,j=1}^{mn} \lambda_{ij}\, \varphi_i(x)\varphi_j(y)\ , \qquad F_{mn} = \sum_{i=1,j=1}^{mn} F_{ij}\, \varphi_i(x)\varphi_j(y)\ ,$$

where $F_{ij} = \int_R \varphi_j\varphi_i F dR$, $F = \nabla\cdot\mathbf{v}^\delta$, $\varphi_i = \sqrt{2/x_M} \sin\omega_i x$, $\varphi_j = \sqrt{2/y_M} \sin\omega_j y$, $\omega_i = i\pi/x_M$, $\omega_j = j\pi/y_M$.

Example 1. Consider the nondivergent field $\mathrm{v} = \mathrm{u}\boldsymbol{i} + \mathrm{v}\boldsymbol{j}$ with $\mathrm{u} = ax + (b+\omega)y$, $\mathrm{v} = (b+\omega)\mathrm{x} - ay$, a=-10^{-5}, b=-a, $\omega = a$ (s^{-1}). This flow has a hyperbolic structure and is perturbed by $\mathbf{u}^\delta = \delta(x\boldsymbol{i} + v\boldsymbol{j})$ with $\delta = -a$ which yields $\mathbf{v}^\delta = \mathbf{v} + \mathbf{u}^\delta$ with an elliptic structure [4]. The VMCM $\mathbf{v}^a$ is computed with m=n=200. Figures 1a, 1b, 1c, show the fields $\mathbf{v}$, $\mathbf{v}^\delta$, $\mathbf{v}^a$. We see that $\mathbf{v}^a$ is almost equal to the desired nondivergent field $\mathbf{v}$ and, mainly, $\mathbf{v}^a$ recovers the hyperbolic structure of $\mathbf{v}$. The main difference between $\mathbf{v}$ and $\mathbf{v}^a$ is in a vicinity of the boundary Γ because of the DBC (3) [7].

IOP Conf. Series: Journal of Physics: Conf. Series **792** (2017) 012034 doi:10.1088/1742-6596/792/1/012034

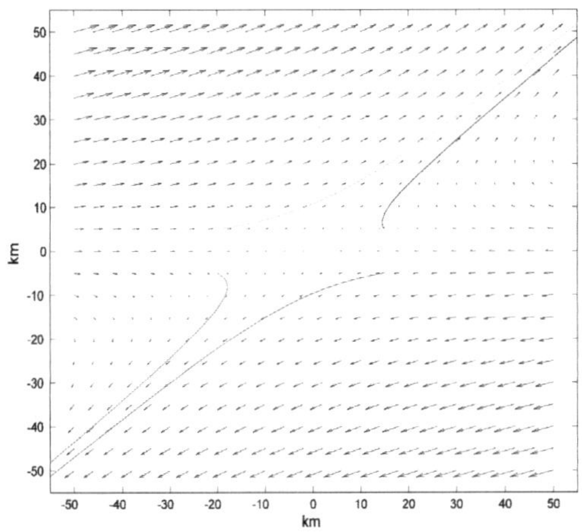

Figure 1a. Nondivergent hyperbolic field **v** and some trajectories.

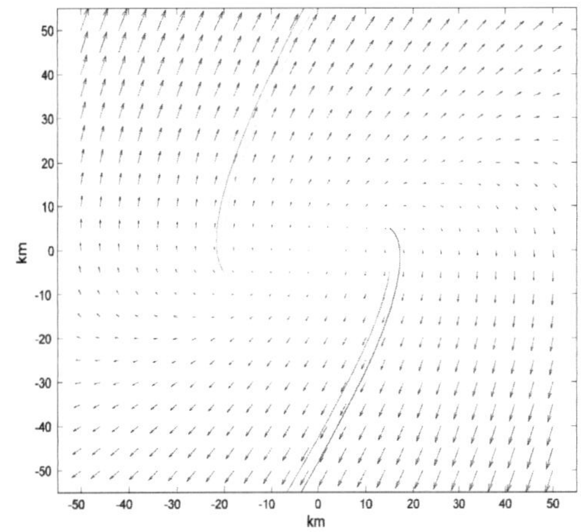

Figure 1b. Perturbed field $\mathbf{v}^{\delta} = \mathbf{v} + \mathbf{u}^{\delta}$ with an elliptic structure.

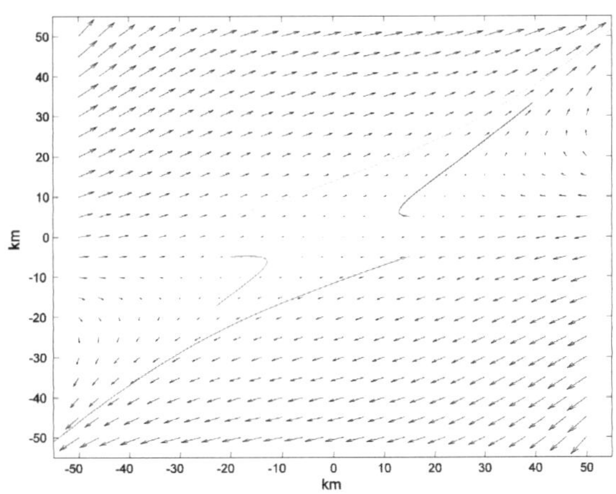

Figure 1c. Adjusted field $\mathbf{v}^{a}_{mn}$ obtained by projection of the perturbed field in Figure 1b.

Variational mass-consistent models have been used mainly as diagnostic wind models at a given time *t* or for climatological studies [5,6] but the next example shows the capability of these models for data assimilation of time-dependent flows and for estimating trajectory models.

Example 2. Consider the nondivergent and nonstationary flow

$$u^0 = V_0 \pi \cos \pi y \,(\sin 2\pi x + \alpha \sin \pi x \cos \omega t), \quad v^0 = -V_0 \pi \sin \pi y \,(2\cos 2\pi x + \alpha \cos \pi x \cos \omega t)$$

where V_0=10km/h, $\alpha = 5$, $\omega = 6$ 1/h, which is modified with the time dependent perturbation $\mathbf{u}^\delta = (1 - \delta x \cos \omega t)\boldsymbol{i} + (1 - \delta y \cos \omega t)\boldsymbol{j}$, δ=5/h. The perturbed flow $\mathbf{v}^\delta = \mathbf{v} + \mathbf{u}^\delta$ has the divergence $\nabla \cdot \mathbf{v}^\delta = 2\delta \cos \omega t$. The adjusted field $\mathbf{v}_{mn}$ at a given time t is computed by solving problem (2), (3), in a sequence of time steps $t_0 = 0, \dots t_n$, used to compute some trajectories with the fields $\mathbf{v}$, $\mathbf{v}^\delta$,, and a fourth-order Runge-Kutta Routine. Figures 2a 2b, 2c, shows the fields **v**, $\mathbf{v}^\delta$, $\mathbf{v}_{mn}$, at t=6 hr, and some trajectories. The perturbation of **v** yields $\mathbf{v}^\delta$ with substantially different trajectories because of the magnitude and time dependence of $\mathbf{u}^\delta$. Nevertheless, Figure 2c shows that the adjusted field $\mathbf{v}_{mn}$ recovers the structure of **v** with high accuracy since the trajectories of $\mathbf{v}_{mn}$ are very similar to those of **v**. To best of our knowledge these time-dependent results have not been reported by the literature devoted to the estimation of trajectory models and mass-consistent models.

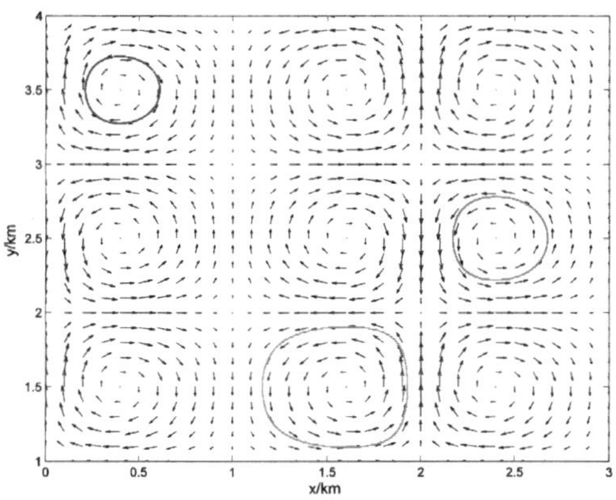

Figure 2a. Nondivergent time-dependent field **v** and some of its trajectories at *t*=6 hr.

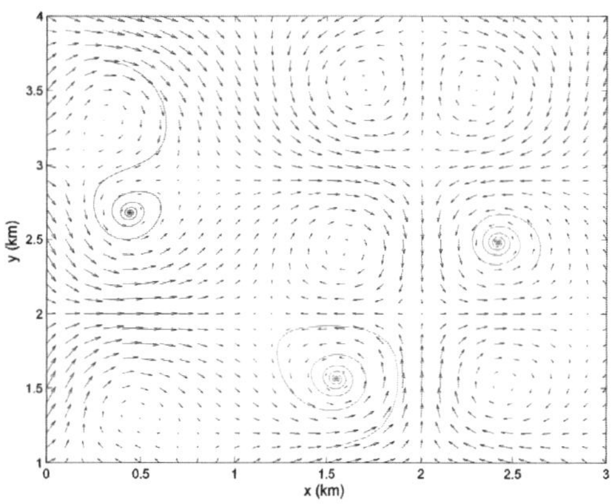

Figure 2b. Perturbed divergent $\mathbf{v}^{\delta} = \mathbf{v} + \mathbf{u}^{\delta}$ and trajectories with the initial conditions of those in Figure 2a.

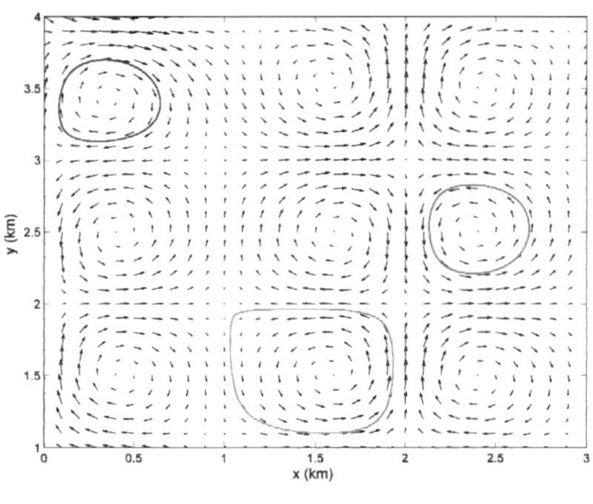

Figure 2c. Adjusted field $\mathbf{v}^{a}_{mn}$ obtained by projection of the perturbed field in Figure 2b at *t*= 6hr and some trajectories with the initial conditions of those in Figure 2a .

3. Conclusions

The results of this work show that VMCM's can be used to improve significantly trajectories models, which are structurally unstable under perturbations of the mass balance, The projection character of VMCM's imply that such models can yield a mass-consistent wind field even when the perturbation of the mass balance is large and time-dependent. Almost all the literature devoted to compute trajectories models has ignored VMCM's [1,3], in part, because such models have been considered as diagnostic models [2,5,6]. However, the Example 2 of this work shows that VMCM's can yield reliable trajectories of real time-dependent atmospheric flows, in other words, VMCM's can be used for four-dimensional data assimilation of velocity data.

Acknowledgments
One of us (R. M.) wants to thank to the Universidad Autónoma Metropolitana for a scholarship.

4. References

[1] Sthol A 1998 Computation, accuracy and applications of trajectories-A review and bibliography *Atmos. Environ.* **32** 947.

[2] Atmospheric Dispersion Modelling Liaison Committee. Annual Report 2005-2006. ISBN 0-85951-587-7.

[3] Bowman K P, Lin J C, Stohl A, Draxler R, Konopka P, Andrews A and Brunner D. 2013 *Input data requirements for Lagrangian trajectory models. Bull. Am. Meteorol. Soc.* **94**, 1051–1058.

[4] Núñez M A and Mendoza R 2015 *Journal of Physics: Conference Series* **582** 012015.

[5] Ratto C F, Festa R, Romeo C, Frumento O A and Galluzzi 1994, Mass-consistent models for wind fields over complex terrain: The state of the art *Environ. Software* **9** 247-268.

[6] Homicz G F 2002 *Three-Dimensional Wind Field Modeling: A Review. Sandia National Laboratories*, Albuquerque, New Mexico.

[7] Núñez M A 2102, *Improving variational mass-consistent models of hydrodynamic flows via boundary conditions, Eur. Phys. J. Plus* **127** 40.

[8] Girault V and Raviart P A 1986 *Finite Element Methods for Navier-Stokes Equation, Theory and Algorithms* Springer, Berlin

IOP Conf. Series: Journal of Physics: Conf. Series **792** (2017) 012035 doi:10.1088/1742-6596/792/1/012035

An approach to get thermodynamic properties from speed of sound

M A Núñez and L A Medina
Departamento de Física, Universidad Autónoma Metropolitana Iztapalapa, A. P. 55-534, C.P. 09340, D. F., México. e-mail: manp@xanum.uam.mx

E-mail: manp@xanum.uam.mx

Abstract. An approach for estimating thermodynamic properties of gases from the speed of sound u, is proposed. The square u^2, the compression factor Z and the molar heat capacity at constant volume C_V are connected by two coupled nonlinear partial differential equations. Previous approaches to solving this system differ in the conditions used on the range of temperature values $[T_{min},T_{max}]$. In this work we propose the use of Dirichlet boundary conditions at T_{min}, T_{max}. The virial series of the compression factor $Z = 1+B\rho+C\rho^2+\ldots$ and other properties leads the problem to the solution of a recursive set of linear ordinary differential equations for the B, C. Analytic solutions of the B equation for Argon are used to study the stability of our approach and previous ones under perturbation errors of the input data. The results show that the approach yields B with a relative error bounded basically by that of the boundary values and the error of other approaches can be some orders of magnitude lager.

1. Introduction

The knowledge of thermodynamic properties of gases is of interest in molecular physics and its industrial applications. Generally, heat capacities are measured with uncertainties which can be orders of magnitude larger than those of thermal properties. The speed of sound u yields a way to estimate caloric properties with an accuracy exceeding that of direct measurements [1]. In this work an approach to estimate thermodynamic properties from speed of sound data, is proposed. The stability of this approach under perturbations is studied together with that of two other approaches [2]-[6], by means of the virial series for gases. Accurate estimation of virial expansions is of interest yield a connection between microscopic and macroscopic gas theories, e. g., to constraint multiparametric equations of state and develop models of intermolecular interactions [7], [8]. The approach and the stability analysis are given in section 2 and section 3 is devoted to some concluding remarks.

Content from this work may be used under the terms of the Creative Commons Attribution 3.0 licence. Any further distribution of this work must maintain attribution to the author(s) and the title of the work, journal citation and DOI.
Published under licence by IOP Publishing Ltd

2. The new approach and its stability

Hereafter partial derivatives will be indicated as follows $\partial Z/\partial T=Z_T$, $\partial^2 Z/\partial T\partial\rho=Z_{T\rho}$. A closed system of equations that connect the speed of sound u with the compression factor $Z=p/R T\rho$ and the molar heat-capacity at constant volume C_V is

$$\rho Z_\rho = - Z - (R/C_V)(Z + TZ_T) + F \,, \quad F = Mu^2/RT \,, \tag{1}$$

$$\rho C_{v,\rho} = - R(T^2 Z_{TT} + 2TZ_T) \,,$$

where T, R, ρ, M, p, are the temperature, the universal gas constant, the molar density and mass, and the pressure. Several approaches have been proposed to solving this system in a rectangular region $\{T_{min} \le T \le T_{max},\ 0 \le \rho \le \rho_{max}\}$ where the speed-of-sound data u are known. Since this is a system of first order in ρ, it is usually solved with initial conditions given by the perfect-gas limit, namely,

$$Z=1, \quad C_V = C_V\,??\,(T) \quad \text{at} \quad \rho_{min}=0 \,. \tag{2}$$

The main difference between previous approaches lies in the conditions used on the interval of temperature values $[T_{min}, T_{max}]$. A first approach, which will be referred to as the initial value method (IVM) [5], has used initial conditions in the temperature

$$Z=Z^{min}(\rho) \,, \quad Z_T=Z_T^{min}(\rho) \quad \text{at } T_{min} \,. \tag{3a}$$

Although this approach can provide accurate thermodynamic properties some inaccuracies in the surface of C_V have been reported and it was pointed out that a source of error is the estimation of the derivative Z_T at T_{min}, whose relative error is one order of magnitude larger than that of Z at T_{min} [5]. In order to avoid the estimation of Z_T in [6] it was proposed to solve the problem with Dirichlet boundary conditions (DBC's) on two isotherms in the lowest range of temperature values T_{min}, T_b,

$$Z=Z^{min}(\rho) \text{ at } T_{min}, \quad Z=Z_b(\rho) \quad \text{at } T_b \,, \tag{4a}$$

e. g, in a range with $T_{max} - T_{min} = 140$ K it was used $T_b \sim T_{min}+5$ K. In this work we propose the use of the initial conditions (2) and the following DBC's

$$Z=Z^{min}(\rho) \quad \text{at } T_{min}, \quad Z=Z^{max}(\rho) \text{ at } T_{max} \,. \tag{5a}$$

A way of studying the stability of these formulations is given by the virial series

$$\begin{aligned} Z &= 1 + B\rho + C\rho^2 + D\rho^3 + \dots \\ C_V &= C_V\,??\,(T) + C_{V,1}\rho + C_{V,2}\rho^2 + \dots \\ F &= \gamma_0(u^2/u_0^2) = \gamma_0 + \gamma_0\beta_a\rho + \dots \,, \qquad u_0^2 = (\gamma_0 RT)/M \,, \end{aligned}$$

which lead to linear ordinary differential equations for B, C, ... The equation for B is

$$B_{xx} + (1+2C_0)\, B_x + 2(C_0+1)C_0\, B = C_0^2\gamma_0\,\beta_a \tag{6}$$

where we set Co= C v ??/*R*, x = ln τ, τ = T/T_r, and T_r is a reference value. To get analytic solutions of equation (6) we consider the gaseous Argon with C v ?? = 3/2*R*, γo = 5/3, and reference values of B are obtained from the equation of state (eos) reported in [8]. Experimental speed-of-sound data u with T between T_{min}=110 and T_{max}=450 yield T_r=280 K yield β_a values reported in [9] with which we get the least-squares estimation

$$\beta_a = -117.7019\tau^{-2} + 210.5188\tau^{-1} - 666.2851 + 1174.822\tau - 839.8723\tau^2 + 314.1316\tau^3 - 48.3716373\tau^4 \quad (7)$$

Hereafter values of B and β are in cm^3mol^{-1}. The conditions corresponding to (3a), (4a), (5a), are

$$B = B^{min}, \quad B_x = B_x^{min} \text{ at } T_{min}, \quad (3b)$$
$$B = B^{min} \text{ at } T_{min}, \quad B = B_b \text{ at } T_b, \quad (4b)$$
$$B = B^{min} \text{ at } T_{min}, \quad B = B^{max} \text{ at } T_{max}. \quad (5b)$$

The eos [8] yields B^{min} = -153.47835, B_x^{min}=280.7384, B^{max}=3.59194, with which the equation (6) is solved analytically. The solution of equation (6) with the conditions (3b), (4b), (5b), will be denoted by B^{IVM}, $B^{DBC(4)}$, $B^{DBC(5)}$, respectively.

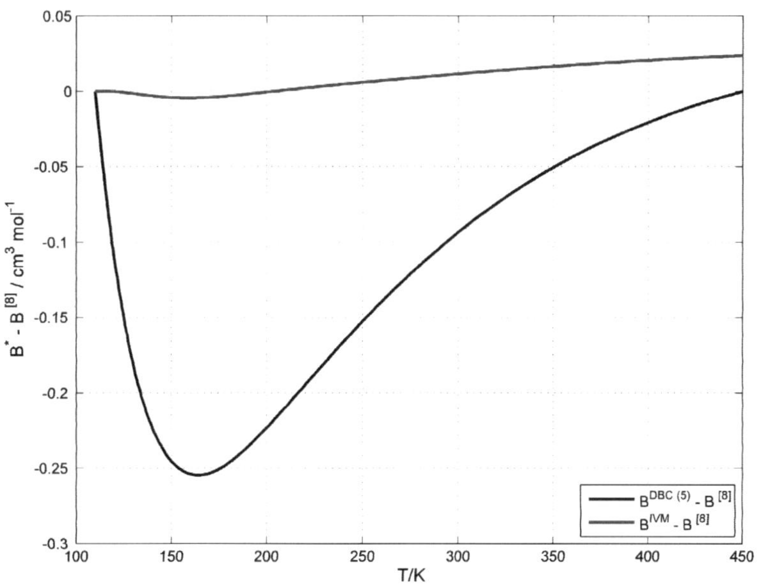

Figure 1. Deviations with respect to the eos for Argon of Ref. [8].

The plot of the difference B^{IVM}-$B^{[8]}$, $B^{DBC(5)}$-$B^{[8]}$, v.s. T is given in figure 1, where $B^{[8]}$ are the values from the eos [8]. The IVM yields B^{IVM} a closer $B^{[8]}$ than the $B^{DBC(5)}$ given by our approach, although $B^{DBC(5)}$-$B^{[8]}$ is of order 0.25, the uncertainty of B values reported in Table 16 of [8]. To study the stability under perturbations of the input data we solve analytically the equation

$$B_{xx}^{e} + (1+2C_0)\, B_{x}^{e} + [(1+k)C_0+2]C_0\, B^{e} = C_0^2\gamma_0\, \beta_a^{e}$$ (8)

where the superscript *e* indicates a solution with errors in the input data. The errors e_j are defined by the following expressions

$$B^e(T_{min}) = (1+10^{-2}e_1)\, B(T_{min}), \quad B_x^e(T_{min}) = (1+10^{-2}e_2)\, B_x(T_{min}), \quad (9)$$
$$B^e(T_{max}) = (1+10^{-2}e_3)\, B(T_{max}), \quad \beta_a^e(T) = (1+10^{-2}e_4 \sin 9x)\, \beta_a(T).$$

Thus e_j yields the percentage error $e_j = 100(q^e - q^{e=0})/q^{e=0}$. The factor $\sin 9x$ introduces a nonuniform error distribution in β_a (7). The effect of an error e_j in the perturbed solution B^e is measured by means of the percentage error

$$\%\Delta B^e = 100\times[(B^e - B^{e=0})/(B^{e=0} - 7.5)] \quad (10)$$

where the denominator $B^{e=0}$-7.5 is used to avoid the singular value B=0 at T~408 K and $B^{e=0}$ denotes the virial coefficient obtained by without input data errors. The relative uncertainty of the velocity data $\Delta u/u \sim 10^{-4}$ yields e_4 of order 0.3%, so that the errors in B are dominated by the error in complementary conditions (3), (4), (5).

The graph of $\%\Delta B^{IVM,e}$ with $e_1=e_4=0.3$ and increasing values of the error e_2=0.3, 1, 5, in B_x, is given in figure 2 and shows that the monotonic growing of the absolute value of $\%\Delta B^{IVM,e}$ is faster as the error in the derivative increases. In contrast, figure 3 shows that the error $\%\Delta B^{DBC(5),e}$ is bounded basically by e_j. The second DBC in (4a) is defined with T_b=120 K and $B^{DBC(5),e=0}$= -130.8880 at T_b. The perturbation of this value is

$$B^{DBC(4),e}(T_b) = (1+10^{-2}e_b)\, B^{DBC(5),e=0}(T_b).$$

For $e_b=e_j$ the absolute value of the error $\%\Delta B^{DBC(5),e}$ is bounded by that of e_j only in the interval $[T_{min},T_b]$ and grows monotonically as T increases in the complementary region $[T_b,T_{max}]$. Figure 4 shows that this absolute error can grow rapidly when the sign of e_1 and e_b is different. Hence, the DBC's (4a) can yield a very unstable solution.

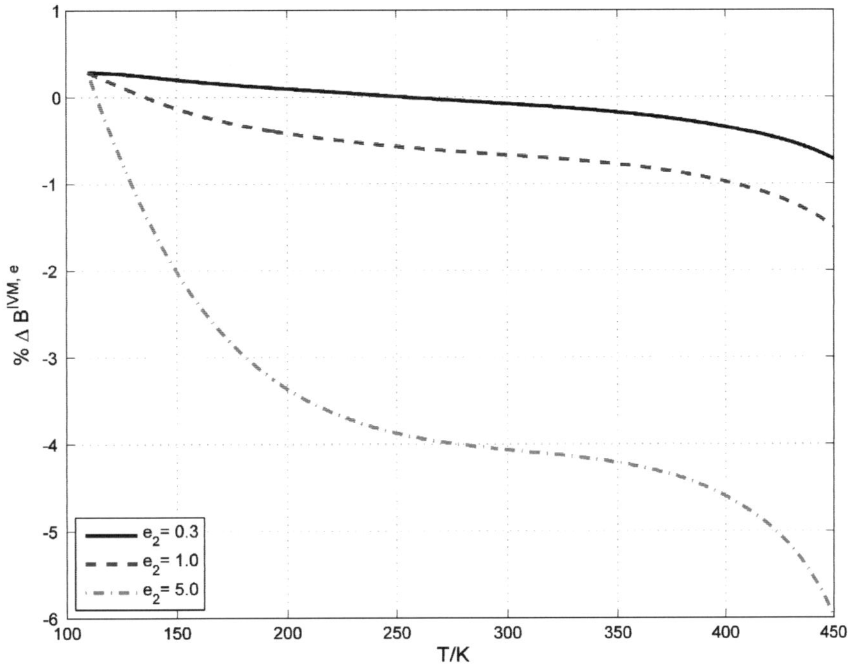

Figure 2. Relative error with e_1=e_4=0.3 and increasing values in the error e_2 of the derivative B_x at T_{min}.

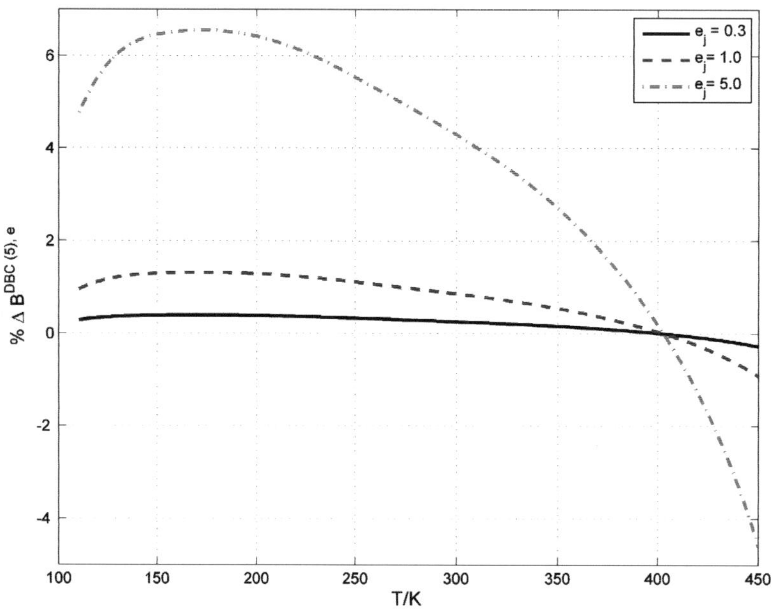

Figure 3. Relative error of $B^{DBC(5)}$ with increasing values of $e_1 = e_3 = e_4$.

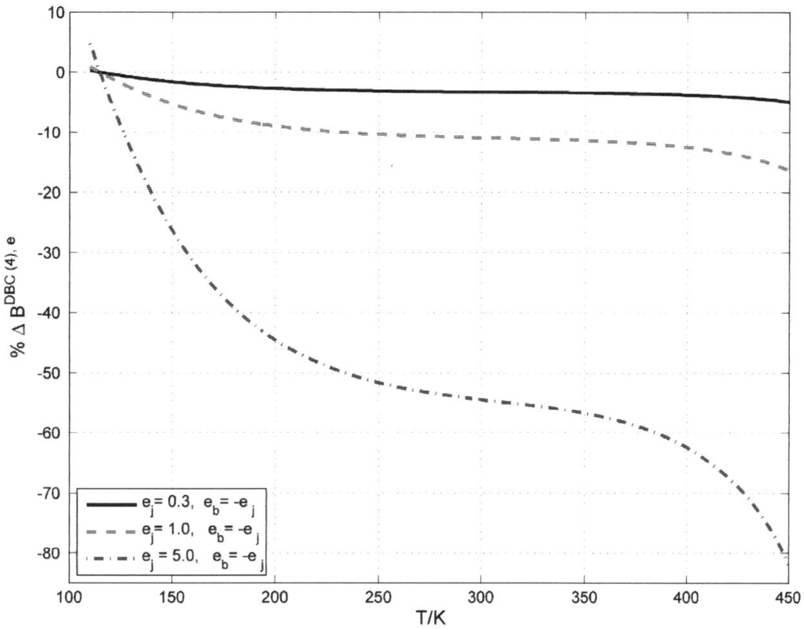

Figure 4. Relative error of $B^{DBC(4)}$ determined by DBC's (4a).

3. Conclusions

This work proposes an approach to compute thermodynamic properties from speed of sound data by solving equations like (1) by means of DBC's (5). Analytic calculations of the virial coefficient B show that these boundary conditions yield stable calculations with an error bounded approximately by that of the DBC's, in contrast with the other approaches whose error can grow rapidly as T increases.

4. References

[1] Trusler J P M 1991 *Physical Acoustics and Metrology of Fluids* (Adam Hilger, Bristol).
[2] Trusler J P M and Zarari M J 1992 *J. Chem. Thermodynamics* **24** 973.
[3] Dayton T C, Beyerlein S W and Goodwin A R H 1999 *J. Chem. Thermodynamics* **31** 847.
[4] Estrada-Alexanders A F, Trusler J P M and Zarari, M P 1995 *Int. J. Thermophys.* **16** 663.
[5] Estrada-Alexanders A F and Trusler J P M 1996 *International Journal of Thermophysics* **17** 6.
[6] Bijedi´c M and Neimarlija N 2013 *Latin American Applied Research* **43** 393.
[7] Maitland G C, Rigby M, Smith E B and Wakeham W A 1981 *Intermolecular Forces* (Clarendon Press, Oxford).
[8] Tegeler C, Span R and Wagner W 1999 *J. Phys. Chem. Ref. Data* **28** 779.
[9] Estrada-Alexanders A F and Trusler J P M 1995 *J. Chem. Thermodyn.* **27** 1075.

IOP Conf. Series: Journal of Physics: Conf. Series **792** (2017) 012036 doi:10.1088/1742-6596/792/1/012036

Design, construction, characterization, and operation of a hybrid cosmic rays detector based on an electron gas

F J Rosas-Torres, M A Hernández-Morquecho, J Sánchez-Sánchez and J Félix

Laboratorio de Partículas Elementales, Departamento de Física, Campus León, Universidad de Guanajuato, Loma del Bosque 103, León 37150, Guanajuato, México

E-mail: felix@fisica.ugto.mx

Abstract. There are several sources that produce very energetic cosmic rays that interact with the Earth's atmosphere and create new particles. To detect them there are different methods such as the ionization of a material and Cerenkov radiation, among others. In this work a hybrid cosmic ray detector of 6 channels was designed, built, tested and operated. Being hybrid is possible to validate the signal with the two detection methods. Three Copper bars were used as detection material, each with an ionization and a Cerenkov radiation detection channel. To detect the Cerenkov radiation, Hamamatsu silicon photodiodes were used, and for the ionization channels an RC circuit was developed to measure the signal. The number of signals was counted using discriminator boards, which digitize the signal. With the counts the cosmic rays flux can be measured. The six channels were tested simultaneously. Data collections and analysis were performed. Details of the design, characterization, testing, operation, data analysis and preliminary results are presented.

1. Introduction

Cosmic rays consist of particles that come from the outer space, large numbers of them interact with the atmosphere of the earth. Such particles are very energetic, of a few GeV´s. These particles are classified into primary and secondary. The primary particles are those created in an astrophysical source. The secondary ones are those that are generated by the interaction between primary cosmic rays and the atmosphere of the planet [1].

Ionization takes place when a particle has enough energy to ionize an atom of a material, which results in an electron and an ion generating a flow of current. This flow of current is amplified if it is immersed in an electric field.

It is thought that Cerenkov radiation only occurs in a dielectric when a particle travels faster than the electromagnetic radiation in the corresponding medium [2].

The Cerenkov radiation detectors are generally developed using materials that are transparent to visible light. On the other hand, detectors that work by ionization normally use a gas. In spite of that, a prototype with an opaque detection material, and that can be ionized, a solid metal that due to the large number of electrons we consider it as an electron gas, was developed.

Content from this work may be used under the terms of the Creative Commons Attribution 3.0 licence. Any further distribution of this work must maintain attribution to the author(s) and the title of the work, journal citation and DOI.
Published under licence by IOP Publishing Ltd

IOP Conf. Series: Journal of Physics: Conf. Series **792** (2017) 012036 doi:10.1088/1742-6596/792/1/012036

To measure the flux of cosmic rays, a hybrid detector was developed using the two techniques mentioned (ionization and Cerenkov radiation). Using both techniques at the same time it is possible to validate the obtained signal. The identification of incident particles is not considered in planning this detector, it is not of interest in this project. Normally, for the detection of Cerenkov radiation a photodiode [3] is used. A photodiode is a device that transforms light signal into voltage signal or electrical current. For the detection by ionization we use an RC circuit that operates when a pulse is produced by a particle interacting with the material immersed in an electric field. To achieve the data analysis, digital signals are used, however the output of the RC circuit is an analog signal. To transform it from analog to digital a discriminator card is used. When the discriminator receives a voltage higher than the control voltage, which is adjusted in the board using a variable resistor, a digital one is obtained, that is, 5 volts, otherwise a 0 is produced.

2. Motivation

To build a detector using ionization detection and Cerenkov radiation, Aluminum was first used. After some tests the material was changed to Copper, since it is approximately three times denser (Al=2.699 g/ml, Cu=8.960 g/ml) [4]. So it is more likely that the particles interact with the material. The ionization energy of the aluminum is less than that of Copper, however, that is not considered a drawback. Cosmic rays energy spectrum is very wide.

Taking advantage that our device is a hybrid detector, the information is validated using two different detection channels in the same bar. An array of three bars, one on top of the other is also used to validate the information. In bar 2 when bar 1 and 3 have signal as shown in Figure 1. With this arrangement it´s possible to know if the particle travels from top to bottom or vice versa, depending on the time of detection in each channel.

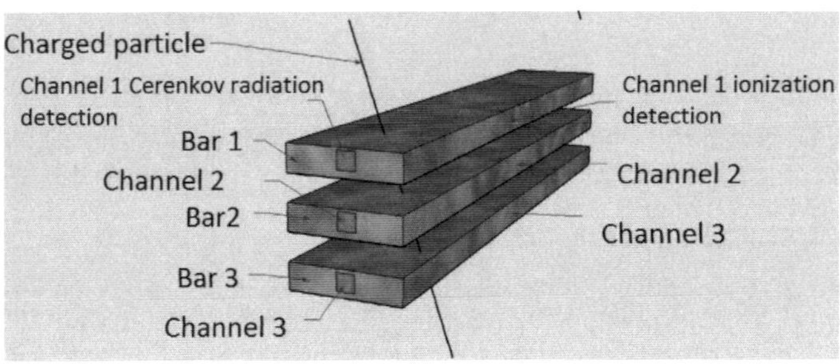

Figure 1. Corroboration system scheme.

3. Design

The design for the prototype was developed using the commercial software SketchUp 2013. The design consists of three Copper bars, one on top of the other, separated by an acrylic bar, to isolate the Copper electrically, as shown in Figure 2. Copper collectors were placed on the sides of the Copper bars that collect the signal and transport it to the RC circuit. As it can be seen in Figure 3. The circuit was fixed to an Aluminum base shaped as a bracket that is screwed to the base system.

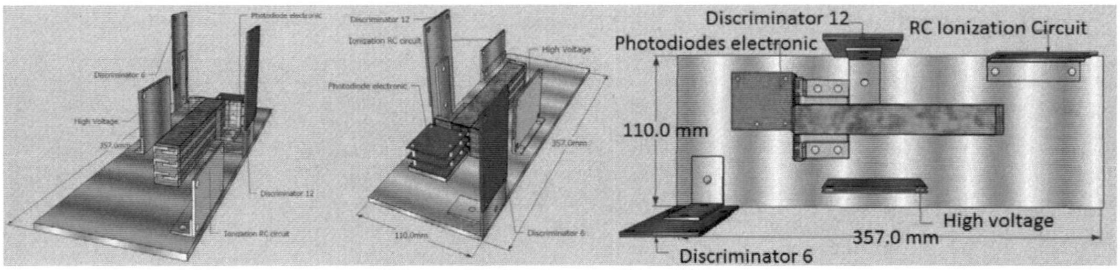

Figure 2. Design on SketchUp.

The electric field is produced using a high voltage source Ultravolt *3M24-P1 M SERIES* [5] with the configuration specified in Figure 4. To fix the electronic board to the Ultravolt voltage source, we used an Aluminum bracket. Two acrylic brackets screwed to the base, were used to maintain the Copper bars in place, and to facilitate the high voltage connection.

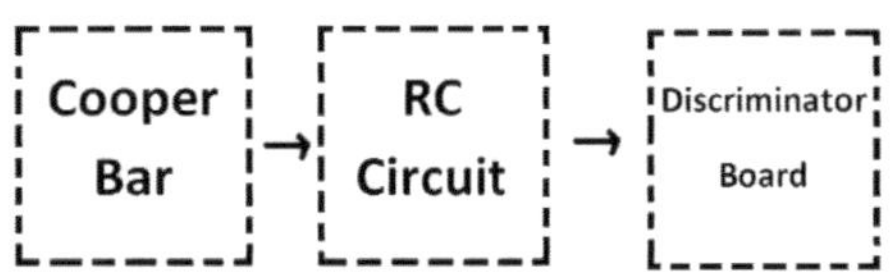

Figure 3. RC circuit scheme.

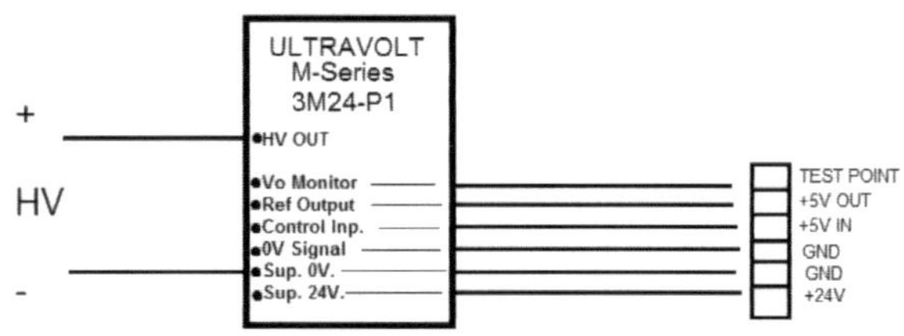

Figure 4. Configuration of high voltage board.

The Hamamatsu photodiode *S12572-100P* [6] was fixed to the Copper bars, making contact with the surface, welded to a phenolic board that has the pins that connect the photodiode to the voltage source and amplification board. To keep in place the photodiode, a small Aluminum rectangle with an orifice the size of the optical sensitive area was made. Electrical insulating tape was used to isolate the photodiodes electrically and Aluminum tape was placed on top.

The voltage source and amplification board that were used in the *hamamatsu photodiode electronic*, to activate the photodiodes and couple the signal, so that the discriminator can process it. To adjust the level of the *hamamatsu photodiode electronic*, we used an Aluminum base glued to the general base and board separators, so that they could be connected without cables to the photodiodes pins. To fix the discriminator to the whole system, Aluminum brackets were used.

4. Construction

As already mentioned, a prototype using Aluminum was first made using a 5 cm x 5 cm x 1 cm Aluminum bar. The bar was polished and cleaned. An RC circuit was constructed so that it gives a signal with exponential decay, with the shortest decay time possible.

For the Copper detector a Copper plate was used. The plate was cut with a waterjet device into three bars with 22 mm x 6 mm x 151 mm bars, then they were polished using sand paper and cleaned.

Three bars of transparent acrylic were cut with the same dimensions as the Copper bars, sanded on the edges and cleaned. They were glued with instant transparent glue in an interleaved way with the Copper bars, with an acrylic bar in the bottom and a Copper bar on the top.

The RC circuit for the ionization channels [Figure 5] was constructed as follows: The high voltage source UltraVolt 3M24-P1 M SERIES was weld to a phenolic board, as seen in Figure 6.

Figure 5. Ionization RC circuit.

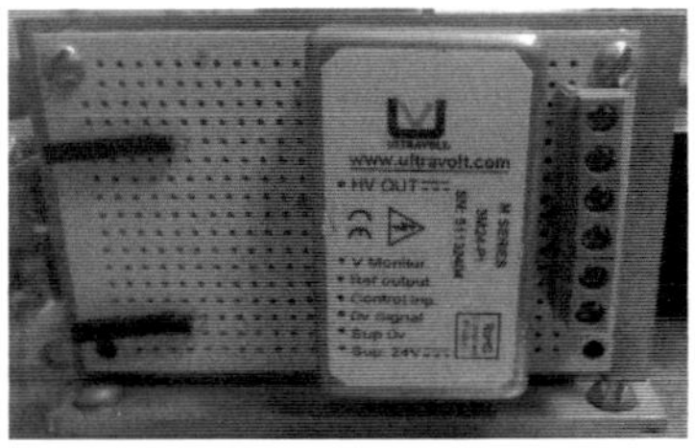

Figure 6. Source of high voltage (UltraVolt).

Figure 7. Frontal view of optical channels.

The RC circuit for the ionization channels [Figure 5] was constructed as follows: The high voltage source UltraVolt 3M24-P1 M SERIES was weld to a phenolic board, as seen in Figure 6.

For the optical channels, a phenolic plate of 6 x 22 mm was cut and two guides were drawn to weld two pins and the photodiode. Three Aluminum plates, with a square orifice at the center, the size of the photodiode, were made. At the edges of the orifice a black liquid electrical insulator was applied. These plates were glued with instant transparent glue to the bare Copper, on the opposite side to the RC circuit. Subsequently the photodiodes were placed and electrical tape was used to isolate the photodiodes electrically and from visible light. To isolate from other kinds of electromagnetic radiation, Aluminum tape was placed on top as shown in Figure 7. Electrical insulating tape was used to avoid electric arcs with the electrodes. The only thing left exposed are the pins.

To place the cards that activate the photodiode at the right height, five Aluminum plates were cut, perforated, cleaned and glued together to the base. Board spacers were used to match the cards with the photodiode pins. For the RC circuits, the two discriminatory cards and the high-voltage source, L-shaped Aluminum pieces were made as a vertical support for the circuits and screwed to the base.

5. Operation

We verified the correct operation of the Copper detector (Figure 8), starting with the optical channels. Next, we connected the ionization channels, the high applied voltage was approximately 2933 V. The expected signal was visualized on the oscilloscope, obtaining the pulses shown in Figure 9.

Figure 8. Detector connected to the CompactRIO.

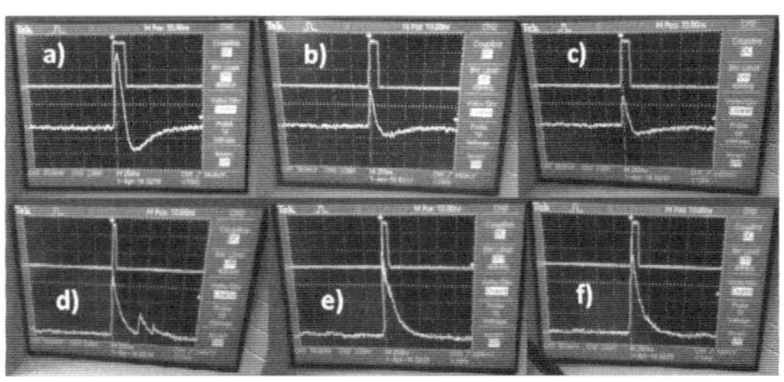

Figure 9. Signal on oscilloscope. Vertical axis voltage, horizontal axis time. Ionization channels: 1 → a), 2 → b), 3 → a); Optical channels: 1 → d), 2 → e), 3 → f).

Having the six channels functioning properly at the same time, the output signal was disabled and connected to a 32 channels data acquisition system (CompactRIO), which works with a 40MHz clock. When the six channels were connected, the output signal was enabled and the reference voltages, that was adjusted for each channel in the discriminator board, to about 3 times the noise level. The obtained values can be seen in Table 1.

6. Characterization

The detector characterization consisted of modifying the voltage supply value for the photodiode in the optical channels and the high voltage for the ionization channels, making runs of 10 minutes to obtain data using the CompactRIO. The ionization voltage was modified from 1222.22 V (2 V in the source) to 3055.55 V (5 V in the source) with increments of 0.2 V in the source. The voltage of the optical channels was changed from 60 V to 100V with increments of 5 V in each run.

The graphs obtained with the characterization data are shown in Figure 10. As it can be seen, each channel has a different voltage with which a region with stable counts is obtained, these voltage values can be observed in the Table 1.

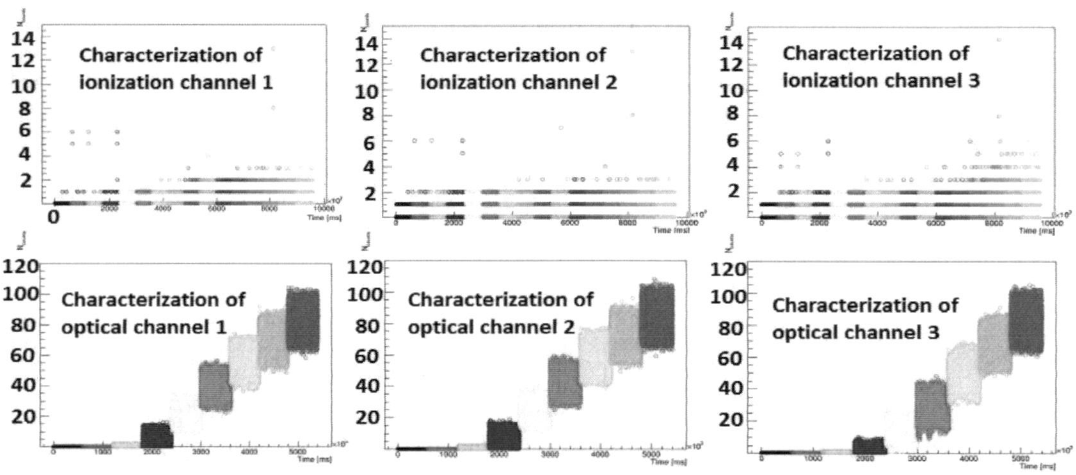

Figure 10. Graphics of characterization, vertical axis counts, horizontal axis time (each color represents 10 minutes).

Table 1. Operation and trigger voltage.

Channel	Trigger Voltage (mV)	Operation Voltage (V)
1 Ionization	50.0	1700 to 1800
2 Ionization	60.0	1700 to 1800
3 Ionization	80.0	1700 to 1800
1 Optical	50.0	70 to 75
2 Optical	50.0	70 to 75
3 Optical	50.0	70 to 75

IOP Conf. Series: Journal of Physics: Conf. Series **792** (2017) 012036 doi:10.1088/1742-6596/792/1/012036

7. Results

With operating voltages of 2750 V and 75 V, a run of 30 minutes was performed and the following distributions of frequencies for each channel were obtained: Figures 11 (a, b, c, d, e, f).

The signal of channel 2 was validated in two different ways, first using the 3 optical channels, by having signal in bar 1 and 3 the signal is validated in bar 2, obtaining what is shown in figure 11 (j). The other validation was done using the ionization and optical channel of bar 2, where the optical signal is validated if a signal is detected in the ionization channel (figure 11 (i)). Channel 1 and 3 were validated in the same way (figures 11 (g, h, k, l)).

8. Conclusion

It was possible to design, build and characterize a detector with 6 channels functioning at the same time, and obtaining signal simultaneously. A similarity between the 3 optical channels is observed when plotting the frequency distribution. The same happens between the 3 ionization channels. This makes us assume that the 6 channels are functioning properly. The calibration of the detector will be performed using a known radiation source, this will allow us to measure the flux of cosmic rays.

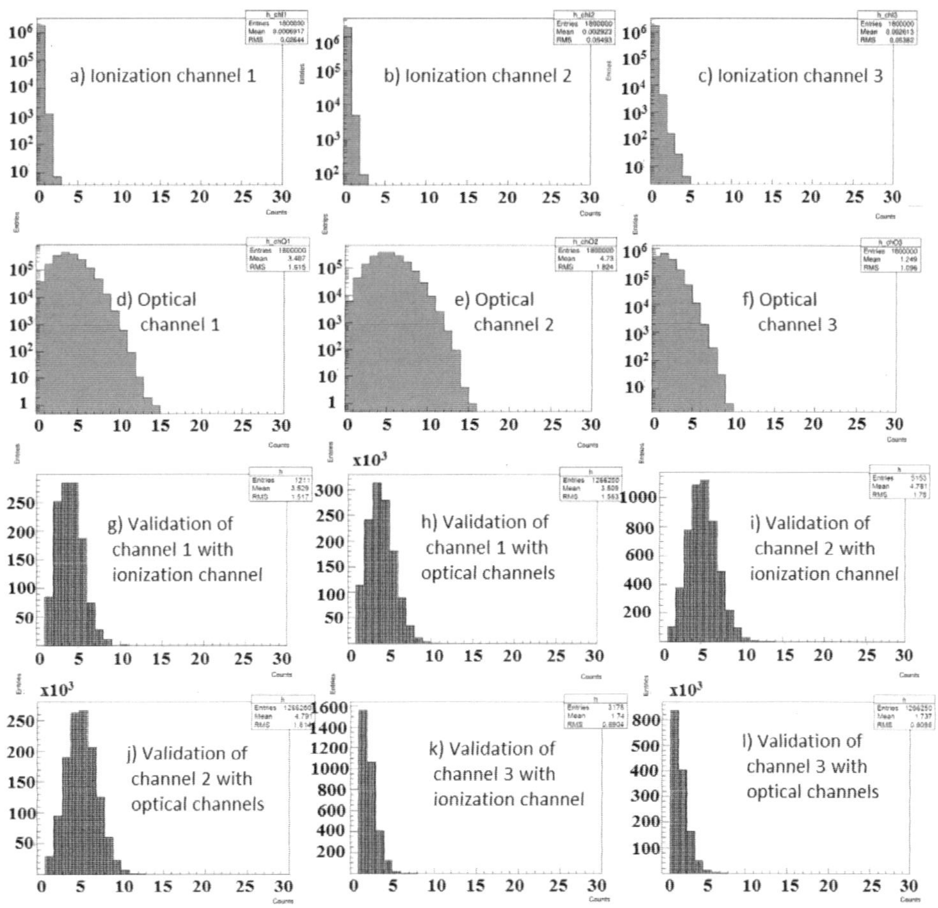

Figure 11. Histograms, vertical axis frequency, horizontal axis counts.

References

[1] Beatt J J, Matthews J and Wakely S P. *Cosmic Ray* [Internet]. http://pdg.lbl.gov/2015/reviews/rpp2015rev-cosmic-rays.pdf (accessed 21 June 2016).

[2] Jin-Kyu So, Jong-Hyo Won, Sattorov M A, Seung-Ho Bak, Kyu-Ha Jang, Gun-Sik Park, Kim D S and Garcia-Vidal F J *Cerenkov radiation in metallic metamaterials 2010* Applied Physics Letters; 97(15). http://scitation.aip.org/content/aip/journal/apl/97/15/10.1063/1.3492846 (accessed 21 june 2016).

[3] Hamamatsu phototonics K.K. and "Opto-semiconductor Handbook" Editorial Committee. *Opto-semiconductor Handbook.* : Hamamatsu phototonics K.K. and Solid State Divition; 2014. https://www.hamamatsu-news.de/hamamatsu_optosemiconductor_handbook/#362/z (accessed 21 June 2016).

[4] *Atomic and nuclear properties of materials.* K.A. Olive et al. (Particle Data Group), Chin. Phys. C, 38, 090001 (2014) and 2015 update.

[5] Advanced Energy Industries, Inc. Ultravolt M Series Minature, *Micro-Sized High Voltage Biasing Supplies* [Internet]. http://www.ultravolt.com/uv_docs/MSeries.pdf (accessed 21 June 2016).

[6] MPPC S12572-100P | *Hamamatsu Photonics:MPPC (SiPM) S12572-100P* [Internet]. Hamamatsu.com. 2016 (cited 22 June 2016). Available from: http://www.hamamatsu.com/jp/en/S12572-100P.html

[7] *The elements of the periodic table sorted by ionization energy* [Internet]. Lenntech.com. 2016 (cited 21 June 2016). Available from: http://www.lenntech.com/periodic-chart-elements/ionization-energy.htm

IOP Conf. Series: Journal of Physics: Conf. Series **792** (2017) 012037 doi:10.1088/1742-6596/792/1/012037

Position-Dependent Mass Schrödinger Equation for the Morse Potential

G Ovando[1*], J J Peña[1], J Morales[1] and J López-Bonilla[2]

1. Universidad Autónoma Metropolitana - Azcapotzalco, CBI - Area de Física Atómica Molecular Aplicada, Av. San Pablo 180, 02200 México City, MEXICO.

2. Instituto Politécnico Nacional, Edif. 5, 1er. Piso, ESIME-Zacatenco, Col. Lindavista, 07738 Mexico City, MEXICO.

*). E-mail: gaoz@azc.uam.mx

Abstract. The position dependent mass Schrödinger equation (PDMSE) has a wide range of quantum applications such as the study of semiconductors, quantum wells, quantum dots and impurities in crystals, among many others. On the other hand, the Morse potential is one of the most important potential models used to study the electronic properties of diatomic molecules. In this work, the solution of the effective mass one-dimensional Schrödinger equation for the Morse potential is presented. This is done by means of the canonical transformation method in algebraic form. The PDMSE is solved for any model of the proposed kinetic energy operators as for example the BenDaniel-Duke, Gora-Williams, Zhu-Kroemer or Li-Kuhn. Also, in order to solve the PDMSE with Morse potential, we consider a superpotential leading to a special form of the exactly solvable Schrödinger equation of constant mass for a class of multiparameter exponential-type potential along with a proper mass distribution. The proposed approach is general and can be applied in the search of new potentials suitable on science of materials by looking into the viable choices of the mass function.

1. Introduction

The Morse potential [1] has been an adequate model for the potential energy function of diatomic molecules, which is the reason for its relevance in chemistry and molecular physics. Although there have been proposed several alternative models of potential functions in order to improve experimental data [2]-[8], Morse potential has maintained its relevance until nowadays [9]. In addition, alternative mentioned potentials are equivalent between them [10], a fact that contributes to maintaining the importance of the Morse potential. On the other hand, the position-dependent mass Schrödinger equation (PDMSE) has been of recent interest for its applications in fields such as condensed matter theory [11]-[15], hetero-structures [16]-[19], nuclear clusters [20]-[21], density functional theory [22]-[26] and related problems. At this regard, different proposals of the kinetic energy operator in the

Content from this work may be used under the terms of the Creative Commons Attribution 3.0 licence. Any further distribution of this work must maintain attribution to the author(s) and the title of the work, journal citation and DOI.
Published under licence by IOP Publishing Ltd

PDMSE Hamiltonian [27] have been considered as useful operators for applications, as for example BenDaniel and Duke [28], Gora-Williams [29], Zhu-Kroemer [30], Li-Kuhn [31], and so on.

In this work we present exact solutions of the Morse potential for the PDMSE and analyze the influence of the position-dependent mass (PDM) in the energy spectra. To solve the PDMSE is used the point canonical transformation scheme [32], which algebraic presentation is known [33]-[34]. In this context it is revealed the potential function of a constant mass Schrödinger equation (CMSE) involved in the resolution. Our exposition considers obtain solutions for any kinetic energy operator of the PDMSE. The general ordering ambiguity has also been treated by means of the Nikiforov-Urvanov method [35]. To perform our task, we use the exact solution of the multiparameter exponential-type potential in the constant mass Schrödinger equation [36]. In this way, the next section outlines the algebraic relation between the PDMSE and the CMSE. Section 3 is devoted to show the use of the multiparameter exponential-type potential in order to solve the required Morse potential. Finally, we study the influence of the position-dependent mass on the energy spectra, which means comparing the CMSE energy spectrum and the PDMSE energy spectrum for the same Morse potential.

2. Algebraic approach to the PDMSE

Let us consider the position-dependent mass Schrödinger equation (PDMSE)

$$-\frac{d}{dx}\left(\frac{1}{2m(x)}\frac{d}{dx}\psi(x)\right)+\left(V(x)+U_{\alpha\gamma}(x)\right)\,\psi(x)=E\,\psi(x), \tag{1}$$

which comes from the von Roos's Hamiltonian $H_{\alpha\gamma}=T_{\alpha\gamma}+V(x)$, where the position-dependent mass kinetic energy operator $T_{\alpha\gamma}$ is [27]

$$T_{\alpha\gamma}=\frac{1}{4}\left[m^{\gamma}\,pm^{\beta}\,pm^{\alpha}+m^{\alpha}\,pm^{\beta}\,pm^{\gamma}\right], \tag{2}$$

$m=m_0 m(x)$ is the mass operator, m_0 is the mass of the involved particle, $p=-i\hbar d/dx$ is the linear momentum operator and the parameters α, β, γ satisfy the constraint $\alpha+\beta+\gamma=-1$. We have selected the natural unit system with $\hbar=m_0=1$. The Eq. (1) follows after using the identity

$$T_{\alpha\gamma}=-\frac{d}{dx}\frac{1}{2m(x)}\frac{d}{dx}+U_{\alpha\gamma}(x), \tag{3}$$

with

$$U_{\alpha\gamma}(x)=\frac{1}{2}\left(\alpha\gamma+\alpha+\gamma\right)\frac{m'^2(x)}{m^3(x)}-\frac{1}{4}\left(\alpha+\gamma\right)\frac{m''(x)}{m^2(x)}. \tag{4}$$

and the expression $V(x)+U_{\alpha\gamma}(x)$ is usually named the effective potential $U_{\mathbf{eff}}(x)$. In [33] we proposed to use instead of Eq. (3) the one given by

$$T_{\alpha\gamma}=A_0^{+}\,A_0^{-}+\tilde{U}_{\alpha\gamma}(x) \tag{5}$$

where $A_0^{\pm} = \mp i\eta p\eta = \mp \eta \frac{d}{dx}\eta$, $\eta = (2m)^{-1/4}$ and

$$\widehat{U}_{\alpha\gamma}(x) = \frac{1}{2}\left(\alpha\gamma + \alpha + \gamma + \frac{7}{16}\right)\frac{m'^2}{m^3} - \frac{1}{4}\left(\alpha + \gamma + \frac{1}{2}\right)\frac{m''}{m^2}, \tag{6}$$

for which Eq. (1) turns to be

$$\left(A_0^+ A_0^- + \widehat{U}_{\alpha\gamma}(x) + V(x)\right)\psi(x) = E\,\psi(x), \tag{7}$$

with the advantage that this form allows the factorization of the PDMSE

$$A^+ A^- \psi(x) = E\,\psi(x), \tag{8}$$

with factorization operators A^+ and A^- of the form

$$A^{\pm} = A_0^{\pm} + \widehat{W}(x). \tag{9}$$

By introducing the main potential

$$\widehat{V}(x) = \widehat{U}_{\alpha\gamma}(x) + V(x) \tag{10}$$

it results that function $\widehat{W}(x)$ satisfies

$$\widehat{V}(x) = \widehat{W}(x)^2 - \eta^2 \frac{d}{dx}\widehat{W}(x). \tag{11}$$

Then, the point canonical transformation defined by

$$\frac{d}{du} = \eta^2 \frac{d}{dx}, \tag{12}$$

which means $u(x) = \int^x \eta^{-2}(x)dx$ and $x = x(u)$ its inverse function, and the functions of the new variable $\tilde{w}(u) = \widehat{W}(x(u))$, $\tilde{v}(u) = \widehat{V}(x(u))$ allows to get from Eq. (11)

$$\tilde{v}(u) = \tilde{w}(u)^2 - \frac{d}{du}\tilde{w}(u). \tag{13}$$

This last equation is identified as the Riccati equation connecting a potential $\tilde{v}(u)$ with its superpotential $\tilde{w}(u)$ in the SUSYQM context. In fact, the factorization operators for a constant mass Schrödinger equation (CMSE) corresponding to this superpotential $a^{\pm} = \mp\frac{d}{du} + \tilde{w}(u)$ satisfy the operator identity

$$a^{\pm} = \eta A^{\pm}\eta^{-1}, \tag{14}$$

meaning that the factorized PDMSE Eq. (8) written in the form

$$(\eta A^{+}\eta^{-1})(\eta A^{-}\eta^{-1})(\eta\,\psi(x)) = E(\eta\,\psi(x)), \tag{15}$$

becomes the CMSE

$$a^{+}\,a^{-}\,\phi(u) = E\,\phi(u), \tag{16}$$

with CMSE eigenfunctions $\phi(u)$ connected with the PDMSE eigenfunctions by $\phi(u) = \eta(x(u))\,\psi(x(u))$.

Previous formalism shows that we can use any solvable potential $\widetilde{v}(u)$ of the CMSE to get a solvable main potential

$$\widetilde{V}(x) = \widetilde{v}(u(x)) \tag{17}$$

of the PDMSE Eq. (8), with solutions given by

$$\psi(x) = \eta^{-1}(x)\,\phi(u(x)). \tag{18}$$

The ground-state energy of CMSE is set to zero (good SUSYQM) and the ground-state eigenfunction is obtained from $a^{-}\phi_0(u) = 0$, which is $\phi_0(u) = \exp(-\int^{u}\widetilde{w}(u)du)$, this implies the ground state of the PDMSE in the form

$$\psi_0(x) = \eta^{-1}e^{-\int^{u(x)}\frac{\widehat{W}(x)}{\eta^2}dx}, \tag{19}$$

although it can also be obtained from $A^{-}\psi_0(x) = 0$.

The main potential Eq. (10) corresponds to the PDMSE with kinetic energy operator determined by selection of operators Eq. (8) which means $\alpha = \gamma = -1/4$. The solvable potentials for other kinetic energy operators come from Eq. (10), that is

$$V(x) = \widetilde{V}(x) - \widetilde{U}_{\alpha\gamma}(x). \tag{20}$$

with $\widetilde{U}_{\alpha\gamma}(x)$ given in Eq. (6) or alternatively by

$$\widetilde{U}_{\alpha\gamma}(x) = (4\alpha+1)(4\gamma+1)\,\eta^{2}\eta'^{2} + \left(\alpha+\gamma+\frac{1}{2}\right)\eta^{2}(\eta^{2})'', \tag{21}$$

where the apostrophe indicates derivative with respect to variable x. Table I lists some solvable PDMSE for particular choices of α and γ.

IOP Conf. Series: Journal of Physics: Conf. Series **792** (2017) 012037 doi:10.1088/1742-6596/792/1/012037

Table I. Parameters and solved potentials of typical Hamiltonians $H_{\alpha\gamma}$ for the PDMSE.		
$\alpha,\ \gamma$	$\tilde{U}_{\alpha\gamma}(x)$	$V(x)=\tilde{V}(x)-\tilde{U}_{\alpha\gamma}(x).$
$\alpha=\gamma=0$	0	$\tilde{V}(x)$
$\alpha=\gamma=0$	$\eta^2(\eta')^2+\frac{1}{2}\eta^2(\eta^2)''$	$V^{BDD}(x)$, BenDaniel-Duke
$\alpha=-1,\ \gamma=0$	$-3\eta^2(\eta')^2-\frac{1}{2}\eta^2(\eta^2)''$	$V^{GW}(x)$, Gora-Williams
$\alpha=\gamma=-1/2$	$\eta^2(\eta')^2-\frac{1}{2}\eta^2(\eta^2)''$	$V^{ZK}(x)$, Zhu-Kroemer
$\alpha=0,\ \gamma=-1/2$	$-\eta^2(\eta')^2$	$V^{LK}(x)$, Li-Kuhn

3. Exact solutions to the Morse potential

Now in order to solve PDMSE with Morse potential we will resort to the CMSE solutions of exponential type, given in ref. [36], and which are obtained from the superpotential

$$\tilde{W}(u)=\frac{K^2+A_0-B_0}{2K}+K\frac{e^{-u}}{1-e^{-u}}, \tag{22}$$

that allows to write the CMSE potential in the form

$$\tilde{V}(u)=\left(\frac{K^2+A_0-B_0}{2K}\right)^2+\frac{A_0e^{-u}}{1-e^{-u}}+\frac{B_0e^{-2u}}{\left(1-e^{-u}\right)^2}, \tag{23}$$

so that A_0, B_0 are the weights of the exponential functions involved and the value $K=-\left(\sqrt{1+4B_0}+1\right)/2$ is selected to acomplish the required eigenvalue $E_0=0$. There is a total of $n_{\max}+1$ eigenstates which eigenfunctions are

$$\begin{aligned}\phi_n(u)&=N_n\left(\exp(-u)\right)^{(c-1)/2}\left(1-\exp(-u)\right)^{(-n+b-c+1)/2}\\&\times\ {}_2F_1\left(-n,b,c,\exp(-u)\right),\\N_n&=\sqrt{\frac{(c-1)\ \Gamma(n+c)\ \Gamma(b)\ \Gamma(n+b)}{n!(n+b+1-c)\ \Gamma^2(c)\ \Gamma(b+1-c)}},\end{aligned} \tag{24}$$

for $n=0,1,2\ldots n_{\max}$. The involved parameters are given by

$$
\begin{aligned}
c &= \frac{-2n(h+n)-2A_0}{h+2n+1}\\
b &= \frac{(h+1)(h+n)-2A_0}{h+2n+1}\\
h &= \sqrt{1+4B_0}\\
0 &< n_{\max} < \sqrt{B_0-A_0}+K
\end{aligned}
\tag{25}
$$

and the eigenvalues are

$$E_n = \left(\frac{K^2+A_0-B_0}{2K}\right)^2 - \left(\frac{(n-K)^2+A_0-B_0}{2(n-K)}\right)^2. \tag{26}$$

Let us consider the mass function

$$m(x) = \frac{1}{2}\left(\frac{a}{1+\lambda e^{-ax}}\right)^2, \tag{27}$$

associated with the η function

$$\eta = \sqrt{\frac{1+\lambda e^{-ax}}{a}}, \tag{28}$$

and the variable change

$$\lambda e^{-ax} = \frac{e^{-u}}{(1-e^{-u})}. \tag{29}$$

Eq. (23) expressed in variable X gives the solvable potential of the main PDMSE, that is, corresponding to $\alpha=\gamma=-1/4$, which is

$$\tilde{V}(x) = \left(\frac{K^2+A_0-B_0}{2K}\right)^2 + \lambda A_0 e^{-ax} + \lambda^2 B_0 e^{-2ax} \tag{30}$$

it being a Morse potential for the main PDMSE. By using Eq. (18) the eigenfunctions will be given by

$$
\begin{aligned}
\psi_n(x) &= \eta^{-1}\phi_n(u(x))\\
&N_n\sqrt{\frac{a}{1+\lambda e^{-ax}}}\left(\frac{\lambda e^{-ax}}{1+\lambda e^{-ax}}\right)^{(c-1)/2}\left(\frac{1}{1+\lambda e^{-ax}}\right)^{(-n+b-c+1)/2}\\
&\times {}_2F_1\left(-n,b,c,\frac{\lambda e^{-ax}}{1+\lambda e^{-ax}}\right)
\end{aligned}
\tag{31}
$$

A property of mass function Eq. (27) is that the corresponding solvable potential for any $T_{\alpha\gamma}$ of the PDMSE, is a generalized Morse potential, in particular, one obtains for the cases listed in Table I, the following solvable cases

$$V^{BDD}(x)=\left(\frac{K^2+A_0-B_0}{2K}\right)^2+\lambda\left(A_0-\frac{1}{2}\right)e^{-ax}+\lambda^2\left(B_0-\frac{3}{4}\right)e^{-2ax}, \tag{32}$$

$$V^{GW}(x)=\left(\frac{K^2+A_0-B_0}{2K}\right)^2+\lambda\left(A_0+\frac{1}{2}\right)e^{-ax}+\lambda^2\left(B_0+\frac{5}{4}\right)e^{-2ax}, \tag{33}$$

$$V^{ZK}(x)=\left(\frac{K^2+A_0-B_0}{2K}\right)^2+\lambda\left(A_0+\frac{1}{2}\right)e^{-ax}+\lambda^2\left(B_0+\frac{1}{4}\right)e^{-2ax}, \tag{34}$$

$$V^{LK}(x)=\left(\frac{K^2+A_0-B_0}{2K}\right)^2+\lambda A_0e^{-ax}+\lambda^2\left(B_0+\frac{1}{4}\right)e^{-2ax}, \tag{35}$$

In Figure 1 we have drawn these potentials and also the main potential $\tilde{V}(x)$ by taking a value of

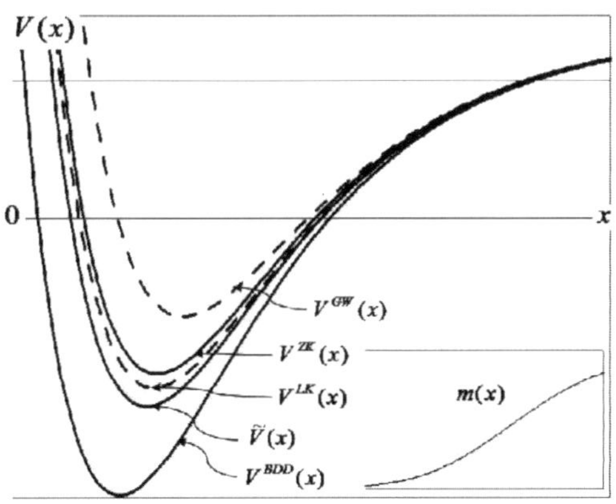

Figure 1. Solvable PDMSE Morse potentials *V*(*x*) for the most important Hamiltonians listed in Table I. The graph of PDM. Mass parameter is λ= 2.5. Horizontal lines represent E_0 and E_1.

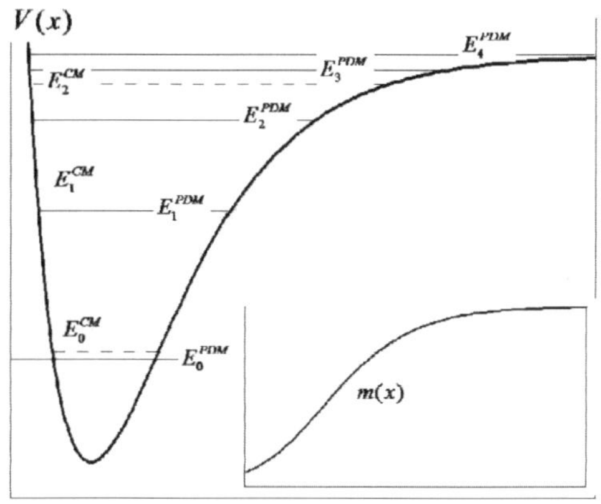

Figure 2. The PDMSE potential $\tilde{V}(x)$ given in Eq. (30) and the Morse CMSE potential of Eq. (36) are made to coincide thus we can see the changes in the eigenvalue spectra. PDM with λ = 0.5.

parameter λ which allows to elucidate a difference between the potentials up to the order of the first gap of the energies.

On the other hand, for small values of λ or $\lambda \approx 1$, the difference between the solved potentials is negligible and we can say that the reported solutions of the potential $\tilde{V}(x)$ correspond to any PDMSE with any value of α and γ. Figure 2 contains the graphs of the five reported potentials which are practically the same potential. Besides, we are able to compare the energies as given in Eq. (26) for the PDMSE, with those that come from considering the CMSE Morse potential written through its depth $D=\lambda^2B_0=-\lambda A_0/2$ in the form

$$\tilde{V}(x,\lambda) = F^2(D,\lambda) + D\left(e^{-2ax} - 2e^{-ax}\right), \tag{36}$$

where

$$F^2(D,\lambda) = \left(\frac{\left(K(D,\lambda)\right)^2 - D\left(\frac{1+2\lambda}{\lambda^2}\right)}{2K(D,\lambda)}\right)^2, \tag{37}$$

and

$$K(D,\lambda) = -\left(\sqrt{1+\frac{4D}{\lambda^2}}+1\right)/2. \tag{38}$$

We have selected a depth D in the potential of Eq. (36) that gives three eigenvalues of the constant mass case and determined the eigenenergies of the PDM case. One can see a difference between the number and value of the eigenenergies. The PDM eigenenergies are deeper and the number of eigenstates is greater as compared with those of the CMSE. We refer to the ref. [34] for similar reports on the effect of the PDM in the energy spectra.

Concluding Remarks

The point canonical transformation method to solve the PDMSE starting from exactly-solvable potentias of the CMSE is presented in an algebraic form. It consists of a unified treatment to the quantum position-dependent mass problems that contains, as particular cases, the kinetic energy operators of various authors such as BenDaniel-Duke, Gora-Williams, Zhu-Kroemer and Li-Kuhn, among others. We have considered the CMSE with multiparameter exponential-type potential to find solvable PDMSE with Morse potential. After, we have studied the influence of the position-dependent mass on the energy spectra. Our results show that PDM distribution affects the eigenvalues, allowing a greater number of eigenstates for a potential with a specific deep. Also, it is found that in a certain range of the mass parameter, there is no difference with respect to the kinetic energy operator considered for the PDMSE. Finally, the algebraic proposal to solve the PDMSE is general and can easily be extended to other potential models and/or position-dependent mass distributions.

IOP Conf. Series: Journal of Physics: Conf. Series **792** (2017) 012037 doi:10.1088/1742-6596/792/1/012037

References

[1] Morse P M 1929 *Phys. Rev.* **34** 57
[2] Pöschl G and Teller E 1933 *Z. Phys.* **83** 143
[3] Manning M F and Rosen N 1933 *Phys. Rev.* **44** 953
[4] Lippincott E R 1953 *J. Chem. Phys.* **21** 2070
[5] Varshni Y P 1957 *Rev. Mod. Phys.* **29** 664
[6] Tietz T 1963 *J. Chem. Phys.* **38** 3036
[7] Deng Z H and Fan Y P 1957 *Shandong Univ. J.* **7** 162
[8] Schiöberg D 1986 *Mol. Phys.* **59** 1123
[9] Wang P Q, Shang L H, Jia C S and Liu J Y 2012 *J. Mol. Spect.* **274** 5
[10] Ovando G, Peña J J and Morales J 2016 *Theor. chem. Acc.* **135** 62
[11] Geller M R and Kohn W 1993 *Phys. Rev. Lett.* **70** 3103
[12] Serra L and Lipparini E 1997 *Europhys. Lett.* **40** 667
[13] Barranco M, Hernández E S and Navarro J 1996 *J.: Phys. Rev.* B **54** 7394
[14] Harrison P 2000 *Quantum Wells, Wires and Dots* (NewYork: Wiley)
[15] Li Y M, Lu H M, Voskoboynikov O, Lee C P and Sze S M 2003 *Surf. Sci.* **532** 811
[16] Bastard G 1988 *Wave Mechanics Applied to Semiconductor Heterostructures* (Les Ulis: Editions de Physique)
[17] Einevoll G T, Hemmer P C and Thomsen J 1990 *Phys. Rev.* B **42** 3485
[18] Weisbuch C and Vinter B 1993 *Quantum Semiconductor Heterostructures* (New York: Academic Press)
[19] Renan R, Pacheco M H and Almeida, C A S 2000 *J. Phys. A: Math. Gen.* **33** L509
[20] Boztosun I, Bonatsos D and Inci I 2008 *Phys. Rev.* C **77** 044302
[21] Bonatsos D, Georgoudis P E, Lenis D, Minkov N and Quesne C 2011 *Phys. Rev.* C **83** 044321
[22] Ring P and Schuck P 1980 *The Nuclear Many Body Problem* (New York: Springer-Verlag) p. 211
[23] Foulkes W M C and Schluter M 1990 *Phys. Rev.* B **42** 11 505
[24] Arias de Saavedra F, Boronat J, Polls A and Fabrocini A 1994 *Phys. Rev.* B **50** 4248
[25] Puente A, Serra L and Casas M 1994 *Z. Phys.* D **31** 283
[26] Barranco M, Pi M, Gatica S M, Hernandez E S and Navarro J 1997 *Phys. Rev.* B **56** 8997
[27] von Roos O 1983 *Phys. Rev.* B **27** 7547
[28] BenDaniel D J and Duke C B 1966 *Phys. Rev.* B **152** 683
[29] Gora T and Williams F 1969 *Phys. Rev.* **177** 1179
[30] Zhu Q G and Kroemer H 1983 *Phys. Rev.* B **27** 3519
[31] Li T and Kuhn K J 1993 *Phys. Rev.* B **47** 12760
[32] Peña J J, Ovando G, Morales J, García-Ravelo J and Pacheco-García C 2008 *Int. J. Quant. Chem.* **108** 2906
[33] Ovando G, Peña J J and Morales J 2014 *J. Phys.: Conf. Ser.* **490** 012201
[34] Quesne C 2009 *SIGMA* **5** 046
[35] Ikhdair S M 2012 *Mol. Phys.* **110** 1415
[36] García-Martínez J, García-Ravelo J, Morales J and Peña J J 2012 *Int. J. Quant. Chem.* **112** 195

Computer vision syndrome (CVS) – Thermographic Analysis

L E Llamosa-Rincón[1], J M Jaime-Díaz[2] and D F Ruiz- Cardona[3]
[1] Universidad Tecnológica de Pereira. Profesor titular - Facultad de Ciencias Básicas, Pereira, Colombia
[2,3] Universidad Tecnológica de Pereira/ Ingeniera Física, Joven Investigadora 2015-2016 - COLCIENCIAS, Pereira, Colombia

E-mail: [1]lellamo@utp.edu.co

Abstract. The use of computers has reported an exponential growth in the last decades, the possibility of carrying out several tasks for both professional and leisure purposes has contributed to the great acceptance by the users. The consequences and impact of uninterrupted tasks with computers screens or displays on the visual health, have grabbed researcher's attention. When spending long periods of time in front of a computer screen, human eyes are subjected to great efforts, which in turn triggers a set of symptoms known as Computer Vision Syndrome (CVS). Most common of them are: blurred vision, visual fatigue and Dry Eye Syndrome (DES) due to unappropriate lubrication of ocular surface when blinking decreases. An experimental protocol was de-signed and implemented to perform thermographic studies on healthy human eyes during exposure to dis-plays of computers, with the main purpose of comparing the existing differences in temperature variations of healthy ocular surfaces.

1. Introduction

Technological advances have given rise to important changes in the visual needs of people, i.e. increase of short-distance vision. The use of computers is one of the principal cause of this visual stress since it requires a different set of visual abilities to achieve eye movements of scanning, visual fixation, focus shifts among others that allow interaction with displays and screens, keyboards and sometimes physical documents to transcript. Therefore, it is advisable a visual assessment that not only considers the refractive state but also the lubrication of the ocular surface [1]. S A. Castillo Estepa y A. Iguti [2] reviewed the state of art in the last 30 years and selected 32 articles that fulfill inclusion criteria since they had used the SVC categorization for different types of symptoms, each one associated to several visual diagnosis including alterations of the ocular surface such as DES. Terminology used to mention symptoms and the assessment of visual alterations is quite diverse, thus hampering an appropriate analysis of diagnosis and symptoms related to SVC. The authors concluded that SVC visual symptomatology reveals problems of binocular vision and accommodation reflex triggered by demands of short-distant vision. DES is considered the principal symptom of the SVC category, but it is not present in all the cases; whereas in other cases, eye dryness somatology is simply caused by extreme environmental conditions that increase vaporization of tears. The combination of bad habits during work with computers and unfavorable environmental conditions in the workplace may risk the visual and ocular health of computer users. Another study classified 200 employees randomly selected into three categories according to daily hours in front of computers that they had previously reported in a survey. The resulting categories were: less than 2 hours, from 2 to 6 hours, and more than 6 rs. Afterwards, they underwent ophthalmic evaluation to assess the visual health of partici-pants. Tests results revealed that computer users who spend more than 6 hours suffer from visual fatigue, and secondly from itching and irritation. Participants worked for institutions and departments from the universities of Teerthanker Mahaveer, Moradabad and Uttar Pradesh [3]. All these alterations of

Content from this work may be used under the terms of the Creative Commons Attribution 3.0 licence. Any further distribution of this work must maintain attribution to the author(s) and the title of the work, journal citation and DOI.
Published under licence by IOP Publishing Ltd

the tear film due to long periods in front of displays and screens may be accompanied by thermal changes. Computer Vision Syndrome (CVS) describes a series of problems related to the use of computers during prolonged periods of time resulting in some alterations in the visual health, most common conditions are visual fatigue, head-aches, blurred vision and DES [4]. Some causes of CVS could stem from features of the visualization such as poor lighting, excessive brightness and quality of screens and displays, upgrading and radiation frequencies, as well as the combination of some of the aforementioned factors with inadequate distances, workplace arrangements, level of toxicity in the environment, appropriate room temperature and humidity, working angle etc. [5,6] American Academy of Optometry published a report where they stated that 70% of employees who need a data processor to carry out tasks at work, suffer from problems in the muscle-skeletal system [7] and visual fatigue due to non-interrupted use of computers. Ophthalmologist refer to this new pathology as CVS. In many cases, eye care professionals ascribe this disease to issues related to workplace arrangement, poorly lighting and uncomfortable contrasts of screens. Many symptoms experienced by users disappear when tasks on computers are interrupted. Nevertheless, some people undergo reduction in their visual capabilities, even after finishing labors with computers [8].

2. Materials and Methods

The dependent variable of this study is the variation of temperature on ocular surface in visual fixation with blinking, which was measured in three different stages: before, during and after an hour of working in front of a computer, they will be detailed on the section Measurement Protocol. Dryness degree of human eyes is the independent variable expected to be evaluated using the designed method based on infrared photography in order to analyze determined time intervals, as well as the behavior of average temperature on ocular surfaces during visual fixation with blinking, before and after working with computers. Both measuring devices and experimental setup can be seen in Figure 1, with participants in the study.

2.1Measuring Devices: Fluke Ti32 is a thermal image processor that captures high quality images, its temperature measurement range is -20°C to +600°C, and manufacturer claims a temperature measurement accuracy of ±2°C or 2% and temperature resolution of 0,01 (°F o °C) [9,10].

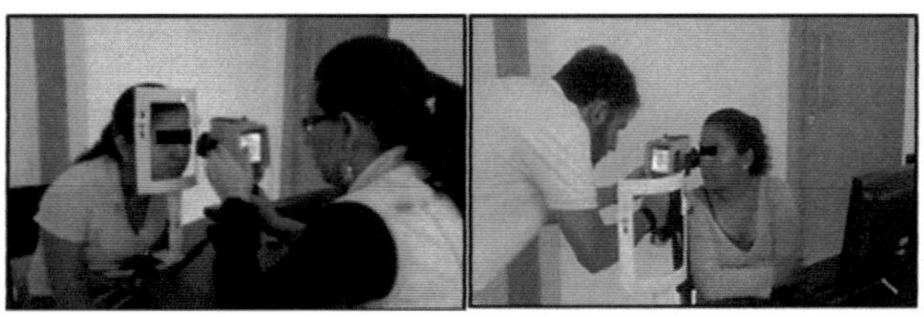

Figure 1. Photographs depicting the experimental protocol.

2.2 Measurement Protocol: Participants were 37 students from Universidad Tecnológica de Pereira (UTP), age range 18 to 28 years old. The study excluded individuals who have suffered from fever or migraine as well as people with systemic vascular diseases such as hypertension and diabetes, these people usually report higher thermal increase in the eye due to blood flow rate on the retina and choroid [11]. To avoid changes in skin emissivity while taking thermal captures, female students were asked to remove their makeups [12, 13]. Although IR thermography applied herein is an noninvasive technique, we abode by bioethical guidelines provided by 1964 Helsinki declaration and 008430 resolution of 1993 by Colombian Ministry of Health; and accordingly, all participants were informed previously about the protocol, so they could give their informed consents [14, 15].

Arrangements, monitoring and controls took place to guarantee appropriate conditions for capturing thermal images in the Electrophysiology Laboratory of UTP. Factors like temperature, relative humidity, air currents, lighting and pollution were taken into account [13];consequently, to obtain measurements from all the participants under the same conditions, constant monitoring and control was per-formed on comfort relative humidity (50%, - 10% to + 20%) and comfort room temperature 22 °C, -3 °C to + 3 °C) [16], finally light intensity was fixed to recommend levels for simple visual task (200 – 300 Lux) [17]. Each participant waited in the laboratory 10 minutes before the thermal images capturing; thereby contributing to not only participant's adaptation to temperature and relative humidity inside the testing room, but also to temperature stabilization of the ocular surfaces [16, 18, 19, 20]. Thermal images capturing procedures always began with the right eyes and ended with the left eyes. The procedure of image capturing was divided into three stages. During the first stage (previous to one hour of work in front of a computer), thermal images of ocular surfaces were taken during 60 seconds with participants holding a visual fixation with blinking, staring at the lens of the thermal images. 10 to 15 images from each participant were captured in first stage. Second stage of measurements consisted in taking thermal images during one hour of work with a computer aiming to determine temperature variation on the ocular surfaces. Participants sat 30 to 50 cm away from the screen. From the beginning of this stage and every 5 minutes thermal images of the ocular surface were taken at a distance of 10 cm. Third stage of measurements took place immediately after the one-hour period in front of the computer and repeated the same protocol of the first stage.

Table 1, Timing of measurements with Ti32 thermal imager.

Stages	Description	Time
First stage	Holding a visual fixation (Open eyes with flicker)	60 seconds
Second stage	Test during the work in front of a computer screen	60 minutes
Third stage	Holding a visual fixation (Open eyes with flicker)	60 seconds

For implementation of the designed experimental protocol, every thermal image was previously analyzed with the built-in software of Ti32 thermal camera: SmartView 3.5, which offers several options of markers (point-shaped, linear, rectangular, elliptical and polygonal) to focus on areas of interest within the image to analyze. Horizontal and vertical profiles, from data based on measurements during stages 1 and 3, were studied to achieve a qualitative analysis using the corresponding horizontal and vertical markers featured in SmartView. The horizontal profiles consisted of the whole area between the nasal and temporal conjunctiva, whereas the vertical ones included the area between upper and lower eyelids. Additionally, a quantitative study focused on variation temperature on the cornea after each blinking by means of the circular marker provided by SmartView, which assisted to determine aver-age temperature of this area of the ocular surface. This quantitative study using the circular marker was the only one performed for stage 2 of measurements corresponding to an hour in front of a computer, with the purpose of determining average temperature of cornea in every thermal image and thereby proceeding to the analysis of temperature variation throughout the task with the data processor. The statistical analysis of the protocol started with running normality tests on the three types of measurements to determine if the data sets were well-modeled by a normal distribution. Since normality tests yielded negative results for data corresponding to stages 1 and 3 (temperature variation experienced during visual fixation with blinking, before and after using the computer), these data were subjected to Mann Whitney U-tests, a non-parametric test that identifies statistically significant differences between populations based on the analysis of two independent samples. Conversely, normality tests yielded positive results for data obtained in stage 2. Therefore, these set of data were subjected to a t-student test, a parametric test to compare the median of a sample against a theoretical median to determine statistical differences between them.

3. Discussions and Results

The principal result of the development of the project was the design of an experimental protocol based on a thermography approach to analyze the thermal behavior of healthy ocular surface when it is exposed to computers. The results obtained from data gathering showed that the proposed methodology is effective for the study of the temperature of healthy ocular surface. The population sample of this study consisted of 37 participants, 19 female individuals (51.35%) and 18 male individuals (48,65), age range 18 to 28 years. As stated in the measurement protocol section, qualitative analysis was performed on data obtained from stages 1 and 3 based on horizontal and vertical profiles of the whole ocular surface. According to the horizontal marker, there is a communality in the behavior of the entire population sample concerning the temperature of the three ocular areas studied herein (temporal conjunctiva, nasal conjunctiva and cornea). The lowest temperatures in average are reported for the cornea surface in first place, followed by the temporal conjunctiva, and the nasal conjunctiva in last place. The greatest difference (1°C) was found between the cornea and the nasal conjunctiva. The vertical profile focused on the intermediate zones be-tween the cornea area and both eyelids (upper and lower). According to the vertical marker, the temperature on the center of the cornea was nearly 0.6 °C less than that recorded for the aforementioned intermediate zone. Thus, coming to the conclusion that ocular surface corresponding to cornea reports lower temperatures. Stages 1 and 3 also included analysis of thermal images using the circular marker to determine the average temperature of the cornea. Temperature variation ranges (initial to final temperature) for the whole population during visual fixation with blinking were the following: in stage1, before working with computer, from -0,79 to -0,01 °C; and in stage 3, after working with computer, from 0,96 to 0,78 °C. Results revealed a considerable drop in temperature in stage 1, while for stage 3 there were many cases where the temperature remained constant, some with a slow drop in tempera-ture and a few where the temperature kept increasing continually.

Table 2. Temperature variation during visual fixation with blinking for stages 1 and 3 (before and after the use of computer).

Test No.	First stage ΔT (T_{FINAL}-$T_{INICIAL}$)[°C]		Third stage ΔT (T_{FINAL}-$T_{INICIAL}$)[°C]	
	Right eye	Left eye	Right eye	Left eye
1	-0,13	-0,20	-0,04	0,19
2	-0,14	-0,11	-0,16	-0,24
3	-0,17	-0,21	-0,03	-0,15
4	-0,39	-0,35	-0,28	0,08
5	-0,12	-0,21	-0,96	-0,26
6	-0,20	-0,18	-0,30	0,04
7	-0,60	-0,60	0,00	0,33
8	-0,47	-0,24	-0,45	-0,06
9	-0,06	-0,11	-0,09	-0,25
10	-0,12	-0,04	-0,14	-0,27
11	-0,26	-0,05	-0,15	-0,20
12	-0,10	-0,11	-0,05	-0,01
13	-0,10	-0,15	0,20	0,06
14	-0,18	-0,07	-0,85	0,09
15	-0,04	-0,01	0,01	-0,37
16	-0,24	-0,33	-0,05	-0,06
17	-0,05	-0,01	-0,20	-0,04
18	-0,18	-0,02	-0,03	-0,04
19	-0,18	-0,02	-0,03	-0,04
20	-0,07	-0,08	0,01	0,10
21	-0,09	-0,01	-0,02	0,17
22	-0,66	-0,47	-0,18	-0,29
23	-0,09	-0,11	-0,06	-0,06
24	-0,16	-0,01	0,78	-0,04
25	-0,07	-0,08	-0,08	-0,20
26	-0,42	-0,16	0,13	-0,04
27	-0,40	-0,20	-0,35	-0,06
28	-0,20	-0,51	-0,74	-0,10
29	-0,45	-0,21	-0,01	-0,01
30	-0,25	-0,38	-0,24	0,06
31	-0,13	-0,06	-0,29	-0,37
32	-0,37	-0,10	-0,30	0,05
33	-0,79	-0,55	-0,18	-0,11
34	-0,07	-0,26	-0,30	-0,18
35	-0,18	-0,12	-0,02	-0,29
36	-0,01	-0.07	-0,17	-0,05
37	-0,16	-0,40	-0,20	-0,31

Table 3, Temperature variation during the use of computer (Stage 2).

Test No.	Second stage ΔT (T_{FINAL}-$T_{INICIAL}$)[°C]	
	Right eye	Left eye
1	0,54	0,37
2	1,01	1,14
3	0,32	0,19
4	0,46	0,56
5	0,25	0,40
6	0,98	0,91
7	0,51	0,57
8	0,16	0,42
9	0,49	0,08
10	0,37	0,61
11	0,08	0,03
12	0,74	0,96
13	0,40	0,31
14	1,07	0,67
15	0,40	0,44
16	1,06	1,29
17	0,24	0,13
18	0,36	0,43
19	0,79	0,87
20	0,71	0,33
21	0,41	0,65
22	0,48	0,70
23	0,85	0,62
24	0,48	0,41
25	0,19	0,30
26	0,45	0,35
27	0,76	0,43
28	0,52	0,21
29	0,95	0,74
30	0,38	0,06
31	0,98	1,22
32	1,11	0,68
33	0,83	1,00
34	1,11	0,68
35	0,99	1,20
36	0,76	0,57
37	0,60	0,81

As stated in the protocol measurement section regarding the statistical analysis; since results of normality test for the data on temperature variation gathered from both stage 1 and 3 yielded negative results, these data was subjected to Mann Whitney U-test with the purpose of determining if there is Statistically significant difference between them. The comparison reported a Z-value of -2.66993 and a p-value of ≤ 0.00733, thereby confirming that distributions of the two set of data were significantly different. By means of the circular marker featured by SmartView, increase of temperature was analyzed based on thermal images corresponding to cornea surface during stage 2 (60 minutes working with a computer). Increases ranges re-ported were from 0.08 to 1.11 °C for right eyes and from 0.03 to 1.29 °C for left eyes and thereby corroborating that under the influence of a computer screen, temperature of both eyes tends to increase considerably. As stated in the protocol measurement section, the normal distribution of the data from stage 2 was confirmed by statistical normality test. Therefore, a t-student test was performed to compare this sample with a theoretical median and determine statistical difference between them. Results from the test yielded a t-statistic = 16.19305, DF=73 y and p-value ≤7.20 X 10-26. The previous determines a confidence level of 90%, which in turn means that the sample of this study is significantly different from the theoretical median which has a confidence level of 99.9%. A record must be kept for the monitoring and control of room conditions during measurement stages (luminosity, temperature and relative humidity). Along these lines, it is important to locate a place where those variables can be controlled [22]. Infrared thermographic techniques applied to human beings are not invasive because there is no direct manipulation of individuals at any time. Despite that, the protocol was designed in accordance with bio-ethical specifications. The affectivity of the experimental protocol proposed for the study of thermal behavior of healthy human ocular surface was proved by the results from measurement stages and the statistical analysis with the significant statistical differences reported.

Acknowledgment
To the Vice Chancelor Office for Research, Innovation, and Outreach of Universidad Tecnológica de Pereira, for the support in the project "Computer Vision Syndrome (CVS) - Thermographic and Electromagnetical Analyses".

References

[1] Casillas Casillas E 2010 Sindrome de fatiga visual por uso de la computadora *Revista Panamericana de Lentes de Contacto* **2** 9-22
[2] Castillo Estepa A and Iguti A 2013 Síndrome de la visión del computador: diagnósticos asociados y causas *Ciencia &Tecnología para la salud visual y ocular*. **11** 97-109
[3] Agarwal S, Goel D and Sharma A 2012 Evaluation of the factors which contribute to ocular complaints in computer users *J Clin. Diagn. Res.* **7**:331-335
[4] Optica Central: Optica En Ibague, Disenio De Lentes Y Monturas, Lentes Progresivos, Venta De Gafas En Ibague, Optometria En Ibague". *Opticacentralltda.com*. N.p., 2016. Web. 4 Oct. 2016
[5] Blehm C, Vishnu S, Khattak A, Mitra S and Yee R 2014 Computer vision syndrome: a review, *Surv. Ophthalmol.* **50** 253-262
[6] Pérez Tejada A, Acuña Pardo A and Rúa Martínez R 2008 *Repercusión visual del uso de las computadoras sobre la salud* (Cuba: Revista Cubana Sobre Salud Pública vol 34) pp 4
[7] Del Río Martínez J and González Videgaray M 2007 Trabajo prolongado con computadoras: consecuencias sobre la vista y fatiga cervical *IX Congreso Internacional de Ergonomia Sociedad de Ergonomistas de México (Ciudad de México)* vol 1
[8] Eroski consumer. *http://www.consumer.es/web/essalud/2003/05/15/61438.php*. N. p., 2014
[9] 2009 *Introducción a los principios de la Termografía* Fluke Corporation and the Snell Group (Netherlands: American Technical Publishers)
[10] 2013 *Manual de Uso Thermal Imager Ti32* (USA: Fluke Corporation)
[11] Buccella C, Satis V, and Feliziani M 2007 Prediction of temperature increase in human eye due to RF sources *IEEE Trans. Electromagn. Compat* **49** 825-833
[12] Lahiri B, Bagavathiappan S, Joyakumar T and Philip J. 2012 Medical applications of infrared thermography: A review *Infrared phys techn* **55** 221-235
[13] Testo A G. Termografía, guía de bolsillo. Teoría, Aplicación Práctica, Consejos y Trucos. 2008.
[14] Asociación Médica Mundial. Declaración de Helsinki. Principios éticos para las investigaciones médicas en seres humanos. Junio 1964, enmendada 2008. [Online] Available: http://www.wma.net/es/30publications/10policies/b3/ 17c_es.pdf.[Accessed: Marzo 2015].
[15] Ministerio de Salud. República de Colombia. Resolución N° 008430 de 1993, Titulo II, Capitulo 1, de los aspectos éticos de la investigación en seres humanos. Octubre de 1993. [Online] Available: http://www.unisabana.edu.co/fileadmin/Documentos/ Investigacion/comite_de_etica/Res__8430_1993_-_Salud.pdf. [Accessed: Marzo 2015]
[16] Kessel L, Johnson L, Arvidsson H and Larsen M 2010 The relationship between body and ambient temperature and corneal temperature *Invest Ophth Vis Sci* **51** 6593-6597
[17] Julio Garavito Protocolo de iluminación laboratorio de producción 2008 Facultad de ingeniería industrial (Colombia: Escuela Colombiana de Ingeniería)

IOP Conf. Series: Journal of Physics: Conf. Series **792** (2017) 012039 doi:10.1088/1742-6596/792/1/012039

[18] Tan L, Cai Z and Lai N 2009 Accuracy and sensitivity of dynamic ocular thermography and inter-subjects ocular surface temperature (OST) in Chinese young adults. *Cont. Lens Anterior Eye* **32** 78-83
[19] Morgan P, Tullot A and Efron N 1996 Ocular surface cooling in dry eye - A pilot study. *Cont Lens Anterior Eye* **19** 7-10
[20] Morgan P, Soh M and Efron N 1993 Potential Applications of Ocular Thermography *Optom. Vis. Sci.* **70** 568-576
[21] Díaz J M 2014 *Protocolo de diagnóstico del síndrome del ojo seco basado en la fotografía termográfica* (Pereira: Universidad Tecnológica de Pereira)
[22] González M 2008 *Alteración de la superficie ocular tras la exposición a un ambiente adverso creado en una cámara de ambiente controlado (CAC)* (Instituto universitario de oftalmo-biología, Colombia: Universidad de Valladolid)

Optoelectronic polarimeter controlled by a graphical user interface of Matlab

J M Vilardy[1], C J Jimenez[2], R Torres[1]

[1] Grupo de óptica e informática. Department of electronic engineering. Universidad Popular del Cesar, Valledupar (Cesar), Colombia

[2] Grupo GIFES. Faculty of engineering. Universidad de La Guajira, Riohacha (La Guajira), Colombia

E-mail: vilardy.juan@unicesar.edu.co

Abstract. We show the design and implementation of an optical polarimeter using electronic control. The polarimeter has a software with a graphical user interface (GUI) that controls the optoelectronic setup and captures the optical intensity measurement, and finally, this software evaluates the Stokes vector of a state of polarization (SOP) by means of the synchronous detection of optical waves. The proposed optoelectronic polarimeter can determine the Stokes vector of a SOP in a rapid and efficient way. Using the polarimeter proposed in this paper, the students will be able to observe (in an optical bench) and understand the different interactions of the SOP when the optical waves pass through to the linear polarizers and retarder waves plates. The polarimeter prototype could be used as a main tool for the students in order to learn the theory and experimental aspects of the SOP for optical waves via the Stokes vector measurement. The proposed polarimeter controlled by a GUI of Matlab is more attractive and suitable to teach and to learn the polarization of optical waves.

1. Introduction

In general, the state of polarization (SOP) for optical waves is described by using the Stokes vectors [1, 2]. These SOP for optical waves are: unpolarized, partially polarized and completely polarized (linear, circular and elliptical) [1]. The optical polarization is very important in several areas of photonics, because a lot of optical devices, such as Faraday rotator, retarder plate, optical fiber sensors, fiber optic gyroscope, etc, are based on optical polarization. In order to use very well the previous optical devices, it is necessary to understand very well the theoretical concepts and experimentation related to the SOP for optical waves. Several practical applications are based on optical polarization, such as diffractive optical elements applied to ophthalmology [3], contrast microscopy [4], biomedicine [5], image processing [6, 7], among other.

In this paper, we present an optical polarimeter that it is controlled by using electronics and a graphical user interface (GUI) of MATLAB. The optical setup is divided in two system. The first and the second systems are used to generate and measure the SOP, respectively. The generation of the SOP is based on the Muller matrices of the linear polarizer (LP) and the quarter wave plate (QWP). The measurement of the SOP is based on the synchronous detection of optical waves in order to compute the Stoke vector. The GUI implemented in MATLAB can control the processes of the generation and measurement of the SOP. This GUI is more attractive and suitable to teach the polarization of optical waves. The proposed optical polarimeter could be used for optical sensing and testing.

Content from this work may be used under the terms of the Creative Commons Attribution 3.0 licence. Any further distribution of this work must maintain attribution to the author(s) and the title of the work, journal citation and DOI.
Published under licence by IOP Publishing Ltd

The rest of the paper is organized as follow. Section 2 presents the optoelectronic devices that will be used to generate and measure the SOP. In sections 3, it is determined the transmission axis orientation of a LP. The fast axis orientation of a QWP is evaluated in section 4. Section 5 shows the generation and measurement of the SOP by computing the Stokes vector and using the synchronous detection of optical waves. Finally, the conclusions are outlined in section 6.

2. Optical setup to generate and measure the SOP

Figure 1 shows the experimental setup that it is used to generate and measure the SOP. The generation system of the SOP is composed by: one laser (with a wavelength of λ = 633 nm and linear vertical polarization), one LP and one QWP. The measurement system of the SOP is composed by: one QWP, one LP and one photodetector. The two LPs are mounted in two different mechanical rotation stages. We implement a GUI in MATLAB to control the generation and measurement systems of the SOP. The two QWPs are mounted in two different mechatronic rotation stages; these two stages are controlled by the GUI implemented in MATLAB. The acquisition process of the optical intensity values obtained from the photodetector is also controlled by the GUI implemented in MATLAB. The electronic control of the two mechatronic rotation stages and the acquisition process of the photodetector is implemented using a serial communication interface between the GUI, the data acquisition board connected to the photodetector, and the physical controller of the two mechatronic rotation stages. Before performing the generation and measurement of the SOP, it is necessary to determine the transmission and fast axis orientations of a LP and a QWP, respectively.

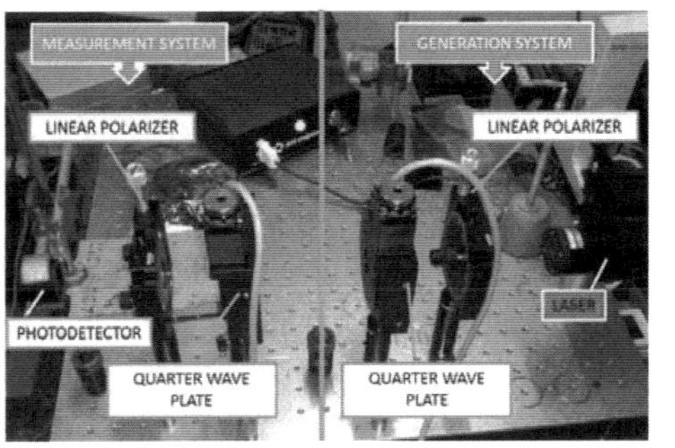

Figure 1. The experimental setup of the optoelectronic polarimeter.

3. Determination of the transmission axis orientation for a LP

In order to determine the transmission axis orientation for a LP with respect to the vertical direction of the laboratory, we perform the following experimental steps: the laser beam with a linear vertical polarization is propagated through to the LP, this LP is rotated by an angle θ_r with respect to the vertical direction of the laboratory. We do not know the transmission axis orientation of the LP. Therefore, the LP will be rotated by an angle $\theta_r + \theta_0$, where θ_0 denotes the angle between the transmission axis of the LP and the vertical direction of the laboratory. At last, we capture the optical intensity value $I(\theta_r + \theta_0)$ produced by the emergent laser beam of the LP using the photodetector. The emergent SOP $\mathbf{S}_{pol}$ from the LP is given by the matrix multiplication between the Mueller matrix of a LP with its transmission axis rotated by an angle $\theta_r + \theta_0$ with respect to the vertical direction of the laboratory and the Stoke vector of a laser beam with a linear vertical polarization [1]

$$\mathbf{S}_{pol} = \frac{1}{2}\begin{pmatrix} 1 & \cos(2\{\theta_r+\theta_0\}) & \sin(2\{\theta_r+\theta_0\}) & 0 \\ \cos(2\{\theta_r+\theta_0\}) & \cos^2(2\{\theta_r+\theta_0\}) & \sin(2\{\theta_r+\theta_0\})\cos(2\{\theta_r+\theta_0\}) & 0 \\ \sin(2\{\theta_r+\theta_0\}) & \sin(2\{\theta_r+\theta_0\})\cos(2\{\theta_r+\theta_0\}) & \sin^2(2\{\theta_r+\theta_0\}) & 0 \\ 0 & 0 & 0 & 0 \end{pmatrix}\begin{pmatrix} 1 \\ 1 \\ 0 \\ 0 \end{pmatrix}. \quad (1)$$

The optical intensity captured by the photodetector $I(\theta_r + \theta_0)$ will be equal to the parameter S_0 of $\mathbf{S}_{pol}$

$$I(\theta_r+\theta_0) = S_0 = \frac{1}{2}\left[1+\cos(2\{\theta_r+\theta_0\})\right] = \frac{1}{2}\left[1+\cos(2\theta_0)\cos(2\theta_r)-\sin(2\theta_0)\sin(2\theta_r)\right]. \quad (2)$$

The transmission axis orientation of a LP given by the angle θ_0, it can be determined by using the synchronous detection of optical waves. We captured the optical intensity $I(\theta_r + \theta_0)$ for the following angles: $\theta_r = 2\pi(r-1)/N$ with $r = 1, \ldots, N$. Then, if every $I(\theta_r + \theta_0)$ is multiplied by the terms $\sin(2\theta_r)$ or $\cos(2\theta_r)$, and finally, when these N products are added, we obtain

$$2\sum_{r=1}^{N} I(\theta_r+\theta_0)\sin(2\theta_r) = -\frac{N}{2}\sin(2\theta_0), \qquad 2\sum_{r=1}^{N} I(\theta_r+\theta_0)\cos(2\theta_r) = \frac{N}{2}\cos(2\theta_0). \quad (3)$$

Using the equation (3), the angle θ_0 is

$$\theta_0 = \frac{1}{2}\arctan\left(\frac{-\sum_{r=1}^{N} I(\theta_r+\theta_0)\sin(2\theta_r)}{\sum_{r=1}^{N} I(\theta_r+\theta_0)\cos(2\theta_r)}\right). \quad (4)$$

The experimental procedure denotes by equation (4) was applied to each LP of the generation and measurement systems of figure 1. The optical intensity measurements $I(\theta_r + \theta_0)$ were captured for the angles θ_r between 0° and 355° with an incremental step of 5°. The GUI of figures 2(a) and 2(b) show that the transmission axis orientations of the LPs for the generation and measurement systems of the SOP with respect to the vertical direction of the laboratory are: $\theta_{01} \approx 17^\circ$ and $\theta_{02} \approx -11^\circ$, respectively.

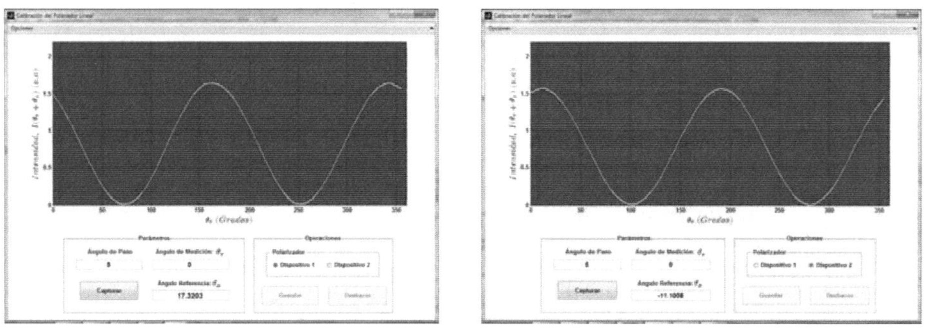

(a) (b)

Figure 2. Determination of the angle θ_0 for the LP located in the system of: (a) generation and (b) measurement, of the SOP.

4. Determination of the fast axis orientation for a QWP

We use an optical system composed by two LP and one QWP, in order to evaluate the fast axis orientation for a QWP. This optical system is called $LP_1 - QWP - LP_2$, and this name represents the location of the two LPs and the QWP. The transmission axis of the LPs LP_1 and LP_2 are oriented along the vertical direction of the laboratory. The emergent SOP from the LP LP_1 is linear vertically polarized, this SOP is propagated through to the rotated QWP with an angle θ_r with respect to the vertical direction of the laboratory. We do not know the fast axis orientation of the QWP. Therefore, the QWP will be rotated by an angle $\theta_r + \theta_l$, where θ_l represents the angle between the fast axis of the QWP and the vertical direction of the laboratory. Using the Mueller matrix of a QWP with its fast axis rotated by an angle $\theta_r + \theta_l$ with respect to the vertical direction of the laboratory and the Stoke vector of the emergent SOP from the LP LP_1, which is linear vertically polarized, the emergent SOP $\mathbf{S}_{LR}$ from the QWP will be [1]

$$\mathbf{S}_{LR} = \begin{pmatrix} 1 & 0 & 0 & 0 \\ 0 & \cos^2(2\{\theta_r+\theta_l\}) & \sin(2\{\theta_r+\theta_l\})\cos(2\{\theta_r+\theta_l\}) & -\sin(2\{\theta_r+\theta_l\}) \\ 0 & \sin(2\{\theta_r+\theta_l\})\cos(2\{\theta_r+\theta_l\}) & \sin^2(2\{\theta_r+\theta_l\}) & \cos(2\{\theta_r+\theta_l\}) \\ 0 & \sin(2\{\theta_r+\theta_l\}) & -\cos(2\{\theta_r+\theta_l\}) & 0 \end{pmatrix} \begin{pmatrix} 1 \\ 1 \\ 0 \\ 0 \end{pmatrix}. \quad (5)$$

The emergent SOP $\mathbf{S}_{P2}$ form the system LP_1–QWP–LP_2 is given by the matrix multiplication between the Stoke vector $\mathbf{S}_{LR}$ and the Mueller matrix of the LP LP_2 with its transmission axis oriented along the vertical direction of the laboratory

$$\mathbf{S}_{P2} = \frac{1}{2}\begin{pmatrix} 1 & 1 & 0 & 0 \\ 1 & 1 & 0 & 0 \\ 0 & 0 & 0 & 0 \\ 0 & 0 & 0 & 0 \end{pmatrix}\mathbf{S}_{LR}. \quad (6)$$

The optical intensity captured by the photodetector $I(\theta_r + \theta_l)$ will be equal to the parameter S_0 of $\mathbf{S}_{P2}$

$$I(\theta_r + \theta_l) = S_0 = \frac{1}{2}\left[1 + \cos^2(2\{\theta_r + \theta_l\})\right] = \frac{1}{4}\left[3 + \cos(4\theta_l)\cos(4\theta_r) - \sin(4\theta_l)\sin(4\theta_r)\right]. \quad (7)$$

The fast axis orientation of a QWP given by the angle θ_l, it can be evaluated by using the synchronous detection of optical waves. We captured the optical intensity $I(\theta_r + \theta_l)$ for the following angles: $\theta_r = 2\pi(r-1)/N$ with $r = 1, \ldots, N$. Then, if every $I(\theta_r + \theta_l)$ is multiplied by the terms $\sin(4\theta_r)$ or $\cos(4\theta_r)$, and finally, when these N products are added, we obtain

$$4\sum_{r=1}^{N} I(\theta_r + \theta_l)\sin(4\theta_r) = -\frac{N}{2}\sin(4\theta_l), \qquad 4\sum_{r=1}^{N} I(\theta_r + \theta_l)\cos(4\theta_r) = \frac{N}{2}\cos(4\theta_l). \quad (8)$$

Using the equation (8), the angle θ_l is

$$\theta_l = \frac{1}{4}\arctan\left(\frac{-\sum_{r=1}^{N} I(\theta_r + \theta_l)\sin(4\theta_r)}{\sum_{r=1}^{N} I(\theta_r + \theta_l)\cos(4\theta_r)}\right). \quad (9)$$

The experimental procedure given by equation (9) was applied to each QWP of the generation and measurement systems of figure 1. The optical intensity measurements $I(\theta_r + \theta_l)$ were captured for the angles θ_r between 0° and 355° with an incremental step of 5°; these measurements and rotations of the QWPs are controlled by the GUI. Figures 3(a) and 3(b) show that the fast axis orientations of the QWPs for the generation and measurement systems of the SOP with respect to the vertical direction of the laboratory are: $\theta_{l1} \approx -24^\circ$ and $\theta_{l2} \approx -31^\circ$, respectively.

5. Generation and measurement of the SOP

The transmission and fast axis orientations of the LPs and the QWPs, respectively, of figure 1 are oriented along the vertical direction of the laboratory. The transmission axis orientations of the LPs of figure 1 are always fixed. The fast axis orientations of the QWPs of figure 1 can be rotated with the purpose of generating or measuring the SOP. The emergent SOP $\mathbf{S}_{gen}$ from the generation system of figure 1 can be described by [1]

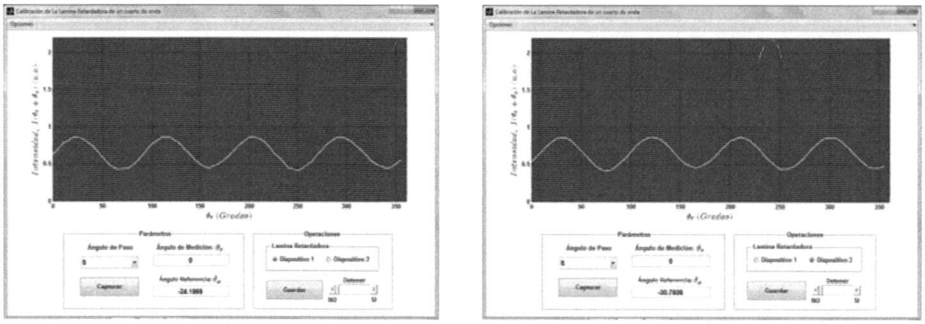

(a) (b)

Figure 3. Determination of the angle θ_l for the QWP located in the system of: (a) generation and (b) measurement, of the SOP.

$$\mathbf{S}_{gen} = \begin{pmatrix} S_0 \\ S_1 \\ S_2 \\ S_3 \end{pmatrix} = \begin{pmatrix} 1 \\ \cos^2(2\theta) \\ \sin(2\theta)\cos(2\theta) \\ \sin(2\theta) \end{pmatrix}, \tag{10}$$

where θ is the rotation angle of the QWP located in the generation system of the SOP and its possible values are between $-90^\circ < \theta < 90^\circ$. The generated SOP for the different angle θ are: linear vertical for $\theta = 0^\circ$, right-circular and left-circular when $\theta = 45^\circ$ and $\theta = -45^\circ$, respectively, and elliptical for all other cases of θ.

In the measurement system, we evaluate the incident SOP by computing its Stoke vector (S_0, S_1, S_2 and S_3). The emergent SOP $\mathbf{S}_{out}$ from the measurement system of figure 1 is given by [1]

$$\mathbf{S}_{out} = \frac{1}{2}\begin{pmatrix} 1 & 1 & 0 & 0 \\ 1 & 1 & 0 & 0 \\ 0 & 0 & 0 & 0 \\ 0 & 0 & 0 & 0 \end{pmatrix}\begin{pmatrix} 1 & 0 & 0 & 0 \\ 0 & \cos^2(2\theta_r) & \sin(2\theta_r)\cos(2\theta_r) & -\sin(2\theta_r) \\ 0 & \sin(2\theta_r)\cos(2\theta_r) & \sin^2(2\theta_r) & \cos(2\theta_r) \\ 0 & \sin(2\theta_r) & -\cos(2\theta_r) & 0 \end{pmatrix}\begin{pmatrix} S_0 \\ S_1 \\ S_2 \\ S_3 \end{pmatrix}, \tag{11}$$

where θ_r is the rotation angle of the QWP located in the measurement system of the SOP. The optical intensity captured by the photodetector $I(\theta_r)$ will be equal to the parameter S_0 of $\mathbf{S}_{out}$

$$\begin{aligned} I(\theta_r) = S_0 &= \frac{1}{2}\left[S_0 + S_1\cos^2(2\theta_r) + S_2\sin(2\theta_r)\cos(2\theta_r) - S_3\sin(2\theta_r)\right] \\ &= \frac{1}{4}\left[2S_0 + S_1 + S_1\cos(4\theta_r) + S_2\sin(4\theta_r) - 2S_3\sin(2\theta_r)\right]. \end{aligned} \tag{12}$$

We apply the synchronous detection of optical waves. We captured the optical intensity $I(\theta_r)$ for the following angles: $\theta_r = 2\pi(r-1)/N$ with $r = 1, \ldots, N$. Then, if every $I(\theta_r)$ is added and if every $I(\theta_r)$ is multiplied by the terms $\sin(2\theta_r)$, $\sin(4\theta_r)$ or $\cos(4\theta_r)$, and finally, when these N products are added, we obtain the four parameters of the Stoke vector to be measured

$$\mathbf{S} = \begin{pmatrix} S_0 \\ S_1 \\ S_2 \\ S_3 \end{pmatrix} = \frac{1}{N}\begin{pmatrix} 2\sum_{r=1}^{N} I(\theta_r) - 4\sum_{r=1}^{N} I(\theta_r)\cos(4\theta_r) \\ 8\sum_{r=1}^{N} I(\theta_r)\cos(4\theta_r) \\ 8\sum_{r=1}^{N} I(\theta_r)\sin(4\theta_r) \\ -4\sum_{r=1}^{N} I(\theta_r)\sin(2\theta_r) \end{pmatrix}. \tag{13}$$

The optical intensity measurements $I(\theta_r)$ were captured for the angles θ_r between 0° and 356° with an incremental step of 4°. Figures 4(a) and 4(b) show the generation and measurement of the following SOP: linear vertical ($\theta = 0^\circ$) and left-circular ($\theta = -45^\circ$), respectively.

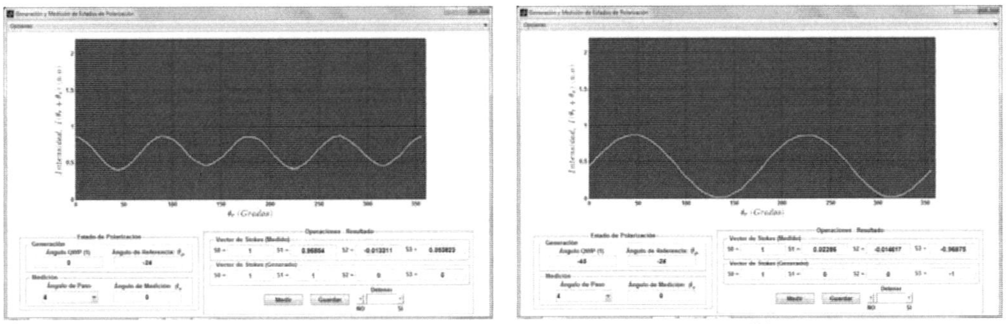

(a) (b)

Figure 4. Generation and measurement of the SOP: (a) linear vertical and, (b) left-circular.

6. Conclusions

The design and implementation of an optoelectronic polarimeter has been presented. We have determined theoretically and experimentally the transmission and fast axis orientations for a LP and a QWP, respectively, by using the synchronous detection of optical waves. We generated completely polarized light in a controlled manner by means of a GUI implemented in MATLAB. We evaluated theoretically and experimentally the Stokes vector of a SOP by using the synchronous detection of optical waves and an optoelectronic system controlled by a GUI. Finally, the proposed optoelectronic polarimeter can be used to teach and to learn the polarization of optical waves, and to determine the Stokes vector of a SOP in a quick and efficient manner.

Acknowledgments

This research has been funded by the Universidad Popular del Cesar and the Universidad de La Guajira.

References

[1] Kumar A and Ghatak A 2011 *Polarization of Light with Applications in Optical Fibers* (SPIE)
[2] Yariv A and Yeh P 1984 *Optical Waves in Crystals* (Wiley: New York)
[3] Ambs P and Bigué L 2001 *SPIE Critical Review* **CR81** 365–393
[4] Plata A and Chacón C 2011 *Memorias XII Encuentro Nacional de Óptica, III Conferencia Andina y del Caribe en Óptica y sus Aplicaciones* (*Barranquilla–Colombia*)
[5] Argel J, Mausa M and Arrieta E 2011 *Memorias XII Encuentro Nacional de Óptica, III Conferencia Andina y del Caribe en Óptica y sus Aplicaciones* (*Barranquilla–Colombia*)
[6] Patel H 2010 *Proceedings of the IEEE National Aerospace and Electronics Conference* (*Dayton–USA*)
[7] Lee J-S and Pottier E 2009 *Polarimetric Radar Imaging* (CRC Press)

Characterisation of a Tungsten X-pinch dense plasma

A Robledo-Martinez, A Villarreal-Garcia, L Miranda, and M R Hernández

Departamento de Ciencias Básicas, Universidad Autónoma Metropolitana, Av. San Pablo 180, Azcapotzalco, 02200 Mexico DF, Mexico.

E-mail: arm@azc.uam.mx

Abstract. An X-pinch experiment using a load made of two 25-µm Tungsten wires was investigated. The X-pinch is a plasma produced by a powerful (~100 kA) current pulse that is applied to two or more slanted wires joined at the centre. The diagnostics employed included PIN diode detection, voltage and current measurement and pinhole X-ray photography. The results show that the X-rays emitted are initially thermal in nature and have energy ~7.7 keV and afterwards there is harder, non-isotropic radiation that can be attributed to a plasma interruption. The radiography shows that there is formation of a jet of plasma at the moment of maximum compression.

1. Introduction

The Z-pinch is a dense, magnetized plasma produced in gases and vapors of metallic fibers that experiences a strong compression due to its own magnetic field. Densities of 10^{17} to 10^{20} cm^{-3} and temperatures in the range 10^2-10^5 eV are typical of a Z-pinch experiment [1]. The high density and temperatures thus produced are useful in the investigation of fusion plasmas and the physics of stellar interior [2]. In a multi-wire Z-pinch experiments a powerful electric-current pulse of the order of 0.1-1 MA is applied to a set of parallel wires surrounded by high vacuum. The electric pulse vaporizes the wires and a plasma is created in the metallic vapor. Then the plasma is compressed by the magnetic pressure and at the moment of maximum compression an X-ray pulse, lasting tens of nanosecond, is produced. A variant of this technique employs a set of crossed wires joined at the center, hence the name: X-pinch [3-6].

A Marx bank is a device used to generate high voltage, high current pulses with power supplied by a moderate-voltage low-current source. Its principle of operation is based in charging capacitors (slowly) in parallel and then discharging them in series. Typically, the capacitors are charged in about 1 minute and discharged in less than 1 µs. The charging current is of the order of tens of milliampere while the discharge current peaks at about hundred kA.

Fast pulsed-power generators comprising a Marx bank and a pulse-forming line were originally used as electron beam sources [7]. In this type of devices the Marx charges the line on a time scale of microseconds and then the line discharges onto a vacuum diode in order to apply an accelerating pulse lasting tens of nanosecond to the electrons. Later these generators were adapted to implode arrays of metallic wires for production of intense X-ray pulses. When the individual wire plasmas merge and implode under the effect of the magnetic pressure produced by the massive current pulse, a dense plasma is created which emits X-rays with energies of the order of 1-10 keV [8].

In pulsed power one of the configurations more often employed consists of a Marx generator feeding a transmission line for pulse forming purposes. In this configuration it is the line the one that feeds the load, not the Marx bank. The main advantage of this set-up is that the Marx is decoupled from the load allowing faster fronts to be employed. Several types of transmission lines are employed, the one using de-ionised water as insulation being one of the most popular [9].

Content from this work may be used under the terms of the Creative Commons Attribution 3.0 licence. Any further distribution of this work must maintain attribution to the author(s) and the title of the work, journal citation and DOI.
Published under licence by IOP Publishing Ltd

The aim of this work is to produce fast pulses of X-rays using an X-pinch configuration of 25-μm Tungsten wires. The time resolved detection of the x-radiation using different filters helps to estimate the energy of the emitted radiation. The time-integrated X-ray photographs helps to visualize the plasma and learn how successful the pinching process was.

2. Experimental set-up

Figure 1 shows the schematics of the Z-pinch generator layout. The Marx bank is housed in a container made of cold-rolled 0.25-inch steel sheet with dimensions 1.2×1.0×0.7 m. The pulse-forming line (PFL), the transfer switch and the transfer line (TL) are inside flanged 19-inch stainless steel tubes. These are joined at the flanges by insulating diaphragms 2 inches thick. The total footprint of the ensemble Marx-line was 1.92 m^2. The vacuum chamber has 8 viewing ports, evenly spaced around its periphery, to incorporate different diagnostics. The inner conductor of both the PFL and the TL are cylindrical tubes with an outer diameter of 10 inches.

The spark gaps of Marx bank were pressurised with dry air to a working pressure of 3.0-3.7 bar. The spark gaps are triggered by applying an electrical impulse to the central trigger electrode; this distorts the field and causes breakdown between the two main electrodes. Once the gap has broken down, the voltage of the first capacitor is applied onto the next stage over-volting it and thus initiating a cascade of spark-gap operations that erect the Marx. When the pulse reaches the transfer switch, this self-triggers and transfers the current to the next section of line (TL) that feeds the load. A one-loop Rogowski coil protruding from the upper lid of the vacuum chamber was used to measure the load current.

In the present work the main parameters of the generator were as follows: the PFL and TL both had an impedance of 4.3 Ω, the Marx bank had 8 stages with a 1.6 μF each and maximum charging voltage was 55 kV. With these parameters a short circuit current of 140 kA was obtained in the preliminary tests at a charging voltage of 55 kV.

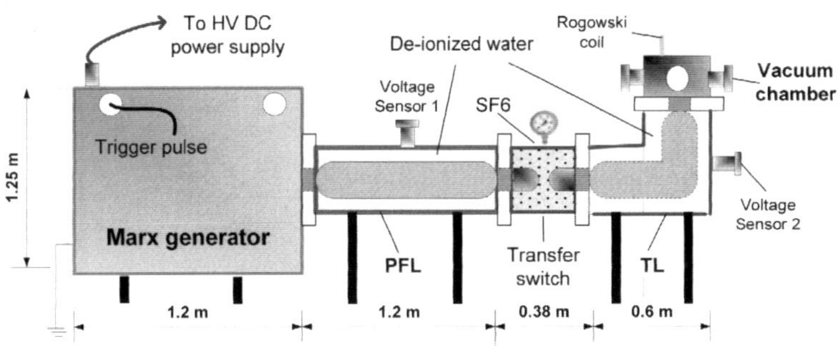

Figure 1 Schematics of the generator and vacuum chamber. Key: PFL=pulse-forming line, TL=transfer line. Not to scale.

In order to detect the X-rays emitted by the plasma two PIN diodes were fitted to different viewports in the vacuum chamber, with identical radial distance from the center. The PIN diodes (Quantrad N-series) had a 250-μm deep intrinsic layer and were operated under a 300 V bias voltage. Each diode was fitted with a different transmission filter. These consisted of thin metallic films: in one case it was 125 μm aluminum and in the other a Titanium foil 12.5 μm thick plus a layer of aluminum 5 μm thick added to avoid fluorescence.

Figure 2 shows the combined response of the filter and the diodes as a function of incident photon energy. The response is obtained from the convolution of the absorbance of the silicon in the PIN with the transmission of the filter. Figure 2 shows that the Titanium-fitted diode is very sensitive in the photon energy interval 2.1-4.94 keV where the aluminum-fitted diode is not. In the interval 5.6-100 keV both are capable of detecting the photons within this energy range but will do that with different amplitudes in their output signals.

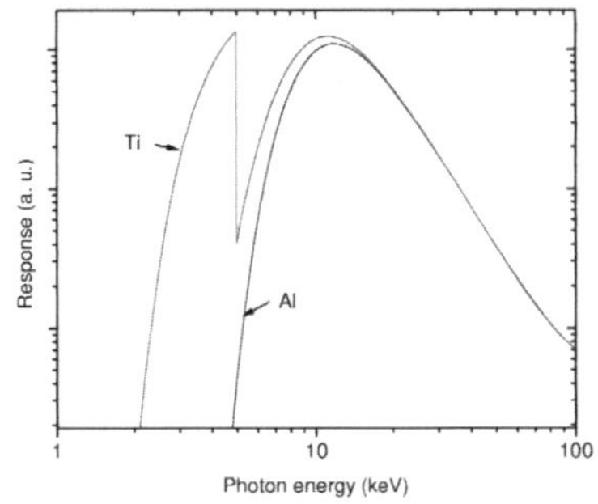

Figure 2 Combined response of the PIN diodes and the filters employed.

An X-ray pinhole camera was placed at one of the viewports of the vacuum chamber in order to take time-integrated radiographs of the plasma. It had a length of 15 cm which, when taking into account the radial distance (30 cm) from the center of the chamber to the flange it was fitted to, give a magnification ratio of 2:1. The pinhole had a diameter of 100 μm and was covered with an aluminum film 5 μm thick in order to prevent the light of the plasma from reaching the film. The film employed was conventional Kodak E-speed dental film.

3. Results

Figure 3 shows the signals obtained in a typical experiment X-pinch using two crossed 25-μm Tungsten wires. The graph shows the voltage, the current and the PIN-diode signals obtained in one of several shots realized in the lab. In the figure t=0 corresponds to the Marx's activation.

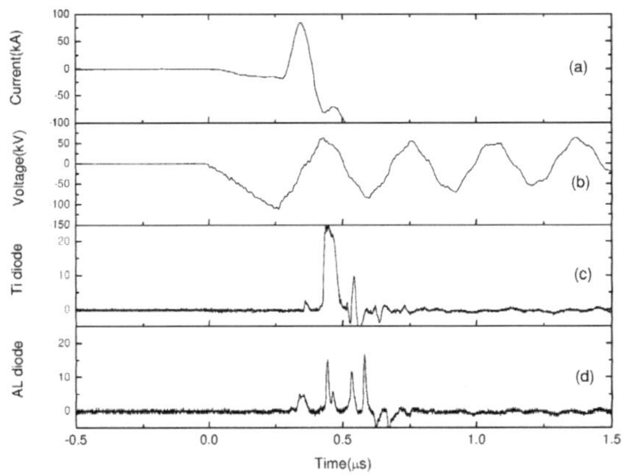

Figure 3 Electrical and X-ray signals produced by the plasma. a) load current, b) PFL voltage, c) titanium-fitted PIN diode, d) aluminum-fitted PIN diode.

The trace in panel (b) of Figure 3 (from voltage probe #1 in Figure 1) shows that when the Marx bank is fired at t=0, the PFL voltage begins to grow steadily in time and afterward experiences several oscillations due to the pulse travelling back and forth along the line. During the time interval t=0-0.28 μs the pulse induces a capacitive pulse of current at the load which can be clearly seen in the current signal. Then at t=0.28 μs the transfer switch closes and consequently the plasma current pulse rises very quickly with a rise-time of 75 ns and a peak value of 84 kA. When the current peaks, the plasma begins to emit X-rays as can be seen from the lower two panels in Figure 3. The difference observed in the signals is due to the fact that the otherwise identical diodes were fitted with different filters.

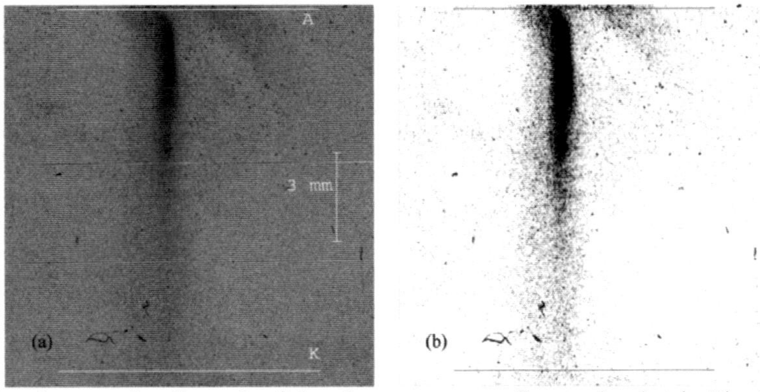

Figure 4 Time-integrated radiograph of the plasma corresponding to the shot in Fig. 3. a) original radiograph, b) enhanced image. Key: *A* anode, *K* cathode.

4. Discussion and conclusions

The X-pinch is a very dynamic plasma: it undergoes dramatic changes in its behaviour in the span of a few nanoseconds. Under normal circumstances, the response curves in figure 2 dictate that the signal detected by the Ti-fitted diode should always be larger than the signal of the other diode. This is the case of the larger peak of the Ti diode at t=0.45 μs where it can be seen that the amplitude of the Ti diode is 1.6 times larger than that of the aluminum one. A comparison with the curves in figure 2 shows that this event matches photons having an energy of about 7.7 keV.

However, there are other peaks in the aluminum-diode's signal that the titanium one does not detect. These we believe are produced by harder, anisotropic radiation produced by breaks in the plasma column. It is a well-known fact that the plasma column sometimes breaks and forms a mini-diode [10]. When this is formed the accelerating voltage for the electrons becomes the voltage of the transfer line that is of the order of 440 kV. This applied voltage can accelerate the electrons toward the anode where they emit X-rays. But if the mini-diode is not parallel to the axis of the electrodes then the emission can be strongly asymmetrical and one of the diodes detects more radiation than the other.

This is consistent with the results of Hass and co-workers [11] who found that in a tungsten X-wire experiment the X-ray high power peak observed was consistent with a Planckian-type spectrum with energy ~5 keV that is purely thermal in nature. This is followed by a break in the plasma and the gap created generates an e-beam that produces harder radiation: above 8 keV.

The radiograph in panel (a) of figure 4 shows that the plasma forms a jet extending to both sides of the point where the two wires join but also that it is more intense on the anode side. In fact, the processed image on panel (b) shows that the plasma has two fragments: one at the center, where the wires merge, and the other is the jet extending from the center to the anode. These type of jets have been observed by other researchers when the coronal plasma from both wires converges at the center [12, 13]. In the present case it is evident that the jet is stronger in the anode side of the A-K gap probably due to an electron beam that was emitted from the center of the pinch in the direction of the anode. The e-beam radiates due to the beam-target mechanism. This type of jet is of interest to astrophysicists as it can be used to simulate in the lab the ejection of stellar material.

References

[1] Haines M G 2011 *Plasma Phys. Control. Fusion* **53** 093001
[2] Remington B A 2006 *Rev. Mod. Phys.* **78** 755
[3] Ampleford D J et al 2015 *IEEE Trans. Plasma Sc.* **43** 3344
[4] Shrestha I et al 2010 *High Energy Density Phys.* **6** 113
[5] Pikuz S A, Shelkovenko TA and Hammer DA 2015 *Plasma Phys. Rep.* **41** 291
[6] Pikuz S A, Shelkovenko TA and Hammer DA 2015 *Plasma Phys. Rep.* **41** 445
[7] Nation J A 1983 High Power Relativistic Electron Beam Sources *Applied Charged Particle Optics* ed A. Septier, (N. York: Academic Press) pp. 171-206

[8] Pereira N R and Davis J 1988 *J. Appl. Phys.* **64** R1
[9] Mankowski J and Kristiansen M 2000 *IEEE Trans. Plasma Sc.* **28** 102
[10] Shelkovenko T A, Sinars D B, Pikuz S A and D A Hammer 2001 *Phys. Plasmas* **8** 1305
[11] Haas D, Bott SC, Ueda U, Eshaq Y and Beg F N 2007 *Proc. Pulsed Power Plasma Conf.* 639
[12] Beg F N, Ciardi A, Ross I, Zhu Y, Dangor AE and Krushelnick K 2006 *IEEE Trans. Plasma Sci.* **34** 2325
[13] Haas D M et al 2011 *Astrophys Space Sci.* ***336*** 33

A simplification of the fractional Hartley transform applied to image security system in phase

Carlos J. Jimenez, Juan M. Vilardy, Ronal Perez

Grupo GIFES. Faculty of engineering. Universidad de La Guajira, Riohacha (La Guajira), Colombia

E-mail: carlosj114@gmail.com, jmvilardy@uniguajira.edu.co

Abstract. In this work we develop a new encryption system for encoded image in phase using the fractional Hartley transform (FrHT), truncation operations and random phase masks (RPMs). We introduce a simplification of the FrHT with the purpose of computing this transform in an efficient and fast way. The security of the encryption system is increased by using nonlinear operations, such as the phase encoding and the truncation operations. The image to encrypt (original image) is encoded in phase and the truncation operations applied in the encryption-decryption system are the amplitude and phase truncations. The encrypted image is protected by six keys, which are the two fractional orders of the FrHTs, the two RPMs and the two pseudorandom code images generated by the amplitude and phase truncation operations. All these keys have to be correct for a proper recovery of the original image in the decryption system. We present digital results that confirm our approach.

1. Introduction

The image encryption techniques use some mathematical tools related to the signal processing, such as the Fourier, Fresnel, fractional Fourier, Hartley, Gyrator, radial Hilbert and Wavelet transforms, among other transforms, in order to encode the image to encrypt into a random noise [1-22]. There are image encryption systems based on the fractional Fourier transform (FrFT) [7-12] or the fractional Hartley transform (FrHT) [14, 15], these fractional transforms improve the security of the encryption systems due to the fractional orders of these transforms represent new additional keys [1]. The FrFT [23] and the FrHT [24] are generalizations of the Fourier and Hartley transforms, respectively, that adding a new degree of freedom given by the fractional order of these fractional transforms. The FrFT and the FrHT are used in applications related to the analysis and processing of signals and images.

The encryption system that encoding the image to encrypt (original image) in phase, was proposed in ref. [25]. This phase encoding is a nonlinear operation that increases the security of the security system in comparison with the encryption systems that encoding the original image in amplitude. In other works, the Hartley transform [13] and the FrHT [14, 15] have been used in encryption systems in order to obtain a real-valued encrypted image and increasing the security of the encryption systems, respectively. The amplitude and phase truncation operations can introduce nonlinear effects in the encryption systems, allowing improve the security of these encryption systems [26].

In this work, we propose an image encryption-decryption system using the FrHT, phase encoding, random phase masks (RPMs) and truncation operations. The original image is encoding in phase because this encoding allows a better protection of the encrypted image against security attacks [1, 8,

Content from this work may be used under the terms of the Creative Commons Attribution 3.0 licence. Any further distribution of this work must maintain attribution to the author(s) and the title of the work, journal citation and DOI.
Published under licence by IOP Publishing Ltd

9, 14, 17]. We introduce a simplification for the computation of the FrHT in order to perform this transform in an efficient and fast way. We use two fractional orders of the FrHTs as key of the security system. The RPMs have random images which are used as keys of the encryption system. The amplitude and phase truncation operations are nonlinear mathematical tools, these truncation operations allow to generate two new keys for the encryption-decryption system. Finally, the main aim of the proposed security system in this paper is to provide a digital images encryption-decryption system with a high security for the protection of the encrypted images.

The rest of the paper is organized as follows: the section 2 shows the definition of the FrHT and introduces a simplification for the computation of this transform. The amplitude and phase truncation operations are presented in section 3. The mathematical description and the numerical simulations of the image encryption and decryption systems are described in section 4. Finally, conclusions are summarized in section 5.

2. A simplification of the fractional Hartley transform

The fractional Hartley transform (FrHT) can be defined using the fractional Fourier transform (FrFT) [24]. The FrFT at order p, is a linear integral transformation that maps a given function $f(x)$ (written in one-dimensional for the sake of simplicity) onto a function $F^p(u)$, by [23]

$$F^p(u) = \mathcal{F}^p\{f(x)\} = \int_{-\infty}^{+\infty} K_p(u,x) f(x)\mathrm{d}x, \tag{1}$$

with

$$K_p(u,x) = C_p e^{i\pi\left[(u^2+x^2)\cot\alpha - 2ux\csc\alpha\right]}, \quad C_p = \frac{e^{-i\left(\frac{\pi}{4}\mathrm{sgn}(\alpha)-\frac{\alpha}{2}\right)}}{\sqrt{|\sin\alpha|}}, \quad \alpha = \frac{\pi}{2}p, \quad -\pi < \alpha \quad \pi, \tag{2}$$

where K_p is the fractional Fourier kernel and sgn is the sign function. For $p = 0$, it corresponds to the identity transform. For $p = 1$, it reduces to the direct standard Fourier transform. For $p = 2$, the reverse transform is obtained. For $p = -1$, it corresponds to the inverse standard Fourier transform. The inverse FrFT corresponds to the FrFT at fractional order $-p$. The FrFT operator is additive with respect to the fractional order, $F^{p1}F^{p2} = F^{p1+p2}$. The FrFT has a periodicity of 4 with respect to the fractional order p.

The FrHT at fractional order p of a function $f(x)$ is defined as [24]

$$f_p(u) = \mathcal{H}^p\{f(x)\} = \frac{1+e^{i\pi p/2}}{2} F^p(u) + \frac{1-e^{i\pi p/2}}{2} F^p(-u), \tag{3}$$

where $\mathcal{H}^p$ is the fractional Hartley operator, $f_p(u)$ is the FrHT at fractional order p of $f(x)$, $F^p(u)$ is the FrFT at fractional order p of $f(x)$. Due to the FrHT can be defined in terms of the FrFT, the FrHT verifies all properties of FrFT [24]. When $p = 1$, the FrHT reduces to the conventional Hartley transform. The FrHT has a periodicity of 2 with respect to the fractional order p. The computation of the FrHT by using the Eq. (3) uses the computation of two FrFTs. We propose the following simplification of the FrHT with the purpose of computing this transform in an efficient and fast way

$$\begin{aligned}\mathcal{H}^p\{f(x)\} &= \frac{1+e^{i\pi p/2}}{2} F^p(u) + \frac{1-e^{i\pi p/2}}{2} F^p(-u) = \frac{1+e^{i\pi p/2}}{2} \mathcal{F}^p\{f(x)\} + \frac{1-e^{i\pi p/2}}{2} \mathcal{F}^p\{f(-x)\} \\ &= \mathcal{F}^p\left\{\frac{1+e^{i\pi p/2}}{2} f(x) + \frac{1-e^{i\pi p/2}}{2} f(-x)\right\}.\end{aligned} \tag{4}$$

Therefore, the computation of the FrHT by using the Eq. (4) uses the computation of only one FrFT. In this way, the computation of the FrHT is more efficient and faster in comparison to the computation of the FrHT used in refs. [14, 15].

3. Amplitude and phase truncation operations

The truncation operations are nonlinear mathematical tools that can be applied to a complex-valued image [26]. Let $f(x) = a(x)\exp\{i2\pi\varphi(x)\}$ be a complex-valued function, where $a(x)$ and $\varphi(x)$ represent the amplitude and the phase of the function $f(x)$, respectively. The amplitude $a(x)$ is a positive real-valued function and the phase $\varphi(x)$ is a real-valued function with positive and/or negative values.

The amplitude truncation (AT) allows to select the phase function $\varphi(x)$ from the complex-valued function $f(x)$. Therefore, the result of the AT when is applied to the complex-valued function $f(x)$ is

$$AT\{f(x)\} = AT\{a(x)\exp\{i2\pi\phi(x)\}\} = \phi(x). \tag{5}$$

The phase truncation (PT) allows to select the amplitude function $a(x)$ from the complex-valued function $f(x)$. When the PT is applied to the complex-valued function $f(x)$, we obtain

$$PT\{f(x)\} = PT\{a(x)\exp\{i2\pi\phi(x)\}\} = a(x). \tag{6}$$

The complex-valued function $f(x)$ can be expressed in terms of the AT and PT operations as

$$f(x) = PT\{f(x)\}\exp\{i2\pi AT\{f(x)\}\}. \tag{7}$$

4. Image encryption and decryption systems

Let $g(x)$ be the real-valued image to encrypt (original image) with values in the interval [0, 1], and let $r(x)$ and $t(u)$ be two random phase masks (RPMs) given by

$$r(x) = \exp\{i2\pi s(x)\}, \quad t(u) = \exp\{i2\pi n(u)\}, \tag{8}$$

where x and u represent the coordinates for the spatial domain and the fractional Hartley domain (FrHD), respectively, $s(x)$ and $n(u)$ are normalized positive functions randomly generated, statistically independent and uniformly distributed in the interval [0, 1].

The image to encrypt $g(x)$ is encoding in phase $g_{ph}(x) = \exp\{i2\pi g(x)\}$ [1, 8, 9, 14, 17]. Then, the image $g_{ph}(x)$ bonded to the RPM $r(x)$ is transformed using the FrHT at order $p1$

$$h_{p1}(u) = \mathrm{H}^{p1}\{g_{ph}(x)r(x)\} = q_{p1}(u)\exp\{i2\pi\phi_{p1}(u)\}. \tag{9}$$

The data distributions $q_{p1}(u)$ and $\varphi_{p1}(u)$ denote the amplitude and the phase of the complex-valued image $h_{p1}(u)$, respectively. When the amplitude and phase truncation operations are applied to the image $h_{p1}(u)$, we obtain the following functions

$$q_{p1}(u) = PT\{h_{p1}(u)\}, \quad \phi_{p1}(u) = AT\{h_{p1}(u)\}. \tag{10}$$

The image $q_{p1}(u)$ bonded to the RPM $t(u)$ is transformed using the FrHT at order $p2$

$$s_{p2}(v) = \mathrm{H}^{p2}\{q_{p2}(u)t(u)\} = e(v)\exp\{i2\pi\theta_{p2}(v)\}, \tag{11}$$

where the functions $e(v)$ and $\theta_{p2}(v)$ represent the amplitude and the phase of the complex-valued image $s_{p2}(v)$, respectively. Finally, the amplitude and phase truncation operations are applied to $s_{p2}(v)$

$$e(v) = PT\{s_{p2}(v)\}, \quad \theta_{p2}(v) = AT\{s_{p2}(v)\}. \tag{12}$$

The real-valued data distribution $e(v)$ is the encrypted image. The six key of the encryption system are given by the two fractional orders ($p1$ and $p2$) of the FrHTs, the two pseudorandom images $\varphi_{p1}(u)$ and $\theta_{p2}(v)$ and the two RPMs $r(x)$ and $t(u)$. The RPMs $r(x)$ and $t(u)$ are used to spread the information content of the original image $g(x)$ onto the encrypted distribution $e(v)$. The data distribution keys $\varphi_{p1}(u)$ and $\theta_{p2}(v)$ are generated by the amplitude truncation operations applied in the encryption system. The keys $\varphi_{p1}(u)$ and $\theta_{p2}(v)$ are not truly random because these images are dependent on the values of the image $g_{ph}(x)$, the two fractional orders ($p1$ and $p2$) of the FrHTs and the RPMs $r(x)$ and $t(u)$, used in Eqs. (8)-(12).

The retrieval of the original image $g(x)$ is possible at the output of the decryption system from the encrypted image $e(v)$ when the correct values keys are used. The decrypted image $d(x)$ can be obtained using the following equations

$$q_{p1}(u) = t^*(u)\mathrm{H}^{-p2}\{e(v)\exp\{i2\pi\theta_{p2}(v)\}\}, \tag{13}$$

$$g_{ph}(x) = r^*(x)\mathrm{H}^{-p1}\{q_{p1}(u)\exp\{i2\pi\phi_{p1}(u)\}\}, \tag{14}$$

$$d(x) = AT\left\{g_{ph}(x)\right\} = g(x). \tag{15}$$

The original image $g(x)$ can be correctly recovered at the output of the decryption system whenever the values of the keys used in the encryption and decryption systems are the same.

Figure 1 shows an example to illustrate the simulation results obtained when we evaluate the proposed image encryption and decryption systems presented in this section. All image used in the numerical simulations have 512 × 512 pixels in grayscale. The original image $g(x)$ and the random distribution code $n(u)$ of the RPM $t(u)$ are depicted in figures 1(a) and 1(b), respectively. The real-valued encrypted image $e(v)$ for the fractional orders $p1$ = 1.77 and $p2$ = 0.43 of the FrHTs is shown in figure 1(c). This encrypted image is a noisy data distribution which does not disclose any information of the original image $g(x)$. The two obtained pseudorandom images $\varphi_{p1}(u)$ and $\theta_{p2}(v)$ in the encryption process, are presented in figures 1(d) and 1(e), respectively. These images have a noisy appearance very similar to the random code $n(u)$ of figure 1(b), but the two images $\varphi_{p1}(u)$ and $\theta_{p2}(v)$ are pseudorandom data distributions. The obtained decrypted image $d(x)$ from the encrypted image $e(v)$ and using the correct security keys is shown in figure 1(f).

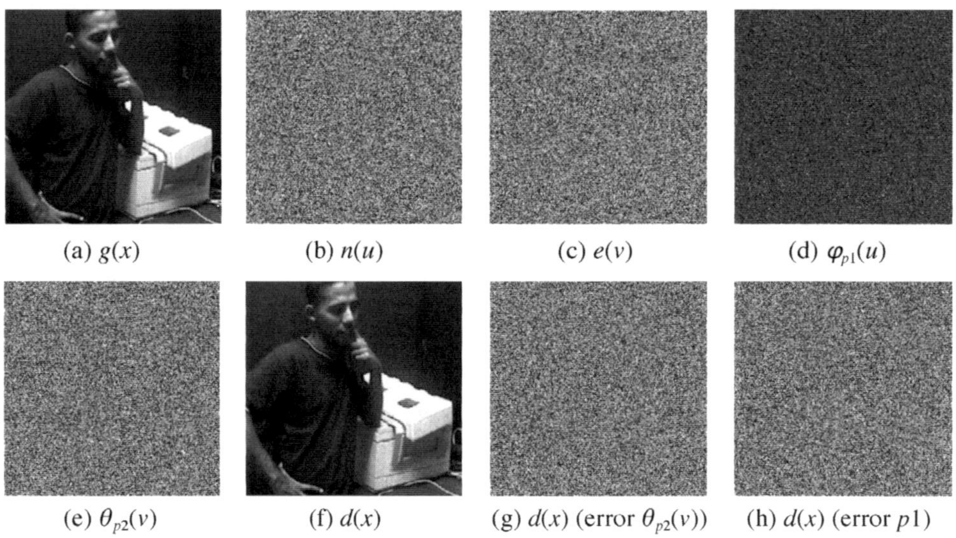

Figure 1. (a) Image to encrypt $g(x)$. (b) Random distribution code $n(u)$ of the RPM $t(u)$. (c) Encrypted image $e(v)$ for the fractional orders $p1$ = 1.77 and $p2$ = 0.43 of the FrHTs. Pseudorandom images: (d) $\varphi_{p1}(u)$, and (e) $\theta_{p2}(v)$. (f) Decrypted image $d(x)$ when the six correct security keys ($p1$, $p2$, $r(x)$, $t(u)$, $\varphi_{p1}(u)$, and $\theta_{p2}(v)$) are used. Decrypted images for the following incorrect keys: (g) the pseudorandom image $\theta_{p2}(v)$, and (h) the fractional order $p1$ = 1.775.

To evaluate the quality of the decrypted images, we use the root mean square error (RMSE) between the decrypted images $d(x)$ and the original image $g(x)$ [3]

$$\mathrm{RMSE} = \left\{ \frac{\sum_{x=1}^{M} [g(x) - d(x)]^2}{\sum_{x=1}^{M} [g(x)]^2} \right\}^{\frac{1}{2}}. \tag{16}$$

The values of the RMSE metric for evaluating image quality are between 0 and 1; when the value of the RMSE is close or equal to 0, this metric indicates an excellent quality image for the retrieval of the decrypted image at the output of the decryption system whereas the values of the RMSE close or equal to 1 represent a worse quality image. The RMSE between the original image of figure 1(a) and the decrypted image of figure 1(f) is 7×10^{-14}.

The noisy decrypted images displayed in figures 1(g) and 1(h), correspond to the retrieved images in the decryption system when the key of the pseudorandom image $\theta_{p2}(v)$ or the value of the fractional order $p1$ are wrong, respectively. The RMSEs between the original image of figure 1(a) and the decrypted images of figures 1(g) and 1(h) are 0.88 and 0.87, respectively. If the values of the fractional order $p2$ of the FrHT, the RPMs $r(x)$ and $t(u)$ or the pseudorandom image $\varphi_{p1}(u)$ used in the decryption system are not equal to the values used in the encryption system, the decrypted image is a noisy pattern very similar to figure 1(g). Therefore, the provided result prove that the all keys (the two fractional orders $p1$ and $p2$ of the FrHTs, the two pseudorandom images $\varphi_{p1}(u)$ and $\theta_{p2}(v)$ and the two RPMs $r(x)$ and $t(u)$) are required in the decryption system for correct retrieval of the original image.

We recall that the FrHT has a periodicity of 2 with respect to its fractional order. We evaluate the sensitivity on the fractional orders of the FrHTs by introducing small errors in these and leaving fixed the other four security keys in the decryption system. The RMSE is used to measure the level of protection on the encrypted image. For this deviation test of the fractional orders on the correct values for the decryption process, it is introduced an small error that varies between -1×10^{-3} and 1×10^{-3}, then for each variation is calculated the RMSE, the results for $p1$ and $p2$ are presented in figure 2. From figure 2, it was found that the fractional orders of the FrHTs are sensitive to a variation of 1×10^{-4}.

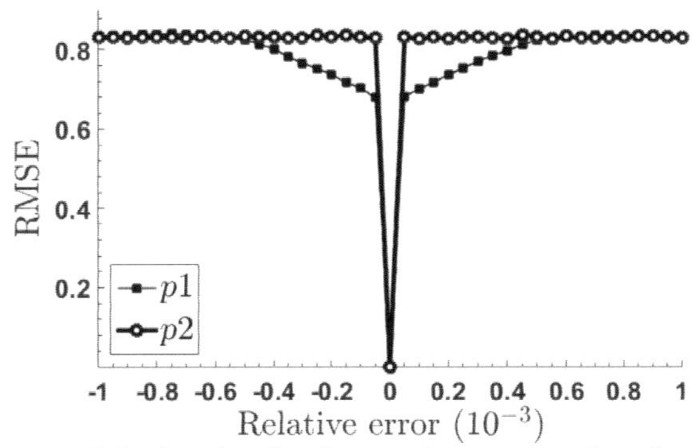

Figure 2. Deviation test of the fractional orders on the correct values for the decryption process.

Using the obtained sensitivity of the fractional orders $p1$ and $p2$, the key space for the fractional orders of the FrHTs is 4×10^8. The two RPMs $r(x, y)$ and $h(x, y)$, and the two pseudorandom images $\varphi_{p1}(u)$ and $\theta_{p2}(v)$ have a size of 512×512 pixels and each pixel has 256 possible values. The number of attempts required to retrieve both RPMs and pseudorandom images is of the order of $256^{4(512)(512)}$. The total key space of the proposed encryption process is $(4\times10^8)\cdot256^{1048576}$, this value for the key space is greater than the results obtained in refs. [2-16].

From numerical simulation developed for this paper, the time computing for the digital FrHT by implementing the Eq. (4) is reduced by half in comparison to the time computing of the Eq. (3). Therefore, the proposed computation for the FrHT is faster and efficient in comparison to the computation presented in refs. [14, 15].

5. Conclusions

We have presented an image encryption system based on the phase encoding, the FrHTs, the RPMs and the amplitude and phase truncation operations. We have introduced a simplification in the computation of the FrHT, this simplification allowed the computation of the FrHT in a more efficient and faster way. The encrypted image is a real-valued random distribution which is very well protected against security attacks because the very larger key space of the proposed encryption-decryption system. The security of the encrypted image is due to the nonlinear operations applied in the encryption-decryption system, these nonlinear operation are the phase encoding and the amplitude and phase truncation operations. The results of the numerical simulations have been shown that the retrieval of the original image in the decryption system is very sensitive to the changes on the six keys of the proposed security system in this work. The two pseudorandom images keys change when the image to encrypt, the two RPMs or the two fractional orders of the FrHTs are modified. Finally, in order to retrieve the original image with a high quality and free of noise at the output of the decryption system, the values of the six keys used in the encryption and decryption systems have to be the same.

Acknowledgments
This research has been funded by the Universidad de La Guajira.

References
[1] Millán M S and Pérez-Cabré E 2011 *Optical and Digital Image Processing: Fundamentals and Applications* ed G Cristóbal, P Schelkens and H Thienpont (Weinheim: Wiley) chapter 33 pp 739–67
[2] Réfrégier P and Javidi B 1995 *Opt. Lett.* **20** 767–9
[3] Vilardy J, Millán M S and Pérez-Cabré E 2013 *J. Opt.* **15** 025401
[4] Situ G and Zhang J 2004 *Opt. Lett.* **29** 1584–6
[5] Vilardy J, Millán M S and Pérez-Cabré E 2014 *Appl. Opt.* **53**, 1674–82
[6] Vilardy J, Millán M S and Pérez-Cabré E 2013 *Proc. of SPIE* **8785**, 87853J
[7] Unnikrishnan G, Joseph J and Singh K 2000 *Opt. Lett.* **25** 887–9
[8] Vilardy J, Torres C and Mattos L 2008 *AIP Conf. Proc.* **992** 1067–72
[9] Vilardy J, Torres C and Mattos L 2008 *Revista Colombiana de Física* **40** 143–6
[10] Vilardy J, Torres C and Mattos L 2011 *Revista Colombiana de Física* **43** 523–7
[11] Vilardy J, Torres Y, Millán M S and Pérez-Cabré E 2014 *J. Opt.* **16** 125405
[12] Vilardy J, Millán M S and Pérez-Cabré E 2014 *Opt. Pura y Apl.* **47** 35–41
[13] Chen L and Zhao D 2006 *Opt. Lett* **31** 3438–40
[14] Vilardy J, Torres C and Jimenez C 2013 *Proc. of SPIE* **8785** 87851R
[15] Zhao D, Li X and Chen L 2008 *Opt. Commun.* **281** 5326-9
[16] Rodrigo J A, Alieva T and Calvo M L 2007 *Opt. Commun.* **278** 279–84
[17] Vilardy J, Millán M S and Pérez-Cabré E 2016 Nonlinear image encryption using a fully phase nonzero-order joint transform correlator in the Gyrator domain *Opt. Laser Eng.* (*Preprint* http://dx.doi.org/10.1016/j.optlaseng.2016.02.013)
[18] Vilardy J, Torres C and Mattos L 2013 *Proc. of SPIE* **8785** 87851Q
[19] Vilardy J, Millán M S and Pérez-Cabré E 2015 *Proc. 20th Symposium on Signal Processing, Images and Computer Vision (STSIVA)* (*Bogotá*) vol 1 (Washington: IEEE) pp 1–6
[20] Morales Y, Díaz L and Torres C 2015 *J. Phys.: Conf. Series* **582** 012063
[21] Vilardy J, Torres C, Useche J and Mattos L 2011 *J. Phys.: Conf. Series* **274** 012047
[22] Vilardy J, Torres C, Useche J and Mattos L 2011 *Revista Colombiana de Física* **43** 170–4
[23] Ozaktas H M, Zalevsky Z and Kutay M A 2001 *The Fractional Fourier Transform: with Applications in Optics and Signal Processing* (Weinheim: Wiley)
[24] Pei S C and Ding J J 2002 *IEEE Trans. Signal Process.* **50** 1661–80
[25] Towghi N, Javidi B and Luo Z 1999 *J. Opt. Soc. Am.* A **16** 1915–27
[26] Qin W and Peng X 2010 *Opt. Lett.* **35** 118–20

The scale of the Fourier transform: a point of view of the fractional Fourier transform

C J Jimenez[1], J M Vilardy[2], S Salinas[3], L Mattos[4], C O Torres[4]

[1] Grupo GIMA. Universidad de La Guajira, Riohacha (La Guajira), Colombia
[2] Grupo GIFES. Faculty of engineering. Universidad de La Guajira, Riohacha (La Guajira), Colombia
[3] Centro de Investigación de Matemática Aplicada (CIMA). Universidad del Zulia, Maracaibo, Venezuela
[4] Grupo de óptica e informática. Department of electronic engineering. Universidad Popular del Cesar, Valledupar (Cesar), Colombia

E-mail: carlosj114@gmail.com, jmvilardy@uniguajira.edu.co

Abstract. In this paper using the Fourier transform of order fractional, the ray transfer matrix for the symmetrical optical systems type ABCD and the formulae by Collins for the diffraction, we obtain explicitly the expression for scaled Fourier transform conventional; this result is the great importance in optical signal processing because it offers the possibility of scaling the size of output the Fourier distribution of the system, only by manipulating the distance of the diffraction object toward the thin lens, this research also emphasizes on practical limits when a finite spherical converging lens aperture is used. Digital simulation was carried out using the numerical platform of Matlab 7.1.

1. Introduction

The Fractional Fourier transform (FrFT) is a generalization of the classical or conventional Fourier Transform, which has been introduced into the mathematical literature by Namias in 1980 [1]. Later, it was used for optical applications by several authors [2, 3]. Solution of differential equations, quantum mechanics, electromagnetic theory of propagation, interference and diffraction of light, optical system design, image processing, multiple space-variant filtering and study of space or time-frequency distributions are some of the applications of the FrFT.

The purpose of this paper is to find a direct relationship between the Fractional Fourier Transform (FrFT), the Collins diffraction formula and the scaled Fourier transform for the complex amplitude distributions on spherical surfaces. As result, we present the digital computation of the scaled Fourier transform with different values of scaling. The two-dimensional fractional Fourier transform (FrFT) at order α, is a linear integral transformation that maps a given function $f(x,y)$ onto a function $f_\alpha(u,v)$, by [2]

$$f_\alpha(u,v) = \mathrm{F}^{\alpha}\{f(x,y)\} = \int_{-\infty}^{+\infty}\int_{-\infty}^{+\infty} K_\alpha(u,v,x,y) f(x,y)\mathrm{d}x\mathrm{d}y, \tag{1}$$

with

$$K_\alpha(u,v,x,y) = C_\alpha e^{i\pi\left[(u^2+v^2+x^2+y^2)\cot\alpha - 2(ux+vy)\csc\alpha\right]}, \quad C_\alpha = \frac{-ie^{i\alpha}}{\sin\alpha}, \quad -\pi < \alpha \le \pi, \tag{2}$$

Content from this work may be used under the terms of the Creative Commons Attribution 3.0 licence. Any further distribution of this work must maintain attribution to the author(s) and the title of the work, journal citation and DOI.
Published under licence by IOP Publishing Ltd

where K_α is the fractional Fourier kernel. For $\alpha = 0$, it corresponds to the identity transform. For $\alpha = \pi/2$, it reduces to the direct standard Fourier transform. For $\alpha = \pi$, the reverse transform is obtained. For $\alpha = -\pi/2$, it corresponds to the inverse standard Fourier transform. The inverse FrFT corresponds to the FrFT at fractional order $-\alpha$. The FrFT operator is additive with respect to the fractional order, $\Phi^\alpha \Phi^\beta = \Phi^{\alpha+\beta}$. The properties of the FrFT have been studied in reference [2].

2. Ray transfer matrix

A paraxial ray in a given cross section of an optical system is characterized by its distance x_1 from the optic axis and the slope y_1; if this slope is assumed small, the ray path through a given structure depends on the optical properties of the structure and the input conditions. In this situation the relation between the input and output parameters of the optical system is given by

$$\begin{pmatrix} x_2 \\ y_2 \end{pmatrix} = \begin{pmatrix} A & B \\ C & D \end{pmatrix} \begin{pmatrix} x_1 \\ y_1 \end{pmatrix}. \tag{3}$$

The ABCD matrix is called the ray transfer matrix with its determinant equal to one [4].

3. The fractional Fourier transform and the Collins formula

The known Collins formula can be rewritten as [5]

$$U_p(u,v) = \frac{-i}{\lambda B} \exp\left\{ \frac{i\pi D(u^2+v^2)}{\lambda B} \right\} \int_{-\infty}^{+\infty}\int_{-\infty}^{+\infty} \exp\left\{ \frac{i\pi A(\xi^2+\eta^2)}{\lambda B} \right\} \exp\left\{ \frac{-i2\pi(u\xi+v\eta)}{\lambda B} \right\} \times U_A(\xi,\eta)\,\mathrm{d}\xi\,\mathrm{d}\eta. \tag{4}$$

Illuminating with a spherical wave of radius R_1/A to the input plane $U_A(\xi,\eta)$, introducing appropriate scale parameters as well as input and output coordinates [3, 5], and using the meaning of the FrFT, the previous equation can be given by

$$U_p(u,v) = \frac{e^{-i\alpha}\sin\alpha}{\lambda B} \exp\left\{ \left(\left(\frac{D}{B} - \frac{\cos^2\alpha}{A\left(1-\frac{B}{R_1}\right)B} \right) \frac{i\pi(u^2+v^2)}{\lambda} \right) \right\} \mathrm{F}^{\alpha}\{U_A(\xi,\eta)\}. \tag{5}$$

An interesting fact is the relationship between a FrFT of order α, the output complex field amplitude $U_p(u,v)$ and the input complex field $U_A(\xi,\eta)$. The phase factor is associated in paraxial approximation with a divergent wave [3, 5-7]. Therefore, the output complex field amplitude $U_p(u,v)$ is over a spherical surface with radius R_2, which is proportional to the FrFT of order α of the input complex field amplitude $U_A(\xi,\eta)$ [8-10]. The equation (5) corresponds to the Collins Formulae in terms of the fractional Fourier transform and it describes the diffraction in an optical system type ABCD, which is characterized for its ray transfer matrix.

4. Optical systems and propagation

Using the previous results, the relationship between the Collins diffraction formula, the ray transfer matrix and the fractional Fourier transform (FrFT), we developed the diffraction formulae for the well-known optical systems, and them we obtain the resultant expression between the complex amplitude distributions of an object and its propagation for an observation plane. In this case, it is considered the proposed setup of figure 1. This system is composed by an illuminating wave that propagates through the surface of the lens placed at point (ξ,η); a pupil is located at point (u,v), and finally, the observation plane is at point (x,y), which is placed in the focal length of the lens; the distance between the pupil and the plane observation is given by d.

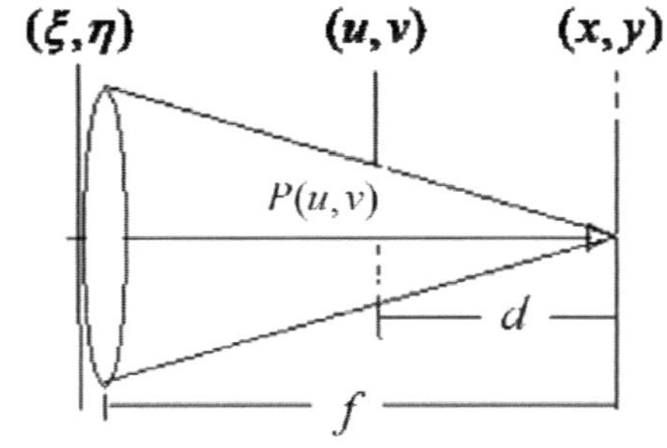

Figure 1. Proposed optical system.

Using a paraxial approach to a spherical wave that illuminates the input, the amplitude of the transmitted wave by the input and which reaches to the pupil can be written as

$$U_0(u,v) = \frac{Af}{d} P\left(u\frac{f}{d}, v\frac{f}{d} \right) \exp\left\{ \frac{-i\pi(u^2+v^2)}{\lambda d} \right\} t_A(u,v). \tag{6}$$

This paraxial approach corresponds to an spherical convergent wave towards the focal plane of radius $R_1 = -d$; applying the FrFT to the wave propagation from the coordinate plane to the point $U_0(u,v)$, where the pupil is found until the coordinate plane to the point $U_S(x,y)$, where the following observation is made and replacing the values of the ray transfer matrix elements for free space propagation ($A = D = 1$, $B = d$, and $C = 0$) in the complex amplitude distribution for the observation plane and $R_1 = -d$, we get

$$U_S(x,y) = \frac{e^{-i\alpha}\sin\alpha}{\lambda d} \exp\left\{ \frac{i\pi(x^2+y^2)}{\lambda R_2} \right\} \mathrm{F}^{\alpha}\left\{ U_0(u,v) \right\}, \tag{7}$$

where

$$R_2 = \frac{d}{1 - \frac{\cos^2\alpha}{2}}. \tag{8}$$

Taking into account that

$$\mathrm{F}^{\alpha}\left\{ U_0(u,v) \right\} = \frac{-ie^{i\alpha}}{\lambda d \sin\alpha} \exp\left\{ \frac{i\pi(x^2+y^2)}{\lambda d \tan\alpha} \right\} \int_{-\infty}^{+\infty}\int_{-\infty}^{+\infty} \exp\left\{ \frac{-i2\pi(xu+yv)}{\lambda d \sin\alpha} \right\} \times \frac{Af}{d} P\left(u\frac{f}{d}, v\frac{f}{d} \right) t_A(u,v)\,\mathrm{d}u\mathrm{d}v. \tag{9}$$

In the particular case if $\alpha = \pi/2$

$$\mathrm{F}^{\frac{\pi}{2}}\left\{ U_0(u,v) \right\} = \frac{1}{\lambda d} \int_{-\infty}^{+\infty}\int_{-\infty}^{+\infty} \frac{Af}{d} P\left(u\frac{f}{d}, v\frac{f}{d} \right) t_A(u,v) \exp\left\{ \frac{-i2\pi(xu+yv)}{\lambda d} \right\} \mathrm{d}u\mathrm{d}v, \tag{10}$$

this result corresponds to the classical Fourier transform $\mathrm{F}^{\frac{\pi}{2}} = \mathrm{F}$

$$\mathrm{F}^{\frac{\pi}{2}}\left\{ U_0(u,v) \right\} = \mathrm{F}\left\{ \frac{Af}{d} P\left(u\frac{f}{d}, v\frac{f}{d} \right) t_A(u,v) \right\}. \tag{11}$$

Finally, replacing the previous result into the equation (7), we obtain

$$U_S(x,y) = \frac{A}{i\lambda d}\frac{f}{d}\exp\left\{\frac{i\pi(x^2+y^2)}{\lambda d}\right\}\mathrm{F}\left\{P\left(u\frac{f}{d},v\frac{f}{d}\right)t_A(u,v)\right\}. \tag{12}$$

As one can notice from the configuration appears the obtained parameter of flexibility described in [4], that corresponds to the scaling factor of the Fourier Transform, which is under control by the researcher. When d increases, it means that the scaling factor in the direct representation (spatial) produces a reduction of the diffraction object; consequently the size of the transform is bigger, which corresponds a zoom of the pupil (where the object of the transmission is) directly to the lens, until the overhead transparency is directly against the lens; when the distance d decreases, it means that the scaling factor in the direct representation produces an expansion of the diffraction object; consequently the size of the transform is smaller, which corresponds a zoom of the pupil in direction of the plane of observation. These considerations are very important in spatial filtering applications.

5. Digital simulations

The results of figure 2 represent how the variety of size of the diffraction object, that is on the pupil, works when a displacement is produced, zooming or moving away the pupil of the surface observation and/or of the pupil. With the purpose of showing the behavior of the proposed optical system of the figure 1, it is used the diffraction object of figure 2(a), which has into its structure a chessboard. The figure 2(b) shows its Fourier transform of the image of figure 2(a) without any scaling; the figure 2(c) shows the Fourier transform when an increase of the diffraction object is produced with a scaling factor of 1.25. The figure 2(d) corresponds to the Fourier transform of a reduced object with a factor 0.5.

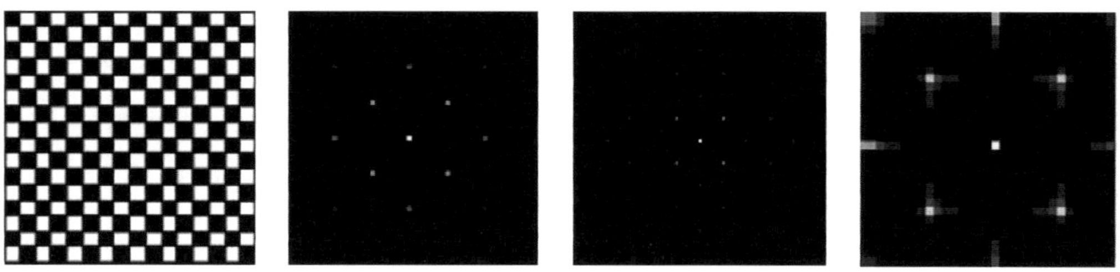

(a) (b) (c) (d)

Figure 2. Fourier transform obtained by digital simulation of the proposed optical system. (a) Diffraction object. (b) Fourier transform of the diffraction object. Fourier transform of the diffraction object with a scaling factor of: (c) 1.25, and (d) 0.5.

6. Conclusions

In the present paper was examined the behavior of the scaled Fourier transform (SFT) for the two-dimensional case, in the particular situation proposed by Goodman when the optical setup corresponds to the case of scaling factor using Fourier transform. We obtain the same result as Goodman, only with the help of the ray transfer matrix, the Collins formula and the fractional Fourier transform. In this analysis we assume that the object is totally illuminated by a spherical beam perfectly convergent and it has been demonstrated that the Fractional Fourier transform allows to interpret the same details of the scaling factors for the diffraction object as well as the conventional Fourier Transform. We proved that the scaling factor obtained is expressed in terms of the distance between the lens position and diffraction object, when the plane of observation is placed in the back focal plane of the lens.

Acknowledgments

This research has been funded by the Universidad de La Guajira and the Universidad Popular del Cesar.

IOP Conf. Series: Journal of Physics: Conf. Series **792** (2017) 012044 doi:10.1088/1742-6596/792/1/012044

References

[1] Namias V 1980 *J. Inst. Maths. Applics.* **25** 241–65
[2] Ozaktas H M, Zalevsky Z and Kutay M A 2001 *The Fractional Fourier Transform: with Applications in Optics and Signal Processing* (Weinheim: Wiley)
[3] Pellat-Finet P and Bonnet G 1994 *Opt. Commun.* **111** 141–54
[4] Goodman J W 1996 *Introduction to Fourier Optics* (Columbus: McGraw-Hill)
[5] Vega B, Peña J and Torres C 2006 *Revista Colombiana de Física* **38** 61–5
[6] Lohmann W 1993 *J. Opt. Soc. Am.* A **10** 2181–6.
[7] Alieva T and Bastiaans M 2003 *Asian J. of Physics* **12** 25–32.
[8] Mendlovic D and Ozaktas H M 1993 *J. Opt. Soc. Am.* A **10** 1875–81
[9] Ozaktas H M and Mendlovic D 1993 *J. Opt. Soc. Am.* A **10** 2522–31
[10] Lohmann A W, Mendlovic D and Zalevsky Z 1998 *Prog. Opt.* **38** 263–342.

Four channel Cosmic Ray detector based on polymaq

K N Herrera-Guzman, R A Gutierrez-Sanchez, J Felix

Department of Physics Engeeniering, University of Guanajuato, Leon, GTO 37150
felix@fisica.ugto.mx

Abstract. The Cherenkov radiation has been widely studied in transparent materials, and applied to detect and identify elementary particles. But it has not been widely studied in opaque materials. A four channels radiation detector has been designed, built, characterized, and operated; based on four polymaq (UHMW-PE) bars of 2.54 cm X 5.08 cm X 25.4 cm, which is an opaque material to visible radiation to the human eye. Silicon photo detectors, Hamamatsu, avalanche type (APD) are used to detect the radiation produced by the passage of particles in the detector blocks. The design, construction, characterization, operation, and preliminary results of this cosmic ray detector details are presented.

1. Introduction

The cosmic rays are particles coming from outer space and can be charged or neutral particles. There are two types of cosmic rays: primary and secondary. The primary are generated by astrophysical sources such as supernovae, stars, pulsars, etc. The secondary are generated by collision of primary cosmic rays with interstellar gas, this means that earth atmosphere is a source of secondary cosmic rays [1].

The Cerenkov radiation is produced when a charged particle travels faster than light in a medium (radiator), this radiation can be detected by photomultipliers [2].

There are several types of cosmic ray detectors using Cerenkov radiation [3] [4] [5], but most of them are based on transparent materials. New radiator materials can allow us to explore different energetic regions of detection and particles.

The aim of this work is to know if Cherenkov radiation is produced in an opaque material like polymaq [6], if so, it must be detectable. Due to that, a prototype with this characteristic (an opaque material as Cherenkov radiation producer) was built tested and characterized.

2. Design

The design consists in four detection channels made of Aluminum tubes with polymaq bars inside, in a vertical arrangement. Two types of polymaq bars are used: White and black. The arrangement is as follow: One tube with a white polymaq bar, on it, a tube with a black polymaq bar and so on up to four.

Each polymaq bar has a cavity in middle of one of its faces and one photodiode is placed in that cavity. The photodiode is supported in its cavity by an electronic support card (PSC) that has two right angle pins for supplying voltage (Figure 1).

Content from this work may be used under the terms of the Creative Commons Attribution 3.0 licence. Any further distribution of this work must maintain attribution to the author(s) and the title of the work, journal citation and DOI.
Published under licence by IOP Publishing Ltd

Figure 1. Photodiode Support Card (PSC).

Two Aluminum caps, for each tube, are used for inclosing completely the polymaq bars. The objective of the Aluminum tubes and caps is isolate as much as possible the photodiode of external radiation. The tubes are between two Aluminum angles and on an Aluminum sheet for supporting them.

The design is made so the electronic cards can be located in one support angle, the left angle. The photodiode electronic reading and power cards (PRPCs) are placed as near to the PSCs as possible. Each PRPC can be used to supply two photodiodes at the same time. The discriminator card is on the same Aluminum angle that the reading and power card, in the other end. Figure 2 shows the detector and its pieces.

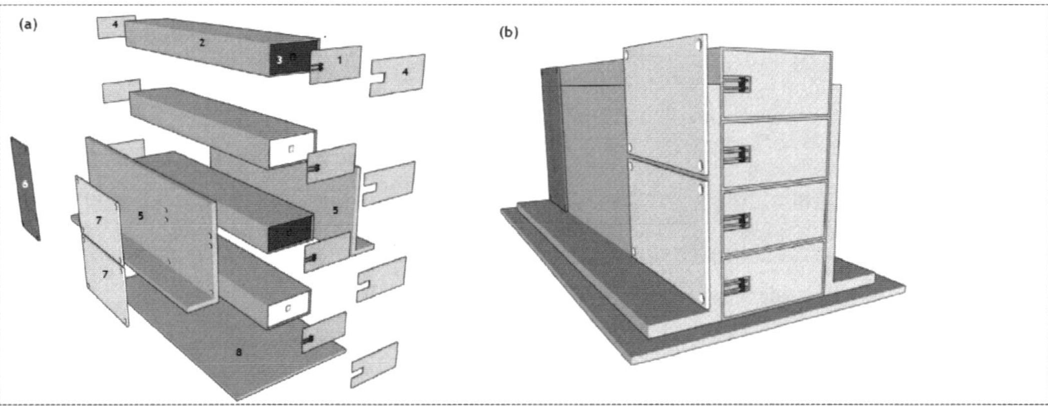

Figure 2. Polymaq detector. a) Parts: PSC (1), Aluminum tube (2), polymaq bar (3), Aluminum caps (4), Aluminum angles (5), discriminator card (6), PRPC (7), base (8). b) Assembled detector.

3. Construction

The Aluminum tubes, which were used in the construction, has the following dimensions: 24.92 cm long, 5.08 cm width (outwards), 2.54 cm height (outwards) and 1 mm thick. The photodiodes used for this detector were the MPPC S12572-100P of Hamamatsu. [7]

First, polymaq bars were manufactured, so that, they fit inside the tubes and a square cavity was made on one of its faces. The dimensions of the cavity are 5 mm, 5 mm and 1.2 mm, being long, width and thick respectively.

Then, the PRPC were made in a phenolic board Steren 405.The Figure 3 shows the necessary circuit to read out the photodiode signal. The values of resistors and capacitors are C1=C2=100 pF, R1=2.2 MΩ, R2=1.2 kΩ and VR1=10 kΩ; these values give the desired output signal. In order to connect the PRPCs and a leading edge discriminator [8] card, SMA connectors were welded.

A measurement of the internal dimensions of the Aluminum tubes was made with a Vernier. The internal width of the tubes were between 48 mm and 49 mm and the heights between 23 mm and 24 mm. It was decided that all the Aluminum caps (8) was going to be cut with 49 mm width and 24 mm height from a 1 mm thick Aluminum sheet.

The Aluminum angles and the base were drilled and the holes sanded so the angles and the base could be screwed.

Due to the size of PRPC (7.8 cm x 4.9 cm, 1.59 mm thick), Aluminum sheet was added to the left Aluminum angle. This sheet is 10.16 cm width and 20.32 cm long. The base was welded to the angle with JB Weld. Eight nuts for placing PRPC were welded on the sheet with JB Weld.

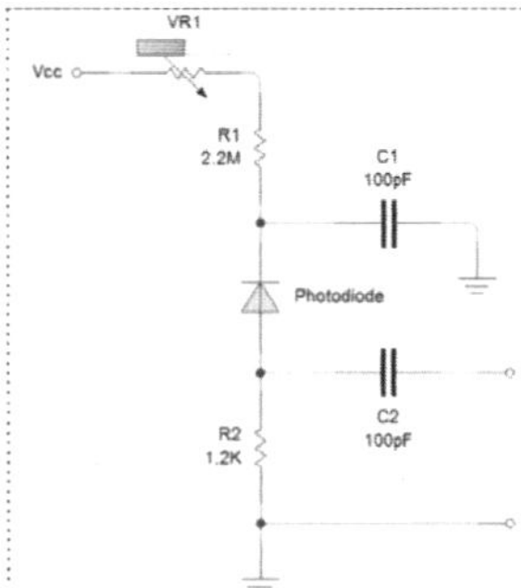

Figure 3. PRPC circuit.

By looking at PSCs it was clear that a second cavity in polymaq bars was required because the right angle pins weld hindered to join PSC to polymaq. That cavities were made with knives.

Before assembling the detector it was necessary to isolate, as possible, the photodiode from external radiation. That was done with Aluminum and electric tape, using electrical tape where Aluminum tape could cause an electroshock.

A second Aluminum sheet was welded to the angle because, when the first sheet was cut, the size of the discriminator card and its needed separation with the PRPC was not considered; the second sheet has the same width and thickness than the first one but is 2 in long. Four nuts were welded to this new sheet, the nuts for supporting discriminator in the angle. Figure 4 shows discriminator card.

Figure 4. Discriminator card. This card has 4 SMA inputs (golden) and 4 BNC outputs (silver).

With aid of the new sheet, the full detector was assembled and fixed the trigger voltage in 100 mV. The tubes in the assembled detector were named CH1, CH2, CH3 and CH4 from bottom to top respectively, showed in Figure 5.

4. Results

4.1. *Photodiode characterization*

For characterization, the discriminator trigger voltage was fixed. Oscilloscope voltage scale was modified to observe the noise of the photodiode analogical signal and it resulted around 20 mV, because of that, trigger voltage was fixed to 70 mV for avoiding noise goes through the discriminator card. With the four discriminator channels, it was possible to count pulses, and due to that, to take data and make a photodiode characterization.

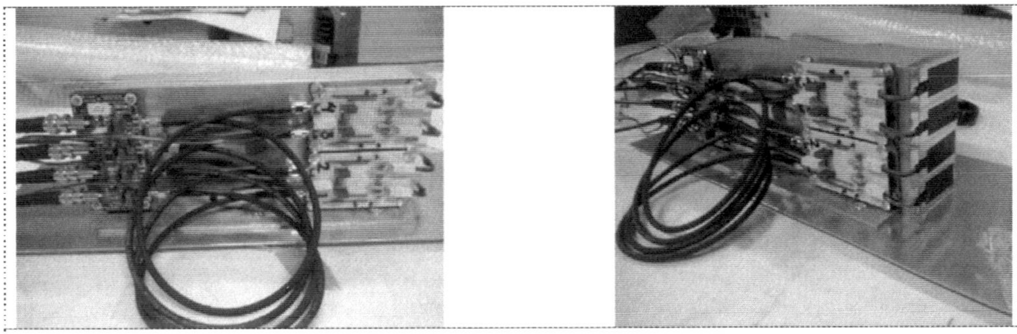

Figure 5. Assembled detector.

The discriminator card outputs were connected to a 32 channels compact-RIO, and the voltage source is connected to the PC. The automatic characterization program (ACP) user have to input the initial and final supply voltage in the program and it takes data for 10 minutes and changes the source output voltage by 5 V, takes another 10 minutes data and so on until finished.

The initial and final values given to ACP for characterization was 60 V and 100 V respectively. The characterization curves obtained are shown in Figures 6, 7, 8 and 9.

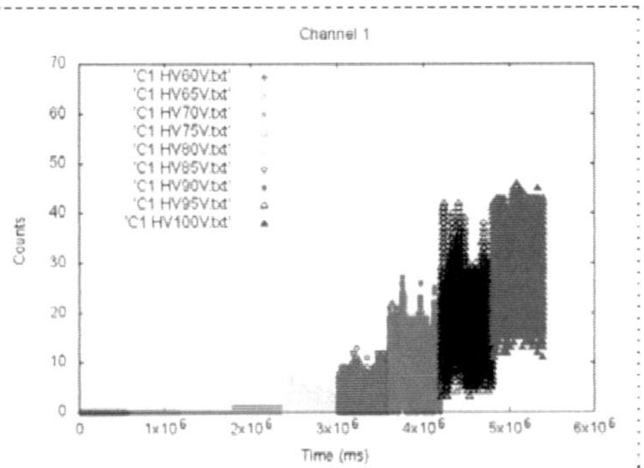

Figure 6. Characterization graph for CH1, corresponding to white polymaq.

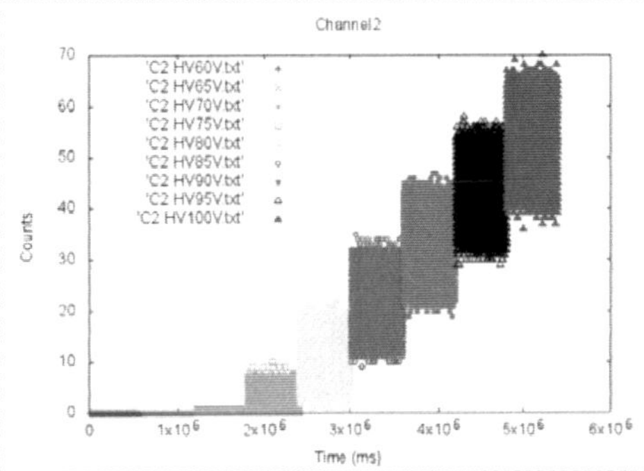

Figure 7. Characterization graph for CH2, corresponding to black polymaq.

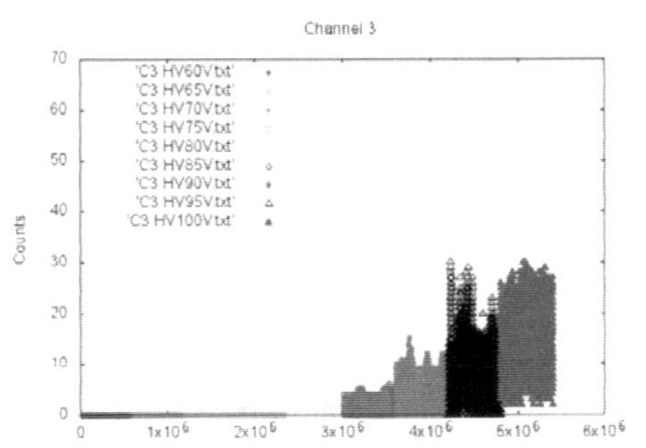

Figure 8. Characterization graph for CH3, corresponding to white polymaq.

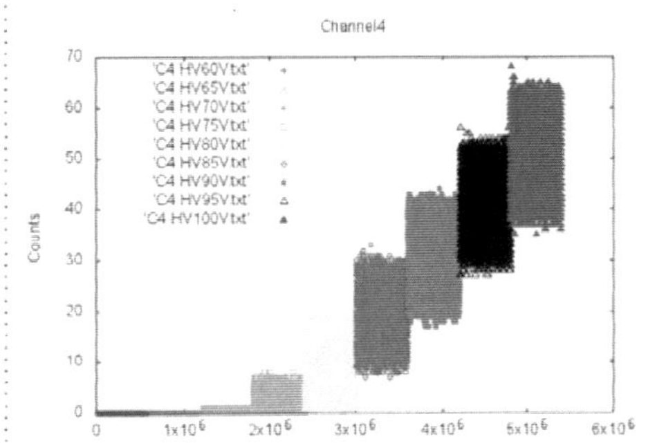

Figure 9. Characterization graph for CH4, corresponding to black polymaq.

4.2. *Data acquisition*

Half an hour of data was taken with cRIO -the program takes data for the time the user indicates and collects them- in 30 minutes files (.txt format). Data histograms are shown in Figures 10, 11, 12 and 13.

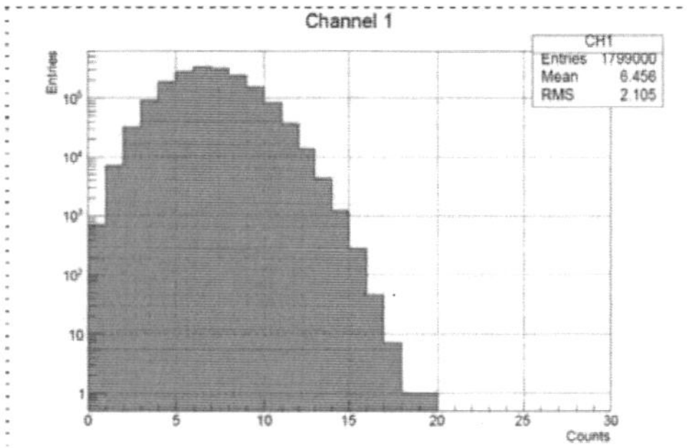

Figure 10. Histogram for CH1.

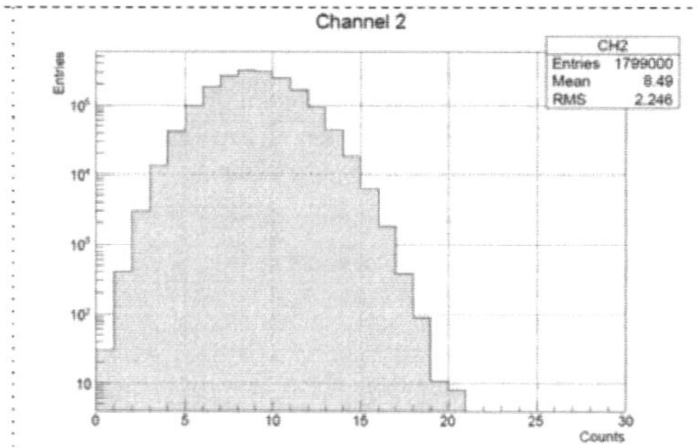

Figure 11. Histogram for CH2.

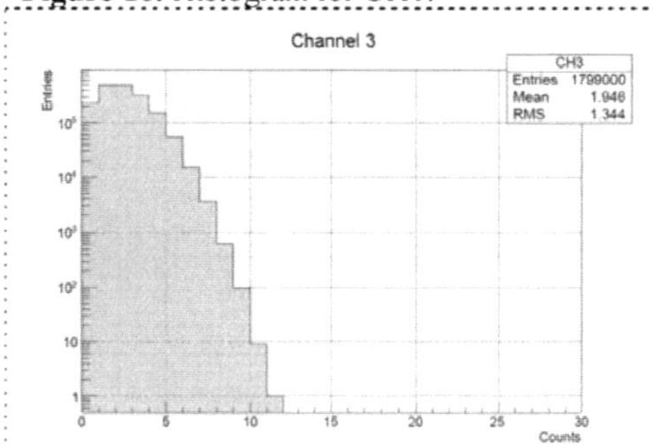

Figure 12. Histogram for CH3.

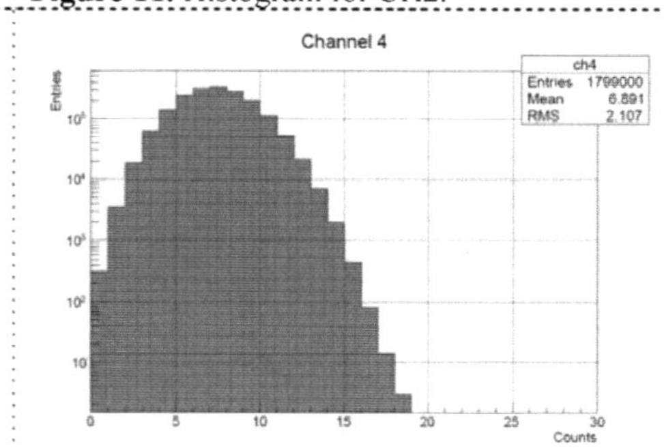

Figure 13. Histogram for CH4.

4.3. *Second Characterization*

A second photodiode characterization was performed. This time the polymaq bars configuration were changed; so that, channels that had black polymaq now have white and vice versa. The characterization curves obtained are shown in Figures 14, 15, 16 and 17.

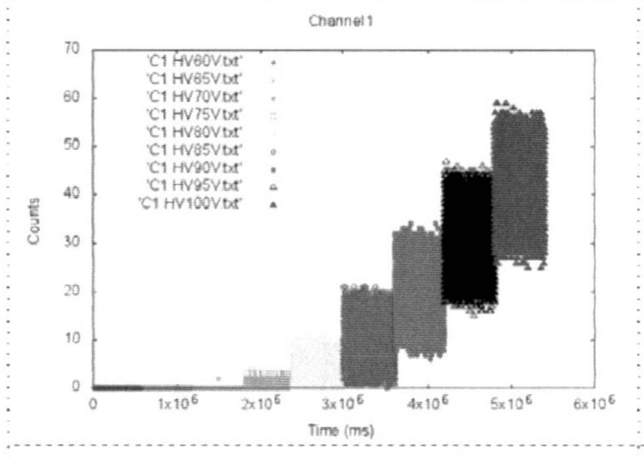

Figure 14. Characterization graph for CH1, corresponding to black polymaq.

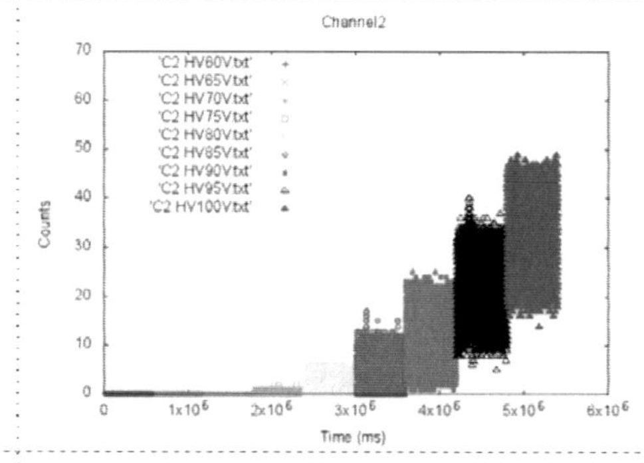

Figure 15. Characterization graph for CH2, corresponding to white polymaq.

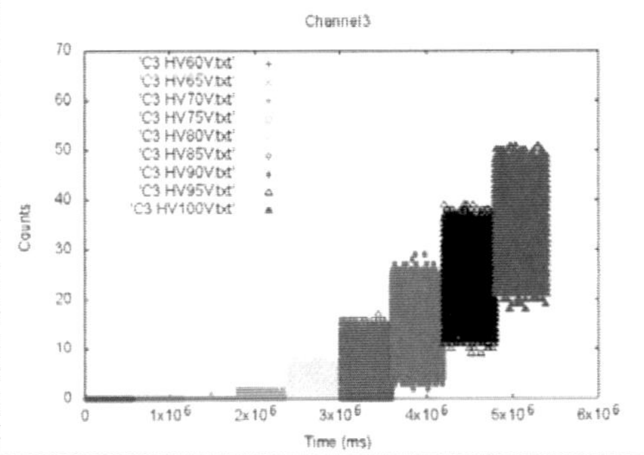

Figure 16. Characterization graph for CH3, corresponding to black polymaq.

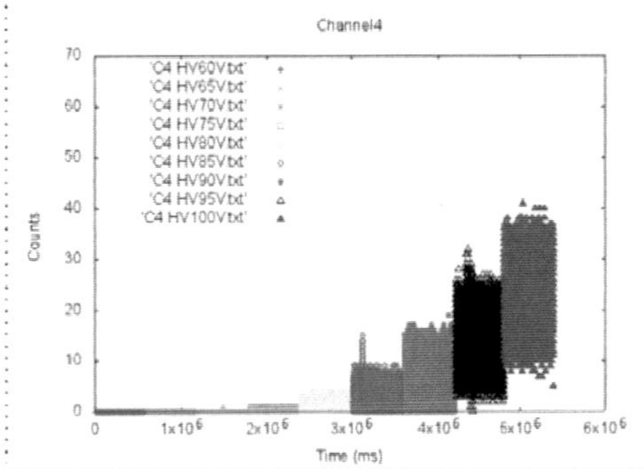

Figure 17. Characterization graph for CH4, corresponding to white polymaq.

5. Conclusions

It can be noted that the characterization graphs of CH1 and CH3 are very similar, CH2 and CH4 too, for both configurations, i.e. characterization graphs of channels with white (or black) polymaq resemble each other and this resemblance is independent of polymaq bars configuration.

From Figures 6, 7, 8 and 9 can be seen that some differences are present between detection channels with white or black polymaq, like that the maximum number of counts in black polymaq is higher than in white for each voltage supply, and these differences can be confirmed from the histograms in the Figures 10, 11, 12 and 13. Besides, it is clear that number of counts increases linearly with voltage.

Figures 14, 15, 16 and 17 show that the maximum number of counts was lower with the second configuration and that may be attributed to the photodiodes. The more important result is that black polymaq still has a bigger number of counts for every supply voltage than the white one. These results indicate that black polymaq is a better detector than the white one.

References

[1] Particle Data Group
http://pdg.lbl.gov/2015/reviews/rpp2015-rev-cosmic-rays.pdf

[2] Mark Chen, Queen's University. Scintillation and Light Sensitive Detectors.
http://neutron.physics.ucsb.edu/docs/scintillation_presentation_info.pdf

[3] Kurtz S M 2004 *The UHMWPE Handbook: Ultra-High Molecular Weight Polyethylene in Total Joint Replacement* (California: Elsevier Academic Press) chapter 12 pp 263-286

[4] Butslov M M, Medvedev M N, Chuvilo I V and M Sheshuno V 1963 *Nucl. Instr. Meth. Phys. Res.* ***20*** *263-266*

[5] Aseev A A, Devitsin E G , Komar A A, Kozlov V A, Hovsepyan Y I, Potashov S Y, Sokolovsky K A and Uvarova T V 1992 Accelerators, spectrometers, detectors and associated equipment *Nucl. Instr. Meth. Phys. Res. A* **317** 143-147

[6] Katharina Müller, University of Zurich. Cherenkov radiation.
http://www.physik.uzh.ch/lectures/empp/14/lectures/empp14_KM_cherenkov.pdf

[7] Photodiode. Hamamatsu S12572-100P.
http://www.hamamatsu.com/jp/en/product/category/3100/4004/4113/S12572-100P/index.html

[8] Hoagland T *A Brief Description of Discriminators* (National Superconducting Cyclotron Laboratory)
http://docs.nscl.msu.edu/daq/samples/Discriminators.pdf

IOP Conf. Series: Journal of Physics: Conf. Series **792** (2017) 012046 doi:10.1088/1742-6596/792/1/012046

Nonlinear image encryption system using the Gyrator transform and truncation operations

Ronal Perez, Juan M. Vilardy, Carlos J. Jimenez

Grupo GIFES. Faculty of engineering. Universidad de La Guajira, Riohacha (La Guajira), Colombia

E-mail: rperez@uniguajira.edu.co, jmvilardy@uniguajira.edu.co

Abstract. A new nonlinear system for image encryption using the Gyrator transform (GT) and the truncation operations is proposed in this work. The original image is encoded in phase at the beginning of the encryption process. The encryption-decryption system is based on the double random phase encoding (DRPE) in the Gyrator domain (GD). The rotation angle of the Gyrator transform is a new key that increases the security of the encryption system. The amplitude and phase truncation operations are nonlinear and no unitary, these truncation operations allow to select the information of amplitude or phase from a complex-valued image. We apply the truncation operations in the image encryption-decryption system in order to generate two new keys, convert the system in nonlinear and increase the security of the system. To retrieve the original image without error (noise-free) in the decryption system, it is needed all the correct security keys: the two RPMs, the rotation angle of the GT and the two key generated by the amplitude and phase truncation operations. The feasibility of this nonlinear encryption-decryption system is verified and analyzed by numerical simulations.

1. Introduction

The double random phase encoding (DRPE) is a successful method for optical image encryption [1, 2], the DRPE uses two random phase masks (RPMs) with the purpose of encoding the image to encrypt (original image) into a stationary white noise pattern (encrypted image). The optical DRPE can be implemented using a processor 4*f* [3] or a joint transform correlator [4, 5]. One of the main drawbacks of the image encryption systems based on the initial DRPE is that the security of the system is vulnerable to attacks, this weakness is due to the linear property of the DRPE scheme [1, 2, 5]. The DRPE has been further extended from the Fourier domain to the Fresnel domain [6-8], the fractional Fourier domain [9-15], the Gyrator domain (GD) [16-19] and other domains [20-23], with the purpose of adding more keys and increasing the security of the DRPE system.

The Gyrator transform (GT) is a mathematical tool for analysis and processing of two-dimensional signals [24]. The GT has been used in optics [25], signal processing [16] and image encryption [16-19]. In this paper, we propose a nonlinear image encryption-decryption system based on the DRPE in the GD and truncation operations. We use the GT to improve the security of the original DRPE by adding a new key for the encryption system, which is the rotation angle of the GT; the value of this rotation angle is a new key and it is set by the user of the encryption system. The amplitude and phase truncations operations are nonlinear [26] and we use them in order to convert the proposed security system in nonlinear, generate two new security keys and improve the security of the encrypted image.

Content from this work may be used under the terms of the Creative Commons Attribution 3.0 licence. Any further distribution of this work must maintain attribution to the author(s) and the title of the work, journal citation and DOI.
Published under licence by IOP Publishing Ltd

The outline of the paper is as follows: the definition and properties of the GT are introduced in section 2. The amplitude and phase truncation operations are presented in section 3. The mathematical description of the encryption and decryption systems are described in section 4. In section 5 are presented the computer simulations of the encryption and decryption systems. Finally, the main ideas of the paper are summarized in section 6.

2. The Gyrator transform

The Gyrator transform (GT) at parameter α, called further as a rotation angle, of a two-dimensional function $f_i(x_i, y_i)$ is defined as

$$f_o(x_o, y_o) = \mathrm{G}^{\alpha}\{f_i(x_i, y_i)\}(x_o, y_o) = \int_{-\infty}^{+\infty}\int_{-\infty}^{+\infty} f_i(x_i, y_i) K_{\alpha}(x_i, x_o, y_i, y_o)\mathrm{d}x_i \mathrm{d}y_i, \tag{1}$$

with

$$K_{\alpha}(x_i, x_o, y_i, y_o) = \frac{1}{|\sin(\alpha)|}\exp\left[i2\pi\frac{(x_o y_o + x_i y_i)\cos(\alpha) - (x_i y_o - x_o y_i)}{\sin(\alpha)}\right], \quad -\pi < \alpha \le \pi, \tag{2}$$

where K_{α} is the kernel function of GT. The GT is mathematically defined as a linear canonical integral transform which produces the rotation in position–spatial frequency planes (x, q_y) and (y, q_x) of phase space [24]. For $\alpha = 0$, it corresponds to the identity transform. For $\alpha = \pm\pi/2$ it reduces to the direct/inverse Fourier transform with rotation of the coordinates at $\pi/2$. For $\alpha = \pm\pi$ the reverse transform described by the kernel $\delta(x_i \pm x_o, y_i \pm y_o)$ is obtained. The GT is periodic and additive with respect to rotation angle: the period is 2π and $\mathcal{G}^{\alpha}\mathcal{G}^{\beta} = \mathcal{G}^{\alpha+\beta}$, respectively. The inverse GT corresponds to the GT at angle $-\alpha$.

The main theorems such as scaling, shift, modulation, etc. for the GT have been formulated in ref. [24]. The optical GT can be implemented using the optoelectronic setup described in ref. [25]. The digital GT will be computed using the fast algorithm of discrete GT based on convolution operation [27] for the nonlinear image encryption-decryption system that it will be described in section 4.

3. Amplitude and phase truncation operations

The truncation operations are nonlinear mathematical tools that can be applied to a complex-valued image [26]. Let $f(x, y) = a(x, y)\exp\{i2\pi\varphi(x, y)\}$ be a complex-valued function, where $a(x, y)$ and $\varphi(x, y)$ represent the amplitude and the phase of the function $f(x, y)$, respectively. The amplitude $a(x, y)$ is a positive real-valued function and the phase $\varphi(x, y)$ is a real-valued function with positive and/or negative values.

The amplitude truncation (AT) allows to select the phase function $\varphi(x, y)$ from the complex-valued function $f(x, y)$. Therefore, the result of the AT when is applied to the complex-valued function $f(x, y)$ is

$$AT\{f(x, y)\} = AT\{a(x, y)\exp\{i2\pi\phi(x, y)\}\} = \phi(x, y). \tag{3}$$

The phase truncation (PT) allows to select the amplitude function $a(x, y)$ from the complex-valued function $f(x, y)$. When the PT is applied to the complex-valued function $f(x, y)$, we obtain

$$PT\{f(x, y)\} = PT\{a(x, y)\exp\{i2\pi\phi(x, y)\}\} = a(x, y). \tag{4}$$

4. Nonlinear image encryption and decryption systems

Let $f(x, y)$ be the real-valued image to encrypt (original image) with values in the interval [0, 1], and let $r(x, y)$ and $h(u, v)$ be two random phase masks (RPMs) given by

$$r(x, y) = \exp\{i2\pi s(x, y)\}, \quad h(u, v) = \exp\{i2\pi n(u, v)\}, \tag{5}$$

where (x, y) and (u, v) represent the coordinates for the spatial domain and the GD, respectively, $s(x, y)$ and $n(u, v)$ are normalized positive functions randomly generated, statistically independent and uniformly distributed in the interval [0, 1]. For the initial step of the encryption system, the original image $f(x, y)$ is encoded in phase $f_{ph}(x, y) = \exp\{i2\pi f(x, y)\}$ [11, 17]. Then, the image $f_{ph}(x, y)$ is multiplied by the RPM $r(x, y)$ and this product is transformed using the GT at parameter α

$$g_\alpha(u,v) = \mathrm{G}^\alpha\{r(x,y)f_{ph}(x,y)\} = q_\alpha(u,v)\exp\{i2\pi\phi_\alpha(u,v)\}, \tag{6}$$

where the functions $q_\alpha(u, v)$ and $\varphi_\alpha(u, v)$ represent the amplitude and the phase of the complex-valued image $g_\alpha(u, v)$, respectively. The real values of the functions $q_\alpha(u, v)$ and $\varphi_\alpha(u, v)$ are dependent on the values of the RPM $r(x, y)$ and the image $f_{ph}(x, y)$. We apply the amplitude and phase truncation operations over the image $g_\alpha(u, v)$ in order to obtain the following real-valued images

$$q_\alpha(u,v) = PT\{g_\alpha(u,v)\}, \quad \phi_\alpha(u,v) = AT\{g_\alpha(u,v)\}. \tag{7}$$

The amplitude image $q_\alpha(u, v)$ is multiplied by the RPM $h(u, v)$ and this product is transformed using the GT at parameter $-\alpha$

$$t(x,y) = \mathrm{G}^{-\alpha}\{q_\alpha(u,v)h(u,v)\} = e(x,y)\exp\{i2\pi\theta(x,y)\}, \tag{8}$$

where the functions $e(x, y)$ and $\theta(x, y)$ denote the amplitude and the phase of the complex-valued image $t(x, y)$, respectively. Finally, the amplitude and phase truncation operations are applied to the image $t(x, y)$

$$e(x,y) = PT\{t(x,y)\}, \quad \theta(x,y) = AT\{t(x,y)\}. \tag{9}$$

The encrypted image is given by the real-valued data distribution $e(x,y)$ and the five security keys of the encryption system are represented by the rotation angle α of the GT, the two RPMs $r(x, y)$ and $h(u, v)$ and the two pseudorandom images $\varphi_\alpha(u,v)$ and $\theta(x,y)$.

The decryption system uses the same process as the encryption system applied in the inverse sense. The inputs of the decryption system are the encrypted image $e(x, y)$ and the five security keys used in the encryption system. The output image of the decryption system is the decrypted image $d(x, y)$. When the security keys used in the encryption and decryption systems are the same, the decrypted image $d(x, y)$ is a replica of the original image $f(x, y)$.

The original image $f(x, y)$ can be retrieved at the output of the decryption system from the encrypted image $e(x, y)$ using the five correct security keys. The decrypted image $d(x, y)$ can be obtained using the following equations

$$q_\alpha(u,v) = h^*(u,v)\mathrm{G}^\alpha\{e(x,y)\exp\{i2\pi\theta(x,y)\}\}, \tag{10}$$

$$f_{ph}(x,y) = r^*(x,y)\mathrm{G}^{-\alpha}\{q_\alpha(u,v)\exp\{i2\pi\phi_\alpha(u,v)\}\}, \tag{11}$$

$$d(x,y) = AT\left\{f_{ph}(x,y)\right\} = f(x,y). \tag{12}$$

The original image $f(x, y)$ can be correctly recovered at the output of the decryption system whenever the values of the security keys (the rotation angle α of the GT, the two RPMs $r(x, y)$ and $h(u, v)$ and the two pseudorandom images $\varphi_\alpha(u, v)$ and $\theta(x, y)$) used in the encryption and decryption systems are the same.

5. Computer simulations

In this section, we present the results of numerical simulations to examine the validity and robustness of our proposed encryption-decryption system. All image used in the encryption and decryption systems have 512×512 pixels in grayscale. The original image $f(x, y)$ and the random distribution code $s(x, y)$ of the RPM $r(x, y)$ are presented in figures 1(a) and 1(b), respectively. The rotation angle of the

GT used in the encryption system is equal to $\alpha = 0.3547\pi$. The two obtained pseudorandom images $\varphi_\alpha(u, v)$ and $\theta(x, y)$ in the encryption process, are depicted in figures 1(c) and 1(d), respectively. These images have a noisy appearance very similar to the random code $s(x, y)$ of figure 1(b), but the two images $\varphi_\alpha(u, v)$ and $\theta(x, y)$ are pseudorandom data distributions because these images are dependent on the values of the image $f_{ph}(x, y)$, the rotation angle α of the GT and the RPMs $r(x, y)$ and $h(u, v)$.

The real-valued encrypted image $e(x, y)$ for the rotation angle $\alpha = 0.3547\pi$ of the GT is shown in figure 1(e). This encrypted image is a noisy data distribution which does not disclose any information of the original image $f(x, y)$. When the decryption process is done with the encrypted image $e(x, y)$ and the correct security keys (the rotation angle α of the GT, the two RPMs $r(x,y)$ and $h(u,v)$ and the two pseudorandom images $\varphi_\alpha(u, v)$ and $\theta(x, y)$), the original image $f(x, y)$ is recovered at the output of the decryption system with loses not visible to the human eye. The decrypted image $d(x, y)$ obtained from the encrypted image $e(x, y)$ and the correct security keys is shown in figure 1(f).

To evaluate the quality of the decrypted images, we use the root mean square error (RMSE) and the signal-to-noise ratio (SNR) between the decrypted images $d(x, y)$ and the original image $f(x, y)$ [5]

$$\mathrm{RMSE} = \left(\frac{\sum_{x=1}^{M}\sum_{y=1}^{N}[f(x,y)-d(x,y)]^2}{\sum_{x=1}^{M}\sum_{y=1}^{N}[f(x,y)]^2} \right)^{\frac{1}{2}}, \qquad \mathrm{SNR} = 10\log_{10}\left[\frac{\sum_{x=1}^{M}\sum_{y=1}^{N}[f(x,y)]^2}{\sum_{x=1}^{M}\sum_{y=1}^{N}[f(x,y)-d(x,y)]^2} \right]. \tag{13}$$

The values of the RMSE metric are between 0 and 1; when the value of the RMSE is close or equal to 0, this metric indicates an excellent quality image for the retrieval of the decrypted image whereas the values of the RMSE close or equal to 1 represent a worse quality image. The SNR is inversely proportional to RMSE and it is given in decibel (dB). The RMSE and the SNR between the original image of figure 1(a) and the decrypted image of figure 1(f) are 4.7×10^{-10} and 186.5 dB, respectively.

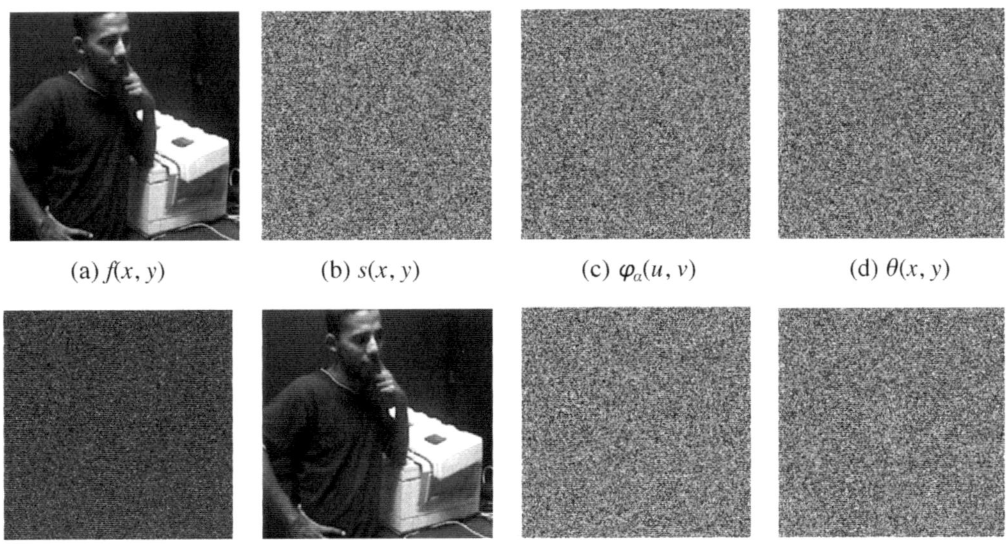

(a) $f(x, y)$ (b) $s(x, y)$ (c) $\varphi_\alpha(u, v)$ (d) $\theta(x, y)$

(e) $e(x, y)$ (f) $d(x, y)$ (g) $d(x, y)$ (error $r(x, y)$) (h) $d(x, y)$ (error $\varphi_\alpha(u, v)$)

Figure 1. (a) Original image to encrypt $f(x, y)$. (b) Random distribution code $s(x, y)$ of the RPM $r(x, y)$. Pseudorandom images for the rotation angle $\alpha = 0.3547\pi$: (c) $\varphi_\alpha(u, v)$, and (d) $\theta(x, y)$. (e) Encrypted image $e(x, y)$ for the rotation angle $\alpha = 0.3547\pi$ of the GT. (f) Decrypted image $d(x, y)$ using the five correct security keys (α, $r(x, y)$, $h_\alpha(u, v)$, $\varphi_\alpha(u, v)$, and $\theta(x, y)$). Decrypted images for the following wrong security keys: (g) the RPM $r(x, y)$, and (h) the pseudorandom image $\varphi_\alpha(u, v)$.

The obtained low RMSE and high SNR values between the original image of figure 1(a) and the decrypted image of figure 1(f) correspond to a good quality and excellent retrieval of the original image at the output of the decryption system. The obtained decrypted images from the encrypted image of figure 1(e) using a wrong RPM $r(x, y)$ or an incorrect pseudorandom image $\varphi_\alpha(u, v)$, are displayed in figures 1(g) and 1(h), respectively. The RMSEs and the SNRs between the original image of figure 1(a) and the decrypted images of figures 1(g) and 1(h) are 0.87 and 0.91, 1.2 dB and 0.82 dB, respectively. If the values of the rotation angle α of the GT, the RPM $h(u, v)$ and the pseudorandom image $\theta(x, y)$ used in the decryption method are not equal to the values used in the encryption method, the decrypted image will be a noisy pattern. Therefore, all the five security keys whit their correct values are needed in the decryption system for the correct retrieval of the original image.

The GT has a period equal 2π with respect to the rotation angle α. The rotation angle can be expressed as $\alpha = p\pi/2$, where p has a period equal 4. We evaluate the sensitivity on the rotation angle p of the GT by introducing small errors in this and leaving fixed the other four security keys in the decryption system. The RMSE is used to measure the level of protection on the encrypted image. For this deviation test of the rotation angle p on the correct values for the decryption process, it is introduced a small error that varies between -1×10^{-6} and 1×10^{-6}, then for each variation is calculated the RMSE, the results for p is presented in figure 2. From computational experiments it was found that rotation angle is sensitive to a variation of 1×10^{-7}.

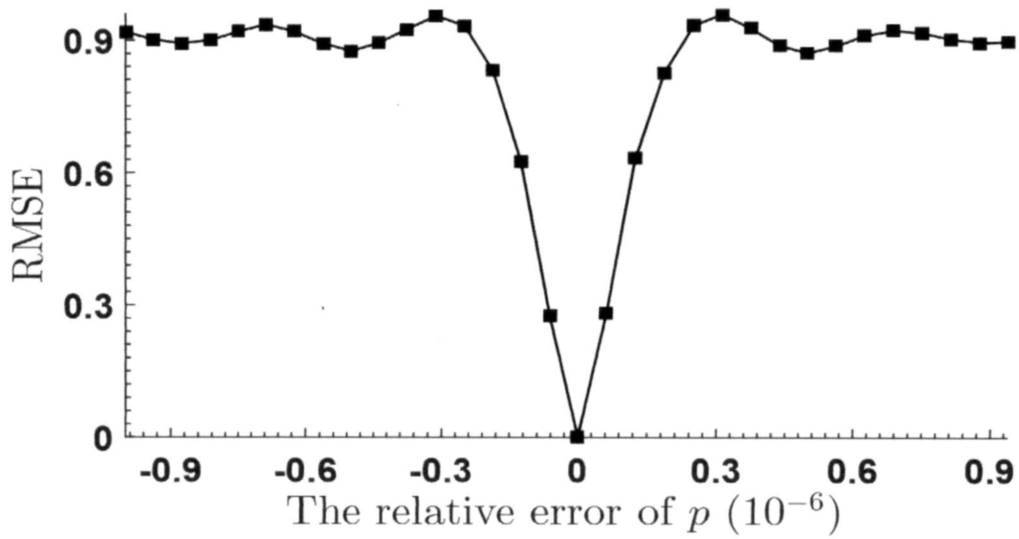

Figure 2. Deviation test of the rotation angle p on the correct values for the decryption process.

For the key space analysis of the proposed encryption system, it is considered every possible combination of keys: the rotation angle p of the GT, the two RPMs $r(x, y)$ and $h(u, v)$, and the two pseudorandom images $\varphi_\alpha(u, v)$ and $\theta(x, y)$. Using the obtained sensitivity of the rotation angle p, the key space for the rotation angle p of the GT is 4×10^7. The two RPMs $r(x, y)$ and $h(x, y)$, and the two pseudorandom images $\varphi_\alpha(u, v)$ and $\theta(x, y)$ have a size of 512×512 pixels and each pixel has L possible values. The number of attempts required to retrieve both RPMs and pseudorandom images is of the order of $L^{4(512)^2}$. For $L = 256$ gray levels, then the number of RPMs and pseudorandom images to try would be $256^{1048576}$. Therefore, the brute force attacks are intractable just considering every possibility of the two RPMs and the two pseudorandom images [17]. The total key space of the nonlinear encryption process proposed in this paper is: $(4\times10^7)\cdot256^{1048576}$, this number represents a very larger key space.

6. Conclusions

A new nonlinear security system has been proposed for image encryption-decryption involving the use of the Gyrator transform and the amplitude and phase truncation operations. The Gyrator transform increases the security on the encrypted image due to the addition of a new key, given by the rotation angle α. The simulation results show that the retrieval of the original image in the decryption scheme is very sensitive to the changes on the security key of the rotation angle. Apart from this, the two obtained pseudorandom images by the amplitude truncation increase much more the security for any cryptanalyst who tries to decrypt the digital image without authorization. The keys of the proposed nonlinear encryption-decryption system are represented by one rotation angle of the GT, two RPMs and two pseudorandom images. These five key allow to obtain a very larger key space for the proposed security system. The simulation results have shown that the encryption-decryption systems are more secure because the decrypted images are very sensitive to the changes on the five security keys.

Acknowledgments
This research has been funded by the Universidad de La Guajira.

References

[1] Millán M S and Pérez-Cabré E 2011 *Optical and Digital Image Processing: Fundamentals and Applications* ed G Cristóbal, P Schelkens and H Thienpont (Weinheim: Wiley) chapter 33 pp 739–67
[2] Réfrégier P and Javidi B 1995 *Opt. Lett.* **20** 767–9
[3] Goodman J W 1996 *Introduction to Fourier Optics* (Columbus: McGraw-Hill)
[4] Nomura T and Javidi B 2000 *Opt. Eng.* **39** 2031–5
[5] Vilardy J, Millán M S and Pérez-Cabré E 2013 *J. Opt.* **15** 025401
[6] Situ G and Zhang J 2004 *Opt. Lett.* **29** 1584–6
[7] Vilardy J, Millán M S and Pérez-Cabré E 2014 *Appl. Opt.* **53**, 1674–82
[8] Vilardy J, Millán M S and Pérez-Cabré E 2013 *Proc. of SPIE* **8785**, 87853J
[9] Unnikrishnan G, Joseph J and Singh K 2000 *Opt. Lett.* **25** 887–9
[10] Vilardy J, Torres C and Mattos L 2008 *AIP Conf. Proc.* **992** 1067–72
[11] Vilardy J, Torres C and Mattos L 2008 *Revista Colombiana de Física* **40** 143–6
[12] Vilardy J, Torres C and Mattos L 2011 *Revista Colombiana de Física* **43** 523–7
[13] Vilardy J, Torres Y, Millán M S and Pérez-Cabré E 2014 *J. Opt.* **16** 125405
[14] Vilardy J, Millán M S and Pérez-Cabré E 2014 *Opt. Pura y Apl.* **47** 35–41
[15] Ozaktas H M, Zalevsky Z and Kutay M A 2001 *The Fractional Fourier Transform: with Applications in Optics and Signal Processing* (Weinheim: Wiley)
[16] Rodrigo J A, Alieva T and Calvo M L 2007 *Opt. Commun.* **278** 279–84
[17] Vilardy J, Millán M S and Pérez-Cabré E 2016 Nonlinear image encryption using a fully phase nonzero-order joint transform correlator in the Gyrator domain *Opt. Laser Eng.* (*Preprint* http://dx.doi.org/10.1016/j.optlaseng.2016.02.013)
[18] Vilardy J, Torres C and Mattos L 2013 *Proc. of SPIE* **8785** 87851Q
[19] Vilardy J, Millán M S and Pérez-Cabré E 2015 *Proc. 20th Symposium on Signal Processing, Images and Computer Vision (STSIVA) (Bogotá)* vol 1 (Washington: IEEE) pp 1–6
[20] Morales Y, Díaz L and Torres C 2015 *J. Phys.: Conf. Series* **582** 012063
[21] Vilardy J, Torres C and Jimenez C 2013 *Proc. of SPIE* **8785** 87851R
[22] Vilardy J, Torres C, Useche J and Mattos L 2011 *J. Phys.: Conf. Series* **274** 012047
[23] Vilardy J, Torres C, Useche J and Mattos L 2011 *Revista Colombiana de Física* **43** 170–4
[24] Rodrigo J A, Alieva T and Calvo M L 2007 *Opt. Express* **15** 2190–203

[25] Rodrigo J A, Alieva T and Calvo M L 2007 *J. Opt. Soc. Am.* A **24** 3135–9
[26] Qin W and Peng X 2010 *Opt. Lett.* **35** 118–20
[27] Liu Z, Chen D, Ma J, Wei S, Zhang Y, Dai J and Liu S 2011 *Optik* **122** 864–7

Mathematical description of the two-dimensional Gabor transform. Application to image encryption

Ronal Perez[1], **Juan M. Vilardy**[1], **Cesar O. Torres**[2]

[1] Grupo GIFES. Faculty of engineering. Universidad de La Guajira, Riohacha (La Guajira), Colombia

[2] Grupo de óptica e informática. Department of electronic engineering. Universidad Popular del Cesar, Valledupar (Cesar), Colombia

E-mail: rperez@uniguajira.edu.co, jmvilardy@uniguajira.edu.co

Abstract. Information security with optical processing, such as the double random phase encoding and the Gabor transform (GT) has been investigated by various researchers. We present a two-dimensional (2-D) generalization of the one-dimensional GT. This 2-D GT is applied to encrypt digital images in this paper. The scaling factors of the GT can be used as new keys, providing a new encryption system with a high security characteristics. This method can encrypt and protect the information of the digital images with a high security for information processing systems.

1. Introduction

The Gabor transform (GT) has been used in texture segmentation [1-4] and defect detection [2]. The GT can be tuned to a specific frequency and orientation and achieve both localization in the time domain and frequency domain; two-dimensional (2-D) Gabor filters are multichannel filters with a multi-resolution decomposition process, which is similar to wavelet analysis. Time-frequency analysis, such as the GT, plays an important role in many signal processing applications [5-7].

In recent years, with the interests of people all around the world growing, optical image encryption techniques have shown great potential in the field of optical information processing, and various optical image encryption techniques have been proposed for optical image encryption by virtue of the arbitrary parameter selection, high computation speed, and extensive application. Some are done by Fourier transform (FT) [8-10], Fresnel transform (FrT) [11-13], or fractional Fourier transform (FrFT) [14-19], which is the generalization of the FT, and compared with it, the fractional orders are the additional key when the original image is encrypted. Since Réfrégier and Javidi proposed the double random phase encryption method in 1995 [8], the method has been effectively and extensively used. Nowadays, most of these above mentioned encryption systems often encrypt information with double random phase in the way of transforming the information into noise-like pictures, and only when the phase random keys and others system parameters are correctly used in decryption, will be recovered the original image. For instance, Situ and Zhang presented a novel encryption technique based on double random phase encoding in Fresnel domain by removing the lenses that are used in Fourier

Content from this work may be used under the terms of the Creative Commons Attribution 3.0 licence. Any further distribution of this work must maintain attribution to the author(s) and the title of the work, journal citation and DOI.
Published under licence by IOP Publishing Ltd

domain [8]. Unnikrishnan et al. proposed double random phase encoding technique in the FrFT domain [14].

In this paper, we suggest a 2-D generalization of the 1-D GT. Like many well-known transforms, the GT also has a number of potential applications. The GT is better in comparison to the conventional FT with respect to the information processing [4]. It is possible that GT will become a useful information-processing tool in the future. Then, we propose a new encryption method based on the GT. Besides, the 2-D GT is a reversible transform analogous to FT. The use of the GT allows to increase the security of the encryption system due to the addition of new keys, these keys are represented by the scaling factors of the GT.

The outline of the paper is as follows: the mathematical description of the 2-D GT is introduced in section 2. The application of the 2-D GT for the digital images encryption and decryption processes are described in section 3. In section 4 are presented the computer simulations of the encryption and decryption methods based on the GT. Finally, the main ideas of the paper are summarized in section 5.

2. Definition of the Gabor transform (GT)

The one-dimensional (1-D) description of the GT is presented, most of the results of the 1D GT can be conveniently generalized to the case 2-D. The normalized 1-D Gabor filter is [4]

$$G_f(x_0,u) = \mathrm{GT}\{g(x)\} = \int_{-\infty}^{+\infty} \exp\left\{-\frac{(x-x_0)^2}{2}\right\}\exp\left\{-i2\pi u\left(x-\frac{x_0}{2}\right)\right\}g(x)\mathrm{d}x, \qquad (1)$$

where u is the frequency coordinate and x_0 is a shift parameter in the spatial domain. The 1-D Fourier transform (FT) is

$$G(u) = \mathsf{F}\ \{g(x)\} = \int_{-\infty}^{+\infty} e^{-i2\pi ux} g(x)\mathrm{d}x. \qquad (2)$$

2.1. 2-D GT

The 2-D is given by

$$G_f(x_0,y_0,u,v) = \int_{-\infty}^{+\infty}\int_{-\infty}^{+\infty} \exp\left\{-\frac{(x-x_0)^2}{2}\right\}\exp\left\{-\frac{(y-y_0)^2}{2}\right\} \times\exp\left\{-i2\pi u\left(x-\frac{x_0}{2}\right)\right\}\exp\left\{-i2\pi v\left(y-\frac{y_0}{2}\right)\right\}g(x,y)\mathrm{d}x\mathrm{d}y. \qquad (3)$$

The 1-D GT of equation (1) can be described by

$$G_f(x_0,u) = \mathrm{GT}\{g(x)\} = \int_{-\infty}^{+\infty} e^{-i2\pi ux} g_r(x)\mathrm{d}x, \qquad (4)$$

where

$$g_r(x) = \exp\left\{-\frac{(x-x_0)^2}{2}\right\}\exp\{i\pi u x_0\}\,g(x). \qquad (5)$$

Taking into account that

$$Gauss\left(\frac{x-x_0}{b}\right) = \exp\left\{-\pi\left(\frac{x-x_0}{b}\right)^2\right\}. \qquad (6)$$

Scaling equation (5) with $x \to \sqrt{\pi}x$ and $x_0 \to \sqrt{\pi}x_0$, we get

$$g_r\left(\sqrt{\pi}x\right) = Gauss\left(\frac{x-x_0}{2}\right)\exp\left\{i\pi\sqrt{\pi}ux_0\right\}g\left(\sqrt{\pi}x\right). \quad (7)$$

Scaling equation (4) with $x \to \sqrt{\pi}x$, $x_0 \to \sqrt{\pi}x_0$ and $u \to \frac{u}{\sqrt{\pi}}$, we obtain

$$\begin{aligned} G_f\left(\sqrt{\pi}x_0, \frac{u}{\sqrt{\pi}}\right) &= \sqrt{\pi}\int_{-\infty}^{+\infty} e^{-i2\pi ux} g_r\left(\sqrt{\pi}x\right)\mathrm{d}x \\ &= \sqrt{\pi}\int_{-\infty}^{+\infty} e^{-i2\pi ux} Gauss\left(\frac{x-x_0}{2}\right)\exp\left\{i\pi ux_0\right\}g\left(\sqrt{\pi}x\right)\mathrm{d}x \quad (8) \\ &= \sqrt{\pi}\mathsf{F}\left\{Gauss\left(\frac{x-x_0}{2}\right)\exp\left\{i\pi ux_0\right\}g\left(\sqrt{\pi}x\right)\right\}. \end{aligned}$$

The Gabor transform can be considered as an inner product of the signal $g\left(\sqrt{\pi}x\right)$ and a shifted and modulated version of the analysis window.

Using the result of equation (8), the 2-D GT can be expressed as

$$\begin{aligned} G_f\left(\sqrt{\pi}x_0, \sqrt{\pi}y_0, \frac{u}{\sqrt{\pi}}, \frac{v}{\sqrt{\pi}}\right) = \pi\mathsf{F}\Big\{&Gauss\left(\frac{x-x_0}{2}\right)Gauss\left(\frac{y-y_0}{2}\right) \\ &\times\exp\left\{i\pi(ux_0+vy_0)\right\}g(\sqrt{\pi}x,\sqrt{\pi}y)\Big\}. \end{aligned} \quad (9)$$

This is a Fourier transform of a product of two functions. Furthermore, if we use the following scaling factors $x \to \frac{x}{a}$, $x_0 \to \frac{x_0}{a}$, $y \to \frac{y}{b}$, $y_0 \to \frac{y_0}{b}$ and coordinate inversions $u \to au$ and $v \to bv$ in previous equation, we obtain

$$\begin{aligned} G_f\left(\sqrt{\pi}\frac{x_0}{a}, \sqrt{\pi}\frac{y_0}{b}, \frac{au}{\sqrt{\pi}}, \frac{bv}{\sqrt{\pi}}\right) &= \pi\mathsf{F}\left\{Gauss\left(\frac{x-x_0}{2a}\right)Gauss\left(\frac{y-y_0}{2b}\right)\right. \\ &\left.\times\exp\left\{i\pi(ux_0+vy_0)\right\}g\left(\sqrt{\pi}\frac{x}{a},\sqrt{\pi}\frac{y}{b}\right)\right\} \quad (10) \\ &= \mathrm{GT}\left\{g\left(\sqrt{\pi}\frac{x}{a},\sqrt{\pi}\frac{y}{b}\right)\right\}. \end{aligned}$$

The previous definition and implementation of the 2-D GT will be applied for the digital images encryption and decryption processes in the next section.

3. Mathematical formulation of the image encryption and decryption

The real image to encrypt is represented by $g\left(\sqrt{\pi}\frac{x}{a},\sqrt{\pi}\frac{y}{b}\right)$. The image encryption uses the 2-D GT and two random phase masks (RPMs) [6], given by $P(x,y)$ and $Q(u,v)$. The implemented 2-D GT of the previous section introduces the scaling factor as new keys of the encryption system, this feature increases the security of the encryption system. The encrypted image is described by

$$e(s,t) = \mathrm{GT}\left\{Q(u,v)\mathrm{GT}\left\{P(x,y)g\left(\sqrt{\pi}\frac{x}{a},\sqrt{\pi}\frac{y}{b}\right)\right\}\right\}. \tag{11}$$

The decrypted image is given by

$$g\left(\sqrt{\pi}\frac{x}{a},\sqrt{\pi}\frac{y}{b}\right) = \left|\mathrm{GT}^{-1}\left\{Q^{*}(u,v)\mathrm{GT}^{-1}\left\{e(s,t)\right\}\right\}\right|, \tag{12}$$

where GT^{-1} represent the inverse GT and $Q^{*}(u,v)$ is the complex conjugate of $Q(u,v)$. The RPM $P(x,y)$ is used to spread the information content of the original image $g\left(\sqrt{\pi}\frac{x}{a},\sqrt{\pi}\frac{y}{b}\right)$ onto the encrypted distribution $e(s,t)$. The security keys of the encryption system are the RPM $Q(u,v)$ and the scaling factor a and b.

4. Computer simulation of the encryption and decryption techniques

The digital simulation for the encryption and decryption techniques was implemented on the platform of Matlab v.7.7, due to their facilities in handling matrices (images).

All image used in the computer simulations have 512 × 512 pixels. The encrypted images are grayscale images with 8 bits of quantization; the values of the pixels are integers between 0 and 255 (different shades of gray), the test image (original image) is depicted in figure 1.

After digitally implement the equations (10) and (11), using a digital algorithm for computing the discrete GT [20, 21], it was possible to encrypt the original image, obtaining an image with a completely random distribution, as shown in figures 1(a) and 1(b), the distribution of encrypted image (both its magnitude and phase) varies by changing the keys (the RPM $Q(u,v)$ and the scaling factor a and b).

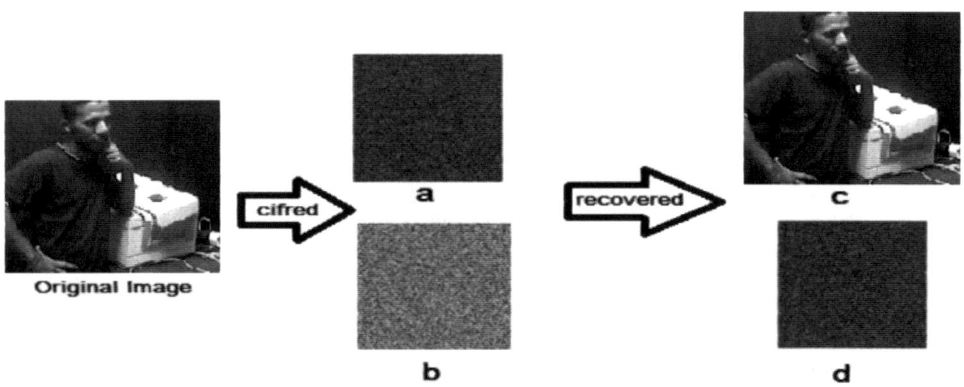

Figure 1. Digital simulation of the image encryption and decryption systems.

When performing the decryption process with the correct keys, we were able to recover the original image without any loss of information, as it is shown in fig. 1(c). If the keys used in the decryption process are not equal to the keys used in the encryption process, we obtain the random pattern image of the figure 1(d) at the output of the decryption system.

The RPM key $Q(u, v)$ have a size of 512x512 pixels and each pixel has 256 possible values. The number of attempts required to retrieve the RPM key $Q(u, v)$ image is of the order of $256^{(512)(512)}$. The possible values of the scaling factors a and b are considered between 0.5 and 1.5. The sensitivity to changes in values of the scaling factors is of the order of 1×10^{-6}. The total key space of the proposed encryption process is $(1\times10^{12})\cdot256^{262144}$, this value for the key space is larger than the results obtained in refs. [10-19].

From numerical simulation, the time computing for obtaining the encrypted image with a size of 512×512 pixels is about 3.14 seconds and the time computing of the decrypted image is 2.92 seconds. These results of the time computing for the encryption and decryption processes are similar to the results presented in the encryption-decryption systems of the refs. [8-10, 11, 14-17].

5. Conclusions

We have described another mathematical expression for the 2-D GT. We have presented the relationship between the GT and the Fourier Transform; which makes possible a computational implementation of an images encryption-decryption system without loss of information in the retrieval of the original image. The 2-D GT was applied to the digital images encryption and decryption processes. A double random phase encoding based technique for image encryption using the GT and RPMs has been proposed. Digital simulations were presented to demonstrate the good performance of the presented encryption-decryption system. The scaling factors of the 2-D GT increased the security of the security system because these factors represent new keys. The three key of the encryption system (one RPM and two scaling factors) allow to obtain a very larger key space for the proposed security system.

Acknowledgments

This research has been funded by the Universidad de La Guajira and the Universidad Popular del Cesar.

References

[1] Okombi-Diba B, Miyamichi J and Shoji K 2000 *IAPR Workshop on machine vision applications (Tokyo)* pp 267–70
[2] Dunn D and Higgins W 1995 *IEEE Trans. Image Process.* **4** 947–64
[3] Lee T 1996 *IEEE Trans. On Pattern Analysis and Machine Intelligence* **18** 959–71
[4] Gabor D 1946 *J. IEE* **93** 429–59
[5] Movellan *J Tutorial on Gabor filters* (Open Source Document)
[6] Redondo R, Šroubek F, Fischer S and Cristóbal G 2009 *Information Fusion* **10** 163–71
[7] Soo-Hwan C, Gilsoo J and Sae-Hyuk K 2010 *IEEE Transactions on Power Delivery* **25** 494–9
[8] Réfrégier P and Javidi B 1995 *Opt. Lett.* **20** 767–9
[9] Nomura T and Javidi B 2000 *Opt. Eng.* **39** 2031–5
[10] Vilardy J, Millán M S and Pérez-Cabré E 2013 *J. Opt.* **15** 025401
[11] Situ G and Zhang J 2004 *Opt. Lett.* **29** 1584–6
[12] Vilardy J, Millán M S and Pérez-Cabré E 2014 *Appl. Opt.* **53**, 1674–82
[13] Vilardy J, Millán M S and Pérez-Cabré E 2013 *Proc. of SPIE* **8785**, 87853J
[14] Unnikrishnan G, Joseph J and Singh K 2000 *Opt. Lett.* **25** 887–9
[15] Vilardy J, Torres C and Mattos L 2008 *AIP Conf. Proc.* **992** 1067–72
[16] Vilardy J, Torres C and Mattos L 2008 *Revista Colombiana de Física* **40** 143–6
[17] Vilardy J, Torres C and Mattos L 2011 *Revista Colombiana de Física* **43** 523–7
[18] Vilardy J, Torres Y, Millán M S and Pérez-Cabré E 2014 *J. Opt.* **16** 125405
[19] Vilardy J, Millán M S and Pérez-Cabré E 2014 *Opt. Pura y Apl.* **47** 35–41
[20] Qian S and Chen D 1993 *IEEE Transactions on Signal Processing* **41** 2429–38
[21] Sondergaard P 2007 *Finite discrete Gabor analysis* (Academic dissertation: Ph.D thesis, Technical University of Denmark)

Respiration monitoring by Electrical Bioimpedance (EBI) Technique in a group of healthy males. Calibration equations.

M Balleza[1], M Vargas[1], S Kashina[2], M R Huerta[3], I Delgadillo[1] and G Moreno[4]

[1] Physics Engineering Department, Division of Science and Engineering, Campus León, University of Guanajuato, León, Mexico.
[2] Department of Chemistry, Division of Natural Sciences, Campus Guanajuato, University of Guanajuato, León, Mexico.
[3] Applied Sciences to Work Department, Division of Health Sciences, Campus León, University of Guanajuato, León, Mexico.
[4] Engineering Department, Universidad Iberoamericana, Campus León.

E-mail: marco.balleza@fisica.ugto.mx

Abstract. Several research groups have proposed the electrical impedance tomography (EIT) in order to analyse lung ventilation. With the use of 16 electrodes, the EIT is capable to obtain a set of transversal section images of thorax. In previous works, we have obtained an alternating signal in terms of impedance corresponding to respiration from EIT images. Then, in order to transform those impedance changes into a measurable volume signal a set of calibration equations has been obtained. However, EIT technique is still expensive to attend outpatients in basics hospitals. For that reason, we propose the use of electrical bioimpedance (EBI) technique to monitor respiration behaviour. The aim of this study was to obtain a set of calibration equations to transform EBI impedance changes determined at 4 different frequencies into a measurable volume signal. In this study a group of 8 healthy males was assessed. From obtained results, a high mathematical adjustment in the group calibrations equations was evidenced. Then, the volume determinations obtained by EBI were compared with those obtained by our gold standard. Therefore, despite EBI does not provide a complete information about impedance vectors of lung compared with EIT, it is possible to monitor the respiration.

1. Introduction

Several research groups have proposed different methods and techniques to monitor and measure the lung tidal volume in a non-invasive way. Nowadays, the gold standard to perform these tasks is the pneumotachometer. However, it is not well tolerated by patients due to the use of a mouth piece and clip nose. This fact cause overestimation of the volume determinations between a 15% and 30% [1-3]. There are other devices such as the respiratory inductive plethysmography (RIP) that allow to perform a qualitative analysis of lung ventilation [4].

One of the most used non-invasive technique for this propose is electrical impedance tomography (EIT). This technique is aimed to image a cross section of thorax and obtain a set of dynamic frames through 16 electrodes placed around the thoracic box. With EIT it is possible to obtain an alternating signal by quantifying all pixels of each EIT image. Therefore, a signal in terms of impedance corresponding to respiration is obtained. In previous studies, our research group has obtained a calibrations equation in order to transform the impedances changes into a measurable volume signal in a group of males, females and chronic obstructive pulmonary disease (COPD) patients. From the obtained results, it was evidenced that the adjustment of calibration equations obtained in males was

Content from this work may be used under the terms of the Creative Commons Attribution 3.0 licence. Any further distribution of this work must maintain attribution to the author(s) and the title of the work, journal citation and DOI.
Published under licence by IOP Publishing Ltd

better than in females [5]. In the case of COPD patients, the calibration equations were corrected by adjusting factor with promising good results [6]. Subsequently, from the main 16-electrodes arrangement, we have chosen the most significant 4 electrodes in order to monitor tidal respiration in a group of healthy males. The most significant results revealed that the mathematical adjustment was as good as that obtained in previous study [7]. Due to previous, our research group proposes the use of 4-electrodes impedance device to monitor and quantify the tidal volume in a group of healthy males. Therefore, the main objective of this research is to obtain a set of calibration equations determined by 4 different frequencies of current injection.

2. Material and methods

2.1 Pneumotachometer
The pneumotachometer was a BIOPAC TSD107B highly linear. It is a wide-range airflow transducer and it interfaces with the BIOPAC DAC100C (general-propose transducer amplifier). The TSD107B is factory calibrated to nominally satisfy the scaling factor: 1 mVolt output = 11.1 liters/sec flow rate. When it is connected to the DA100C with Gain =1,000, the calibration factor is: 1 Volt = 11.1 liters/sec.

2.2 Impedance Device
The impedance device used in this study was a BIOPAC EBI100C. The EBI100C measures both impedance magnitude and phase simultaneously. Impedance can be recorded at four different measurement frequencies (12.5 kHz, 25 kHz, 50 kHz and 100 kHz). This device injects a small current of 400 μA. This device interface with the MP150 data acquisition and analysis platform and Acq*Knowledge* software.

2.3 Volunteers
In this study, a group of 8 healthy males was analysed. We have chosen this sample because this study is the pilot one. So, in this group we have assessed new anthropometric parameters that describe fat mass, muscle mass and other different perimeters of body. All subjects were non-smokers. All test were performed between 9:00 am and noon in the Laboratory of Bioimpedance of the Department of Physics Engineering of the University of Guanajuato, Campus León. All volunteers consented to participate in the study singing an informed consent document.

2.4 Procedure
The pneumotachometer (TSD107B) and the impedance device (EBI100C) were connected simultaneously to each volunteer. The impedance and volume signal were recordered at the same time by the BIOPAC software (Acq*Knowledges*) for the frequencies of 12.5 kHz, 25 kHz, 50 kHz and 100 kHz. Single respiratory manoeuvre for each frequency was performed. In the group of volunteers, we used a general equation that allows to adjust impedance changes into a measurable volume signal. The equation is expressed as following:

$$\Delta Z = A \times \Delta V \tag{1}$$

where ΔZ is the impedance changes obtained with BIOPAC impedance amplifier, ΔV is the volume changes obtained with the pneumotachometer and A is the proportionality coefficient or calibration coefficient. This latter parameter is defined by different anthropometric parameters (A=[age, weight, height, BMI, etc.]).

The respiration of each volunteer was recorded at standing position during periods of 40 seconds, with 3 minutes rest between measures. Each respiratory test was registered numerically and graphically. Prior to measure of the respiration pattern, we have recorded anthropometric parameters: age, weight, height and body mass index. Also, different skinfold measures (lateral chest, subscapular, triceps, biceps, suprailiac, pectoral, abdominal, the skinfold of both thigh and calves) have been recorded. We have measured the diameters corresponding to both humerus and femurs. Also, we have registered some other body perimeters as wrist, arm, chest, waist, abdomen, hip, both thighs and calves.

2.5 Statistical analysis

All data were expressed in terms of mean ± standard deviation. The test of Kolmogorov-Smirnov was used for assessing the normal distribution of data ($p > 0.05$). The calibration equations were obtained by multivariate linear regression of anthropometric parameters and by the proportionality coefficients (A). This parameter is estimated by the quotient of impedance mean value and tidal volume mean value, see equation 1. So, for each frequency of current injection, a calibration mathematical model was obtained. The comparison of the volume determinations obtained by both devices was performed by a Statistical test for paired data. All volume differences were assessed by a set of graphics of Bland & Altman [8].

3. Results

All data series have shown a normal distribution (K-S test $p > 0.05$). Therefore, all the statistical tests are parametric. The main values of anthropometric parameters corresponding to the male sample were: an age of 25 ± 6 years, a height 1,72 ± 0.06 m, a weight of 73.8 ± 6.8 kg and a body mass index (BMI) of 24.9 ± 1.6 kg/m^2. The main values of body skinfolds were: for the lateral chest 18.8 ± 7.4 mm, for the subscapular 19.3 ± 3.2, for triceps 15.0 ± 2.3 mm, for biceps 7.0 ± 1.2 mm, for suprailiac 19 ± 4.2 mm, for the pectoral 18.5 ± 4.9 mm and for the abdominal 22.3 ± 4.0 mm. The mean values of skinfold for the left and right thigh were of 17.1 ± 3.4 and 17.7 ± 2.7 mm, respectively. The mean values of skinfolds for the left and right calf were of 13.3 ± 3.3 and 12.4 ± 3.7 mm, respectively.

The mean values of the diameters of the left and right humerus were of 6.4 ± 0.4 cm, respectively. The mean values of the diameters of the left and right femur were of 9.3 ± 0.3 cm, respectively.

The mean values of the of body perimeters were: for the wrist 16.4 ± 1 cm, for the arm 29.9 ± 1.9 cm, for the chest 93.8 ± 3.9 cm, for the waist 83.9 ± 3.5 cm, for the abdomen 88.2 ± 3.7 cm and for the hip 99.4 ± 3.9 cm. The perimeters of the left and right thigh were of 50.7 ± 2.0 cm and 51.1 ± 2.5 cm, respectively. The perimeters of the left and right calf were of 37.0 ± 1.1 cm and 37.0 ± 1.3 cm, respectively.

The mean value of volume determinations obtained by the pneumotachometer at the frequencies of 12.5 kHz, 25 kHz, 50 kHz and 100 kHz were 0.579 ± 0.176 ml, 0.579 ± 0.225 ml, 0.572 ± 0.192 ml and 0.556 ± 0.211 ml, respectively.

The calibration equations obtained to each frequency are expressed as following:

$$A_{12\text{kHz}} = 28.23 - 2.13 \times \text{DLHum} + 0.111 \times \text{SkSuprail} - 0.092 \times \text{PAbdom} - 0.134 \times \text{PLCalf} \qquad R^2 = 0.999\ (p < 0.01) \tag{2}$$

$$A_{25\text{kHz}} = 7.94 - 0.02 \times \text{Age} + 0.014 \times \text{SkLChest} + 0.165 \times \text{SkRCalf} - 0.019 \times \text{DRHum} - 2.48 \times \text{DLHum} + 0.087 \times \text{DLFem} + 0.094 \times \text{PChest} \qquad R^2 = 0.999\ (p < 0.01) \tag{3}$$

$$A_{50\text{kHz}} = 11.57 + 0.27 \times \text{SkRCalf} - 0.21 \times \text{SkLCalf} + 0.027 \times \text{SkRThigh} - 1.69 \times \text{DRHum} + 0.24 \times \text{DRFem} - 0.004 \times \text{PWaist} + \alpha_{14} \times \text{PAbdom} - 0.005 \times \text{PLCalf} \qquad R^2 = 0.999\ (p < 0.01) \tag{4}$$

$$A_{100\text{kHz}} = 13.54 + 0.17 \times \text{SkRCalf} - 0.083 \times \text{SkLThigh} - 1.71 \times \text{DRHum} \qquad R^2 = 0.974\ (p < 0.01) \tag{5}$$

Where the A_i stands for the calibration factor at 12 kHz, 25 kHz, 50 kHz and 100 kHz, SkSuprail is the suprailiac skinfold, the SkRCalf and SkLCalf are the right and left calf skinfolds, the SkRThigh and

SkLThigh are the right and left thigh skinfold, DRHum and DLHum are the diameters of right and left humerus, DRFem and DLFen are the diameters of right and left femurs, PChest is the perimeter of chest, PWaist is the perimeter of waist, PAbdom is the perimeter of abdomen and PLCalf is the perimeter of the left calf.

The volume values obtained with pneumotachometer and impedance device, using the calibration equations, and its differences are show in table 1.

Table 1. Volume determinations obtained by pneumotachometer and impedance device using the calibration equations.

Frequency at 12 kHz (using equation 1)			**Frequency at 25 kHz (using equation 2)**		
V_Pneumotach[a] (liters)	**V_Impedance[b] (liters)**	**Differences (liters)**	**V_Pneumotach[a] (liters)**	**V_Impedance[b] (liters)**	**Differences (liters)**
0.604	0.603	0.002	0.574	0.579	-0.005
0.425	0.426	-0.001	0.427	0.431	-0.004
0.397	0.395	0.002	0.421	0.425	-0.004
0.380	0.379	0.002	0.276	0.278	-0.002
0.881	0.869	0.013	0.949	0.964	-0.015
0.669	0.667	0.002	0.824	0.835	-0.011
0.721	0.729	-0.008	0.674	0.685	-0.010
0.557	0.555	0.002	0.486	0.492	-0.006
0.579 ± 0.176	0.578 ± 0.174	0.002 ± 0.006	0.579 ± 0.225	0.586 ± 0.229	-0.007±0.004
		Error: 0.96%			Error: 0.76%
Frequency at 50 kHz (using equation 3)			**Frequency at 100 kHz (using equation 4)**		
V_Pneumotach[a] (liters)	**V_Impedance[b] (liters)**	**Differences (liters)**	**V_Pneumotach[a] (liters)**	**V_Impedance[b] (liters)**	**Differences (liters)**
0.591	0.595	-0.004	0.600	0.614	-0.013
0.390	0.393	-0.003	0.393	0.397	-0.004
0.382	0.385	-0.003	0.371	0.346	0.025
0.376	0.379	-0.003	0.335	0.340	-0.004
0.830	0.842	-0.012	0.904	0.845	0.059
0.850	0.859	-0.009	0.793	0.812	-0.018
0.637	0.644	-0.007	0.623	0.638	-0.016
0.516	0.522	-0.006	0.426	0.451	-0.026
0.572 ± 0.192	0.577 ± 0.196	-0.006± 0.003	0.556 ± 0.211	0.555 ± 0.202	0.0003±0.028
		Error: 0.58%			Error: 5%

[a] Volume determinations obtained by pneumotachometer.
[b] Volume determinations obtained by impedance device using calibration equations.

The analysis of volume differences was performed by a Bland and Altman graphics (see figure 1). From the t-test statistical analysis of the volume differences obtained by both monitoring devices at each frequency, statistically significant differences in the volume determinations obtained at 25 kHz ($p = 0.003$) and 50 kHz ($p = 0.002$) were evidenced. However, there was no significant differences for the frequencies 12.5 kHz ($p = 0.499$) and 100 kHz (0.981). The Pearson's correlation between volume determinations obtained by pneumotachometer and impedance device at four frequencies was over 0.99.

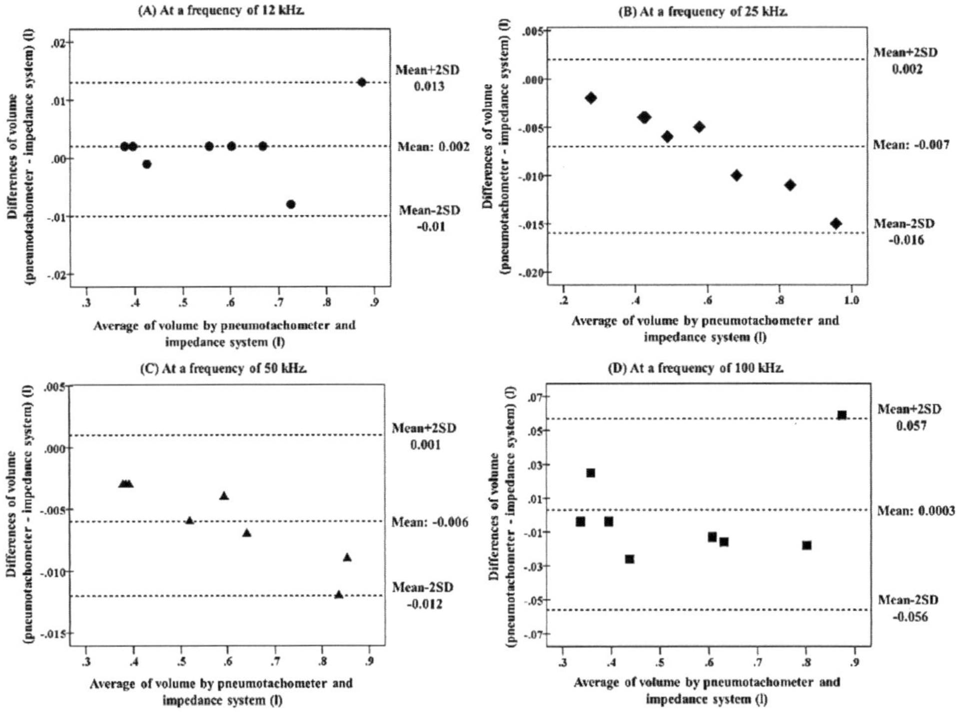

Figure 1. Analysis of volume differences by a Statistical analysis of Bland and Altman. Graphic (A) shows the differences of volume determinations at 12.5 kHz. Graphic (B) shows the differences of volume determinations at 25 kHz. Graphic (C) shows the differences of volume determinations at 50 kHz. Graphic (D) shows the differences of volume determinations at 100 kHz.

4. Discussion

The first objective of this research was to obtain a set of calibration equations to adjust a 4 electrodes impedance signal at four different frequencies (corresponding to ventilation) into a measurable volume signal. In order to accomplish this aim, we have performed a statistical multivariate analysis of anthropometric parameters and calibration factors (A). This factor is unique for each subject. From the resulted analysis we have obtained 4 calibration equations with a high determinations coefficients. All those equations involved parameters that describe the anthropometry of lower extremities. It seems that the physical activity and the possible development of lower extremities have a high relation.

In previous studies [5-7], we have measured those anthropometrical parameters that describe the anthropometry of thorax as thoracic skinfolds, diameters of chest at rest, maximal expiration and maximal inspiration. The obtained calibration equations evidenced a high coefficient of determination. However, the errors obtained in each obtaining group were between 5% and 7%. Comparing these results with the obtained in this study, we have improved the mathematical adjustment at a 20% just in the obtaining group. This fact will affect the estimation of volume in a different group of volunteers.

In the case of those results obtained from the statistical comparison, we evidenced that the volume differences at 25 kHz and 50 kHz showed statistical significant differences despite to obtain lower volume differences. This fact can be attributed to the small sample. Probably, the calibration equation does not adjust correctly the impedance determinations due to the anthropometrical parameters of one subject. In graphics A, C and D, it is possible to evidence one point around the upper and lower boundaries. Therefore, we need to increase the sample in order to obtain more reliable results. It is important to mention that this study is a first attempt to calibrate a set of tetrapolar impedance determination in order to obtain a volume signal.

5. Conclusion

All these obtained results are limited to a sample of healthy males. Therefore, the assessed calibration equation might not be suitable for patient. All volume determinations are restricted to a range of 300 ml to 900 ml. So, all conclusion are restricted to this range. The main conclusion of this study is that the impedance determinations can be adjusted by using the parameters that describe the anthropometry of lower extremities. This fact can be related with the physical activities and therefore, the lung conditioning.

Acknowledgments

The authors thank FOSSIS (CONACYT) for financial support for performing of this researching work (Reference number: 261277).

References

[1] Askanazi J., Silverberg P A, Foster R J, Hyman A I 1980 *J Appl Physiol* **48** 577-80.
[2] Perez W and Tobin J 1985 *J Appl Physiol* **59** 1515-20.
[3] Paek D and McCool D (1992) J Appl Physiol 73 887-93.
[4] Whyte K F, Gugger M, Gould G A, Molloy J, Wraith P K, Douglas N J 1991 *J Appl Physiol* **71** 1866-71.
[5] Balleza M, Fornos J, Calaf N, Feixas T, Gonzalez M, Antón D, Riu P J, Casan P 2007 *Arch. Bronconeumol* **43** 300-3.
[6] Balleza M, Calaf N, Feixas T, Gonzalez M, Antón D, Riu P J, Casan P 2009 *Arch Bronconeumol* **45** 320-24.
[7] Balleza-Ordaz M, Alday-Perez E, Vargas-Luna M, Kashina S, Huerta-Franco M R, Torres-González L A, Riu-Costa P J 2016 *Biomed Signal Process Control* **27** 68-76.
[8] Bland J M, Altman D 1986 *The lancet* **327** 307-10.

IMPLEMENTATION AN IMAGE PROCESSING TECHNIQUE FOR VIDEO MOTION ANALYSIS DURING THE GAIT CYCLE CANINE

G López[1] and J O Hernández[1]
[1]INSTITUTO POLITÉCNICO NACIONAL, Escuela Superior de Ingeniería Mecánica y Eléctrica, Sección de Estudios de Posgrado e Investigación, Unidad Profesional "Adolfo López Mateos" Zacatenco, Edif. 5, 2do. Piso, Col. Lindavista, C.P. 07738, Ciudad de México, México.

E-mail: gpe.lopez.contreras@gmail.com, johm0906@hotmail.com

Abstract. Nowadays the analyses of human movement, more specifically of the gait have ceased to be a priority for our species. Technological advances and implementations engineering have joined to obtain data and information regarding the gait cycle in another animal species. The aim of this paper is to analyze the canine gait in order to get results that describe the behavior of the limbs during the gait cycle. The research was performed by: 1. Dog training, where it is developed the step of adaptation and trust; 2. Filming gait cycle; 3. Data acquisition, in order to obtain values that describe the motion cycle canine and 4. Results, obtaining the kinematics variables involved in the march. Which are essential to determine the behavior of the limbs, as well as for the development of prosthetic or orthotic. This project was carried out with conventional equipment and using computational tools easily accessible.

1. Introduction

The walk is considered as one of the activities most commonly performed by quadrupeds and bipeds. This kinematic analysis describes the movements without considering the forces that produce them [1]. The development of this analysis only considers the sagittal plane.

Through gait, bipedal and quadrupeds may be moved from one place to another at different speeds. For these movements are carried out, quadrupeds (for this case) use the locomotor system. It includes musculoskeletal structures of various parts of the body. Highlighting the forelimbs and hindlimbs [2]

Quadrupeds are able to perform various types of gait; walk, trot and canter. In each case, the limbs perform a cycle of step. It is considered a step cycle the period from the instant that a limb is detached from the floor until it is touch again. Each of these cycles of step it is divided into two phases; advance and support [3].

In walking, the limbs are in contact with the ground more than 50% of the time. Each movement is independently, they are not synchronized. Usually two or three limbs support the weight of the animal [4]. The analysis is limited to gait dog when is walking. Because is the simplest and natural activity that performed by the dog.

---

1 gpe.lopez.contreras@gmail.com

Content from this work may be used under the terms of the Creative Commons Attribution 3.0 licence. Any further distribution of this work must maintain attribution to the author(s) and the title of the work, journal citation and DOI.
Published under licence by IOP Publishing Ltd

2. Methodology

This section describes the steps for the project and the equipment that it is used. The characteristic of the test subject are: Australian Shepherd, female, approximate age 4years, height 50cm, length 57cm, body mass: 18kg.

The height of a dog is measured from the cross to the ground. Where the cross is located at the base of the neck and the highest points of the shoulder blades form it. The length of the dog is measured from the shoulder to the ischial tuberosity [5]. The estimated height for males is 51-58 cm, for females is 46 to 53 cm [6].

2.1. Training

To perform the analysis, the animal had training for walking comfortably on the electric treadmill. The training time was 3 or 4 months because the dog has a nervous character that way the dog could perform the assigned activity. To achieve the goal, it is used the following equipment: Chain 2 m long, Food rewards (positive reinforcement) and Dog Toys.

With the help of positive reinforcement in other words motivate the dog with positive experiences and awards to achieve the goal successfully without the dog suffer any physical injury and/or trauma, it taken precautions and all kinds of care.

2.1.1 Shooting gait cicle

For the gait cycle shooting, it is adapts a large stage for placing equipment well as lighting sufficient. Where gait recordings were made in the electric treadmill (in horizontal position without declivities without inclination) at a constant speed (0.1 km / h). The digital camera is located to 1.2 m from the electric treadmill and a height of 25 cm as it is shows in Figure 1. Table 1 describes the devices used and the requirements.

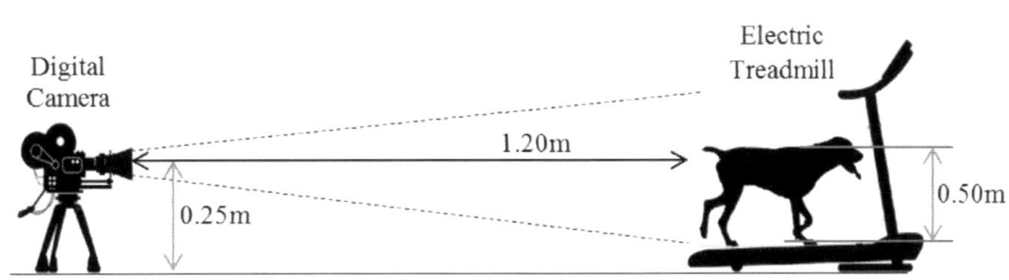

Figure 1. Distances of equipment for the filming.

Table 1. Elements and requirements.

Element	Description
Digital video camera	GoPro Hero 4 Silver Resolution 1280x720 pixels
Electric treadmill	Flat base, Model E318N, variable speed, dimensions 1.02 x 0.33 m
Circular markers	Diameter 30 mm Forelimb: Blue, Hindlimb: Gold

As it is mentioned above, to walk is the most natural movement of a living being to move from one place to another. That is why this paper focuses only on the gait cycle to walk. It is considered the minimum set speed on the electric treadmill, in order to have a better appreciation of the cycle.

2.1.2 Data acquisition

The canine gait analysis has the intended to indicate the clinical condition of the patient or specimen to be treated. With the joints movement in each limb. They can be obtained kinematic variables involved in driving tests. With the joints, movement in each limb can be obtained kinematic variables involved in gait tests.

Before shooting, markers are placed on the joints of the dog. The order of placement for markers is shows in Table 2. Then Figure 2 shows the markers placed in joints of dog. These are placed in all four limbs.

Table 2. Marker location.

Forelimb	Hindlimb
Shoulder joint (AH)	Hip joint (AC)
Elbow flexion (FC)	Knee flexion (FR)
Carpal bone (HC)	Flexing of tarsus (FT)
Metacarpal bones (HM)	Metatarsal bone (HMT)

Figure 2. Location of markers in forelimbs (blue) and hindlimbs (gold), right side view.

The videos obtained from the canine gait. One for hindlims and one for forelimbs in right and left side. They are digitized to be processed by the mathematical program.

Upon detecting the marker positions represented in a virtual way (in two dimensions), it can see the trajectory of biomechanical model of segment for each markers.

For better quality of video, a setting of 1280 x 7000 pixels at 60 fps (frames per second) is used. This is the best option in terms of size for the generated file. At a higher frequency, better video quality is obtained but requires more computational resources to process video file.

The results are obtained by the mathematical software, which calculates the position values, performing arithmetic difference of the angles generated.

3. Results

In Figure 3, it observed the forelimb results and Figures 4, 5 and 6 show the corresponding graphs of results for the kinematic variables; linear velocity of shoulder, angular velocity of elbow. Also the kinetic variable is shows; carpus joint angle.

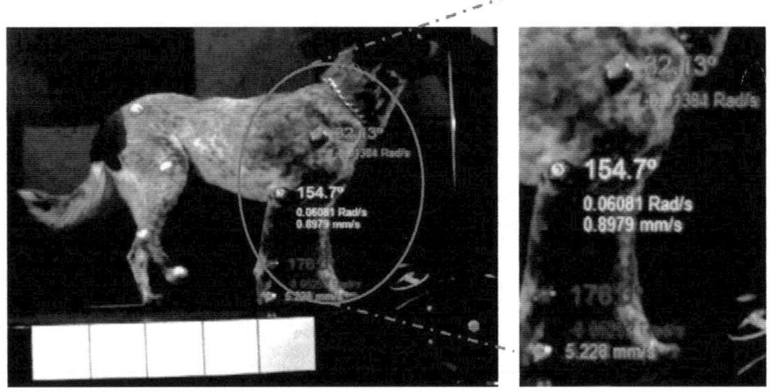

Figure 3. Results obtained in forelimbs

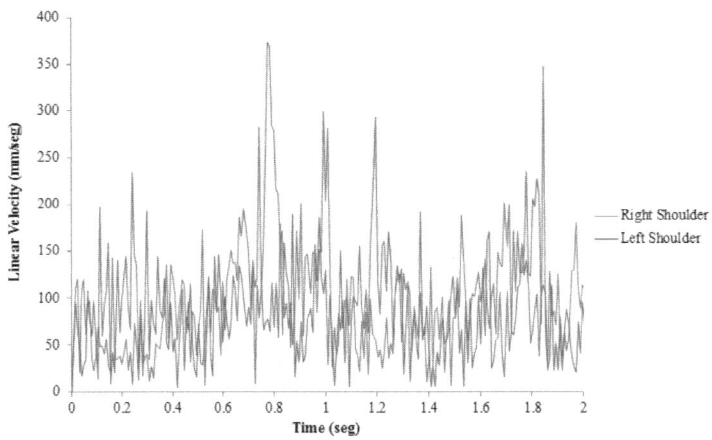

Figure 4. Linear velocity of shoulder.

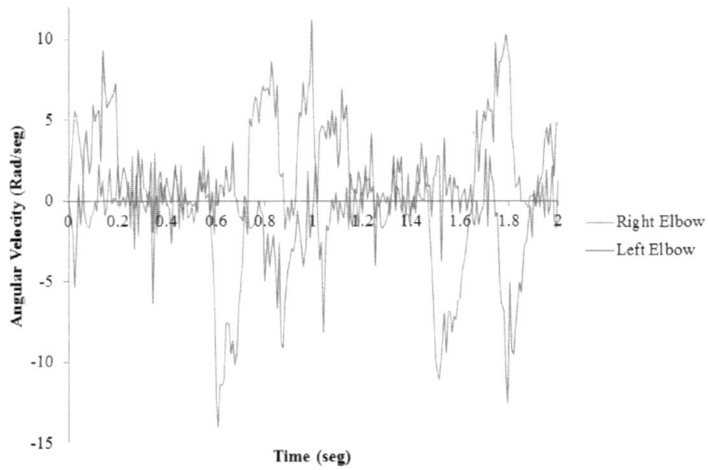

Figure 5. Angular velocity of elbow.

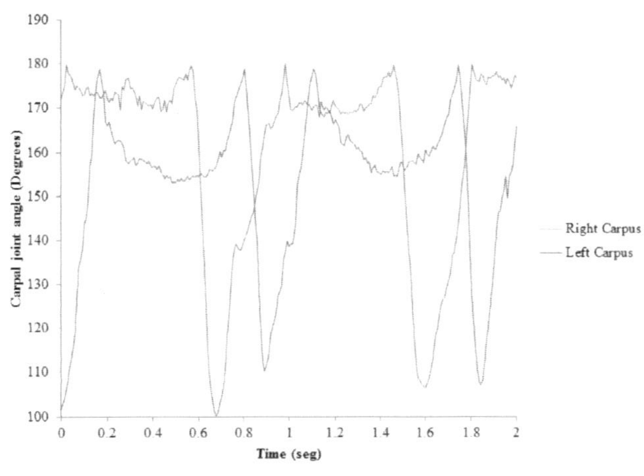

Figure 6. Angle of the joint carpus

4. Conclusions

With this technique, was achieved implement a system that allows viewing kinematic variables as: linear velocity and angular velocity. In addition, the kinetics variable: angles of articulations, which they were detected in the cycle of canine gait.

The analysis developed in this work, is a simple procedure economic and reliable. Is a tool with that it can obtain various variables (kinematics and kinetics) of the joints of limbs of a quadruped (in this case dog). With the help of a conventional camera of low cost and a computational tool, it can improve the data acquisition.

The idea of developing this technique is justified due to the high demand that today presents this type of analysis. There are certainly various sophisticated equipments of those can get the same results. With the research work that is presented here, it exposes another perspective of the various equipment with those can perform this type of studies. In such a way that any person interested in the area of biomechanics or other related areas, can obtain reliable results of this analysis.

It should be noted that, with the results of this analysis were taken decisions and clarify expectations about potential designs and manufacturing of prototypes of prostheses for hindlimb and forelimb in canine.

5. References

[1] Unda J and García-De Jalon J 1985 Análisis cinemático y dinámico de sistemas mecánicos formados por varios sólidos rígidos *Vet. J.*

[2] Fischer S Anders A, Nolte I and Schilling N 2013 Compensatory load redistribution in walking and trotting dogs with hind limb lameness *Vet. J.* **187**

[3] Goldner B, Fuchs A, Nolte I and Schilling N 2015 Kinematic adaptations to tripedal locomotion in dogs *Vet. J.* **204**

[4] Griffin T, Main R and Farley C 2014 Biomechanics of quadrupedal walking: how do four-legged animals achieve inverted pendulum-like movements? *J. Exp. Biol.* **207**

[6] Federation Cynologique Internationale (AISBL) 2009 *Australian Shepherd*

IOP Conf. Series: Journal of Physics: Conf. Series **792** (2017) 012050 doi:10.1088/1742-6596/792/1/012050

Sensitivity of a PMMA polymer capillary microresonator for measuring relative humidity.

D A Avila Padilla [1], C O Torres Moreno[1] and Cristiano M B Cordeiro[2]
[1]Department of Physics, Laboratorio de óptica e informatica, Universidad Popular del Cesar UPC, Colombia [2]Institute of Physics, Gleb Wataghin, University of Campinas UNICAMP, Brazil

E-mail: duberavilap@unicesar.edu.co

Abstract. This paper studied experimentally the behavior of a capillary microresonator made from PMMA polymer as a sensor for measuring relative humidity. In the manufactured device, the WGMs modes within the microcavity are excited by the proximity of an optical fiber Taper made from the stretching of a standard optical fiber of silica by the method of scanning flame with waist of the order of 3-5 μm. In the device, the field from a tunable laser system TLS is coupled into the capillary along its cross section where the resonant modes were observed WGMs into the cavity. When the system is subjected to change of the relative humidity of the external environment, the wavelengths of peaks resonances of WGMS modes of the resonant system experience a spectral shift, so that a sensitivity of the microresonator is observed at changes in humidity of the external environment. During the experiment, it was manufactured capillaries with different diameters and different wall thickness obtaining a sensitivity of the order of 0.07 nm/% RH for a capillary thickness 42.1 μm.

1. Introducción

The optical resonators play an important role in modern optics, being not only important for the manufacture of Laser, but are constituted as indispensable devices for performing accurate measurements, experiments nonlinear optics and other applications. One of the objectives most important in manufacturing such devices, is the miniaturization of optical resonators made from different materials, the most important materials that have been reported, Silice, Silica and some types of polymers, which have enabled the interconnection with integrated optical systems used for optical telecommunications. These devices have great attention due to their ability to confinement of light inside, which has led to breakthroughs in the area of sensors applied to biology, physics, chemistry and the study of nonlinear optical effects.

Currently, these structures are designed with dimensions on the order of microns to adapt easily to dielectric structures of waveguides and high symmetry to take advantage of the features and their properties, including their relative ease for mathematical modeling.

In the manufacture of microcapillaries silica and polymers, has led to the development of sensors for applications in measuring physical variables, chemical and biological [1-4]. Some measurement techniques reported in the literature have used the microcapillary as cavity for confining the field within it through the excitation of the WGMs modes within the cavity using different techniques, such as coupling technique through optical fibers Taper [1-3].

The capillaries thin wall used as optical microcavities are important in the development of sensors to study properties microfluidic due to their small size, low modal volume and relatively high Q factor, so that, due to its structural design can be used as a channel for the transit of small amounts of liquids

[1] duberavilap@unicesar.edu.co

Content from this work may be used under the terms of the Creative Commons Attribution 3.0 licence. Any further distribution of this work must maintain attribution to the author(s) and the title of the work, journal citation and DOI.
Published under licence by IOP Publishing Ltd

and gases, which depending on the type of study being performed can modify the characteristics of the optical field is confined inside.

2. WGMs modes within a cylindrical dielectric cavity.

The dielectric structures having a circular symmetry, can confine the field inside by total internal reflection within its surface, whose propagation modes are commonly called Whispering Gallery Modes WGMS. Indeed, in the literature is reported the excitation of WGMS modes within cavities with rotational symmetry in the form of spheres, cylinders, toroids, rings and discs made from different materials, which constitute a resonant structure where a field external is introduced into the cavity. Therefore, in such structures is of importance to study the wavelengths/frequencies of resonance, the Q factor of the oscillations within the cavity, the free spectral range FSR and the maximum width at half maximum FWHM.

In the case of a resonant structure in the form of capillary, the light is trapped by the outer wall by total internal reflection when this flows along of the circumference of the capillary, where the study of the wavelengths of the resonant modes WGMS allows its use as an optical sensor.

The cylindrical cavities, present a set of analogous resonances the case of structures in the form of spheres, where the main difference is supported by the fact that the cavities have a capillary not support the polar field component having spherical cavities, as it assumes that the field confinement occurs through the cross section of a two-dimensional cylinder. In this sense, the expressions used to determine the resonances in both resonators is the same with the difference that in the case of a cylindrical structure the functions of cylindrical Bessel is used while in a spherical structure Bessel functions spherical used as the solution of Maxwell's equations with boundary conditions.

Some studies report the analytical and experimental study of the optical microcavities as capillaries to very specific practical situations, such as refractometer sensor [3] and hydrostatic pressure sensor in microfluidics [1]. In the case of this research we are interested in the study of the sensitivity of the cavity to changes in relative humidity of the medium, which is determined from shifts in the wavelength of peak resonances obtained experimentally. In such systems a theoretical analysis based on Maxwell's equations with boundary conditions can determine the sensitivity of the devices, the Q factor of the oscillations and the FSR of the cavity.

The excitement of the WGMs modes inside a microcapillary could contain losses curvatures when the diameter of the microcapillary is the order of the micron low and may have differences in index of major refraction between the internal or external environment with the material capillary. Similarly, the resonances obtained in different modes can achieve important factor Q enabling improved signal/noise ratio in measurement applications and sensitivity.

With respect to coupling of the field within this type of microcavities, the power transferred from the optical fiber tapered to the microcavity is optimal when there is significant overlap of the field between the modes of the coupled structures, wherein the coupling length depends on the proximity of the fiber Taper. Therefore, it is possible to achieve a process of efficient power transfer when the optical modes have equal phase, which depends on the geometry of the resonant structure and the type of material used in its manufacturing process. For coupling the field within this type of resonant dielectric structures such as microcavities as spheres, discs, toroids, rings, bottles, bubbles etc., have specifically reported two types of techniques used for coupling of the field from a source coherent light known as coupling prism and optical fiber Taper coupling.

In the case of the technique of coupling prism, this has been used to couple the field into microspheres and microtubes through total internal reflection in the microsphere-prism interface obtaining an efficient coupling of the field into the microcavity, however, this technique has not been widely used because of the complexity of the resonant system related to the effect of displacement Goos-Hanchen and the precision of the instruments used [5]. In the case of the microresonators in the form of toroids, discs, bottles, bubbles and even into bottles and rings they have been reported various studies where the field confinement is achieved through an optical fiber taper, which can reach easily required phase equalization between the modes of the cavity and the optical fiber taper. This technique last coupling of the field has proven efficient results for two fundamentals reasons: The first one, is related to the manufacturing techniques used to design of fiber taper with specific characteristics in the fibers, which are relatively easy and the other reason is related with its ease of interconnection with standard telecommunication systems through optical fibers [1,3,6]. For a possible theoretical study of the problem, it is proposed to study the structure as a system composed of three layers where Maxwell's equations apply with the boundary conditions appropriate to determine the wave equation in the microcapillary which can obtain the wavelengths resonances and the field distribution of the corresponding modes with different radial orders WGMS. To understand the model, Figure (1) illustrates the general scheme of the dielectric structure designed to confine the field inside.

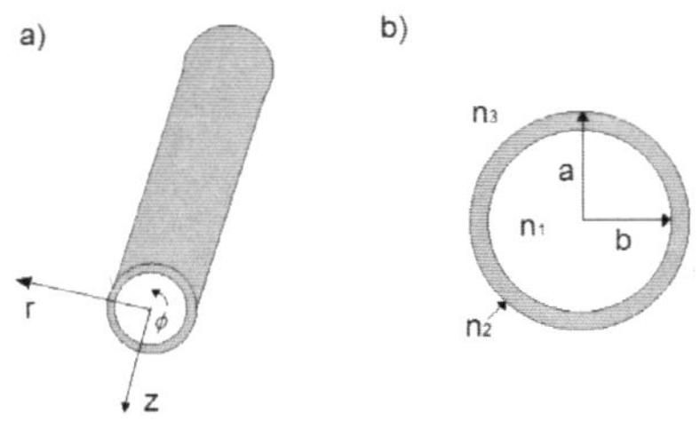

Figure 1. a) General outline of the microcavity for confinement of modes WGMS inside. **b)** three-layer system in a cylindrical shape with internal and external radii ***b*** and ***a*** respectively and with different refractive indices.

For the theoretical analysis, the symmetry of the cavity is exploited and cylindrical coordinates (r, θ, z) whose origin is the center of the capillary are chosen. In the microcavity, the internal radius ***b*** has a refractive index n_1, the cavity wall has a thickness $a-b$, where ***a*** is the outer radius having a refractive index n_2 and the external medium has a refractive index n_3. In the device, the field confinement is realized in the medium with refractive index n_2 which is physically wall structure. In the study of the field distribution of the problem, it is required to know the states of polarization of the field within the cavity. These polarization states are classified according to how electric and magnetic fields propagate with respect to the plane of propagation of the WGMS modes and which are known as TE and TM modes.

3. Experimental setup.

For the resonant structure can be used as an optical measuring system sensitive to changes in humidity, it is required to couple the field inside the microcavity. For this, the technique used is through a taper optical fiber, which allows the excitation of resonant modes WGMS within the cavity. In the manufacturing process of the taper optical fiber, it was used flame brushing technique, which is suitable due to its high flexibility for controlling the movement of the flame, the length and speed of fiber drawing.

During manufacture was used an optical fiber commercial step index SMF 20 Corning, where a small portion of the fiber in the order of millimeters is selected to remove the coating and then subjected to the heat produced by a flame that heats fiber at a temperature of 1500 ° C, while simultaneously the fiber is attached from the ends until achieving stretching and thus obtain the desired profile. During this process, fibers are obtained with neck diameters of 2-5 μm, sufficient to ensure optimal coupling of the field inside the cavity. In the test transmission fiber, is used a tunable laser system TLS with operating wavelengths in the range of 1500-1600 nm and optical spectrum analyzer OSA, mounted BraggMeter FS 2100 model Fibersensing with a resolution of 10 pm. The diagram of experimental setup for obtaining transmission spectrum of manufactured fibers is shown in Figure 2a, while in Figure 2b the spectrum of transmission obtained for a taper fiber of 4 μm in diameter is observed in the range of 1500-1600 nm of wavelength. In the same figure, a close is performed on the spectral response of the fiber in the range of 1562-1572 nm wavelength.

For the manufacture of the microcavity, they were manufactured microcapillaries PMMA by stretching a preform tube from a tower PMMA optical fiber design polymer. For stretching the preform tube, it requires heating the tube to a temperature of 210 °C and then you stretch to reach diameters and the desired wall thickness. In Figure 3, microscope images of some of the samples manufactured are observed. For the experimental setup performed, it was necessary to manufacture different capillary samples with different geometrical parameters.

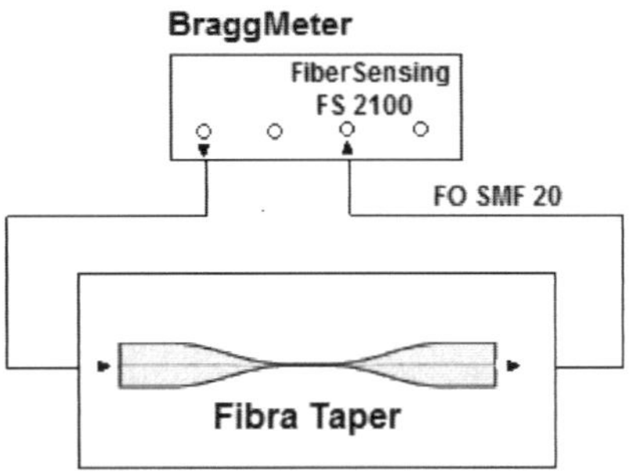

Figure 2. a) Diagram of the experimental setup for obtaining transmission spectrum of the fiber taper fabricated.

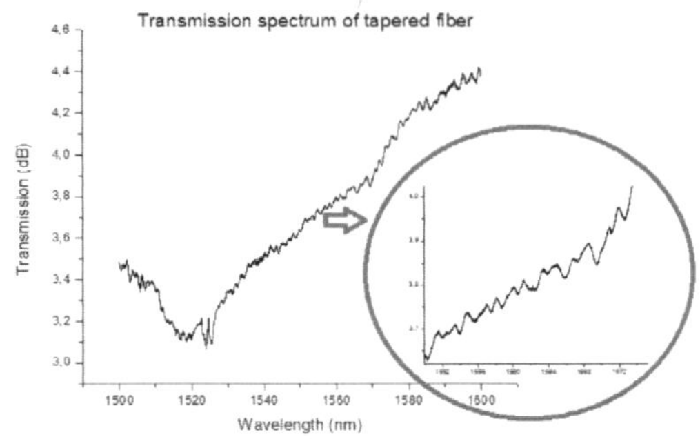

Figure 2. b) Transmission spectrum 4 µm taper diameter. Within the circle an image with increased range of 1562-1572 nm.

For the manufacture of the microcavity, they were manufactured microcapillaries PMMA by stretching a preform tube from a tower PMMA optical fiber design polymer. For stretching the preform tube, it requires heating the tube to a temperature of 210 °C and then you stretch to reach diameters and the desired wall thickness. In Figure 3, microscope images of some of the samples manufactured are observed. For the experimental setup performed, it was necessary to manufacture different capillary samples with different geometrical parameters.

The general diagram of the experimental setup is shown in Figure 4a and the general structure of the system for coupling the field to the cavity shown in Figure 4b.

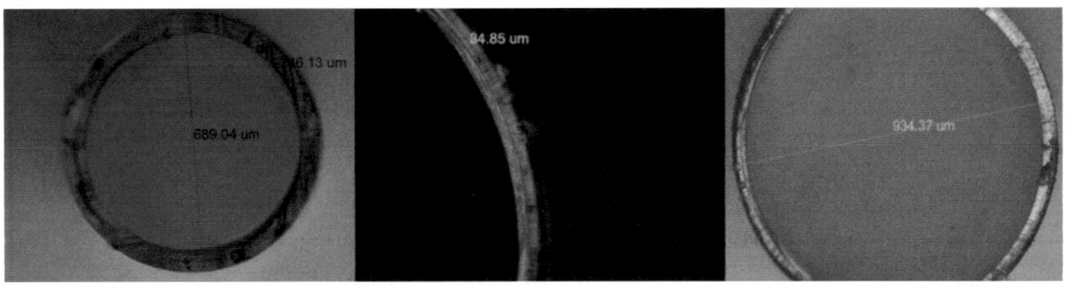

Figure 3. a) Capillary fiber manufactured in the laboratory with an external diameter of 689.04 µm and wall thickness of 46.13 µm. **b)** Image of a capillary fiber thin wall 34.85 µm. **c)** Image of capillary thin wall of 34.85 µm and external diameter of 934.37 µm.

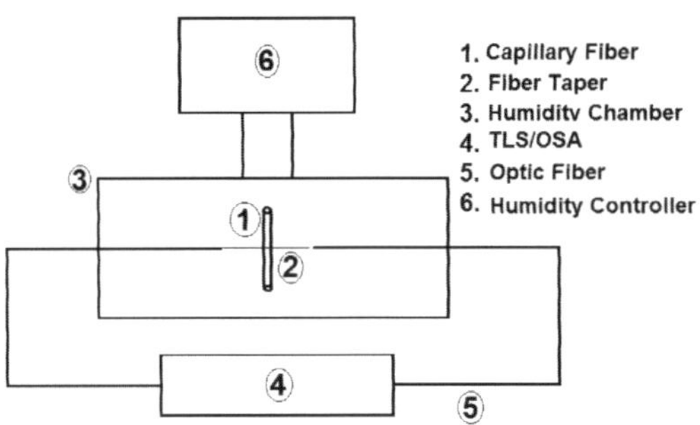

Figure 4. a) General diagram of the experimental setup.

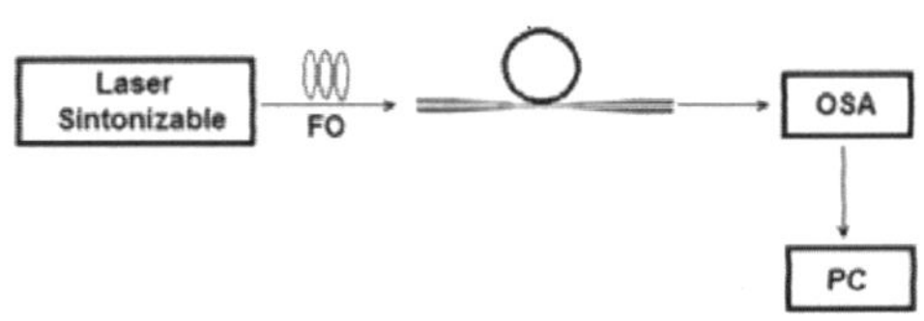

Figure 4. b) Outline of taper fiber coupled to the capillary fiber.

In figure it shows that the taper–capillary fiber system is coupled within the humidity chamber. The humidity inside the chamber is controlled electronically and the tapered fiber is connected to the TLS / OSA for the analysis of the displacement of the peaks of resonances observed at the output of the taper, which verifies the resonant characteristics of the system. In Table 1, the geometrical parameters and experimental sensitivities of each of the samples obtained in the laboratory are observed.

Table 1. Geometrical parameters and sensitivity of the manufactured samples

SAMPLE	External diameter **(μm)**	Wall thickness **(μm)**	***Sensibility*** Humidity up **(nm/% RH)**	Humidity down **(nm/% RH)**
Sample N° 1	2300,22	42,10	Test A: 0,05247 Test B: 0,05265	Test A: 0,07139 Test B: 0,06485
Sample N° 2	306,81	78,00	Test A: 0,03843	Test A: 0,04592
Sample N° 3	1143,59	101,00	Test A: 0,02661	Test A: 0,03336
Sample N° 4	2180,37	50,70	Test A: 0,04548	Test A: 0,0608
Sample N° 5	1231,07	109,73	Test A: 0,02008	Test A: 0,02989

The output of the taper, we allowed to determine the transmission spectrum of each of the samples produced through the OSA, obtaining peaks characteristic resonances of resonant systems. In Figure 5, the transmission spectrum of the capillary sample No. 2 is shown in the range of 1530 to 1535 nm wavelength to a fixed humidity value of 65% RH. With the information provided by the resonance peak of Figure 5b, we obtain a factor $Q \approx 4x10^3$ which is approximately equal to the case of silica capillaries reported in the literature [3].

In order to determine the sensitivity of the device to changes in humidity medium, it was necessary a humidity chamber in an operating range between 53-78% RH. For measuring humidity it was used a PCE-555 hygrometer PCE Intruments with an accuracy of 0.1% RH.

Figure 6 shows the displacement of the resonance peaks for different values of the relative humidity in the range from 60.6 to 64.0% RH. The curves were obtained with relative humidity up to higher values, for which a shift of the resonance peak of black color is initially centered at a wavelength of 1534.3248 nm at a relative humidity of 60,6 % RH is observed, however, by changing the relative humidity of the medium, the central wavelength of the peak is shifted to longer wavelengths.

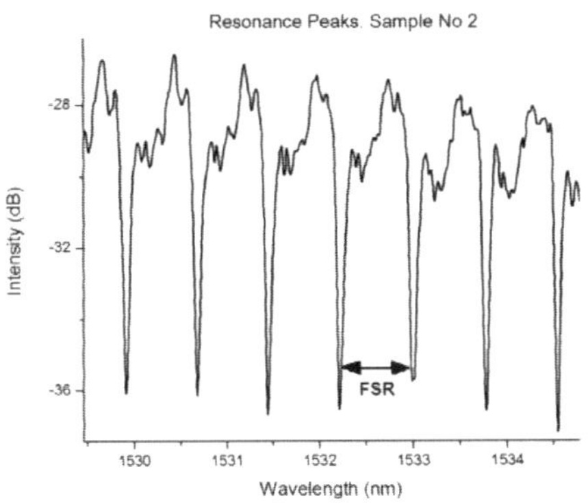

Figure 5. Peaks resonances of the sample No. 2. a) Curves resonances in the range of wavelengths 1530 to 1535 nm.

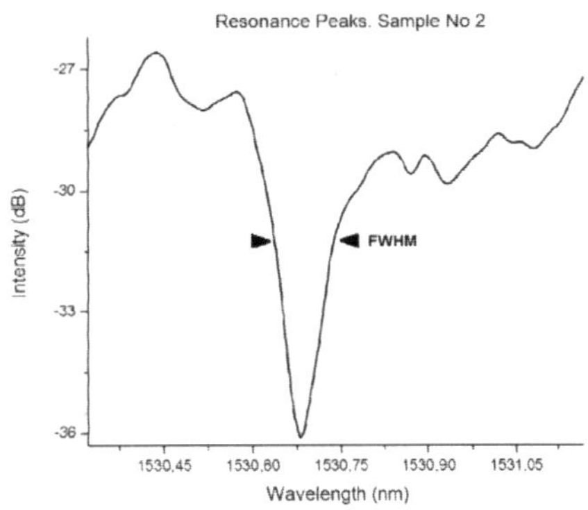

Figure 5. b) Resonance peak centered at 1530.68 nm

Figure 6 shows the displacement of the resonance peaks for different values of the relative humidity in the range from 60.6 to 64.0% RH. The curves were obtained with relative humidity up to higher values, for which a shift of the resonance peak of black color is initially centered at a wavelength of 1534.3248 nm at a relative humidity of 60,6 % RH is observed, however, by changing the relative humidity of the medium, the central wavelength of the peak is shifted to longer wavelengths.

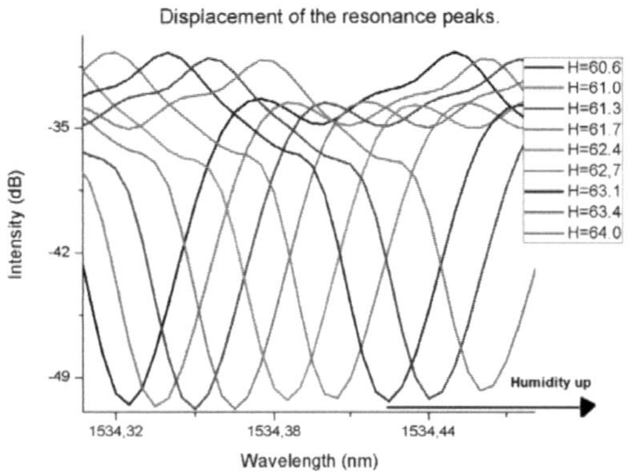

Figure 6. Graph of intensity Vs wavelength peak resonances at different values of relative humidity.

The peak shifts of resonances in Figure 6, corresponds to the capillary sample No. 1, whose geometric parameters are summarized in Table 1. During the experiment, was given the experimental sensitivity of the sensors manufactured for several times to determine the reproducibility of the experiment.

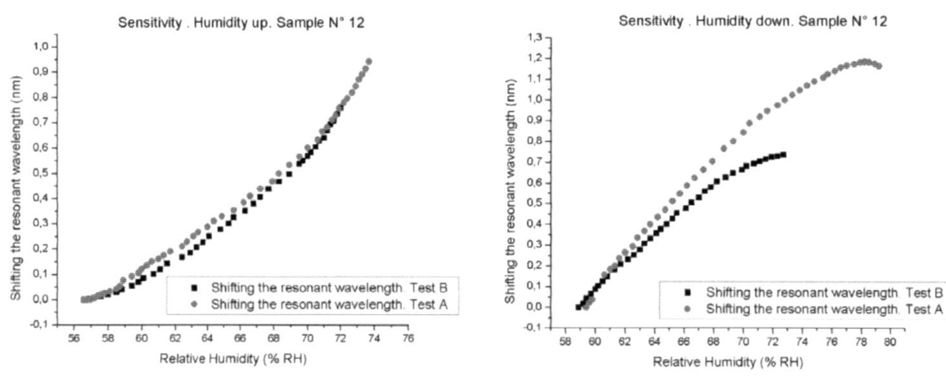

Figure 7. Graph of sensitivity for sample No. 1 obtained from two different tests (Test A and Test B) with increasing and decreasing humidity.

Figure 7 shows the results of the reproducibility in the experiment for the sensitivity of sample No. 1. The experiment was initially developed with humidity ascending and then descending in a controlled manner. In the results, it was analyzed hysteresis in sensitivity measures for changing a relative humidity of up and down manner, which results are shown in Figure 8.

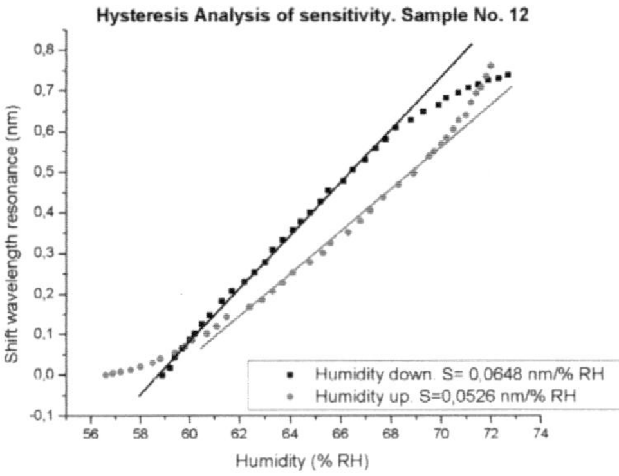

Figure 8. Analysis of Hysteresis for sample No. 1. The red curve corresponds to the measure of sensitivity to humidity increasing, while the black curve corresponds to decreasing humidity.

4. Analysis and conclusions

Hysteresis in the analysis performed, it is evidenced that the sensitivity of the different samples analyzed for humidity measurement is not completely linear. However, there are ranges where humidity sensor response can be considered linear. During the hysteresis test, it is remarkable to note that there are slight differences in the sensitivity of the device to measure humidity when this increases or decreases gradually. During the experimental results it can be seen that the sensitivity is slightly higher when the humidity is down for all samples tested. The measurement sensitivity experimentally for each of the samples analyzed is summarized in Table 1. During the analysis of the results can be determined that the best sensitivity was obtained for the sample No. 1 with a sensitivity 0.071 nm /% RH, whose sample had a wall thickness of 42.1 μm. The analysis of samples, show an increase in sensitivity when the samples analyzed lower wall thickness.

As a result of the experiments, we found that the microcavity as manufactured cylinder from PMMA has certain sensitivity to humidity changes in the medium, unlike those made from the silica as it was found in some experiments above. The sensitivity depends on the wall thickness of the microcapillary, which improvement microcapillaries with thin wall. In our experiment the sensitivity was achieved in the range 0,027-0,071 nm/% RH measure through different microcapillaries with different wall thickness.

5. References

[1] Gouveia M A, Duber-Avila P, Marques T H R, Cesar Torres M. C and Cordeiro C M B 2015 Morphology dependent polymeric capillary optical resonator hydrostatic pressure sensor *Opt. Express* **23** pp 10643-10652

[2] Avila D A, Gouveia M A, Marques T H R, Torres C O and Cordeiro C M B 2014 Polymeric Capillary Optical Resonator in Latin America Sensors *J. Opt. Soc. Am.* LTu4A.38

[3] Zamora V, Díez A, Andrés M V, and Gimeno B 2007 Refractometric sensor based on whispering-gallery modes of thin capillaries *Opt. Express* **15**, 12011-12016

[4] Matthew S. Luchansky and Ryan C. 2012 Bailey High-*Q* Optical Sensors for Chemical and Biological Analysis *Anal. Chem.* **84** (2) pp 793–821

[5] Tao Ling and L. Jay Guo A. 2007 Unique resonance mode observed in a prism-coupled micro-tube resonator sensor with superior index sensitivity *Opt. Express* **15** 17424-17432

[6] Yang Y, Saurabh S, Ward J, and Chormaic S N, 2015 Coupled-mode-induced transparency in aerostatically tuned microbubble whispering-gallery resonators *Opt. Lett.* **40**, 1834-1837

IOP Conf. Series: Journal of Physics: Conf. Series **792** (2017) 012051 doi:10.1088/1742-6596/792/1/012051

Influence of void fraction in the power distribution for a GE-12 fuel assembly

S J Castillo[1], G A Vargas[2], E del Valle Gallegos[1, 3]

[1]Instituto Politécnico Nacional Avenida Instituto Politécnico Nacional S/N, Gustavo A. Madero, San Pedro Zacatenco, 07738 Ciudad de México, D.F. [2]Instituto Nacional de Investigaciones Nucleares Carretera Mexico-Toluca S/N, La Marquesa, 52750 Ocoyoacac, Méx.[3]On sabbatical leave at Instituto Nacional de Investigaciones Nucleares.

sid.jc1@gmail.com, gustavo.alonso@inin.gob.mx, evalle@ipn.mx

Abstract. Analysis of the influence of void fraction distribution is very important to learn about the fission process and heat produced by the fuel assembly, here in this study several void fraction (VF) values along different burnup values have been considered in order to observe their influence in power distribution, uranium consumption, neutron flux and k_{inf} behaviour for a GE-12 fuel assembly. For this study, burnups up to 60 MWd/kg and VF values up to 0.8 were considered setting the uranium enrichment at 3.5 weight percent at the start of every VF scenario, results show that higher void fractions reduce the thermal flux decreasing thermal fission and limiting heat production.

1. Introduction

Fuel assembly design plays a very important role in reactor core performance. A nuclear reactor is designed for heat production in the core. The heat is extracted through the coolant by forced convection. Core thermal design seeks to extract the highest amount of fuel heat within temperature limits of the cladding maximizing the fuel-coolant contact area (Rust, 1979). This implies a large number of fuel rods in the fuel assembly. Nowadays, some fuel assembly designs are built in a ten by ten rods array basis, like the GE-12 fuel assembly shown in Fig. 1 (Dalrymple, 2012).

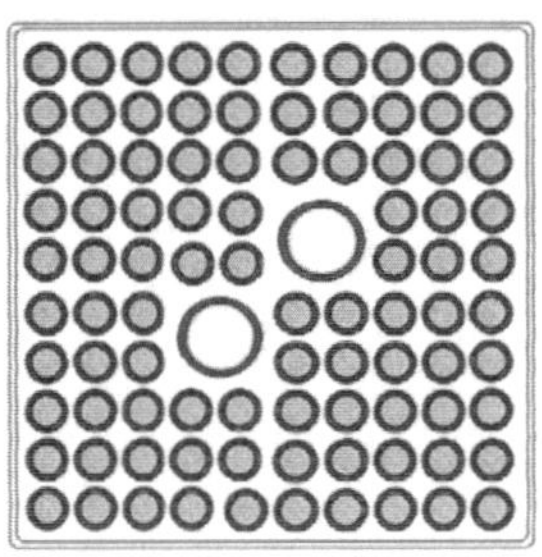

Figure 1. GE-12 fuel assembly radial cut array.

As the fuel heat is being extracted from the rod, the amount of bubbles in the water surrounding the fuel rod increases and accumulates, going through different flow regimes. In a BWR, typical VF values go from 0.38 to 0.75, reaching an annular flow regime where the liquid flows as a film along the cladding while the steam goes through among the fuel rods.

Content from this work may be used under the terms of the Creative Commons Attribution 3.0 licence. Any further distribution of this work must maintain attribution to the author(s) and the title of the work, journal citation and DOI.
Published under licence by IOP Publishing Ltd

On the other hand, current fuel assembly designs use inner water rods to provide a more homogeneous burnup in the fuel. The GE-12 design has two inner water rods occupying the space of 8 fuel rods.

2. Methodology

In order to perform the neutronic analysis of the GE-12 fuel assembly, the CASMO-4 (Rhodes, 2007) code from Studsvik was used. CASMO-4 is a multigroup two-dimensional transport theory code capable of performing burnup calculations for BWR and PWR lattices or simple pin cells by solving the transport equation using the characteristic method.

The code handles a geometry consisting of cylindrical fuel rods of varying composition in a square pitch array that allows for fuel rods loaded with gadolinium, erbium, integral fuel burnable absorber (IFBA), burnable absorber rods, cluster control rods, in-core instrument channels, water gaps, and cruciform control rods in the regions separating the fuel assemblies.

In general, a GE-12 fuel assembly has up to 100 pins. In particular, in this study 92 fuel rods with 3.5% enrichment and 8 water rods are considered. The input file was run considering an assembly with diagonal symmetry. The initial power density considered was 50.66 KW/lt, with a fuel and moderator temperature of 739 K and 559 K, respectively. Every input file was run for burnups up to 60 MWd/kg in every VF scenario considered.

3. Results and discussion

In the present study, three VF values are evaluated for a GE-12 fuel assembly. They are of 0.2, 0.4, and 0.8 VF values. Also, three burnup values are considered namely 20, 40, and 60 MWd/kg.

Figures 2-4 show the fast neutron flux for burnups over 60 MWd/kg with 0.2, 0.4, and 0.8 VF values in coolant. Figures 5-7 show the thermal neutron flux for the same parameters. The fast neutron flux energy range studied goes from 0.625 eV to 10 MeV, while the thermal neutron flux goes from 0.005 eV to 0.625 eV. Fast neutrons are the first born from fission, so the highest concentration is seen on the fuel rods zone in every scenario studied (Figures 2-4). The thermal neutrons are moderated fast neutrons, so the highest concentration is seen on the water rods and the boundaries of the assembly (Figures 5-7).

As void fraction increases, fewer fast neutrons are moderated, so the quantity of thermal neutrons is lesser in 0.8 than 0.2 VF values. This means that there will be less thermal neutrons colliding with uranium 235, producing less fissions and fast neutrons. The uranium consumption is shown in Figures 8-10.

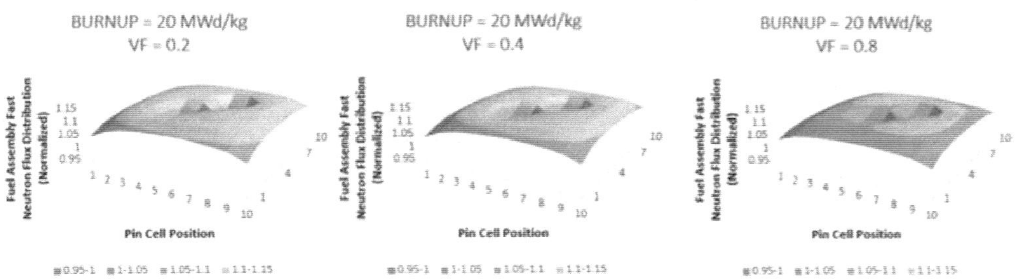

Figure 2. Fast neutron flux for a 20 MWd/kg burnup with different VF values.

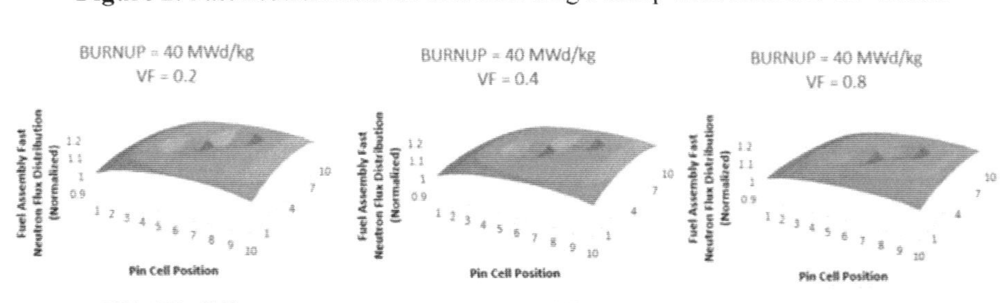

Figure 3. Fast neutron flux for a 40 MWd/kg burnup with different VF values.

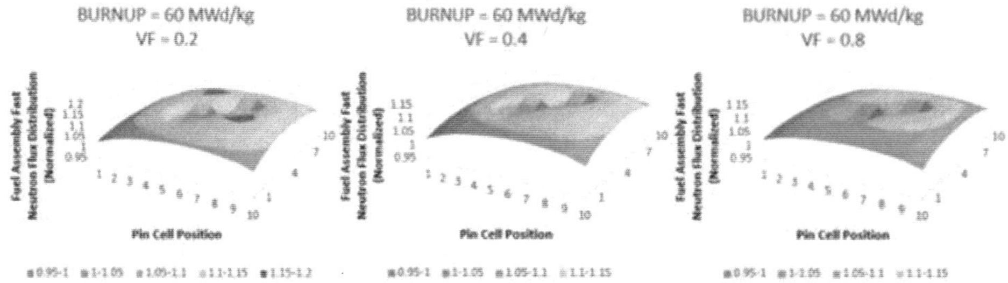

Figure 4. Fast neutron flux for a 60 MWd/kg burnup with different VF values.

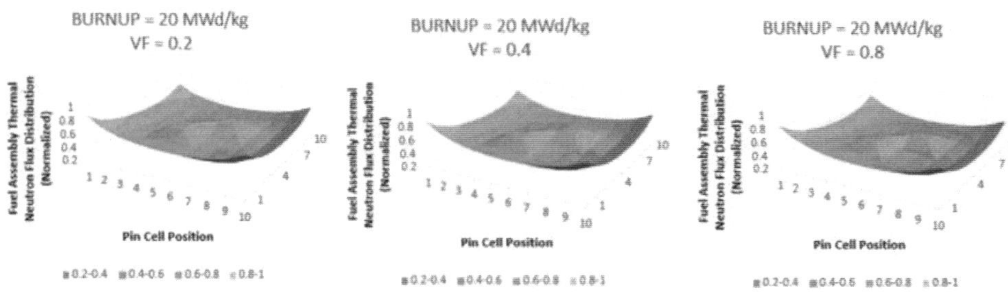

Figure 5. Thermal neutron flux for a 20 MWd/kg burnup with different VF values.

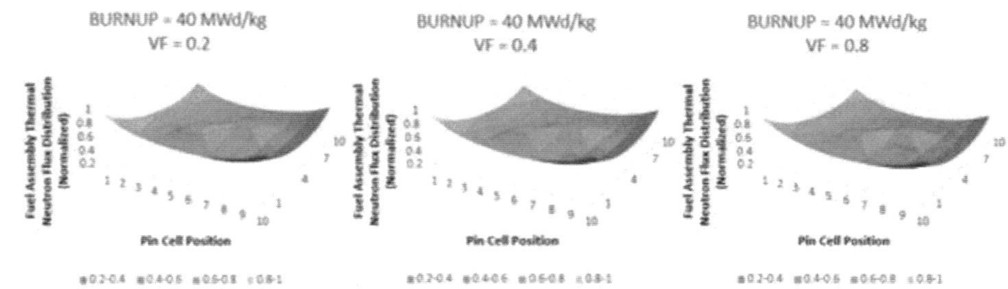

Figure 6. Thermal neutron flux for a 40 MWd/kg burnup with different VF values.

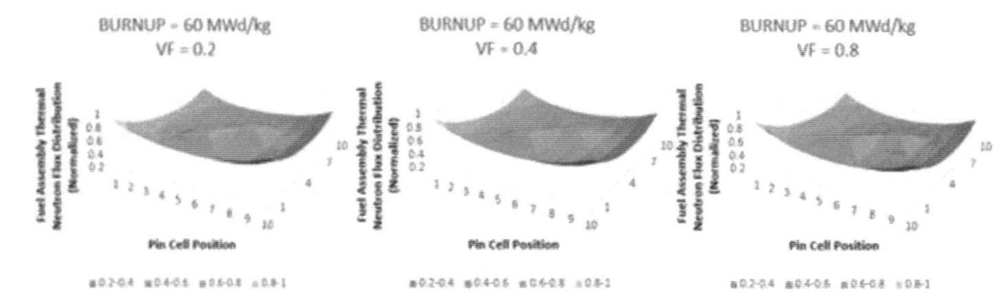

Figure 7. Thermal neutron flux for a 60 MWd/kg burnup with different VF values.

Figures 8-10 show the uranium 235 consumption for the same parameters seen in the neutron flux. At the start of every scenario studied, the uranium enrichment is stated at 3.5 weight percent. As void fraction increases, less uranium 235 is consumed due to thermal neutron flux decrease. However, the uranium 235 consumption at the boundaries is still the highest consumption of the whole assembly. This could be due to neutrons first impact with the external fuel rods, while inner fuel rods suffer from self-shielding. The fast neutron flux does not affect the uranium 235 consumption, but other elements consumption not shown in this study.

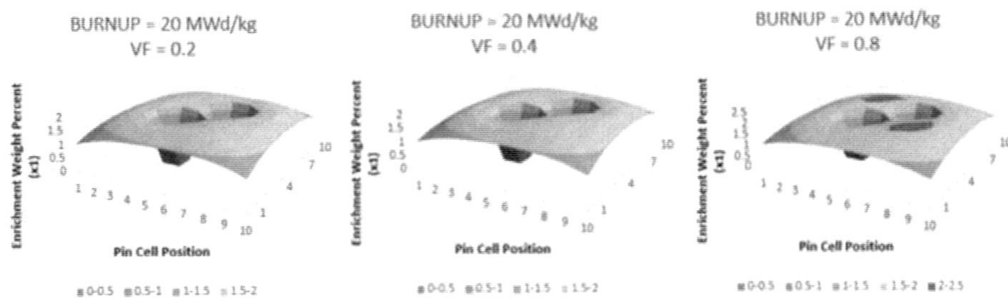

Figure 8. Weight percent of U-235 for a 20 MWd/kg burnup with different VF values.

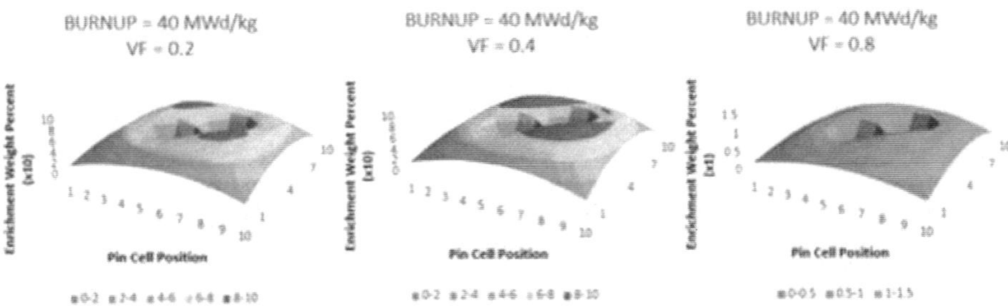

Figure 9. Weight percent of U-235 for a 40 MWd/kg burnup with different VF values.

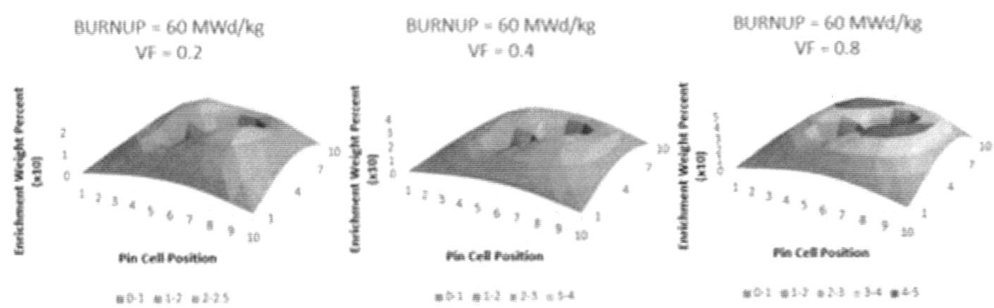

Figure 10. Weight percent of U-235 for a 60 MWd/kg burnup with different VF values.

Figures 11-13 show the normalized power distribution for 20, 40, and 60 MWd/kg burnups with 0.2, 0.4, and 0.8 VF values in coolant. As the VF increases for every scenario studied, the power distribution in the boundaries increases while in the central zone decreases. The highest values of the power distribution in every graphic agree with the lowest values of the weight percent of uranium 235.

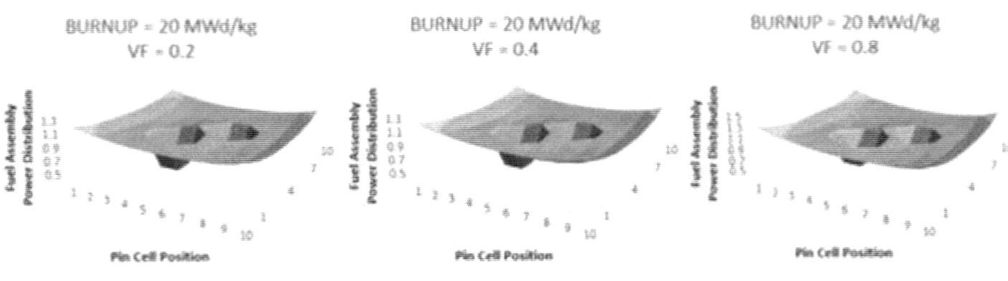

Figure 11. Power distribution for a 20 MWd/kg burnup with different VF values.

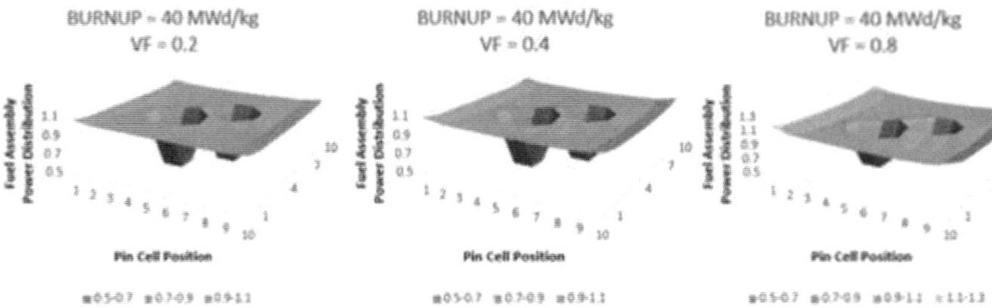

Figure 12. Power distribution for a 40 MWd/kg burnup with different VF values.

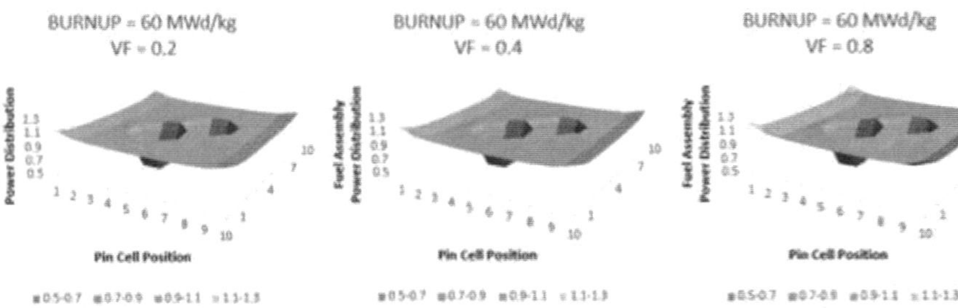

Figure 13. Power distribution for a 60 MWd/kg burnup with different VF values.

The infinite medium multiplication factor (k_{inf}) is the ratio of the number of neutrons produced in reactor and the number of neutrons loss by absorption in reactor with no leakage (Duderstadt, 1976). The behavior of the k_{inf} for the three VF scenarios studied is shown in Fig. 14. As burnup increases, the k_{inf} decreases for the three VF values. At the beginning of the burnup steps, the k_{inf} has a higher value for the VF scenario of 0.2 than the VF scenarios of 0.4 and 0.8. This is because uranium enrichment has the same value at the beginning of the burnup steps for all the scenarios studied, so there are more fissions produced when the moderation is higher for the VF scenario of 0.2 than in the others two scenarios. At the end of the burnup steps this behavior is inverted, so the highest k_{inf} corresponds to the VF scenario of 0.8 while the lowest k_{inf} value corresponds to the VF scenario of 0.2 since the moderation is lower. This means that there are less thermal neutron colliding with the uranium 235 but there are more fast neutrons colliding with another fissionable materials, like uranium 238 which can also produce plutonium 239 when absorbs a thermal neutron.

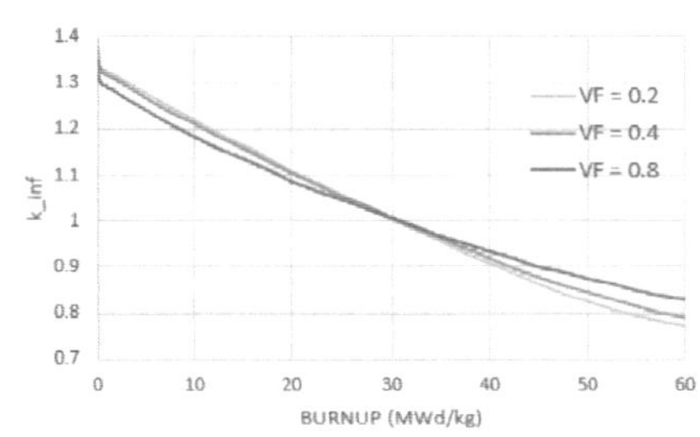

Figure 14. k_{inf} vs burnup for different VF values.

4. Conclusions

A study on the effect of void fraction's change with burnups over 60 MWd/kg on neutron flux, uranium consumption, power distribution and k_{inf} behavior for a GE-12 fuel assembly has been conducted.

For higher VF values in the same burnup scenario studied, fast and thermal neutron fluxes decrease while uranium 235 weight percent increases. The thermal neutron flux shares a similar behavior with the power distribution behavior due to its contribution to the fission production. In all the VF scenarios studied, the k_{inf} value decreases as the burnup increases. The k_{inf} higher decrease is seen for the VF value of 0.2, while the lower decrease is seen for the VF value of 0.8. This might be due to the rate consumption of uranium 235.

Reference

[1] Alonso G, Bilbao S, and del Valle E 2015 Impact of the moderation ratio over the performance of different BWR fuel assemblies *Ann. Nucl. Energy* **85** pp. 670-678
[2] Duderstadt J and Hamilton L 1976 *Nuclear Reactor Analysis*, (John Wiley & Sons Inc.)
[3] Fuel Review/Design Data 2012 *in Nucl. Eng. Int.* **57** pp. 42-54
[4] Rhodes J. And Smith, K. A 2007 *Fuel Assembly Burnup Program* User's Manual Studsvik Scandpower
[5] Rust J 1979 *Nuclear Power Plant Engineering* (Haralson Publishing Company)
[6] Waris A, Su'ud Z, Mehmet Şahin H, Kurt E and Şahin S 2015, Influence of void fraction on BWR spent fuel direct recycling scenario *Int. J. Hydrogen Energy* **40** pp. 15172-15178
[7] Waris A, Su'ud Z, Permana S, and Sekimoto H 2008 Influence of void fraction change on plutonium and minor actinides recycling in BWR with equilibrium burnup *Prog. Nucl. Energy* **50** pp. 295-298

IOP Conf. Series: Journal of Physics: Conf. Series **792** (2017) 012052 doi:10.1088/1742-6596/792/1/012052

Radial distribution function at the particle-bottom interface in granular self-assembly

J J Becerríl-González, F Peñuñuri, M A Zambrano, C Acosta and O Carvente
Departamento de Ingeniería Física, Universidad Autónoma de Yucatán, 97310, Mérida, Yucatán, México.

E-mail: osvaldo.carvente@correo.uady.mx

Abstract. Dynamic self-assembly of confined non-Brownian spheres under gravity is investigated in terms of the statistical behavior of the like-gaseous particles closest to oscillating bottom. For strong vibrations, these particles resemble a quasi-two-dimensional gas (qtd-gas) that supports upper particles. By vibrational annealing, wet particles inside a commensurate rectangular container self-assembling into face-centered-cubic (fcc). From high temporal resolution images we study, by means of the associated Radial Distribution Function (RDF), the evolution in time of spatial correlations in the transition: qtd-gas to dense liquid of the vibrating particles closest to the oscillating bottom during self-assembly. As the vibrational annealing advances, the particles closest to the bottom reach this final position and the associated RDF shows a clear increase of the spatial correlation, which is congruent with the fcc crystallization phase.

1. Introduction

Due to friction forces, gravity and confinement, static granular materials are into a metastable static states. In this sense, a sudden agitation of the container is enough for the particles composing the granulate system to attain a new spatial configuration (a new metastable state). On the other hand, continuous vibrations (usually sinusoidal) *thermalize* the system into a stationary state that resembles, for particular vibrating conditions, like granular gases at low density (feature that exclude many experimental conditions of current interest). For this particular case the Boltzmann equations correctly describe the system [1], but for denser gases the Maxwell-Boltzmann distribution, of the horizontal velocities, fails [2]. Despite granulates are inherently dissipative and have no reaction to thermal fluctuations, some sort of stochastic mobility is provided by continuous external excitations, for this the system resemble and have been used as model system for atomic, molecular, and colloidal ensembles in thermal equilibrium [3, 4]. Into a stationary state some thermodynamic-like variables (density, temperature or pressure) are statistically well defined [5-15]. Because of dissipative collisions, such variables exhibit a non-monotonic behavior [13, 14, 16].

In many situations of the life, including the pharmaceutical, food, mineral or construction industries, the granular materials are prominent. Notwithstanding their importance, distinctive phenomena, such crystallization [17-20], convection [21-25], or compaction [26, 27] where nature guides toward one or the other depending, at first instance, on the oscillating conditions, has been extensively studied without considering the dynamics of particles closest to the oscillating bottom.

Content from this work may be used under the terms of the Creative Commons Attribution 3.0 licence. Any further distribution of this work must maintain attribution to the author(s) and the title of the work, journal citation and DOI.
Published under licence by IOP Publishing Ltd

In this work we study the dynamics of cohesive particles closest to oscillating flat bottom during a granular self-assembly process. It is well known that granulates change their mechanical properties when a small amount of liquid is added to them [28]. Liquid bridges between particles give place to attractive capillary forces and, therefore, granulates behave in a different way. At the beginning of self-assembly process, particles under scrutiny (cohesive ball bearings) behave as a confined (by the vertical walls of rectangular container) quasi-two-dimensional dense gas (qtd-gas), which transfers their kinetic energy by collisions and supports the upper particles (also cohesive). We chose wet beads because an analogous behavior with thermal systems has been established in [29]. As the self-assembly advances, the particles of the qtd-gas slowly evolves to a crystalline phase. Throughout the Radial Distribution Function (RDF), calculated from the analysis of high temporal resolution images of the evolving qtd-gas (a fast camera was mounted underneath of the translucent oscillating bottom, Fig.1), we study the structural evolution of the granular self-assembly. Our results show that spatial correlation of the qtd-gas grows in agreement with the crystalline structure. Hence, the behavior of the whole vibrating granular system can be considered as consequence of the dynamics in the particle-bottom interface.

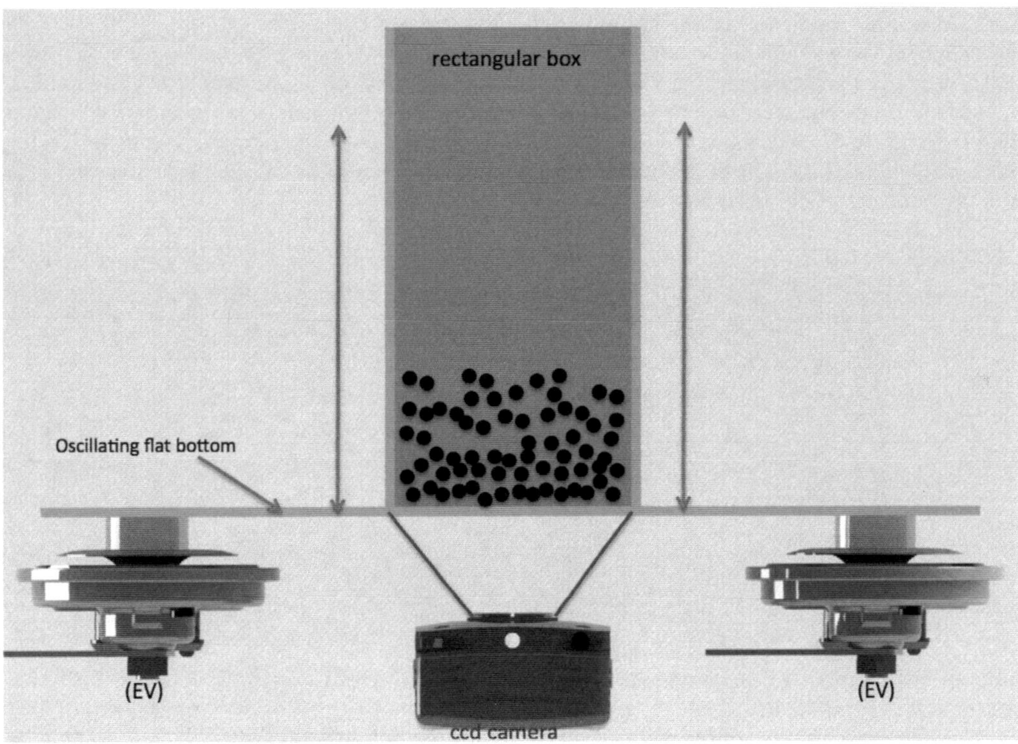

Figure 1. (Color online) Experimental setup. A rectangular box mounted on the oscillating flat bottom. Two fixed electromechanical vibrators (EV) drives the system to vertical oscillations. The fast speed camera (ccd) was mounted below the square box to capture the movement of the particles closest to oscillating flat bottom.

2. Experimental methodology

The granular self-assembly, by vibrational annealing method, consists in shaking the granular system with a sinusoidal oscillation; where the frequency v and the amplitude A is increased and decreased by discrete steps, respectively, in such a way that the dimensionless acceleration $\Gamma = A(2\pi v)^2 / g$, (g is the acceleration of gravity) remains constant. The total self-assembly process is completed in about

200 seconds. The frequency and amplitude were changed in intervals of 5 seconds before. After changing the frequency and amplitude the system reaches a stationary state in 0.5 seconds. At the beginning of the process, when the amplitude is large and the frequency small, most of the assembly is a dense gas. Slowly, the assembly becomes a fluid-like system where the beads reduce their velocity but still move all around the container. Finally, at high frequencies and small amplitudes, the system is a solid where beads only vibrate around their equilibrium positions [30]. The initial conditions of the annealing (amplitude and frequency), and the rate of the cooling program, are given to the computer at the beginning of the process. As long as the annealing rate is not too fast to quench the system in a disordered structure, this rate is not crucial. The experimental setup consists of a vertical shaking system driven by a function generator (HP-33120A), an amplifier for the electromechanical vibrators (EV), and a controlling computer. The particles of the qtd-gas were filmed with a high-speed charge-coupled device (CCD) camera (Photron SA3 60K M1 512 × 512 pixels), mounted underneath of a translucent flat bottom. In order to capture the whole dynamic, the particles were filmed at 1000 frames per second. ImageJ free distribution software was used for the video analysis, this give us the particles position for each one frame. In Fig. 2 we show an instantaneous frame (left) and their corresponding reconstruction from the particles positions given by the ImageJ software (right). Unfocused images are the best technique to study only the particles of the qtd-gas. In this way, the sphere's images were recorded just from the bright spot shining at their center, making it easier for the particle center localization algorithm to process the frames. We use cohesive steel beads (a thin film of oil cover the beads surface), of diameter d=3.18 *mm.*, inside a commensurate rectangular container made of plexyglas, having dimensions of $L_x = L_y = 61.6$ *mm.* and $L_z = 150$ mm. The cohesive force between two particles consists of two parts: the surface tension, acting at the wet perimeter, and a term arising from the capillary pressure in the liquid [31].

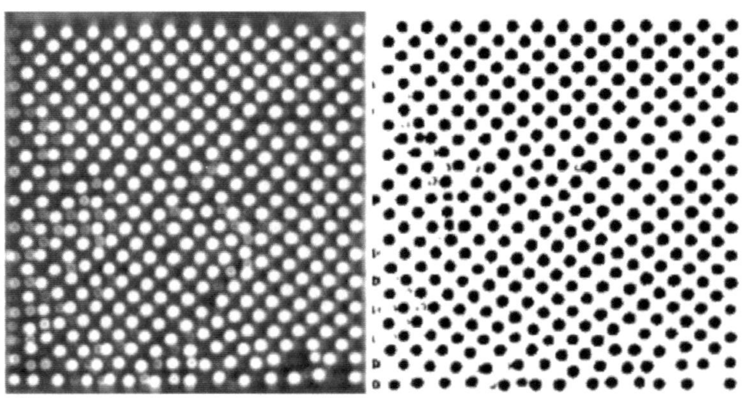

Figure 2. (Color online) Left, the instantaneous image from the fast camera of the particles closest to the bottom. Right, the reconstruction by means of the particles position obtained from the ImageJ software.

Radial Distribution Function: The radial distribution function $g(r)$ is a standard way of describing the average positional structure of particulate systems [32]. $g(r)$ measures the probability that two particle centers are at distance r, regardless of orientation. For hard particles there is zero probability that the particles overlap so $g(r) = 0$ for $r < d$. The construction of the RDF is as follows: consider, for example, a gas of hard particles and chose, in any instant of time, a particle and draw it a series of concentric spheres, set at a small fixed distance (dr) apart. At regular intervals of time the number of particles found in each shell is counted and stored. At the end, the average number of particles in each shell is calculated. Then, this is divided by the volume of each shell and by the average density of particles in the system. The result is the RDF. The mathematical expression for the RDF is given by:

$$g(r) = \frac{n(r)}{\rho 4\pi r^2 dr}, \tag{1}$$

where $n(r)$ is the mean number of particles in a shell of width dr at distance r, ρ is the mean particle density. The method need not be restricted to one particle. All the particles in the system can be treated in this way, leading to an improved determination of the RDF as an average over many particles.

In our experiments, despite the intrinsic effects of the limiting walls of the container and the cohesion forces, we calculate the RDF for particles of the qtd-gas at each one of the stationary state during the self-assembly. For this, we first consider the particle closest to the center (Fig. 3) to calculate the mean number of particles in a ring of radius r and thickness dr, thereafter we extend the count to the five particles closest to the center and go one for each one of the 5000 frames belonging to some particular stationary state, at the end we averages over time and the five particles to obtain $g(r)$. We choose dr enough small compared with the dimensions of the system; hence the particles outside the longest circle do not contribute to the RDF.

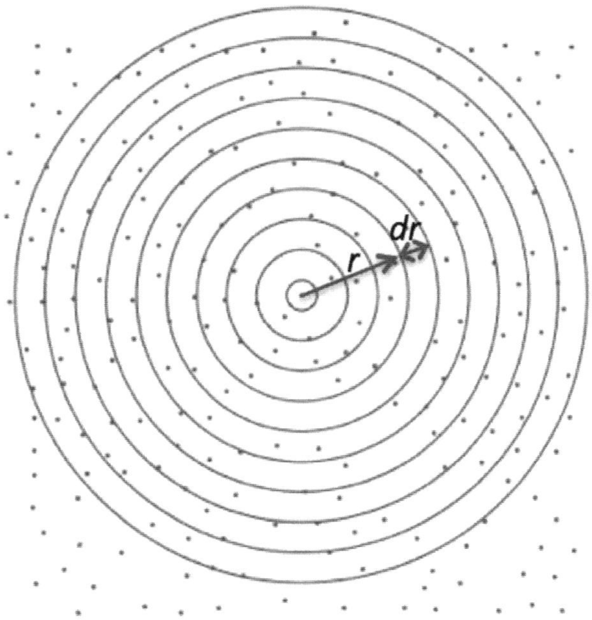

Figure 3. (Color online) Each dot represents the center of a particle.

3. Discussion and Results

The annealing process starts with v = 22 Hz and ρ = 5.5. For these oscillating conditions the system behaves as a dense fluid and the closest particles to the oscillating bottom move erratically throughout the container. Due to periodic oscillations, the number of these particles is not constant, so the packing fraction associated takes values φ = (0.54, 0.64). From one frame to the next, the particles of the qtd-gas could be different; nevertheless the RDF captures correctly the positional correlation of granular gases in a stationary state [33, 34], being these in out of equilibrium but stationary ensembles. In Fig. 4a we show the RDF at the beginning of the self-assembly, this is consistent with lead particles (lesser restitution coefficient, μ) in two-dimensional granular gases at low density (φ = 0.35) [35]. It is interesting to note that particles with bigger restitution coefficient and packing fraction (our system) behave as a gas of particles with minus restitution coefficient and packing fraction. Then, μ and φ must be considered for a correct description. The annealing goes one and the positional correlation increase. For example, at v = 28 Hz the RDF (Fig. 4b) shows strong correlation, and several peaks indicates that the closest particles to the bottom vibrates as a very dense fluid. At this time the system has been suffered a structural phase transition of first order [36], where dense fluid particles support a more dilute phase in the upper part of the vibrating system. The RDF at v = 40 Hz and v = 28 Hz looks very similar, with a first peak at $\sqrt{2}r/d$ and the second principal at $2r/d$. This indicates that a cubic lattice (face centered cubic) is formed over the oscillating bottom. Before the second principal peak a little peak emerges at v = 40 Hz (Fig. 4c) around the vertical line. We think that this small peak is due to the periodic oscillations. At some particular time the particles collides with the oscillating bottom, and after they leave it, so the system undergoes a small periodic expansion-contraction, this produces an overlap that gives place to the small peak. Despite the dissipative collisions, the effects of the walls, and the increase of the packing fraction, the RDF captures the correlations as the annealing advances and the crystalline phase begin to appear.

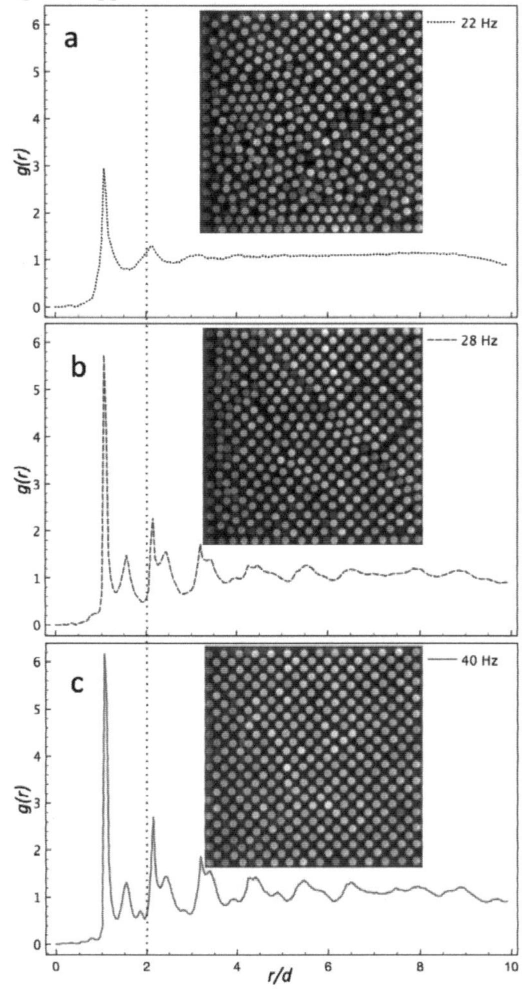

Figure 4. (Color online) Radial Distribution Function at different stationary states, characterized by the oscillating frequency. With the aim to appreciate, we show only the RDF for three stationary states.

4. Conclusions

We have presented an experimental study of the spatial correlations associated with the closest particles at the particle-bottom interface in a granular self-assembly process. The radial distribution functions of these particles were obtained when the system was in a stationary state, characterized by the frequency of vibration ν. Despite the effects of the confining walls and the relatively high values considered for ρ ($0.54 < \rho < 0.75$), the corresponding RDF's for each stationary state, shows congruency with the conformation of the FCC structure. Our work could be used as the basis for calculating effective potentials (via the Orstein-Zenike equations with the Percus-Yevick closure) at the particle-bottom interface.

References

[1] Brilliantov N V and Pöshel T 2004 *Kinetic Theory of Granular Gases* (Oxford U. Press, New York)
[2] Rouyer F and Menon N 2000 *Phys. Rev. Lett.* **85** 3676
[3] Jaeger H.M, Nagel S.R and Behringer R.P 1996 *Rev. Mod.Phys.* **68** 1259
[4] Tata B V R, Rajamani P V, Chakrabarti J, Nikolov A and Wasan D T 2000 *Phys. Rev. Lett.* 84, 3626
[5] Knight J B 1995 *Phys. Rev.* **51** 3957-3963
[6] Nowak E. R. 1998 *Phys. Rev.* **57** 1971-1982
[7] Olafsen J S and Urbach J. S. 1999 *Phys. Rev.* **60** R2468R2471
[8] Warr S and Hansen J P 1996 *Europhys. Lett.* **36**
[9] Rouyer F and Menom N 2000 *Phys. Rev. Lett.* **85**
[10] Windows-Yule C R K and Parker D J 2013 *Phys. Rev. E.* **87**
[11] Anna G D' 2003 *Nat. Phys.* **424**
[12] Shinbrot T, Nature c, Vol. **9**, 263264 (2013).
[13] Ramírez R and Soto R. 2003 *Physica A.* **322** 372-380
[14] Falcon E 1999 *Eur. J. Phys* **9** 183-186 (1999).
[15] Goldhirsch I *Powder Technol.* **182** 130-136
[16] Bodrova A 2012 *Phys. Rev. Lett.* **109** 178001-178005
[17] J-C Tsai, G A Voth and J P Gollub 2003 *Phys. Rev. Lett.* **91** 064301
[18] Daniels K E and Behringer R P *Phys. Rev. Lett.* **94** 168001 (2005).
[19] O. Carvente and J C Ruiz-Surez 2005 *Phys. Rev. Lett.* **95** 018001
[20] Reis P M, Ingale R A and Shattuck M D 2006 *Phys. Rev. Lett.* **96** 258001
[21] Ehrichs E E, Jaeger M H, Karczmar G S, Knight J B, Kuperman V Yu and Nagel S R 1995 *Science* **267** 1632-1634
[22] Bourzutschky M and Miller J 1995 *Phys. Rev. Lett.* **75** 4154
[23] Hsiau S S and Chen C H 2000 *Powd. Tech.* **111** 210217
[24] Huerta D A and Ruiz-Surez J C 2004 *Phys. Rev. Lett.* **93** 069901
[25] Eshuis P, van der Meer D, Alam M, van Gerner H J, van der Weele K and Lohse D 2010 *Phys. Rev. Lett.* **104** 038001
[26] Ben-Naim E, Knight J B, Nowak E R, Jaegerd H M and Nagel S R 1998 Physica D 1998 **123**, 380
[27] P. Richard, M. Nicodemi, R. Delannay, P. Ribiére and D. Bideau 2005 Nature Materiales **4** 121
[28] Mason T G, Levine A J, Ertas D, and Halsey T C 1999 *Phys. Rev. E* **60** R5044
[29] Carvente O and Ruiz-Suárez J C 2008 *Phys. Rev. E.* **78** 011302
[30] Carvente O, Peraza-Mues G G, Salazar J M and Ruiz-Suárez J C 2012 *Granular Matter* **14** 303
[31] Albert R, Albert I, Hornbaker D, Schiffer P and Barabasi A L 1997 *Phys. Rev. E* **56** R6271
[32] B. Bernal 1964 *Proc. Roy. Soc. A*, 280, 299

[33] Tata B V R, Rajamani P V, Chakrabarti J, Nikolov AA and Wasan D T 2000, *Phys. Rev. Lett.* **84**, 3626
[34] Olafsen J S and Urbach J S 1998 *Phys. Rev. Lett.* **81** 4369
[35] Bordallo-Favela R A, Ramírez-Saíto A, Pacheco-Molina C A, Perera-Burgos J A, Nahmad-Molinari Y and G. Pérez 2009 *Eur. Phys. J. E*
[36] Carvente O, Salazar-Cruz M, Peñuñuri F and Ruiz Suárez J.C 2016 *Phys. Rev. E.* **93**, 020902(R)

IOP Conf. Series: Journal of Physics: Conf. Series **792** (2017) 012053 doi:10.1088/1742-6596/792/1/012053

Numerical simulation of Flow Pressure Drop and Friction Factor of Water in 2D channel

H Aya Baquero[1] and L H Camargo Casallas
Facultad de Ingeniería, Universidad Distrital Francisco José de Caldas Bogotá, Colombia.

E-mail: haya@udistrital.edu.co

Abstract. The paper presents the results obtained from the numerical study of the dynamic properties of a straight channel 50 mm long and 780 μm wide on a 2D model. Numerical simulations were performed by using Navier-Stokes equation. The results showed a good agreement with experiments and other models. Pressure drop and friction factor of water in the channel in the studied ranges of Reynolds number are due to viscosity effects.

1. Introduction

The development on the manufacture of new devices at a micro and nano scale allows the construction of microchannels of different sizes and shapes that are used in a wide variety of ways, such as heat transfer, biologic material analysis: DNA, cells, embryos, chemical reagents among others. In the present situation, understand the microchannel dynamic characteristics will be followed for a design improvement. Nowadays, device design with integrated electronic microchannels has growth in the industry and research, to be used in medicine, biology and environmental sectors, for instance. Numerical simulation plays an important role study and design of microfluidic devices [1,2,3].

2. Theory

Under the framework of a continuous model, a Newtonian liquid dynamic is represented according the Navier-Stokes incompressible equation [4].

$$\rho\left[\frac{\partial \boldsymbol{u}}{\partial t} + (\boldsymbol{u}\nabla)\boldsymbol{u}\right] = -\nabla p + \mu\nabla^2\boldsymbol{u} + \boldsymbol{f}, \tag{1}$$

$$\nabla . \boldsymbol{u} = 0, \tag{2}$$

where $\boldsymbol{u}$ and μ are the speed and the dynamic viscosity of the liquid, p is the pressure, ρ is the fluid density, $\boldsymbol{f}$ is the volumetric density of external forces acting on the liquid. The characteristic dimensions in systems such as water-filled microchannels or electrolytic aqueous solutions, with a D diameter, fulfil the conditions for a Reynolds number $Re = \frac{\rho u D}{\mu} \ll 1$. This means that for most of the stationary problems, in equation (1) can be used the Stokes approximation

$$-\nabla p + \mu\nabla^2\boldsymbol{u} + \boldsymbol{f} = 0 \tag{3}$$

E-mail: haya@udistrital.edu.co

Content from this work may be used under the terms of the Creative Commons Attribution 3.0 licence. Any further distribution of this work must maintain attribution to the author(s) and the title of the work, journal citation and DOI.
Published under licence by IOP Publishing Ltd

Nevertheless, It is necessary to know the channel behaviour, including values of Reynolds number $R \geq 1$. In a laminar flow, the drop of pressure depends linearly on the fluid median velocity. Then, it also depends of the Reynolds number

$$\Delta p = K\frac{\mu u}{D}, \tag{4}$$

where K is the dimensionless coefficient of pressure drop. The channel friction factor f is calculated with the following expression

$$f = \frac{\Delta p D}{2L\rho u^2} \tag{5}$$

From equation (5), it is deduced that the friction factor is inversely proportional to the Reynolds number.

3. Numeric model and results

The purpose of the present work is evaluate the pressure drop and the friction factor of a 2D channel 50 mm long and 780 μm wide, where water flows at 18 ^{0}C. The pressure drop is assessed along the channel for different values of the Reynolds number under no-slip boundary condition. To calculate the pressure drop COMSOL models are used at different mean velocities. In this configuration, the pressure drop is only due to the friction of the bounding walls and no other effect influences in the channel hydraulic performance.

In figure 1, it is shown the flow velocity inside a channel when mean speed is 2 mm/s. As it is observed, no-slip boundary condition is accomplished (**u**=0) in the channel bounding walls. It is calculated the velocity profile along the channel transverse arc length (figure 2).

Figure 1. Velocity inside the selected channel [mm/s].

The absolute pressure diminishes along a channel, in which sliding is not present between the fluid and the bounding walls, mainly due to viscosity. Therefore, the loss in pressure depends proportionally to the dynamic viscosity of the fluid. Figure 3 shows the variation in the pressure along the channel for different values of the Reynolds number.

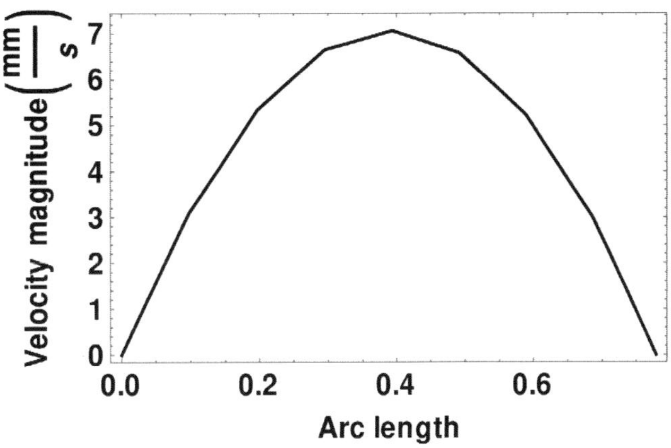

Figure 2. Velocity magnitude along an arch length for a mean velocity of 4.5 mm/s.

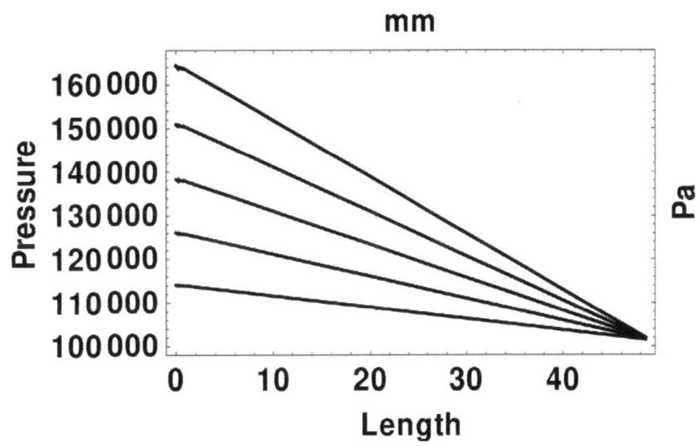

Figure 3. Pressure along the channel for Re values of 7.4, 14.8, 22.2, 29.6 y 37.0 (bottom to top).

The pressure drop is evaluated along the channel according the Reynolds number. Figure 4 shows the percentage of the pressure drop along the channel determined by the Reynolds number.

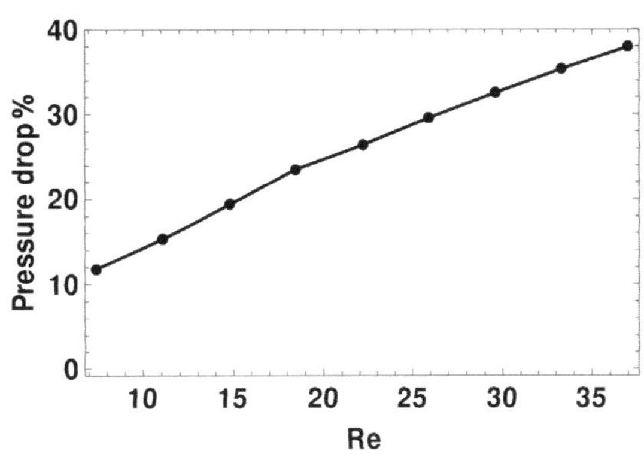

Figure 4. Pressure drop according the Reynolds number.

Lastly, the friction factor for the fluid inside the channel is obtained using the equation (5). Figure 5 shows the friction factor diminishing inversely proportional to the Reynolds number value.

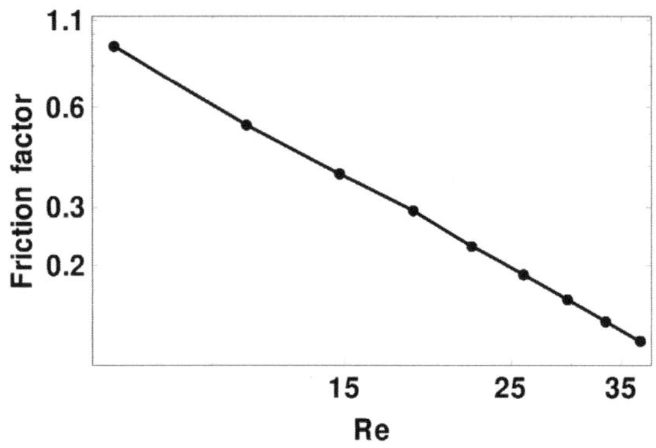

Figure 5. Friction factor determined by the Reynolds number.

4. Conclusions

The data obtained in this study satisfactory agrees with other studied models and theory literature. The development and design of devices at a micro and nano scale is possible with the advancement in the knowledge of the hydraulic characteristics, including how different scale values affect the performance. Recently, new approaches have appeared in this field with a series of novel proposals, technical and experimental developments. For instance, new geometric designs and materials to the improvement of dragging inside of a superhydrophobic channel, new designs of mixers and separators, heat transfer, among others [5,6,7,8]. This project is partially founded by the Engineering Faculty of the Universidad Distrital Francisco José de Caldas de Bogotá, inside the Proyecto Curricular de Ingeniería Electrónica.

5. References

[1] Niklas M, Favre-Marinet M and Asendrich D 2005 Bulletin of Polish Academic of Sciences **53** 4 351-359

[2] Mirmanto M 2013 Journal of Mechanics Engineering and Automation **3** 641-649

[3] Federspie W J and Valenti I 2012 *Open J. Appl. Sci.* **2** 28-34

[4] Bejan A 1995 Convection Heat Transfer John Wiley & Sons New York USA

[5] Akbari M, Sinton D and Bahrami M 2009 *J. Fluids Eng.* **131** 041202-1-8

[6] Zhao H, Liu Zh, Zhang Ch, Guan N and Zhao H 2016 *Exp. Therm. Fluid Sci.* *71* 57-69

[7] Chai L, Xia G D and Wang H Sh 2016 *Int. J. Heat Mass Transfer* **97** 1081-1090

[8] Chong D T, Liu J P and Yan J J 2011 *Energy Convers. Manage* **52** 2272-2281

Some real, periodic stationary solutions of the one-dimensional nonlinear Schrödinger equation for constant potentials

G Torres-Vega
Physics Department, Cinvestav, Apdo. postal 14-740, 07000 Mexico city, Mexico.

E-mail: gabino@fis.cinvestav.mx

Abstract. We introduce a nonsymmetric version of Jacobian elliptic functions. With these functions, we are able to obtain linear superposition solutions of the non-linear Schrödinger equation for a free condensate. These functions are more general than that of Jacobi and contain them as a special case. The eigenfunctions for a condensate in a box are considered.

1. Introduction

Since the Gross-Pitaevskii equation (GPE)[1,2] is widely accepted as a valid model for the dynamics of the Bose-Einstein condensate at $T{\sim}0\$$ K, as well as for other systems, the properties of this equation are of interest, in particular the finding of its eigenstates.

For a free condensate, besides the ground state[3] and Zakharov's solutions[4], the solutions are known in terms of Jacobi's elliptic functions[5,6] and of hyperbolic functions and they have been used in order to find the eigenstates of the particle in free space, inside a box and in a ring[7,8].

It was also found a way to linearly superpose Jacobi's elliptic functions by adding constant terms to their arguments[9].

In this paper, we introduce a three parameters elliptic-type functions which allows us to deal with the linear superposition of the usual Jacobi's elliptic functions, and then find additional steady solutions for nonlinear quantum systems.

In Jacobi's elliptic function sn(u), the nonlinearity is found as the assignment sn(u)=sin(θ) with the relation between the variables u and θ given as $u = \int_0^\theta dx/\sqrt{1 - m\, sin^2(x)}$. In this paper we explore the use of other change of variable. In section 2, we use another relationship between u and θ and obtain additional solutions to the nonlinear Schrödinger equation, real, periodic ones and which contain the usual Jacobi's functions as special cases.

With these functions, one can find additional eigenstates for some quantum systems. In Sec. 3, we use the functions introduced in this paper in order to find some eigenstates for the nonlinear quantum condensate in the infinite line.

2. A set of nonlinear functions

Let us consider the change of variable from θ to u defined by the Jacobian

$$dna(u) := \frac{d\theta}{du} = \sqrt{1 + \frac{\alpha}{2}(A^2 - B^2)\cos(2\theta) + \alpha\, A\, B \sin(2\theta)}, \tag{1}$$

Where $\alpha, A, B \in$ are real, $|\alpha| < 4|AB|/(A^2 + B^2)^2$. Thus, the relationship between θ and u is

Content from this work may be used under the terms of the Creative Commons Attribution 3.0 licence. Any further distribution of this work must maintain attribution to the author(s) and the title of the work, journal citation and DOI.
Published under licence by IOP Publishing Ltd

IOP Conf. Series: Journal of Physics: Conf. Series **792** (2017) 012054 doi:10.1088/1742-6596/792/1/012054

$$u = \int_0^\theta \frac{d\theta}{\sqrt{1+\frac{\alpha}{2}(A^2-B^2)\cos(2\theta)+\alpha\, A\, B \sin(2\theta)}}, \tag{2}$$

and we define the nonlinear functions

$$sna(u) := A\sin(\theta) - B\,cos(\theta), \quad cna(u) := A\cos(\theta) + B\,sin(\theta). \tag{3}$$

As with the trigonometric and elliptic functions, let us also consider the ratios between the above functions,

$$osa(u) := \frac{1}{sna(u)}, \quad oca(u) := \frac{1}{cna(u)}, \quad oda(u) := \frac{1}{dna(u)}, \tag{4}$$
$$csa(u) := \frac{cna(u)}{sna(u)}, \quad sca(u) := \frac{sna(u)}{cna(u)}, \quad dsa(u) := \frac{dna(u)}{sna(u)}, \tag{5}$$
$$dca(u) := \frac{dna(u)}{cna(u)}, \quad sda(u) := \frac{sna(u)}{dna(u)}, \quad cda(u) := \frac{cna(u)}{dna(u)}, \tag{6}$$

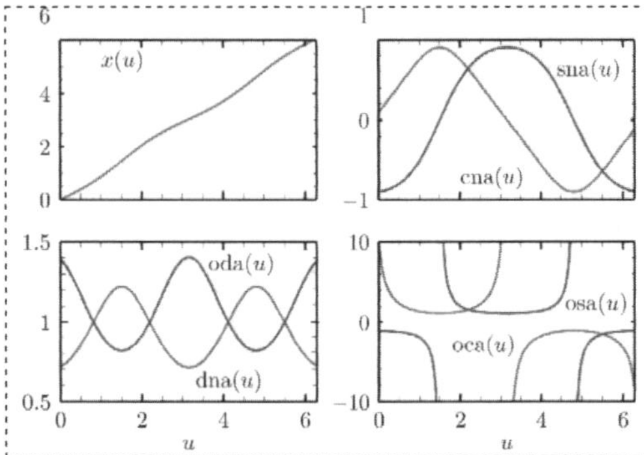

Figure 1. Plots of the nonlinear functions for A=0.1, B=0.9 and α=1.2. Note that the functions cna and sna have different shapes and, thus, they are not just the other shifted by some amount.

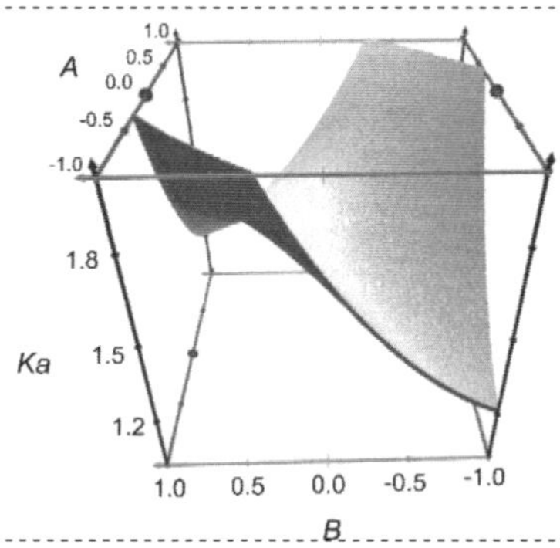

Figure 2. Some of the values of nonlinear quarter period $Ka(\alpha, A, B)$, for $\alpha = 1.2$.

Some plots of these functions are shown in Fig. 1. Note that the $cna(u)$ function is not just the shifed version of $sna(u)$, and viceversa, as is the case of the trigonometric functions $cos(u)$ and $sin(u)$ functions which have the same shape but are shifted by $\pi/2$.

Quarter period of these functions is defined as

$$Ka(\alpha, A, B) = \int_0^{\pi/2} \frac{dt}{\sqrt{1+\frac{\alpha}{2}(A^2-B^2)\cos(2t)+\alpha\, A\, B \sin(2t)}}, \tag{7}$$

A plot of $Ka(\alpha, A, B)$ can be found in Fig. 2 for $\alpha = 1.2$.

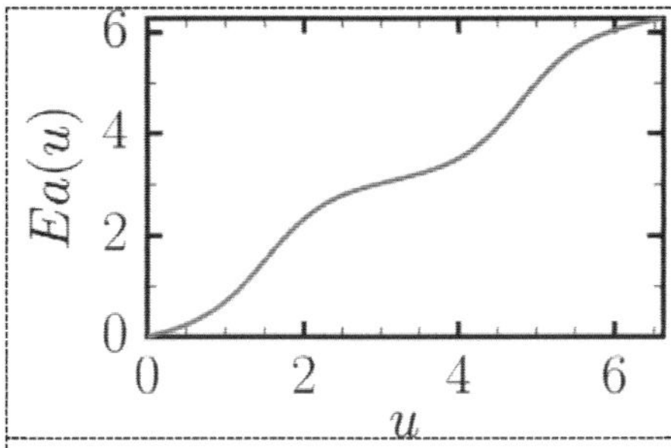

Figure 3. Plot of $Ea(u)$ for A=0.1, B=0.9 and α=1.2.

We also introduce the integral

$$Ea(u) = \int_0^u dw\, dna^2(w)$$

which resembles Jacobi's Elliptic integral of the second kind. This function is shown in Fig. 3 for a set of values of the parameters.

In next subsections, some algebraic properties of these functions are derived.

2.1. Algebraic properties

Some algebraic relationships between the above functions are

$$A^2 + B^2 = sna^2(u) + cna^2(u), \quad dna^2(u) = 1 - \frac{\alpha}{2}(sna^2(u) - cna^2(u)), \tag{8}$$

$$dna^2(u) = 1 - \frac{\alpha}{2}(A^2 + B^2) + \alpha\, cna^2(u) = 1 + \frac{\alpha}{2}(A^2 + B^2) - \alpha\, sna^2(u), \tag{9}$$

$$sda^2(u) = (A^2 + B^2)\, oda^2(u) - cda^2(u), \tag{10}$$

$$1 - oda^2(u) = \frac{\alpha}{2}[\, cda^2(u) - sda^2(u)], \tag{11}$$

$$1 + \alpha\, sda^2(u) = [1 + \frac{\alpha}{2}(A^2 + B^2)]oda^2(u), \tag{12}$$

$$1 - \alpha\, cda^2(u) = [1 - \frac{\alpha}{2}(A^2 + B^2)]oda^2(u), \tag{13}$$

$$sca^2(u) = (A^2 + B^2)\, oca^2(u) - 1, \quad dca^2(u) = \left[1 - \frac{\alpha}{2}(A^2 + B^2)\right] oca^2(u) + \alpha, \tag{14}$$

$$csa^2(u) = (A^2 + B^2)\, osa^2(u) - 1. \tag{15}$$

2.2. Derivatives

We now find the derivatives of the functions introduced earlier.

$$sna'(u) = cna(u)\, dna(u), \quad cna'(u) = -sna(u)\, dna(u), \tag{16}$$

$$dna'(u) = -\alpha\, cna(u)\, sna(u), \quad osa'(u) = -cna(u)\, dna(u) osa^2(u), \tag{17}$$

$$oca'(u) = sna(u)\, dna(u) oca^2(u), \quad oda'(u) = \alpha\, cna(u)\, sna(u) oda^2(u). \tag{18}$$

2.3. Eliminant equations

The eliminant equations, also known as energy or Liapunov functions are

$$[sna'(u)]^2 + \left[1 + \frac{3}{2}\alpha(A^2 + B^2)\right] sna^2(u) - \alpha\, sna^4(u) = (A^2 + B^2)\left[1 + \frac{\alpha}{2}(A^2 + B^2)\right], \tag{19}$$

$$[cna'(u)]^2 + \left[1 - \frac{3}{2}\alpha(A^2 + B^2)\right] cna^2(u) + \alpha\, cna^4(u) = (A^2 + B^2)\left[1 - \frac{\alpha}{2}(A^2 + B^2)\right], \tag{20}$$

$$[dna'(u)]^2 - 2\, dna^2(u) + dna^4(u) = \frac{\alpha^2}{4}(A^2 + B^2)^2 - 1, \tag{21}$$

$$[osa'(u)]^2 + \left[1 + \frac{3}{2}\alpha(A^2 + B^2)\right] osa^2(u) - (A^2 + B^2)\left[1 + \frac{\alpha}{2}(A^2 + B^2)\right] osa^4(u) = \alpha, \tag{22}$$

$$[oca'(u)]^2 + \left[1 - \frac{3}{2}\alpha(A^2 + B^2)\right] oca^2(u) - (A^2 + B^2)\left[1 - \frac{\alpha}{2}(A^2 + B^2)\right] oca^4(u) = -\alpha, \tag{23}$$

$$[oda'(u)]^2 - 2\, oda^2(u) + \left[1 - \frac{\alpha^2}{4}(A^2 + B^2)^2\right] oda^4(u) = -1. \tag{24}$$

2.4. Differential equations

Second derivatives of the functions lead to the differential equations for them. For sna, cna and dna we have that

$$sna''(u) + \left[1 + \frac{3}{2}\alpha(A^2 + B^2)\right] sna(u) - 2\alpha\, sna^3(u) = 0, \tag{25}$$

$$cna''(u) + \left[1 - \frac{3}{2}\alpha(A^2 + B^2)\right] cna(u) + 2\alpha\, cna^3(u) = 0, \tag{26}$$

$$dna''(u) + 2\, dna(u)[dna^2(u) - 1] = 0, \tag{27}$$

The other functions, osa, oca and oda, are also solutions of similar differential equations, according to the following lemma,

Lemma. If $f(u)$ is a function which satisfy the nonlinear differential equation

$$f''(u) + E\, f(u) - 2C\, f^3(u) = 0, \tag{28}$$

and the derivative of $f(u)$ is such that

$$[f'(u)^2] - C\, f^4(u) + \frac{1}{2}(E - A)f^2(u) - \frac{B}{2} = 0, \tag{29}$$

with E and C real constants, then $1/f(u)$ also satisfy a similar nonlinear equation,

$$\frac{d^2}{du^2}\frac{1}{f(u)} - \frac{A}{f(u)} - \frac{B}{f^3(u)} = 0. \tag{30}$$

Then, the other functions, namely osa, oca and oda will also satisfy a similar differential equation,

$$osa''(u) + \left[1 + \frac{3}{2}\alpha(A^2 + B^2)\right] osa(u) - 2(A^2 + B^2)\left[1 + \frac{\alpha}{2}(A^2 + B^2)\right] osa^3(u) = 0, \tag{31}$$

$$oca''(u) + \left[1 - \frac{3}{2}\alpha(A^2 + B^2)\right] oca(u) - 2(A^2 + B^2)\left[1 - \frac{\alpha}{2}(A^2 + B^2)\right] oca^3(u) = 0, \tag{32}$$

$$oda''(u) - 2\, oda(u) + 2\left[1 - \frac{\alpha^2}{4}(A^2 + B^2)^2\right] oda^3(u) = 0, \tag{33}$$

The derivatives of the inverse functions are

$$\frac{d}{dy}sna^{-1}(y) = \frac{\pm 1}{\sqrt{(A^2+B^2-y^2)\left(1+\frac{\alpha}{2}(A^2+B^2)-\alpha y^2\right)}}, \tag{34}$$

$$\frac{d}{dy}cna^{-1}(y) = \frac{\pm 1}{\sqrt{(A^2+B^2-y^2)\left(1-\frac{\alpha}{2}(A^2+B^2)+\alpha y^2\right)}}, \tag{35}$$

$$\frac{d}{dy}dna^{-1}(y) = \frac{\pm 1}{\sqrt{\left[1+\frac{\alpha}{2}(A^2+B^2)-y^2\right]\left[y^2-1+\frac{\alpha}{2}(A^2+B^2)\right]}}, \tag{36}$$

$$\frac{d}{dy}osa^{-1}(y) = \frac{\pm 1}{\sqrt{[(A^2+B^2)y^2-1]\left[\left(1+\frac{\alpha}{2}(A^2+B^2)\right)y^2-\alpha\right]}}, \tag{37}$$

$$\frac{d}{dy}oca^{-1}(y) = \frac{\pm 1}{\sqrt{[(A^2+B^2)y^2-1]\left[\left(1-\frac{\alpha}{2}(A^2+B^2)\right)y^2+\alpha\right]}}, \tag{38}$$

$$\frac{d}{dy}oda^{-1}(y) = \frac{\pm 1}{\sqrt{\left[\left(1+\frac{\alpha}{2}(A^2+B^2)\right)y^2-1\right]\left[1-\left(1-\frac{\alpha}{2}(A^2+B^2)\right)y^2\right]}}, \tag{39}$$

Then, as expected, we can see that these functions also invert the same integrals that Jacobi's functions invert.

This is the minimum set of properties which will allow us to find some eigenstates of the one-dimensional nonlinear Schrödinger equation for a free condensate.

3. Condensate in an infinite or periodic medium

As an application of the use of the functions introduced in the previous section, we find some real, periodic solutions of the non-linear Schrödinger equation for a condensate in an infinite medium and a constant potential.

The one-dimensional nonlinear time-dependent Schrödinger equation for a condensate in a constant potential is written as

$$-2i\frac{\partial\, \psi(u;t)}{\partial u} = \frac{\partial^2\, \psi(u;t)}{\partial u^2} - 2(V + \alpha|\psi(u;t)|^2)\psi(u;t), \tag{40}$$

where t, u, $\psi(u;t)$ and $\psi(u;t)$ are dimensionless quantities, with scaling factors $t_s = ML^2/\hbar$ for time, $V_s = \hbar^2/ML^2$ for energy. The length is scaled by L, a characteristic length of the system of interest, and the normalization factor for the wave function is A_s. $\psi(u;t)$ is the wave function for the Bose-Einstein condensate (BEC), $\alpha = ML^2A^2NU_0/\hbar^2$, M is the mass of a single atom, N is the number of atoms in the condensate, $U_0 = 4\pi\hbar^2 a/M$ characterizes the atom-atom interaction, and a is the scattering length.

The functions introduced in this work can be used to find stationary states for the infinite line or for a system which is periodic like a particle moving on a ring. For the ring, the condition is that the length of the ring should be a multiple of the period of the nonsolitonc functions which is $4Ka(\alpha, A, B)$ For a solitonic function we can have symmetry-breaking eigenstates.[4,5]

As with the other solutions,[7] there are states which can travel without distortion; the soliton-type states. Their time dependency is given by

$$\psi(u;t) = k\, e^{\pm iu\sqrt{-k^2(1+3\alpha(A^2+B^2)/2)-2V}} sna\left[k\left(u \mp t\sqrt{-k^2\left(1 + \frac{3\alpha(A^2+B^2)}{2}\right) - 2V}\right)\right], \tag{41}$$

for $V < -k^2(1 + 3\alpha(A^2 + B^2)/2)/2$. Another soliton-type solution is

$$\psi(u;t) = \frac{k}{\sqrt{-\alpha}}\, e^{i\theta \pm iu\sqrt{2(k^2-V)}} dna\left[k(u \mp t\sqrt{2(k^2 - V)})\right], \tag{42}$$

valid when $k^2 > V$. A third soliton-type solution is provided by the oca function,

$$\psi(u;t) = \frac{k}{\sqrt{\alpha}}\sqrt{(A^2 + B^2)\left(1 - \frac{\alpha}{2}(A^2 + B^2)\right)}\; e^{\pm iu\sqrt{-k^2\left(1-\frac{3\alpha(A^2+B^2)}{2}\right)-2V}}$$
$$\times\; oca\left[k\left(u \mp t\sqrt{-k^2\left(1 - \frac{3\alpha(A^2+B^2)}{2}\right) - 2V}\right)\right], \tag{43}$$

with $\alpha > 0$, $2 > \alpha(A^2 + B^2)$, and $2V < -k^4(1 - 3\alpha(A^2 + B^2)/2)$. There is also a soliton type solution involving the function osa,

$$\psi(u;t) = \frac{k}{\sqrt{\alpha}}\sqrt{(A^2 + B^2)\left(1 + \frac{\alpha}{2}(A^2 + B^2)\right)}\; e^{\pm iu\sqrt{-k^2(1+3\alpha(A^2+B^2)/2)-2V}}$$
$$\times\; osa\left[k\left(u \mp t\sqrt{-k^2\left(1 + \frac{3\alpha(A^2+B^2)}{2}\right) - 2V}\right)\right], \tag{44}$$

This solution is valid if $\alpha < -2(k^2 + 2V)/3k^2(A^2 + B^2)$. And there is a last soliton-type solution given by

$$\psi(u;t) = \pm\frac{k}{\sqrt{\alpha}}\sqrt{\frac{\alpha^2}{4}(A^2+B^2)^2-1}\,e^{\pm iu\sqrt{2(k^2-V)}}\,oda\big[k\big(u\mp t\sqrt{2(k^2-V)}\big)\big], \tag{45}$$

where $\sqrt{-k^2(1+3\alpha(A^2+B^2)/2)-2V}$, $\sqrt{2(k^2-V)}$, $\sqrt{-k^2(1-3\alpha(A^2+B^2)/2)-2V}$, $\sqrt{-k^2(1+3\alpha(A^2+B^2)/2)-2V}$ and $\sqrt{2(k^2-V)}$ are the corresponding velocities of these waves.

Remarks

We have found that one can obtain additional solutions of the nonlinear Schrödinger equation, with a constant potential, by using other change of coordinate than the one used in Jacobi's elliptic functions.

An advantaje of these functions is that they facilitate the solving of boundary problems for a free Bose-Einstein condensate. Then, we can have solutions for the same type of systems that are found in the linear case.

As is the case with Jacobi's functions, the set of functions we have introduced are interpolations functions between the linear solutions (trigonometric functions) and purely nonlinear (hyperbolic and trigonometric times a complex factor) functions.

An inconvenience of these solutions is that they do not carry any momentum at all. Another one is that they do not describe the decay of the wave function. These behaviours are found in the linear system and we expect that something similar can be found in the nonlinear case. We are currently looking for the appropriate changes of variables for those situations, as well as for solutions in higher dimensions.

References

[1] Pitaevskii L 1961 *Sov. Phys. JETP* **13** 451
[2] Gross E 1961 *Nuovo Cimento* **20** 454
[3] Sulem C and Sulem PL 1999 *The Nonlinear Schrödinger Equation. Self-focusing and Wave Collapse (Applied Mathematical Sciences* vol 139*)* (New York: Springer)
[4] Zakharov VE and Shabat AB 1972 *Sov. Phys. JETP* **34** 62
[5] Cayley A 1961 *An elementary treatise on Elliptic functions. Second edition* (New York: Dover Publications, Inc.)
[6] Abramowitz M and Stegun IA 1964 *Handbook of Mathematical Functions* (New York: Dover Publications, Inc.)
[7] Carr LD, Charles WC and Reinhardt WP 2000 *Phys. Rev. A* **62** 063610
[8] Carr LD, Charles WC and Reinhardt WP 2000 *Phys. Rev. A* **62** 063611

Design and Construction of an ultrasound transmit pulser system for non-destructive testing

E Alonso[1], S Solis-Najera[2], F Vazquez[2] and L Medina[2]
[1]Posgrado en Ingeniería, Universidad Nacional Autónoma de México
[2] Depto. de Física, Facultad de Ciencias, Universidad Nacional Autónoma de México

E-mail: lucia.medina@ciencias.unam.mx

Abstract. The design and construction of an ultrasonic pulser is part of a multichannel ultrasonic system, for the analysis, monitoring and characterization of heterogeneous and highly dispersive materials. The pulser has been built on the FPGA platform and allowed to modify the pulse repetition rate, the pulse width and the number of pulses of a burst. This open architecture is a basic module for a more complex ultrasonic pulser, such as arbitrary pulse generator. The results show the ultrasonic transducer responses under the above parameters.

1. Introduction

The acoustic wave propagation throughout the heterogeneous and highly dispersive media is very complex, due to scattering, dispersion and mode change of the energy, causing the phase velocity and attenuation coefficient becomes frequency dependent [1-3]. In a non destructive testing, the received ultrasonic signals, that contain information about microstructure and embedded defects of the tested media, are an attenuated, slower down and shifted version of the transmit pulse. In order to enhance the defect information from the microstructural noise it is necessary to design the transmit pulse that able to work with the 6-dB bandwidth of the transducer, and also with the power of the excitation pulse. These can be done by varying the rectangular pulse width and designing a train of pulses, respectively.

The ultrasonic pulser proposed in this work is the basic excitation module of a multichannel system for non-destructive testing. It is based on FPGA architecture, is a single channel with configurable parameters such as pulse width, train of pulses (burst), pulse repetition rate, to avoid second-fire echoes, and the delay of the excitation pulse. The latter is for the multichannel implementation of beamforming, to focus the energy to a certain region depending on the user needs. The design of this open architecture is based on rectangular pulses as many commercial systems work, such as US Ultratek [4], Advanced OEM Solutions [5], Ultrasonix [6], among o

[1] lucia.medina@ciencias.unam.mx

Content from this work may be used under the terms of the Creative Commons Attribution 3.0 licence. Any further distribution of this work must maintain attribution to the author(s) and the title of the work, journal citation and DOI.
Published under licence by IOP Publishing Ltd

IOP Conf. Series: Journal of Physics: Conf. Series **792** (2017) 012055 doi:10.1088/1742-6596/792/1/012055

2. The ultrasonic pulse generator system

A single channel ultrasonic pulser generation system is shown in Figure 1. The system is composed of the main pulse generation module, voltage driver and ultrasonic transducer. The pulse generation module is responsible for rectangular excitation signals and it is controlled using USB connection. It has been designed and built on a FPGA Xilinx Spartan-6 (model LX45) architecture, the main structure of the Digilent Atlys development module [7].

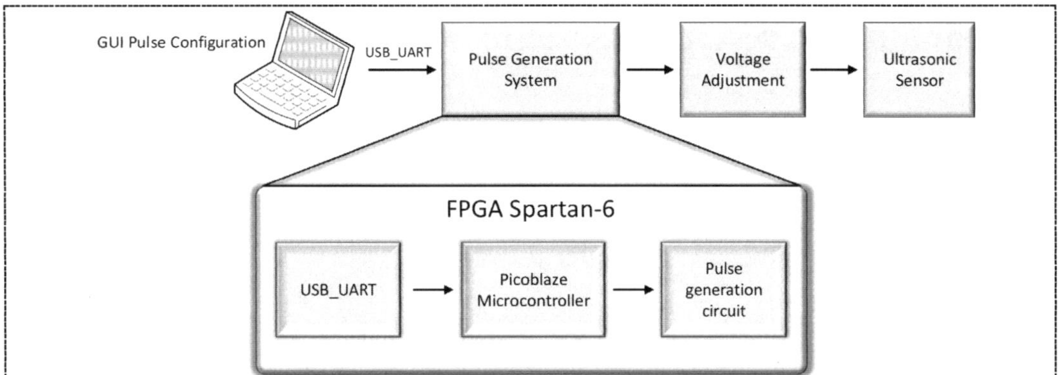

Figure 1 Ultrasonic pulse generation system

The FPGA Spartan 6 has an embedded Picoblaze microcontroller that is in charge of reading the input data, from the user via the UART and sending to the generation pulse module. Once a specific transmit pulse is built and voltage adjusted is fitted for the ultrasonic transducer to generate the transmit pulse.

2.1 Pulse Generation Module

This module is composed by the control architecture and pulse generation circuit. The control architecture sends the bytes coming for the microcontroller to each of the internal configuration registers of the pulse generation circuit through an 8-bit bus. This operation has been done in assembly language based on the flow chart shown in Figure 2.

2.1.1 Pulse Generation Circuit

The circuit has been developed on a Finite State Machine with Datapath (FSMD). This circuit has five configuration registers; one register for the time delay, other for the operational frequency, a third is for the number of pulses, the fourth and fifth are dedicated to the burst and the pulse repetition rate, respectively. The FSMD load the initial parameters values in the registers, and wait for the enable signal. If a time delay is greater than zero, the machine is on hold until the value of the parameter is reached, then a logic "1" is generated in the circuit output followed by a logic "0". The time taken between this to finite machine states define the frequency given as initial parameter. This process is repeated depending on the input value respect to the number of pulses. Once the pulses are generated, the FSMD must halt long enough to fulfill the requirement of the pulse repetition rate, as shown in Figure 3. This can be an interactive process depending on the initial parameters given by the user.

2.2 Voltage Adjustment

The amplitude of the FPGA output signal is 3.3 V, that for non-destructive testing of heterogeneous media is very low. Therefore, the digital output data of the pulse generation circuit is voltage adjusted by an optocoupler 6N137, which amplified the signal up to 5V (TTL level), and then it is transfer to a voltage driver SN75472 which produces an amplified ultrasonic pulse in the range of 12-50 V. (see Figure 4). The excitation amplitude is limited by the transducer construction and the electronics capabilities. At this stage this procedure is manually operated, however we are working on automatic amplification to control the gain of the output voltage.

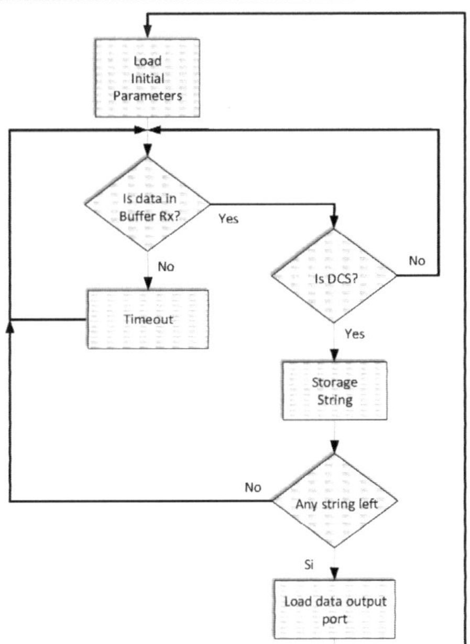

Figure 2. Flow of the input parameters from the Picoblaze microcontroller to the pulse generation circuit

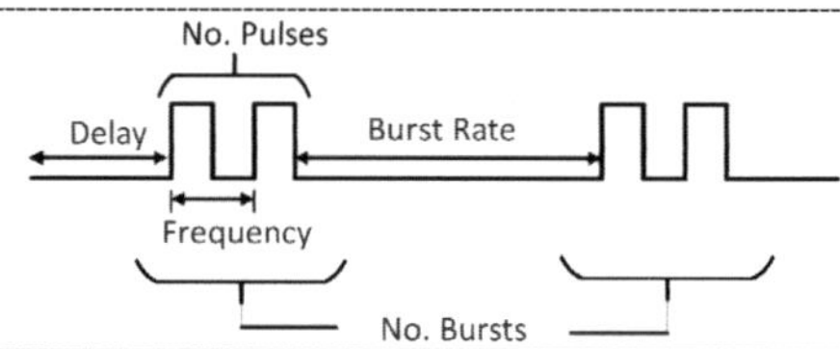

Figure 3. Ultrasonic pulse parameters.

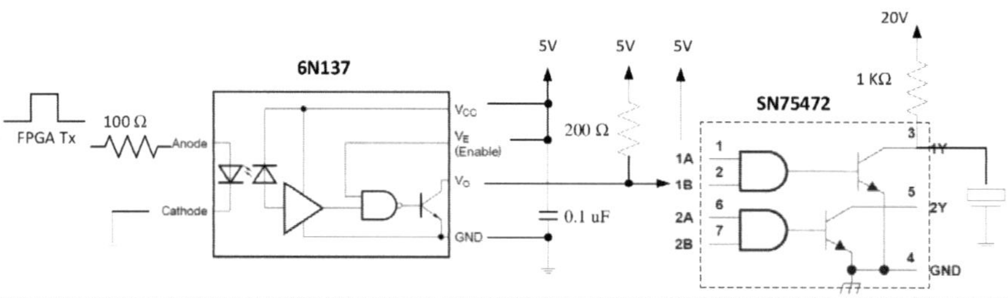

Figure 4. Voltage adjustment circuit: optocoupler 6N137 and voltage driver SN75472.

3. Control software

The user Graphic Interface (GUI), developed in Python platform, allow the input of the transmit pulse parameters by the user, as shown in Figure 5. As you can see it is not restricted to the operating frequency of the transducer, the user can choose any frequency of interest depending on the application, also this open architecture able to decide whether to use a single pulse or a train of pulses and the pulse repetition rate, which depend on the tested material elastic parameters and dimensions.

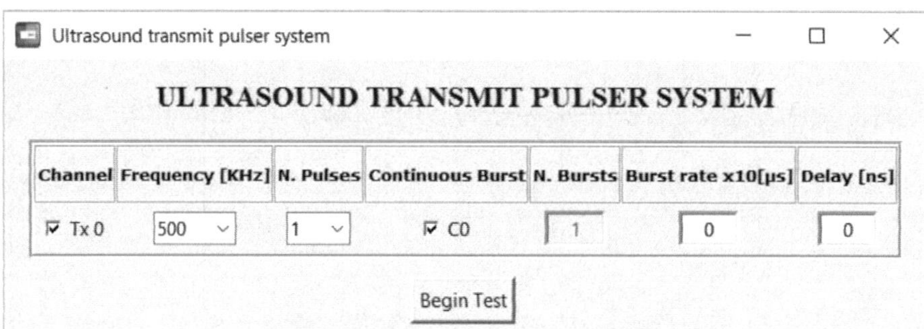

Figure 5. Single-channel GUI.

4. Results and future work

The ultrasonic transmission module has been implemented and experimentally validated. The experimental setup in Figure 6 is an immersion through transmission ultrasound, the distance between the transmitter and the receiver is 45 cm. This configuration has been chosen under the following considerations: the slowest tested media is water, therefore the time taken for the wave to travel is the longest time we going to study, that give us the maximum pulse repetition rate, the maximum dimension of the tested material is 30 cm, this can be related to the minimum power needed of the transmit pulse.

The results, shown from Figure 7a to Figure 9b, highlight the responses of the transmitter and receiver transducers of several pulses configuration.

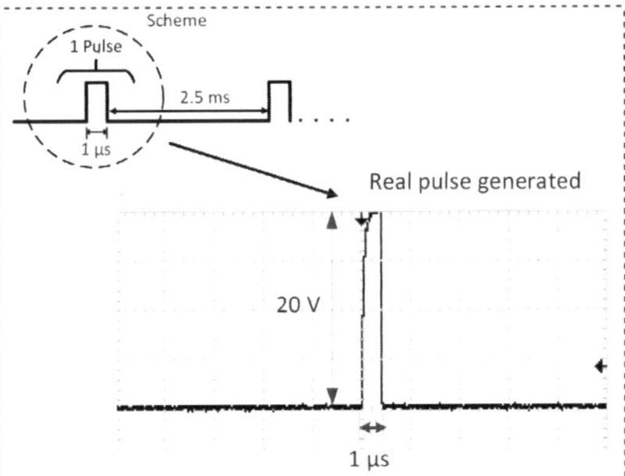

Figure 7. (a) Pulse parameters, given by the user (upper), and the transmitter response (lower).

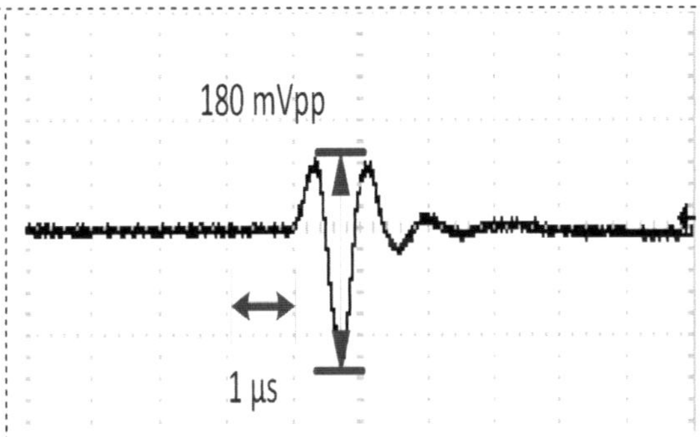

Figure 8. (b) The received signal for a single emitted pulse.

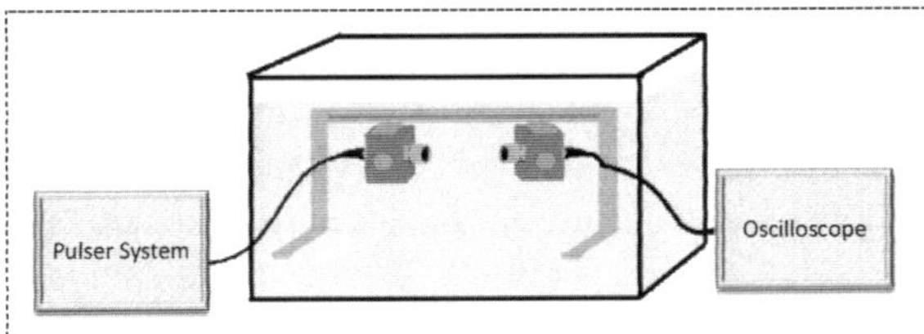

Figure 9. Immersion through transmission testing. The transmit system is connected to a transmitter and an oscilloscope acquired the received transducer response.

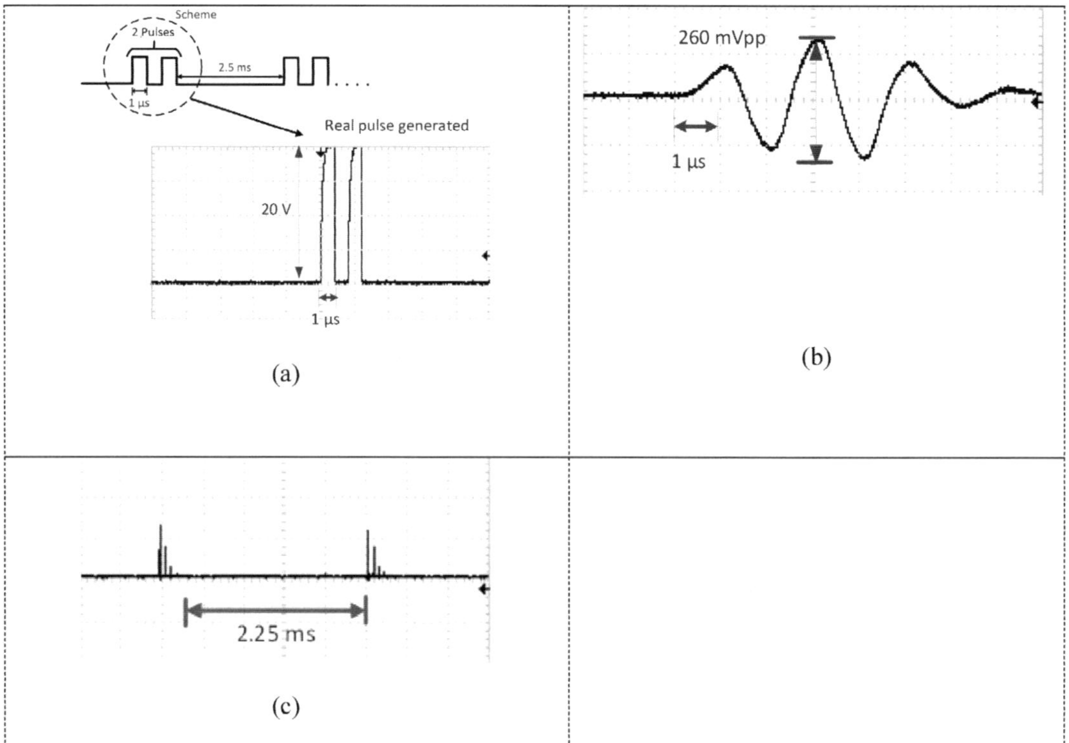

Figure 8. Two pulses generation for a given width and pulse repetition rate. (a) Pulse parameters, given by the user (upper), and the transmitter response (lower). (b) The received signal for a two consecutive emitted pulses. (c) Output of two burst fired a 2.25ms apart.

As can be seen from the results a single pulse produces a narrow ultrasonic signal with low amplitude, as the number of pulses increases the width of the detected pulse and its power increases.
This open system give us the possibility of study an heterogeneous material from several points of view, for a narrow pulses at different frequencies able to highlight microfractures embedded in granular noise, on the other hand broader transmit pulses can give information about the effective elastic properties of the tested media.

It is worth to mention that the ultrasonic pulser is not restricted to the immersion through transmission experimental setup, it can also be used for contact through transmission system. Also the lower and maximum frequencies are 127 KHz and 5 MHz, respectively. This single channel pulser is the basis for a multichannel ultrasonic system, the future work leads to have an 8 independent transmitter channels and 8-receivers to generate a 2D image of the tested material, including an arbitrary signal generator.

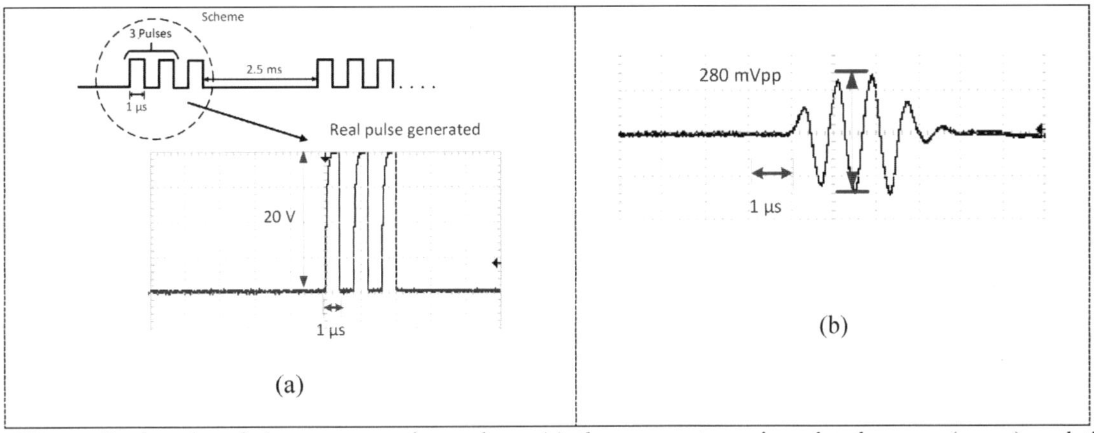

Figure 9. A train of three consecutive pulses: (a) the parameters given by the user (upper) and the transmitter response (lower). (b) The received signal due to a burst under the characteristics of (a).

5. Reference

[1] Waterman P C and Truell R 1961 Multiple scattering of waves *J. Math. Phys.* 2(4) 512-537

[2] Yang R B A 2003 Dynamic generalized self-consistent model for wave propagation in particulate composites. *J. Appl. Mech.* 70(4), 575-582, 2003.

[3] Kim J Y 2010 Models for wave propagation in two-dimensional random composites: A comparative study. The Journal of the Acoustical Society of America 127(4) 2201-2209

[4] Holger J H, Welsch H J, Fonfara H and Tretbar S H 2012 Highly scalable and flexible FPGA based platform for advanced ultrasound research 2012 IEEE International Ultrasonics Symposium Proceedings pp. 2075-2080

Acknowledgments

AcknowledgmentThe system was design and built under the financial support of DGAPA-UNAM (IT102116) and E. Alonso would like to thank CONACyT for the Ph.D. stipend.

IOP Conf. Series: Journal of Physics: Conf. Series **792** (2017) 012056 doi:10.1088/1742-6596/792/1/012056

Numerical and semi-analytical solutions of the compound layer growth kinetics in cylindrical surfaces during plasma nitriding of pure iron

E M Hernández, F Castillo, R D Santiago, R Martínez, J A Otero, J E Oseguera and A Jiménez

Tecnológico de Monterrey, Campus Estado de México, Atizapán de Zaragoza, 52926 Edo. de Mex., Mexico.

E-mail: francast@itesm.mx

Abstract. Numerical and approximate analytical solutions for the compact layer growth kinetics in a pure iron solid cylinder, during a plasma nitriding process, are obtained. The numerical simulations of the model are performed by using a front tracking finite difference scheme and the heat balance integral method. We propose a model where the main assumption is to consider a diffusion zone of constant thickness, which is solved for cylindrical symmetries. We present results for the time evolution of the compound layer, where the obtained solutions from both methods are consistent with the expected behaviour in the asymptotic time limit and experimental data from other authors.

1. Introduction

Nitriding thermochemical treatments in different kind of steels are broadly used in industry because of their numerous applications [1, 2]. Nitriding can be performed with gas containing ammonia or cyanide salts and also through weakly ionized plasma [3]. The nitrogen flux from the surface into the bulk, carries allotropic transformations, which notably enhance the mechanical and chemical properties of piece. The combination of a compact nitride layer, followed by a diffusion zone in ferrite, produces an important improvement in wear resistance, a significant increase in hardness and in many cases, progression of fatigue resistance. Plasma assisted thermochemical treatments allow diffusion to occur at low temperatures compared to other processes [1-3], so that very low distortions are developed in pieces. In addition, these treatments produce no pollutants.

Modeling the concomitant growth of compact layers of nitrides in the Fe-N_2 thermodynamic system has been targeted by several authors [4-7]. An important aspect of the mathematical modelling of nitriding processes, is the geometry of the piece under treatment [8, 9]. Many papers are devoted only to the modeling of layer growth kinetics over plane surfaces; therefore, this work is focused on the study of nitride layer growth on cylindrical symmetries. Considering mass balance at each interface, we have performed a numerical simulation for the concomitant growth of compact nitride layers during plasma nitriding within a solid cylinder of pure iron. We will show that for these kind of geometries, the solutions obtained from the finite difference scheme (FDS) and heat balance integral method (HBIM), are able to capture the physical behavior of the process.

Content from this work may be used under the terms of the Creative Commons Attribution 3.0 licence. Any further distribution of this work must maintain attribution to the author(s) and the title of the work, journal citation and DOI.
Published under licence by IOP Publishing Ltd

2. Problem statement

2.1 Mathematical model

We will consider an iron sample with a cylindrical cross section where nitrogen gas has penetrated in such a way that at time t_0, the ε, γ' and α phases are already formed. The surface concentration is constant and equal to $C_\varepsilon^{\max}$ and the boundary conditions at each interface are determined through the Fe-N_2 phase diagram. The surface concentration of nitrogen located at $x = R$ is denoted as $C_\varepsilon^{\max}$, where R is the radius of the sample. Therefore, the boundary conditions in cylindrical coordinates are given by

$$C_1(R,t) = C_\varepsilon^{\max},\ C_1(R-\xi_\varepsilon,t) = C_\varepsilon^{\min}, \tag{1}$$

at the ε phase, where ξ_ε is the thickness of the ε layer. Furthermore, with $\xi_{\gamma'}$ and ξ_α defined as the γ' and diffusion zone thicknesses respectively, the boundary conditions for each layer are given by

$$C_2(R-\xi_\varepsilon,t) = C_{\gamma'}^{\max},\ C_2(R-\xi_\varepsilon-\xi_{\gamma'},t) = C_{\gamma'}^{\min}, \tag{2}$$

and

$$C_3(R-\xi_\varepsilon-\xi_{\gamma'},t) = C_\alpha^{\max},\ C_3(R-\xi_\varepsilon-\xi_{\gamma'}-\xi_\alpha,t) = C_\alpha^{\min}. \tag{3}$$

Nitrogen transport within iron is assumed to be of Fickian nature. Consequently, classical diffusion equations are used to describe mass transport phenomena, and due to the symmetry of the system, the diffusion equations are solved in cylindrical coordinates. Interface motion is driven by the flux jump through the following Stefan condition

$$\bar{C}_{i,i+1}\frac{dL_i}{dt} = D_{i+1}\frac{\partial C_{i+1}(x,t)}{\partial x}\Big|_{x=L_i} - D_i\frac{\partial C_i(x,t)}{\partial x}\Big|_{x=L_i},\ i = 1,2 \tag{4}$$

where the constants $\bar{C}_{i,i+1}$ represent the concentration jump at each interface, in such a way that for $i=1$, $\bar{C}_{1,2} = C_\varepsilon^{\min} - C_{\gamma'}^{\max}$, $D_1 = D_\varepsilon$ and $D_2 = D_{\gamma'}$. For $i=2$, $\bar{C}_{2,3} = C_{\gamma'}^{\min} - C_\alpha^{\max}$ and $D_3 = D_\alpha$. Moreover, for $i=1$, L_i represents the epsilon layer thickness, as $L_1 = \xi_\varepsilon$; for $i=2$, L_i represents the compound layer thickness $L_2 = \xi_\varepsilon + \xi_{\gamma'}$, and for $i=3$, L_i represents the compound layer and diffusion zone widths $L_3 = \xi_\varepsilon + \xi_{\gamma'} + \xi_\alpha$. The main idea of this proposal, is the assumption of a diffusion zone with constant thickness. Based on this assumption, the kinetics of the interface between the α zone and pure iron matrix is only governed by the motion of the ε/γ' and γ'/α interfaces. As a result of this, the diffusion zone is being pushed by the compound layer.

2.2 Numerical solutions

A second order FDS and the HBIM are used to solve the diffusion equations, along with the Stefan condition at the ε/γ' and γ'/α interfaces given by equation (4). In order to obtain a physically viable solution from the HBIM, we test the type of concentration profiles that have been proposed for cylindrical configurations in heat conduction problems [11], such as a linear-log and parabolic-log profile. The initial parabolic profiles used to solve the diffusion equations and the Stefan condition are given by

$$C_i(x,t) = C_i^{\min} + a_i(t)(R-L_i-x) + b_i(t)\ln\left(\frac{x}{R-L_i}\right), \tag{5}$$

for $i = 1,2$. Finally, the proposed concentration profile for the diffusion zone is written as

$$C_3(x,t) = C_\alpha^{\max} + a_3(t)(R - L_2 - x) + b_3(t)\ln\left(\frac{x}{R - L_2}\right) \tag{6}$$

The initial concentration profiles are obtained by calculating the functions $b_i(t)$ from the boundary conditions described in section 2.1, and assuming an initial logarithmic profile. It was shown in [10] that for this model, the concentration profile is linear for large time values, and does not depend on the shape of the initial profile. It will be shown here, that for a cylindrical geometry this holds as well.

3. Results and Discussion

The results obtained with the FDS and the HBIM described in section 2 will be shown in this section. For the numerical simulations, the nitriding temperature used is $793\,\mathrm{K}$, effective diffusion constants at each layer are $D_\varepsilon = 5.3367\times10^{-15}\,\mathrm{m}^2/\mathrm{s}$, $D_{\gamma'} = 5.8555\times10^{-14}\,\mathrm{m}^2/\mathrm{s}$ and $D_\alpha = 4.9343\times10^{-12}\,\mathrm{m}^2/\mathrm{s}$. The concentrations at each interface are taken from the Fe-N_2 phase diagram as $C_\varepsilon^{\max} = 7.68$, $C_\varepsilon^{\min} = 7.5204$, $C_{\gamma'}^{\max} = 5.8849$, $C_{\gamma'}^{\min} = 5.7738$, $C_\alpha^{\max} = 0.0627$ and $C_\alpha^{\min} = 0.0$ in $\%\ \mathrm{wt}$. Finally, the initial thickness of each layer is assumed to be $\xi_\varepsilon = 0.0480\,\mu\mathrm{m}$, $\xi_{\gamma'} = 0.3739\,\mu\mathrm{m}$ and $\xi_\alpha = 80\,\mu\mathrm{m}$, as in [10].

All figures below, show the results obtained for each layer, by using the FDS and HBIM described in section 2. In figure 1, we show the solutions from both methods for a sample of radius $R = 0.10$ m and a diffusion zone $80\,\mu\mathrm{m}$ thick. An important consequence of this model is that for large time values, the jump in the net flux through each interface approaches zero; therefore

$$D_\varepsilon(C_\varepsilon^{\max} - C_\varepsilon^{\min})/\xi_\varepsilon = D_{\gamma'}(C_{\gamma'}^{\max} - C_{\gamma'}^{\min})/\xi_{\gamma'}, \quad D_{\gamma'}(C_{\gamma'}^{\max} - C_{\gamma'}^{\min})/\xi_{\gamma'} = D_\alpha(C_\alpha^{\max} - C_\alpha^{\min})/\xi_\alpha. \tag{7}$$

From equation (7) it is easy to obtain the thickness of each layer in the asymptotic time limit [10], as

$$\xi_{\gamma'\text{-lim}} = \frac{D_{\gamma'}\left(C_{\gamma'}^{\max} - C_{\gamma'}^{\min}\right)}{D_\alpha\left(C_\alpha^{\max} - C_\alpha^{\min}\right)}\xi_\alpha, \quad \xi_{\varepsilon\text{-lim}} = \frac{D_\varepsilon\left(C_\varepsilon^{\max} - C_\varepsilon^{\min}\right)}{D_{\gamma'}\left(C_{\gamma'}^{\max} - C_{\gamma'}^{\min}\right)}\xi_{\gamma'\text{-lim}}. \tag{8}$$

In this time limit, the net flux of mass through a point x within any layer is zero; therefore

$$\frac{D_i\left(C_i^{\max} - C_i(x)\right)}{R - x - L_{i-1}} = \frac{D_i\left(C_i(x) - C_i^{\min}\right)}{x - (R - L_i)},$$

which can be solved for the concentration at an arbitrary point within the compound layer or diffusion zone as

$$C_i(x) = \frac{C_i^{\max} - C_i^{\min}}{(L_i - L_{i-1})}(x - R) + \frac{C_i^{\max}L_i - C_i^{\min}L_{i-1}}{(L_i - L_{i-1})}, \text{ for } R - L_i < x < R - L_{i-1} \tag{9}$$

On each figure we use equation (8) to check that every solution obtained through any of the proposed methods, captures the predicted behaviour for large time values. In figures 2 and 3 the results for several values of the sample radius are shown and compared against the predicted thickness of each layer, in the asymptotic time limit. In figure 2, the results from the FDS and HBIM are shown for the ε layer. As can be observed, there is a weak dependence on the radius of the sample.

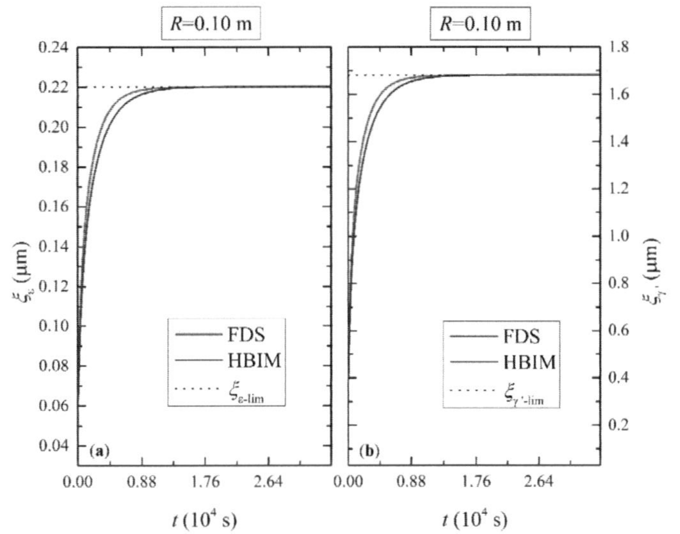

Figure 1. FDS and HBIM solutions for the time evolution of the ε and γ' layer thicknesses.

Finally, similar results for the γ' layer growth kinetics are shown in figure 3. Also in this case, the solutions obtained from the FDS and HBIM, do not depend significantly on the radius of the sample.

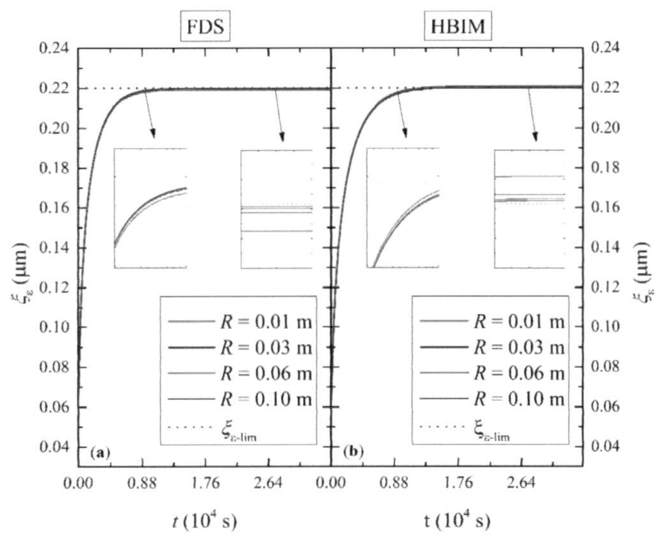

Figure 2. Time evolution of the ε layer for different values of the sample radius. a) FDS and b) HBIM solutions for a diffusion zone of $80\,\mu$m.

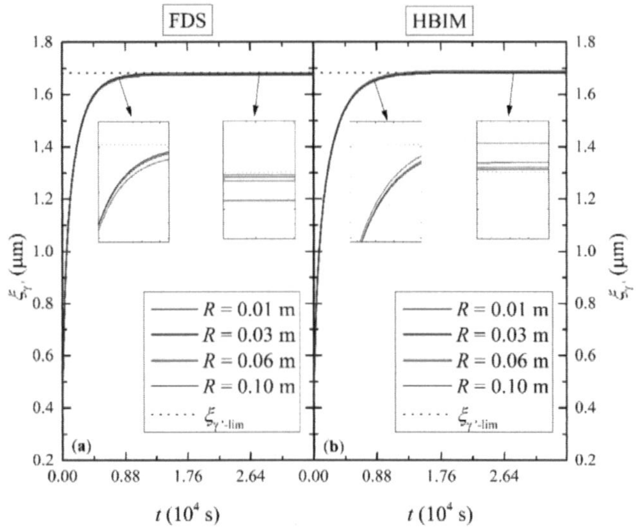

Figure 3. a) FDS and b) HBIM solutions for γ' layer growth with different values of R.

A comparison between the experimental results of [4] which are performed on planar samples and the solutions obtained for a cylindrical geometry is shown in figure 4. Each curve belongs to the solution calculated for a specific thickness of the diffusion zone. As the width of ξ_α is increased, the relative difference between adjacent curves reduces monotonically. The largest value of ξ_α belongs to the continuous line, which is shown in figure 4. For this value, the asymptotic limit is reached at $t \approx 2\times10^{10}$ s, which is 10^6 orders of magnitude larger than the total nitriding time used in the experiment of [4]. For this reason, further increase in ξ_α, produces almost identical results to the curves shown in solid lines, which correspond to the solutions used to compare with the experimental data. These numerical experiments suggest a constant thickness of the diffusion layer. Solutions of this model were obtained for planar geometries [10], which are very close to the numerical solutions obtained here for cylindrical geometries.

For small values of ξ_α, the mass flux through the diffusion zone is of the same order of magnitude than the flux through the γ' layer, which implies that α iron is quickly saturated with Nitrogen, slowing any further growth of the γ' layer. This is illustrated in figure 4, where the dashed line corresponding to the smallest value of ξ_α is observed to approach very rapidly to the asymptotic limit. In fact, once the compound layer has formed, diffusion zones are typically several orders of magnitude thicker than ε and γ' layers, because effective diffusion coefficients for the diffusion zone are two orders of magnitude larger than those corresponding to ε and γ' layers. Thus, for typical values of ξ_α, the flux through this zone is very small compared to the flux within the γ' layer, this is why the influence of mass transport within the diffusion zone is almost negligible. By considering a large diffusion zone of constant thickness, the proposed model is neglecting the effects that mass transport within the diffusion zone may have on the compound layer growth kinetics. This is shown to be a good approximation, once each layer is formed. Therefore, in order to compare the results from the proposed model, the starting point is chosen at time values where the diffusion zone is at least one order of magnitude thicker than any layer. For example, the numerical results shown in figure 4, are obtained by starting at $t = 3001.22$s, where the first data point is reported in [4] and every layer is already formed.

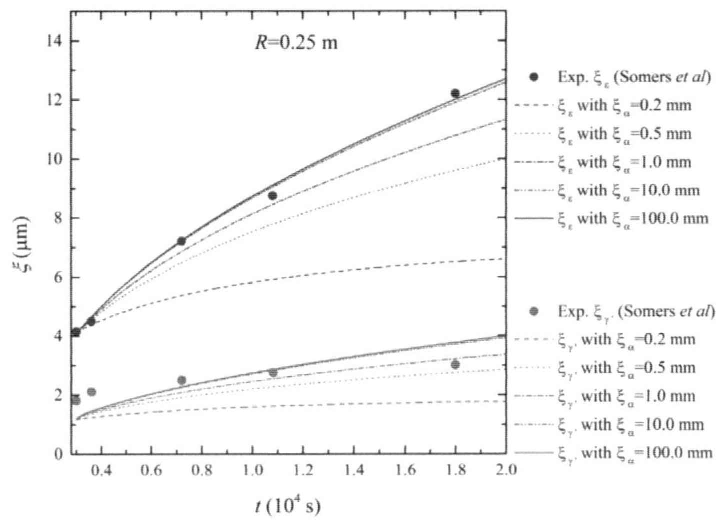

Figure 4 Solid lines correspond to the solutions obtained with the HBIM and data points from the experiment in [4] are shown in blue and red circles for each layer. The maximum relative difference between the numerical results and the experimental data is observed to be of 6.32%.

4. Conclusion

This work presents a correctly stated mathematical model of growth kinetics for compact concomitant layers of nitrides during plasma nitriding of pure iron in cylindrical surfaces. The introduced model is consistent with the mechanism of mass transfer (Fick's Second Law and mass balance at the interfaces) and relevant available information on the problem (diffusion coefficients and nitrogen solubility limits). The obtained solutions, employing data from other authors and information produced in our analysis, yield congruent results between the numerical and semi-analytical methods. Finally, our solutions are observed to be in excellent agreement with the values predicted by equation (8), and no significant dependence on the shape of the sample is observed, as can be expected from the predicted thickness of each layer, in the asymptotic time limit.

5. References

[1] Pye D 2003 *Practical Nitriding and Ferritic Nitrocarburizing* (ASM International Materials Park, Ohio)

[2] Bell T et al 1977 *Controlled Nitriding in Ammonia-Hydrogen Mixtures* (Source Book on Nitriding, American Society for Metals)

[3] Totten G E 2004 *Steel Heat Treatment: Metalurgy and Technologies* (CRC Press, Boca Raton)

[4] Somers M A J and Mittemeijer E J 1995 *Metall. Mater. Trans. A*, **26A** p 57

[5] Jaoul C, Belmonte T, Czerwiec T, Ablitzer D, Ricard A and Mitchell H 2006 *Thin Solid Films* **163** 506

[6] Christiansen T, Dahl K V and Somers M A J 2008 *Mater. Sci. Tech.* **24** p 159

[7] Marciniak A 1985 *Surface Eng.* **1** p 283

[8] Henry J, Topolsky J and Abramczuk M 1992 *Crankshaft Durability Prediction- A new 3-D Approach* (SAE Technical Paper No. 920087)

[9] Zoroufi M and Fatemi A 2005 *A literature review on durability evaluation of cranksshafts including comparison of competing manufacturing processes and costs analysis* (Proceedings of the 26thForging Industry Technical Conference, Chicago)

[10] Hernandez E, Jimenez A, Otero J A, Santiago R D, Martinez R, Castillo F and Oseguera J 2016 *Sci. Res. Essays* **11(13)** pp 135-146

[11] Driscoll R J 2010 *Convective and Radiative Heat Transfer in a Thick Cylinder Using the Heat Balance Integral Method* (46th AIAA/ASME/SAE/ASEE Joint Propulsion Conference and Exhibit, Nashville)

Floating Gate sensor for in-vivo dosimetry in radiation therapies. Design and first characterization.

A Faigon*, I Martinez Vazquez, S Carbonetto and M García Inza, G
RedinLaboratorio de Física de Dispositivos-Microelectrónica. Facultad de Ingeniería. Universidad de Buenos Aires.

E-mail address: afaigon@gmail.com

Abstract. A floating gate dosimeter was designed and fabricated in a standard CMOS technology. The design guides and characterization are presented. The characterization included the controlled charging by tunneling of the floating gate, and its discharging under irradiation while measuring the transistor drain current whose change is the measure of the absorbed dose. The resolution of the obtained device is close to 1 cGy satisfying the requirements for most radiation therapies dosimetry. Pending statistical proofs, the dosimeter is a potential candidate for wide in-vivo control of radiotherapy treatments.

1. Introduction

"Adequate in-vivo dosimetry would prevent most accidental exposures" states the International Commission for Radio-Protection (ICRP) in its Publ. 86 aiming to assist in the prevention of accidental exposures involving patients undergoing cancer radiotherapy from external beam or solid brachytherapy sources[1]. Similar advice is found in reports from the American Association of Physicist in Medicine (AAPM) and by the European Society for Therapeutic Radiology and Oncology (ESTRO) dealing with the use of Thermo-Luminescent Dosimeters (TLD), Radiochromic films or Diodes for in-vivo dosimetry [2], [3]. The importance of in-vivo dosimetry in radiotherapy treatments for the clinical efficacy and for the safety of the patients has been emphasized last years by the regulation agencies. It is aimed to verify that the delivered dose does not deviate from the planned treatment in more than 5%. In spite of all these recommendations, the implementation of in-vivo dosimetry is slow and limited to a few treatments (Total Body Irradiation for example) or reduced to a complementary procedure in the Quality Assurance (QA) program of the radiotherapy centers. Different detectors has been used for in-vivo dosimetry. Luminescent devices TLD, Opto-Stimulated Luminescent (OSL); Electronic devices: diode, MOSFET; Films: Radiochromic films.

In the last years a relative increase in the use of MOSFET sensors has been observed. Their main advantages are: very small volume due to the close to 20000 times the sensitivity of a ionization chamber of the same volume, robust for their manipulation, easy electronic real time reading, produced in CMOS technology allowing thus easy integration with processing electronics, suitable for intracavitary positioning and for intra-operatory and brachytherapy treatments [4].

Within the MOS sensors family, the floating gate sensor introduced a decade ago has been the subject of several studies and improvements [5-7]. Its main advantage lies in its ability to sustain an electric field in the oxide which increases the net radiation induced pair generation without the need to be externally biased. In this work we present the design, fabrication and first characterization of a floating gate device for its use in in-vivo dosimetry for radiation therapies.[*]

[*1] afaigon@gmail.com
This work was supported by Universidad de Buenos Aires, grant UBACyT Q025 and by ANPCyT, PICT 1812.
Adrian Faigon , Mariano Garcia-Inza, and Sebastian Carbonetto are also with CONICET.

Content from this work may be used under the terms of the Creative Commons Attribution 3.0 licence. Any further distribution of this work must maintain attribution to the author(s) and the title of the work, journal citation and DOI.
Published under licence by IOP Publishing Ltd

IOP Conf. Series: Journal of Physics: Conf. Series **792** (2017) 012057 doi:10.1088/1742-6596/792/1/012057

2. Materials and methods

2.1 Structure and Principle of operation

The floating gate (FG) dosimeter is a MOS transistor (labeled "Reader" in Fig. 1) with an isolated/floating polysilicon gate whose potential, given by its stored charge, controls the channel current. Before measurement, charge is injected through the thin gate oxide of a second PMOS (labeled "injector"). During irradiation, charges are created in a thick field oxide under a large area of the floating gate. The electric field set up by the injected charge moves electrons and holes in opposite directions, discharging the floating gate. The reader's drain current ID-VDS characteristics vary with FG charge and therefore with absorbed dose.

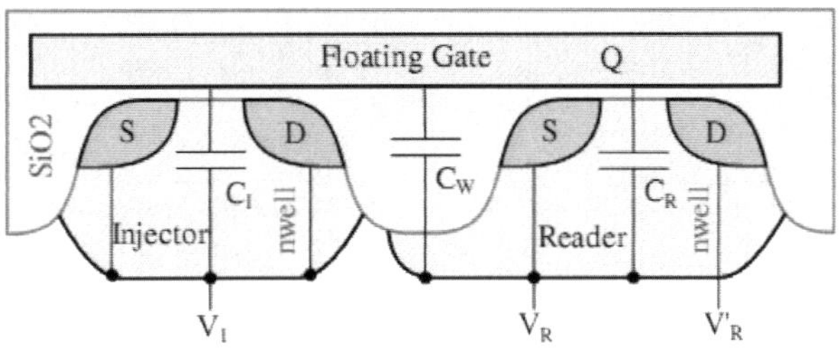

Fig. 1. Floating Gate capacitance model. CW is the capacitance between FG and the reader's N well.

2.2 Design considerations

Charging the floating gate occurs by Fowler-Nordheim tunnelling from the injector [8]. The expression for the tunnelling current is

$$J_{FN} = AF_{ox}^2 \exp(-B/F_{ox}) \tag{1}$$

with A and B constants. The oxide field Fox is given by

$$V_{FG} - V_I = F_{ox}t_{ox} + \psi_s + V_{FB}, \tag{2}$$

where the left hand side is the potential difference across the tunnel oxide, tox the field oxide thickness and ψ_s the band bending at the substrate surface. VFB is the flat band voltage. In order to negative charge the floating gate, the injector is negatively biased (VI < 0) causing the silicon surface to be accumulated and thus

ψ_s is negligible in last expression. The floating gate potential is given by

$$V_{FG} = \frac{C_I V_I + C_R V_R + Q}{C_I + C_R} \tag{3}$$

being CI and CR the capacitances -shown in Fig. 1- injector-floating and reader-floating, VI and VR the corresponding voltages and Q the floating gate stored charge.

For the injection to be efficient, i.e. for the applied voltage at the injector VI to fall mainly across the tunnelling oxide, VFG should be as close as possible to VR in absence of charge (Q=0). This set the condition CI<<CR. The second design condition is that all the capacitances should be as small as possible in order to maximize the electrical sensitivity, i.e. maximum dVFG/dQ, given

$$Q_{FG} = (C_R + C_W)V_{FG} + C_I(V_{FG} - V_I) \quad (4)$$

Finally, as the radiation induced charge in the floating gate is proportional to the volume of the insulators in these capacitors where it is generated, we have for the radiation sensitivity,

$$S = \frac{dV_{FG}}{dDose} \propto \frac{\sum A_i t_i}{\sum A_j / t_j} \quad (5)$$

where A_i and t_i are the area and thickness of each insulator. The result of the above criteria is synthesized in Fig. 2, where using the thicknesses of the foundry process the sensitivity is represented by the level curves in the plane of the areas ratios.

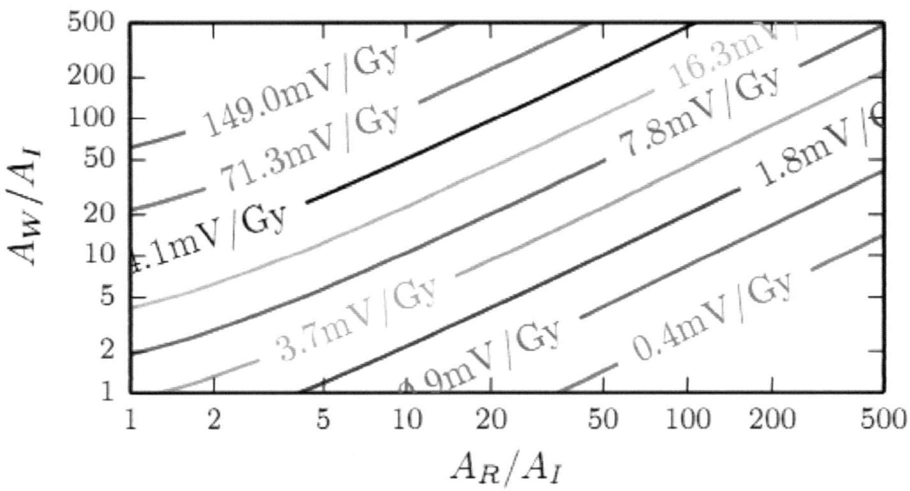

Fig. 2. The level curves of iso-sensitivity in the plane of the capacitors area ratios

Whereas the maximum sensitivity is in the upper left corner, efficient injection requires, as explained below eq.(3) $A_{R/}$ A_I>>1. We choose the ratios $A_I/A_R/A_W$: 1/50/500, being the largest area limited by the cost of the device which is proportional to its footprint on Silicon.

3. Measurements and results

3.1. Charging the floating gate dosimeter

The floating gate was charged forcing a tunnel current across the capacitor Ci provided by a Keithley 220 current source connected between the terminals Vi and Vr of the injector and reader transistors. The drain current at the reader transistor and the injected charge were monitored during the tunnel injection. A schematic of the injection configuration is given in Fig. 3.

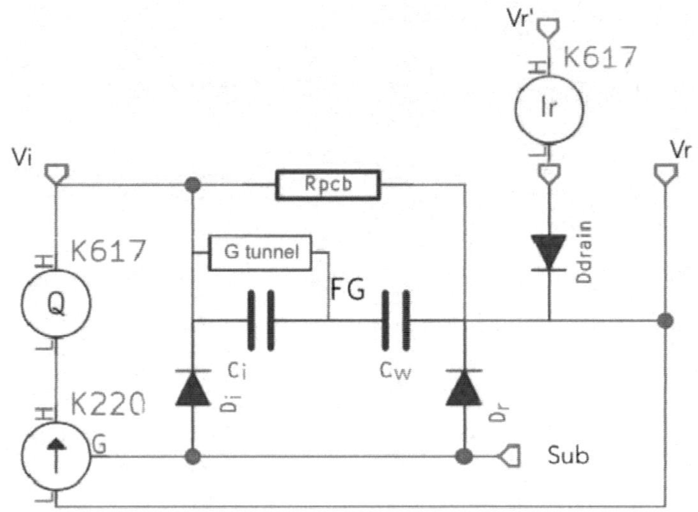

Fig. 3. Details of the injection setup. K220 is the Keithley 220 current source, there are two K617, Keithley 617 electrometer, one measuring the injected charge Q, the other one monitoring the drain current of the reader transistor Ir. G tunnel represents the tunnel conductance of the oxide in the Ci capacitor. The possible leakage paths for the injected current through Rpcb representing the pcb where the chip is mounted, and the diodes Di and Dr are blocked by applying to the substrate the same injecting voltage Vi through the guard terminal G of the K220.

The result of charging the floating gate are shown in Fig. 4. The drain current Id vs drain-source voltage Vds is measured after controlled injection pulses. The control is carried out by measuring the same drain current at fixed Vds and stopping the injection once the current reaches desired levels.

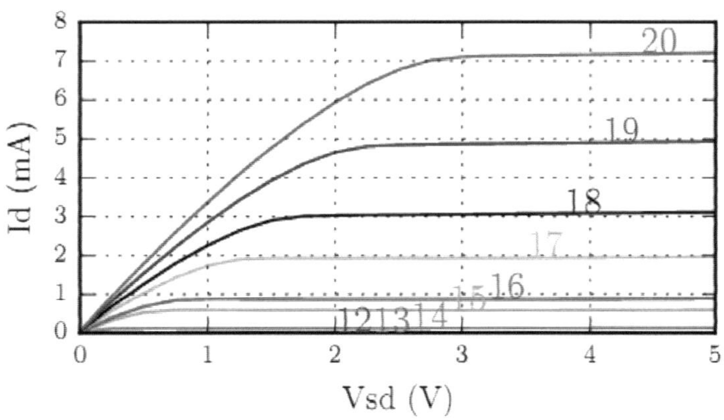

Fig. 4. Drain Current vs. Drain-Source Voltage after successive injection pulses

3.2.- Irradiating the floating gate dosimeter

After the device was charged at the maximum level of the Fig. 4, it was exposed to a 90Sr source with a dose rate of about 0.5 Gy/min measured with a ionization chamber. Along the irradiation the drain current at a fixed drain-source voltage was tracked as shown in Fig. 5.

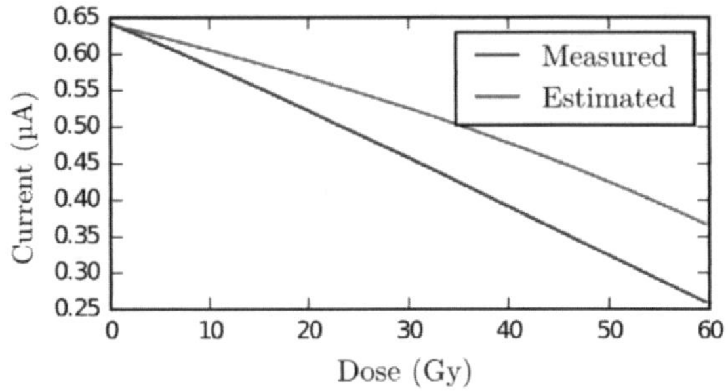

Fig. 6. Drain Current vs. Dose at Vds=0.1 V.

3.3 Sensor resolution

The measured current is almost linear with dose exhibiting the sensor a sensitivity of 6.5 uA/Gy. In order to establish the sensor resolution a figure of noise is needed. It was taken from the statistic of repeated measurements as shown in Fig. 6.

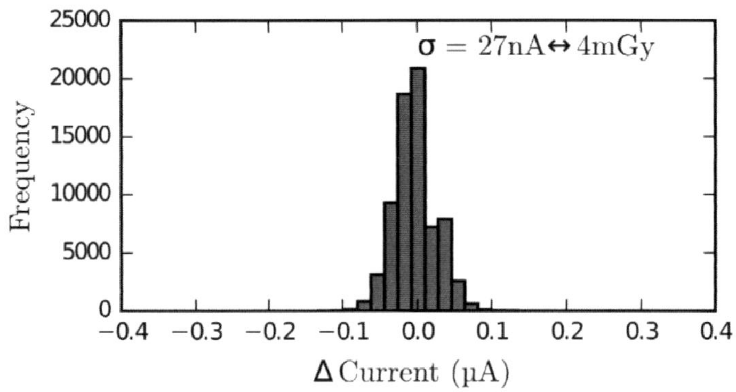

Fig. 6. Estimation of the noise. Distribution of the differences between successive current readings. The standard deviation in the distribution is equivalent to 4 mGy.

Conclusion

A floating gate dosimeter was designed and fabricated in a commercial CMOS process. First characterization of its performance was done. The resolution obtained from the estimation of noise is 4 mGy for one standard deviation. A safe figure of four standard deviations gives a noise equivalent dose of 1.6 cGy. Following the 5% accuracy required by the control agencies, the sensor complies with the requirement for radiotherapy control for doses greater than about 30 cGy. This is a fairly good performance given that in standard treatments doses of the order of 1 Gy per session are delivered.

The presented results are preliminary ones. Pending further tests, the presented device is a potential candidate to be used in in-vivo dosimetry for Radiation Therapies.

IOP Conf. Series: Journal of Physics: Conf. Series **792** (2017) 012057 doi:10.1088/1742-6596/792/1/012057

References

[1] "Preventions of Accidents to Patients undergoing Radiation Therapy". ICRP Publication 86. Ann ICRP 30 (3), 2000
[2] "Diode in vivo dosimetry for patients receiving external beam Radiation Therapy". Report of task group 62 of the Radiation Therapy Committee, AAPM 2005
[3] "Comprehensive QA for radiation oncology", AAPM Report No 46, 1994
[4] A. B. Rosenfed "Electronic dosimetry in radiation therapy", Radiation Measurements, vol 41, pp 134-153, 2006.
Oldham, Timothy R., and F. B. McLean. "Total ionizing dose effects in MOS oxides and devices." (2003).
[5] Tarr, N. Garry, et al. "A sensitive, temperature-compensated, zero-bias floating gate MOSFET dosimeter." *IEEE Transactions on Nuclear Science* 51.3 (2004): 1277-1282.
[6] Cesari, J., et al. "Floating Gate P-MOS Radiation Sensor Charging Cycles Characterization." *2014 IEEE Radiation Effects Data Workshop (REDW)*. IEEE, 2014.
[7] Pavan, Paolo, Luca Larcher, and Andrea Marmiroli. *Floating Gate Devices: Operation and Compact Modeling*. Springer Science & Business Media, 2007.
[8] Lenzlinger, M., and E. H. Snow. "Fowler Nordheim Tunneling into Thermally Grown SiO2." *Journal of Applied physics* 40.1 (1969): 278-283.

IOP Conf. Series: Journal of Physics: Conf. Series **792** (2017) 012058 doi:10.1088/1742-6596/792/1/012058

Substrate temperature effects on the structure and properties of ZnMnO films prepared by pulsed laser deposition

H Riascos[1], J S Duque[1] and S Orozco[1]
[1] Plasma, Láser y Aplicaciones, Universidad Tecnológica de Pereira, Pereira, Colombia.

E-mail: hriascos@utp.edu.co

Abstract. ZnMnO thin films were grown on silicon substrates by pulsed laser deposition (PLD). Pulsed Nd:YAG laser was operated at a wavelength of 1064 nm and 100 mJ. ZnMnO thin films were deposited at the vacuum pressure of 10^{-5} Torr and with substrate temperature from room temperature to 600 °C. The effects of substrate temperature on the structural and Optical properties of ZnMnO thin films have been investigated by X-ray diffraction (XRD), Raman spectroscopy and Uv-vis spectroscopy. From XRD data of the samples, it can be showed that temperature substrate does not change the orientation of ZnMnO thin films. All the films prepared have a hexagonal wurtzite structure, with a dominant (002) peak around 2θ=34.44° and grow mainly along the c-axis orientation. The substrate temperature improved the crystallinity of the deposited films. Uv-vis analysis showed that, the thin films exhibit high transmittance and low absorbance in the visible region. It was found that the energy band to 300 ° C is 3.2 eV, whereas for other temperatures the values were lower. Raman reveals the crystal quality of ZnMnO thin films.

1. Introduction

ZnO can be considered as an "old" semiconductor which has been attracted research attention for a long time because of its large area of applications. Zinc oxide is a direct band gap semiconductor (Eg) with 3.37 eV (at room temperature) having hexagonal (wurtzite) structure with lattice parameters a=0.325 nm and c=0.512 nm [1]. Stoichiometric films are insulating whereas nonstoichiometric and oxygen deficient films are very good as transparent semiconducting oxides.

The transition metals (Ni, Mn, and so on) are used as attractive dopants for ZnO because they are isomorphic to Zn, possess variable oxidation states, and have different acceptor properties in a ZnO matrix, thus affecting the electronic surface band structure of ZnO. Mn incorporation in metal oxide systems is interesting because of the band gap variations based on the percentage of Mn ion introduced and the property of the MZO film is essential for wide-ranging applications. Among the oxide semi conducting materials, ZnO is one of the most important oxide material [2]. Beside the generation of novel magnetic properties, the introduction of Mn^{2+} ions can also tune the optical properties viz., band gap (Eg) of ZnO. Generally band-gap is estimated from UV–vis spectrophotometry. But like origin of magnetism, another debatable issue is the nature of variation of Eg with Mn concentration. Thus transition metal-doped ZnO has the potential to be a highly multifunctional material with coexisting magnetic, semiconducting, electromechanical, and optical properties. A large number of reports are available on the different properties of transition metal doped ZnO [3-5]. The first experiments on Mn-doped ZnO were performed by Fukumura et al [6] and room-temperature ferromagnetism for Mn doped ZnO was first reported by Sharma et al. [7].

Some studies have been carried out to realize Mn doped ZnO as a dilute magnetic semiconductors (DMS), however, studies to understand the other properties of Mn doped ZnO are also necessary.

Content from this work may be used under the terms of the Creative Commons Attribution 3.0 licence. Any further distribution of this work must maintain attribution to the author(s) and the title of the work, journal citation and DOI.
Published under licence by IOP Publishing Ltd

Consequently, the synthesis and the characterization of transition metals doped ZnO-based DMS have become the subject of intense research.

On the other hand, the deposition method and parameters are influencing the structural properties of the films and consequently the electrical and optical ones. Therefore, apart from dopant nature and concentration, the deposition method and parameters must be optimized such that the obtained films to possess the desired properties.

Mn doped zinc oxide thin films can be obtained by various deposition techniques including spray pyrolysis technique [8], sol–gel technique [9], chemical vapor deposition (CVD) [10], DC magnetron sputtering [11,12], molecular beam epitaxy (MBE) [13], and pulsed laser deposition [14-17].

Pulsed Laser Deposition (PLD) method has many advantages and technological possibilities. These advantages are: effectiveness and simplicity of the deposition equipment, high deposition rates, wide spectrum of deposition parameters for the control and the optimization of film properties, accurate control of stoichiometry and film thickness, use of background atmosphere of various gases, and high energy and high reactivity of the ablated material. The sum of all these special features enables the growth of oxide thin films at low temperature substrates with perfect crystallinity and good stoichiometry.

In this paper, ZnMnO films were deposited by pulsed laser deposition. The influence of temperature substrate on the structure and optical properties of the deposited films has been investigated.

2. Experimental Details

Mn-doped ZnO thin films were grown by conventional pulsed laser deposition (PLD) using a stainless steel HV deposition chamber and a pulsed NIR laser (Nd:YAG, 1064 nm wavelength, 9 ns pulse duration). The laser was operated at 10 Hz and the laser beam was incident at an angle of 45 with respect to the target surface. The targets were prepared from high-purity powders of ZnO (99.999%), with MnO (99.999%) serving as the doping agent. The pressed targets were sintered at 1000 °C for 12 h in air. The target was fabricated with a nominal composition of 5 wt.%Mn, and was continuously rotated during laser ablation to renew the irradiated surface and to prevent crater formation. The ablated material was collected onto a Si(100) substrate, over a temperature range of RT to 600°C, placed in front of and at 6 cm from the target. Prior to an experiment, the chamber was evacuated to 7×10^{-4}Pa. During deposition, oxygen (99.999 %) was flown through the chamber and the background pressure was of 6.13 Pa. The laser fluence was kept at (3.0±0.4) J cm^{-2}. The target to substrate distance was 6 cm, and the deposition time was kept at 30 min, the average thickness of the film was approximately 200 nm. Structural analysis of the films was carried out in a Bruker D8 advanced diffractometer by X-ray (XRD), using Cu$K\alpha$ line of 1.5418 Å and scan rate 0.04 2θ/s. The identification of crystalline phases was done using the JCPDS database cards [18]. Optical analysis of the films was carried out in a Jobin Yvon spectrometer.

3. Results and discussions

3.1. XRD analysis

ZnMnO thin films were grown on Si (100) at substrate temperatures from room temperature to 600°C. Characteristic XRD patterns of doped ZnO films are presented in figure 1. The XRD patterns of the studied samples show that the films have a preferential orientation with (002) planes parallel to the substrate. It indicated that the films are well oriented along the c-axis. No characteristic peak of Mn was observed by the means of X-ray diffraction. The absence of Mn XRD peaks indicates that a diffusion of Mn atoms in new formed ZnO lattice takes place.

This indicates that these ZnMnO thin films show a good crystallinity with a wurtzite structure and a c-axis preferred orientation. We can also see the others weak reflection peaks, in the samples grown at substrate temperatures RT, 100°C and 600°C, at diffraction angle $2\theta = 31.8°$, $2\theta = 36.3°$, corresponding to (100) and (101) reflection of the hexagonal ZnO respectively (figure1). The presence of these reflections clearly points to the polycrystalline nature of the ZnMnO.

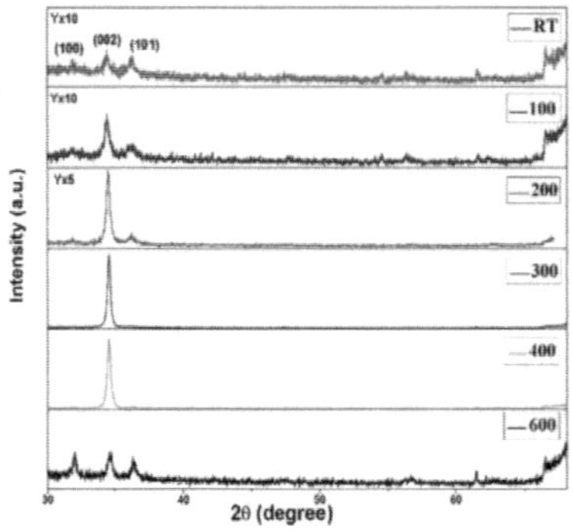

Figure 1. XRD patterns of ZnMnO thin film deposited by PLD with the substrate temperature varying from room temperature to 600 °C. The Mn-doping and oxygen pressure were kept constant, 5 wt.% and 6.13 Pa, respectively.

The increasing of the substrate temperature is in favor for the diffusion of atoms absorbed on the substrate and accelerates the migration of atoms to the energy favorable positions, resulting in the enhancement of the crystallinity and c-axis orientation of film, which can be indicated by increase of (002) peak strength. On the other hand, when substrate temperature is low, the adatoms are rapidly cooled and their diffusion or hopping distance becomes small. Thus, the surface becomes rough and its structure looks like polycrystalline as confirmed by XRD in samples which were grown at RT and 100°C.

The average crystallite size is estimated from Debye-Scherrer equation and the values of lattice parameters a and c for Mn doped ZnO were estimated from the equation (1)

$$\frac{1}{d_{hkl}} = \frac{4}{3}\left[\frac{h^2 + hk + k^2}{a^2}\right] + \frac{l}{c^2} \tag{1}$$

where h, k and l are the Miller indices and dhkl is the interplanar spacing for the plane (h k l) and the film stress along the c- axis, for a hexagonal crystal structure can be calculated using the biaxial strain model [19] based on the 002 peak position and are shown in table 1.

The value of lattice parameters decreases with the substrate temperature, the average crystallite size is around 24.2 nm and all films are under tensile stress as shown in table 1. The decrease in lattice parameters is caused due to zinc vacancies [20]. Manganese can exist in Mn^{2+}, Mn^{3+}, and Mn^{4+}. The ionic radius of Mn^{2+}, Mn^{3+}, and Mn^{4+} are 0.66 Å, 0.58 Å and 0.53 Å respectively. In addition, based on the fact that the ion sizes of Mn are smaller than that of Zn, it is possibly that the Mn ions successfully substitute the Zn sites as increasing growth substrate temperature but there are still interstitially positioned Mn ions in crystal due to smaller ionic radii of Mn ions compared to Zn^{2+} (0.74 Å). Substitution by the smaller manganese atoms decreases the lattice constants due to the intrinsic stress induced in the film [21]. These interstitially positioned ions also may have some contributions to the peak shift of ZnO(0 0 2). We also note that this report have been supported by Dole et al. [22].

Table 1. The XRD (002) peak position, lattice constants, grain size and stress calculated at 6.13 Pa and Mn 5 wt%.

Temp (°C)	(002) Pos. (°)	a (Å)	c (Å)	Grain size (nm)	Stress (10^9 Pa)
RT	34.425	3.188	5.206	24.656	0.029
100	34.491	3.182	5.196	23.106	0.461
200	34.490	3.182	5.197	24.465	0.454
300	34.540	3.178	5.189	25.997	0.781
400	34.550	3.177	5.188	23.696	0.846
600	34.596	3.173	5.181	23.112	1.150

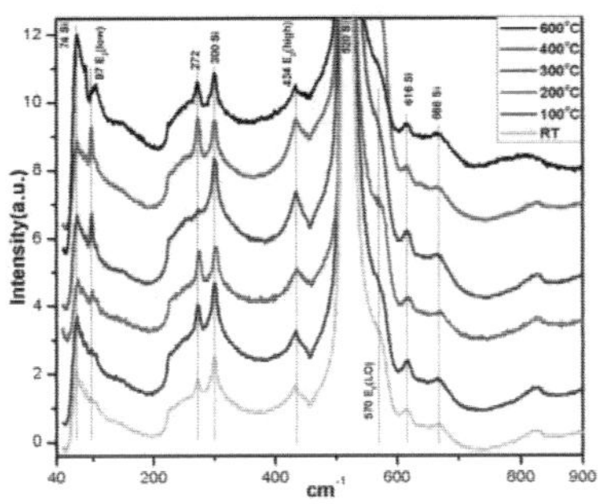

Figure 2. Raman patterns of ZnMnO thin film deposited by PLD with the substrate temperature varying from room temperature to 600°C. The Mn-doping and oxygen pressure were kept constant, 5wt.% Mn and 6.13Pa, respectively.

3.2. Raman Analysis

Raman spectrum can sensitively reveal the changes of microstructure of the materials, leading to the analysis of crystal quality. According to the group theory eight vibrational modes are predicated at the Brillouin zone for wurzite structure. The A1 and E1 modes are polar and split into transverse optical (TO) and longitudinal optical (LO) phonons, and E2 mode consists of two modes of low and high frequency phonons. On the other hand, the B1 modes are not Raman active. To investigate the influence of substrate temperature on structure and vibrational properties further, Raman studies are performed and the results are shown in Figure. 2. The Raman spectra of ZnMnO thin films were analysed in the range 40-900cm^{-1}, with the substrate temperature varying from room temperature to 600°C. Figure 2 show several peaks at 100cm^{-1} and 434cm^{-1} are assigned to the two non-polar optical phonons modes, E2 (low) and E2 (high), respectively. The peak located at 570cm^{-1} is associated to E1(LO) vibration modes. The peaks at about 520cm^{-1} come from the Si(1 0 0) substrates, which can be ignored to be discussed. An additional peak appears at 273cm^{-1} and it indicates host lattice defects in the ZnO structure [23].

The E2 phonon mode centered around 434 cm^{-1} is obviously a good choice in order to understand the stress-induced phenomena in wurzite ZnMnO. An increase in the E2 phonon frequency is generally ascribed to compressive stress, where as a decrease is caused by the tensile stress.

The sharpest peak at about 434cm^{-1} (E2 (high) mode) is the strongest Raman mode in wurtzite ZnO structure, and the values of the full width at half maximum of E2 (high) mode of these samples are less than 10cm^{-1}, which indicates that all the ZnMnO films keep a good ZnO wurtzite structure. E2(low) and E2(high) appear with low line width and higher intensity at 300°C.

In our Raman spectra, no Mn related Raman mode is observed. In a word, from the analysis of Raman spectra, we can deduce that Mn ions, at least partially, substitute for Zn ions in ZnO and microscopic structural disorders are introduced. The Raman spectra also exhibit a weak shoulder at about 570 cm^{-1} (E1(LO)) and has been attributed to the formation of oxygen deficiency, interstitial Zn. The broadening of these bands implies formation of defects electronic states within the band gap [24]. The vibrational modes are not sensitive function of substrate temperature, such behavior was observed and reported before [25].

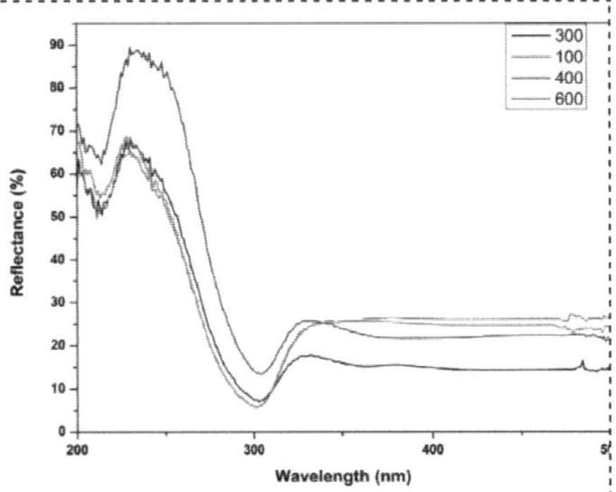

Figure 3. Optical reflectance spectra of ZnMnO thin films deposited by PLD with the substrate temperature varying from 200°C to 600°C. The Mn-doping and oxygen pressure were kept constant, 5wt.% Mn and 6.13Pa, respectively.

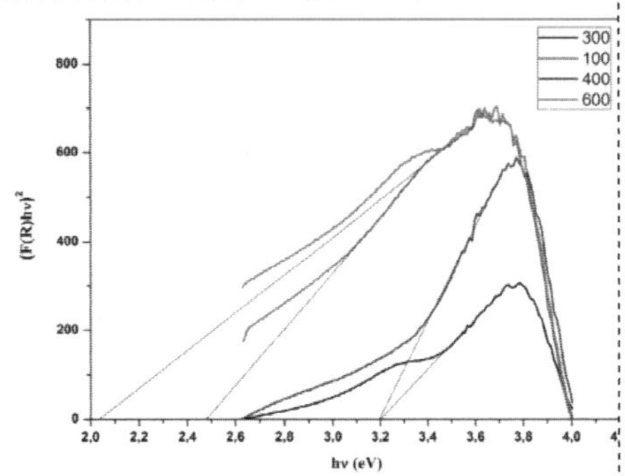

Figure 4. Plot of $(F(R)\ h\nu)^2$ versus hν. Obtaining energy bang values of ZnMnO thin films.

3.3. Uv-vis Analysis

The figure 3 shows the UV–visible diffuse reflectance spectra for ZnMnO, the absorption edge in the UV region depending on film growth temperature. For analysis purposes the diffuse-reflectance, R, of the samples can be related to the Kubelka–Munk function F(R) by the relation $F(R) = (1 - R)^2/2R$ [26]. The energy band gap of the ZnMnO samples was determined from the diffuse-reflectance spectra by plotting the square of the Kubelka–Munk function $(F(R)h\nu)^2$ versus energy and extrapolating the linear part of the curve to $(F(R\ h\nu))^2 = 0$, as shown in figure 4. The optical band gap (Eg) values were determined by the above method to be 2.83, 2.03, 2.89, 3.2, 3.2, and 2.39 eV for the films grown at substrate temperatures RT to 600°C, respectively, indicating that Eg varies irregularly with substrate temperature. The energy gap, Eg, is related to the grain size and carrier concentration in films [27]. In addition, quantum size effects may be involved due to nanofeatures on the films [28]. The maximum energy gap Eg for films grown at 300°C may relate to the smallest grain size. Compared with films grown at 600°C, the Eg of films grown at 300°C show a slight increase. This reveals that high substrate temperature will create more defects such as oxygen vacancies. Furthermore, more oxygen vacancies may cause a carrier concentration increase, thus the energy gap Eg slightly increases although Eg must decrease as result of the increase of grain size.

4. Conclusions

The structure of the ZnMnO films clearly varies with substrate temperature. The vibrational modes appear not to be sensitive function of substrate temperature. The increasing of the substrate temperature is in favor for the diffusion of atoms absorbed on the substrate and accelerates the migration of atoms to the energy favorable positions, resulting in the enhancement of the crystallinity and c-axis orientation of film.

The optical band gap of the films was found to be around 2.03–3.2 eV, which is lower than the theoretical value of 3.37 eV. At 300 °C the band gap is nearest to band gap undoped ZnO.

References

[1] Jagadish C and Pearton S J 2006 Thin Films and Nanonstructures: Processing, properties and aplications, *(Amsterdam: Elsevier)*
[2] Nirmala M and Anukaliani A 2010 *Photon Lett. Pol.* **2** 189
[3] Janisch R, Gopal1 P and Spaldin N A 2005 *J. Phys.: Condens. Matter* **17** R657–R689
[4] Pradhan A K, Hunter D, Lasley-Hunter B, Dadson J B, Zhang K, and Rakhimov R R 2006 *J. Appl. Phys.* **99**, 08M108
[5] Bai H, He S, Liu G, Cao Q, Zhu D, Chen Y, Yan S, Mei L, 2012 *Physica B: Condensed Matter* **407** 2126 [109]
[6] Fukumura T, Jin Z and Kawasaki M 2001 *Appl. Phys. Lett.* **78** 958
[7] Sharma P, Gupta A, Rao K V, Owens F J, Sharma R, Ahuja R, Guillen J M O, Johansson B, Gehring G A, 2003 *Nature Materials* **2** 673
[8] Singh P, Kaushal A and Kaur D 2009 *J. Alloys Compd.* **471** 11
[9] Xue S W, Zu X T, Zhou W L, Deng H X, Xiang X, Zhang L and Deng H, J. Alloys Compd. 2008 *J. Alloys Compd.* **21** 448
[10] Waugh M R, Hyett G and Parkin I P 2008 *Chem. Vap. Dep.* **14** 366
[11] Houng B, Hsi C S, Hou B Y and Fu S L 2008 *J. Alloys. Compd.* **456** 64
[12] Devy F C, Barreau N and Kessler J 2008 *Thin solid Films* **516** 7094
[13] Wang D, Zhang J W, Peng Y P, Bi Z, Bian X M, Zhang X A and Hou X 2009 *J. Alloys. Compd.* **478** 325
[14] Xiao B, Ye Z Z, Zhang Y Z, Zeng Y J, Zhu L P and Zhao B H 2009 *Appl. Surf. Sci.* **253** 895
[15] Chrisey D B and Huber G K (Eds.) 1994 *Pulsed Laser Deposition of Thin Films* (John Wiley & Sons, Inc., New York) p. 613
[16] Kotlyarchuk B and Savchuk V 2003 *Proc. IEEE "CAOL-2003"* **1** 149
[17] Kotlyarchuk B and Savchuk V 2003 *Phys. Chem. Sol. State* **4** 434
[18] Jones B S and Smith R Z 1992 *Joint Committee on Powder Diffraction Standards, Powder Diffraction File* (Philadelphia: ASTM)
[19] Riascos H, Orozco S and J Uzuriaga 2015 *Procedia Mat. Sci.* **9** 523
[20] Khalid M, Ziese M, Setzer A and Esquinazi P 2009 *Physical Review B* **80** 035331
[21] Bhatti K P, Chaudhary S, Pandya D K and Kashyap S C 2005 *Solid State Communications* **136** 384
[22] Dole B N, Mote V D, Huse V R, Purushotham Y, Lande M K, Jadhav K M and Shah S S 2011 *Appl. Phys.* **11 3** 762
[23] Zhaoa Y, Jianga Y and Fang Y 2007 *J. Cryst. Growth* **307** 278
[24] Zeng J N, Low J K, Ren Z M, Liew T and Lu Y F 2002 *Appl. Surf. Sci.* **197–198** 362
[25] Orozco S, Riascos H and Duque S 2016 *J. Phys.: Conf. Ser.* **687** 012036
[26] Kortum G 1969 *Reflectance Spectroscopy* (New York: Springer)
[27] Lin S S and Huang J L 2004 *Surf. Coat. Tech.* **185** 222
[28] Ozgur U, Ya I, Alivov C, Liu A, Teke M A, Reshchikov S, Dogan V, Avrutin S-J, Cho H and Morkoc 2005 *J. Appl. Phys.* **98** 041301

Acknowledgments

This work was be supported by a grant from the Programa Latino-Americano de Física in 2013 (Brazil). We also acknowledge Dr. Pedro Alejandro Ayala with the XRD measurements.

PMMA Solid bottle optical microresonator for measure relative humidity.

D A Avila[1], S D Horta[1] and C O Torres[1]
[1]Department of Physics, Laboratorio de Óptica e Informatica, Universidad Popular del Cesar, Colombia.

E-mail: duberavilap@unicesar.edu.co

Abstract. In this work we studied experimentally the performance of an solid bottle optical resonator made of PMMA (polymethylmethacrylate) for measure the relative humidity of the medium. In the developed device, the WGMs modes within the microcavity are excited by the proximity of an optical fiber taper with an outer diameter of the order of 3-5 microns made from stretching a standard optical fiber of Silica by the flame brushing technique. In the resonant device, the field produced by a laser system tunable TLS is guided through the fiber taper and is coupled into the microcavity by the approach of the fiber taper to the equatorial zone of the microbottle, causing the excitation of the WGMs resonant modes inside the same. When the device is subjected to changes in relative humidity of the medium, the wavelengths of resonance of WGMs modes that have been coupled in the microresonator are shifted spectrally depending on the external humidity, showing an experimental sensitivity in the resonator due to changes in the relative humidity of the medium.
In the experiment, it was possible to produce different samples of optical resonators with a profile shaped bottle with different maximum diameters achieving a maximum sensitivity of 0.032 nm /% RH for a resonator with equatorial diameter of 1250 μm.

1. Introduction

The optical resonators play an important role in modern optics, being not only important in the design and manufacture of Laser, but also are essential devices for performing high-accuracy measurements in interferometry, experiments nonlinear optics, etc

One of the most significant advances in the manufacture of these devices is the miniaturization of optical resonators made from different materials, the most important material reported in the literature, silice, silica and some types of polymers that have allowed an interconnection with systems integrated optics used in the area optical telecommunication. These devices have great attention due to its ability to light confinement within them, which has led to breakthroughs in the area of sensors applied to biology, physics, chemistry and the study of nonlinear optical effects.

Actually, these structures are designed with dimensions on the order of microns to allow coupling easily dielectric structures of waveguides with high symmetry and thus exploit the advantages of the morphological characteristics, its properties and its relative ease for treatment mathematical.

The propagation of Whispering Gallery Modes WGMs in a resonator bottle have been experimentally obtained in 3D structures shaped bottles using materials such as silica [1]-[4]. This type of bottle-shaped microresonators double neck support WGMs modes that are different modes of the resonators of microspheres and WGMs modes microtoroides called equatorial modes.

Currently, the manufacture and its corresponding theoretical study the transmission spectrum of such resonators, made from fibers of silica is reported, however not been reported jobs that relate the application of this type of resonators in some area using a specific polymer. Actually, the manufacture and its corresponding theoretical study the transmission spectrum of such microresonators, made from

Content from this work may be used under the terms of the Creative Commons Attribution 3.0 licence. Any further distribution of this work must maintain attribution to the author(s) and the title of the work, journal citation and DOI.
Published under licence by IOP Publishing Ltd

fibers of silica is reported, however not been reported works that relate the application of this type of resonators in some area using a specific polymer.

During this investigation, were exploited the resonant properties of the optical microresonators in a solid bottle and was studied its application as relative humidity sensor using the physical principle of the displacement of the wavelengths of resonance modes WGMs which are confined to their inside, the same physical principle used in some previous work [5],[6]. Samples of optical microcavities in the form of solid bottles were manufactured from the polymer PMMA, with the intention to increase the sensitivity for measurement of relative humidity, for it was necessary to characterize the spectral response of the samples according to the changes relative humidity.

2. Excitation WGMs modes within a dielectric cavity in a solid bottle.

In the dielectric structures have a circular symmetry, the field can be confined by total internal reflection inside and whose modes of propagation are commonly referred Whispering Gallery Modes WGMs. In the literature the excitation of WGMs modes is reported within cavities with rotational symmetry in the form of spheres, cylinders, toroids, rings, discs, and in some cases shaped structures bottle and bubbles, which have been fabricated from different materials such as silice and silica. These structures show behavior of resonant cavities when an external field is coupled to the cavity. For this reason, in such cavities is important to study the wavelengths/frequencies of resonance, the Q factor of the oscillations within the cavity, the free spectral range FSR and Full width at half maximum FWHM.

According to the theoretical analysis reported in the literature by some authors, when a cavity has the shape of a bottle it is possible to associate a profile that depends on the geometric shape of the cavity, which can be adjusted through various functions such as a function parabolic, cosine or a truncated harmonic oscillator [1]

$$D(z) = \begin{cases} D_b\left[1+(\Delta kz)^2\right]^{-1/2} & |z| \le L_b/2 \\ D_s & |z| > L_b/2 \end{cases} \qquad (1)$$

Where D_s is the diameter of the solid polymer fiber chosen before the heating process, whereas D_b is the maximum diameter reached at the center of the cavity through the heating technique, L_b is the length of the protuberance and Δk is the radius of curvature of the profile. The latter term is considered in many situations a very small amount, for bottles having a minimum variation of fiber diameter or a protuberance in the center of the solid fiber, so that it is necessary that the relation $dD/dz \ll 1$ is satisfied for a smooth protuberance. This type of microcavities of solid bottle, can be used as optical cavities 3D dimensional asymmetrical, topologically equivalent to a sphere-shaped resonator, which have been extensively studied in the literature.

In the literature it has been reported manufacturing optical microcavities silica in the form of bottles with a high Q factor, where the light confinement is caused by total internal reflection on the walls of the cavity. In this regard, it may be noted that the confinement of modes WGMS into the cavity is performed along a thin ring along the Ecuador on the surface of the bottle by total internal reflection experienced by rays. The WGMs excitation of modes within the cavity is performed through coupling of the field by the phase-matching of WGMs modes with external waveguide to achieve important values for the factor Q.

3. Experimental setup.

In order that the resonant structure can be used as an optical sensor for detecting changes in relative humidity it is necessary to couple the field inside the microcavity. To this end, it was used an optical fiber taper, which enables the excitation of resonant modes WGMs within the cavity.

In the fabrication process of the optical fiber taper, it was used the flame brushing technique, which is suitable due to its high flexibility for controlling the movement of the flame, the length of the taper zone and speed control stretching the fiber. In the experiment was used an optical fiber commercial step index SMF 20 type Corning, where a small portion of the fiber of the order of millimeters is selected to remove the coating and then is subjected to a heat source produced by a flame to heat the fiber to a temperature close to 1500 ° C, while simultaneously the fiber is fixed at the ends to achieve the stretch and thus to obtain the desired profile. During the fabrication process of the tapered fibers, it was possible to obtain neck diameter fibers 2-5 microns, sufficient to ensure optimum field coupling cavity.

In the test transmission fiber, a tunable laser system TLS with operating wavelengths in the range of 1500-1600 nm and optical spectrum analyzer OSA available in BraggMeter FS 2100 FiberSensing with a resolution 10 pm was used for measurements.

The diagram of experimental setup for obtaining transmission spectrum of fabricate fibers is shown in Figure 1a, while in Fig 1b the transmission spectrum obtained for a taper 4 μm diameter observed in the range of 1500 -1600 nm wavelength. In the same figure, the spectral response of the fiber in the range of 1562-1572 nm wavelengths is observed.

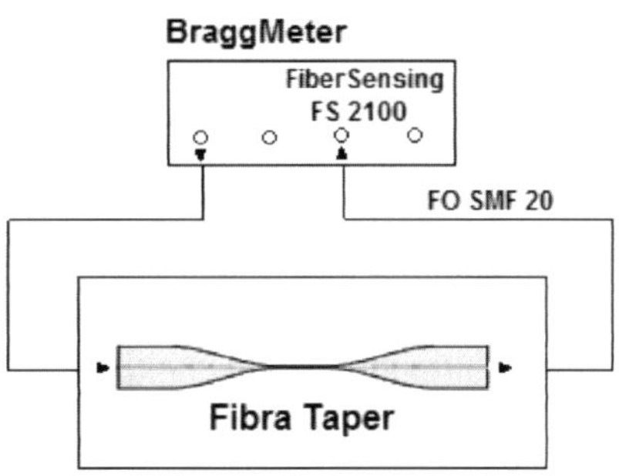

Figure 1. a) Diagram of the experimental setup for obtaining transmission spectrum of the fiber taper fabricated.

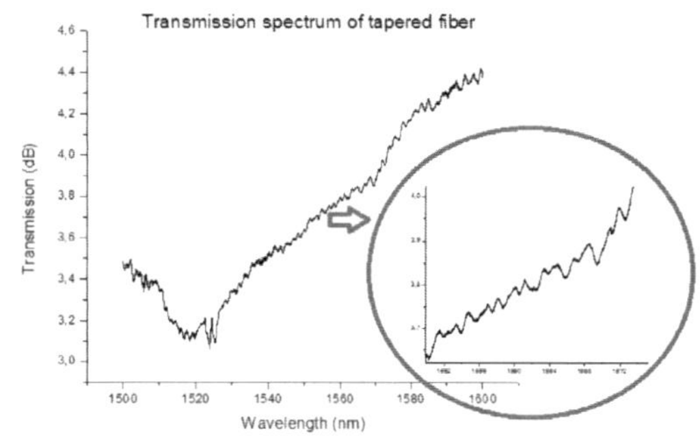

Figure 1. b) Transmission spectrum 4 μm taper diameter. Within the circle an image with increased range of 1562-1572 nm.

BMRs bottles microresonators are solid structures having shape of ellipsoid that can support WGMs nondegenerate modes. These optical cavities are fabricated in the form of hollow and solid bottles from Silica and have been used primarily as sensors optofluidics applications.

In this paper was fabricated a type of solid shaped microresonator bottle made from PMMA for use in optical measurement of the relative humidity and thus, it was possible for the first time report for the first time the fabrication of this type of resonators fabricated from a solid PMMA fiber.

The fabrication process of the cavity began with the use of a PMMA solid optical fiber with diameters in the range of 300-2000 μm. To acquire the desired shape, the solid fiber must be heated to a temperature near 200 ° C. To this it was possible to use a heating technique through an induction heating coil. When the tip of the solid fiber is inserted inside the solenoid during a given time, it is produced a protuberance at the tip of the fiber, which is molded by the gravitational force and the force of surface tension. The exposure time of the fiber to the heat source is of great importance for achieve the symmetrical shape.

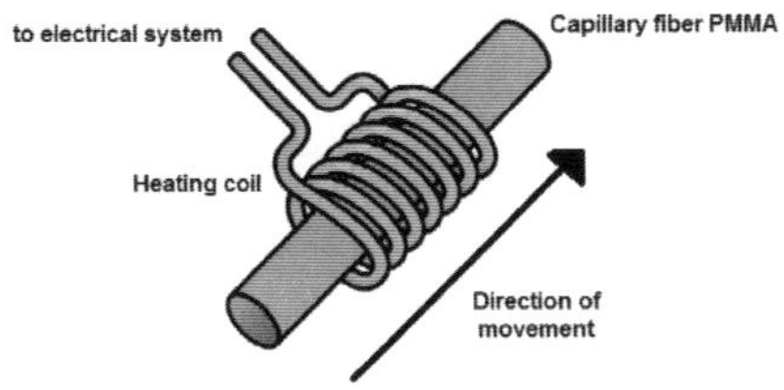

Figura 2. a) The solid fiber made of PMMA introduced into the induction coil.

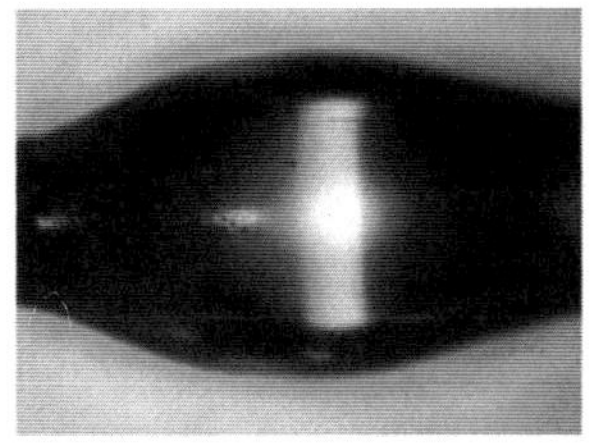

Figura 2. b) Microscope image of bottle made with maximum diameter 1917 μm

The figure 2a shows a solid PMMA fiber within the induction coil, which is heated and softened in the process, while in Figure 2b is observed microscope image of the fabricated samples with the technique. During the experiment, the selective excitation of WGMs modes within the cavity is performed through a Tapers fiber coupled to the resonator along its equatorial zone, which allowed the excitation of WGMs modes and thus was possible analyzing resonance curves obtained by confining the field within the cavity.

In order to determine the sensitivity of the device to changes in relative humidity environment, it was necessary to a humidity chamber that was constructed from the controlled ingress of wet and dry air into a chamber, which presented an operating range between 53-78% RH. For measuring the relative humidity it was used a hygrometer PCE-555 of PCE Instruments with an accuracy of 0.1% RH. In Figure 3a, the coupling of the field in the equatorial region is observed through the Taper fiber. The excitation of the modes into the cavity, is performed using a tunable laser (TLS) in the range of 1500 to 1600 nm and optical spectrum analyzer (OSA) with a maximum resolution of 10 pm (Fiber Sensing FS22).

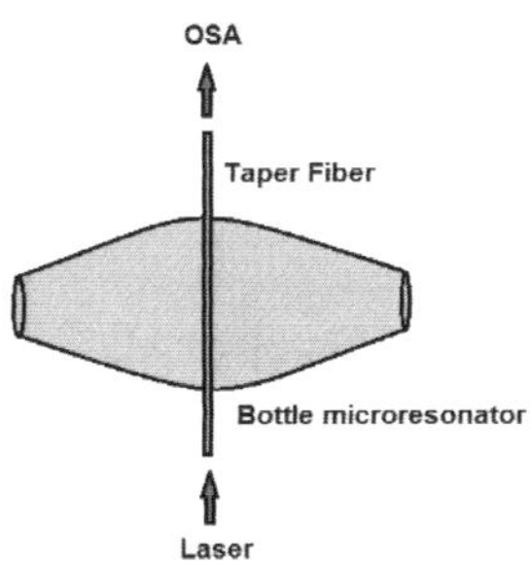

Figura 3. a) Fiber taper coupling to the microcavity.

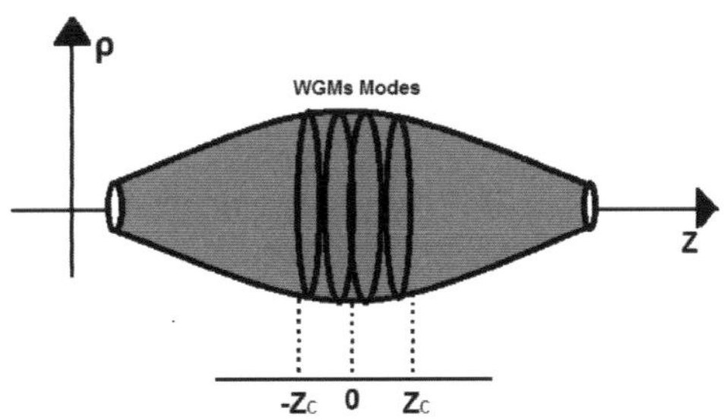

Figura 3. b) Excitation of WGMs modes within the cavity.

In the measurement of the spectral response, the field was collected at the exit of the field transmitted along the fiber taper, showing a very dense spectrum with a Q factor of the order of $3.5x10^4$ when the taper was approached the equatorial zone of the bottle. During the experiment it was necessary to ensure stability of the spectrum in each of the fabricated samples, for it was important to control external variables to avoid possible cross-sensitivity that may experience the sensor. Within these external variables that can cause changes in the spectral response, it was possible to consider external vibrations, which may occur involuntarily during measurements. To do this, once the system is adjusted, a test spectral stability consisted of measuring the spectrum at ten minute intervals under constant temperature and humidity was performed. The result of measurement of the spectral response for the sample No. 1 is shown in Figure 4. The test results show a spectrum with good stability for the selected sample.

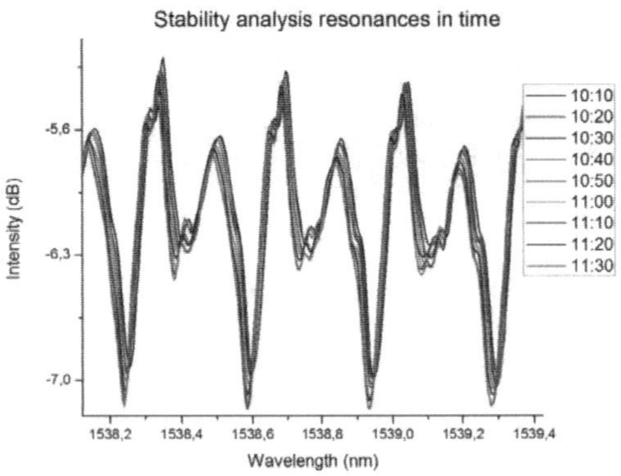

Figure 4. The spectral stability of the sample No. 1 under conditions of humidity and temperature constant. The spectrals measurements were taken every ten minutes.

During the experiment, they were fabricate different samples of bottle-shaped resonators and were selected five different samples for the experiment. The geometric parameters of the samples are summarized in Table 1.

Table 1. Geometrical parameters and relative humidity sensitivity of microcavities fabricated bottle-shaped. In all cases fiber preform PMMA polymer used it had a diameter of $D_s = 850\mu m$.

Sample N°	Maximum diameter D_b (μm)	Sensitivity nm/%
1	1522	0.0117
2	1425	0.0219
3	1250	0.0325
4	1491	0.0161
5	1917	0.0088

In Table 1, the results of experimental measurements of the sensitivity of the samples were manufactured are recorded. In Figure 5a, an analysis of the hysteresis is observed the sample No. 1. The experimental data plotted in red show the displacement of a resonance peak with fixed wavelength in response to changes in relative humidity of the environment when the humidity increases. Likewise, the line plotted in black shows the response of the same cavity when the humidity is decreasing. In this case, the sensitivity measured by increasing or decreasing humidity acquire not significantly important differences, therefore have a better response than the capillaries resonators reported in some papers for humidity measurement [6].

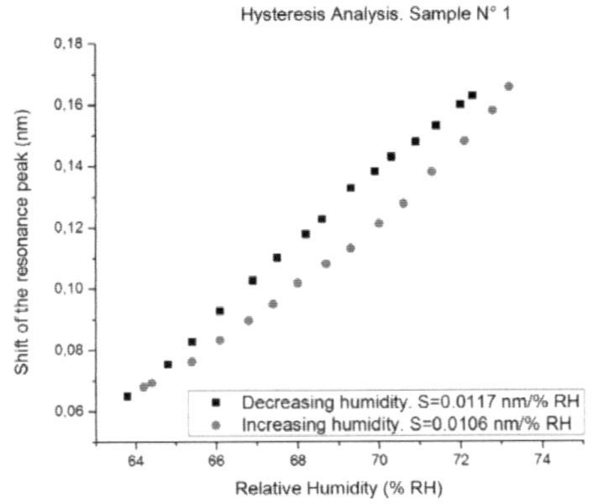

Figure 5. a) Spectral response of the sample No. 1 to changes in medium humidity with humidity increasing and decreasing.

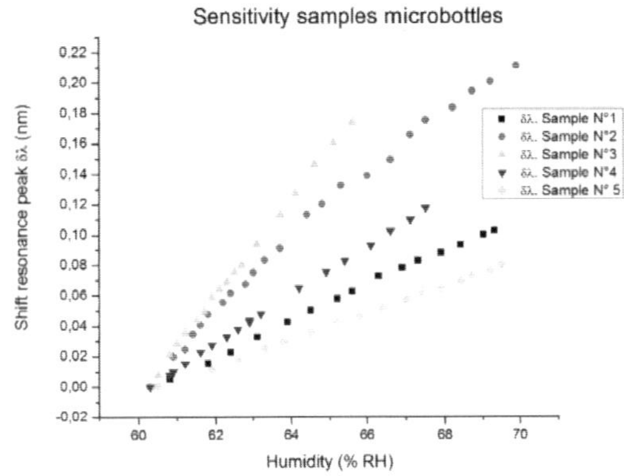

Figure 5. b) Experimental sensitivity of all samples fabricated.

In Figure 5b, the response sensitivity of the samples is observed. Of the analyzed results it is clear that the device with smaller maximum core diameter Db corresponding to sample No. 3 has better sensitivity 0.033 nm/% RH, while the sample with the highest maximum diameter (sample No. 5) has lower sensitivity 0.009 nm/% RH [6].

4. Analysis and conclusions

In this paper we have demonstrated an efficient, simple and practical method for the design and manufacture of microcavities of solid bottle using a solid fiber standard PMMA polymer, which is used as an optical microresonator whose field has been effectively confined by a fiber optical adiabatic taper profile. The experimental Q factor of the oscillations confined within the microcavity is about 3.5 x104, which is slightly lower than those reported for microcavities shaped capillary for the same type of material [5,6]. The fabricated device can measure with good accuracy the relative humidity of the medium when the radius of curvature of the profile is small enough.

During the developed experiments, it was found that the solid shaped microresonators bottle exhibit an improvement in spectral stability for measuring the relative humidity under changing relative humidity ensuring a good hysteresis in the measuring device. However, the experimentally obtained sensitivity are not comparable with those obtained in the capillary fiber microresonators reported in some previous studies, since this type of resonators have a sensitivity ten times higher. The fabrication of these devices allow measurement of relative humidity with good accuracy and at the same time can be incorporated into integrated photonic devices for detecting substances, viruses or bacteria.

5. References

[1] Murugan G S, Wilkinson J S and Zervas M N 2009 Selective excitation of whispering gallery modes in a novel bottle microresonator Opt. Express 17 11916-11925

[2] Murugan G S, Wilkinson J S and Zervas M N 2010 Optical excitation and probing of whispering gallery modes in bottle microresonators: potential for all-fiber add–drop filters *Opt. Lett.* **35** 1893-1895

[3] Murugan G S, Wilkinson J S and Zervas 2010 Novel fiber bottle microresonator add-drop filters Optical Components and Materials VII *Proc. of SPIE* **7598**, 75980K

[4] Pöllinger M, O'Shea D, Warken F and Rauschenbeutel A 2009 Ultrahigh-Q tunable whispering-gallery-mode microresonator *Phys. Rev. Lett.* **103**(5) 053901

[5] Gouveia M A, Avila P D, Marques T H R, Torres M C and Cristiano M. B 2015 Cordeiro, "Morphology dependent polymeric capillary optical resonator hydrostatic pressure sensor *Opt. Express* **23** 10643-10652

[6] Padilla D A A 2014 Polymeric Capillary Optical Resonator Sensors Latin America Optics and Photonics Conference *Opt. Soc. Am., Tech. Dig. Ser.* paper LTu4A.38

From Phonomecanocardiography to Phonocardiography computer aided

J Granados[1], F Tavera[1], G López[2], J M Velázquez[1], R T Hernández[1] and G A López[1]

[1]Universidad Autónoma Metropolitana Azcapotzalco, México.
[2]Instituto de Educación Media Superior del Distrito Federal, México.

Email: jalgras@yahoo.com

Abstract. Due to lack of training doctors to identify many of the disorders in the heart by conventional listening, it is necessary to add an objective and methodological analysis to support this technique. In order to obtain information of the performance of the heart to be able to diagnose heart disease through a simple, cost-effective procedure by means of a data acquisition system, we have obtained Phonocardiograms (PCG), which are images of the sounds emitted by the heart. A program of acoustic, visual and artificial vision recognition was elaborated to interpret them. Based on the results of previous research of cardiologists a code of interpretation of PCG and associated diseases was elaborated. Also a site, within the university campus, of experimental sampling of cardiac data was created. Phonocardiography computer-aided is a viable and low cost procedure which provides additional medical information to make a diagnosis of complex heart diseases. We show some previous results.

1. Introduction

Some years ago, cardiac auscultation was the way in which physicians, based on the sounds emitted by the heart, diagnosed many diseases from simple to complex one. The tool was just the stethoscope. To correctly interpret what they listened to, it was necessary to submit resident's doctors to a training, which lasted two years, after completing their studies of general medicine [1].

On auscultation, a physician should detect to be present, not just the two basic sounds that make up a normal heartbeat, their uniformity and regularity, as well as the existence of some other sounds, clicks, or blows, that would indicate a normal or abnormal functioning of the heart and the parts that constitute it [4,7].

1.1. The Phonomecanocardiography and cardiac physicians training

W. Einthoven, 1894, did the first Phonocardiogram in the world [3]. B. Fishleder [2], in the 1950s, created a device, the Phonomecanocardiograph, that made possible, among other things, to graph in paper the sounds emitted by the heart, therefore it was possible to visualize what was heard with a stethoscope. The closer sounds and their intensities, as the so-called splits in Systole or Diastole, which the human ear cannot tell the difference, are clearly visible in a graphic record. The Phonomecanocardiograph served as a great support for the process of educational training of cardiologist's future. Guadalajara, mentions that the capacity of the resident doctors in cardiology to discriminate different diseases, only using the stethoscope, is 50% [1, 11], that is why the use of the Phonomecanocardiograph was very important.

Content from this work may be used under the terms of the Creative Commons Attribution 3.0 licence. Any further distribution of this work must maintain attribution to the author(s) and the title of the work, journal citation and DOI.
Published under licence by IOP Publishing Ltd

1.2. Other diagnostic technologies

Unfortunately, with the arrival of new technologies such as the electrocardiogram, echocardiogram, computerized tomography and the magnetic resonance imaging, the use of the Phonomecanocardiograph gradually reduce until its disappearance losing a valuable diagnostic tool.

On the other hand, these new technologies are not only more expensive, they require special equipment and trained staff to use them, moreover their availability is only found in clinics and hospitals in big cities.

2. The interpretation of the PCG and associated diseases

Fishleder, Guadalajara, in Mexico [1, 2], and several other cardiology researchers from different countries, left a legacy of very important information since they related graphs or images of various heart sounds emitted with different heart pathologies [4-10], [12-17]. This is the basis on which can now make the interpretation of the phonocardiograms.

The acoustic analysis of a PCG seeks to respond several aspects: Periodicity. The captured wave recurs on a regular basis? If it is regular, how is it formed? What is the frequency? What additional peaks presented and where are they? What cardiac pathologies can be inferred?

In healthy subjects, the phonocardiogram represents only two basic sounds in a completely regular form, usually called S1 and S2, simulated in Figure 1, [20]. The frequency of the wave is called heart rate or number of pulses per minute. But in non-healthy subjects may appear other sounds added visualized as peaks situated in several places of the wave form.

Taking into account the symbology used to describe phonocardiograms [20], it is possible to make a first classification of 15 possible alterations that may be visible in the graphic wave, and Table I outlined some of the associated pathologies. This is not an exhaustive list.

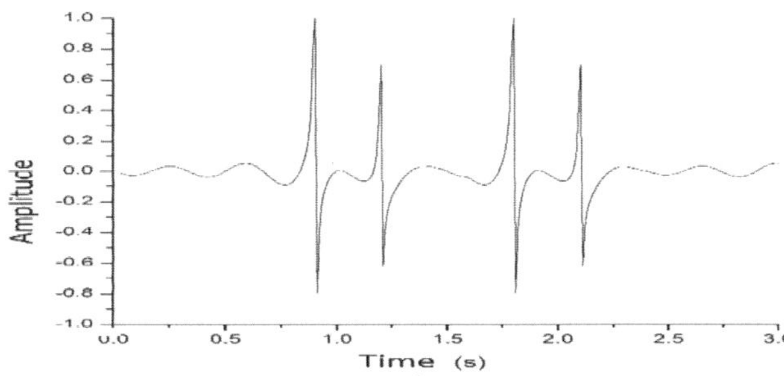

Figure 1. Simulation of two beats in a healthy person. The peaks indicate S1 and S2 sounds [20].

1. The S1peak split, in two tightly together peaks called M and T.
2. The S2 peak split in two tightly together peaks called A and P.
3. If a third sound S3, complex form, after S2, appears.
4. If a fourth sound S4, a narrow peak, before S1, appears.
5. If a click or an additional narrow peak after S2, appears.
6. If a peak of sound after S1, appears.
7. If an intermediate sound between S1 and S2, appears.
8. If high-frequency sinusoidal waves between S1 and S2 are seen.
9. If appears high frequency sinusoidal waves cushioned exponentially immediately after S1.
10. If see high frequency sine wave between S1 and S2, growing exponentially.
11. If see high-frequency sinusoidal waves between S1 and S2, increasing and decreasing exponentially to form a diamond.
12. If high frequency sinusoidal waves appear near to S2, cushioned exponentially
13. If waves are sinusoidal decreasing exponentially low amplitude and long decay time, between S2 and S1.

14. If high frequency sine waves type immediately before S1, growing exponentially, are seen.
15. If waves are increasing and decreasing high-frequency sinusoidal exponentially forming a diamond, in the time of systole and diastole, absence of S1 and S2.

Table 1. Although the possible associated cardiac pathologies to waveform alterations listed are so many, some are mentioned.

Case	Name	Possible Pathology
1	Split in S1	Myxoma in Atria, atrial septal defect
2	Split in S2	Aortic stenosis. Hypertrophic cardiomyopathy
3	Third heart sound	Mitral regurgitation. Ischemic cardiomyopathy
4	Fourth heart sound	Aortic stenosis. Pulmonary Arterial hypertension
5	Opening Snap	Mitral stenosis. Tricuspid stenosis.
6	Ejection Sound	Aortic stenosis. Pulmonary stenosis
7	Click Mid Systolic	Hypertrophic cardiomyopathy. Mitral Valve prolapse
8	Holosystolic Murmur	Mitral and Tricuspid Regurgitation. Severe pulmonary
9	Early systolic Murmur	stenosis
10	Late Systolic Murmur	Hypertrophic cardiomyopathy. Pulmonary stenosis.
11	Mid Systolic Murmur	Mitral regurgitation.
		Hypertrophic cardiomyopathy. Coarctation of the aorta
		Aortic and pulmonary stenosis
12	Early Diastolic Murmur	Aortic and pulmonary regurgitation
13	Mid Diastolic Murmur	Aortic and Mitral regurgitation. Atrial Septal Defect
14	Presystolic Murmur	Mitral stenosis and aortic Tricuspid regurgitation
15	Continuous Murmur	Cameral coronary fistula. Pulmonary embolism

3. The Phonocardiograms and data acquisition technology

Due to technological advances in computing and scientific instrumentation, using an electronic stethoscope, like sound sensor, software and a laptop computer now is possible to obtain the PCG of any subject.

As a research project, Phonocardiography, which currently takes place in the Optics Laboratory of the Universidad Autónoma Metropolitana Azcapotzalco, a digital process of cardiac data collection: blood pressure, heart rate, oxygenation and heart sound, in which also has been made phonocardiograms of people in the University community, healthy and diagnosed patients, was implemented. In aleatory form, subjects were selected among pupils, Teachers or Administrative staff.

Using a digital meter of blood pressure and heart rate, Omron model HEM - 7114, the Oximeter Hergom MD300, Littmann Electronic Stethoscope Model 3200, Littman StethAssist and LabVIEW 2013 software, the heart information from the subjects was obtained.

Through a program of computation, data acquisition instruction is given to a laptop that the acoustic signal emitted by the heart and detected by the electronic stethoscope, to be transferred, via Bluetooth. That wave is exported in format .wav. With LabVIEW software, built a program that converts the signal to .lvm format, which allows us to show it as an amplitude vs. time graphic on the PC screen, which can be extended or compressed for display in different time intervals.

4. Image Recognition

Using new computer technology and developing an artificial vision algorithm for pattern recognition, in LabVIEW software, built a program that consists of three routines, which are described below.

The first routine consists of reading of a .wav file, previously obtained with the electronic stethoscope. The program displays the sound wave on the screen and computes the following parameters: the main frequency of the sound, the period with which each peak is repeated times, the number of pulses per minute, the amplitude of the maximum peaks for the characterization of regular and strong or weak sound hearts, the time that it takes to pass the systole and diastole in each cycle and the differences between these times.

The Next routine, which converts the .wav to an image format .jpg, the image is analyzed using artificial vision to identify the distance between each wave peak in pixels, thus giving us a relationship with the time elapsed between consecutive pulses.

In the same way as with acoustic analysis, can determine the existence of some type of arrhythmia or tachycardia if the repetition period is not constant over time.

This method can also be used to identify the presence, and time of occurrence, of peaks with abnormal amplitude between pulses. These peaks suggest the presence of an intense sound, which diagnose a possible blow and its type. The algorithm can also measure the vertical distance of the wave, the amplitude of the sound.

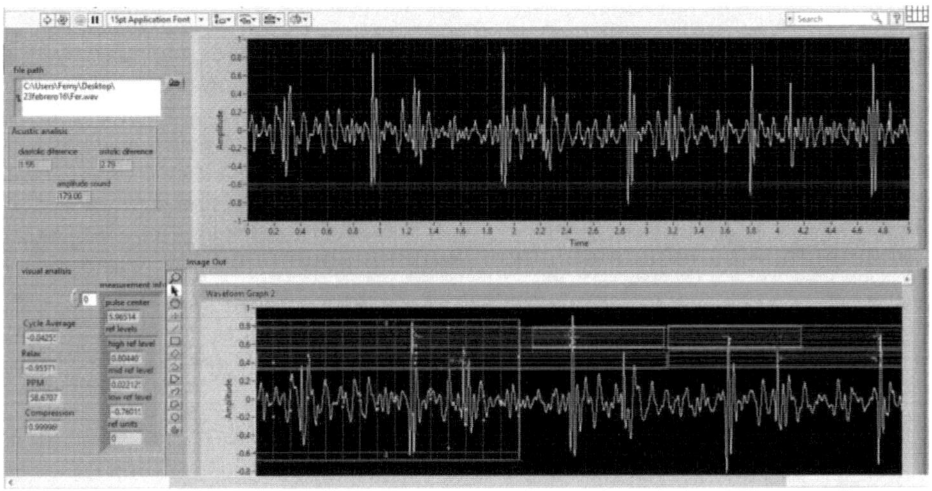

Figure 2. Front panel of the LabVIEW program of acoustic and visual Phonocardiograms Diagnostics

5. Results

We have performed about 62 Phonocardiograms of people, randomly chosen, into university community, of several ages, physical conditions, sex, healthy and non-healthy, diagnosed and not.

Important heart functioning irregularities in different ages, apparently healthy individuals, were detected, with ease, without cost, without special facilities and expensive equipment. In another paper will present in detail the respective analysis. In Fig. 3 to 6, shows four cases of PCG displayed in 10 seconds, identifying possible diseases associated to variations of waveforms obtained in comparison with the Table I. The PCG of student 1, Fig. 3 shows the first sound S1, too weak, that means a possible Atrioventricular Block. In the case of the student 2, Fig. 5 the PCG shows a regular waveform, the student is healthy. For the student 3, Fig. 6 the PCG shows a split in S2, which means a possible Aortic Stenosis. And for the student 4 the graph shows split in S1 and split in S2 which means possible Myxoma in Atria and Aortic Stenosis.

Table 2. Data from volunteers who underwent phonocardiograms

Voluntary	Gender	Age (year)	Weight (kg)	Height (m)	Blood pressure	Heart rate	blood oxygenation	Diseases
1	M	23	64	1.7	108/66	64	98	Nothing
2	F	20	48.5	1.57	97/65	72	98	Nothing
3	M	28	63	1.69	119/71	70	96	Nothing
4	F	24	54	1.54	105/72	70	97	Lung diseases

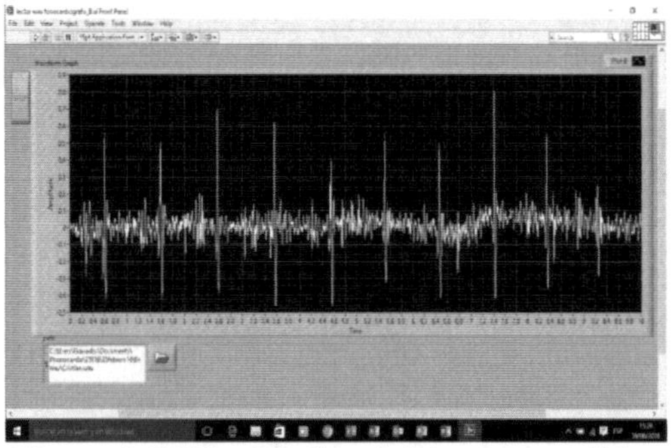

Figure 3. PCG of student 1. Means a possible Atrioventricular Block.

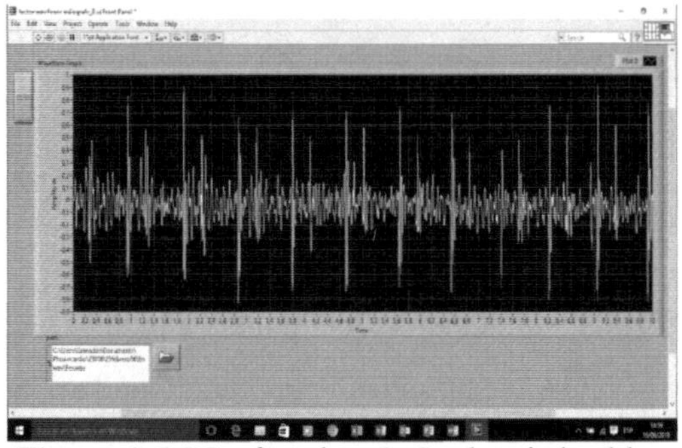

Figure 4. PCG of student 2. Student is healthy.

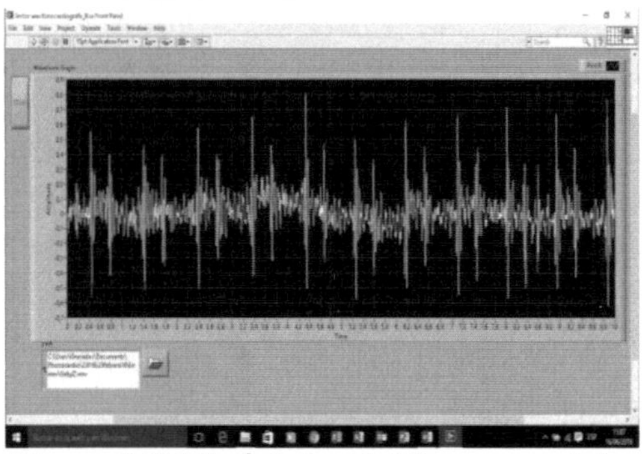

Figure 5. PCG of student 3. Means a possible Aortic Stenosis.

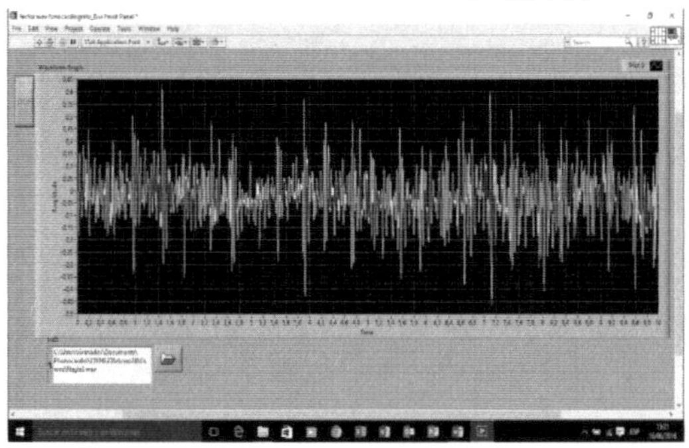

Figure 6. PCG of student 4. Means possible Myxoma in Atria and Aortic Stenosis.

6. Conclusions

Looking for an objective method for quickly, by means of sound that emits the heart, obtain important information of its operation, with the aim to develop a procedure to help in diagnostic of hearth pathologies, we have developed the technique called Phonocardiography Computer Aided. In it, the sound generated by the heart of people is captured in a laptop computer. Subsequently turning sound into image, it analyzes the image looking for alterations in the waveform, relating them with pathologies.

The research project, Phonocardiography, which currently takes place in the UAM-A has achieved: Implement a digital process of cardiac data acquisition: blood pressure, cardiac frequency, heart sound and oxygenation. Present the sound wave as image in format .lvm, which can be extended or compressed for display in different time intervals. Mathematically simulate the heart sound wave. Build an image recognition program to detect critical areas in the image of the heart sound.

Phonocardiography is an alternative method to diagnose cardiac pathologies. Phonocardiography Computer Aided is a feasible technique, very low cost, which gives immediate objective results and which can be applied in remote populations with limited economic resources.

Thanks to the application of image recognition and acoustic visual techniques, is easy to identify alterations in the sound wave and correlate them with diseases.

7. References

[1] Guadalajara B 2015 *Gaceta Médica* (México) **151** 260-5

[2] Fishleder B 1966 *Exploración cardiovascular y fonomecanocardiografía clínica* (México: La Prensa Médica Mexicana)

[3] Einthoven W and Geluk M 1894 *Die Registrierung, Der Herztöne. Pflügers Arch Ges Physiol* pp 57-617

[4] Leatham A 1958 *Auscultation of the Heart* (London, Lancet) **2** 7049 703-8
[5] Suton G and Leadam A 1968 *Second Heart Sound in pulmonary Hipertension* (Br Heart) **30** 6 743-56
[6] Stel I 1952 *The murmur of high pressure in the pulmonary artery Med Chron* (Manchester) **9** 1888-9
[7] Harvey W 1968 Some newer and poorly recognized findings with clinical auscultation (I) *Mod. Concepts Cardiovasc. Dis.* **37**(3) 85-8
[8] Adolf R 1970 The second heart sound: a screening test for heart disease *Mod. Concepts Cardiovasc. Dis. **39**(4) 91-6*
[9] Perloff J 1964 Clinical recognition of aortic stenosis *Prog. Cardiovasc. Dis.* **10** 323-9
[10] Leatham A 1987 *Auscultation and phonocardiography: a personal view of the past 40 years* (Br Heart) **57**(5) 397-403
[11] Mangione S, Nieman L, Gracely E and Kaye D 1993 The teaching and practice of cardiac auscultation during internal medicine and cardiology training *Ann. Intern. Med.* **119**(1) pp 47-54
[12] Shaver J, Salerni R and Reddy P 1985 *Normal and abnormal heart sounds in cardiac diagnosis, Part I: Systolic sounds Current Problems in Cardiology* **10**(3) pp 1-68
[13] Reddy P, Salerni R, and Shaver J 1985 *Normal and abnormal heart sounds in cardiac diagnosis, Part II: Diastolic sounds Current Problems in Cardiology* **10**(4) pp 1-55
[14] Akay A, Semmlow J, Welkowitz W, Bauer M and Kostis J 1990 Detection of coronary occlusions using autoregressive modeling of diastolic heart sounds *IEEE Trans. Biomed. Eng.* **37**(4) pp 366-373
[15] Stein P and Sabbah H 1980 Intensity of the second heart sound: Relation of physical, physiological and anatomic factors to auscultatory evaluation *Henry Ford Hosp. Med. J.* **28**(4) pp 205-209
[16] Wang J, Tie B, Welkowitz W, Semmlow J and Kostis J 1990 Modeling sound generation in stenosed coronary arteries *IEEE Trans. Biomed. Eng.* **37**(11) pp 1087-1094
[17] Fredberg J 1977 Origin and character of vascular murmurs: Model studies *J. Acoust. Soc. Am.* **61**(4) pp 1077-1085
[18] Granados J, Rodríguez L and Velázquez J 2010 Phonocardiograph signals acquisition and analysis system *Proceedings of the 1st. International Congress on Instrumentation and Applied Sciences* (México)
[19] Granados J, Tavera F, Velázquez J, López G, Hernández R and Morales A 2014 Fonocardiografía asistida por computadora *Memorias del V Congreso Nacional de Tecnología aplicada a ciencias de la salud* (México)
[20] Granados J, Tavera F, Velázquez J, López G, Hernández R and Morales A 2015 Acoustic heart. Interpretation of Phonocardiograms by Computer *J. Phys.: Conf. Ser.* **582** 012057

1020 steel coated with Ti/TiN by Cathodic Arc and Ion Implantation

F Bermeo[1], J P Quintana[2,3], A Kleiman[2,3], F Sequeda[4] and A Márquez[2,3]
[1] Universidad Santiago de Cali, Cali, Colombia
[2] Universidad de Buenos Aires, Facultad de Ciencias Exactas y Naturales, Departamento de Física, Buenos Aires, Argentina
[3] Universidad de Buenos Aires, Consejo Nacional de Investigaciones Científicas y Técnicas, Instituto de Física del Plasma (INFIP), Facultad de Ciencias Exactas y Naturales, Buenos Aires, Argentina
[4] Universidad del Valle, Cali, Colombia

E-mail: frank@usc.edu.co

Abstract. TiN coatings have been widely studied in order to improve mechanical properties of steels. In this work, thin Ti/TiN films were prepared by plasma based immersion ion implantation and deposition (PBII&D) with a cathodic arc on AISI 1020 steel substrates. Substrates were exposed to the discharge during 1 min in vacuum for the deposition of a Tiunderlayer with the aim of improving the adhesion to the substrate. Then, a TiN layer was deposited during 6 min in a nitrogen environment at a pressure of $3x10^{-4}$ mbar. Samples were obtained at room temperature and at 300 °C, and with or without ion implantation in order to analyze differences between the effects of each treatment on the tribological properties. The mechanical and tribological properties of the films were characterized. The coatings deposited by PBII&D at 300 °C presented the highest hardness and young modulus, the best wear resistance and corrosion performance.

1. Introduction

Cathodic arc coatings have been extensively studied and used due to their high deposition rates. In cathodic arc discharges a metallic plasma jet is ejected from the cathode surface and it is deposited onto a substrate. The substrate conditions used during the deposition process -temperature and bias respect to the plasma potential- influence on the structure, morphology and adhesion of the film [1,2].

Nanostructured Ti thin films have been obtained using cathodic arcs and also by combining plasma immersion ion implantation and deposition (PIII&D) [3, 4]. Polycrystalline titanium thin films have been widely employed as interlayer between the substrate and different coatings in order to improve adhesion strength, corrosion resistance and wear performance, as well as to promote the growth of crystalline phases of the coating.

Studies on the properties of TiN coatings obtained using PIII&D revealed significant improvement of tribological performance with the application of pulsed bias [5]. There are also studies on the tribological properties of TiN coatings grown with cathodic arc on stainless steel including Ti interlayers [6].

In this work, thin Ti/TiN films were prepared by PIII&D with a cathodic arc on AISI 1020 steel substrates. Coatings were obtained at different temperature and with or without ion implantation in order to analyze differences between the effects of each treatment on the tribological properties.

Content from this work may be used under the terms of the Creative Commons Attribution 3.0 licence. Any further distribution of this work must maintain attribution to the author(s) and the title of the work, journal citation and DOI.
Published under licence by IOP Publishing Ltd

2. Experiment

The experiment was carried out in a d-c vacuum arc system, which is schematically shown in figure 1. The discharge was produced between a cylindrical Ti cathode (60 mm in diameter) and a grounded annular Cu anode (80 mm in diameter). Both electrodes were mounted on an insulating piece that set an electrode separation of about 15 mm. The cathode was surrounded by a floating shield. The discharge circuit consisted in a current supply (18 kW, 150 A) in parallel with a capacitor bank (165 mF) connected to the electrodes through a series inductor (2.8 mH) in order to provide arc stability. A high voltage generator applying pulses of 8 kV (30 μs, 200 Hz) was employed for ion implantation. The chamber was an electrically insulated stainless steel cylinder (25 cm long, 10 cm diameter). The base pressure of the system was lower than 10^{-5} mbar.

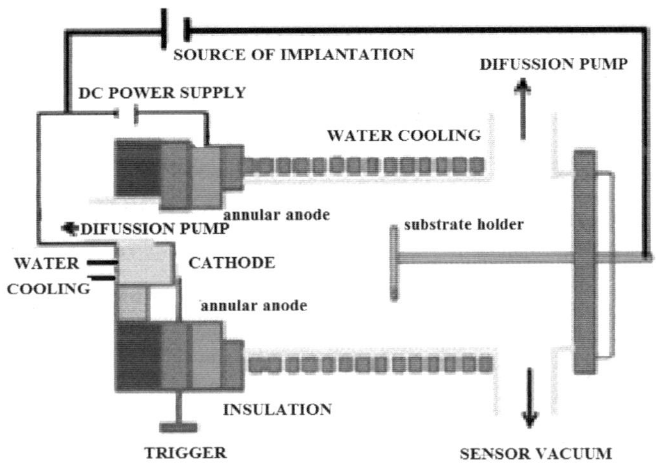

Figure 1. Schematic diagram of the d-c vacuum arc system.

The employed substrates were AISI 1020 steel sheets (22 mm long, 17 mm wide and 3 mm thick). They were placed at 20 cm from the cathode surface on a heater. The surfaces were previously cleaned with alcohol in an ultrasound bath for 30 min. Substrates were exposed to the discharge during 1 min in vacuum for the deposition of a Ti underlayer, and during 6 min in a nitrogen environment at a pressure of 3x10-4 mbar for TiN deposition. Samples were obtained at room temperature and at (300 ± 10) °C, with and without implantation.

Film thickness was determined by using a Calotester, the coating was eroded with a steel ball (diameter 22.2 mm) employing standard metallurgical diamond lapping paste (1/4 μm). The tribological characterization was performed using a brand tribometer (CSEM Instruments SA) with an alumina ball at a speed of 5 cm/s with a load of 1 N for a distance of 400 m, in air at 18 °C and 43% relative humidity. Nanohardness was measured with a nanoindenter Nanovea using the model of Oliver and Pharr and a maximum load of 5 mN. Corrosion rate measurements were performed by a potentiostat galvanostat PG-TEKCORR USB 4.1 with a brine solution (H2O + NaCl 3.5%) and Ag/AgCl electrodes. The roughness was studied with a KLA Tencor ALPHA STEP 120 profilometer, using a diamond-tipped stylus, with a resolution of 10 Å.

3. Results and discussion

To identify samples prepared in each condition, coatings deposited at room temperature with cathodic arc are denoted as A, with cathodic arc and ion implantation as AI, and coatings obtained with arc and implantation at 300 °C are named AIT.

Coatings obtained in different growth conditions were examined by a calotester and the thickness resulted in (740 ± 50) nm for all the samples. This value corresponds to the whole film (Ti + TiN), the thickness of each layer being not able to assess. Surface roughness results are shown in table 1. Roughness is very similar in all the films and is clearly lower than that of the substrate. Typical hardness measurements by the method of Oliver and Pharr are shown in figure 2. The figure shows measurements at different places along an AI sample. It can be noticed an elastic/plastic deformation and that the signals are continuous, indicating the absence of film cracking during the indentation. Results obtained for hardness and elastic modulus from these measurements for different deposition conditions are listed in table 1.The substrate hardness measured with the Brinell hardness test was 130 HB (equal to 1,4 GPa) and the elastic modulus for the AISI 102 steel was ~ 200 GPa, both values were included in table 1 in order to compare them with the obtained results. It can be seen that the hardness and elastic modulus increased with the coatings, the obtained values being similar for different deposition methods. It is worth mentioning that the hardness measurement can be influenced by the hardness of the substrate, given that the depth is higher than 10% of the thickness.

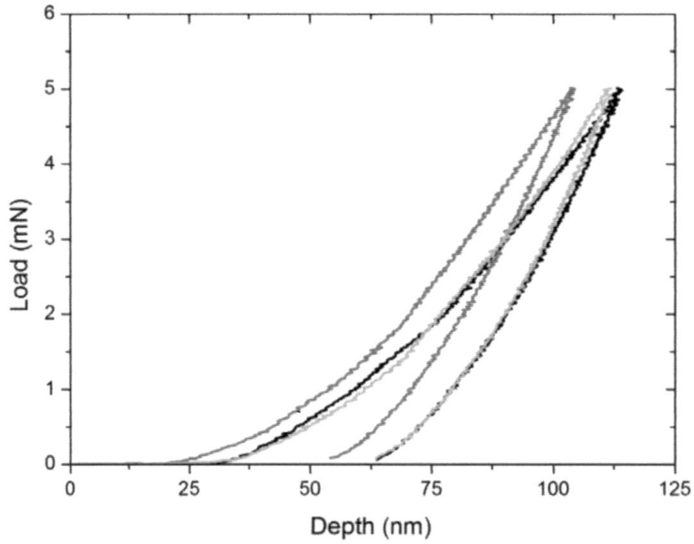

Figure 2. Hardness measurements at different places of an AI sample.

Table 1. Roughness, hardness, elastic modulus and wear rate for different growth conditions

Growth conditions	Substrate	A	AI	AIT
Roughness (± 2 nm)	624	189	194	202
Hardness (± 2 GPa)	1.4[a]	27	26	28
Elastic modulus (± 40 GPa)	200[a]	290	320	400
Wear rate ($\mu m^3/Nm$)	0.84	0.22	0.27	0.003

[a] Hardness and elastic modulus of the substrate are indicated as reference

Figure 3 shows wear measurements corresponding to an AIT sample. In the post wear scarring, no detachment or delamination can be observed. The wear rate obtained from the evaluation of the worn volume for different deposition conditions is presented in table 1. The wear rate decreases significantly, respect to the substrate, in samples A and AI, and even more for samples AIT.

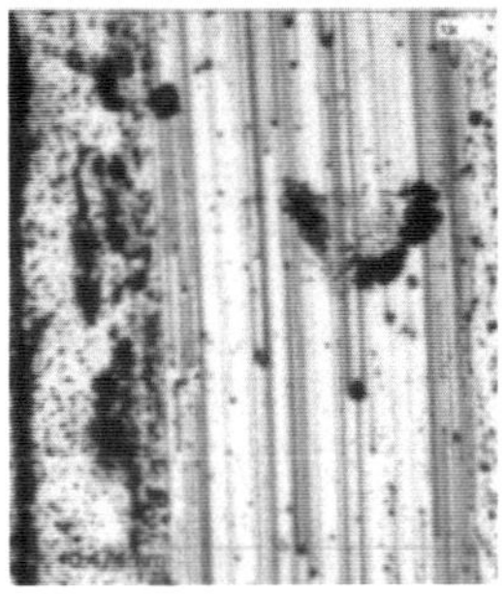

Figure 3. Wearscarring from an AIT coating.

Friction coefficient for Ti/TiN coating deposited in different conditions is plotted in figure 4. Signals corresponding to the coatings were registered up to 400 m distance, while that of the substrate was measured to 100 m. Samples prepared in the three conditions presented a similar behaviour, the friction coefficient being initially in the range 0.2 – 0.3 and finally reaching the substrate value, around 0.8, indicating an abrasive process. It can be seen that the transition from the initial to the final value for the friction coefficient takes places at longer distance in the cases of the samples AIT and A.

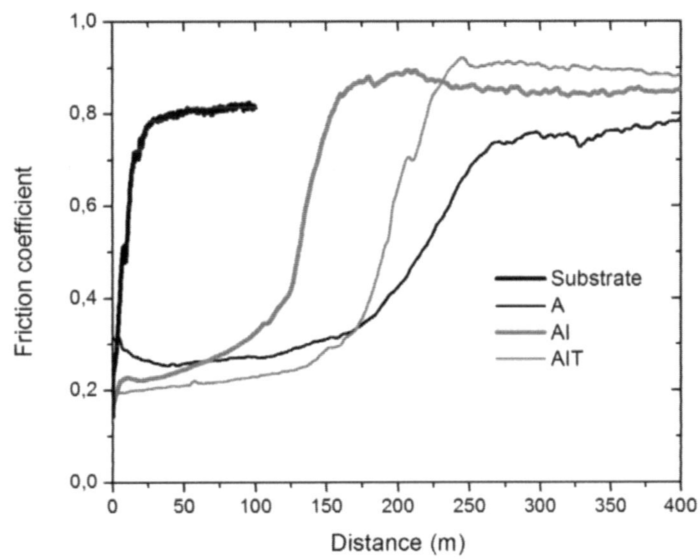

Figure 4. Friction coefficient for Ti/TiN samples deposited in different conditions.

Figure 5 shows the corrosion potential as a function of the current density for samples deposited in different conditions. It can be clearly noticed that the sample prepared with arc and ion implantation at 300 °C (AIT) exhibited the lowest corrosion potential, while samples grown in other conditions presented similar values to that of the substrate. Results obtained from corrosion measurements are summarized in table 2. The AI sample presented similar corrosion behaviour to that of the substrate, regarding all the parameters. On the other hand, the AIT coating showed the lowest corrosion velocity, and then the best corrosion resistance performance.

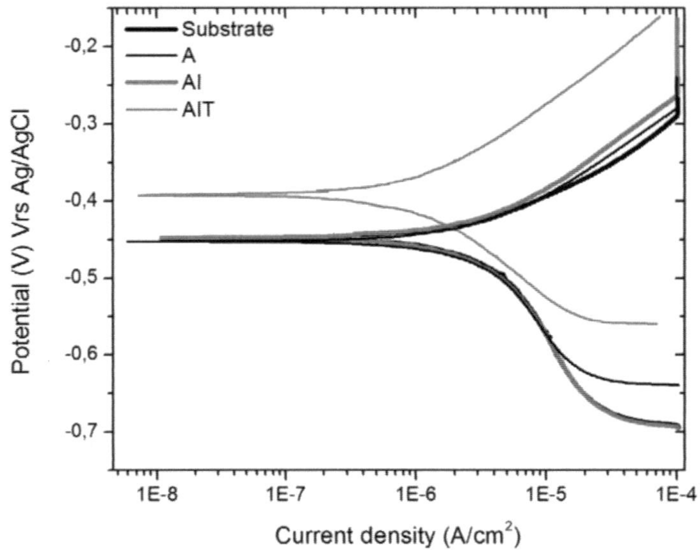

Figure 5. Corrosion potential as a function of the current density for Ti/TiN samples deposited in different conditions.

Table 2. Corrosion parameters obtained from Ti/TiN samples deposited in different conditions.

Growth conditions	Corrosion velocity (mm/y)	Current density (nA/cm^2)	Corrosion potential (mV)
Substrate	164	320	-448
A	128	249	-453
AI	163	318	-448
AIT	119	231	-393

4. Summary

Ti/TiN coatings on stainless steel substrates were obtained by cathodic arc deposition and ion implantation. The obtained hardness and friction coefficient was similar for all the samples. Surface hardness of the coated samples was one order of magnitude greater than that of the substrate. Despite the final friction coefficients were similar for the different coating conditions, for A and AIT samples the transition from the initial low value to the final value for the friction coefficient takes places at longer distance.

The best corrosion and wear properties were obtained for AIT samples. The corrosion velocity for AIT samples was ~ 30% lower than that of the substrate, while the reduction in the corrosion velocity for A samples was ~ 20%. The wear rate in all deposition conditions was significantly reduced respect to the substrate, the value corresponding to AIT samples being two orders of magnitude lower than those of the other conditions.

Therefore, deposition by cathodic arc combined with ion implantation and heating the substrate led to the best overall performance and is promising for industrial applications.

5. Acknowledgments

This work was supported by grants from Universidad de Buenos Aires, CONICET, Hard Coatings and Industrial Applications Laboratory (RDAI, Universidad del Valle) and Santiago de Cali University. The authors would like to thank Carolina Ortega for the collaboration in the characterization of samples.

IOP Conf. Series: Journal of Physics: Conf. Series **792** (2017) 012061 doi:10.1088/1742-6596/792/1/012061

6. References

[1] Boxman R L, Sanders D M and Martin P J (Eds.) 1995 *Handbook of Vacuum Arc Science and Technology, Fundamentals and Applications* (Park Ridge, NJ: Noyes)
[2] Lieberman M and Lichtenberg A 2005 *Principles of Plasma Discharges and Material Processing* (Hoboken, NJ: Wiley) 2nd Edition pp 152-154
[3] Giuliani L, Bermeo F, Lamas D, Grondona D, Kelly H and Márquez A 2009 *Rev. Latin. Am. Metal. Mater.* S **1** 1073-79
[4] Fazio M, Kleiman A, Lamas D G, Grondona D and Márquez A 2014 *J. Phys. Conf. Series* **511** 012069
[5] Akkaya S S, Vasyliev V V, Reshetnyak E N, Kazmanlı K, Solak N, Strel'nitskij V E and Ürgena M 2013 *Surf. Coat. Technol.* **236** 332-40
[6] Lain G C, Cemin F, Menezes C M, Aguzzoli C, Baumvol I J R, Tomiello S S and Figueroa C A 2016 *Surf. Eng.* **32** 279-83

IOP Conf. Series: Journal of Physics: Conf. Series **792** (2017) 012062 doi:10.1088/1742-6596/792/1/012062

Sample entropy applied to the analysis of synthetic time series and tachograms

A Muñoz-Diosdado[1], G G Gálvez-Coyt and E Solís-Montufar
[1]Basic Sciences Department, UPIBI, Instituto Politécnico Nacional, Mexico City, Mexico

E-mail: amunozdiosdado@gmail.com

Abstract. Entropy is a method of non-linear analysis that allows an estimate of the irregularity of a system, however, there are different types of computational entropy that were considered and tested in order to obtain one that would give an index of signals complexity taking into account the data number of the analysed time series, the computational resources demanded by the method, and the accuracy of the calculation. An algorithm for the generation of fractal time-series with a certain value of β was used for the characterization of the different entropy algorithms. We obtained a significant variation for most of the algorithms in terms of the series size, which could result counterproductive for the study of real signals of different lengths. The chosen method was sample entropy, which shows great independence of the series size. With this method, time series of heart interbeat intervals or tachograms of healthy subjects and patients with congestive heart failure were analysed. The calculation of sample entropy was carried out for 24-hour tachograms and time subseries of 6-hours for sleepiness and wakefulness. The comparison between the two populations shows a significant difference that is accentuated when the patient is sleeping.

1. Introduction

Currently, cardiovascular diseases are the leading cause of death worldwide and more people die for this reason than for any other reason [1, 2]. Factors such as obesity, the consumption of tobacco, physical inactivity, high blood pressure, hyperlipidemia, diabetes, and inadequate diet, are risk factors for cardiovascular disease, even more when these factors arise together [1]. Age is a factor that affects the heart and vessels, operating on the structures and systems of the body in general, so that it favours the emergence of cardiovascular diseases.

The use of non-linear analysis for the study of complex systems with fractal properties, allows more accurate descriptions of their behavior. The human body consists of different complex systems that have fractal and multifractal behaviours in healthy conditions [3], and it has been reported a decrease in these fractal trends when the systems are in a pathological condition [5].

Congestive Heart failure (CHF) is a syndrome in which patients typically have lack of air both at rest and during exercise; they show signs of fluid retention, and objective evidence of a structural or functional cardiac disturbance at rest [5]. The heartbeat interval time series obtained from ECG records of patients with CHF have been studied with various techniques of Nonlinear Dynamics, especially fractal and multifractal analysis [3, 4] and found significant differences between healthy and CHF subjects. Complexity of these signals seems to be different, so we propose that computational sample entropy could help to differentiate the two types of series, and thus be able to support the diagnosis of patients with heart failure.

Content from this work may be used under the terms of the Creative Commons Attribution 3.0 licence. Any further distribution of this work must maintain attribution to the author(s) and the title of the work, journal citation and DOI.
Published under licence by IOP Publishing Ltd

2. Methodology

In many cases, the information obtained from the measuring devices provides temporary signals $x(i)$, where $i = 1,..., N$. The temporary registration of data obtained from measurements allows predicting the behaviour of the systems [6]. Time series can be characterized by four basic types of variation, which produce the changes observed in the series in a certain time interval, and give the series an erratic behaviour. These components are: secular trend, seasonal variation, cyclical variation and irregular variations, which are short term variations that are unpredictable and non-recurrent.

Frequently, it seems to be that a time series does not have useful information, but a more careful analysis can show the existence of correlations within the series, that is, the existence of a relation between the value at a moment i with the value at the moment j if j > i. When there is no relation between the value at the moment i with the value at the moment j (i ≠j) it is said that the time series is a succession of independent samples, this sort of time series is called white noise. In the case of white noise time series, a random ordering of its components would be again a white noise.

On the other hand, when the value of the variable i influences the future values j > i, it is said that there is dependence and therefore there is correlation. For example, a time series obtained from the accumulation of independent events of white noise is called Brownian noise; this type of noise corresponds to short range correlations [7].

We used a database of physiological signals from the web page Physionet [8] to obtain beat to beat intervals time series of 24 hours, this kind of time series is called tachogram. It was used the Normal Sinus Rhythm RR Interval Database, consisting of a population of 54 healthy individuals (30 men aged between 28 and 76) and 24 women aged between 58 and 73. It was also used two populations of CHF patients, one of them corresponding to the BIDMC Congestive Heart Failure RR Interval Database, which consists of 15 patients (11 men aged between 22 and 71) and 4 women aged between 54 and 63 classified in the NYHA scale 3-4, and a last database named Congestive Heart Failure RR Interval, with 29 patients (which includes 8 men and two women aged 34 to 79 and other 17 individuals of unknown gender) of different NYHA classifications [9]. The NYHA index was proposed by the Nueva York Heart Association as a method to classify CHF patients, the value IV corresponds to the patients in worst conditions.

The 24-hour signals were segmented into series of 6 hours for wakefulness and 6 hours for sleep intervals. Discrimination was possible thanks to the significant reduction in heart rate that individuals suffer while they are asleep, and the contrast with the beat to beat interval of the awake individual, which demands a greater flow of blood and this is reflected in the time between R and R [10] in the electrocardiogram signal.

To apply computational entropy analysis it is required to ensure that the used method provide values that do not depend on the series size, which is a problem with this type of methodologies, also is required a proper calibration of the method. Time series generated with a suitable methodology were used for this. For this reason were used time series with a certain value of β spectral power were generated, these series were obtained using the method proposed by Gálvez-Coyt et al. [7]. In this method a Gaussian white noise with N data is generated, and then the discrete transform of Fourier is applied. The result is the flat spectrum corresponding to a $\beta = 0$ spectral power, where the amplitude $|X(m)|$ is the same for the coefficients with $m = 1, 2, 3, \ldots, N$.

Subsequently new coefficients of Fourier transform $X'(m)$ were built, from those obtained from the Fourier transform of the Gaussian white noise:

$$X'(m) = \left(\frac{m}{N}\right)^{-\beta/2} X(m) \quad (1)$$

With $m = 1, 2, 3, \ldots, N$. Now the magnitude of the filtered data is taken $|X'(m)|$ and a phase θ_m uniform and random is generated for each one in the $[0,\ 2\pi]$ interval. Then we get the Fourier coefficients for the magnitudes and phases previously calculated by using:

$$X''(m) = |X'(m)|(\cos\theta_m + i\sin\theta_m), \tag{2}$$

$n = 1, 2, 3, \ldots, N$. The inverse Fourier transform of $|X''(m)|$ is obtained,

$$x(n) = \frac{1}{N\delta}\sum_{n=1}^{N} X''(m) e^{2\pi i n m / N}, \tag{3}$$

With the purpose of obtaining a noise Gaussian with a certain value of β, the real part of the series [19] is considered. In Figures 1, 2 and 3 are shown the self-affine time series with N = 10000 data for β values 0, 1, and 2.

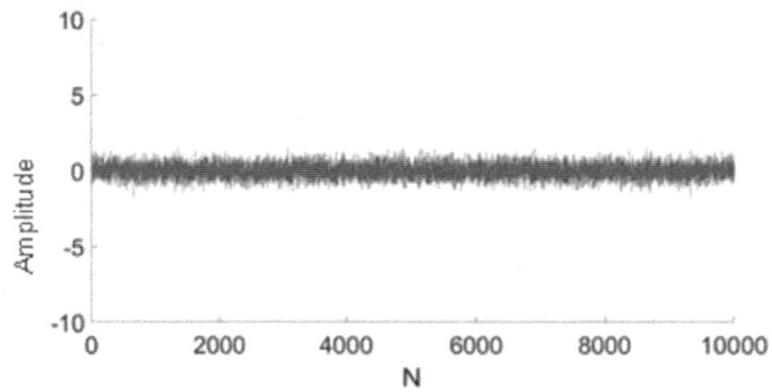

Figure 1. Graph obtained for a time series generated by the algorithm described in the text with power spectra exponent β = 0.

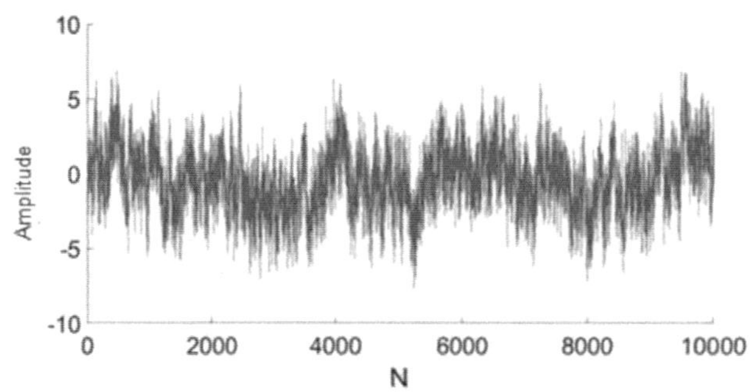

Figure 2. Graph obtained for a time series with power spectra exponent β = 1.

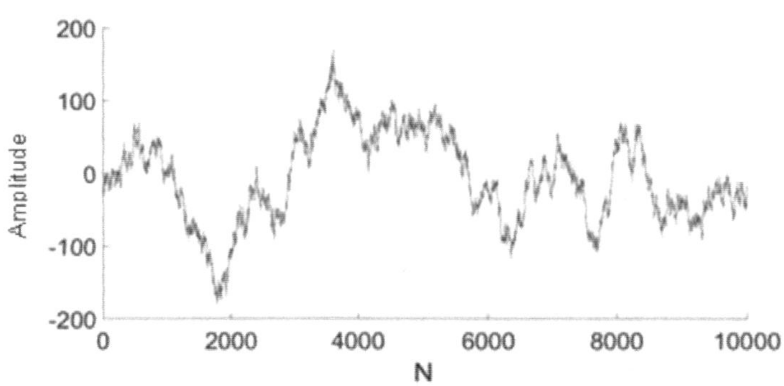

Figure 3. Graph obtained for a time series with power spectra exponent β = 2.

The method was implemented to generate a group of signals with N = 10000 data with known values of β, with β ranging from β = 0 to β = 2 with 0.1 increments between each series. Then we proceed to evaluate different types of entropy, but ultimately only the sample entropy was used because the best results were obtained for this type of entropy. The method proposed by Richman and Moorman was used for the calculation of the sample entropy (SampEn) [11]. Taking into account the time series, vectors of length m are defined by: $X_m(i) = \{x(i), X(i+1), \ldots, x(i+m-1)\}$, for $1 \leq i \leq N - m + 1$. These vectors represent m consecutive values of x starting at point i, subsequently the distance between the vectors Xm (i) and Xm (j) is defined as the absolute value of the maximum difference between its scalar components:

$$d_m[X_m(i), X_m(j)] = \max[X_m(i+k) - X_m(j+k)], 0 \leq k \leq m-1 \quad (4)$$

Subsequently, for a given $X_m(i)$, the number of $j(1 \leq j \leq N-m, j \neq 1)$ is counted, such that the distance between $X_m(i)$ and $X_m(j)$ is less than or equal to r. Taking into account that $1 \leq i \leq N-m$:

$$B_i^m(r) = \frac{1}{N-m-1} B_i \quad (5)$$

$B^m(r)$ is defined as:

$$B^m(r) = \frac{1}{N-m} \sum_{i=1}^{N-m} B_i^m(r) \quad (6)$$

Then m is changed to m + 1 and $A_i^m(r)$ is calculated as:

$$A(r) = \frac{1}{N-m} \sum_{i=1}^{N-m} A_i^m(r) \quad (7)$$

As a result, $B^m(r)$ is defined as the probability that two sequences match in m points, and thus $A^m(r)$ corresponds to the probability that two sequences match on m + 1. So the sample entropy can be defined as:

$$SampEn(m, r, N) = -\ln\left[\frac{A^m(r)}{B^m(r)}\right] \quad (8)$$

The values of *r* and *m* are decisive for the precision of the method. In this case, values of $r = 0.2$ and $m = 3$ were chosen, which proved to be suitable for this type of time series analysis, allowing a greater efficiency in the method and reducing the loss of information [10].

3. Results

For the characterization of the sample entropy method, it was applied to the theoretical time series previously generated, calculating the value of entropy for series of N = 10000 data and values of power spectral exponent ranging from β = 0 to β = 2 with increments of 0.1 between each series. Figure 4 shows the graph obtained for this calculation, being the abscissas axis the corresponding β values and the ordinates axis the entropy value. The method shows a linear trend extending from the values of *β = 0.7* to *β = 2.0*, the human beat to beat intervals time series belongs to this interval.

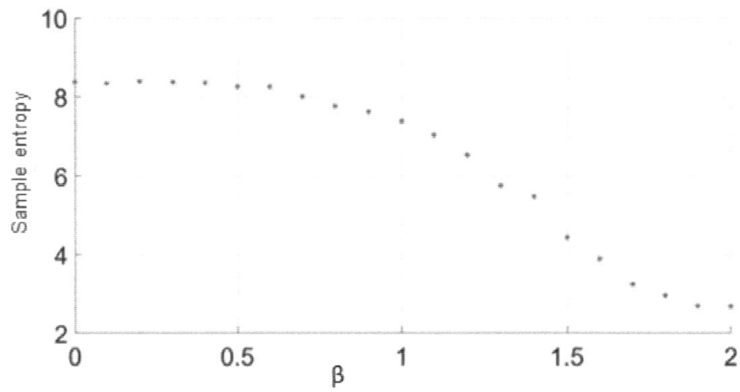

Figure 4. Sample entropy for different values of β.

Once characterized the method with time series of known values of β, it followed the analysis of the time series of the CHF database and healthy people for the three types of series (total 24 hours, sleep 6 hours and wake 6 hours). We proceeded to perform the analysis with the method of sample entropy. In the following figures, the graphic representation of the values sample entropy for each one of the samples, and the average value of both populations is represented as continuous lines. On dispersion graphs shown below, blue colour corresponds to healthy people, and red colour to heart failure patients.

Likewise, the statistical test for each type of study, i.e., complete series, series of 6 hours of sleep and series of 6 hours of wakefulness was made by using T-Student with a significance value of 0.5 and the hypothesis that were used for the previous study: the null hypothesis was $H_0: SE\ healthy = SE\ CHF$ and the alternative hypothesis $H_1: SE\ healthy \neq SE\ CHF$ (SE is sample entropy).

In the comparison carried out when they are sleeping, it was found a considerable difference between both averages, this difference is statistically significant. Whereas, for the series in which the individuals are awake, it can be seen a very small gap between the two average values, and even the average entropy of healthy people has a lower value than the patients with CHF, which does not correspond with the values obtained for the sleep period.

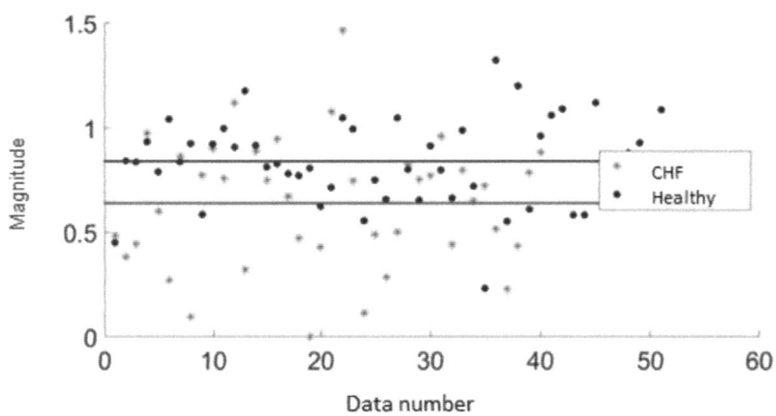

Figure 5. Sample entropy for healthy persons and CHF patients in time series of 6 hours when they were sleeping.

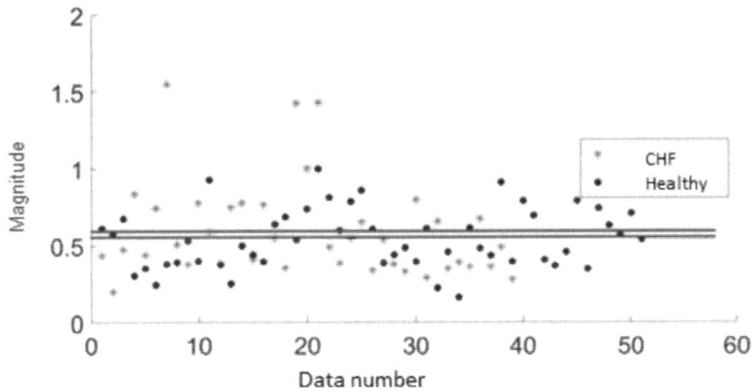

Figure 6. Sample entropy of patients and healthy people in 6 hour time series in the awaken phase.

4. Conclusions

Unlike other methods for evaluating the entropy, sample entropy does not present alterations in their results by the length of the series, however, the processing time increases exponentially as the size of series increases. The use of the $r = 0.2$ value gave reliable results in the calculation of the sampling entropy. We recommend the use of value m = 3 which was taken from the literature, which is recognized as a value that allows a satisfactory study of entropy and bringing precision to the method, without losing information. Applying the method to the tachogram analysis of healthy people and CHF patients it was possible to differentiate both groups, but the best results were obtained when people are sleeping, it is believed that this is because the condition of patients with CHF is aggravated by night.

5. References

[1] Murray C J L and Lopez A D 1997 *Lancet* **349** 1269
[2] Alberti K G M, Eckel R H, Grundy S M, Zimmet P Z, Cleeman J I, Donato K A, Fruchart J C, James W P T, Loria C M and Smith S C 2009 *Circulation* **120** 1640
[3] Stanley H E, Amaral L A N, Goldberger A L, Havlin S, Ivanov P Ch and Peng C K 1999 *Physica A* **270** 309
[4] Muñoz-Diosdado A, Angulo-Brown F and Del Río-Correa J L 2005 *Proceedings IEEE EMBS Annual International Conference* **27** 6981
5] McKee PA, Castelli W P, McNamara P M and Kannel W B 1971 *N. Engl. J. Med.* **285** 1441
[6] Falk M 2012 *A First Course on Time Series Analysis* (Germany: University of Würzburg)
[7] Gálvez-Coyt G, Muñoz-Diosdado A, Del Río-Correa J L and Angulo-Brown F 2010 *Fractals* **18** 235
[8] Goldberger A L, Amaral L A N, Glass L, Hausdorff J M, Ivanov P Ch, Mark R G, Mietus J E, Moody G B, Peng C K and Stanley H E 2000 *Circulation* **101** e215
[9] Bennett J A, Riegel B, Bittner V, Nichols J (2002) *Heart and Lung: Journal of Acute and Critical Care* **31** 262
[10] Muñoz-Diosdado A, Ramírez-Hernández L, Aguilar-Molina A M, Zamora-Justo J A, Gutiérrez-Calleja R A and Virgilio-González C D 2014 *AIP Conf. Proc.* **1626** 155
[11] Richman J S, Moorman J R 2000 Heart and Circulatory Physiology *Am. J. Physiol* **278** 2039
[12] Wessel R, Koch C and Gabbiani F 1996 Coding of time-varying electric field amplitude modulations in a wave-type electric fish *J. Neurophysiol.* **75** 2280–93

Reducing hydrogen permeation in 304 stainless steel by compound layers of Al, Zr and Ti oxides films

R T Hernández L[1], V Cortes S[1], J Granados S[1] and S Orozco S[2]
[1]División de Ciencias Básicas e Ingeniería, UAM-Azcapotzalco, Av. San Pablo 180, Azcapotzalco, Cd. de México, 02200, México
[2]Departamento de Física, Facultad de Ciencias, UNAM, Av. Universidad 3000, UNAM, Coyoacán, Cd. México, 04510, México

E-mail: hlrt@correo.azc.uam.mx

Abstract. A single and double layer formed by thin films coatings of aluminium oxide, zirconium oxide and titanium oxide were deposited over 304 stainless steel surface by ultrasonic spray pyrolysis technique. The steel samples were conformed for tensile tests. The purpose of these layers is to reduce hydrogen embrittlement effect in steel. An electrochemical cell was used in hydrogen charged, where a low concentration of sulfuric acid is utilized like electrolytic solution. Tension trials show the change the fracture type in samples with or without coating after hydrogen charged. The embrittlement percent factor and SEM micrographs indicate a reduction of hydrogen permeation for coated samples with double layer.

1. Introduction

In many cases the metals and alloys are in environments with extensive possibilities to generate hydrogen by the electrolytic natural phenomena, by example the different steels used in pipelines of the petrochemical industries or like structural materials in different devices and vehicles [1]. In these conditions hydrogen can adhere to the surface metal and diffuse into, this process is possible to ambient temperature [2, 3]. Then to occur the formation in the metal of hydrogen-rich zones, which will is accumulating more and more thereby causing cracking and embrittlement of the metal [4-6]. This can lead to catastrophic consequences. Is necessary to protect metals in rich ambient of hydrogen because can occur embrittlement [7, 8]. To prevent or reduce hydrogen diffusion it been proposed ceramic coatings like barriers [9].

J. P. Hirht [3], propose a parameter to evaluate the hydrogen embrittlement effect (%HE) in terms of the area reduction percent for samples without (%ARsH) and with (%ARcH) hydrogen charged after tensile test,

$$\%HE = \frac{\%AR_{sH} - \%AR_{cH}}{\%AR_{sH}} * 100 \qquad (1)$$

Content from this work may be used under the terms of the Creative Commons Attribution 3.0 licence. Any further distribution of this work must maintain attribution to the author(s) and the title of the work, journal citation and DOI.
Published under licence by IOP Publishing Ltd

In this work, single and double layers formed by aluminium (Al), zirconium (Zr) and titanium (Ti) oxides thin films are used to reduce the hydrogen diffusion in the 304 stainless steel. The charged of hydrogen is realized with an electrochemical cell.

2. Experimental details

Fig. 1 has the drawing and photograph of the tensile test piece of commercial purity 304 stainless steel. The steel has the nominal composition (weight %): 0.08 C (max.), 1.00 Si (max.), 2.00 Mn (max), 18.00-20.00 Cr, 8.00-12.00 Ni, Fe (balance). The steel pieces were stress relieved treatment at 650°C by 60 min and sensitized by 30 hr at 1050°C.

Coated films over tensile samples were deposited by ultrasonic spray pyrolysis technique from a solutions of Al, Zr and Ti acetyl-acetonates dissolved in N,N-dimethylformamide, with the respective concentrations; 0.32M, 0.38M and 0.10M. In only test piece one-side films were deposited. The deposit of a single layer was performed in the following order; first, the Al oxide film at 550°C by 50s, after the Zr oxide film at 550°C by 50s and finally the Ti oxide film at 500°C by 50s (Fig.2). The second layer is obtained in the same form. The coated morphology is analysed with optical (OM) and electronic (SEM) microscopies.

The charged of hydrogen was realized with an electrochemical cell, a 0.5M concentration of acid sulfuric was used like solutions media and 0.45 mA/cm2 current density applied during 96 hours.

The tensile test was doing with the mechanical essay universal machine, (Instron, mod. 5500R). Test was to 0.16mm/s mobile head velocity. Mechanical parameters; yield stress, tensile strength, elongation % and reduction area % are evaluated from the stress vs. strain curve and SEM micrographs. The parameters values between specimen with and without coated are compared. And the fracture surface morphology was evaluated with OM and SEM.

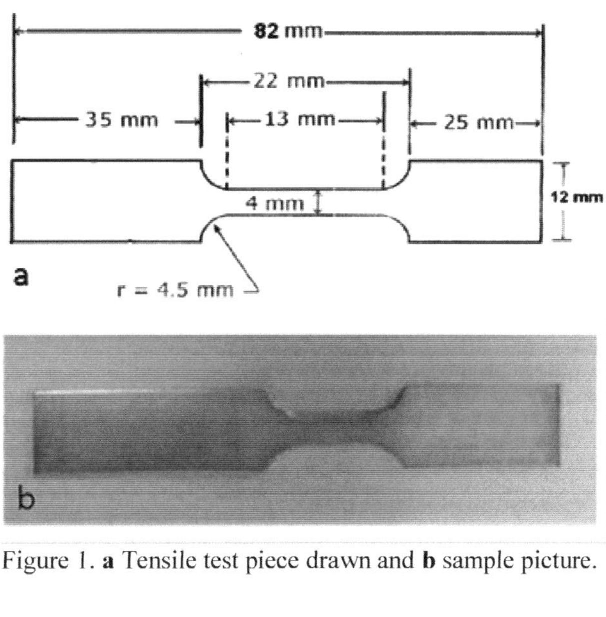

Figure 1. **a** Tensile test piece drawn and **b** sample picture.

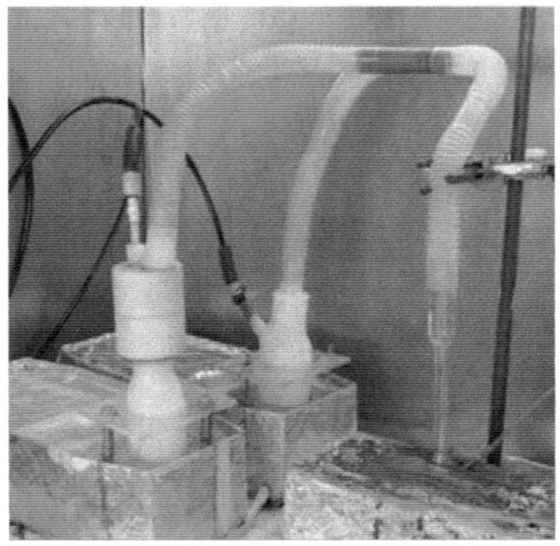

Figure 2. Section of films deposit equipment.

3. Results and discussion

OM and SEM micrographs show a smooth and uniform coating with 3μm for single-layer and 5.4μm of double-layer of thickness average (Fig. 3b). The Fig. 3a show grains in form of twins and many precipitates in the grains limit in the sensitized steel surface. Initially in hydrogen charged process a transitory stage the hydrogen diffusion process begins with the accumulation of hydrogen in the surface and proper diffusion follows.

The Fig. 4 shows the characteristics stress-strain curves for tensile testing until fracture for the specimens in different conditions: testing (1 curve), single layer coating and hydrogen charged (3 curve), with double-layer and hydrogen charged (2 curve) and only hydrogen charged (4 curve). The curves present a similar form. The maximum elongation occurs for the testing samples and minimum in the sample only with charged. The yield stress, elongation %, tensile strength and area reduction % are shown in the Table 1. The control sample and the sample charged double layer and hydrogen have similar mechanical properties, it appears to have no effect of the presence of ceramic films.

It is remarkable the difference in value of the maximum stress or tensile steel hydrogen only charged relative to the other samples and lower % elongation. This behaviour is associated with greater hydrogen diffusion and therefore steel hardening. The mechanical behaviour of sample with double layer is nearing to testing sample, which is indicative of lower hydrogen diffusion; the layer acts as a barrier for hydrogen movement to the metal interior.

The embrittlement percent factor (% HE) is evaluated in the fracture surfaces of tension test. Minor % HE values obtained for the samples with ceramic layer, 6.2 and 8.4, with respect to the sample only hydrogen charged of 17.2, indicate a reduction in the hydrogen permeation. That is, ceramic layers decrease the intensity of hydrogen diffusion in the steel.

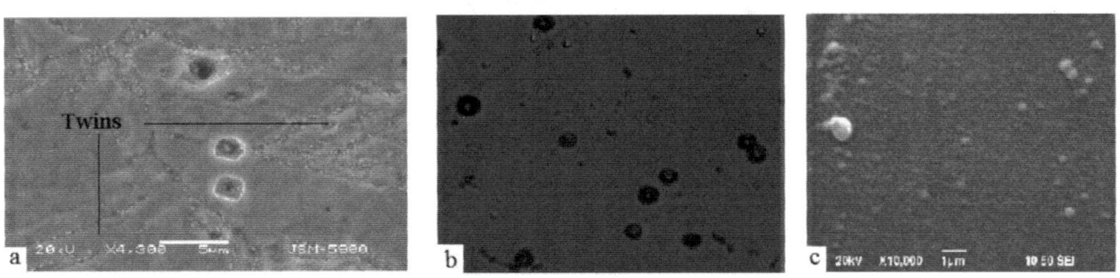

Figure 3. Test samples surfaces micrographs; **a** only sensitized SEM, **b** with single-layer coating by OM to 100X and **c** double-layer coating by SEM.

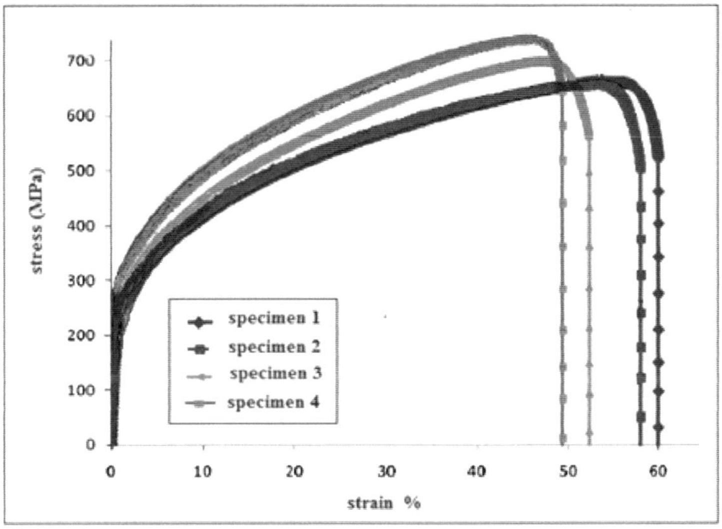

Figure 4. Stress-strain curves; 1 testing, 2 double-layer coating, 3 single-layer and hydrogen charged and 4 only charged.

The comparison of the morphologies of fracture surfaces shows different fracture surface (Fig. 5). The testing sample has a typical morphology of ductile fracture; neck formation and void coalescence (Figs. 4 and 5). The sample only hydrogen charged (control piece) and samples with coating and hydrogen charged don´t form neck, this result indicates that a brittle fracture. However, the fracture mechanism is different, in the first case samples show a quasi-cleavage fracture type and in the other case the surface shows big voids, similar morphology to a ductile material. Brittle fracture is related to hardening of the material associated with the presence of hydrogen-rich zones and fractures.

Table 1. Tensile test parameters and embrittlement factor of characteristics samples.

Sample	**Yield Stress**	**Tensile strength**	**Elongation**	**Area reduction**	**Embrittlement factor**
1(testing)	258.4 MPa	633.3 MPa	58.1 %	53.6 %	----
2(double-layer)	249.1 MPa	624.9 MPa	60,0 %	50.3 %	6.2 %
3(single-layer)	282.4 MPa	637.2 MPa	52.3 %	49.1 %	8.4 %
4(only charged)	270.5 MPa	698.1MPa	40.7 %	44.4 %	17.2%

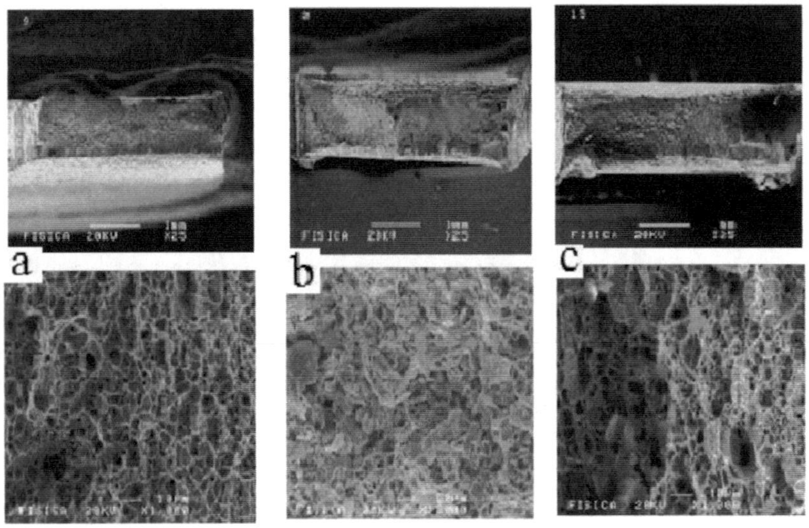

Figure 5. Surface fracture micrographics; testing (a), only hydrogen charged (b) and with double-layer coating charged (c).

4. Conclusions

Uniform coating of one and two layers of aluminium, zirconium and titanium oxides films were deposited on 304 stainless steel by the ultrasonic spray pyrolysis technique. The results of the mechanical tests for charged and uncharged samples with hydrogen have the effect on the steel mechanical properties. Embrittlement factor (% HE) resulted largest for the sample with double layer, this suggests that the layer is reducing diffusion of hydrogen. Also, the fracture surface micrograph shows big voids, morphology characteristics of ductile material.

5. References

[1] Majciejny A 2000 *Jour. Mats. Process Tech.* **106** pp 184-190
[2] Wojciechowski S 2000 *Jour. Mats. Process Tech.* **106** pp 230-235
[3] Hirth J P 1984 *Hydrogen Embrittlement and Stress Corrosion Cracking* ed R. Gibala and R.F. Hohemann (Ohio: ASM) pp 29-41
[4] Thompson A W 1976 *Effect of Hydrogen on Behavior of Materials* (Merced, California: AIMM) pp 467-479
[5] Thompson A W 1979 *Environment-Sensitive Fracture of Engineering Materials* ed Z.A. Foroulis (Englewood, Colorado: Conf. Proceed. Metall. Soc. AIME) pp 379-410
[6] Cheng P 1981 *Corrosion* **37** pp 39-50
[7] Davnahan M V A and Stachruski Z 1964 *J. Electrochem. Soc.* **111** pp 619-623
[8] Hirth J P 1980 *Met. Trans.* **11A** pp 861-890
[9] Yen S K, Huang I B and Yen Z S 2000 *NACE Int.* **56** 10 pp 120-127

MEGARA Optics: stain removal in PBM2Y prisms

D Aguirre-Aguirre[1,2], R Izazaga-Pérez[3], B Villalobos-Mendoza[2,4], E Carrasco[3], A Gil de Paz[5], J Gallego[5] and J Iglesias[6]
[1]Centro de Ciencias Aplicadas y Desarrollo Tecnológico, Universidad Nacional Autónoma de México, Cd. Universitaria, Apdo. Postal 70-186, C.P. 04510, Cd. Mx., México.
[2]Polo Universitario de Tecnología Avanzada, Universidad Nacional Autónoma de México, Vía de la Innovación No. 410, Autopista Monterrey-Aeropuerto Km. 10, PIIT C.P. 66629 Apodaca N. L., México.
[3]Instituto Nacional de Astrofísica, Óptica y Electrónica, Luis Enrique Erro No.1, Sta. María Tonantzintla, C.P. 72840 Puebla, México.
[4]Centro de Investigación y de Estudios Avanzados, Unidad Monterrey, Apodaca N.L. 66629, México
[5]Universidad Complutense de Madrid, España.
[6]Instituto de Astrofísica de Andalucía-CSIC, España.

E-mail: daniel.aguirre@ccadet.unam.mx

Abstract. MEGARA is the new integral-field and multi-object optical spectrograph for the GTC. For medium and high resolution, the dispersive elements are volume phase holographic gratings, sandwiched between two flat windows and two prisms of high optical precision. The prisms are made of Ohara PBM2Y optical glass. After the prisms polishing process, some stains appeared on the surfaces. For this, in this work is shown the comparative study of five different products (muriatic acid, paint remover, sodium hydroxide, aqua regia and rare earth liquid polish) used for trying to eliminate the stains of the HR MEGARA prisms. It was found that by polishing with the hands the affected area, and using a towel like a kind of pad, and polish during five minutes using rare earth, the stains disappear completely affecting only a 5% the *rms* of the surface quality. Not so the use of the other products that did not show any apparent result.

1. Introduction

MEGARA is the acronym of Multi-Espectrógrafo en GTC de Alta Resolución para Astronomía. It is the new generation optical spectrograph of the 10.4m Gran Telescopio CANARIAS. The instrument is being built by a consortium of public institutions lead by the Universidad Complutense de Madrid (UCM, Spain) and that includes the Instituto Nacional de Astrofísica, Óptica y Electrónica (INAOE, México), the Instituto de Astrofísica de Andalucía (IAA, Spain), and the Universidad Politécnica de Madrid (UPM, Spain) [1-6].

To provide the 18 spectral configurations 60 optical elements are required, 36 windows: 12 for low resolution (LR), 20 for medium resolution (MR), 4 for high resolution (HR) and 24 prisms: 20 for MR and 4 for HR. These 73 optical elements are being built at INAOE optical workshop in Mexico. The diameters of the lenses vary between 110 mm and 272 mm. In the case of the pupil elements, the dimensions of the of VPH's windows are 220 mm x 180 mm x 30 mm for LR, and 240 mm x 190 mm x 25 mm for MR and HR, with similar dimensions for the MR and HR prisms apertures [1-5].

A pair of prisms is obtained from an Ohara PBM2Y blank [7], with dimensions of 247 mm x 149 mm x 195 mm, figure 1a.

Content from this work may be used under the terms of the Creative Commons Attribution 3.0 licence. Any further distribution of this work must maintain attribution to the author(s) and the title of the work, journal citation and DOI.
Published under licence by IOP Publishing Ltd

An overview of the MEGARA project is presented by Gil de Paz et al. [1]. The manufacturing process is described by Carrasco et al. [2]. Specific processes in the optical fabrication are briefly described by Izazaga et al. [6].

In figure 1b, are shown the prisms mounting used during the polishing process. In the instrument design, the prisms have two polished surfaces. The required quality of the prisms HR are, for the first surface, 1 frg at the used aperture (180mm x 140 mm); and 0.5 frg at the central aperture (160 mm x 150 mm). For the second surface, the required quality is 2 frg at the used aperture (180 mm x 200 mm); and 1 frg at the central aperture (160mm x 200 mm). The first surface was designed to have an antireflective (AR) coating while the second surface will be cemented with a fused silica window.

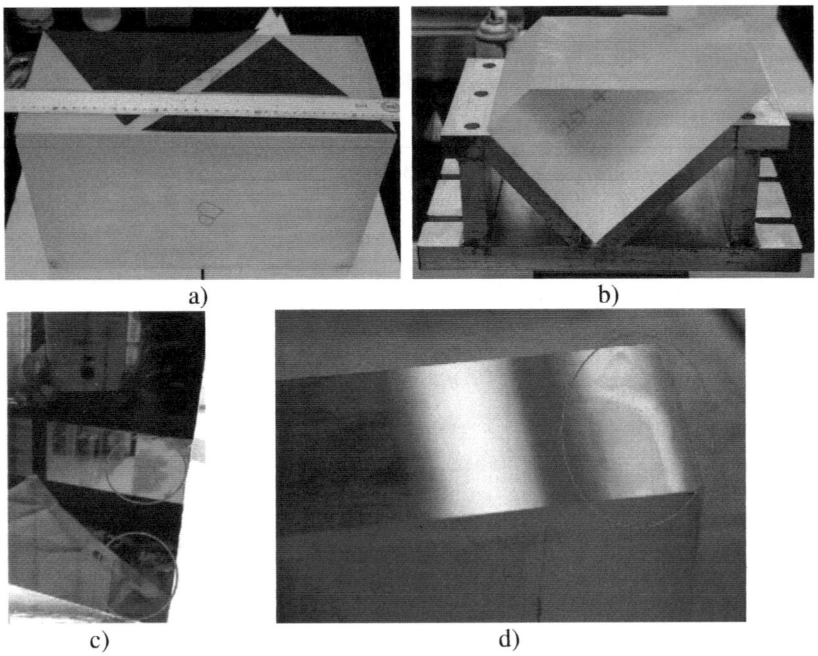

Figure 1. a) Ohara PBM2Y blank, b) HR prism mounted in a cast iron base, c) and d) stains on the polished surface.

During the polishing process, the prism is mounted in a cast iron base made in a CNC machine [7]. To clamp the prism in the base blocking wax was used. When the second surface of the prism is polished, the first surface is covered with a protecting coating to prevent scratches or digs. Then the prism is stuck to the base with wax. The stains problem appeared on the first surface when the polishing of the second surface was ended and the prism was removed from the mounting and it was cleaned [7], see figure 1c. It is believed that the stains were produced by the combination of the wax, the protected coating and the high temperature used to melt the blocking wax (≈70ºC). In figure 1d the stains are evident on the surface by using the reflection of a fluorescent lamp; we can compare the left side of the surface that it is free of stains. It is important to say that these stains appears only on a polished surface and are present in an optical glass with high acid resistance value.

These kind of stains on MEGARA prisms could produce diffused images, optical quality decreasing and maybe some problems with the AR coating or in the cementing with fused silica windows [1]. Thus, it is important that this optics do not have any kind of stains to avoid damages and guarantee the instrument performance.

From the above, the chemical products used are presented and also the techniques used for the stain removal at the INAOE´s optical workshop.

2. Tests for stain removal

Due to the stain problem that appeared during the polishing process of the second surface of the HR prism, it was necessary to make some tests to eliminate these stains. Polished pieces of the prism, obtained from the cutting process of the PBM2Y blank, were used for these tests, figure 1a. These pieces of glass have the same stains problem as the principal prism on the polished surface (large basis of the prism, figure 2).

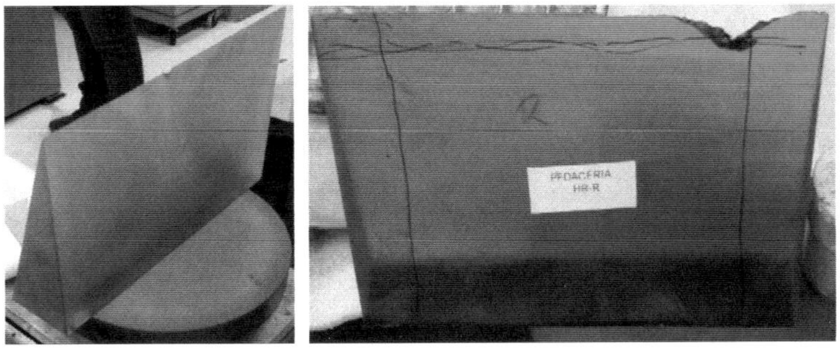

Figure 2. PBM2Y polished piece used for the cleaning tests.

The piece of glass was divided into three parts and the cleaning test was made on the ends of each surface. Figure 3 shows how the glass was divided.

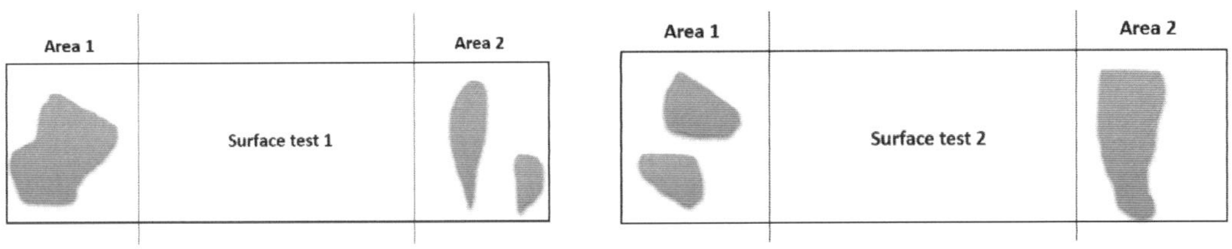

Figure 3. Scheme of the pieces used for the cleaning tests.

In Table 1 are shown the thermal and chemical properties of the glass.

Table 1. Properties of Ohara PBM2Y optical glass.

Thermal properties		*Chemical properties*	
Strain Point StP (°C)	385	Water Resistance(Powder) Group RW(P)	2
Annealing Point AP (°C)	418	Acid Resistance(Powder) Group RA(P)	1
Transformation Temperature Tg (°C)	436	Weathering Resistance(Surface) Group W(S)	1
Yield Point At (°C)	470	Acid Resistance(Surface) Group SR	1

2.1 Commercial products

The first tests were made with two different commercial products. The first product was Hydrochloric acid (muriatic acid), and the second product was paint remover (water + benzyl alcohol +titanium dioxide). These kinds of products were used because it was thought that the stains could be produced by a kind of thin film deposited on the surface because the chemical evaporation produced by the protecting coating and the blocking wax. Also these products are easy to get in any hardware store.

The first product was spread it above the stains of the surface test 1, area 1 figure 4a; the second product was spread it above the test surface 2, area 1 figure 4b. Both products were left act during 20 minutes but the results obtained were negatives and none of the products produced any perceptible change removing the stains.

a) b)

Figure 4. Tests with a) hydrochloric acid, and b) paint remover.

2.2 High corrosive chemical products

Due to the results obtained in the previous test, it was decided to make some tests with products used to remove some kinds of thin films deposited on optical glasses, i.e. to remove the aluminum deposited on optical mirrors.

The first of these products was the sodium hydroxide (caustic soda), and the second product was aqua regia (nitric acid + hydrochloric acid). The sodium hydroxide is highly corrosive and is used in the soap industry and for the cleaning and whitening in the textile industry. As solid, the sodium hydroxide has a white color. The aqua regia is also a highly corrosive solution and is composed with a ratio of one-to-three parts of nitric acid and hydrochloric acid mixture that has a yellow color.

The first surface test was left soaking in aqua regia, figure 5a, and the second test surface was left soaking in sodium hydroxide, figure 5b. This procedure was done similar to the process that is followed when a thin film of aluminum coating is removed from optical mirrors.

The two products were left act during 4 hours and the areas under test were checked every 20 minutes to see the progress. During that time was observed that the sodium hydroxide did not cause any apparent change on the surface stains. Whereas that, a little change or vanishing on the surface stains after two hours was observed with the aqua regia, but the stain did not disappear completely.

a) b)

Figure 5. Tests glasses into a) aqua regia and b) sodium hydroxide.

2.3 Cerium oxide polishing test

Because the results obtained in the last section, it was decided to use the solution that is applied in the majority of the cases when these kinds of troubles appear in optical fabrication, which is re-polish the damaged surface, with the polish machine. This alternative has some problems, one of them is that the grinding and polishing process has to be re-done, but this is time-consuming. Another problem is that, since the stain problem is not controlled, to polish the first surface implies to cover with a protecting coating the second surface, stick it to the base with blocking wax and heating again the material. As we know, the stains appeared when the prism was removed from the base, so risk exists with stains hat could appear on the other surface making this a cycle without end. To avoid this problem, it was decided to hand polish the surface with stains. These tests were done using cerium oxide (Universal Photonics rare-earth polishing compound).

In order to follow the figure behavior of the surface during the hand polishing we did an interferometric analysis of the surface irregularity. This analysis was done by using a Newton interferometer with a wavelength of 546.1 nm and an Edmund Optics lambda/10 reference flat. An interferogram was taken before and after the hand polishing. The interferogram was analyzed with the Durango interferometry software (Diffraction International Inc.). With this analysis, we can know the influence of the hand polishing on the surface irregularity in terms of PV and *rms* values. The results obtained in these tests are shown next.

2.3.1 Results for surface 1

Before polishing the area 2 of the first surface, the interferogram of figure 6a was captured. The values obtained when this interferogram was analyzed were 0.9158 frg PV and 0.1190 frg *rms*. The polishing process was done as follow: the area of interest was polished following a path of circles and using a smooth cotton towel similar to a polishing pad, this was done until the stains disappeared completely. This process took about 5 minutes.

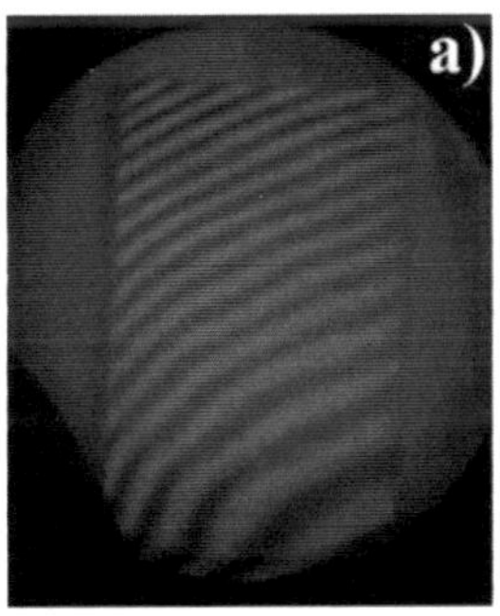

Statistics

Statistic	Value
P-V:	0.9158 fringes
RMS:	0.1190 fringes
Irregularity:	0.9216 fringes
RMSi:	0.1189 fringes

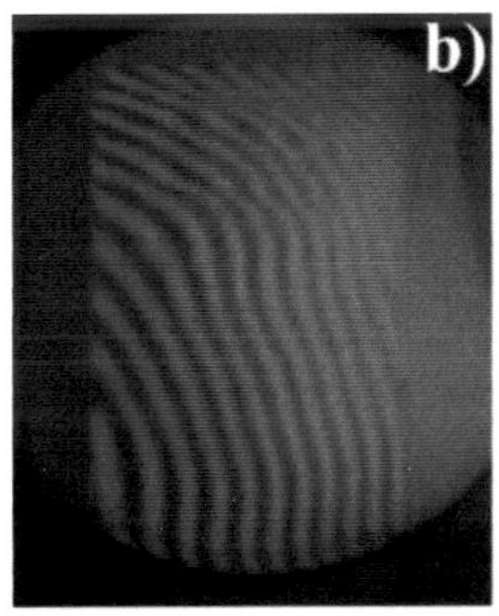

Statistics

Statistic	Value
P-V:	1.427 fringes
RMS:	0.2310 fringes
Irregularity:	1.325 fringes
RMSi:	0.2201 fringes

Figure 6. Data and interferogram a) before and b) after polishing surface 1.

Figure 6b shows the surface interferogram after hand polishing. As can be seen, the values of PV and rms were 1.427 frg and 0.2310 frg, respectively. It is observed that two values increased more or less 55% and 94%, respectively. This can be corroborated by observing the interferograms of figure 6b where the fringes are irregular compared with the interferogram fringes of figure 6a.

Although the stains almost disappeared, the problem of scratch and digs appeared on the surface. This effect is attributed to the aqua regia which caused damage to the surface due to its highly corrosive nature.

2.3.2 Results for surface 2

The interferogram that is shown in figure 7a was taken before polishing the area 2 of surface 2. In this figure it can be seen a horizontal gray shadow in the fringe pattern. This shadow corresponds to one of the stains on the surface. The values obtained from the analysis of this interferogram were 1.069 frg PV and 0.2125 frg *rms*.

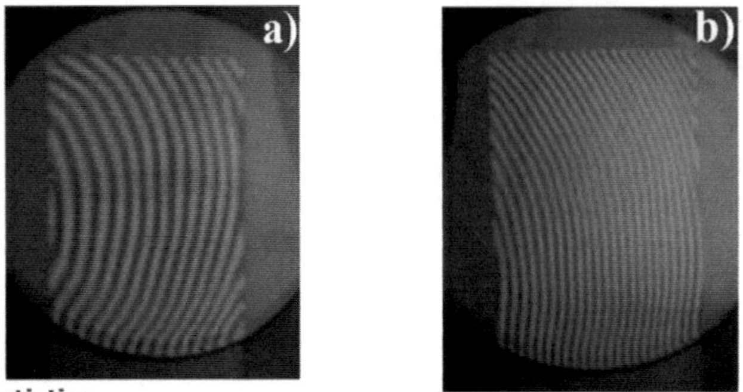

Statistics

Statistic	Value
P-V:	1.069 fringes
RMS:	0.2125 fringes
Irregularity:	1.047 fringes
RMSi:	0.2119 fringes

Statistics

Statistic	Value
P-V:	1.069 fringes
RMS:	0.2235 fringes
Irregularity:	1.074 fringes
RMSi:	0.2235 fringes

Figure 7. Data and interferogram a) before and b) after polishing surface 2.

The hand polishing time consuming for surface 2 to get stains vanishing was 5 minutes, the same of the last test. Comparing these results against the results for surface 1, this test the scratch and dig defects did not appear because the caustic soda is not more aggressive than the aqua regia. After the hand polishing, the values obtained for the interferogram shown in figure 7b were 1.069 frg PV and 0.2235 *rms*. In this test, the variation in PV and *rms* was 0% and 5%, respectively. Also, it can be seen in figure 7b that the shadow of figure 7a disappeared completely without affecting the irregularity of the interference pattern.

3. Conclusions

In this work was shown the comparative study of different commercial products used for the stain removal in the HR prisms for the MEGARA spectrograph. The results with hydrochloric acid, paint remover, and caustic soda were not satisfactory. The results obtained with aqua regia after 2 hours of treatment were not perceptible, besides this substance produced serious damages on the surface (scratch and digs). On the other hand, the use of cerium oxide for hand polishing worked very well. After 5 minutes of polish all the stains disappeared completely without produce significant changes in the quality of the surface (*rms* 5% and 0% PV, approx.). It is important to remark that these stains appears only in a polished surface, and are present in an optical glass with high acid resistance value.

Due the results obtained in this work, the hand polishing is used at the INAOE´s optical workshop for the stains removal in the MEGARA prisms. Besides this technique is used for the stain removal in the COLL-D2 and COLL-D3 lenses, which are part of the collimator system of the instrument.

With this study we can guarantee the instrument performance and avoid damages, i.e., diffused images, optical quality decreasing and maybe some problems with the AR coating or in the cementing process for fused silica windows.

4. References

[1] Gil de Paz A et al. 2016 *Proc. SPIE* **9908**, 9908-57
[2] Carrasco E et al. 2014 *Proc. SPIE* **9147** 914771
[3] Gil de Paz A et al. 2015 *Proc. SPIE/ASP* **9147** 91470O
[4] Maldonado-Medina M et al. 2012 *Proc. SPIE* **8446** 844655
[5] Aguirre-Aguirre D et al. 2015 *Proc. SPIE/APOMA* **9633** 96331P
[6] Izazaga-Perez R et al. 2016 *Proc. SPIE* **9912** 99120D
[7] http://www.oharacorp.com/iline.html
[8] Izazaga-Perez R et al. 2015 *Proc. SPIE/APOMA* **9633** 963324

IOP Conf. Series: Journal of Physics: Conf. Series **792** (2017) 012065 doi:10.1088/1742-6596/792/1/012065

Industrial wastes as alternative raw materials to produce eco-friendly fired bricks

V Jiménez-Quero[1], O T Maza-Ignacio[2], J Guerrero-Paz[3] and K Campos-Venegas[4]
[1]Instituto Politécnico Nacional, CIIDIR Unidad Oaxaca. Hornos 1003 Sta. Cruz Xoxoxcotlán Oaxaca, México. C. P. 71230
[2]M.Sc. Student. Instituto Politécnico Nacional, CIIDIR Unidad Oaxaca, Hornos 1003. Sta. Cruz Xoxocotlán Oaxaca, México C. P. 71230
[3]Universidad Autónoma del Estado de Hidalgo. Carr. Pachuca - Actopan Km. 4.5, Colonia Campo de Tiro, Pachuca de Soto, México C.P. 42039
[4]Centro de Investigación en Materiales Avanzados S.C. Av. Miguel de Cervantes 120, Complejo Industrial Chihuahua, Chihuahua, México C.P. 31109

E-mail: ¿?

Abstract. This work focuses on the incorporation of sugarcane bagasse ash (SCBA) and silica fume (SF) wastes as an alternative raw material into clay bricks, replacing clay by up to 40 wt.%. Fly ash (FA) was used as reference. The plasticity of the batches was determined by Atterberg´s consistency limits. Bricks were produced by uniaxial pressing and fired a 900 and 1000°C. Physical properties (fired shrinkage, water absorption, apparent porosity and Initial water absorption rate) and mechanical properties (compressive strength and flexural strength) as function of the firing temperature and type waste were investigated. The results showed that wastes into clay body increase its global plasticity. 80%Clay-20%SCBA mixture has the lower linear shrinkage. After firing process, the brick produced with clay-SCBA show the higher water absorption and apparent porosity, regardless of the firing temperature. The brick produced with 60%clay-40%SF show the water absorption and apparent porosity similar to control bricks. The SCBA waste additions tend to decrease the mechanical strength of the clay bricks, therefore amounts of 40% SCBA waste should be avoided because it reduce the mechanical strength of the red fired bricks. The fired bricks with 40% SF, firing a 900°C show mechanical properties similar control bricks.

1. Introduction

The fired brick is one of the main materials in construction due to the availability of raw materials for manufacturing [1]. The current world-wide production of construction bricks is about 1,391 trillion units per year, and demand for bricks is expected to continue rising [2]. The conventional bricks are produced from natural clay and fired by high-temperature kiln. Besides clay, which confers cohesion and plasticity, ceramic pastes can also include inert materials that provide structural support that helps to retain shape during drying and firing. Quartz (SiO2) is the most commonly used inert material. Depending on the source and the calcination temperatures some waste ash can contain high levels of SiO_2. Have been carried out research on the reuse of different ash in order develop better quality bricks and reducing waste [3]. Ash wastes such as fly ash, blast furnace slag, wood sawdust and rice husk ash are also being used as raw materials for ceramic products [4].

Flay ash (FA) is the solid waste produced from thermal power station, this waste has been shown that might improve the compressive strength of bricks and durable [5]. In some research it has been utilized FA as raw material to make bricks; the volume ratio used is about 10-30% by volume [6].

Content from this work may be used under the terms of the Creative Commons Attribution 3.0 licence. Any further distribution of this work must maintain attribution to the author(s) and the title of the work, journal citation and DOI.
Published under licence by IOP Publishing Ltd

Sugarcane bagasse ash (SCBA) is a residue resulting from the burning of bagasse in boilers in the sugarcane/alcohol industry. A comparison of the chemical compositions of SCBA reported in the literature [7] shows that they are variables due to differences in the soil where the sugarcane was grown. In recent years, different studies are indicating the feasibility of using SCBA in Portland cement based materials, but research on the incorporation of SCBA in the clay ceramics are scarce [8]. In spite of the fact that SCBA waste generated in large amounts in Mexico, insufficient attention was devoted to the incorporation of SCBA waste in the production of fired clay bricks. There are no researches on the effects of incorporating silica fume (SF) in the manufacture of fired clay bricks. SF also known as condensed silica fume or microsilica is by-product from furnaces used in the production of ferrosilicon and silicon metals. In some studies have reported benefits in the mechanical properties of bricks made with clay and ashes, however these results were obtained by post-treatment to which the ashes were submitted.The implementation of several post-treatments to improve the properties of the ash includes re calcination at 500 ° C [9] or grindings for 35 h. Most of the post-treatments used require a significant amount of energy to operate causing additional emissions of pollutants to the atmosphere. In others studies the generation of ashes in the laboratory, by controlling calcination processes is reported.

The main of this work is to study the effect on the physical and mechanical properties of eco-friendly fired bricks produced by the incorporation of SCBA, SF and FA wastes into clay. The SCBA was only sieved using the 75 mm ASTM sieve for four minutes, was used “practically as was collected”.

2. Experimental program

2.1. Materials and Methods

The clay used in the present study was collected from a local brick manufacturing plant located in San Pablo Huitzo, Oaxaca México. The fly ash (FA) used throughout the investigation is commercially available in the market as a building material (Admix Tech®). The Sugar cane bagasse ash (SCBA) was collected from the Constancia sugarcane mill, located in Tezonapa, Veracruz Mexico. The silica fume (SF) used was supplied by EUCO® México Corporation. The clay was oven-dried (110°C), ground in a ball mill using porcelain balls (40 minutes) and sieved below 150 μm. The SCBA was sieved below 75 μm sieve only for four minutes. The FA and SF were used as received.

2.2. Clay and ash characterization

The chemical composition of the clay and ash was determined by the X-ray fluorescence (XRF) method with an Epsilon 3 XL energy dispersive X-Ray spectrometer. The clay and ash crystalline phases were identified by X-ray diffraction using a Bruker D8 Advance® diffractometer. The morphology of clay and ash was examined by scanning electron microscopy (HITACHI®, model SU3500).

Batches with 0, 20 y 40 wt. % ash, which formulations are show in Table 1, were studied. The plasticity of the batches was determined by Atterberg´s consistency limits (ASTM D4318 – 10).

Table 1. Clayey formulation (wt. %)

Batches	Clay	FA	SCBA	SF
B1	100	0	0	0
B2	80	20		
B3	60	40		
B4	80		20	
B5	60		40	
B6	80			20
B7	60			40

FA: fly ash, **SCBA**: sugarcane bagasse ash, **SF**: silica fume

2.3. Samples preparation
Based on the results obtained in the consistency limits, four mixtures (clay-waste) for the preparation of samples (bricks) were selected. The raw materials were mixed and homogenized by using a cylindrical mixer during 20 min. The moisture content was adjusted to plastic limit of each batch. Prismatic probes (140 mm x 50 mm x 25 mm) were produced by uniaxial pressing. The probes were dried for 72 h at room temperature and then fired (900°C y 1000°C) [10] in a laboratory gas kiln at a heating rate of 10°C/min.

A series of tests were performed to determine physical properties: linear shrinkage, water absorption, apparent porosity (ASTM C373-88) and mechanical properties: compressive strength and flexural strength (ASTM C-67). The compressive strength was estimated using an ELVEC® model E659-5 compression machine. The tensile strength was determined by using a Geotest Instrument Corp®. Model S5830 multi-load machine. Initial water absorption rate was estimated (ASTM C-67).

3. Results

3.1. Chemical characterization
The chemical composition of the clay and ash used in this study are summarized in Table 2.

Table 2. Chemical analysis of the used materials, %

Element/Compound	Clay	FA	SCBA	SF
Aluminium Oxide (Al_2O_3)	18.25	20.58	14.61	0.18
Calcium Oxide (CaO)	3.89	4.83	2.36	0.72
Iron Oxide (Fe_2O_3)	2.94	2.78	5.04	0.05
Potasium Oxide (K_2O)	1.95	1.30	3.29	0.87
Magnesium Oxide (MgO)	1.10	1.20	1.43	0.36
Maganesum Oxide (MnO)	0.06	0.05	0.18	0.01
Sodium Oxide (Na_2O)	3.42	1.28	1.57	0.12
Phosphorous Oxide (P_2O_5)	0.40	0.22	0.85	0.13
Silicon Dioxide (SiO_2)	61.19	61.10	56.37	92.74
Titanium Oxide (TiO_2)	0.94	0.72	0.96	-
LOI at 1000 °C	4.67	3.69	10.53	3.84

The clay showed a typical composition of clay minerals of the kaolinite group, rich in SiO2, Al2O3 CaO, Na2O and Fe2O3, with minor amounts of K, Mg, Ti, P and Mn, oxides. The content of iron oxide (2.94%) is high, and characterizes the clay as red firing clay. The sum of the major oxides for the FA (SiO2+Al2O3+Fe2O3) was 89.55 % with a CaO content of 4% and 2.6% of LOI, therefore is classified as a Type F pozzolan (ASTM 618-05). SCBA show similar composition at clay, the mainly difference are LOI. Loss on ignition (LOI) implied in high weight loss of 10.53 % and is mainly attribute to the presence of organic matter in waste. As expected, silica fume has mainly SiO_2.

The XRD pattern of clay shows kaolinite as the dominant mineral. Accessory minerals quartz, gibbsite, goethite, and potash feldspar were identified. The XRD pattern for the FA shows quartz and mullite as the two main crystalline phases. The XRD pattern of the SCBA show quartz, cristobalite, mullite and hematite were identified. The XRD pattern of SF shows cristobalite as main crystalline phase.

Figure 1 show the SEM micrographs of the used materials: a) Clay particles to be granular in shape, b) FA particles to be spherical in shape, c) the SCBA particles to be elongated and flake in shape. Further, d) SF particles to be spherical in shape.

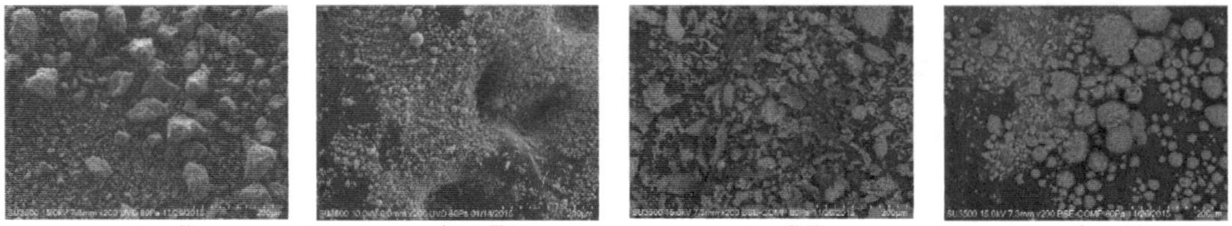

a) Clay b) FA c) SCBA d) SF

Figure 1. SEM micrographs

Atterberg consistency limits results are presented in Figures 2 and 3.

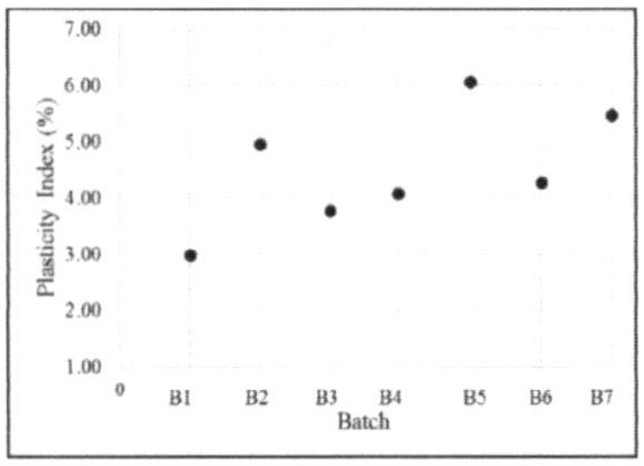

Figure 2. Plasticity Index

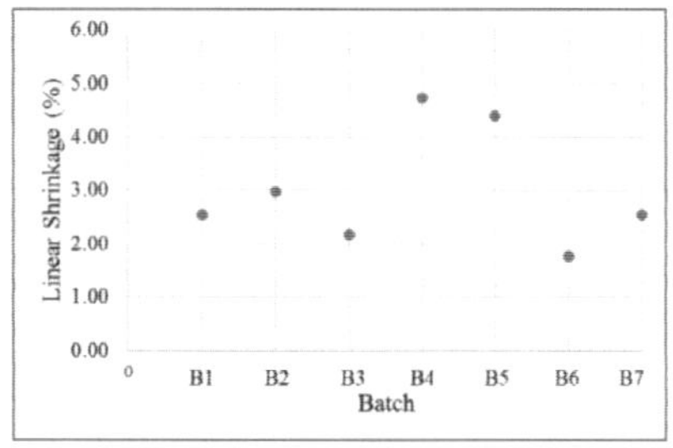

Figure 3. Linear Shrinkage

The incorporation of ash waste into clay body increases its global plasticity. B5 batch has the highest plastic index. This effect is attributed to the shape of the particle morphology of SF. All values are within the adequate range for industrial production of clay-based products [11]. The results demonstrate that no significant differences in the linear shrinkage between B1, B3 and B7 batches. B4 mixture has the highs linear shrinkage. The incorporation of the FA and SBCA waste could improve the dimensional control of the red ceramic pieces.

Table 3. Shows the water absorption and apparent porosity of the fired bricks.

Batches	Firing temperature		Firing temperature	
	900 °C	1000°C	900 °C	1000°C
	Water absorption %		Apparent porosity %	
B1	21.12	21.02	37.96	37.73
B3	26.91	27.29	40.70	42.42
B5	22.60	21.40	38.30	36.49
B7	44.82	46.57	53.64	55.57

The water absorption is related to the microstructure of the fired ceramic matrix, and determines the volume fraction of open pores. In the 900 - 1000 °C temperature range, the water absorption presented only a small variation on B1 and B5 batches. The water absorption is greater on B3 and B7 batches. The sugarcane bagasse ash waste bearing clay bricks presented acceptable values of water absorption for clay brick industrial production. At 1000 °C the apparent porosity decreased (1.81%) on B5. Indicating higher densification at this temperature. This result is in accordance with the absorption.

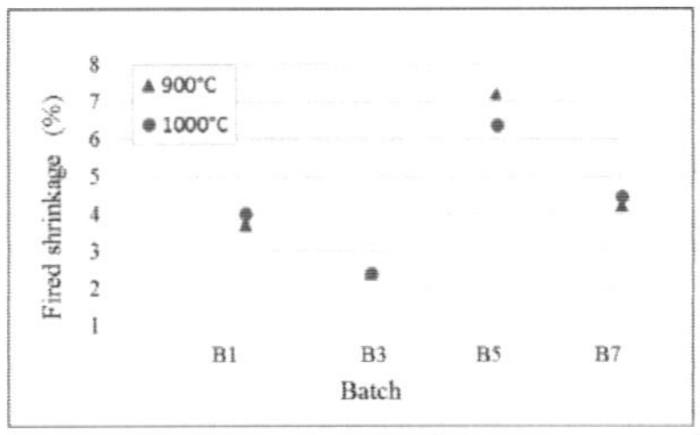

Figure 4. The fired shrinkage of the red fired brick at different temperatures.

This property is very important for the dimensional control of the finished red ceramic products. The higher linear shrinkage was presented in the bricks made whit B5 and the lower fired shrinkage was presented in the bricks made whit B3, this result is in accordance with Plasticity index. The increase firing temperature only affects the bricks made with B5 batch. According with the literature [12] for temperatures up to 1000 °C, the red ceramic pieces are sintered predominantly via solid state sintering mechanism. This explains the lower values of linear shrinkage on the bricks made with B7 (similar at control bricks) resulting in good dimensional control. This effect is related to the SCBA composition, rich in silica, which is a non-plastic component and, as such, behaves as a filler material and decreases the plasticity of the clay/ash waste mixes.

Mechanical properties bricks are show in Figures 5 and 6.

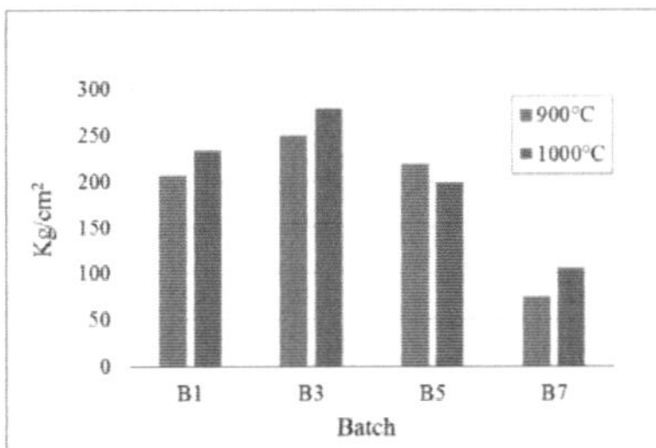

Figure 5. Compressive strength

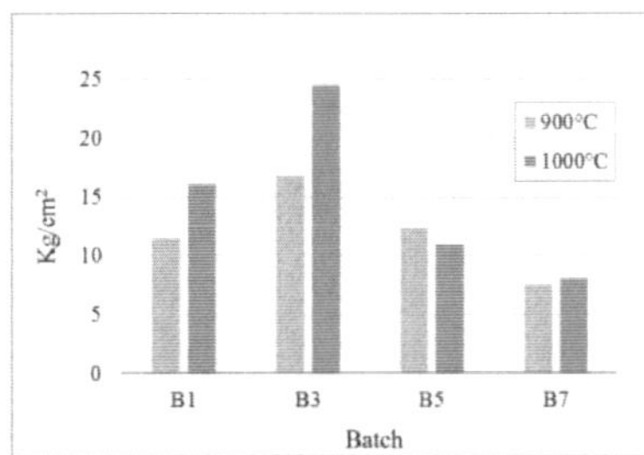

Figure 6. Rupture module

The results indicate that the bricks compressive strength increase with increase firing temperature, except in bricks made with B5 batch (Figure 5). SCBA addition reduces the compressive strength of the bricks principally ascribed to the enhancement of porosity which is known to have a marked influence on the mechanical strength of the ceramic.

Figure 6 shows the flexural strength of the red fired brick. As expected, the effect of ash and the firing temperature in the flexural strength of fired bricks, was the same as obtained in the compressive strength. This behavior is in accordance with the densification behavior during sintering. Such behavior is mainly related to the following factors: i) decomposition of organic matter from the waste sample, thus generating pores in the fired structure; and ii) presence of a high content of crystalline silica particles in the waste sample that tends to induce flaws in the fired ceramic matrix [13]. Specimens of B2 showed the highest flexural strength, at 1000 °C. Figure 7 shows the results of water absorption test for different fired bricks. The bricks that showed rates lower water absorption were made with B1 and B5. The bricks that showed rates higher water absorption were made with B7.

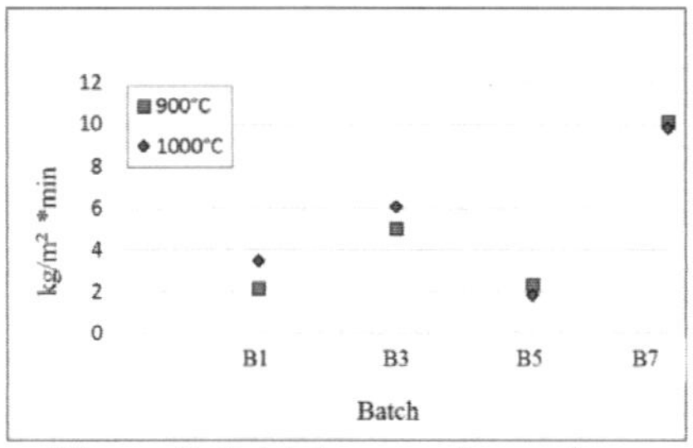

Figure 7. Initial water absorption rate

This fact is expected for the high LOI present in the SCBA, which is eliminated during the thermal process and it leads to an increase of the open porosity of the ceramics bodies.

4. Conclusions

The incorporation of sugarcane bagasse ash wastes as raw material into clay bricks, increase its plasticity, water absorption and apparent porosity. Although confers dimensional stability, the amounts of 40% SCBA waste should be avoided because tends to reduce the mechanical strength of the red fired bricks, regardless of the firing temperature.

The use of silica fume as raw material into clay bricks, no effect physical neither mechanical properties of fired bricks, when amounts 40% is used and fired to 900°C. This options shows highly positive results in terms of environmental protection, is an effectively measure of saving clay.

5. Acknowledgments

The authors would like to thank to Mr. Juan Ruiz (ICAPET Oaxaca, México) for the contribution in carrying out firing bricks and EUCO® México Corporation by supplying of silica fume.

6. References

[1] Sutas J, Mana A and Pitak L 2012 Effect of Rice Husk and Rice Husk Ash to Properties of Bricks *Procedia Eng.* **32** pp 1061–1067
[2] Zhang L 2014 Production of bricks from waste materials – a review *Constr. Build. Mater.* **47** pp 643–55
[3] Lin KL 2006 Feasibility study of using brick made from municipal solid waste incinerator fly ash slag *J. Hazard. Mater. B* **137** pp 1810–6
[4] Little M, Adell V, Boccaccini A and Cheesman C 2008 Production of novel ceramic materials from coal fly ash and metal finishing wastes *Resour. Conservat. Recycl.* **52**(11) pp 1329-1335
[5] Lingling X, Wei G, Tao W and Nanru Y 2005 Study on fired bricks with replacing clay by fly ash in high volume ratio *Constr. Build. Mater.* **19** pp 243–7
[6] Guler R, Patla P and Hess TR 1995 Properties of fly ash bricks produced for environmental applications *Environ. Sci. Health.* **30**(3) pp 505–24
[7] Frías M, Villar E and Savastano H 2011 Brazilian sugar cane bagasse ashes from the cogeneration industry as active pozzolans for cement manufacture *Cem. Conc. Comp.* **33** pp 490-496
[8] Borlini M C, Lopes C R, Monteiro S N and Vieira C M F 2005 Influence of Sugarcane Bagasse Ash in Substitution for the Fluxes of a Vitrified Ceramic Tile *49th Brazilian Ceramic Congress* São Paulo, Brazil
[9] Andreola F, Martínc M I, Ferrari A M, Lancellotti I, Bondioli F, Rincon J M, Romero M and Barbieri L 2013 Technological properties of glass-ceramic tiles obtained using rice husk ash as silica precursor *Ceram. Int.* **39** pp 5427–5435
[10] Mancuso A 2009 Degree thesis. University of Modena and Reggio Emilia. Tutor: Lancellotti I and Andreola F
[11] Dondi M 2003 *Int. Ceram. J.* pp 55-59
[12] Milheiro F A C, Freire M N, Silva A G P and Holanda J N F *Ceram. Int.* **31** pp 757-763
[13] Faria K C P, Gurgel R F and Holanda J N F 2012 Recycling of sugarcane bagasse ash waste in the production of clay bricks *J. Environ. Manage.* **101** pp 7-12

IOP Conf. Series: Journal of Physics: Conf. Series **792** (2017) 012066 doi:10.1088/1742-6596/792/1/012066

Erbium Doped Fiber Optic Gravimeter

G G Pérez-Sánchez[1], J R Pérez-Torres[2], J A Flores-Bravo[3], J A Álvarez-Chávez[3] and F Martínez-Piñón[3]

[1]Universidad Autónoma Metropolitana, Av. San Pablo 180, Reynosa Tamaulipas, Azc. México City 02200.

[2]Tecnológico de Estudios Superiores de Coacalco Av. 16 de Septiembre 54, Cabecera Municipal, Coacalco de Berriozábal, México. 55700.

[3]Centro de Investigación e Innovación Tecnológica-IPN, Cerrada de Cecati S/N, Santa Catarina, Azc. México City. 02250.

Email: ggps@correo.azc.uam.mx

Abstract. Gravimeters are devices that can be used in a wide range of applications, such as mining, seismology, geodesy, archeology, geophysics and many others. These devices have great sensibility, which makes them susceptible to external vibrations like electromagnetic waves.

There are several technologies regarding gravimeters that are of use in industrial metrology. Optical fiber is immune to electromagnetic interference, and together with long period gratings can form high sensibility sensors of small size, offering advantages over other systems with different technologies.

This paper shows the development of an optical fiber gravimeter doped with Erbium that was characterized optically for loads going from 1 to 10 kg in a bandwidth between 1590nm to 1960nm, displaying a weight linear response against power. Later on this paper, the experimental results show that the previous described behavior can be modeled as characteristic function of the sensor.

1. Introduction

The use of optical fiber has been proposed for several applications due to its features such as small size, lightweight, and immunity to electromagnetic interference; furthermore, in some cases optical fiber can operate as both, a sensor and the transmission media, hence avoiding the optoelectronic conversion. Optical fiber sensors have a significant importance since they offer a great sensitivity on detecting changes in temperature, pressure, strain or polarization.

Fiber Bragg gratings have been used on developing different optical fiber sensors such as long period fiber gratings (LPFG) and mechanically induced long period gratings [1-4].

In [5], an erbium doped fiber with LPFGs in cascade is used to generate an interference pattern around 1550 nm spectrum.

Gravimeters are a type of sensor intended for measuring in several knowledge areas. Such devices use different technologies as direct mass measurement, strain gage load cell, frequency shift technology, force restoration, etc. [6]. This work proposes developing a gravimeter, taking advantage of optical fiber properties.

---

[1] Grethell Georgina Pérez Sánchez, Depto Ciencias Básicas, Universidad Autónoma Metropolitana, Av. San Pablo 180, Reynosa Tamaulipas, Azc. Mexico City 02200.

Content from this work may be used under the terms of the Creative Commons Attribution 3.0 licence. Any further distribution of this work must maintain attribution to the author(s) and the title of the work, journal citation and DOI.
Published under licence by IOP Publishing Ltd

Gravimeters are prone to be affected by factors such as local gravity, air buoyancy, tidal effects, moisture condensation, among others. Due to its properties, optical fiber offers an interesting choice when it comes to deal with the previously described factors. This work shows the experimental results of an erbium doped fiber gravimeter with three LPFGs in cascade.

2. Amplified Spontaneous Emission in Erbium doped fiber

The change on the pump and signal powers of augmented radiation along Erbium doped fiber scan be described with the following equations:

$$\frac{dP_p(z)}{dz} = -\gamma_p(z)P_p(z) \tag{1}$$

$$\frac{P_s^{\pm}(z,\lambda_i)}{dz} = \pm\{G_e(z,\lambda_i)[P_s^{\pm}(z,\lambda_i) + P_0] - G_a(z,\lambda_i)P_s^{\pm}(z,\lambda_i)\} \tag{2}$$

Where $P_p(z)$ refers to the pump power propagation towards z, and $P_s^{\pm}(z,\lambda_i)$ represents the output power in both directions, backward and forward. The absorption value for the pump is represented by $\gamma_p(z)$, and $G_e(z,\lambda_i)$ and $G_a(z,\lambda_i)$ stand for amplification factor and the absorption coefficient of spontaneous emission respectively. $P_0$is the equivalent of the input noise power, represented by:

$$P_0 = 2hv_s\Delta v \tag{3}$$

Solving equations 1 and 2 considering the ASE as a weak-signal[7] the signal power solution is.

$$P_s^{+}(z,\lambda_i) = \left(\frac{G_e(z,\lambda_i)}{G_b(z,\lambda_i)}P_0e^{zG_b(z,\lambda_i)} - \frac{G_e(z,\lambda_i)}{G_b(z,\lambda_i)}\right) \tag{4}$$

$$P_s^{-}(z,\lambda_i) = \left(\frac{G_e(z,\lambda_i)}{G_b(z,\lambda_i)}P_0e^{(L-z)G_b(z,\lambda_i)} - \frac{G_e(z,\lambda_i)}{G_b(z,\lambda_i)}\right) \tag{5}$$

Where:

$$G_b(z,\lambda_i) = G_e(z,\lambda_i) - G_a(z,\lambda_i) \tag{6}$$

3. Long Period Fiber Gratings

A long period grating can attach light between the fundamental core mode and the propagation modes within the fibercladding. As a result of such attachment, a signal filtering is displayed as a power loss around a frequency. The wavelength of the filter is denominated as resonance wavelength λ_m, defined by the phase attachment condition $\lambda_m = (n_{eff1} - n_{eff2}^m)\Lambda$, where Λ is the periodicity of the grating, n_{eff1} and n_{eff2}^m are the effective refraction indices of the core and the cladding respectively. The coupling mode equations are:

$$\frac{dA}{dz} = \kappa B e^{i\Gamma z} \tag{7}$$

$$\frac{dB}{dz} = \kappa A e^{-i\Gamma z} \tag{8}$$

$$\Gamma = \beta_1 - \beta_1 - K \tag{9}$$

Where A is the fundamental mode propagated in the fiber core, and B is the propagation mode of the cladding, Γ represents the propagation coefficient. $\kappa$is the coupling coefficient, and the period of the grating is defined as $\Lambda = \frac{2\pi}{\beta_1-\beta_2}$, where $\beta_1 = \frac{2\pi n_{eff\,1}}{\lambda_0}$ and $\beta_2 = \frac{2\pi n_{eff\,2}}{\lambda_0}$are the coefficients of the core and the cladding propagation respectively [5,8].

When light is passed from the first grating towards the second one, being A_0 y B_0 for the former and A_1 y $B_1$for the latter, where A and B are initial cladding and core powers, respectively, the 7, 8 and 9 equations are solved. The relationship between the Interferometers length and wavelength selective separation can be embodied as:

$$I(\lambda_m) = I_{core} + I_{clad} + 2\sqrt{I_{core} I_{clad}} \cos\phi \qquad (10)$$

In this equation Icoreand Iclad correspond to the core and the cladding-dominant mode magnitudes, respectively. $\phi = 2\pi\Delta n_{eff} L/\lambda$ is the phase difference between the core and cladding modes, λ stands for the propagating light wavelength, L is the length of the MZI, and neff is the variation of the effective refractive indices between the core and cladding modes that propagate light. When $\phi = (2m+1)\pi$ the interference signal reaches its minimum. The wavelength of the mth order attenuation peak is represented by:

$$\lambda_m = \frac{2\Delta n_{eff} L}{2m+1} \qquad (11)$$

Wavelength spacing between two interference minima can be approximated as:

$$\Delta\lambda_m = \frac{4n_{eff} L}{(2m+1)(2m-1)} \approx \frac{\lambda_m^2}{\Delta n_{eff} L} \qquad (12)$$

Taking into account the variation of environment of the neff or the length L.

4. Optical fiber gravimeter development

In figure 1 the optical fiber gravimeter setup is proposed. The gravimeter consists of a 980 nm fiber laser connected to an erbium-doped fibersection, which acts as a superluminiscent source, that is an optical source with a broad spectrum that is compatible with single-mode fibers on the end side of the doped fiber an SMF28 fiber section is spliced. The output power signal is inspected in an optical spectrum analyzer. Three mechanically induced long period fiber gratings are fixed in series on the SMF28 section. Experiments with different controlled loads going from 0 kg to 10 kg in steps of 1 kg are placed on top of a plate on the gratings. For such range of weights, the power spectrum variation is well defined between 1590 nm and 1610 nm.

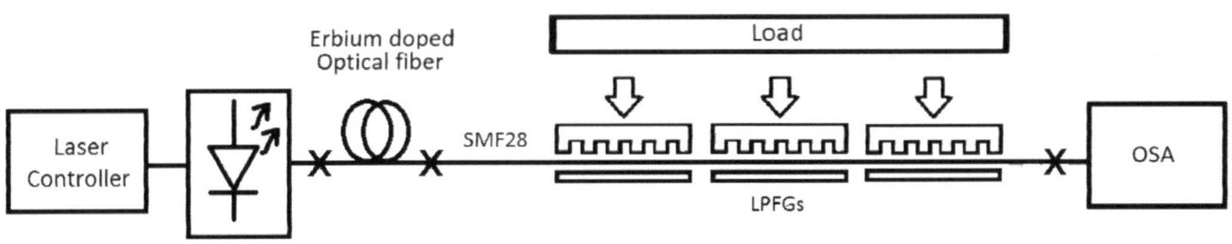

Figure 1.Er^{3+} doped fiber and three long period gravimeter setup.

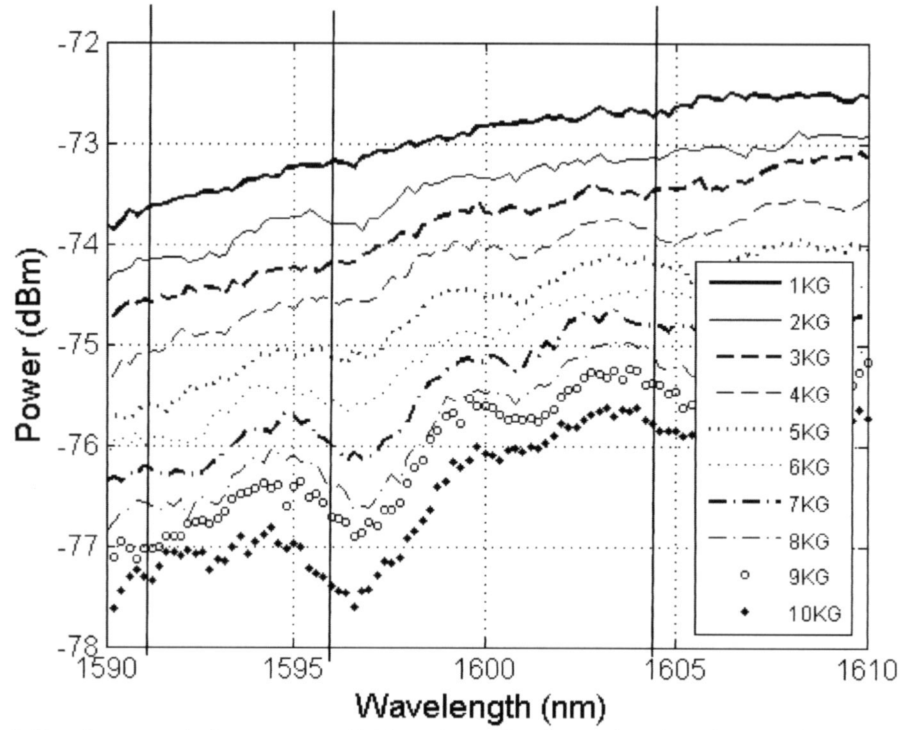

Figure 2. Gravimeter emission spectrum Erbium doped implementing three long period gratings with changing weight.

Figure 2 shows the spectral response of the gravimeter in the 1590 nm – 1610 nm band. It is noted that power declines as the weight increases. Particularly, for the 1597 nm wavelength there is a power loss of -4.42 dB for a load increase from 1 kg to 10 kg. We selected three different wavelengths for the analysis .1580 nm, 1597 nm and 1605 nm. Figure 3 presents the optical power variation as a function of the applied weight. We can see that the curve corresponding to the wavelength of 1597 nm behaves more linearly and it is also the wavelength at which a depression appears because of the LPG mechanism. We propose the use of this wavelength for the integration of a fiber optics gravimeter sensor with an Erbium doped superluminiscent fiber laser.

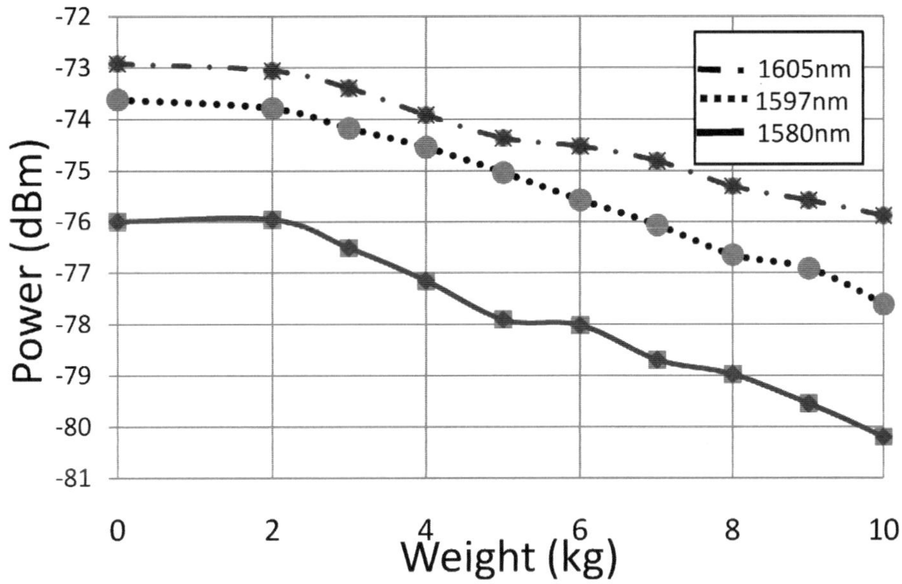

Figure 3. Optical Power variation for three different wavelengths in the fiber optic gravimeter.

Figure 4 shows the behaviour of the function relating weight versus optical power received at the 1597 nm wavelength. We observe that the function is not exactly linear. In fact, if we make a polinomial fitting, a parabolic model is more appropiate giving a maximum error of 5.2 % in the range of 3 kg to 8 kg (with a minimum error of 0.45%).

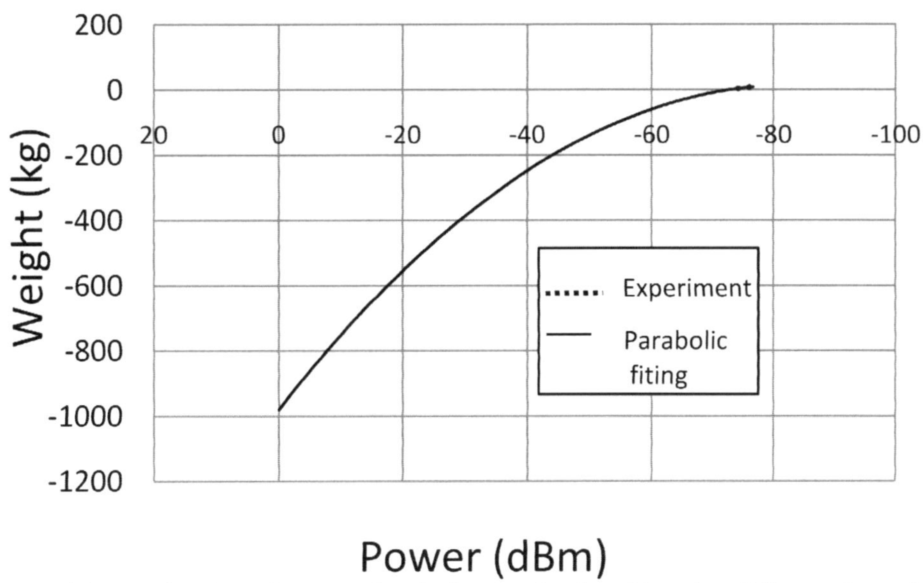

Figure 4. Weight as a function of measured optical power for the erbium doped fiber optic gravimeter.

The parabolic model obtained is:

$$W(p) = a_0 + a_1 p + a_2 p^2 \tag{13}$$

$a_0 = -979.7108$
$a_1 = -24.1395$
$a_2 = -0.1468$

$W = \text{Weight in kg.}$
$p = \text{Optical power in dBm's.}$

The dynamic range of this fiber optic gravimeter is then for measurements from 3 to 10 kg. The resolution of this sensor is related to optical power resolution of the OSA. This resolution is about 0.01 dBm that corresponds to approximately 22 grams of weight resolution. Outside this dynamic range the error in weight increases as for example for measuring 2 kg, the error is almost 10% and in 9 kg the error is 5.7%.

The calibration for this sensor is performed by making an initial measurement with a 3 kg load and setting the measurement system to give the optical power of -74.18 dBm´s.

5. Conclusions

In summary this paper presents the results of a gravimeter optical fiber doped with erbium. The erbium section acting as a superluminiscent source and the test section formed by a standard SMF-28 single-mode fiber with three long period gratings coupled mechanicaly to a load from 0 to 10 kg. It was observed an excelent behavior of the sensor with an error less than 5.2% from 3 to 8 kg with a resolution of 22 grams. The results obtained by the developed device presents a parabolic decreasing behavior of the optical power as the load increases on the sensor. We have selected a 1597 nm wavelength test as this produces a relatively smooth variation which coincides with the depression wavelenght peak. We believe that this optical fiber gravimeter will provide a useful instrument for the measurement of mass and weight in may scientific and industrial applications as for example in sismology , geophysics, geodesic.

6. References

[1] Yun-Jiang R 1997 *In-fibre Bragg grating sensors Measurement Science and Technology* **8**
[2] Bhatia V V and Vengsarkar A M 1996 Optical Fiber Long-Period Grating Sensors *Opt. Lett.* **21** 9.
[3] Jiang Y, Zhao J, Yang D and Tang D 2010 High-Sensitivity Pressure Sensors Based on Mechanically Induced Long-Period Fiber Gratings and Fiber Loop Ring-Down *Opt. Commun.* **283** pp 3945-3948.
[4] Mizunami T, Ishida Y and Yoshikura H 2008 Long-Period Fiber-Grating Temperature Sensors In Ge–B-Codoped Fibers with Temperature/Strain Discrimination *Jpn. J. Appl. Phys.* **47** 8 pp 6833–6837
[5] Krcmarik D, Slavik R, Karasek M and Kulishov M 2009 Theoretical and experimental analysis of long-period fiber gratings made directly into Er- doped active fibers *J. Lightwave Technol.* **27** 13 pp 2335-2342
[6] Boynton R 2001 Precise Measurement of Mass *SAWE PAPER* **3147**
[7] Pérez-Sánchez G G et al. 2013 Rare-earth-doped fiber designs for superluminescent sources *Opt. Eng.* **52** 8
[8] Savin S, Digonnet M J F, Kino G S and Shaw H J 2000 Tunable mechanically induced long-period fiber gratings *Opt. Lett.* **25** 10 pp 71-712

Electrical model of dielectric barrier discharge homogenous and filamentary modes

J A López-Fernández[1,2,*], R Peña-Eguiluz[1,+], R López-Callejas[1], A Mercado-Cabrera[1], R Valencia-Alvarado[1], A Muñoz-Castro[1] and B G Rodríguez-Méndez[1]
[1] Instituto Nacional de Investigaciones Nucleares, Plasma Physics Laboratory, AP 18-1027, 11801 México City, Mexico
[2]Tecnológico de Estudios Superiores de Tianguistenco.
52600 Tianguistenco, Estado de México, Mexico

+e-mail address: rosendo.eguiluz@inin.gob.mx
*e-mail address: j.a.lopezfernandez.test@gmail.com

Abstract. This work proposes an electrical model that combines homogeneous and filamentary modes of an atmospheric pressure dielectric barrier discharge cell. A voltage controlled electric current source has been utilized to implement the power law equation that represents the homogeneous discharge mode, which starts when the gas breakdown voltage is reached. The filamentary mode implies the emergence of electric current conducting channels (microdischarges), to add this phenomenon an RC circuit commutated by an ideal switch has been proposed. The switch activation occurs at a higher voltage level than the gas breakdown voltage because it is necessary to impose a huge electric field that contributes to the appearance of streamers. The model allows the estimation of several electric parameters inside the reactor that cannot be measured. Also, it is possible to appreciate the modes of the DBD depending on the applied voltage magnitude. Finally, it has been recognized a good agreement between simulation outcomes and experimental results.

1. Introduction

Dielectric barrier discharge (DBD) has been utilized in many research and industrial fields as an interesting technology due it combines the characteristics of the non-thermal plasma with the advantage of operating at atmospheric pressure, besides that small research reactors can be scaled to industrial reactors. This type of discharge has been used successfully in applications such as pollution control, polymer processing, medicine, fluorescent lamps, ultraviolet lamps, etc. Figure 1(a) shows an argon atmospheric pressure DBD established in a coaxial reactor. While its applied voltage and resulting electric current waveforms are depicted in Figure 1(b), where it can be distinguished the homogeneous and filamentary modes in the current curve.

The DBD can be modeled using computer numerical methods based on continuity equations coupled to the Poisson equation and boundary conditions. However, it should be necessary to make several considerations related to the gas ionization to simulate the physical process. The aim of these models is to study the physical-chemical process discharge, mainly the phenomena associated with the electron conservation, the spatiotemporal evolution of chemical species, electric field along the axis discharge, etc. One of the disadvantages of these computer models is that they are complex, require great computing power and consume considerable time [1]. Instead, there are DBD models based on the use of electric circuits [2], [3], [4], [5] which are mainly composed by a high voltage power supply coupled to a reactor or cell implemented by passive components and the discharge process emulated by switches, dependent current sources and passive elements also. This class of models allows obtaining the dynamic relationships between external parameters (applied voltage, total current, flow and gas type, dielectric materials and reactor configuration).

Content from this work may be used under the terms of the Creative Commons Attribution 3.0 licence. Any further distribution of this work must maintain attribution to the author(s) and the title of the work, journal citation and DOI.
Published under licence by IOP Publishing Ltd

There are several models of electric discharge DBD for example: Kogelschatz [6], uses a variable resistor, which represents the channels of microdischarges, Valdivia-Barrientos [7], proposes a method in which superimposes the microdischarges to the displacement current, Flores-Fuentes [2] and Chen [8] used a concept of power law, Pal [9] implements a model that uses multiple switches to reproduce the current peaks of the DBD, Kudryavtsev [10] uses a diode bridge model and a source of DC voltage, Naude [11] uses two Zener diodes and an RC circuit, among others.

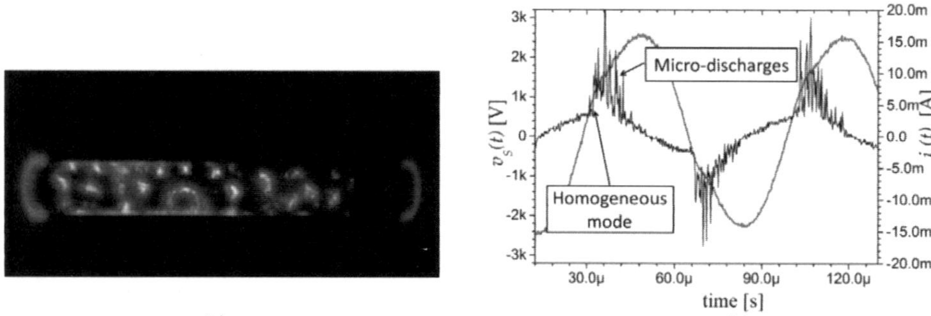

(a) (b)

Figure 1. Dielectric barrier discharge, (a) cylindrical reactor photography, (b) current and voltage waveforms.

2. Proposed Model

The proposed model does not represent the physical conditions of a DBD but aims to reproduce the electrical signals to serve as support in the optimal design of electric power converters. Besides the model allows obtaining relevant electrical functioning parameters such as voltage discharge, power efficiency, etc. which are useful for scaling of even other applications. When there are many filaments in the discharge process, the plasma is a mixture of glow and filamentary discharges. When the whole system is properly adjusted the filamentary discharge can be suppressed, since the visible streamers and filaments disappear, as a consequence, the discharge becomes more homogeneous. The waveform of the current can remain with few peaks as a result of instabilities caused by irregularities in the electrode edges, dielectric imperfections, water vapor and oxygen species in the plasma [8].

A double dielectric barrier parallel plate reactor with a space gap for containing the gas is powered by an alternating current to generate the DBD discharge. Equivalent electric circuit considers the dielectric plate as a capacitive element C_d and the gas inside the gap corresponds to a capacitive component C_g. When the DBD develops, the proposed model operation is divided into two stages: off and ignition. When voltage applied to the reactor is less than the breakdown gas voltage the reactor impedance is mostly capacitive and can be modeled by the dielectric capacitance C_d in series with the gas capacitance C_g, as it can be seen in Figure 2(a), since the applied voltage is a sinusoidal waveform the current maintains the same form but phase shifted by 90°. The ignition phase is represented by the circuit of figure 2(b), where the reactor current can be defined by equation (1), where the current reactor $i_{DBD}(t)$ is constituted by the addition of three components: current in the gas background $i_g(t)$, current due to homogeneous mode $i_h(t)$ and filamentary current due to microdischarges $i_f(t)$.

$$i_{DBD}(t) = i_g(t) + i_h(t) + i_f(t) \tag{1}$$

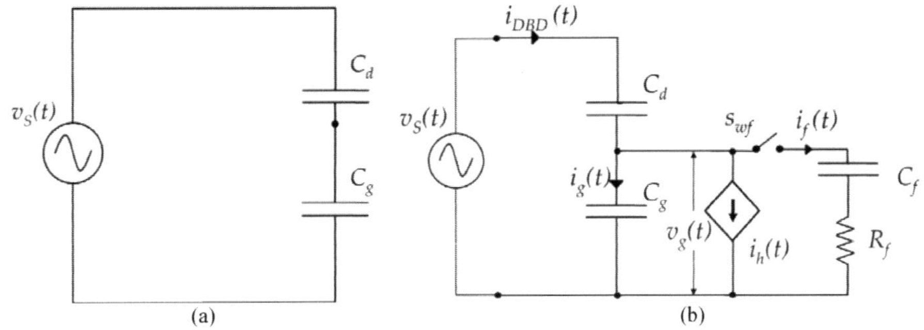

Figure 2. DBD electric model, (a) off stage circuit, (b) ignition stage circuit.

2.1. DBD homogenous mode

The ignition stage begins when the applied voltage inside the gap $v_g(t)$, surpasses the gas breakdown voltage magnitude V_b, at this moment the current $i_h(t)$, turns in an exponential form due to the changes in gas conductivity. The high voltage involves an intense electric field that produces ionization and transportation of charges. Thus this phenomenon is modelled by a power law proposed by Roth [12]:

$$i_h(t) = i_o \left(\frac{v_g(t)}{V_b}\right)^{\alpha} \tag{2}$$

The electric current $i_h(t)$ is implemented in the electric circuit model by a current source controlled by the gap voltage $v_g(t)$ and α is the rate of change of current with respect to time di_h/dt which takes values between 2 and 12, and i_o is the current density due to the continuous emission of electrons from the cathode.

2.2. DBD filamentary mode

When the dielectric has a non-uniform surface, there are areas where the electric field becomes more intense, leading to the development of microdischarges. First, a primary electron by inelastic collisions develops an electrons avalanche and consequently a weakly ionized conductor channel, trough the gap in a few nanoseconds [13], [14], [15]. Then electron accumulation attracted by the positive electrode on the dielectric surface causes the extinction of the local electric field and by the way the ending of the current associated with the electronic avalanche. The number of microdischarges depends fundamentally on the power density, a typical amount of is about 10^6 per cm^2 per second [6]. The proposed model represents the current due to the appearance of microdischarges by the commutation effect of an *RC* circuit, in which a switch connects and disconnects both the capacitor C_f and resistor R_f, as can be seen in Figure 3(a). The control circuit of s_{wf} is shown in Figure 3(b), in order to activate s_{wf} a pulse generator with a nanoseconds period defines the main frequency, this signal is applied to an "and" logic gate with a signal generated if and only if the gap voltage $v_g(t)$ surpasses the gas breakdown voltage level V_b. When s_{wf} is closed, the voltage $v_g(t)$ is supplied to C_f and R_f, applying the laws of Kirchhoff is established the following equation:

$$v_g(t) = V_{Cf} + V_{Rf} \tag{3}$$

Whereas the width of a microdischarge is on the order of nanoseconds while the period of signals that energize the reactor are on the order of microseconds, it can be assumed $v_g(t)$ constant, therefore the above equation can be rewritten as:

$$v_g(t) = \int i_f(t)\,dt - V_{Cf}(0) + R_f \cdot i_f(t) \tag{4}$$

Solving the above equation for the filamentary current $i_f(t)$ and considering the initial voltage of $C_f = 0$ is obtained:

$$i_f(t) = \left(\frac{V_g}{R_f}\right) e^{-\frac{t}{R_f C_f}} \tag{5}$$

Microdischarges can reach peak currents of the order of 0.1 A, considering the proposed circuit, the maximum amplitude of $i_f(t)$ is obtained at time $t = 0$, then, the amplitude depends on $I_{fmax} = V_g / R_f$ and solving equation for R_f:

$$R_f = \frac{V_g}{I_{fmax}} \tag{6}$$

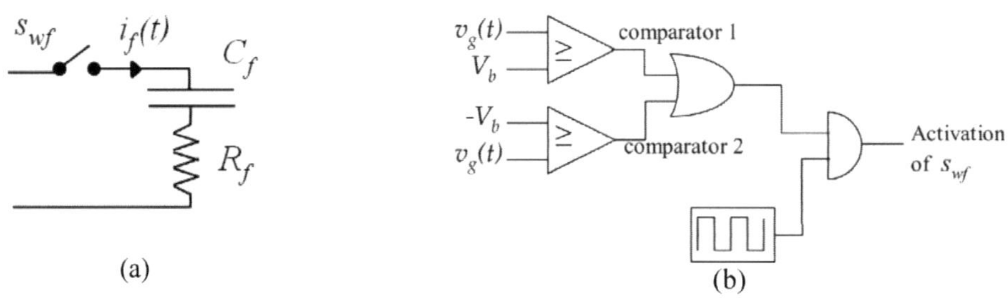

Figure 3. (a) Microdischarge proposed model implemented by an RC circuit, (b) the control circuit of s_{wf}.

3. Simulation and Experimental Results

The experimental setup consists of a circle parallel plate reactor, where both electrodes are stainless steel with 18 cm diameter, each electrode is covered with dielectric tempered glass with a thickness of 2 mm and a space gap of 2 mm to contain the gas. The working gas is N_2 with a flow of 1 L/min. The reactor is energized by a sinusoidal signal with an amplitude of 16 kV peak to peak at a frequency of 16 kHz. The model has been implemented in Simulink/Matlab® taking into account the physical dimensions of the reactor (C_g= 112 pF, C_d= 253 pF) and the voltage specified in experimentation. The electrical signals obtained by simulation are shown in Figure 4(a) whereas experimental waveforms can be appreciated in Figure 4(b).

Through the proposed model, the curve in Figure 5 was obtained, considering the graph it can be seen six different points denoted by letters A to F. The DBD starts at the point A, where the current remains almost constant until the voltage reaches a value of 3 kV, from this point current, starts to increase exponentially to the point B, reaching its maximum, while the voltage remains nearly constant, then the current decreases from point C until point D, where the current crosses by zero and becomes negative, υ_g remains positive due to the phase shift that occurs between current intensity and voltage due to a predominantly capacitive load, until the voltage reaches zero volt. From point D voltage becomes negative until the point E where it stabilizes while the current has an exponential behavior, then decrease to zero crossing at point F where the current becomes positive while the voltage is negative. Finally, the current becomes positive while the voltage decreases to return to the starting point A.

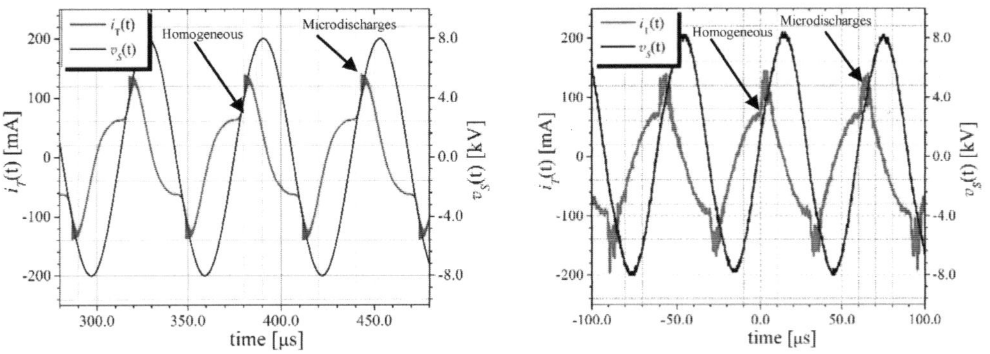

Figure 4. Reactor current and voltage waveforms, (a) simulation outcomes, (b) experimental results.

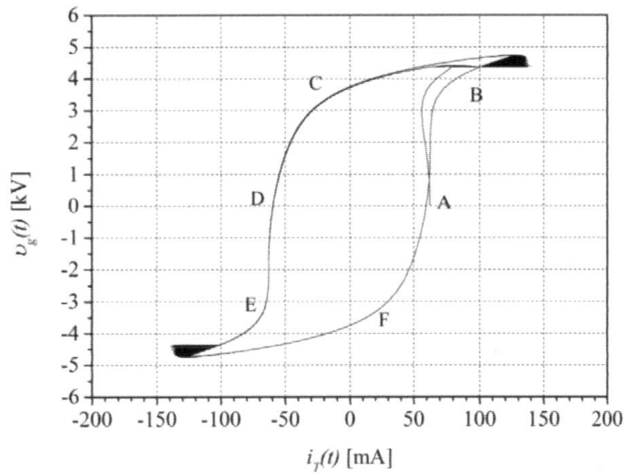

Figure 5. Estimated voltage-current discharge curve resulting from the proposed model.

Signals currents and voltages within the reactor are shown in Figure 6 since it can be seen that the space gap voltage $v_g(t)$ is phase shifted relative to the applied reactor voltage $v_s(t)$, the homogeneous mode starts at point B when $i_h(t)$ increases exponentially. Finally, during the filamentary mode, it has been observed a higher voltage level than that corresponding to the start of the homogeneous mode.

4. Conclusion

The proposed model includes the electrical behavior of both modes of a DBD: homogeneous and filamentary. From the model it has been observed that the homogenous mode starts at lesser voltage level than the filamentary mode voltage level, this could be due to the effect of the increased electric field which begins to produce ionization by changing the gas conductivity and generating displacement of electrical charges, the last are associated with the electric current increase during the homogeneous mode. If the voltage level continues increasing the electric field will become even more intense, and the protrusions of dielectric could present the tip effect, forming conducting current channels (microdischarges). Moreover, the presence of humidity in certain regions of the space gap could produce that the local conductivity is increased favoring the appearance of microdischarges.

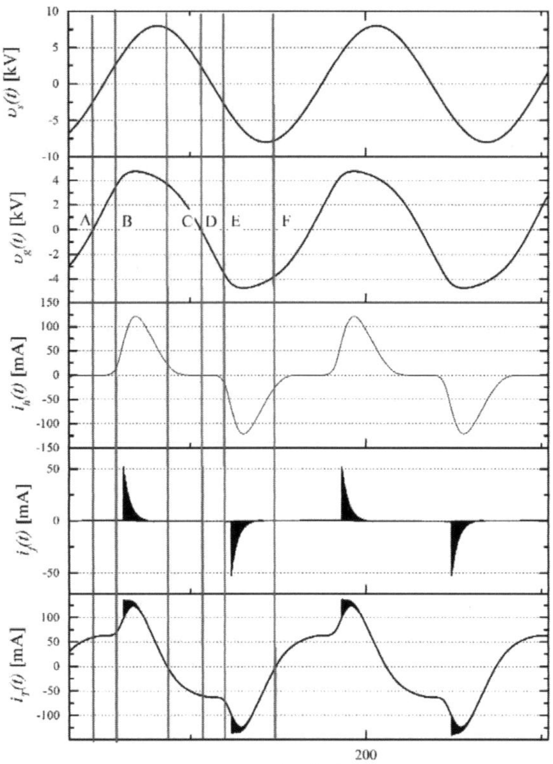

Figure 6. Waveforms inside the reactor obtained from the modelled DBD.

5. References

[1] Fang Z, Qiu Y, Zhang C, and Kuffel E 2007 *J. Phys. D: Appl. Phys.* **40** 1401
[2] Flores-Fuentes A, Peña-Eguiluz R, López-Callejas R, Mercado-Cabrera A, Valencia-Alvarado R, Barocio-Delgado S and de la Piedad-Beneitez A 2009 *IEEE Trans. Plasma Sci.* **37** 128
[3] Fang Z, Ji S, Pan J, Shao T and Zhang C 2012 *IEEE Trans. Plasma Sci.* **40** 883
[4] Pal U N, Gulati P, Kumar N, Prakash R and Srivastava V 2012 *IEEE Trans. Plasma Sci* .**40** 1356
[5] Yao Y, Zhang Z T, Yu Z, Xu S J, Yu Q X and Zhao J S 2013 *J. Phys. Conf. Ser.* **418** 012023
[6] Kogelschatz U. 2003 *Plasma Chem. Plasma P.* **23** 1
[7] Valdivia-Barrientos R, Pacheco-Sotelo J, Pacheco-Pacheco M, Benítez-Read J S and López-Callejas R 2006 *Plasma Sourses Sci. Technol.* **15** 237
[8] Chen Z Y 2003 *IEEE Trans. Plasma Sci.* **31** 511
[9] Pal U N, Gulati P, Kumar N, Prakash R and Srivastava V 2012 *IEEE Trans. Plasma Sci.* **40** 1356
[10] Koudriavtsev O, Wang S, Konishi Y and Nakaoka M 2002 *IEEE Trans. Ind. Appl.* **38** 369
[11] Naudé N, Cambronne J P, Gherardi N and Massines F 2005 *J. Phys. D: Appl. Phys.* **38** 530
[12] Roth J R 1995 *Industrial Plasma Engineering: Volume 1: Principles* (London: Taylor & Francis)
[13] Meek J M 1940 *Phys. Rev.* **57** 722
[14] Xu X and Kushner M J 1998 *J. Appl. Phys.* **84** 4153
[15] D. Braun, V. Gibalov and G. Pietsch 1992 *Plasma Sources Sci. Technol.* **1** 166

Shape-control of Zinc Oxide nanoparticles: enhancing photocatalytic activity under UV irradiation

M Montero-Muñoz[1], J E Ramos-Ibarra[1], J E Rodríguez-Páez[3], A Ramirez[2] and J A Huamaní-Coaquira[1]
[1]Institute of Physics, University of Brasilia, 70910-900 Brasília - DF, Brazil
[2]Department of Chemistry, University of Cauca, Popayán - Cauca, Colombia.
[3]Department of Physics, University of Cauca, Popayán - Cauca, Colombia

E-mail: *marlymontero@gmail.com*

Abstract: Zinc oxide (ZnO) nanostructures with different sizes and morphologies were synthesized using the Controlled Precipitation Method. It follows a standard process, but with different synthesis and washing solvents to modify the features related to the photocatalytic activity. The solid phase evolution during aging step was followed using Infrared Spectroscopy (FTIR) and the solids obtained, after the washing process, were characterized using X-ray diffraction (XRD). The Rietveld refinement indicates a Wurtzite phase (space group P63mc) as majority phase with lattice parameters a = 3.2530 Å and c = 5.2125 Å. Scanning electron microscopy (SEM) image shows a sponge-like morphology for the sample synthesized with ethylene glycol as solvent, acidified with nitric acid and washed with water. The sample synthesized and washed with water shows a needle-like morphology; and the sample synthesized in acetic acid and washed with water shows particles with undefined morphology. The optical properties of the as-prepared ZnO samples were investigated by UV-vis absorption spectroscopy. Finally, the photocatalytic activity of ZnO powders was studied from the initial rate of decomposition of H_2O_2 in aqueous solution. The best results were obtained with samples synthesized and washed with water; the influence of all the solvents on the morphology of ZnO samples and the effect of the morphologies on the photocatalytic activity are discussed.

1. Introduction.

Zinc oxide semiconductor, a wide direct band gap (3.37 eV) with large excitation binding energy (60 meV), is an excellent material with multifunctional properties and applications in electronics, photoelectronics, sensors and photocatalysts [1-2]. The optical and electronic properties of ZnO particles can be modulated by changing their size and morphology. Then, the control over these properties will represent a great challenge to realize novel functional devices. Up to now, a large quantity of ZnO nano and microstructures with various specific shapes have been successfully synthesized by different techniques; chemical methods such as: sol-gel, hydrothermal, precursor polymeric (Pechinni) and controlled precipitation [3-4] and physical methods such as: thermal evaporation, chemical vapor deposition (CVD), and molecular beam epitaxy (MBE). Specifically, N. Morales et al. [5], fabricated ZnO nanostructures with different morphologies through ultrasound-assisted hydrolysis of zinc acetate at room temperature. They study the effects of morphology and other parameters like specific surface area, defect content, and surface contamination on photocatalytic activity, using methylene blue (MB) degradation under UV light, evidencing that all the above mentioned parameters seen to affect the photocatalytic performance of ZnO nanostructures. Linping Xu et al. synthesized ZnO materials with different morphologies via solvothermal method, with different solvents, exhibiting different activities to phenoldegradation. They indicate that there was no

Content from this work may be used under the terms of the Creative Commons Attribution 3.0 licence. Any further distribution of this work must maintain attribution to the author(s) and the title of the work, journal citation and DOI.
Published under licence by IOP Publishing Ltd

IOP Conf. Series: Journal of Physics: Conf. Series **792** (2017) 012068 doi:10.1088/1742-6596/792/1/012068

correlation between the surface areas and the catalytic activity data for the studied samples suggesting that there are other more important factors that govern the photocatalytic activity, such as crystal habits [6].

In this work, ZnO particles were synthesized by the controlled precipitation method using zinc acetate and different synthesis and washing solvents. The powder obtained was washed (or chemically attacked) to modify and control the initial features such as morphology, particle size and surface area in order to study and improve the photocatalytic activity of ZnO structures.

2. Experimental

2.1 Materials

Zinc acetate dehydrate ($Zn(CH_3COOH)_2$) (ACS reagent ≥ 98%, Sigma Aldrich), acetic acid (ACS reagent ≥ 99.7%, Sigma Aldrich), Ethilene glycol (anhidrous ≥99.8%, Sigma-Aldrich), nitric acid (ACS 70%, Sigma-Aldrich), and ammonium hydroxide solution (28-30% NH_3 basis, Sigma-Aldrich) were commercially available and used as received. Deionized water (18.2 MΩ.cm) was obtained from a Milli-Q water purification system (Millipore, Billerica, MA, USA)

2.2. Synthesis

ZnO nanoparticles were synthesized by the controlled precipitation method (controlled precipitation method: formation mechanics of ZnO nanoparticles), for this purpose, zinc acetate was dissolved in different solvents: 200 mL of water, ethylene glycol acidified with nitric acid and acetic acid in a concentration of 0.3M. The mixture was stirred for 20 minutes until reach a completely transparent solution. Then the NH_4OH solution was added at a rate of 1ml/min, in constant stirring, until the previously determinate pH value with the titration curve; pH 8.8 for the sample synthesizedwith water, pH 9.0 for the sample synthesized with ethylene glycol acidified with nitric acid and pH 7.0 for the sample synthesized with acetic acid.

The wet solid obtained was washed (or chemically attacked). In this process was used 100 mL of water and re-dispersed using a high shear equipment (Ultraturrax-T50) for 10 minutes at 10000 rpm, then the system was allowed to stand for 24 hours and proceeded to remove the liquid phase by decanting.This process was repeated by four times. After the fourth wash, the liquid phase was removed and the solid dried at a temperature of 100 °C.The resulting product was macerated in an agate mortar until a fine powder which was subjected for 2 hours heat treatment at temperature of 300 °C, respectively, using an oven (Termolyne).

3. Results and Discussion

XRD was used to investigate the crystalline structure of the as-prepared ZnO particles, as shown in Figure 1(a). From the XRD patterns, it has been observed that the ZnO structures exhibit single-phase Wurtzite structure with average crystallite size of about 7, 36 and 16 nm for the sample synthesized with ethylene glycol acidified with nitric acid, water and acetic acid, respectively. The average crystallite size has been calculated using the Scherrer formula. Figure 1(b) shows FTIR spectra of ZnO samples within 400–4000 cm^{-1}. These results revealed that the peak at 3437 cm^{-1} comes from the stretching mode vibrations of OH, while the peak at 1639 cm^{-1} is attributed to the bending vibrations of adsorbed H_2O molecules. The small peak at 1397 cm^{-1} is due to asymmetric stretching modes COO^- which probably comes from the residues of preparation processes. It indicates the presence of small amount of organic residuesabsorbed on the surface of nanoparticlesin the samples, especially in the synthesized with ethylene glycol with nitric acid. The strong peak at 525 and 463 cm^{-1} can be ascribed to the vibration of Zn–O bond as shown in the literature (J.E. Rodriguez – Páez, 2001) [7-8]. The displacement of these absorption bands can be justified based on the morphology of the particles.

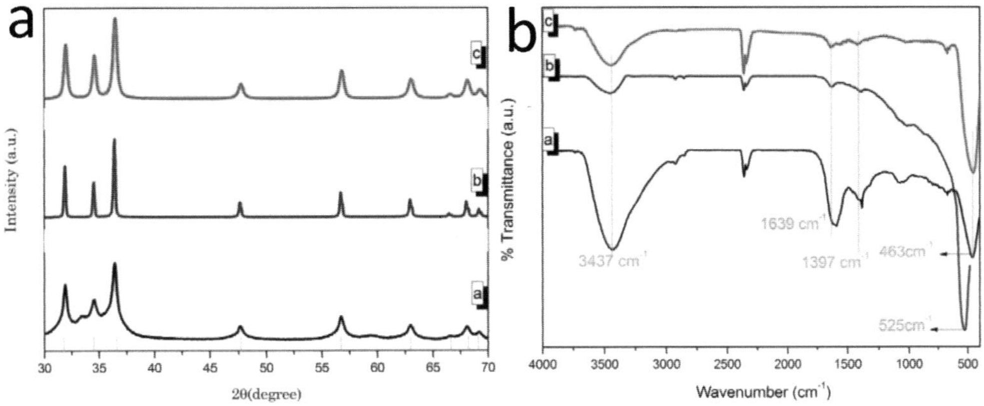

Figure 1. XRD and FTIR patterns of the as-prepared samples; (a) ethylene glycol acidified with nitric acid, (b) water and (c) acetic acid.

The optical characterization of the samples was recorded on UV-Vis absorption spectrophotometer. Figure 2(a) shows the UV-Visible absorption spectra of ZnO particles as a function of wavelength. The presence of a single slope in the plot suggests that the nanoparticles have direct and allowed transition for the sample synthesized with water and acetic acid, attributed to the band edge absorption of wurtzite hexagonal ZnO; it is clearly seen that the absorbance decreases with an increase in wavelength, and a sharp decrease in absorbancenear the band edge (390 nm) indicating the better crystallinity of the samples.

Energy band gap studies of these materials have been reported using Tauc plot, this intrinsic band gap is given as the intercept with the energy axis of a linear fit to the square of the absorption data in a energy interval slightly above the band gap, where the data is linear. An illustration of this procedure is shown in figure 2(b). The band gap values of ZnO are 3.19 eV for the samples sinthesised with water and acetic acid and 2.72 eV for the sample synthesized with ethylene glycol.

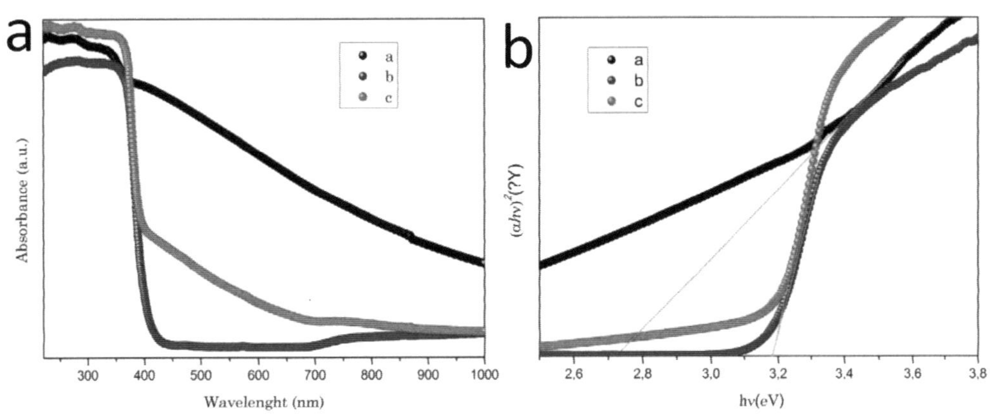

Figure 2.UV-visible spectra of as-synthesized ZnO samples at different synthesis solvents; (a) ethylene glycol acidified with nitric acid, (b) water and (c) acetic acid.

The morphology of samples was examined by Scanning electron microscopy (SEM) and Transmission Electronic Microscopy (TEM) as shown in Figure 3.

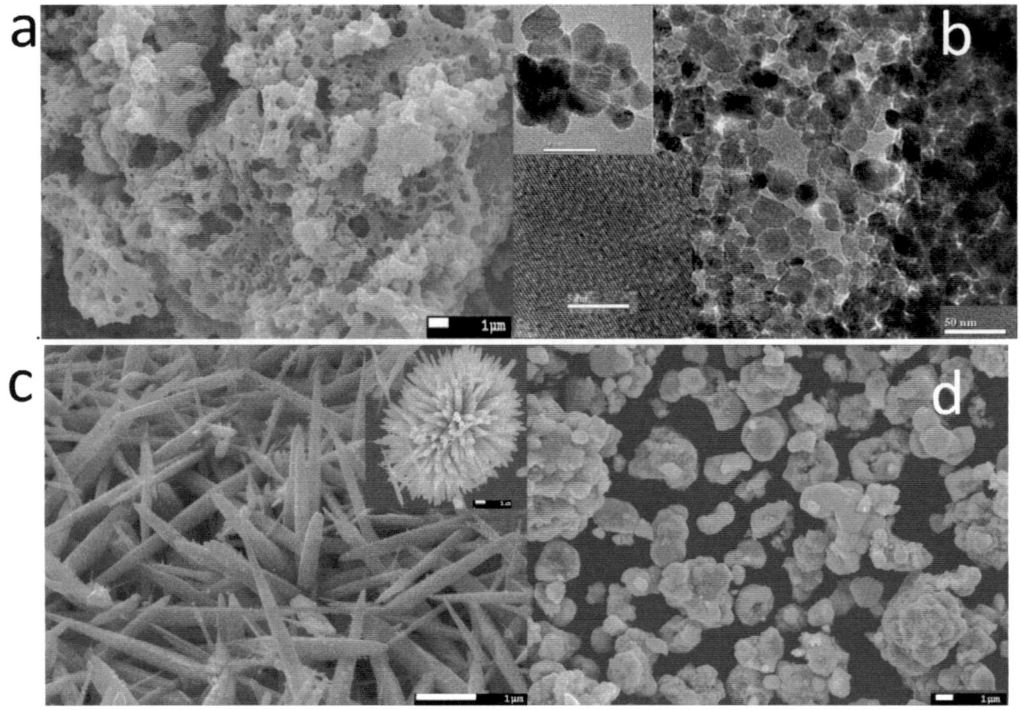

Figure 3. SEM and TEM images of as-synthesized ZnO samples at different synthesis solvents; (a,b) ethylene glycol acidified with nitric acid, (c) water and (d) acetic acid.

SEM images show a sponge-like morphology for the sample synthesized with ethylene glycol solvent, acidified with nitric acid and washed with water (Figure 3a). It composed of primary particles of nanometric sizes that make up the secondary particle sponge (Figure 3b).The sample synthesized and washed with water shows a nanoneedles morphology (Figure 3c), with hexagonal cross section, forming nanoflowers and the sample synthesized in acetic acid and washed with water shows particles with undefined morphology (Figure 3d). The morphology of ZnO particles would be governed by the balance of the nucleation and crystal growth determined by the reaction route. Experimental data show that the variation of the synthesis conditions significantly affects the specific surface area (BET) and pore size distribution as shown in Figure 4a. According to the IUPAC classification, the hysteresis loops observed in samplesynthesized withethylene glycol acidified with nitric acid and the sample synthesized with and washing with water can be attributed to H3-type mesoporous material and isotherm of the sample synthesized with water corresponds to a microporous solid.

The catalytic and photocatalytic activities of ZnO samples were studied from the initial rate of decomposition of H2O2 in aqueous solution. The kinetic study was followed by the production of O2 via gas pressure monitoring (O2 – monitored method [9], under UV-visible light irradiation ($\lambda \geq 400$ nm), at 20 °C. Experimental data showed that the rate of H2O2 decomposition, under ultraviolet visible light irradiation, obeys the first order kinetic law. Under visible light irradiation the H2O2 decomposition rate achieved a maximal value of 9.5 mmol g−1 s−1 for ZnO2 sample, 5.8 mmol g-1s-1 for sample ZnO3 and 3.74 mmol g-1s-1 for sample ZnO1.

IOP Conf. Series: Journal of Physics: Conf. Series **792** (2017) 012068 doi:10.1088/1742-6596/792/1/012068

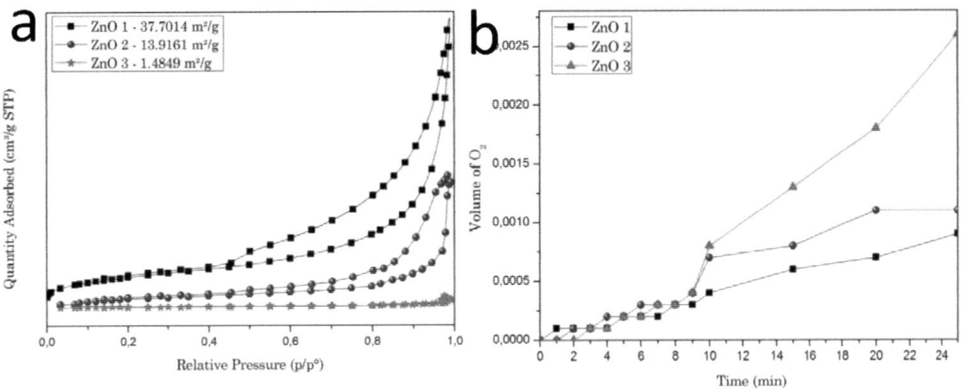

Figure 4. (a) Typical N_2 adsorption–desorption isotherms for ZnO nanostructures, (b) Photodecomposition of H_2O_2 catalyzed by ZnO compounds.

These results show that nanoparticles with well-defined morphology exhibit large activity in the photocatalytic photo-decomposition of H_2O_2 (Figure 4b) regardless of specific surface area. This indicates that a high surface reactivity of ZnO samples is not necessarily related to a high surface area. It seems that the abundance of specific sites, which strongly depends on the preparation route, is the most important feature for the active of the ZnO. Results indicate that ZnO nanoflowers showed significantly higher photocatalytic activity than others ZnO samples. The photocatalytic activities were found to be associated with their content of oxygen vacancy. The catalytic activity is related to defects created in the structure which allow a greater or lesser degree of vacancies affecting the electron-hole pair recombination. Additional measurements and studies are being developed in order to understand better the photocatalytic activity of these samples.

4. Conclusion

ZnO samples with diverse sizes and morphologies were prepared by the controlled precipitation method using different solvents and modifying the washing process. Comparing the photochemical test with the morphology and surface area results, it seems that the abundance of specific sites, directly related to the morphology of ZnO samples which strongly depends on the preparation route, is the most important feature for the active of the ZnO.

5. Acknowledgments

Authors would like to thanks CAPES, CNPQ and FAP – DF for the financial support .

6. References

[1] Morkoc H 2007 *Zinc oxide: fundamentals, materials and device technology* (Weinhein Germany:Willey-VCH)
[2] Malato S *Proc. IOCD/CYTED Symp. VIII-G* **4** pp 79-101
[3] Wang Y et al. 2008 *Separation and Purification Technology* **62** pp 727–732
[4] Pardeshi S K and Patil A B 2009 *J. Mol. Catal. A: Chem.* **308** pp 32–40
[5] Morales-Flores N, Pal U, Galeazzia R and Sandovalb A 2014 *RSC Adv.* **4** pp 41099–41110
[6] Xu L, Hu Y, Pelligra C, Chen C, Jin L, Huang H, Sithambaram S, Aindow M, Joesten R and Suib S L 2009 *Chem. Mater.* **21** 13
[7] Rodriguez–Paez J E, Moure C, Duran P and Fernández J F 1998 *Bol. Soc. Esp. Cerám. Vidrio* **37** 2-3 pp 187-192
[8] Rodríguez–Páez J E 2001 *J. Eur. Ceram. Soc.* **21** pp 925-930
[7] Páez C A 2011 *Catal. Commun.* **15** pp 132–136

Changes in the reflectivity of a lithium niobate crystal decorated with a graphene layer

O. Salas[1,2], E. Garcés[1], F. L. Castillo[1] and L.F. Magaña[2*]
[1] Escuela Superior de Físico-Matemáticas, Instituto Politécnico Nacional. Avenida Instituto Politécnico Nacional S/N, Edificio 9, Gustavo A. Madero, San Pedro Zacatenco, 07738 Ciudad de México, México.
[2]Instituto de Física, Universidad Nacional Autónoma de México. Apartado Postal 20-364, Código Postal 01000, Ciudad de México. México.

*Email:fernando@fisica.unam.mx

Abstract. Density functional theory and molecular dynamics were used to study the interaction of a graphene layer with the surface of lithium niobate. The simulations were performed at atmospheric pressure and 300K. We found that the graphene layer is physisorbed with an adsorption energy of -0.8205 eV/C-atom. Subsequently, the optical absorption of the graphene-(lithium niobate) system was calculated and compared with that of graphene solo and lithium niobate alone, respectively. The calculations were performed using the Quantum Espresso code with the GGA approximation and Vdw-DF2 (which includes long-range correlation effects as Van der Waals interactions).

1. Introduction

The photorefractive crystals motivate a special interest because of their prominent and highly advanced applications such as holographic memories, phase conjugators, dynamic interferometry and pattern recognition among many more. Since the first publication on the explanation of their physical properties [1, 2], these materials have revolutionized and given life to many applications of nonlinear optics.

Lithium niobate ($LiNbO_3$) is an anisotropic crystalline material. This means that the refractive indices depend upon the direction of crystal axes. It is a photorefractive material. It is a strongly asymmetric crystal, and it can change the polarization state of the light passing through, with the variation of an electric field applied externally. This change is due to an electro-optical effect, which can be explained as follows. An external electric field applied across the crystal changes the refractive index n_e (special) and n_o (ordinary), according to the crystal orientation and applied field. This material is used for making modulators intensity, optical components in advanced technology and in many other devices.

The structure of $LiNbO_3$ belongs to the rhombohedral space group (trigonal) R3c, with a point group 3m, when it is grown under controlled temperature. For crystal growth at temperatures above the phase transition temperature (≈1483 K), the crystal structure is center-symmetrical of the group R3m. There are three options in the selection of the axes of $LiNbO_3$ cells. The first one corresponds to the rhombohedral. The second corresponds to the hexagonal, and the last to the hexagonal-ortho hexagonal. The first two are suitable for crystallographic purposes and for determining its structure. For most applications, the preferred setting is orthogonal and tensor components give properties with respect to these axes.

The $LiNbO_3$ crystal (CLN) is very commercial, and is usually grown with the mixture of ([Li]/[Nb] = 48.6/51.4) by the Czochralsk method. The CLN is grown easily with a homogeneous

Content from this work may be used under the terms of the Creative Commons Attribution 3.0 licence. Any further distribution of this work must maintain attribution to the author(s) and the title of the work, journal citation and DOI.
Published under licence by IOP Publishing Ltd

composition and great optical quality. However, there may be some intrinsic defects within the crystal due to lack of Li. At present, the CLN can be obtained as mono crystal with excellent quality [3].

On the other hand, graphene, a sheet of a single layer of graphite, is a semiconductor [4, 5] with a forbidden zero band which has been of great interests in theoretical and experimental studies. Graphene has unique electronic properties resulting from its hexagonal honeycomb lattice structure, which causes electrons to behave as massless relativistic fermions satisfying the Dirac equation [6]. The Hall Effect [7] has been observed in graphene. Studies have also shown that there is spin-orbit coupling in this material [8-11]. These and other unusual electronic properties of graphene make it a promising material for electronic devices. However, the lack of a band gap in graphene establishes limitations on their practical applications. It is very important to find ways to effectively tune the band gap of graphene to expand its applications in nano electronics and optoelectronics.

In this work, density functional theory and molecular dynamics were used to study the interaction of a graphene layer with the surface of lithium niobate. The simulations were performed at atmospheric pressure and 300K. Then we calculated the reflectivity of the graphene-(lithium niobate) system. For our simulations, we considered a lithium niobate slab (figure 1) with a thickness of 5.14 Å.

2. Method

The density functional theory was used, with the general gradient approximation (GGA) and Vdw-DF2 (which includes long-range correlation effects as Van der Waals interactions), and molecular dynamics (MD) within the Born-Oppenheimer approximation with the Quantum Espresso code [12]. For exchange-correlation energies, we used the Perdew-Burke-Ernzerhof (PBE) approximation for GGA [13]. We used the norm conserving Martins -Troullier and Goedecker- Hartwigsen-Hutter-Teter [14] pseudopotentials.

We considered as valence electronic states for carbon $2s^2p^2$, for lithium: $2s^1$, for niobium: $4d^45s^2$, and for oxygen: $2s^22p^4$. We performed non-relativistic, and non-spin polarized calculations. The cut-off energy was 748 eV, and we took 50 k points within the Monkhorst-Pack scheme [15]. The threshold energy convergence was 10^{-8} Ry. To validate our pseudopotentials we obtained by energy minimization, in the graphene layer, for the value of the C-C bond length 1. 42 Å (the experimental values are 1. 42 Å). For the lithium niobate crystal with rhombohedral unit cell, the lattice parameters were **a** = 5.4940 Å, **c** = 13.8625 Å and ∝=55.867° (experimental values **a** = 5.494 Å, **c** = 13.825 Å, and ∝=55.867°, respectively [16]).

The graphene-$LiNbO_3$ system was represented using a hexagonal cell with **a** = 5.1479 Å and **c** = 39.9804 Å with periodic conditions. It contains eight carbon atoms, eighteen oxygen atoms, six niobium atoms, and six lithium atoms (see figure 1). In the **c** direction, we took a large enough separation to ensure that there is no interaction between adjacent decorated lithium niobate slabs.

The adsorption energy of the graphene layer onto the surface of lithium niobate was obtained from:

$$\Delta E = E(graphene - LiNbO_3) - \left[E(graphene) + E(LiNbO_3)\right]$$

Where *E (Graphene-LiNbO3)* is the energy of the final configuration; *[E (Graphene) + E ($LiNbO_3$)]* is the energy of the initial system, which is the graphene alone plus the $LiNbO_3$ crystal energy alone, with no interaction between them. It was found that graphene is physisorbed on the $LiNbO_3$ surface, and the adsorption energy is -0.8205899 eV/(C-atom), when using GGA, and -0.8154758 eV/(C-atom) GGA when we used Vdw-DF2 (which includes long-range correlation effects as Van der Waals interactions).

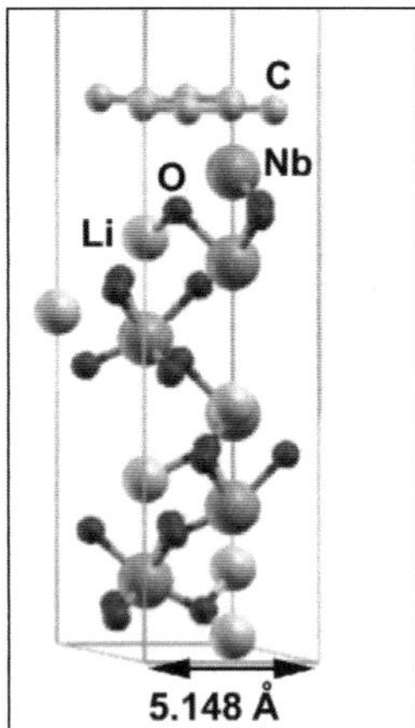

Figure 1. Optimized unit cell of the new system.

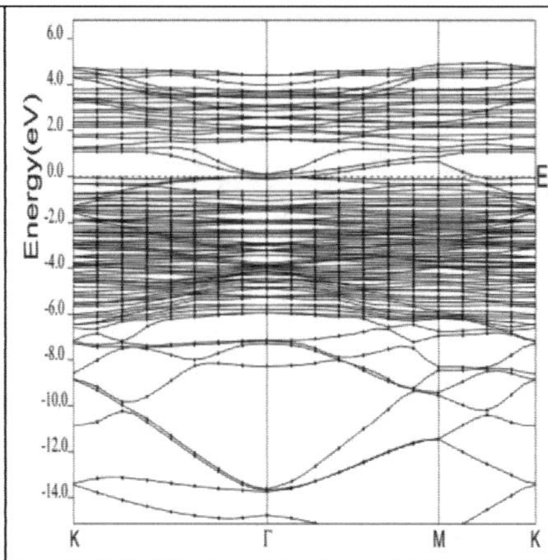

Figure 2. Calculated energy band structure of our system.

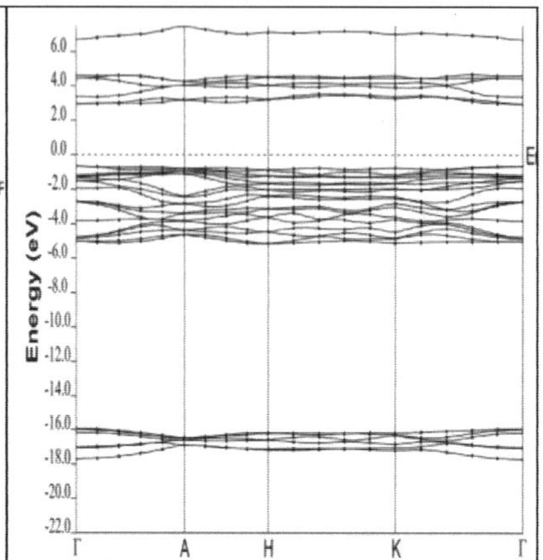

Figure 3. Calculated energy band structure of lithium niobate.

After the optimization of the graphene-($LiNbO_3$) system, we calculated the band energy structure for the new system, using the method of the modified tetrahedron Blöch for integration in the first Brillouin zone [17]. The energy gap of the $LiNbO_3$ crystal was, $Eg = 3.6\ eV$ (the experimental value is $Eg = 3.78\ eV$ [18]), the energy gap for the graphene-$LiNbO_3$ system is $Eg = 0.148\ eV$.

The imaginary component of the dielectric tensor can be calculated by knowing the structure of electronic bands in the limit of the linear optics, and in the formalism of the random phase approximation (RPA) where the expression is obtained [19]:

$$Im\varepsilon_{\alpha\beta}(w)=\frac{4ne^2}{m^2w^2}\sum_{c,v}\int dkc_k|p^0|v_kv_k|p^\beta|c_k\delta(\varepsilon_{ck}-\varepsilon_{vk}-w) \tag{1}$$

The real component of the dielectric tensor is derived from the Equation (1) by the Kramers-Kronig relations [19, 20]:

$$Re\varepsilon_{\alpha\beta}(w)=\delta_{\alpha\beta}+\frac{2}{n}p\int_0^\infty\frac{w'Im\varepsilon_{\alpha\beta}(w')}{w'^2-w^2}dw' \tag{2}$$

The reflectivity at normal incidence is

$$R_{ii}(w)=\frac{(n-1)^2+k^2}{(n+1)^2+k^2} \tag{3}$$

Where n and k are the real and imaginary part of the complex refractive index (refractive index and extinction coefficient, respectively). The components of n and k are given in equations (4) and (5).

$$n_{ii}(w)=\sqrt{\frac{|\varepsilon_{ii}(w)|+Re\varepsilon_{ii}(w)}{2}} \tag{4}$$

$$k_{ii}(w)=\sqrt{\frac{|\varepsilon_{ii}(w)|-Re\varepsilon_{ii}(w)}{2}} \tag{5}$$

Results and discussion

After optimization of our system (see figure 1), we calculated the adsorption energy. We found that the graphene layer is physisorbed with an adsorption energy of -0.8205 eV per carbon atom. The result for the corresponding energy band structure is shown in figure 2. For comparison, in figure 3 we show the energy band structure of lithium niobate alone. We can see big differences from the first case. Afterwards, we calculated the complex dielectric tensor for the equilibrium configuration in an energy range of 0-30 eV (which corresponds to 0-240000 cm^{-1}).

In Figure 4, we show the resulting reflectivity for the graphene-($LiNbO_3$) system. In this figure, we also show the reflectivity of the graphene alone and the corresponding one of the lithium niobate alone, for comparison.

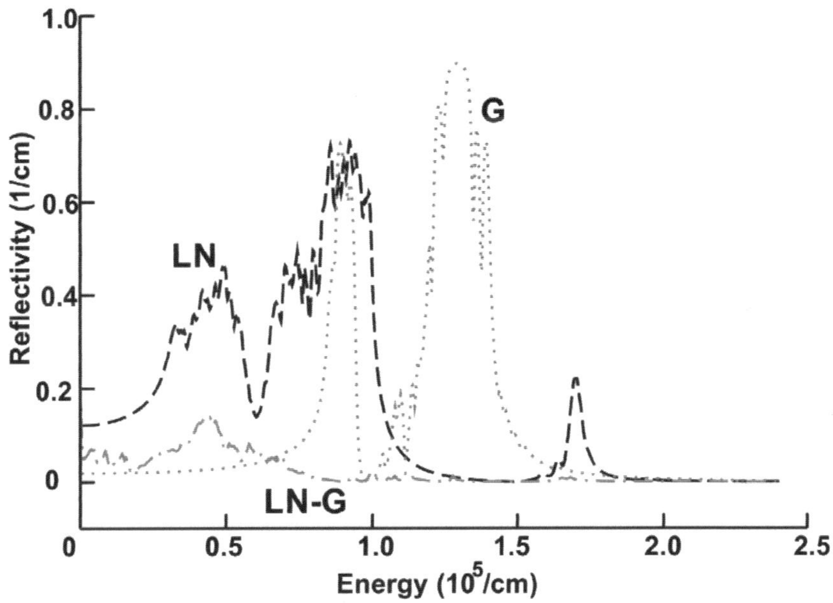

Figure 4. Calculated reflectivities of our system, and those of graphene alone and LiNbO3 alone.

Notice that the reflectivity of lithium niobate alone has several peaks, with the maximum around 10^5 cm^{-1}. After that, the reflectivity decreases with increasing energy, and then it flattens up to around $1.7X10^5$ cm^{-1}, when the last peak appears.

On the other hand, the reflectivity of graphene also has several peaks. The first one is similar in position and in magnitude to the corresponding reflectivity of lithium niobate, a little bit before 10^5 cm^{-1}. The largest peak is around $1.3X10^5$ cm^{-1}. After this peak, the graphene reflectivity goes to zero.

Finally, in the reflectivity of the lithium niobate decorated with graphene, we have a smaller variation with changes in energy, than in the two previous cases. Furthermore, the largest value of the reflectivity occurs around $0.45X10^5$ cm^{-1} (around 5.5 eV), in a similar position of the first peak of the reflectivity of lithium niobate. However, the maximum value of the reflectivity is about 0.14 cm^{-1}. This value is 31% of the value of the reflectivity of lithium niobate at the same position. There is a small peak at around $1.6X10^5$ cm^{-1}, where the reflectivity of lithium niobate presents a peak. However, the value of the reflectivity is 16% of the corresponding value of the reflectivity of graphene.

Clearly, the effect of the graphene layer on the reflectivity of lithium niobate is strong. Besides, the resulting reflectivity of the decorated lithium niobate crystal is also very different from the reflectivity of the graphene alone.

4. Conclusions

It was found that graphene is physisorbed on the lithium niobate surface at atmospheric pressure and 300K. There are large differences of the energy band structure of lithium niobate alone with respect to graphene - (lithium niobate).

The resulting reflectivity of lithium niobate decorated with graphene is very different from the reflectivity of graphene alone, and from single lithium niobate.

The reflectivity of the lithium niobate decorated with graphene, shows a much smaller variation with changes in energy, than any of graphene and lithium niobate decorated with graphene.

Acknowledgements

We thank Direccion General de Asuntos del Personal Academico de la Universidad Nacional Autonoma de Mexico, partial financial support by Grant IN-106514 and we also thank UNAM-Super-Computing Center the technical assistance.

References

[1] Kukhtarev N Kiev Semenyuk A and Nauka Dumka 1971 *Radiation Physics of Nonmetallic Crystals* Vol.3, Chap. 1
[2] Kukhtarev Sov N *Tech. Phys*. Lett., 1, 155 1975. Vinetskii V, Kukhtarev N and Semenyuk Sov A *Phys. Semicond*., 6(18), **879** (1972) Vinetskii V and Kukhtarev Sov N *Phys. Solid State Phys*., 16, 3714 (1974).
[3] Volk T and Wöhlecke M 2008 *Lithium Niobate: Defects, Photorefraction and Ferroelectric Switching. Springer series in materials science* **115**.
[4] Springer-Verlag, Berlin K S, Novoselov, Geim A K, Morozov S V, Jiang D, Zhang Y, Dubonos S V, Grigorieva I and Vand Firsov A A 2004 *Science* **306** 666
[5] Novoselov K S, Jiang D, Schedin F, Booth T J, Khotkevich V V, Morozov S V and Geim A K 2005 Proc. Natl. Acad. Sci. U.S.A. **102**, 10451
[6] Novoselov K S, Geim A K, Morozov S V, Jiang D, Katsnelson M I, Grigorieva I V, Dubonos S V and Firsov A A, 2005 *Nature* (London) **438** 197
[7] Zhang Y B, Tan Y W, Stormer H L and Kim P 2005 *Nature* (London) **438** 201
[8] Kane C L and Mele E J, 2005 *Phys. Rev*. Lett. **95** 226801
[9] Huertas-Hernando D, Guinea F and Brataas A, 2006 *Phys. Rev*. B **74** 155426
[10] Min H, Hill J E, Sinitsyn N A, Sahu B R, Kleinman L and A H MacDonald 2006 *Phys. Rev*. B **74**, 165310
[11] Yao Y, Ye F, Qi X L, Zhang S C and Fang Z 2005 *Phys. Rev. B* **75** 041401(R)
[12] Giannozi P , Baroni S, Bonini N, Calandra M, Car R, Cavazzoni C, Ceresoli D, Chiarotti , Cococcioni L M J 2009 *Phys. Condens. Matter* **21** 395502.
[13] Perdew J P, Burke K and Ernzerhof M 1996 *Phys. Rev*. Lett. **77** 3865.
[14] Troullier N, Martins J L and Perdew J P 1993 *Phys. Rev*. B **43**.
[15] Monkhorst H J and Pack J D 1976 *Phys. Rev*. B. **13** 5188.
[16] Weis R S and Gaylord T K *Lithium Niobate: Summary of Physical Properties and Crystal Structure* USA School of Electrical Engineering, Georgia Institute of Technology, Atlanta, GA 30332
[17] Blöchl, Jepsen P E and O. & Andersen OK 1994 Improved tetrahedron method for Brillouin-zone integrations. *Phys. Rev*. B **49**(23), 16223-33.
[18] Dhar A and Mansingh A 1976 *J. Appl. Phys.* **68** 5804.
[19] Bohm, D. & Pines and D A 1953 Collective description of electron interactions: III. Coulomb interactions in a degenerate electron gas. *Phys. Rev*. **92**(3), 609-625.
[20] Kittel and C 1966 *Introduction to Solid State Phys*. 3rd edition. New York, John Wiley p 648

Inverse problem of HIV cell dynamics using Genetic Algorithms

J A González, F S Guzmán
Laboratorio de Inteligencia Artificial y Supercómputo. Instituto de Física y Matemáticas, Universidad Michoacana de San Nicolás de Hidalgo, Morelia México

E-mail: gonzalez@ifm.umich.mx, guzman@ifm.umich.mx

Abstract. In order to describe the cell dynamics of T-cells in a patient infected with HIV, we use a flavour of Perelson's model. This is a non-linear system of Ordinary Differential Equations that describes the evolution of healthy, latently infected, infected T-cell concentrations and the free viral cells. Different parameters in the equations give different dynamics. Considering the concentration of these types of cells is known for a particular patient, the inverse problem consists in estimating the parameters in the model. We solve this inverse problem using a Genetic Algorithm (GA) that minimizes the error between the solutions of the model and the data from the patient. These errors depend on the parameters of the GA, like mutation rate and population, although a detailed analysis of this dependence will be described elsewhere.

1. Introduction

Modelling the dynamics of HIV and AIDS is important because this disease represents a main concern in public health policies, as infected people are counted in dozens of millions around the world [1]. There are two important concerns: prevention and treatment. The treatment involves a constant monitoring of each patient, specifically the dynamics of infected cell concentrations in the blood and is the subject we focus on in this manuscript.

From the clinical point of view, what is required for the present model is that an essential component of the immune system are the Lymphocytes which destroy invaders. Lymphocytes are of two types B-cells and T-cells, being B-cells antibody factories producing antibodies as fast as they can and also clone themselves, whereas T-cells either direct the activity of B-cells (called CD4+T-cells) or act as suppressors (called CD8+T-cells) destroying infected cells and thus damp out the activity of the immune system. The AIDS has three stages, one including the initial infection, a second one of latency and the third one corresponding to a runaway destruction of the immune system. During the latency lapse healthy T-cells are infected although its number remains high. When the concentration of T-cells decreases and that of HIV virus cells increases, it is the AIDS stage.

In practice, as an example, cytometry is a procedure useful to count the concentration of healthy, latently infected and infected T-cells in the blood stream, for instance using the dispersion of laser light, that depends on the enzymes covering the cells [2]. Knowing the concentration of CD4+T-cells it is possible to wonder about modelling the evolution of the different cell populations. Among the various models trying to describe the dynamics of CD4+T-cells (see for instance [3,4]), a simple one describing the phases of latency and the destruction of the immune system is Perelson's model [5,6,7].

In general, a model is expected to describe clinical data and make predictions. Starting from patients data we define the inverse problem of determining the parameters of the model. Knowing the parameters allows one to determine the evolution of the concentration of healthy and infected CD4+T-cells, so as possibly infer the conditions during an early stage of the disease. .

Assuming Perelson's model, we intend to solve the inverse problem and reconstruct the coefficients of the evolution model for the cell populations, out of a time series that in a realistic case might correspond to the data taken from a particular patient. This means, we assume that there is a Perelson system of equations for each patient with different parameters. In this sense, what we propose is a personalized set of equations that can be used to diagnose and decide on the treatment of each particular patient. This paper is about testing a method to estimate the parameters characterizing a

Content from this work may be used under the terms of the Creative Commons Attribution 3.0 licence. Any further distribution of this work must maintain attribution to the author(s) and the title of the work, journal citation and DOI.
Published under licence by IOP Publishing Ltd

IOP Conf. Series: Journal of Physics: Conf. Series **792** (2017) 012070 doi:10.1088/1742-6596/792/1/012070

given patient. Our method to solve this inverse problem uses Genetic Algorithms (GAs) to find the best combination of parameters that minimizes the difference between the experimental data (in our case generated numerically) and the numerical solution to the Perelson's model.

The paper is organized as follows. In Section 2 we describe the particular model of T-cell populations, whereas in Section 3 we present the GAs applied to estimate the parameters of the ODE system associated to the model. In Section 4 we present some results and in 5 draw some final comments.

2. The T-cell dynamics model

Among the various models describing the evolution of T-cell in the blood stream in patients, in order to show how our method works, we use the basic Perelson model [5]. Following [8], the variables that evolve under this approach are the healthy T-cells (X), latently infected T-cells (Y), actively infected T-cells (Z) and free viral cells (W), which obey the following set of non-linear coupled ODEs

$$\begin{aligned} \frac{dX}{dt} &= s + rX\left(1 - \frac{X+Y+Z}{x_{\max}}\right) - \mu_X X - k_1 XW, \\ \frac{dY}{dt} &= k_1 XW - \mu_Y Y - k_2 Y, \\ \frac{dZ}{dt} &= k_2 Y - \mu_Z Z, \\ \frac{dW}{dt} &= N\mu_Z Z - k_1 XW - \mu_W W \end{aligned} \qquad (1)$$

Here s is the rate at which T-cells enter the blood stream; r grades the logistic growth rate in the number of CD4+T cells, and their evolution as a function of actively infected and latently infected cells; the concentration of destroyed T-cells is proportional to the population of infected cells with clearance k_1; the coefficient $-\mu_X$ is based on how patients recover after therapy; $x_{\max}$ is the maximum possible CD4+T concentration; μ_Y is the death rate of latently infected CD4+T cells; μ_Z is the death rate of actively infected CD4+T population, μ_W is the death rate of free virus cells, k_1 is the rate of infection of CD4+T by the free virus, k_2 is the rate at which latently infected CD4+T cells convert to actively infected cells and N is the number of free virus produced by lysing a CD4+T cell.

Solving system (1) is straightforward provided initial conditions and specific values of the coefficients/parameters. We solve these equations using a 4th order accurate Runge-Kutta integrator and graphically the solution is shown in Figure 1 for the parameters in Table 1, that we set following [8]. In this case the solution approaches asymptotic values in all variables.

In order to test our approach, this solution will work as our experimental data, consisting of the four time series of $(\tilde{X}, \tilde{Y}, \tilde{Z}, \tilde{W})$ which in a realistic case would be the actual measurements from a patient's blood stream.

variable	value	units	variable	value	units
s	10	$mm^{-3}day^{-1}$	r	0.03	day^{-1}
$x_{\max}$	1500	mm^{-3}	μ_X	0.02	day^{-1}
μ_Y	0.02	day^{-1}	μ_Z	0.24	day^{-1}
μ_W	2.4	day^{-1}	k_1	.000024	$mm^{-3}day^{-1}$
k_2	0.003	day^{-1}	N	500	
X_0	500	mm^{-3}	Y_0	0	mm^{-3}
Z_0	0	mm^{-3}	W_0	0.001	mm^{-3}

Tabla 1. Parameters used to construct an illustrative solution. Such solution will also be considered as data from a patient, being these parameters the target to be found by the GA. X_0, Y_0, Z_0, W_0 are the initial conditions of the four variables.

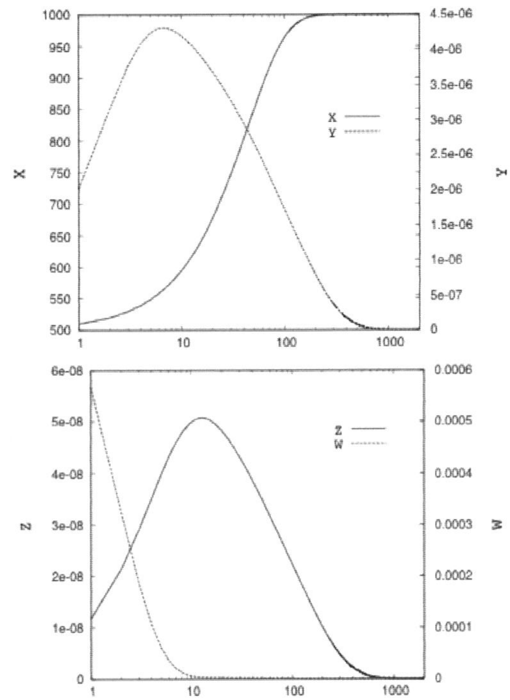

Figure 1. Integration of the system during 2000 days for the parameters in Table 1. All the variables reach a stationary state within this time domain. The horizontal axes are in days.

3. Parameter estimates using Genetic Algorithms

The initial conditions and the various parameters of system (1) are patient dependent. Counts of the four variables before or during treatment work as experimental data that should be fit by the solutions of the system. Thus, the parameters in these equations are particular of each patient submitted to his own treatment. In order to foresee the evolution of the patient as a function of experimental data it is necessary to estimate the whole set of coefficients in system (1). In this manuscript we present an efficient method that estimates these parameters using GAs.

As mentioned before, instead of using realistic data from patients to test our method, we consider the solution of the system for the parameters in Table 1 as our time series data. The advantage of this approach is that we know the values of the parameters and we can measure the accuracy in the parameter estimate of our GAs.

Inverse problem strategy. We assume each patient is described by a Perelson system of equations with unknown coefficients. Therefore we set the inverse problem as an optimization problem where the variables are the coefficients of system (1). In order to solve this problem we use a Genetic Algorithm described as follows. We assume that the array of numbers $\{s, r, x_{\max}, \mu_X, \mu_Y, \mu_Z, \mu_W, k_1, k_2, N\}$ is the DNA of an organism, and each entrance is a gene. With this assumption the GA contains the following elements

- Initial Population. We define an initial population of $j = 1, \ldots, N_{org}$ organisms with random entrances with the values of the parameters of the data, but randomly shifted within a given range $0 < \beta < 1$. For example, we chose the parameter s to be $s = 10(1 \pm \beta\varepsilon)$, with ε a random number between 0 and 1. N_{org} is the size of the population of organisms we start the GA with.

- For each organism j, a particular system of equations of type (1) is defined and is solved numerically.
- The resulting solution for each organism is compared with the data $(\tilde{X}, \tilde{Y}, \tilde{Z}, \tilde{W})$.
- We calculate a weighted $L_1(E_{t,j})$ norm of the error between the solution and the time-series for organism j. Considering each of the variables X, Y, Z, W has values within different scales, the error is measured as follows:

$$L_1(E_{t,j}) = \sum_i \left(\frac{|X_j - \tilde{X}_j|}{\tilde{X}_j} + \frac{|Y_j - \tilde{Y}_j|}{\tilde{Y}_j} + \frac{|Z_j - \tilde{Z}_j|}{\tilde{Z}_j} + \frac{|W_j - \tilde{W}_j|}{\tilde{W}_j} \right)$$

where i labels a summation over the total number of days in the time series. We define the fitness of the organism j as $F_j = 1 / L_1(E_{t,j})$ and classify the organisms from the best to the worst fitted.

- The selection: Using a roulette selection algorithm [9,10], we choose a couple of organisms, one called the mother and the other the father, with the only restriction that they cannot be the same. Each couple will generate a new organism using the crossover mechanism described below. We select $N_{org} - 2$ couples to create the same number of children that will populate the next generation. The remaining 2 organisms missing to complete the next generation will be the copies of the 2 best fitted DNAs of the current generation.
- The crossover: We choose randomly 5 genes out of the 10 of the mother. Then complete the DNA of a child with the 5 genes from the father not chosen from the mother.
- Mutation: Every individual has 40% probability to mutate. Every gene of a selected individual is forced to mutate within $\alpha\%$ of its current value.

This process is repeated in a non-standard way. Consider that t^k labels the time in the time series (in days) and that $\tilde{X}^k, \tilde{Y}^k, \tilde{Z}^k, \tilde{W}^k$ are the data at t^k. The GA starts at t^0 and evolves the necessary number of generations to find the DNA that produces a numerical solution X^1, Y^1, Z^1, W^1 with coefficients $\{s_1, r_1, x_{\max,1}, \mu_{X,1}, \mu_{Y,1}, \mu_{Z,1}, \mu_{W,1}, k_{1,1}, k_{2,1}, N_1\}$, with an error within a given tolerance θ from the data. These coefficients are then used as the initial guess to evolve another number of generations to fit the variables at t^2 and so on and so forth.

4. Results

We thus use the GA to evolve the initially randomly generated organisms within an initial departure β. In Fig. 2 we show the solution using the DNA of the fittest organism found after a number of generation using $\theta = 0.1$ and two different values of $\beta = 0.1, 1$. Notice that the values of β correspond to parameters possibly at 10 and 100% distant from the parameters used to generate the data.

IOP Conf. Series: Journal of Physics: Conf. Series **792** (2017) 012070 doi:10.1088/1742-6596/792/1/012070

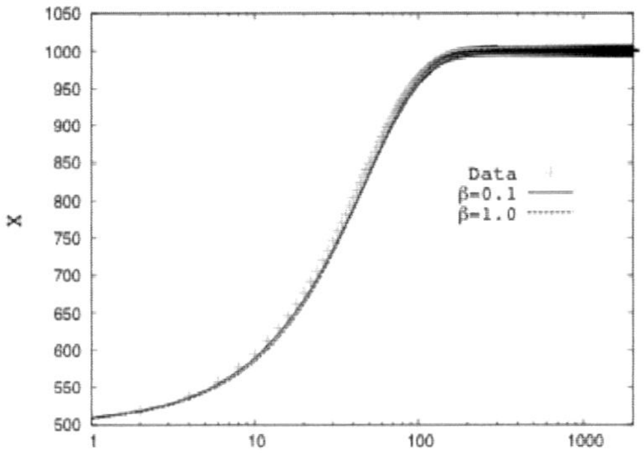

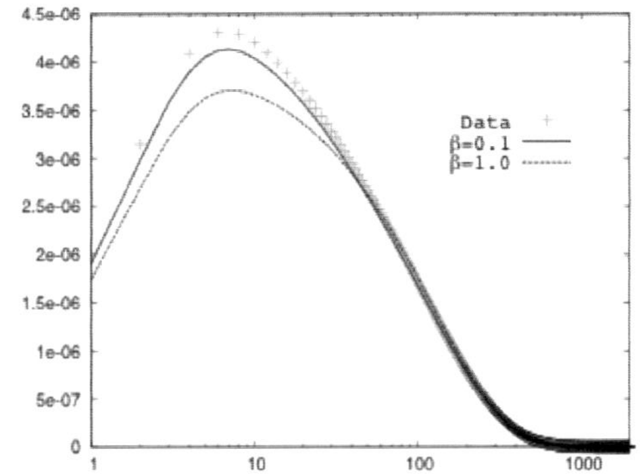

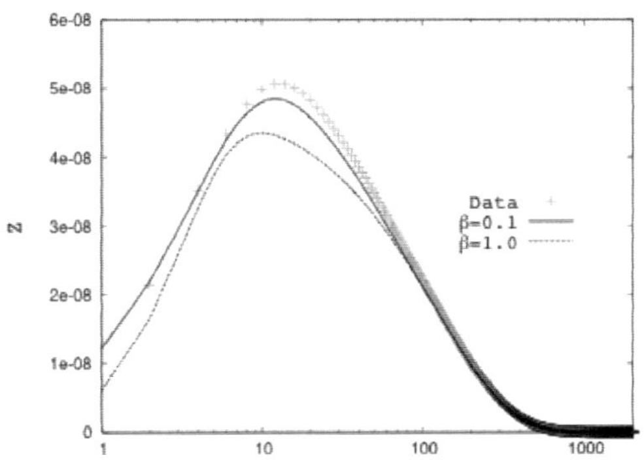

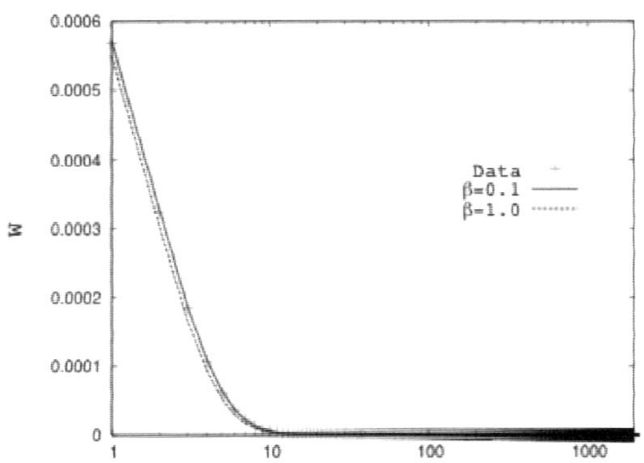

Figure 2. Solution calculated using the coefficients corresponding to the best fitted organism of a given generation with $\beta = 0.1, 1$ and tolerance $\theta = 1$. These are compared with the data from Figure 1. The horizontal axes are in days.

5. Final comments and conclusions

We implemented an application that uses GAs to estimate the parameters of a Perelson's model. This means we solve the inverse problem which in practice implies that, within a given tolerance, we have a method to assign a particular set of equations to a particular patient, based on the T-cell concentration measurements. For this we do not use patient's real data, but a numerical solution, and the GA tracks down the parameters these data were constructed with.

A simple improvement of our method can include the initial conditions of the four variables as genes of the DNA of our organisms. This would allow to estimate X, Y, Z, W during the latency period and possibly bound (going backward until the initial conditions) a lapse of infection. It is believed that early phases of HIV contain crucial information about the further immune response and viral dynamics and is expected to influence the progression of AIDS [11].

Our method works for this non-linear system of equations and is expected to work for straightforward generalization of Perelson's model as well. The complexity of this simple model will allow us to engage the parameter estimation of more elaborate problems, for instance the patient's parameter reconstruction based on the response to treatment considering interrupted therapy [11,12,13,14].

6. Acknowledgments

This research is partly supported by grants CIC-UMSNH-4.9, CIC-UMSNH-4.23 and CONACyT 258726 (Fondo Sectorial de Investigaci6n para la Educaci6n).

7. References

[1] S Armstrong, C Fontaine and A Wilson, Report on the global AIDS epidemic UNAIDS/Joint *United Nations Programme on HIV/AIDS*, (Geneva, Switzerland. Online: http://www.unaids.org)
[2] M G Ormerod 2008 *Flow Cytometry* (Garland Pub.)
[3] J D Murray 2002 *Mathematical Biology I. An Introduction.* (Ed. Springer)
[4] A S Perelson and G Weisbuch 1997 Immunology for physicists. *Rev. Mod. Phys.* **69** 1219-1267
[5] A S Perelson, D E Kirschner, R De Boer 1993 Dynamics of HIV infection of CD4+ T cells. *Mathematical Biosciences* **114** 81-125
[6] A S Perelson and P W Nelson 1999 Mathematical analysis of HIV-1 dynamics in vivo. *SIAM review* **41** 3-44
[7] A S Perelson 2002 Modelling viral and immune system dynamics *Nature Reviews Immunology* **2** 28-36
[8] P A Isihara 2005 Immunological and Epidemiological HIV/AIDS modelling. *THE UMAP JOURNAL* ***26*** *49-90*
[9] J H Holland 1975 *Adaptation in Natural and Artificial Systems* (University of Michigan Press. 2nd Ed. MIT press)
[10] M Mitchell 1998 *An introduction to Genetic Algorithms.* (MIT Press)
[11] B M Adams, H T Banks, M Davidian, E S Rosenberg 2007 Estimation and Prediction with HIV-Treatment Interruption Data. *Bulletin of Mathematical Biology* **69** 563-584
[12] B M Adams, H T Banks, M Davidian, H D Kwon, H T Tran, S N Wynne, E S Rosenberg 2005 HIV Dynamics: Modeling data analysis and optimal treatment protocols *J. Comput. Appl. Math.* **184** 10-49
[13] Y Yang and X Xiao 2010 Threshold dynamics for an HIV model in periodic environments. *J. Math Analysis and Applications*, **361** 59-68
[14] A Jarne, D Commenges, M Prague, Y Levy, R Thiebaut Modeling CD4+T cells dynamics in HIV infected patients receiving repeated cycles of exogenous Interleukin 7 arXiv:1602.05399.

IOP Conf. Series: Journal of Physics: Conf. Series **792** (2017) 012071 doi:10.1088/1742-6596/792/1/012071

Estimation of the Reynolds number in a Poiseuille flow using artificial neural networks

M. Carrillo[1], J.A. Gónzalez[2] and U. Que[3]

Laboratorio de Inteligencia Artificial y Supercómputo, Instituto de Física y Matemáticas, Universidad Michoacana de San Nicolás de Hidalgo. Edificio C-3, Cd. Universitaria, 58040, Morelia, Michoacán, México.

Email:, [1]mcarrillo@ifm.umich.mx, [2]gonzalez@ifm.umich.mx, [3]qsalinas@ifm.umich.mx

Abstract. In this work the estimation of Reynolds number in a 2-dimensional Poiseuille flow is explored employing artificial neural networks (ANNs). The velocity fields of the fluids were generated evaluating the Hage-Poiseuille equation for different Reynolds (*Re*) from 20 to 2000. The velocity profile obtained for each case is used as input data for the ANNs, which is then trained to predict the *Re*. The results show an accuracy of at least of 99.5% in all prediction cases. This analysis is the first step towards the construction of a Machine Learning algorithm capable of computing physical parameters in more general scenarios.

1. Motivation

The study of flows in incompressible fluids at moderate and low Reynolds (*Re*) number has applications in many relevant physical problems, such as laminar flows through channels and pipes, for which, extensive studies have been done in literature [1,2].

Nowadays, many of these applications are present in the field of engineering , involving viscous incompressible flows through ducts, filters and other devices. With that motivation and given its complexity, several studies have been performed in order to obtain accurate solutions to the Navier-Stokes equations describing the behavior of the fluid.

In complicated problems, where the solutions have to be computed using numerical methods, the estimation of some physical parameters is not an easy task. For instance, if we want to estimate the Reynolds number for a 2-dimensional flow around an obstacle, we need a method that using the vector fields obtained from the simulation (for example, the velocity field) can predict the *Re* associated to that experiment. As a first step to implement it, in this paper we present a simpler configuration: a Poiseuille flow and propose a method to estimate Reynolds numbers in the range from 20 to 2000.

To perform the estimation of the Reynolds number we use a Machine Learning method: artificial neural networks (ANNs), that have been very useful in pattern recognition, in particular they have been used to estimate and predict parameters in problems involving fluids. For example in [3] the authors compute the velocity and diffusion coefficient of a fluid that creates a flow around an obstacle in two dimensions using ANNs. In [4] an implementation that predicts the power ratio and torque of a vertical axis wind turbine training ANNs with experimental data were done. In [5] it is predicted the friction factor of an open channel flow. In [6] the flow patterns of different Reynolds numbers are obtained using ANNs, and in [7] the authors estimate the Reynolds number and the dimensions of the enclosure from the knowledge of the centreline velocity field using an inverse method. The structure of the paper is as follows: In section 2 we briefly, describe the Poiseuille flow problem, in section 3

Content from this work may be used under the terms of the Creative Commons Attribution 3.0 licence. Any further distribution of this work must maintain attribution to the author(s) and the title of the work, journal citation and DOI.
Published under licence by IOP Publishing Ltd

the concepts of ANNs, in section 4 the methodology to estimate the Reynolds number and in section 5 we present the results and in section 6 we conclude.

2. Two dimensional Poiseuille flow

The Poiseuille equation, is a physical law that allows to describe the presssure drop in a Newtonian incompressible fluid with a laminar flow through a cylindrical pipe of constant cross section. It is assumed that the diameter of the cylinder is negligible compared with its length, and no acceleration of the fluid within the pipe is considered.

In classical hydrodynamics, the flow velocity at a given point in the space, can be obtained by solving the momentum equation of Navier-Stokes [1]. For a Poiseuille flow in two dimensions the analytical solution for the time-dependent behaviour is given by

$$v_x(y,t)=\frac{F}{2\nu}y(y-L)+\sum_{n=0}^{\infty}\frac{4FL^2}{\nu\pi^3(2n+1)^3}\sin\left(\frac{\pi y}{L}(2n+1)\right)\exp\left(-\frac{(2n+1)^2\pi^2\nu}{L^2}t\right), \quad (1)$$

where F is the magnitude of the body force in the direction of the flow, L is the diameter of the cylinder, and ν is the kinematic viscosity. The second term in the right hand side of equation (1) computes the transitional behavior of the solution. Thus, the stationary solution for the Poiseuille flow is

$$v_x(y,t)=\frac{F}{2\nu}y(y-L). \quad (2)$$

For a Poiseuille flow, the Reynolds number is

$$Re=\frac{v_{\max}L}{\nu}, \quad (3)$$

where $v_{\max}$, is the maximum velocity characteristic achieved in its steady state, related to the magnitude of the body force F by

$$v_{\max}=\frac{FL^2}{\nu}. \quad (4)$$

In this paper we propose a method to estimate Re as a function of the x component of the velocity of the y-axis, i.e., $v_x(y)$, using the ANNs machinery. In the next section, details about ANNs' concepts and operations are described.

3. Artificial neural networks concepts

ANNs are a Machine Learning paradigm based on biological systems, commonly used in regression, pattern recognition and classification problems. ANNs are fed with some variables or data on a particular problem, and after a series of calculations made by the composition of non linear functions, give an outcome related with the information they were trained. The most common ANN is the Multilayer Perceptron (MLP), derived from the simple Perceptron [6], which consists of interconnected layers of nodes or neurons: an input layer, one or several middle or hidden layers and an output layer. In a MLP information goes from input to hidden layers and later from hidden to the output one. For neurons in hidden and output layers the connections are regulated by the weight coefficients, which are adjusted in the training phase.

In order to illustrate a MLP, or ANN in this case, in organization and operation, consider a 3 layer network with n inputs, m hidden neurons and l outputs. The relation between and input vector $\vec{x}=\{x_1,x_2,\ldots,x_n\}$ and the k-the output neuron is determined by the expression

$$y_k=F\left(\tilde{\omega}_{0k}+\sum_{j=1}^{m}\tilde{\omega}_{jk}*G\left(\sum_{i=1}^{n}\omega_{ij}*x_i+\omega_{0j}\right)\right), \quad (5)$$

where ω_{ij} and $\tilde{\omega}_{jk}$ are the connection weights between the neurons in input-hidden and hidden-output layers respectively; ω_{0j} and $\tilde{\omega}_{0k}$ are some extra weights called bias, operating as thresholds; F and G are the named neurons' activations functions on each layer, commonly the logistic or hyperbolic tangent functions are employed.

In order to obtain the proper outcomes the weights must be adjusted properly. One way to do this is through a supervised training process, minimizing a cost function from a collection of examples with already known results or targets. Thus, the training phase consists of running a set of coupled inputs and targets $\left(\tilde{x}_i^p, \tilde{y}_k^p\right)$, with $i = 1,\ldots,n;\ k = 1,\ldots,l;\ p = 1,\ldots,N$, where N is the number of example pairs, adjusting the weights such that minimize the cost function

$$E(\vec{\omega}) = \frac{1}{N}\sum_{p=1}^{N}\sum_{k=1}^{l}\frac{1}{2}\left(\tilde{y}_k^p - y_k^p\right), \tag{6}$$

where $\vec{\omega}$ are the ANN's weights defined as before in a vector representation and y_k^p are the ANN's outcomes for the $\tilde{x}_i^p$ inputs. In this work, the minimization of this cost function is made by a gradient descent algorithm called *backpropagation* [7]. This method searches the directions where the error diminishes more on each time step t, and updates the value of each ω_{ij} at the next time step $t+1$ by the rule

$$\omega_{ij}(t+1) = \omega_{ij}(t) - \gamma\frac{\partial E(t,\vec{\omega})}{\partial\omega_{ij}(t)} + \alpha\left(\omega_{ij}(t) - \omega_{ij}(t-1)\right), \tag{7}$$

with $0 < \gamma < 1$ the parameter known as the learning rate, determined by the user; $\alpha\left(\omega_{ij}(t) - \omega_{ij}(t-1)\right)$ is called the momentum term, added for preventing that the minimization algorithm gets trapped in a local minimum, and $0 < \alpha < 1$ is another parameter to be adjusted by the user. The number of iterations for training depends on the learning parameter, the random weights initialization and the training set. The resulting weights after this process are used to compute the values in the prediction set.

4. Methodology

As mentioned in section 1, the main idea of this work is to use the velocity profile as input for the ANN, so it learns its associated Reynolds number giving some particular characteristics of the fluid and the cylinder. As a first step, we generate the velocity profiles using the analytic solution given by equation (2). It was assumed a fluid with a kinematic viscosity $\upsilon = 0.001\mathrm{m}^2/\mathrm{s}$ flowing through a cylinder of 0.5m, considering the magnitude of the body force F as the independent variable to construct 100 representative cases, in a range from 20 to 2000 *Re* with steps of $\Delta = 20$. On each case, 100 values of the velocity of the fluid $v_x(y)$ inside the cylinder have been extracted over a line transverse to the flow direction to be the inputs of the ANN. In figure 1, we present some of these profiles.

As the rule of thumb suggests, 80% of data should be used as training set, 10% as validation set and the remaining as the prediction set. The validation set is completely unknown to the ANN in the weight adjustment process, and is considered to check the performance of the ANN. In general, training is continued until the error in validation set gets into a local or global minimum, with the intention to avoid over-fitting on the training pairs. Sets were selected as follows:

- Prediction set: Values of the x component of the velocity, i.e., $v_x(y)$, for the simulation with *Re* = 300, 600, 900, 1200, 1500, 1800, 1940, 1980, 2000, some of them are shown in figure 1.
- Validation set: $v_x(t)$ values and their corresponding *Re* for 20, 40, 120, 280, 480, 1280, 1520, 1760, 1880 and 1920.

- Training set: The $v_x(y)$ values for the remaining cases for *Re* starting from 60 to 1900. Leaving the cases with *Re* = 1940, 1960, 1980, 2000 as a test to explore the extrapolation capabilities of the algorithm.

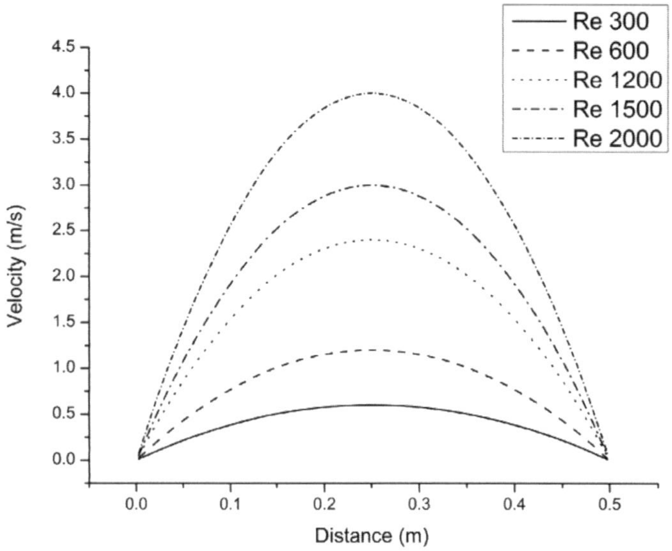

Figure 1. Velocity profiles inside the cylinder with a 0.5m diameter, for different *Re*. These curves were obtained modifying *F* in equation (2).

ANN's weights are randomly initialized. To show consistency on the results independently of their initial condition, ten ANNs have been constructed with the same structure, to calculate an average of the *Re* predicted. Every network has the following structure:

- Input layer with 100 inputs consisting of the x component of velocity, i.e., $v_x(y)$.
- Hidden layer with ten neurons, using the hyperbolic tangent as activation function.
- Output layer with a single neuron equipped with a linear function, giving the ANN's *Re* prediction.

Scaling data in ML is required in almost all cases. In this particular one, data seems to be in an appropriate range, however the number of inputs might overload the neurons' activation functions, in this case the hyperbolic tangent function is used in the hidden layer. We scaled the input values by a factor of 10 and the targets by a factor of 100, such that the ANNs produce the best results compared with other approaches such as max-min, mean-standard normalization in inputs; or logarithmic rescaling at the targets. The ANN's outcomes were rescaled back to obtain the physical results.

5. Results

Using a learning rate $\gamma = 0.05$ and $\alpha = 0.5$ together with backpropagation algorithm, the ANNs required in general from one thousand to five thousand iterations and a couple of minutes to reach a minimum on the validation set, time at which a signal to stop the training set was triggered. On figure 2, we observe one of the ANN's root mean square error (RMSE) on the training phase. The errors in training and validation decrease initially together, but in this case, after 1173 iterations the validation error starts growing slowly. This is the moment when the training algorithm was stopped saving those weights for the prediction phase.

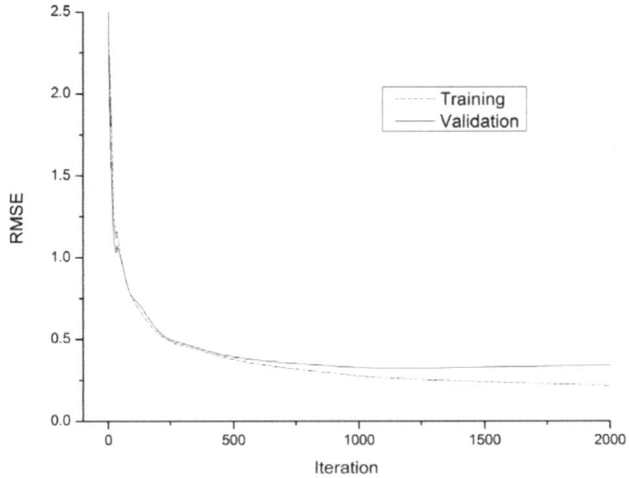

Figure 2. RMSE of training and validation phase. The ANN's training is stopped when the validation error gets a local or global minimum. In this case it was stop at iteration 1173, after which the error increases smoothly.

In table 1, we present the averaged values computed by the 10 ANNs and the relative error for their predictions. The network reached an accuracy of 100%, considering an error less than 0.5% over all sets, this means that the network was capable to classify all unknown elements with an error less than 0.5%. In the extrapolation regime (*Re* = 1940, 1960, 1980, 2000), the relative errors are bigger than in the interpolation regime since the ANNs were trained with *Re* = 1900 as the top training case, figure 3. The best results are on the middle region of data, for *Re* from 600 to 1800, those samples lie within the interpolation regimen and nearly other samples in the validation set. This shows the sensitivity of the ANN to the selection of the training and validation sets.

6. Conclusions

In this work, we presented a method to predict the Reynolds number in a two dimensional Poiseuille flow with an accuracy of 99.5 % for unknown flow patterns, this performance is reached based on the simplicity of the problem. The paper shows the first step towards the construction of a numerical method capable to predict physical constants analyzing only a snapshot of a vector field (in this case the x component of the velocity). Considering the good performance shown in here we are planning to consider more complicated scenarios. In particular, we would like to implement an ANN that predicts the Reynolds number of an incompressible fluid flowing around a cylindrical obstacle, expecting that the complexity of the problem could decrease the ANN performance. These results were obtained evaluating a formula. However, in more complicated scenarios, the inputs for the ANN cannot be generated using analytic solutions, and it will be necessary to use data directly from experiments or obtained using numerical simulations, nevertheless once trained the results can be obtain faster than other approaches.

Acknowledgments
This research is supported by grant CIC-UMSNH 4.23

Table 1. Predictions and errors made by the ANNs in the prediction set. All error are less than 0.5%. In the cases out of the training range (*Re* = 1940, 1960, 1980, 2000) the error increases as expected.

Reynolds	Prediction	Relative Error (%)
300	300.414±0.355	0.138
600	600.107±0.192	0.018
900	900.232±0.295	0.026
1200	1200.028±0.124	0.002

1500	1499.702±0.348	0.020
1800	1800.922±0.101	0.051
1940	1937.841±0.596	0.111
1960	1956.434±0.842	0.182
1980	1974.664±1.140	0.270
2000	1992.493±1.490	0.375

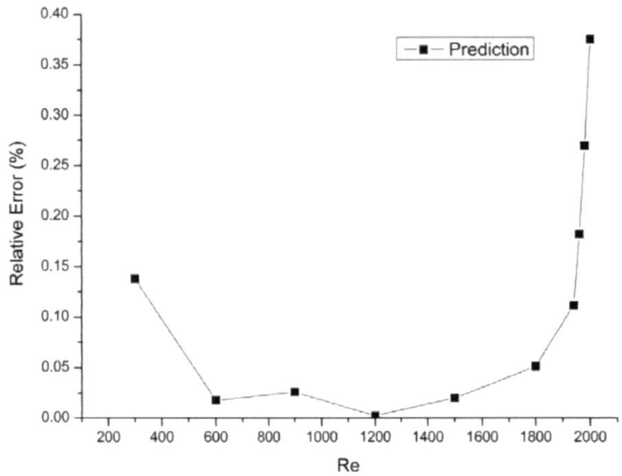

Figure 3. Relative errors between the real *Re* and the ANNs' averaged outcomes in the prediction set. All errors are less than 0.5%, however those outside the interpolation regime increase.

References

[1] Batchelor G K 1997 *An Introduction to Fluid Dynamics* (Cambridge: London Cambride University Press)
[2] Landau L D and Lifshitz E M 1987 *Fluids Mechanics* (Oxford: Pergamon Press)
[3] Lauret P, Heymes F, Aprin L, Johannet A and Slagen P 2015 *Chemical Engineering Transactions* **43** 1621-26
[4] Sargozaei J and Kianifar A 2009 *Simulation Modelling Practice and Theory* **17** 1290-98
[5] Yuhong Z and Wenxin H 2009 *Commun Nonlinear Sci Numer Simulat* **14** 2373-78
[6] Zhang L, Akiyama M, Huang K, Sugiyama H and Ninomiya N 1996 *Estimation of Flow Patterns by Applying Artificial Neural Networks* **2** 1358-63
[7] Das R 2012 A *simulated annealing-based inverse computational fluid dynamics model for unknown parameter estimation in fluid flow problem* **26** 499-13
[8] Rosenblatt F 1961 *Principles of Neurodynamics: Perceptrons and the Theory of Brain Mechanisms* (Washington DC: Spartan Books)
[9] Rumelhart D E, Hinton G E and Williams R J *Nature* **323** 533-36

Design of a rib impactor equipment

C R Torres[1], G García[1], L A Aguilar[1] and L Martínez[2].

[1]Instituto Politécnico Nacional, Escuela Superior de Ingeniería Mecánica y Eléctrica, Av. Luis Enrique Erro S/N, Gustavo A. Madero, Zacatenco, 07738 Ciudad de México.

[2]Universidad Politécnica de Madrid, Instituto de Investigación del Automóvil, Campus Sur UPM, Carretera de Valencia (A-3), km7, 28031 Madrid, España.

E-mail: ctorress@ipn.mx, gesellegarcia@hotmail.com, laguilarpe@gmail.com, luis.martinez@upm.es

Abstract. The human ribs must be analyzed as long and as curved bones, due to their physiology. For the development of an experimental equipment that simulate the application of loads, over the rib in the moment of a frontal collision in an automobile with seat belt, it was applied a methodology that constituted in the identification of needs and the variables which led the design of 3D model, from this it was used the technique of fused deposition modeling for the development of the equipment pieces. The supports that hold the rib ends were design with two and three degrees of freedom that allows the simulation of rib movement with the spine and the breastbone in the breathing. For the simulation of the seat belt, it was determined to applied two loads over the front part of the rib from the sagittal and lateral plane respectively, for this it was made a displacement through a lineal actuator with a speed of 4mm/s. The outcomes shown a design of an equipment able to obtain the load parameters required to generate fractures in rib specimens. The equipment may be used for the study of specimens with nearby geometries to the rib taken as a reference.

1. Introduction

The bone is a biologic element that suffers changes during its growth also depending its location it has a complex geometry. There are several investigations with principal interest on determination and analysis of the mechanics proprieties of the ribcage elements with the objective of establish the answer that suffers chest in a traumatic event.

Those investigations usually search to measure the applied load and the damage caused in the body, either because the impact force which was generated, speed, acceleration or applied compression; the rib fractures are considered as an important indicator, because these are caused by serious injuries and are related specific with injuries in organ and a high mortality rate as a result.

In 2009 was realized an analysis of Time-Frequency by [1] in order to get the mechanics proprieties of thorax, for that purpose; specimens with the absence of pre-existing fractures, lesions or other bone pathology were used. The tests were realized in a seat designed to imitate an automobile standard seat. The impacts were applied using a pneumatic impactor; the impact velocity was 3 m/s. The specimen was successively impacted at three levels: shoulder, upper chest and mid chest. After each test, were found fractures for the number 3 ribs and below, each fracture was described as displaced, non-displaced, mono-cortical or bi-cortical.

For the evaluation of stiffness and kinematic of ribcages in 2010 was designed by [2] a process in which was applied a load to the thorax by a plate connected to an actuator, for avoid a fracture by shearing were used spherical segments as an interface between a loading plate and the ribcage. Were

Content from this work may be used under the terms of the Creative Commons Attribution 3.0 licence. Any further distribution of this work must maintain attribution to the author(s) and the title of the work, journal citation and DOI.
Published under licence by IOP Publishing Ltd

IOP Conf. Series: Journal of Physics: Conf. Series **792** (2017) 012072 doi:10.1088/1742-6596/792/1/012072

realized two types of test: unilateral in which was used only one load plate to move a single point and bilateral, in this test two load plates applied a displacement in two points simultaneously.

In 2011 was realized a test by [3] for the analysis and reduction the thoracic injuries in vehicle collisions, for this test it was used a testing machine with a 5 cm wide of nylon belt, which is the width of a standard automobile seatbelt, situated at 40 degrees from a sagittal plane of the body, the load was applied to the thorax by a hydraulic actuator and it distributed by a rigid plate.

The deflection of thorax was measured using a string potentiometer joined to the sternum belt, the force of the loads cells was added for obtaining the reaction force for effective deflection of the thorax under these conditions it was determined for realization of a linear regression of the force versus percent deflection from 0% to 20% of compression.

A study on 2013 realized by [4] reported that the stiffness of the ribs can be evaluated by a horizontal impact test (Fig. 1), for the test, the extremes of the rib were hold to a pair of rotation elements that allow free rotation of the specimen, the test was realized with a horizontal impact by a pendulum with 54.4 kg of mass to a velocity of 1 m/s and 2 m/s. The load was measured with a load cell aligned to the rib center. The test was realized with different sizes of ribs and the conditions of tests are destined for the simulation of a front dynamic impact.

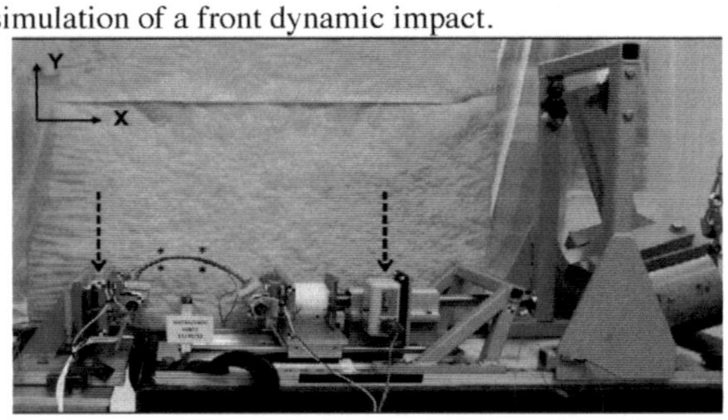

Figure 1. Experimental test of frontal impact by [5].

2. Materials and Methods

2.1. Methodology

For development of equipment design, was used a methodology of design constituted in the identification of needs established by the variables that affects directly the tests to apply on the ribs, which led to the design of a mechanic model on 3D, the construction of the equipment and the performance validation. The parameters of design were:

- Rib´s breaking forces.
- Rib anatomy.
- Load ´s application points on the ribs.
- Freedom degrees in the equipment.
- Load Displacement.
- Load Application velocity

2.2. Design considerations

During a frontal impact automobile, displacements due a seatbelt on the occupant's body are generated, specifically in the thorax and abdomen. Because of the displacements were defined forces that act on the rib in the impact time, so for the test experimental equipment were determinated apply two loads on the front part of the rib, one with sagittal plane relation and posterior with relation to lateral plane of the body respectively for simulate the mechanic of the rib (Fig. 2).

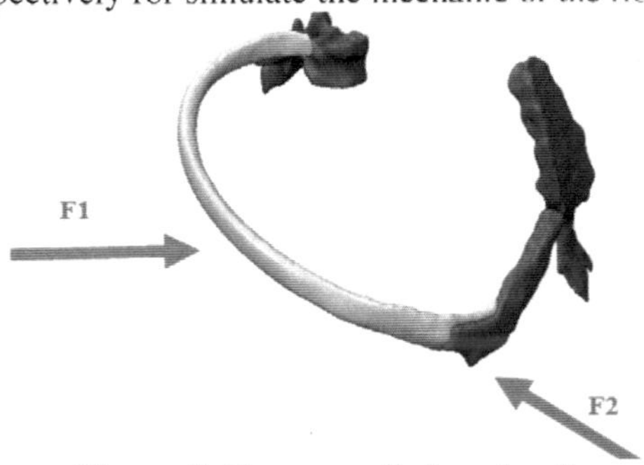

Figure 2. Forces applied on the rib.

The equipment dimentions were defined in relation of the ribs real dimentions; for the specimen; rib number 7 of a male adult person with more than 70 years was taken as a reference.

The natural move of the rib in the moment of breathing was observed to define the freedom degrees of the supports and mechanisms of equipment, for the end of the rib liable to the spine and the torax move up and down in the moment of inspiration and expiration was considered.

For the determination of the force magnitude to apply on the rib structure, was considered a rib with 200 mm of length and the tansversal area with 14 mm, a breaking compression force of the rib with compact structure of 170 N/mm2 and with trabecular structure of 2.2 N/mm2.

It was determined to realized the test in static conditions, the conducted studies by [5-9] define conditions of applied velocity force between 5 and 16 mm/s, on that basis it was established the applied force over the specimens of 4 mm/s.

2.3. Design for brackets of clamping ribs

For the supports, it was designed a worktable in which a pair of riels were assembly; its function is hold the base supports that hold the extremes of the rib.

The restrictions and freedom degrees of the equipment rib supports were defined in specific anatomy of the rib and the relation of the rib move with the spine and sternum in the breathing moment.

The support A was fixed to the worktable for simulate the rib subjection with the spine, was defined with two freedom degrees to cause the rotation generated for the supports on the subjection elements; the support B was not fixed in the structure, for its attributed of three freedom degrees (Fig. 3).

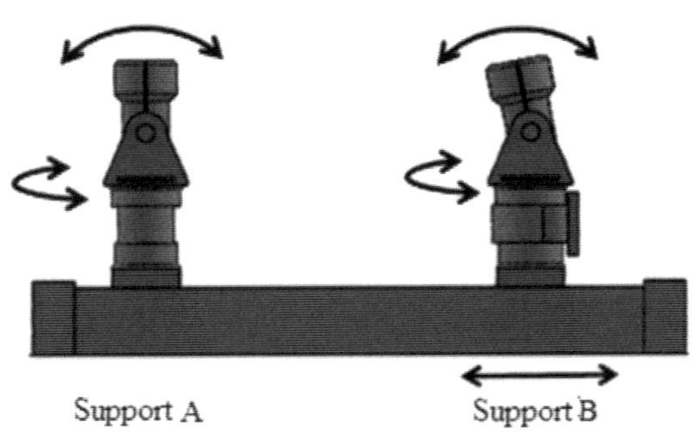

Figure 3. Freedom degrees of the supports

2.4. Movement of mechanisms
With respect to breaking compression force contemplated, was obtained the necessary motor torque for applied the load on the specimen, for that was use the equation (1).

$$\tau = F * d \tag{1}$$

$$\tau = 5950 \frac{N}{mm}$$

The displacement for applying the force on the specimen was defined for a linear actuators located to the top and side of the equipment.

The linear actuator has declared a displacement of 150 mm (L_{max}) because the height of the rib placed in the subjection supports is approximately 140 mm, the displacement of actuator the motor moves to pinion and the wheel to threaded rod were associated, the relation of gears used for transmission of move was 4:1. The step (P) of the threaded rod it's from 2 mm, due to relation of gears (i) the revolution displacement (Lp) was described:

$$L_R = \frac{P}{i} \tag{2}$$

$$L_R = 0.5mm$$

The number of revolutions (R) necessary for specify the displacement for the actuator was defined by:

$$R = \frac{L_{max}}{L_R} \tag{3}$$

$$R = 300\, rev$$

For define the velocity displacement of the actuator, was defined the total time (t) associated to velocity (4 mm/s).

$$t = \frac{L_{max}}{v} \tag{4}$$

$$t = 37.5\, s \approx 38s$$

2.3 Mechanic 3D Model
For realizing the mechanic 3D model of the experimental equipment of test in ribs (Fig. 4) it was used SolidWorks® for design. The retaining brackets were inserted to the base, its structure and freedom degrees allows movement rotation of supports for adjust of rib geometry. Within brackets are the subjection elements with the function of retaining the extremes of rib in the moment of test.

On the top of the equipment was assembly a linear actuator which is coupled a load support, this actuator moves in descending order to apply a point load in two points, on the side of the equipment was assembly a second actuator, the displacement realized by the actuator to the mobile support and allows applying a front loads on the specimen with relation to sagittal plane of the body.

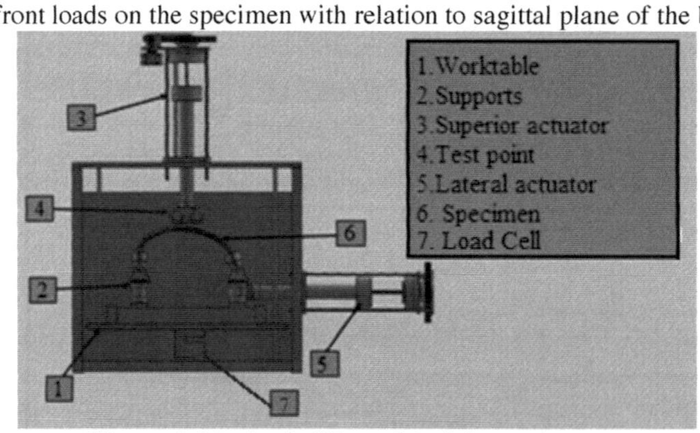

Figure 4. Mechanic 3D model of the equipment.

2.4 Electronic instrumentation

For the function of the equipment were designed electronic circuits (Fig. 5) that altogether perform a specific function with the movement of equipment mechanism for apply the necessary loads on the ribs in the moment of the test and perform the measure of parameter test, through a load cell placed in lower part of equipment.

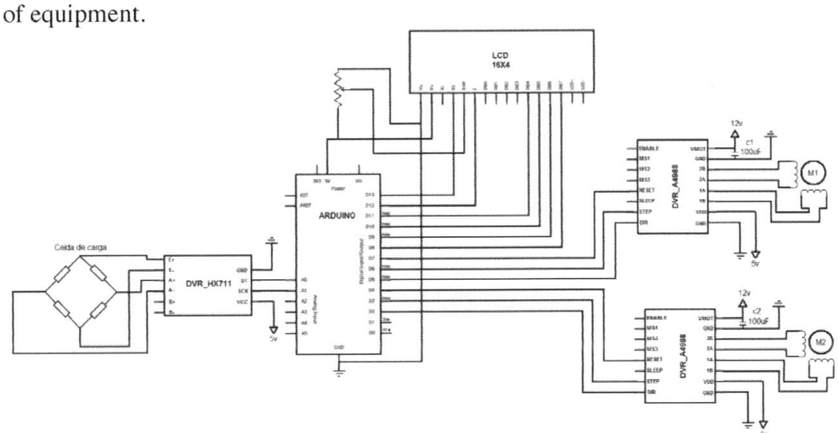

Figure 5. Electronic diagram of the equipment. circuits.

3. Results

With the manufacture of the equipment, it was observed the physical deformation generated on the rib in a situation of a frontal collision with seat belt starting with static force conditions and applying a displacement of 4 mm/s on the specimen.

In the moment of applying the loads on the specimen, it was observed that the displacement of 4 mm passed on by the linear actuator of the superior structure of the equipment generates an increment of flexion than the lateral linear actuator generates on the frontal of the rib (Fig. 6), what caused a displacement of 120 mm before fracture of the rib, concluding that with the point loads that were generated on the front of the rib the experimental test equipment represents a chest compression due to frontal impact.

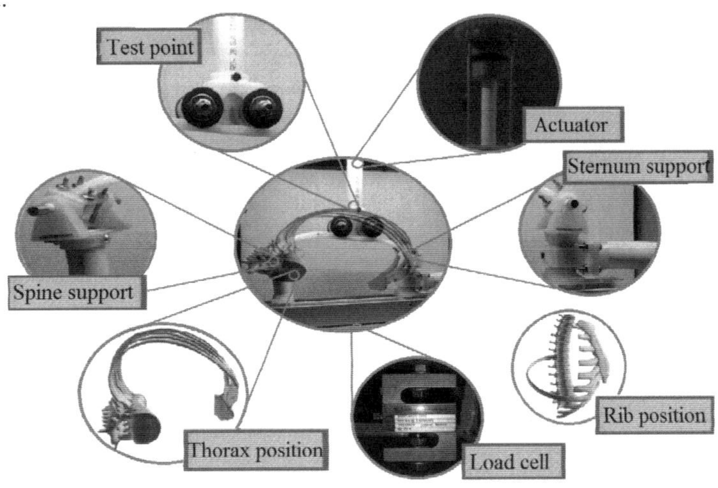

Figure 6 Flexion applied on the rib.

The specimen was obtained by tomography and "fused deposition modeling" technique was used. The specimen used for the test have different properties that a real bone, the modulus of elasticity of the material used is about 10 times lower than the bone, however the design considerations allow apply test in structures that do not exceed the elastic modulus of the bone (170 GPa), it is possible to use similar specimens to those of a human rib having physical properties equal or less magnitude than those established for compact bone physical and geometric properties. Due the structure design and the biofidelity of the freedom degrees of the supports given to the equipment, the experimental test can be used for evaluate cases which implies changes in the bone proprieties, either due to age conditions or bone pathologies.

4. Discussion

The fracture of the specimen submitted to the test was found in the anterolateral region of the rib near to the extreme from fixed support of equipment, location of fracture is coincident with the experimental works reported by [5]; in comparison to mentioned work; that was realized with dynamics experimental conditions present a double fracture in the anterolateral sites of the rib near to the extreme subjects to the supports. The above mentioned presents a validation of the design of experimental equipment able to simulate a frontal impact representing the chest compression.

The development of the impact simulation is a fundamental progress to identify the mechanism of injury in the body due to impacts, however for results are reliable the design of equipment for its study needs to include physical proprieties that are able to represent real behaviour of human body for the biofielity of equipment, which applied the integration of freedom degrees of the supports that simulate the mechanics of the rib.

5. Future Works

Based on the work done, the followings future works are suggested:

- Perform a test machine to study of the behaviour of the ribs under twisting forces.
- Implement a database in the testing machine that allows a comparison of the behaviour of displacement of ribs subjected to various forces.
- Manufacture an experimental model to evaluate the conditions of frontal and lateral load dynamic load conditions.

Acknowledgements

The authors gratefully acknowledge the financial support from the Mexican government by de Consejo Nacional de Ciencia y Tecnología, The National Polytechnic Institute and the participation of the biomechanics group of INSIA that it´s incorporated at the Polytechnic University of Madrid.

References

[1] Gabrielli and F 2009 et al., *Time-frequency analysis to detect bone fracture in impact biomechanics. Application to the thorax*. Medical Engineering & Physics,.

[2] Kindig and M W. et al., *Structural Response of Cadaveric Ribcages under a Localized Loading: Stiffness and Kinematic Trends* Stapp Car Crash Journal, 2010.

[3] Kemper and A R et al., Reducing Chest Injuries in Automobile Collisions: Rib Fracture Timing and Implications for Thoracic Injury Criteria. Annals of Biomedical Engineering, 2011.

[4] Agnew and A M 2013et al., *Age-Related Changes in Stiffness in Human Ribs*, in *IRCOBI Conference*..

[5] Bir and C D. Viano, and A. King 2004 *Development of biomechanical response corridors of the thorax to blunt ballistic impacts*. Jounal of Biomechanics,

[6] Currey and J D *The structure and mechanics of bone*. Springer Science Business Media, 2012.

[7] Kieser, J.A., et al., *Compressive rib fracture: Peri-mortem and post-mortem trauma patterns in a pig model*. Legal Medicine, 2012.

[8] Li and Z 2009et al., *Rib fractures under anterior-posterior dynamic loads: Experimental an finite element study*. Journal of Biomecanics,.

[9] Robin and S *Humos: Human model for safety- a joint effort towards the development of refined human-like car occupant models*.

Proposal by simple design of the lower limb exoskeleton of continuous use, provided of own mobility and body load support. Case: application due to an illness

Rafael Rodriguez-Martinez[1], Julio Alberto Lopez-Amaya[2], Guillermo Urriolagoitia-Sosa[1], Beatriz Romero-Ángeles[1], Guillermo Manuel Urriolagoitia-Calderón[1]

[1]Instituto Politécnico Nacional, Escuela Superior de Ingeniería Mecánica y Eléctrica Unidad Profesional "Adolfo López Mateos", Av. Instituto Politécnico Nacional S/N, Sección de Estudios de Posgrado e Investigación, Departamento de Ingeniería Mecánica en Diseño, Edif. 5, 2° piso, Tel. +52 1 5557296000 ext. 54740

[2]Instituto Politécnico Nacional, Escuela Superior de Ingeniería Mecánica y Eléctrica Unidad Profesional "Adolfo López Mateos", Av. Instituto Politécnico Nacional S/N. Estudiante de 6th semestre de la licenciatura en Igeniería en Control y Automatización.

E-mail: rafarm68@hotmail.com, ljulio.950@hotmail.com, guiurri@hotmail.com, urrio332@hotmail.com

Corresponding Authors: rafarm68@hotmail.com, ljulio.950@hotmail.com

Abstract. In recent times it has established a debate between experts and academics about the social and economic impact of advances in robotics. The robotic exoskeletons mounted as suits on affected parts of the human body, represent one of the most significant examples of which is oriented towards robotics. With recent technological advances have increased the fields of application of these devices widely with respect to the first applications were teleoperation and increase in strength of a human being for various tasks. The aim of this work is to contribute as much as possible, to start a discussion about the vision of offering future developments in socio-economic terms and its impact resulting from the use of robotic exoskeletons, especially with regard to its application in medical rehabilitation of lower member and especially its use permanent, replacing cumbersome devices such as crutches, walkers, canes. All this, focused on the health sector, which is most affected by different diseases cannot have access to these devices. In this paper, only it proposes a design that could be inexpensive and used for various ailments.

Keywords: Exoskeleton, robotics, lower limb, joint, automation.

Content from this work may be used under the terms of the Creative Commons Attribution 3.0 licence. Any further distribution of this work must maintain attribution to the author(s) and the title of the work, journal citation and DOI.
Published under licence by IOP Publishing Ltd

1. Introduction

The Initial studies about the movement of lower limb orthoses assisted by date from the 1960's in the United States [1] and the newly formed Yugoslavia [2] respectively, for military purposes and medical [3]. Since then, orthoses and exoskeletons have been developing prosperously in different types of structures, actuators and interfaces. Today, established from the point of view as tools to ease the hard work of physical therapists, while improving (neurological or orthopaedic) retrieves the effectiveness of patient movement−Loko Mat [4] LOPES [5] they are also the goal of helping paraplegics or quadriplegics people to recover the ability to walk in daily life–ATLAS [6] Re Walk ™ [7], Ekso (Ekso Bionics, US, formerly e Legs [8]).; which they are adopted as systems capacity increase strength in healthy people in order to carry heavy loads–BLEEX [9] Sarcos Exosqueleton [10], MIT Exosqueleton [11]. Furthermore, being placed in the use of exoskeletons of lower limb, as part of aid to the motor independence patients of Mexican health sector is necessary to consider that there are many causes that can affect the functioning of the human locomotor system, leading to the onset of joint disorders and lower limb generation atypical patterns of motion. The importance of research and development in assistive technologies to compensate for the pathological gait have been recognized since the early twentieth anniversary, and many challenges still lie ahead for clinical application a reality.

In this work, the lower limb is presented as an exoskeleton, conceived as a system of compensation and evaluation of the pathological gait, for applications under real conditions as a methodology of assistance and joint assessment of the problems affecting mobility in individuals with neuromotor disorders.

The main technological challenges are discussed with regard to detection, performance and control subsystems. A special emphasis is placed on advances in robotics for lower limb orthoses, and biomechanical requirements, considerations of structural design are analysed and existing approaches to develop robust real-time controllers, to mobile solutions with a common goal, control human motor.

2. Normal and pathologic human walking

Concerning disabilities of walking due to neurological, orthopedic or traumatic conditions, different robotic approaches, and classification of robotic exoskeletons for lower limb, is presented in rehabilitation robots, systems evaluation and monitoring and portable systems functional recovery.

The cyclic process of events during the march known as the gait cycle, and starts and ends at the moment when one foot makes contact with the ground and support phase begins. While leg moving, body moves the other leg acts as a support. So, the state of the lower extremity is divided into two phases depending on their situation with regard to the plant: impulsion phase, and phase support. During walking at normal speeds there is a short period of simultaneous support of both legs. As velocity increases a cycle in which no bipodal support is reached. The march can be characterized by a set of parameters: stride length, step length, rhythm and speed. The complexity of human locomotion involves the study of the cyclical movement which runs, considering the kinematics and kinetics (forces and moments) and work, energy and power consumed in the process. Other hand, to understand entire phenomenon, it is necessary to have a knowledge of neurological principles that control movement, sensory input peripherals or systems involved and the mechanisms that command the musculoskeletal system.

3. Exoskeleton structure

Depending on the type of interface or set the system to the extremity, unilateral or bilateral structure can be used. For example, if the interfaces cannot generate sufficient torque in the transverse plane at the points of connection with lateral bars, a unilateral system is not feasible. Conversely, if the interfaces can generate moments in the transverse plane, it is possible to use a unilateral exoskeleton, whose supports are rigid to stabilization in the sagittal plane. The alignment in a bilateral solution is a critical problem because if the two joints are not parallel, friction and undesired stress will be produced at each step. For the design of the mechanical structure of a unilateral frame it was used, with four interfaces load on soft surfaces, such as the right to test the application of compensation strategies. The configuration adopted exoskeleton consisted of two joints of the knee, which was changed last for more efficient time consisting of a polycentric four-bar mechanism [12] that generates a more accurate instantaneous center trajectory, making this joint, which is the most sensitive for their degrees of freedom is more suitable for the exoskeleton and other ankle included only with one degree of freedom, sidebars attached to the thigh (proximal), the leg (distal) and foot (support and inserted in the shoe insole), and four fastening strips to the tip. To make the design took into account important factors such as compensation for ankle instability, compensation knee instability, the condition of impact with the ground, so that the entire mechanism was able to compensate for these instabilities actuator means mounted on the knee joints and ankle. Figure 1 shown the exoskeleton (without mounted compensating actuators) where you can observe the joint originally proposed, which was replaced by a polycentric four-bar mechanism, fairly stable phases swing in walking and support phase.

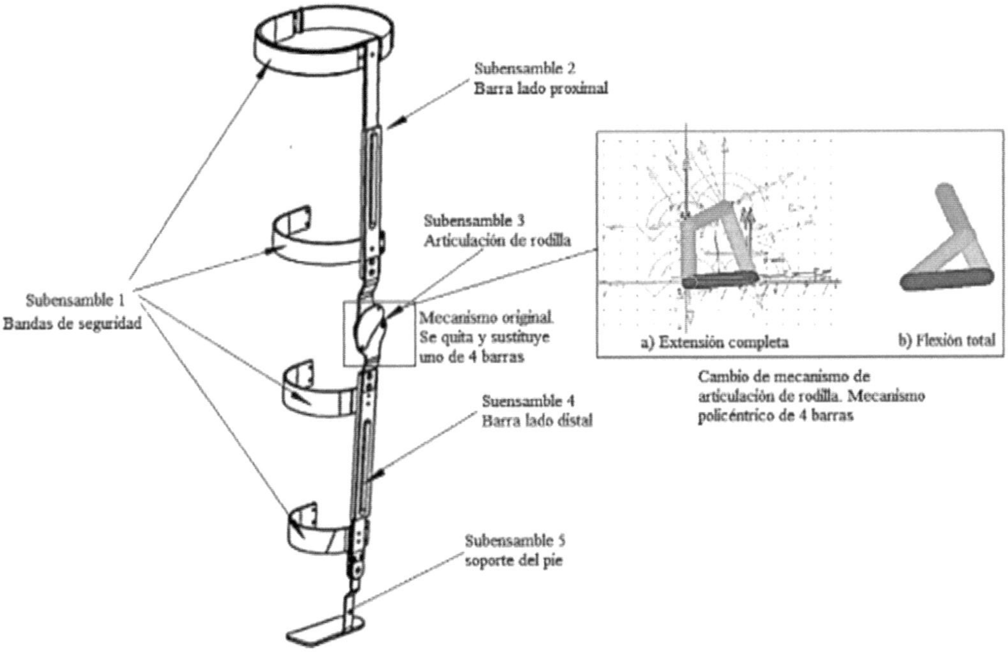

Figure 1. Complete assembly of the exoskeleton, which shows the knee joint that was changed [12].

The main movement of the knee joint is flexion and extension occurs in the sagittal plane; this is not a simple rotational movement, because if so, by rotating the femur over the tibia to the knee would end dislocated [13]. During flexion and extension of the knee, the femur, in addition to the rotary movement, makes a sliding movement on the tibial plateau, which makes the rotation axis is not fixed and therefore there is an instantaneous center of rotation (ICR). The knee joint performs its natural movement through the use of a complex structure of muscles, tendons and ligaments.

3.1. Synthesis of knee mechanism

The synthesis of polycentric knee mechanism was performed using the genetic algorithm developed by Lugo [12] which is easily adapted to the femoral and tibial structure. The general genetic algorithm for synthesis can request the author. Basically is a fairly extensive program developed by Lugo for the synthesis of four-bar linkage including polycentric (instantaneous centers of rotation), such as monocentric like the knee and ankle.

3.2. Balancing the knee joint

Given the linear rotations was found that the performance of gaskets set for some phases of the gait cycle, the operating principle of the actuator knee (Figure 2) is given by: an elastic element K2 is activated to heel contact, with lock in B1 in the phase of initial position, the application of stiffness K2

IOP Conf. Series: Journal of Physics: Conf. Series **792** (2017) 012073 doi:10.1088/1742-6596/792/1/012073

in the knee corresponding to a bending angle maximum longitudinal path restriction. An elastic element K1 (compression spring) is activated in the terminal phase support, by unlocking B1 to release the joint and allow free oscillation, given that K2 >> K1. The element has a maximum longitudinal displacement, corresponding to the restriction of knee flexion. Total lock on B1 to keep the entire leg in extension and provide security when standing or climbing stairs or slopes. The K2 elastic element, configured using disks stacked Belleville, has a maximum trajectory of 13 mm corresponding to 15 degrees of flexion, which enables safe load response and consequent extension before turning. The elastic element has a maximum displacement K1 of 55 mm, which corresponds to a knee flexion 95 degrees. The actuator for balancing the knee joint was selected in the catalog of Festo [14] which conform to the restrictions of movement of flexion and extension of the knee, in the case of flexion to 95° by providing a similar rigidity normal knee extension and leaving almost free but limited very close to zero degrees to prevent extension by the actuator hyperextension, as described in Figure 2.

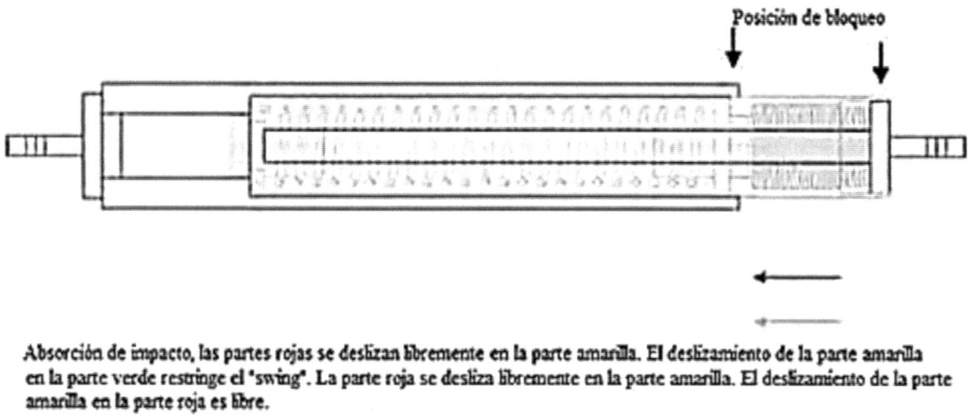

Figure 2. Linear actuator for balancing knee joint

3.3. Balancing the ankle joint

The mechanism ankle is much easier to synthesize, since it is a mechanism articular pivoted only by a bolt, where only the lengths mentioned were selected to determine the damping in flexion and extension as described in this section. The passive system acting on the ankle contains springs which are dimensioned according to the linear relationships given by K3 and K4 which model the operation of the board, the operating principle of the actuator ankle (Figure 3) is given by: A K3 elastic element which is compressed, stored energy and controls the dorsiflexion of the support phase. In compression, K3 has a maximum longitudinal trajectory equivalent to a restriction of ankle dorsiflexion. Energy stored in dorsal flexion recovers towards a path plantarflexion before pivotable extension K3. An elastic member that adds rigidity K4 extension to control plantar flexion posture and prevent foot drop during the swing. The element has a maximum longitudinal trajectory equivalent to a restriction of ankle plantar flexion.

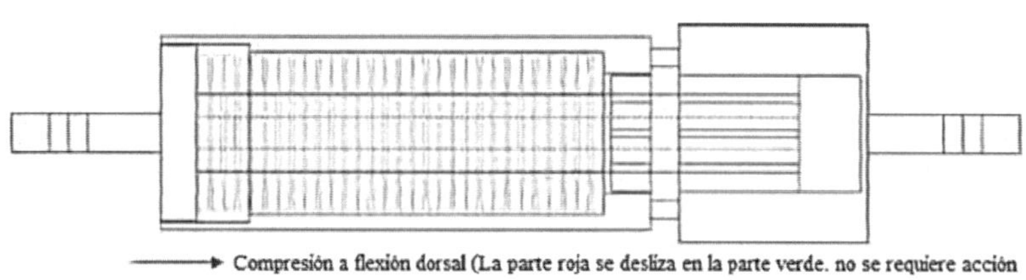

Figure 3. Linear actuator for balancing ankle joint.

Finally, it should be mentioned, in particular, to design the actuator knee integrated into the ankle joint, the elastic element K3 (disks Belleville stacked) is compressed with a history of 15 mm equivalent to 20 degrees of ankle dorsal flexion. The elastic element compression K4 has a maximum displacement of 6.5 mm plantar flexion, equivalent to 15 degrees of flexion. The final design of integrated drive system bi-articular in full exoskeleton structure is shown in Figures 4 and 5.

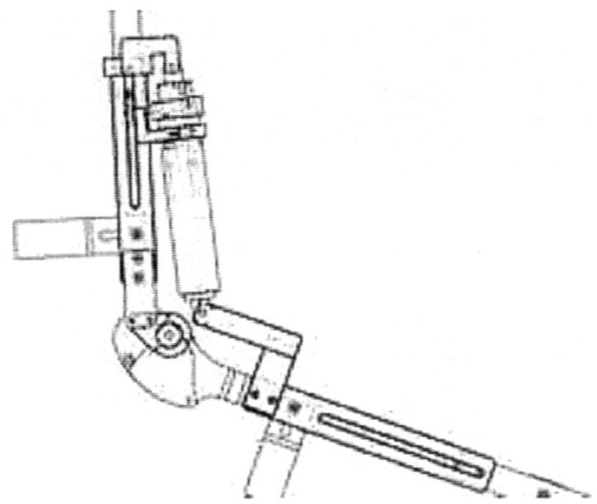

Figure 4. Complete knee joint.

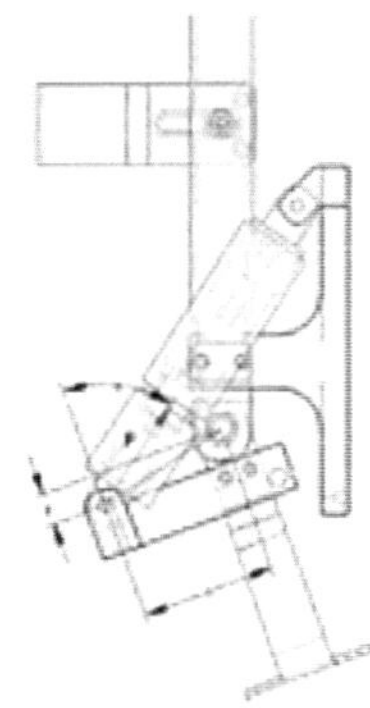

Figure 5. Complete ankle joint

4. Results

The results obtained are specifically shown as follows:

1. The armature is very slender, which does not sacrifice strength, since it is made of duraluminium and because of this it is very light.
2. The cost of the exoskeleton is relatively low, compared to other robotic systems trademarks.
3. Due to the type of fastening the lower member, assembly and disassembly is easy.
4. Exoskeletal system was integrated into a polycentric high efficiency articulated of four-bar linkage and very stable, in most of the flexion-extension movement.

5. Conclusions

Conclusions obtained from this work are very encouraging, since to be manufactured and used in the health sector, could be made experimental tests in different patients, which would throw valuable results to improve aspects of the exoskeleton and optimize for the exclusive use of the sectors of the population who have no access to private medical care systems, This project proposes the design of an exoskeleton , inexpensive and is accessible to low income patients in the Mexican health sector

6. References

[1] Mosher R 1967 "Handyman to hardiman, Technical Report" SAE Technical Paper,

[2] Vhbjgjb M, Vukobratovic, Borovac B, Surla D and D 1990 "Stokic, Biped Locomotion", Springer-Verlag, 349p,

[3] Dollar A and Herr H 2008 "Lower extremity exoskeletons and active orthoses", challenges and state-of-the-art" Robotics IEEE Transactions 24

[4] Colombo G, Joerg M, Schreier R and Dietz V 2000 et al. "Treadmill training of paraplegic patients using a robotic orthosis", Journal of rehabilitation research and development Vol. 37, pp 693-700

[5] Veneman J, Kruidhof R, Hekman E, Ekkelenkamp R, Van Asseldonk and Van Der Kooij E 2007 "Design and evaluation of the lopes exoskeleton robot for interactive gait rehabilitation", Neural *Systems and Rehabilitation Engineering* IEEE Transactions on Vol. 15, pp. 379-386

[6] Sanz-Merodio D, Cestari M, Arevalo J, Carrillo X and Garcia E 2014 "Generation and control of adaptive gaits in lower-limb exoskeletons for motion assistance" *Advanced Robotics* Vol 28, pp. 329–338

[7] Esquenazi A, Talaty M, Packel A and Saulino M 2012 "The rewalk powered exoskeleton to restore ambulatory function to individuals with thoraciclevel motor-complete spinal cord injury", *American Journal of Physical Medicine & Rehabilitation* Vol. 91, pp. 911-921

[8] Strausser K A and Kazerooni H 2011, The development and testing of a human machine interface for a mobile medical exoskeleton, in: Intelligent Robots and Systems (IROS), 2011 IEEE/RSJ International Conference on, IEEE, pp. 4911–4916.

[9] Kazerooni H and Steger R, 2006 "The berkeley lower extremity exoskeleton", Journal of dynamic systems, *measurement, and control* Vol. 128, pp. 14-25

[10] Guizzo E and Goldstein H 2005, "The rise of the body bots [robotic exoskeletons]", Spectrum, IEEE Vol. 42, pp. 50-56

[11] Walsh C, Endo K and Herr H 2007, "A quasi-passive leg exoskeleton for loadcarrying augmentation", *International Journal of Humanoid Robotics* Vol. 4, pp. 487-506

[12] Lugo G 2010 "Diseño de mecanismos utilizando algoritmos genéticos con aplicación en prótesis para miembro inferior", *Tesis de Doctorado, Instituto Politécnico Nacional, Sección de Estudios de Posgrado e Investigación*, ESIME Unidad Profesional Zacatenco, 223p.

[13] Viladot, R, Cohi, O-Clavell and S 1989 "*Órtesis y Prótesis del Aparato Locomotor* Ed. Masson

[14] Catálogo de FESTO™ , obtenido el 10/06/2016, en: https://www.festo.com

Interaction of CO with an hBN surface doped with Ti and Pt: A First Principles Study

J M Ramirez-de-Arellano[1] and L F Magana-Solis[2,3]

[1] Tecnologico de Monterrey, Campus Ciudad de Mexico, Calle del Puente 222, Distrito Federal, Mexico, 14380.
[2] Instituto de Física, Universidad Nacional Autónoma de México, Apartado Postal 20-364, C.P. 01000 México, D.F., México.

E-mail: fernando@fisica.unam.mx

Abstract. We studied the possible catalytic effect that a transition metal atom could have after being adsorbed in a surface of hexagonal boron nitride (hBN). We considered platinum and titanium, performing *ab-initio* calculations, including molecular dynamics at 300K, within the Density Functional Theory. We considered an hBN surface either with a vacancy of a Boron atom, or with a vacancy of a Nitrogen atom, and we found that both titanium and platinum are absorbed at the place of the vacancy for both cases considered. Afterwards, we found that this decoration of the hBN surface indeed has a catalytic effect on the adsorption of a carbon monoxide molecule. Possible desorption was explored, at 800 K. To perform the calculations, the Quantum ESPRESSO package code was used. The generalized gradient corrected Perdew-Burke-Ernzerhof (PBE) approximation was used for the exchange and correlation functional.

1. Introduction

The study and development of surfaces that can absorb air pollutants is of a continuous interest in our society. In particular, carbon monoxide (CO) is a very poisonous gas even at levels of only 0.1% (1000 ppm), and a prolonged exposure can cause death in minutes [1,2]. Various materials, like graphene, have been studied to find ways of absorbing or detecting CO from the air [3,4].

Among those different materials, Boron nitride may be of particular interest, because of its availability and properties. Boron nitride has a structure similar to that of graphene and carbon nanotubes, for instance, Hexagonal boron nitride (hBN) has the same atomic structure than graphene, only with a lattice constant 1.8% bigger than the latter. hBN is also an electrical insulator [5], which can be useful when the electrical conductivity of graphene is undesirable.

Several ab-initio studies have been made on the CO adsorption in BN surfaces. Some have considered the BN sheet with no vacancies, with Stone-Wales defects, or with doping of different atoms, like Aluminum [6]. The interaction between CO and BN-based fullerenes has also been studied [7].

In this work, we considered an hBN surface with a vacancy of either a Nitrogen or a Boron atom. In each case, we studied its interaction with a CO molecule and then we considered the case when the vacancies are filled with a platinum atom and with a titanium atom. A good understanding of the mechanisms of adsorption for these pollutant molecules is relevant in the search of surfaces that can be used as sensors or cleaning devices of air and water.

[3] To whom any correspondence should be addressed.

Content from this work may be used under the terms of the Creative Commons Attribution 3.0 licence. Any further distribution of this work must maintain attribution to the author(s) and the title of the work, journal citation and DOI.
Published under licence by IOP Publishing Ltd

IOP Conf. Series: Journal of Physics: Conf. Series **792** (2017) 012074 doi:10.1088/1742-6596/792/1/012074

2. Method

Ab–initio molecular dynamics calculations were made, at 300K, based on the Density functional Theory (DFT) [8]. The Quantum ESPRESSO package code was used, this package considers a plane-wave expansion for the electronic wave functions, and periodic boundary conditions [9] . The generalized gradient corrected Perdew-Burke-Ernzerhof (PBE) approximation [10] was considered for the exchange and correlation functional. When considered appropriate, some of the calculations were repeated by using the van der Waals functional vdW-DF2 of Lee et al. [11]

The pseudo potentials are norm-conserving Troullier-Martins [12] in the Kleinman-Bylander fully separable form [13]. All calculations are non-relativistic and without polarized spin. The temperature is controlled via velocity rescaling and the adsorption energy is calculated according to the following definition:

$$E_{ads} = E_{system1+system2} - E_{system1} - E_{system2} \tag{1}$$

We considered an adsorption with energy equal or lower than 0.100 eV as physisorption, while any higher energy can be considered as chemisorption. The XCrySDen package was used for crystal structure visualization [14].

3. Results and Discussion

The first case we considered was that of the hBN surface with a vacancy of a Boron atom, or a vacancy of a Nitrogen atom. The MD calculations showed that in both cases, the CO molecule remains unaffected by the hBN surface. There is no adsorption in either case. To include the van der Waals interactions, we repeated the calculations with the vdW-DF2 density functional, and the results were similar, with no adsorption between the CO molecule and the hBN surface with the aforementioned vacancies.

We next considered a platinum atom, and the MD calculation showed that it was strongly absorbed in place of a nitrogen atom and also in place of a boron atom. In both cases, the adsorption energies of the platinum atom in the vacancy are similar (see Table 1). Subsequently we explored the interaction of a CO molecule with the two resulting Pt-doped hBN surfaces. In both cases, the CO molecule is chemisorbed by the platinum atom at the surfaces (Figure 1), with a higher adsorption energy in the case of the platinum in place of a boron atom (see Table 3).

Finally, we considered the catalytic effect of a titanium atom, which –as in the case of the Pt atom– was well absorbed into either a boron vacancy or a nitrogen vacancy (Figure 2). Unlike the platinum adsorption, however, we found that the titanium atom is more strongly absorbed into a boron vacancy than it is into a nitrogen vacancy, with the adsorption energy of the former case being more than three times larger than the one of the latter case (see Table 2). We then found that the CO molecule can be chemisorbed by the titanium atom when the former is either at the boron vacancy or at the nitrogen vacancy (see Table 3).

Our results are consistent with the fact that the hBN surface is usually considered as chemically stable when pristine, but the addition of either the Ti or Pt atoms, facilitates the charge exchange between any of those two and the CO molecule, resulting in adsorption.

We explored the possibility of desorption by performing MD calculations of the four cases of a chemisorbed CO molecule, at 800 K. We found that, at such temperature, the CO remains absorbed into the surface. We could then consider that the CO is chemisorbed by the catalysts, and that the hBN surface acts as a substrate.

Table 1. Energies of adsorption, E_{ads}, (in eV) of Pt into the hBN surface, for the two cases considered.

Surface	E_{ads}	Type of adsorption
hBN with a B vacancy	−6.926	chemisorption
hBN with an N vacancy	−6.813	chemisorption

Table 2. Energies of adsorption, E_{ads}, (in eV) of Ti into the hBN surface, for the two cases considered.

Surface	E_{ads}	Type of adsorption
hBN with a B vacancy	−12.302	chemisorption
hBN with an N vacancy	−3.825	chemisorption

Table 3. Energies of adsorption, E_{ads}, (in eV) of the CO molecule into the hBN surface, for all cases considered.

Surface	E_{ads}	Type of adsorption
hBN with a B vacancy	---	No adsorption
hBN with an N vacancy	---	No adsorption
hBN with Pt in B vacancy	−2.106	chemisorption
hBN with Pt in N vacancy	−1.164	chemisorption
hBN with Ti in B vacancy	−1.530	chemisorption
hBN with Ti in N vacancy	−1.679	chemisorption

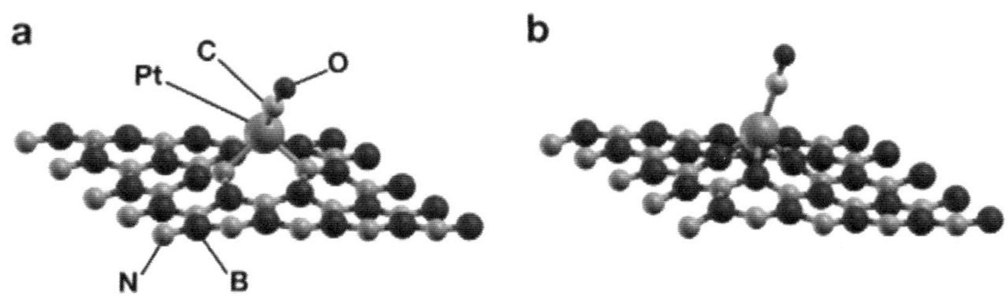

Figure 1. Final configurations for the interaction of a CO molecule atom and an hBN surface where a Pt atom has been previously absorbed in place of (a) a boron vacancy and (b) a nitrogen vacancy. In both cases, the CO molecule is chemisorbed by the system, without being dissociated by it. The first case shows a higher value for the adsorption energy of the CO molecule.

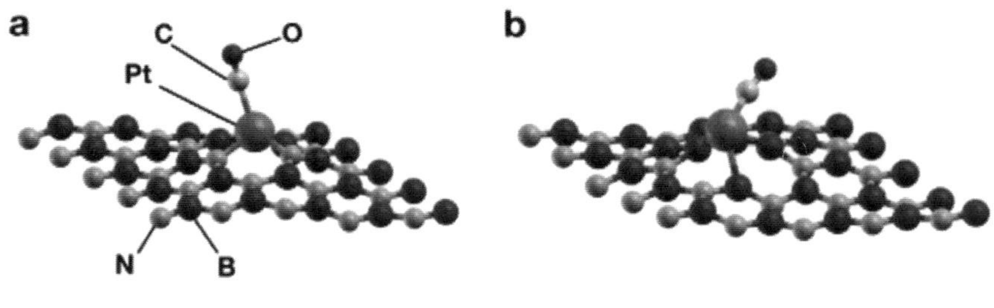

Figure 2. Final configurations for the interaction of a CO molecule atom and an hBN surface where a Ti atom has been previously absorbed in place of (a) a boron vacancy and (b) a nitrogen vacancy. Here, the CO molecule is chemisorbed as well, and the energies of adsorption are similar in value.

4. Conclusions

We found that the hBN surface has no relevant effect in the interaction with CO when pristine, or even when there are vacancies in the surface. The mechanism of adsorption may depend on the charge distribution due to either the platinum or titanium atoms. The catalytic effect of both the Pt and Ti atoms in the interaction between the hBN surface and the CO molecule can be due to their modification of such charge distribution.

Both the Ti and Pt atoms, while occupying vacancies of the hBN surfaces, can absorb the CO molecule in all cases considered. We found that the CO molecule is chemisorbed in the Pt-decorated hBN surface, with adsorption energies of −2.106 eV (B vacancy) and −1.164 eV (N vacancy). The CO molecule is also chemisorbed in the Ti-decorated hBN surface, with adsorption energies of −1.530 eV (B vacancy) and −1.679 eV (N vacancy). The fact that the substrate is a readily available material, such as hBN, instead of the bulk titanium or platinum surfaces, would be potentially useful in technological applications.

Acknowledgements

We thank Direccion General de Asuntos del Personal Academico de la Universidad Nacional Autonoma de Mexico, partial financial support by Grant IN-106514 and we also thank UNAM-Super-Computing Center the technical assistance.

References

[1] Omaye S T 2002 Metabolic modulation of carbon monoxide toxicity *Toxicology* **180** 139–50
[2] Prockop L D and Chichkova R I 2007 Carbon monoxide intoxication: an updated review. *J. Neurol. Sci.* **262** 122–30
[3] Huang M and Fabris S 2008 CO Adsorption and Oxidation on Ceria Surfaces from DFT+U Calculations *J. Phys. Chem. C* **112** 8643–8
[4] Wanno B and Tabtimsai C 2014 A DFT investigation of CO adsorption on VIIIB transition metal-doped graphene sheets *Superlattices Microstruct.* **67** 110–7
[5] Watanabe K, Taniguchi T and Kanda H 2004 Direct-bandgap properties and evidence for ultraviolet lasing of hexagonal boron nitride single crystal. *Nat. Mater.* **3** 404–9
[6] Zhang Y-H, Zhou K-G, Gou X-C, Xie K-F, Zhang H-L and Peng Y 2010 Effects of dopant and defect on the adsorption of carbon monoxide on graphitic boron nitride sheet: A first-principles study *Chem. Phys. Lett.* **484** 266–70
[7] Nigam S and Majumder C 2008 CO oxidation by BN-fullerene cage: effect of impurity on the chemical reactivity. *ACS Nano* **2** 1422–8
[8] Parr R G and Yang W 1989 *Density Functional Theory of atoms and molecules* (Oxford Science Publications)

[9] Giannozzi P, Baroni S, Bonini N, Calandra M, Car R, Cavazzoni C, Ceresoli D, Chiarotti G L, Cococcioni M, Dabo I, Dal Corso A, de Gironcoli S, Fabris S, Fratesi G, Gebauer R, Gerstmann U, Gougoussis C, Kokalj A, Lazzeri M, Martin-Samos L, Marzari N, Mauri F, Mazzarello R, Paolini S, Pasquarello A, Paulatto L, Sbraccia C, Scandolo S, Sclauzero G, Seitsonen A P, Smogunov A, Umari P and Wentzcovitch R M 2009 QUANTUM ESPRESSO: a modular and open-source software project for quantum simulations of materials. *J. Phys. Condens. Matter* **21** 395502
[10] Perdew J P, Burke K and Ernzerhof M 1997 Generalized Gradient Approximation Made Simple [Phys. Rev. Lett. 77, 3865 (1996)] *Phys. Rev. Lett.* **78** 1396–1396
[11] Lee K, Murray É D, Kong L, Lundqvist B I and Langreth D C 2010 Higher-accuracy van der Waals density functional *Phys. Rev. B* **82** 081101
[12] Troullier N and Martins J L 1991 Efficient pseudopotentials for plane-wave calculations *Phys. Rev. B* **43** 1993–2006
[13] Kleinman L and Bylander D 1982 Efficacious Form for Model Pseudopotentials *Phys. Rev. Lett.* **48** 1425–8
[14] Kokalj A 1999 XCrySDen—a new program for displaying crystalline structures and electron densities *J. Mol. Graph. Model.* **17** 176–9

IOP Conf. Series: Journal of Physics: Conf. Series **792** (2017) 012075 doi:10.1088/1742-6596/792/1/012075

Factors that determine the presence of particles in YBCO films grown by PLD

V R Barrales-Guadarrama, E M Rodríguez-Rodríguez, R Barrales-Guadarrama, N Reyes Ayala
Universidad Autónoma Metropolitana - Azcapotzalco, Laboratorio de Sensores y Señales, Av. Sn. Pablo 180, Col. Reynosa-Tamaulipas, México D. F.

E-mail: vrbg@correo.azc.uam.mx

Abstract. The method of growing thin films PLD, is widely used in applications and possesses great potential in thin $YBa_2Cu_3O_{7-\delta}$ films production with outstanding physical properties. However, it is limited in nano and micro technology due to the presence of particles on the surface of the films. This article describes some causes that create these particles. $YBa_2Cu_3O_{7-\delta}$ films have been grown on electrolytic copper used as a variable model the distance target-substrate. The effects are studied through Scanning Electronic Microscopy. It is observed particles with a large variety of shapes and distributions. The results show that ranging the target-substrate distance, the superficial morphology is modified. An evidence of it, is that the evaporation of d_{B-S} = 7 cm, is more coherent that d_{B-S} = 3 cm. Therefore, exist a relation between the morphology and the parameters of growing. Also affect, the structural change that exists among the substrate and the film formation, the substrate preparation and it must not be monocrystalline, these factors define a kinetic and a mechanism of growing that promotes a heterogeneous nucleation.

1. Introduction

Growing thin films with PLD (Pulse Laser Deposition) is the result of forming plasma making fall on laser radiation pulsed upon the surface of a solid target. It is used in a diversity of applications maintaining the capacity of $YBa_2Cu_3O_{7-\delta}$ (YBCO) thin films production with excellent properties [1,2] to be used in the manufacture of low temperatures high performance electronic devices[3,4].

However, its applicability is limited in nano and microelectronic mostly due to two factors: splashing, this is, expel of drops and fragments from the target during the ablation, for the creation of outgrowths upon the surface of the film. Both factors depend on the physical properties of the surface of the target and the substrate that include its temperature, density, optic absorption, chemical composition, steam pressure, texture and the net parameter adjustment and the film in formation; from thermodynamic conditions given for growing parameters, that, at the same time, define the mechanisms and the growing kinetic of films, examples of this are: the laser properties that include the fluency, wavelength, uniformity of the beam profile, repetition rate; gas pressure in the growing chamber; distance target-substrate, growth rate, dimensions and the beam print track size, among others. All these parameters are related, this is, exist a range of value for which reduce the splashing and outgrowths creation. This implies that the phenomenology associated to this interrelation is complex; part of this knowledge still is qualitative despite the effects being measured. These effects refer to physical properties of the grown material, for instance, electric transportation, thermo electronics, and

Content from this work may be used under the terms of the Creative Commons Attribution 3.0 licence. Any further distribution of this work must maintain attribution to the author(s) and the title of the work, journal citation and DOI.
Published under licence by IOP Publishing Ltd

magnetics. Its phenomenology must be understood trough the interrelation between the physical properties of the film and the thermodynamic limited for growing parameters, highlights that the approach to the explanation of the origin of the particles represents the capacity of diminish its density or even remove them, in the meantime, this represents be even closer of the optimum range of one or several interesting physical properties for the obtained film, which implies that the applications where are used these materials have competitive efficiencies with other technologies. This is one of the main reasons that justified the concern of the study of the particles origin upon de surface of YBCO films grown for PLD. The applications refer, since alternating current cables, generators, or engines operating at 77 K[5], until applications in nano and microelectronics[6,7], for example, in the last decades researches have been had great interest for superconducting nanostructures and semiconducting of low dimensionality [8,9], due to its low energy applications and energy harvesting devices[10,11]. Because of the surface effects, the performance of these devices depends of the morphology of the nanocrystals.

Outstanding efforts are being made to understand the origin of the particles and to find techniques to diminish its density upon the films surface. More likely, the best implement better classified, among the desirable and necessary to develop a technology that produces thin film crio electronic devices, based on super conductive materials of high T_c (HTSC) as the YBCO, it is the one that gives the possibility of manipulate locally the superconductive properties as the critical current density, J_c, or the transitional temperature T_c, being possible in a nano metric scale.

An approach to achieve this is the one called "selected" or, more descriptively, "inhibited" growing [12]. The underlying idea is, to use optimized monocrystalline substrates[13,14], for example, YBCO films with a *c* axis oriented perpendicular to the film plane, moreover, with a pre-structure located in the substrate areas where is desirable impede or inhibit entirely the epitaxial growing of the film HTSC[12]. Once more, through this alternative can be obtained high performance applications, it is necessary the lack of existence of particles upon the surface of the film, in other words, having nanometric control of the morphology of the surface of the film. Examples where the procedure has being applied to YBCO films grown are the deposition of Si[15,16], Si_xN_y [17], Ti[18,19], Nb and W[19]. Another important progress regarding YBCO films with good morphology are the development of technologies for conductors covering that have achieved $J_c > 1$ mA/cm^2, which has stimulated greater interest in the comprehension and improvement of the deposition technique of YBCO films by PLD [20,21], being as its high J_c and its intrinsic magnetic properties provide a widely range of solutions to engineering problems [22].

In this work is studied the origin of particles that are created during the deposit of YBCO, grown through the technique on steam phase PLD, considering the variation target-substrate and letting the rest of the growing parameters steady. The superficial morphology was analyzed and it was obtained the chemical composition in different regions of the thin film and a characteristic target using a scanning electron microscope (SEM). It is argued the correlation between the distances target-substrate, the particles density and their composition.

2. Experimental Description

YBCO films were deposited on electrolytic copper substrates with a Nd:YAG laser of $\lambda = 1064$ nm, a 650 mJ, with a pulse duration $\tau \approx 5$-7 ns and a frequency of 10 Hz. So as to, a steady target position, the beam image focused upon the target surface was adjusted experimentally with a round shape with 2 mm diameter and it was focused with a convex lens ($f = 50$ cm). For all the experiments the laser fluency, spot magnitude and the distance between the lens and the target were stable. The deposit time was t = 5 minutes, the distance between the target and substrate was modified in a range that went from $d_{T\text{-}S} = 3$ cm to $d_{T\text{-}S} = 7$ cm, the growing temperature was T = 650°C and the oxygen pressure into the chamber was $P_{O2} = 200$ mTorr. The sole parameter that showed a variance was the distance between target-substrate. Once this distance was defined, all other parameters remain constant during the growing. The samples were exposed to an *in situ* annealing at the end of the growing to T = 450°C y $P_{O2} = 270$ mTorr during two hours.

The microstructural superficial observation and the elemental chemic analysis were realized in a SEM, brand LEO 440 with a capacity of 40 KeV, adapted with a solid state detector of SiLi with 1024 channels and a range of 20 KeV, OXFORD brand. The micro photographs were obtained in an acceleration voltage of 20 KeV and a current density of 200 pA, through a secondary electron detector. The chemical analysis was obtained through X-ray diffraction with energy dispersion application (EDS) electrostatic dispersion, punctual analysis was realized with an acceleration voltage of 20 KeV and currect density 1 nA, at the time of the acquisition of each specter was 60 seconds real time. All the elements that generate X-rays were detected from Beryllium. The calculus of semi quantifications was realized using the correction factor ZAF (atomic number effect, absorption, and fluorescence of the processed sample) al the results returned to normal.

3. Results and Discussion

Figure 1 reveals an expulsion direction of the preferential material, located in the internal part of the periphery of the crater caused by the "feeder" upon de target surface. It is observed, for the topographical characteristics of the surface, that experimented a re-solidification process of mould when it did not receive enough energy to sublimate.

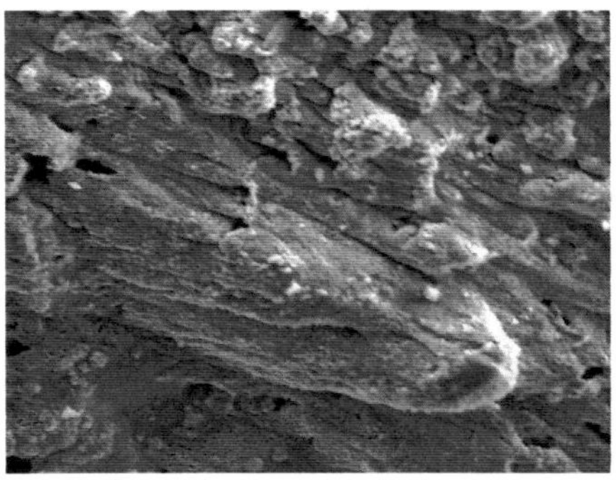

Figure 1. Micrography that shows the preferential direction of the material. Amp: 500x; Esc: 1.7 cm = 500 μm.

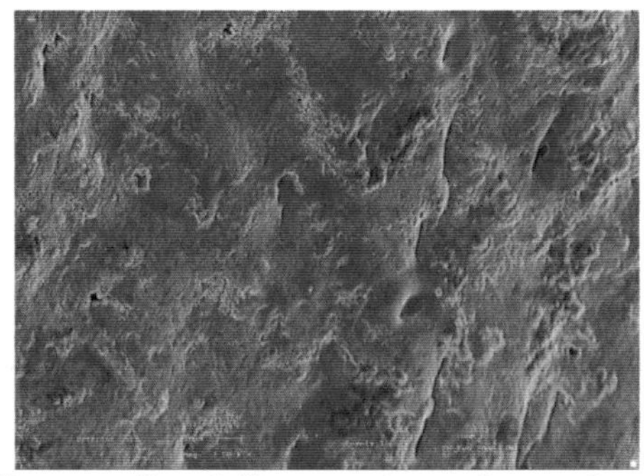

Figure 2. Micrography that shows liquefaction effects and re –solidification of the periphery of the crater.

Figure 2 is a micrography of the external periphery of the crater, where is observed the liquefaction process that the material suffered, later on resolidify, without the influence of the impulsive force that produces the feeder.

Figure 3 shows solidified drops with an elliptic shape, in the periphery of the exterior edge of the crater, randomly distributed, produced for the splashing. The isotropic superficial tension of the liquid drives to the spherical shape equilibrium, it is possible that the spherical shape may not be entirely achieved before the solidification so that they possess that shape, even the biggest axis of all drops have the same direction, towards outside of the crater edge.

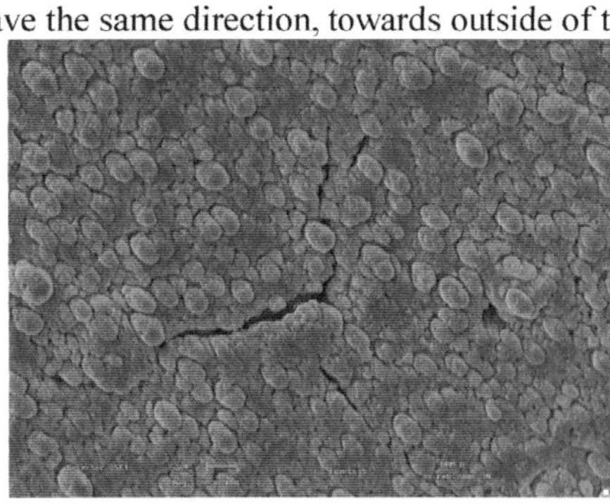

Figure 3. Random distribution of drops with elliptic shape, solidified with a preferential direction

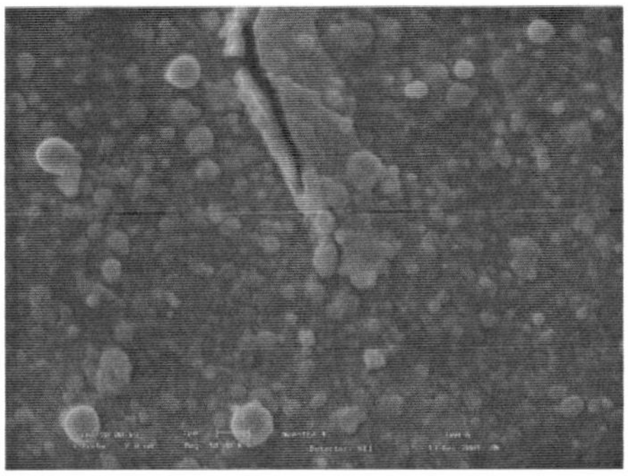

Figure 4. Micrography of a YBCO surface film grown at $d_{T\text{-}S}$ = 3 cm.

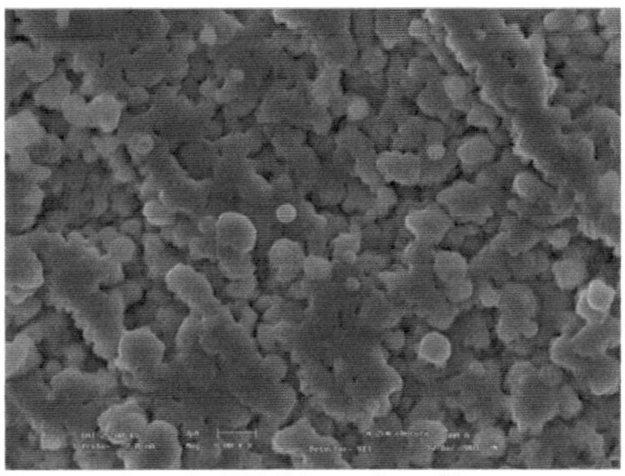

Figure 5. Micrography of a YBCO surface film grown at $d_{T\text{-}S}$ = 7 cm.

In Figure 4, some drops between 0.4 y 1 μm of diameter are observed superficial, meanwhile other are subsequently covered by the deposit. As magnitudes, shapes and chemical composition are similar to observed drops upon the target surface. This implies that are drops of mould resolidified, which are expelled directly since the target. On the upper side of the micrography is observed a fractured outgrowth. Instead, in figure 5 shows that the density of particles has decreased and do not appear any outgrowth, this highlights the splashing and outgrowths reduction for $d_{T\text{-}S}$ = 7 cm.

On the surface of the films, there exist areas with a different composition to phase 123. In other words, there are areas with a phase of CuO, with an abundant concentration of Ba and low concentration of Y. Figure 6 shows the EDS spectrum of a thin film, $d_{T\text{-}S}$ = 3 cm, which is an example of a growth YBCO film.

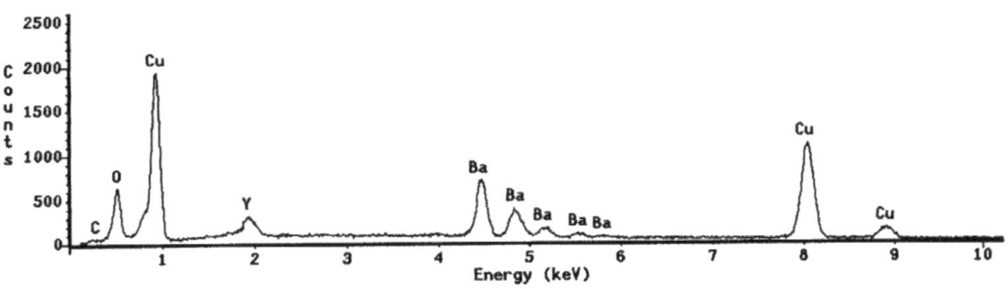

Figure 6. EDS spectrum representative of the YBCO films grown by PLD.

In order to determine the effect of "splashing" in the composition, different areas were chosen to determine the EDS spectrum on the surface of a YBCO film at $d_{T\text{-}S}$ = 3 cm. Before the growing, a screen of 2 mm width and 10 mm length was placed on the substrate surface, with the purpose of avoiding that particles generated in the process of ablation are deposited under this area, but allowing that steam is deposited. After the growing, three areas were identified as: "step", "outgrowth", and "film", as it is shown in Figure 7. The "step" is an area which denotes the height of the film after growing. The "outgrowth" is an area which comprises of a multiphase microstructure and the "film" represents the area of the growth film under the screen which is characterized by deposited steam without splashing. At the same time, four specific subareas were selected, in order to quantify their chemical composition, as it is shown in Figure 7.

Figure 8 shows the quantification of Cu for each subarea. A significant difference of Cu is shown among areas. However, when the atomic weight of Cu and O_2 is calculated for these areas, the stoichiometry agrees with the CuO phase (Tenorite).

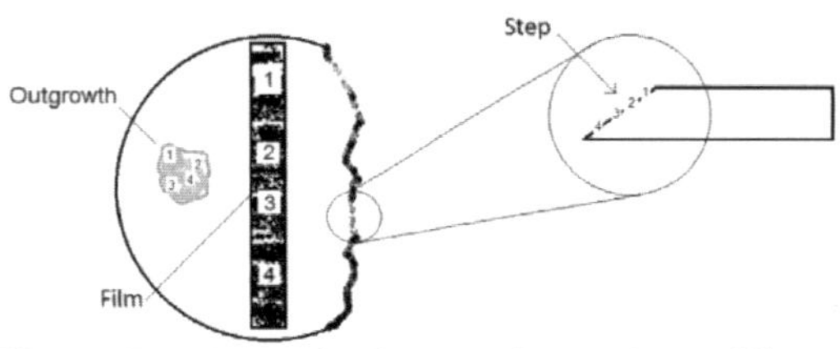

Figura 7. Schematic diagram shows three different areas to quantify their chemical composition.

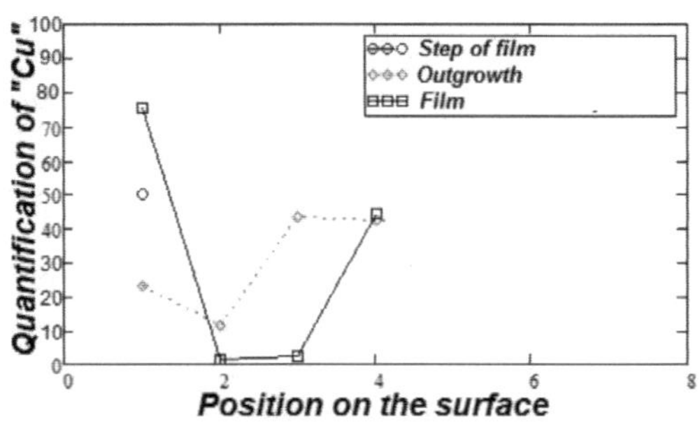

Figure 8. Cu quantification.

In the same way, a quantification of Y and Ba was performed on those areas previously selected. Figures 9 and 10 show this quantification of both elements, Y and Ba, respectively. It is observed a low quantity of Y for all subareas selected and a relative increment in the quantity of Ba.

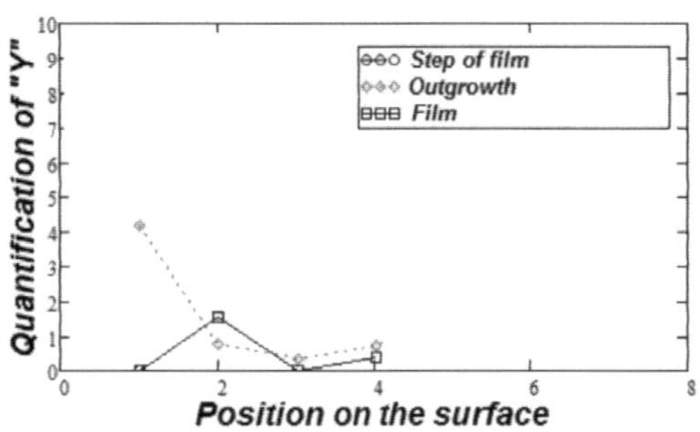

Figure 9. Y quantification.

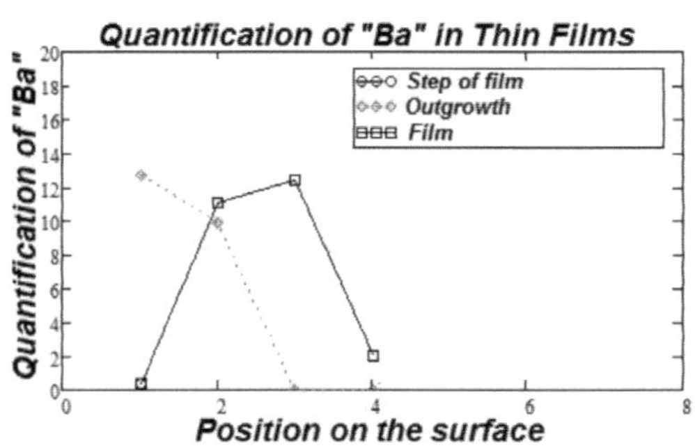

Figure 10. Ba quantification.

An analysis was performed under the area where screen was placed. Figure 11 shows the quantification of elements where steam was deposited and splashing is not present on the surface. Those subareas identified as 3 and 4 show a particular and interesting to be considered as adequate subarea. Table 1 quantifies a regular stoichiometry relationship, where the component 123 of YBCO is 20%.

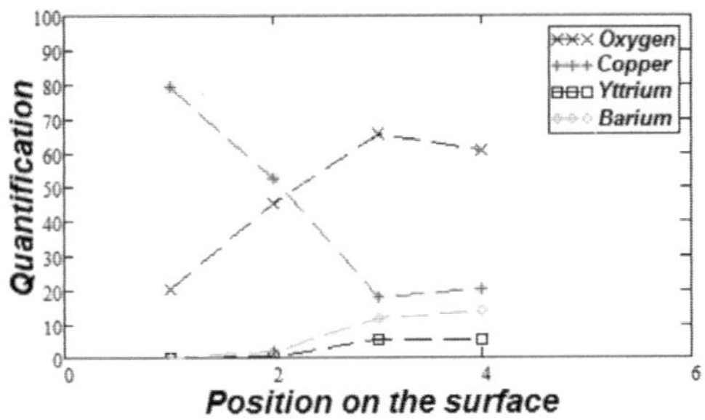

Figure 11. Quantification of YBCO composition.

Table 1.

Atomic species	*Stoichiometric relation between species (average % of atomic weight)*
Y	1
Ba	2.395
Cu	3.602
O	11.978

4. Conclusions

The results show that the morphology of the surface, in thin films grown by PLD, depends mainly on the distance target-substrate. However, it suggests a correlation between the distance and the thermodynamic growing of the thin film. The effects of splashing and outgrowths are decreased at d_{T-S} = 7 cm, which in turns may determine specific characteristics on the structure and morphology for engineering applications on nano scale.

The evaporation process at d_{T-S} = 7 cm is more congruent than d_{T-S} = 3 cm. A homogeneous composition on the thin film is better preserved during the ablation for d_{B-S} = 7. The larger the distance between target-substrate, the lower quantity of undesired particles on the surface of a thin film.

The analysis shows four factors that mainly contribute to the lack of a homogeneous composition and the generation of heterogeneous nucleation: the target and the substrate were very close, increasing the probably of deposited particles on the thin film; the lack of an adequate adjustment between the net parameters target-substrate; the set up in the chemical process in the substrate to eliminate all kind of impurity before the growing; and the substrate must not have a monocrystalline surface.

The use of a screen during the growing is another factor that may prevents the effect of "splashing", which in turns affect an adequate composition. It means that a screen improves both composition and morphology on the surface with a difference of 20% respect to phase 123.

The ablation phenomena and the vaporized material, used as target, are steps that are repeated hundreds of times in a typical deposition, modifying the original target surface since the first laser pulse impacts. After the target-laser interaction, amorphous superficial areas may be observed, more than a crystalline crater upon the film surface. This factor influences the quality in the composition of the thin film.

5. References

[1]Talantsev1 E F and Tallon J L 2015 "Universal self-field critical current for thin-film superconductors", Nature Communications 6, Article number: 7820.
[2] Tang, Xiao 2015 *et al*, "High-Jc $YBa_2Cu_3O_{7-x-Ag}$ superconducting thin films synthesized through a fluorine-free MOD method", *Journal of the European Ceramic Society*, Vol. 35, No. 6, p. 1761–1769.
[3] Gülce Çoker, E; Kolemen, U; Topal and U. 2015 "Physical and Mechanical Properties of B_2O_3 doped YBaCuO Superconductors: Applicability as a Low DC Magnetometer", *Journal of Superconductivity and Novel Magnetism*, , Volumen 28, Número 2.
[4] Galiano 2015 *et al*, "Fast pyroelectric response of semiconducting YBaCuO detectors with high IR sensitivity; development of THz imaging arrays", *40th International Conference on Infrared, Millimeter, and Terahertz waves* (IRMMW-THz).

[5]Bourdillon A and Tan Bourdillon N X 1994., "High Temperature Superconductors: Processing and Science", *Academic Press* (San Diego), CA
[6] Korzec M D 2014 *et al*, "Equilibrium shapes of polycrystalline silicon nanodots" *Journal of Applied Physics* **115** 074304
[7] Thompson C V 2012, "Solid-state dewetting of thin films," *Annu. Rev. Mater. Res.* **42**, 399–434.
[8] Char,Colclough M S,Lee L P, and Zaharchuk G 1991, *Appl.Phys.Lett.* **59** 2177.
[9] Shi L, Hashishin Y, Dong S Y, Zheng J P and Kwok H S 1991 *Appl.Phys.Lett.* **59** 1377.
[10] Law M, Goldberger J and Yang 2004 *Annu. Rev. Mater.* Res. 34
[11] Pearton S J 2008., *et al*, Nanosci J. *Nanotechnol.* **8**, 99
[12] Lengl G, Ziemann P, Banhart F and Walther P2003 "Anomalous behavior of gold nanoislands on top of $SrTiO_3$(001) during their overgrowth by thin YBaCuO films", *Physica* C390 175–184.
[13] Maiorov B2009, *et al*, "Synergetic combination of different types of defect to optimize pinning landscape using $BaZrO_3$-doped $YBa_2Cu_3O_7$", *Nat. Mater.* **8** 398–404
[14] Aytug T 2008 *et al*, "Enhanced flux pinning and critical currents in $YBa_2Cu_3O_{7-\delta}$ films by nanoparticle surface decoration: extension to coated conductor templates", *J. Appl. Phys.* 104
[15]Fork D K 1990, *et al*, Appl. Phys. Lett. **57** 2504.
[16]Ma Q Y, Yang E S, Treyz G V 1989, Appl. *Phys. Lett.* **5** 896.
[17] Kern D P, Lee K Y,Laibowitz R B and Gupta A 1991 *J. Vac. Sci. Technol.* B **9** 2875.
[18] Rossel C, Kaufmann U, Downar H,Schulz R 1991, *Physica C* 255 185–189.
[19] C. A. J. Damen1998, *et al*, *Supercond. Sci. Technol.* 11 (1993)437.
[20] Wu X D 1995, *et al*, Appl. Phys. Lett. 67 (1995) 2397.
[21] Goyal A 1996, *et al*, Appl. Phys.Lett. 69 (1996) 1795.
[22] García-González E,Wagner G, Reedyk M and Haber-Meier H U 1995, J. *Appl. Phys.* 78 (1995) 353

A comparative study of power law scaling in large word-length sequences

J Quezada-González[1], D Aguilar-Velázquez[1], T Contreras-Uribe[1], B Obregón-Quintana[2], L Guzmán-Vargas[1,3]

[1] Unidad Interdisciplinaria en Ingeniería y Tecnologías Avanzadas, Instituto Politécnico Nacional, Av. IPN No. 2580, L. Ticomán, México, D.F., 07340, México
[2] Facultad de Ciencias, Universidad Nacional Autónoma de México, Circuito Exterior s/n, Ciudad Universitaria, Coyoacán, México, D.F., 04510, México

E-mail: lguzmanv@ipn.mx

Abstract. A study of the correlation of lengths of words in large literary texts is presented. We use the statistical tools based on Allan factor and fractal dimension for estimating the fractal indices associated with the presence of correlations in the original sequences. We found that there is a scaling behavior at large scales for Allan factor with an exponent value, which reveals positive correlations. By means of the fractal dimension, we confirmed the presence of memory in the sequences of lengths of words. In addition, when considering a random rearrangement of the lengths of words in the original text, the fractal exponents are consistent with an uncorrelated process.

1. Introduction

The texts have been instrumental throughout the history of mankind to explore and document the progress in the development of human knowledge, from hieroglyphics, scrolls and printed books until the appearance of e-books. Human communication is ruled by grammar and syntax that allow to give meaning to the sentences which are comprised by several words [1, 2]. In 1851, the work of August of Morgan began the study of the frequency in the language. Morgan found that the average length of words (letters per word) was an indicator of individual style. For details of this work and state of the art, see [3]. In the decade of the 40s of the last century, George Kingsley Zipf, established that the frequency of the words shows a power law behavior with the rank, i.e., if $f(r)$ is the frequency of a word and r the rank of the word, then $f(r) \sim 1/r^{\alpha}$, with $\alpha \approx 1$ [4]. These type of studies are widely known and have been applied to other areas, particularly with regard to language are: content of information [5], polarity and information [6], recurrence times [7], correlations [8-10], allometries [11], and many others [2, 5, 12-19].

The work of Piantadosi *et al.*, noted that the length of a word is also important, because the length has a nonlinear relationship with the frequency and information, that is, if the duration or length of a word is longer, it contains more information [5]. Garcia *et al.* established a relationship between the frequency of a word and its positive or negative content, being more frequent positive words, however, the negative ones contain more information [6]. In other studies, the relation between words has been considered taking into account their sound, if two words are pronounced similarly, then there is a relationship between them [16-17]. Moreover, words have a life time or time of use in everyday language

[3] To whom any correspondence should be addressed.

Content from this work may be used under the terms of the Creative Commons Attribution 3.0 licence. Any further distribution of this work must maintain attribution to the author(s) and the title of the work, journal citation and DOI.
Published under licence by IOP Publishing Ltd

where the frequency is not the only factor in determining survival (birth, survival and death) as it was shown in the work of Petersen *et al.* [11]. On the other hand, diverse works have been done on specific texts manipulated as time series to identify characteristics of language considering temporal correlations with respect to the length or frequency of words [8, 20-22].

However, these previous works have not widely studied the interdependence between lengths of words and the relationship between ideas and stories developed in large literary texts, therefore, it is important to study the all these properties and not only the frequency in isolation. In this paper we consider the temporal correlation between the length of the words for several large literary texts. First, we consider the sequence of the length of words as a point process and calculate the Allan factor to determine the scaling regime characterized by the correlation exponent. Next, we use the fractal dimension method to estimate the fractality index for the word-length sequences. We find that both methods show the presence of positive correlations, indicating that word-length time series is not a random process. This paper is organized as follows, in the Section II, we explain the Allan and fractal dimension methods and a description of the texts in our study. Next, in Section III, we present the results, and finally the conclusions are given.

2. Methods and data

We use the point process statistics to determine the presence of scaling properties in word-length sequences. A point process can be described as events (spaces between words in our case) located along a time axis. We use the so-called Allan factor (AF), which is defined as the variance of successive counts for a specific window time, divided by twice the average number of events in that count [23]. For a process where this ratio remains constant and close to the unit for different time windows, it is identified as a Poisson processes (with no memory). In contrast, for a process with a scaling behavior between AF and the time window T, i.e., $A(\mathrm{T})$ behaves like a power law function of the time window T, $A(\mathrm{T}) \sim \mathrm{T}^{\alpha}$, the process is known to exhibit positive correlations and the exponent α characterizes the behavior of the strength of the correlations.

We also use the fractal dimension method (FDM) [24, 25] to estimate the value of the dimension of the set represented by the sequence of lengths of words. The FDM basically consists of the following steps: given the time series $x_1, x_2, \ldots, x_N$, new series are constructed x_m^k, which are defined as $x_m, x_{m+k}, x_{m+2k}, \ldots, x_{\left(m\left[\frac{N-k}{k}\right]k\right)}$, with $m = 1, 2, \ldots, k$; [ ] denotes Gauss notation, i.e., the greatest integer, m and k are integers representing the initial time and the time interval, respectively. The length of the curve is defined as:

$$L_m(k) = \frac{1}{k}\left[\left(\sum_{i=1}^{\left[\frac{N-m}{k}\right]} |x(m+ik) - x(m+(i-1)k)|\right)\frac{N-1}{\left[\frac{N-m}{k}\right]k}\right]$$

where $(N-1)\{[(N-m)/k]k\}^{-1}$, represents a normalization factor. In order to get the total length of the original curve for the time interval k, it is considered the length of each sequence x_m^k. Finally, if $\langle L(k)\rangle \propto k^{-D}$, then the curve is fractal with dimension D [24, 25].

For a Brownian type process resulting from the integration of an uncorrelated process, typically we get $D = 1.5$, whereas for a process with $D < 1.5$, we have positive correlations or persistence and $D > 1.5$ reflects anticorrelations. There is a relationship between the Allan factor exponent α and D, given by $\alpha = 3 - 2D$ [24, 25].

Our database consists of 10 electronic e-books in English language (see table 1), obtained from the Project Gutenberg (http://www.gutenberg.com) and Project Gutenberg Australia (http://gutenberg.net.au). For the selection of titles, we considered books published in different times and from different genres, and in particular that were large enough texts (number of words $> 10^5$) in order evaluate the asymptotic scaling behavior.

Table 1. List of books used in the study. M denotes the number of words and N_U the number of different words

No.	Title and author	*M*	N_U
1	Ulysses, J. Joyce	272,415	29,898
2	The Bible, Ed. King James	884,964	12,806
3	On the Origin of Species, C. Darwin	156,811	7,273
4	Frankenstein, M. Shelley	104,753	8,078
5	Gulliver's Travels, J. Swift	104,797	8,188
6	Hopscotch, J. Cortazar	195,702	15,338
7	Moby Dick, H. Melville	218,704	17,150
8	Pride and Prejudice, J. Austen	122,878	6,450
9	The Koran (Al-Qur'an)	210,967	11,058
10	The Idiot, F. Dostoyevsky	247,952	10,102

3. Results

In our case we are interested in comparing the asymptotic behavior of the scaling exponents for large time scales. First, we apply the Fano factor to word-length sequences. For this purpose, the event of interest is the space between words, and the interevent times is given by the word-lengths. In figure 1 we show the results of analysis of Allan for the 10 books in our study. It is observed that at small scales, $A(\mathrm{T})$ is close to 1, indicating that the process is Poissonian; for intermediate scales, there is a valley with values smaller than 1, reflecting an increase in the regularity. At large scales, a scaling regime can be identified, that is, $A(\mathrm{T}) \sim \mathrm{T}^{\alpha}$, with an average exponent $\alpha = 0.57 \pm 0.09$, denoting the presence of correlations. Our results of exponent values indicate that when word-lengths are observed at large scales (windows), they exhibit persistence, i.e., word-lengths separated by large intervals display memory whereas at short intervals the dynamics is uncorrelated. Moreover, when we build a shuffled version of the word-lengths in the original text, the statistics $A(\mathrm{T})$ behaves differently than the original, with values near and below the unit value (homogeneous Poisson process), indicating that the memory observed in the original sequences is destroyed by the shuffled process (see figure 1).

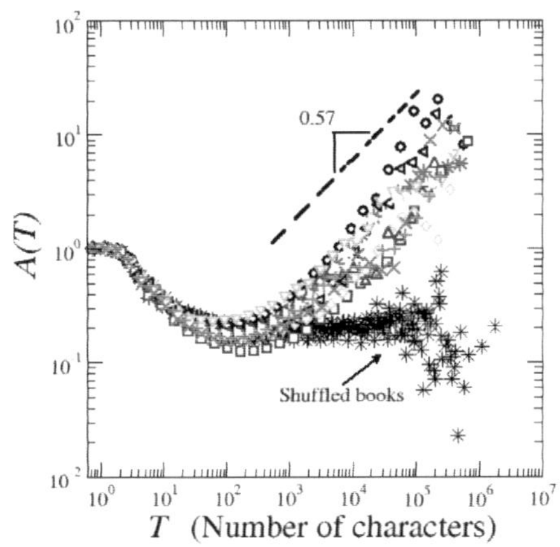

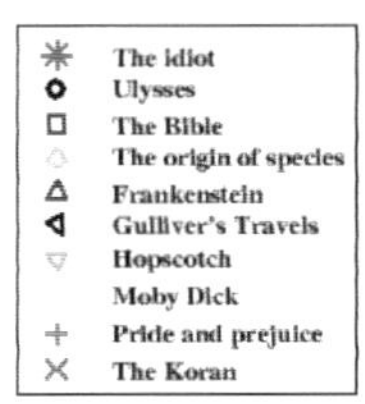

Figure 1. Plot of $A(\mathrm{T})$ $vs.$ T for the books in our study. We identify a scaling regime for the scales $10^3 < \mathrm{T} < 10^6$, with and average exponent close to 0.5. For shuffled data, the values exhibit a flat distribution, indicating that the memory is destroyed.

To further explore the behavior of the power-law scaling at large scales, we proceeded to estimate the value of the local exponent α^l given by the following expression: $\alpha^l = \delta log AF(\mathrm{T})[\delta log(\mathrm{T})]^{-1}$. For a better accuracy in the estimation of the exponent, we use the centered derivative. The results of these calculations are shown in figure 2. We observe a smooth transition from the regime corresponding to the Poisson process, to an average value close to 0.5, confirming that in the asymptotic region the local exponent is consistent with the presence of positive correlations and long-range memory. We also noticed that there is a greater dispersion at large scales, possibly associated with the heterogeneity of our database, a slightly different trend in the level of correlations is identified for texts like the Bible and the Koran (see figure 1), i.e., some texts emphasize the presence of correlations in the long term while others exhibit more oscillations in the statistics.

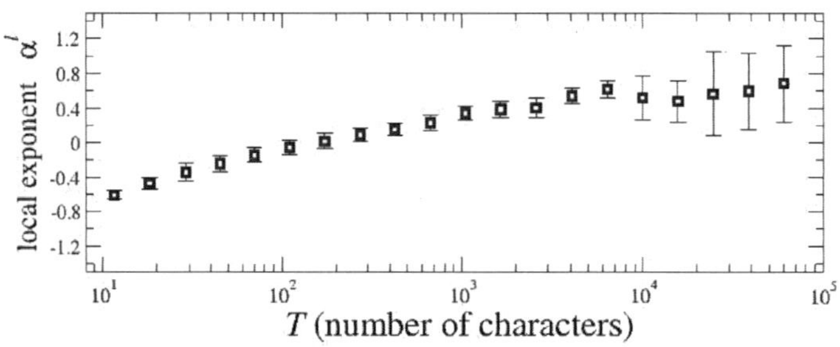

Figure 2. Behavior of the average local exponent α^l in terms of the time scale T. We observe a smooth transition from the Poissonian regime to the correlated one. We show the mean value ± 1 standard deviation.

To complement our study, we use the FDM method for estimating the values of the fractal dimension of the same of word-length sequences. For a better estimation of the fractal exponent D, an integration of the original word-length time series is performed prior the application of the FDM.

Figure 3 shows the results of $\langle L(k) \rangle$ $vs.k$ for the ten books in our dataset. After an inspection, it is possible to identify two trends in the slopes that fit the data, that is, for short time intervals, it is observed that the statistics is characterized by the average exponent value of the exponent $D = 1.45 \pm 0.03$, which corresponds to an uncorrelated process, whereas for large scales, the fractal index is $D = 1.25 \pm 0.04$, confirming that the process is persistent with long-term memory. Additionally, the estimation of the local exponent was performed to quantify the asymptotic behavior of this scale invariant. In Figure 4 we present the results of the mean value of the local exponent D^l, we also observe a smooth transition from the uncorrelated regime ($D = 1.5$) until it stabilizes around $D \approx 1.3$, indicating a good concordance with the global exponent. It is also worth to mention that the obtained results for the fractal dimension are in general agreement with the values for the invariants obtained by means of the Allan factor, i.e., both methods confirm the presence of long-term memory in the word-length sequences. More specifically, if use the relationship $\alpha = 3 - 2D$, for $D \approx 1.25$ we get $\alpha^* \approx 0.5$ which is in good concordance with the value $\alpha = 0.57$ obtained by means of the Allan Factor.

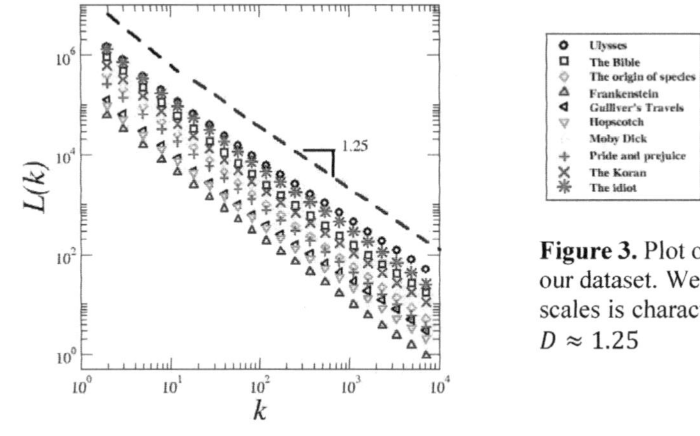

Figure 3. Plot of $L(k)$ $vs.k$ for the ten books in our dataset. We observe that the regime at large scales is characterized by the average exponent $D \approx 1.25$

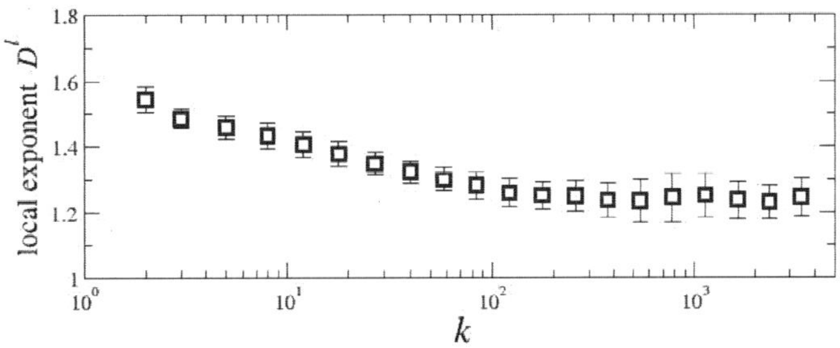

Figure 4. Behavior of the average local exponent D^l in terms of the time interval k. We observe a smooth transition from the uncorrelated regime to the correlated one. We show the mean value ± 1 standard deviation

4. Conclusions

We have presented the Allan factor and fractal dimension analyses for word-length sequences obtained from large literary texts. The interest has focused on the estimation of the stability of the local exponent and its agreement with the global exponent that characterizes the correlations in the long term. It has been shown that the exponent describing the scaling in Allan factor, has an average value $\alpha = 0.57$, which reflects the presence of positive correlations, while the trend of the local index displays a smooth transition from the Poisson regime to the correlated one, achieving stability in the very large scales, but exhibiting a large spread. We also found that the fractal dimension value ($D \approx 1.5$) is consistent with the presence of positive correlations and confirm the results of the Allan statistics. The local behavior of this scaling index at large scales, reflects a stability with a moderate dispersion. This result assigns an advantage to this methodology regarding the estimation of a fractality index for irregular time series. In summary, we have shown that when we observe a word-length sequence as a point process, the Allan factor reflects the presence of scaling with an index that is consistent with the dimensionality of the original series, being more locally stable the exponent extracted by means of the FDM algorithm.

Acknowledgments

This work was partially supported by COFAA-IPN, EDI-IPN and SNI-Conacyt.

References

[1] Nowak M A, Plotkin J B and Jansen V A A 2000 *Nature* **404** 495-498
[2] Solé R V, Corominas-Murtra B, Valverde S and Steels L 2010 *Complexity* **15** 20-26
[3] Grzybek P 2006 History and Methodology of Word Length Studies: The State of the Art *Contributions to the Science of Text and Language (Springer Amsterdam, The Netherlands vol 31)* pp 15-90
[4] Zipf G K 1935 *The Psycho-Biology of Language: An Introduction to Dynamic Philology (MIT Press, Cambridge, MA, USA)*
[5] Piantadosi S T, Tily H and Gibson E 2011 *Proc. Natl. Acad. Sci.* **108** 3526-3529
[6] Garcia D, Garas A and Schweitzer F 2012 *EPJ Data Sci.* **1**:3
[7] Altmann E G, Pierrehumbert J B and Motter A E 2009 *PLoS ONE* **4** e7678
[8] Kosmidis K, Kalampokis A and Argyrakis P 2006 *Physica A* **370** 808-816
[9] Nuñez A, Lacasa L, Valero E, Gómez J P and Luque B 2012 *Int. J. Bifurc. Chaos* **22**
[10] Lacasa L, Nuñez A, Roldán E, Parrondo J and Luque B 2012 *Eur. Phys. J. B* **85**
[11] Petersen A, Tenenbaum J, Havlin S and Stanley H E 2012 *Sci. Rep.* **2**
[12] Kalimeri M, Constantoudis V, Papadimitriou C, Karamanos K, Diakonos F K and Papageorgiou H 2012 *Int. J. Bifurc. Chaos* **22**
[13] Kalimeri M, Constantoudis V, Papadimitriou C, Karamanos K, Diakonos F K and Papageorgiou

H 2015 *J. Quant. Linguist.* **22** 101-118
[14] Rêgo H H A, Braunstein L A, D'Agostino G, Stanley H E and Miyazima S 2014 *PLoS ONE* **9** e110213
[15] Michel J B, Shen Y K, Aiden A P, Veres A, Gray M K, Team T G B, Pickett J P, Hoiberg D, Clancy D, Norvig P *et al.* 2011 *Science* **331** 176-182
[16] Arbesman S, Strogatz S H and Vitevitch M S 2010 *Int. J. Bifurc. Chaos* **20** 679-685
[17] Arbesman S, Strogatz S H and Vitevitch M S 2010 *Entropy* **12** 327-337
[18] Chan K Y and Vitevitch M S 2010 *Cognit. Sci.* **34** 685-697
[19] Samoilenko A, Karimi F, Edler D, Kunegis J and Strohmaier M 2016 *EPJ Data Sci.* **5**:9
[20] Guzmán-Vargas L, Obregón-Quintana B, Aguilar-Velázquez D, Hernández-Pérez R and Liebovitch L 2015 *Entropy* **17** 7798-7810
[21] Rodriguez E, Aguilar-Cornejo M, Femat R and Alvarez-Ramirez J 2014 *Physica A* **414** 378-386
[22] Ausloos M 2012 *Phys. Rev. E* **86** 031108
[23] Allan D W 1966 *Proc. IEEE* **54** 221-230
[24] Higuchi T 1990 *Physica D* **46** 254-264
[25] Guzmán-Vargas L and Angulo-Brown F 2003 *Phys. Rev. E* **67** 052901

Plasmonic resonances in ordered and disordered aluminum nanocavities arrays.

R. G. Campuzano[1], D. Mendoza[2]
Instituto de Investigaciones en Materiales, Universidad Nacional Autónoma de México, 04510, Coyoacán, Ciudad de México.
E-mail: [1]naedra_9999@hotmail.com, [2]doroteo@unam.mx

Abstract. Nanocavities arrays were synthesized by electrochemical anodization of aluminum using oxalic and phosphoric acids as electrolytes. The morphology and topography of these structures were evaluated by SEM and AFM. Plasmonic properties of Al cavities arrays with different ordering and dimensions were analysed based on specular reflectivity. Al cavities arrays fabricated with phosphoric acid dramatically reduced the optical reflectivity as compared with unstructured Al. At the same time pronounced reflectivity dips were detectable in the 300nm-400nm range, which were ascribed to (0,1) plasmonic mode, and also a colored appearance in the samples is noticeably depending on the observation angle. These changes are not observed in samples made with oxalic acid and this fact was explained, based on a theoretical model, in terms that the surface plasmons are excited far in the UV range.

1. Introduction

Nano metallic objects derive their interesting optical properties from an ability to support collective electron excitations known as surface plasmons (SPs). Localized surface plasmon resonance (LSPR) confines light to nanostructures that are smaller than the wavelength of light, thereby enabling a family of novel plasmonic effects, such as extraordinary optical transmission (EOT) in subwavelength hole arrays **[1-4]** caused by a resonant coupling of light and surface plasmons.

In nanoscale systems, the plasmons of silver and gold (the most common plasmonic metals) are only tailored across the visible range, extending into the infrared regions and have their inherent drawbacks limits for extended applications **[5,6]**. Contrastingly, the plasmons of Aluminum are tailored in the ultraviolet–visible–near-infrared (UV–vis–NIR) region of the electromagnetic spectrum. That is why it has been paid attention to this simple metal, because it possesses attractive properties including high natural abundance, low cost, high stability (because of the surface oxide layer); thereby making Al a promising alternative to conventional noble plasmonic materials. Recently, Al plasmonic has been increasingly used in studies and applications for its high performance as biosensors **[7]**, Surface Enhanced Fluorescence (SEF) **[8]**, and Surface Enhanced Raman Scattering (SERS) **[9]**.

In this work we present an electrochemical method **[10,11]** to fabricate regular and irregular arrays of aluminum concaves using oxalic acid and phosphoric acid, and their optical characterization by reflectance measurements. The main results are that plasmon resonances are excited in ordered as well

Content from this work may be used under the terms of the Creative Commons Attribution 3.0 licence. Any further distribution of this work must maintain attribution to the author(s) and the title of the work, journal citation and DOI.
Published under licence by IOP Publishing Ltd

as in disordered array of concaves, and also the existence of colored appearance of the samples; the last result to our knowledge has not been reported yet.

2. Experimental details

High-purity aluminum foil (Sigma-Aldrich 0.25 mm thick, 99.999% purity) was cut in 1cmX2.5cm samples. Before the anodization process the Al foils were mechanically polished (mirror finish) and ultrasonically degreased in deionized water, acetone and ethanol and subsequently electropolished in a 1:4 mixture of 70% $HClO_4$ and 96% CH_3CH_2OH at a constant voltage of 18V for 2 min at 3 ºC. Two kinds of samples using oxalic and phosphoric acids as electrolytes for anodization were used. In one case, 0.3 M oxalic acid solution at 40 V for 7 h and for the second type, 0.1 M phosphoric acid solution at 130V and 160V at ambient temperature, and 150 V and 170 V at 3 ºC for 7 h were employed. A large aluminum plate (99.99% purity) was used as the counter electrode and the distance between both electrodes was kept 1cm apart, and the solution was continuously stirred to homogenize both the electrolyte and its temperature.

After anodizing, the samples were immersed in a mixture of 1.8 wt% CrO_3 and 6 wt% H_3PO_4 at 60 ºC for 12 h to selectively dissolve the anodic aluminum oxide (AAO), with this procedure an array of concaves is observed in the remaining aluminum surface. Morphology of the nanostructured aluminum samples and microanalysis of chemical composition was made using field-emission scanning electron microscope (SEM) JEOL 7600F FESEM equipped with energy dispersive X-ray spectrometer (EDS). Topography analysis was made with an Atomic Force Microscope JEOL Scanning Probe Microscope JSPM-4210.

To obtain geometrical parameters of the fabricated Al concaves Fast Fourier Transforms (FFTs) were generated based on SEM images using WSxM **[12,13]** and Image J software **[14]**. Reflectivity measurement was performed using UV-VIS Spectrophotometer (UV-2600 Shimadzu Corporation). The probe was set at normal angle and reflectance spectra were collected at 190-1400 nm wavelength range at an angle of 8 º. For comparison purposes, electropolished Al foil and Al film obtained by thermal evaporation were also measured.

3. Results and discussion

After anodization and thorough removal of the resulting AAO an array of concaves appear on the Al sample, which is an exact replica of the morphology of the AAO pore bottoms and their size depends on the applied voltage. SEM and AFM images of Al concave arrays fabricated in different electrolytes are shown in Fig. 1. The profiles show that the depths of the concavities made from phosphoric acid at 130V, 150 V, 160 V an 170 V were 210 nm, 250 nm , 310 nm and 360 nm respectively, and for oxalic acid was 20 nm.

To get better statistics on the geometrical parameters of Al concaves, the FFT analysis was performed on SEM images (Fig. 1). Based on FFT, the regularity ratio (RR) is a parameter that is related to the arrangement of concavities, which was estimated according to the following formula **[15]**:

$$RR = \frac{H}{W_{1/2}} \cdot n \qquad (1)$$

where n is the number of concaves on the analyzed image, H the maximal intensity value of the FFT intensity profile, and $W_{1/2}$ is the width of the intensity profile at half of its height. The average interconcave distance (D_c) and pore concave diameter (D_d) were estimated from SEM images for each sample using Image J software. The determined values are presented in Table 1.

IOP Conf. Series: Journal of Physics: Conf. Series **792** (2017) 012077 doi:10.1088/1742-6596/792/1/012077

Table 1. Concave diameter (D_p), interconcave distance (D_c), regularity ratio (RR), and the wavelength where the minimum in reflectance appears (λ) values for the samples anodized with different acids.

Acid	Voltage (V)	Temperature	D_p (nm)	D_c (nm)	RR	λ(nm)
Oxalic	40	ambient	73.3±2.5	114.8±4.4	26.1±0.3	--
Phosphoric	130	ambient	240.9±12.6	357.1±4.4	25.3±0.9	317
Phosphoric	150	3 °C	236.56±89.7	438.0±60.6	13.8±3.5	342
Phosphoric	160	ambient	353.2±27.3	417.1±18.6	25.98±1.6	371
Phosphoric	170	3 °C	358.9±54.8	481.1±61.8	14.3±4.7	395

The FFT images of the samples fabricated at ambient temperature demonstrates six distinct points in the corners of a hexagon, confirming good hexagonal arrangement of concaves. In the case of the samples anodized at 3 °C, the points in the FFT image are more blurred, suggesting lower degree of concaves order. In accordance with **[15,16]** the greater RR a better order in the arrangement of the concavities. As can be observed, the hexagonal concaves arrangement of the samples produced at ambient temperature is much better than the ones fabricated at 3 °C, and this fact is reflected in the corresponding values of RR shown in Table 1.

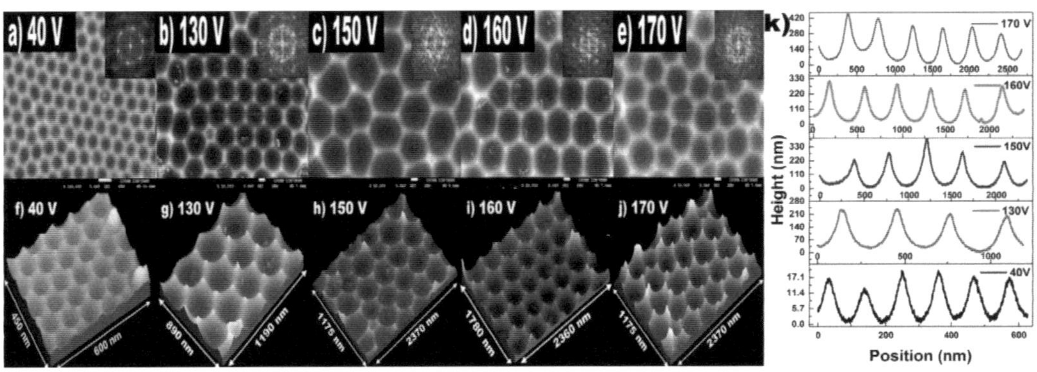

Fig. 1. SEM images of Al concaves fabricated using a) oxalic acid and b)-e) phosphoric acid with their respective FFT images (insets). AFM characterization of aluminum concaves in f) oxalic acid and g)-j) phosphoric acid. k) Concaves profiles obtained by AFM are shown for different acids.

Reflectance spectrum as a function of light wavelength of the Al nanoconcaves are shown in Fig. 2. Results for Al film obtained by thermal evaporation and electropolished aluminum are shown for comparison. A significant dip in the reflectivity intensity at 826 nm (~1.5 eV) is observed in all cases. This feature is related with interband transitions of electrons in the crystalline aluminum **[17,18]**. Additionally, a pronounced dip in the reflectivity in the 300nm-400nm range is observed in all the samples with concaves arrays obtained using phosphoric acid as the electrolyte. Note that the additional dip in the reflectivity of the sample made with oxalic acid does not appear within the worked spectral range.

The minima in UV region is a signature of SPs excitations **[19]**, which means that the incident light couples with surface plasmons located at Al concaves. The concave arrays provide the additional momentum G necessary to fulfill the resonance conditions, giving rise to the dips in reflectivity spectra **[3, 19]**.

The surface plasmon resonances are directly associated with the periodicity of the concave arrays and to the optical properties of Al. At normal incidence, coupling of photons with 2-dimensional hexagonal periodic array gives SP resonances or minimum of the reflectivity at the wavelength given by **[3,20]**:

$$\lambda = \frac{P}{\sqrt{\frac{4}{3}(i^2 + ij + j^2)}}\sqrt{\frac{\varepsilon_m(\lambda)\varepsilon_d}{\varepsilon_m(\lambda) + \varepsilon_d}} \tag{2}$$

where $\boldsymbol{P}$ is the period of the array (in our case interconcave distance D_c), ε_m and ε_d are respectively

the dielectric constants of the metal and the dielectric material in contact with the metal and $\boldsymbol{i}$, $\boldsymbol{j}$ are the scattering orders of the array. Although the eq. (2) was developed for periodic arrays supporting SPPs resonances, to generate the SPPs a long-range order is not necessary; such as are shown in Fig. 2 for samples with disordered array of concaves.

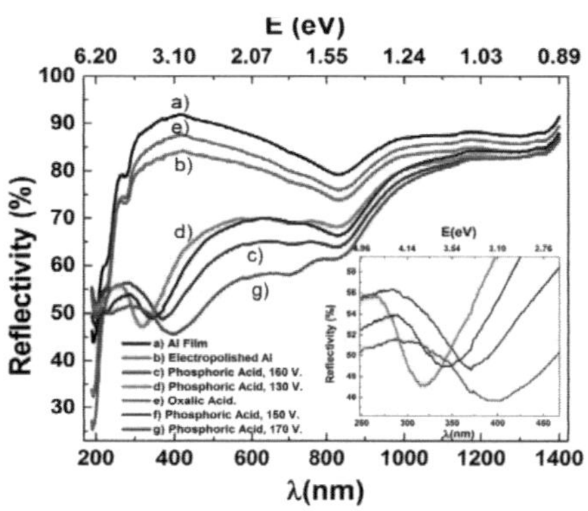

Fig. 2. Reflectivity spectrum of Al concaves fabricated with different acids. Results for electropolished Al foil and Al film obtained by thermal evaporation are shown for comparison.

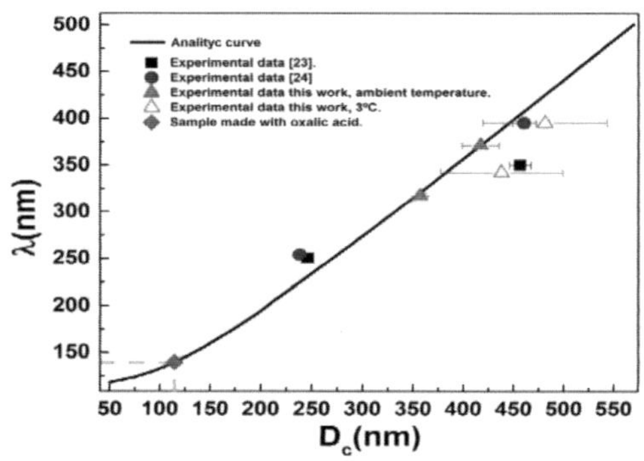

Fig. 3. Comparison of experimental data dip of this work and other works, and analytical curve for the surface plasmon resonance model for the fundamental mode (i, j)=(0,1). The green symbol marks the position for the wavelength where the dip in reflectance would locate for the sample made with oxalic acid.

Fig. 3 shows a comparison of experimental data of the reflectivity dip observed in this work and reported in other works **[23,24]**, and theoretical curve for the surface plasmon resonance model with periodicity described by Eq. (2). For calculations, Eq. 2 was solved using experimental values for the

dielectric function of aluminum $\varepsilon_m(\lambda)$ **[21,22]** and taking $\varepsilon_d = 1$ for air. Experimental data of the samples made with the conditions here presented and also for data reported for other authors fit well with the theoretical curve. Note that the results for the samples with a disordered array of concaves are also well reproduced by the theoretical model, which may be interpreted as that the periodicity is not essential for the excitation of surface plasmons and that the irregularity on the surface is the important characteristic for the momentum conservation rule. As a predicting capability of the model, we can say that for the sample made with oxalic acid with interconcave distance of 114 nm, the plasmonic resonance would locate at a wavelength of around 139 nm; which is out of the experimental range of measurements, and this may be the reason for which no dip is observed for this kind of samples (see Figure 2).

The SPR model as shown in Eq. (2) is limited in that it does not account for variations in metal thickness or hole (concave) size. Resonant optical modes are known to shift to shorter wavelengths and decrease in bandwidth with increasing metal thickness **[4]**. It has also been shown that the resonant optical modes increase in magnitude, shift to longer wavelengths, and increase in bandwidth with an increase in hole size **[25].**

Finally, it is interesting to note that the samples made with phosphoric acid show a colored appearance depending on the observation angle, regardless the order in the concaves array, but not observed in the samples made with oxalic acid (see figure 4). Our hypothesis is that the surface plasmons decay radiatively emitting photons in the visible range in the case of samples made with phosphoric acid, and in the far-UV range for the samples made with oxalic acid; this process because the strong coupling of plasmons and photons through the irregularities on the aluminum surface **[19]**. But this observed phenomenon deserves further careful investigation.

Fig.4. Colored appearance of samples anodized with a) phosphoric acid for different angles and b) oxalic acid.

4. Conclusión

Aluminum concave arrays were prepared by electrochemical anodization using oxalic and phosphoric acid as electrolytes. Geometrical order was determined by FFTs analysis. Plasmonic properties of Al arrays of concaves with different periodicities and dimensions were analyzed based on reflectivity measurements. It was found that, regardless the periodicity, the arrays of Al concaves obtained with phosphoric acid dramatically change the optical reflectivity of the surface compared with flat unstructured Al surface. The modified aluminum surface present a pronounced reflectivity dip as well as a colored appearance. These changes are not observed in samples made with oxalic acid and were explained, based on a theoretical model, in terms that the surface plasmons are excited far in the UV range. The order and geometrical dimensions of the concaves can be simply changed by manipulation of operational parameters, such as voltage, temperature or anodization time.

Acknowledgements. We thank Josue Esau (SEM), Carlos Flores (AFM), M. A. Canseco (UV-Vis) for their valuable assistance and Prof. E. Geffroy for kindly providing us with deionized water, all them at IIM-UNAM.

5. References

[1] Ebbesen T W 1998, et. al. *Nature* **391** 667-9.
[2] Coe J V, Heer J M 2008, et. al. *Annu. Rev.Phys.* Chem. **59** 179–202.
[3] Genet C, Ebbesen T W 2007 Light in tiny holes. *Nature* **445** 39–46.
[4] Martín-Moreno L 2001, et. al. *Phys. Rev.* Lett. **86** 1114–17.
[5] Samuel E and Lohse 2014, et. al. *Chem. Mater.* **26** 34–43.
[6] Matthew N O'Brien 2014, et. al. J. *Am. Chem. Soc.* 136 (21) 7603–06.
[7] Wanbo Li 2016, et. al. *Biosensors and Bioelectronics* 79 500–7.
[8] Mustafa H. Chowdhury 2009, et. al. *Anal. Chem.* 81 (4) 1397–1403
[9] Xiaoyu Zhang 2006. et. al. *J. Am. Chem. Soc.* 128 (31) 10304–09.
[10] Hideki Masuda 1995, et. al. Science **268** 1466-68.
[11] Sousa C T 2014, et. al. *Applied Physics Reviews* **1**, 031102

IOP Conf. Series: Journal of Physics: Conf. Series **792** (2017) 012077 doi:10.1088/1742-6596/792/1/012077

[12] Horcas I 2007, et. al. Rev. Sci. Instrum. 78 8, 013705.
[13] WSxM, http://www.wsxmsolutions.com/
[14] ImageJ, https://imagej.nih.gov/ij/
[15] Wojciech J. 2011 et. al. *Surf. Coat. Technol.* **206** 1416–22.
[16] Wojciech J and Stępniowski.2014 et. al. Materials Characterization. **91** 1–9.
[17] Hughes A. 1969 et. al. *J. Phys. C: Solid State Phys.* **2**, 102.
[18] H. Ehrenreich 1963 et. al. *Phys. Rev.* **132** 1918-28.
[19] Heinz Raether 1988 Surface Plasmons on Smooth and Rough Surfaces and on Gratings. Springer
[20] Tracts in Modern Physics, Vol. 111, Springer Berlin
[20] E.T. Papaioannou, et. al. Opt. Express 19 (2011) 23867-877.
[21] Rakic A D 1998, et. al. *Appl. Opt.* **37** (5271-83.
[22] Weaver J H, et. al. Optical Properties of Metals II: Noble Metals, Aluminum, Lanthanides, and Actinides, 0.1-500 eV, Zentralstelle für Atomkernenergie-Dokumentation, ZAED, 1981, 320 pp.
[23] Małgorzata Norek 2014, et. al. Current Applied Physics **14** 1514-20.
[24] Małgorzata Noreka, 2014 et. al. *Applied Surface Science* **314** 807–14.
[25] Van Der Molen K L 2004, et. al. Appl. Phys. Lett., **85** 4316–18.

Magentite nanoparticle for arsenic remotion.

H Viltres[1, 2], O F Odio [3], R Borja [4], Y Aguilera [2] and E Reguera [1]
[1]Centro de Investigación en Ciencia Aplicada y Tecnología Avanzada-Unidad Legaria, Instituto Politécnico Nacional, Ciudad de México, Distrito Federal, México.
[2]Instituto Superior de Tecnologías y Ciencias Aplicadas, La Habana, Cuba.
[3]Instituto de Ciencia y Tecnología de Materiales, Universidad de La Habana, La Habana, Cuba.
[4]Centro de Nanociencias y Micro-Nanotecnologías, Instituto Politécnico Nacional, Ciudad de México, México.

E-mail: herlysvc.231289@gmail.com

Abstract. Inorganic As (V) and As (III) species are commonly found in groundwater in many countries around the world. It is known that arsenic is highly toxic and carcinogenic, at present exist reports of diverse countries with arsenic concentrations in drinking water higher than those proposed by the World Health Organization (10 μg/L). It has been reported that adsorption strategies using magnetic nanoparticles as magnetite (<20 nm) proved to be very efficient for the removal of arsenic in drinking water. Magnetic nanoparticles (magnetite) were prepared using a co-precipitation method with $FeCl_3$ and $FeCl_2$ as metal source and NaOH aqueous solution as precipitating agent. Magnetite nanoparticles synthesized were put in contact with As_2O_3 and As_2O_5 solutions at room temperature to pH 4 and 7. The nanoparticles were characterized by FT-IR, DRX, UV-vis, and XRF. The results showed that synthesized magnetite had an average diameter of 11 nm and a narrow size distribution. The presence of arsenic on magnetite nanoparticles surface was confirmed, which is more remarkable when As (V) is employed. Besides, it is possible to observe that no significant changes in the band gap values after adsorption of arsenic in the nanoparticles.

1. Introduction

Arsenic poisoning has become one of the major environmental worries worldwide, as millions of people, which have been exposed to high arsenic concentrations (through contaminated drinking water), developed severe health problems [1]. Besides, it ranks twentieth in abundance in the earth's crust, fourteenth in seawater and twelfth in the human body [2]. Both organic and inorganic arsenic exist in natural waters but organic arsenic is of less environmental concern as it undergoes biotransformation and detoxification through methylation. Evidence of chronic arsenicosis has been observed in populations ingesting arsenic contaminated drinking water in many parts of the world [1, 2]. Different reports described that major sources of arsenic are geologic formations such as soil and bedrocks, weathering of rocks, mine tailings, industrial wastes discharge, fertilizers, agricultural use of pesticides, smelting of metals, and burning of fossil fuels [3, 4]. Chronic exposure to arsenic from groundwater has been recognized to cause the largest environmental health disaster in the world, putting more than 100 million people at risk of cancer and other arsenic-related diseases [5].

Varieties of methods have been developed for the removal of arsenic compounds from drinking water including precipitation, adsorption, ion exchange and nanofiltration [6-8]. However, nanomaterials and the nanotechnologies inspire new potential solutions to major environmental issues in actuality; the

Content from this work may be used under the terms of the Creative Commons Attribution 3.0 licence. Any further distribution of this work must maintain attribution to the author(s) and the title of the work, journal citation and DOI.
Published under licence by IOP Publishing Ltd

removal of arsenic by nanoparticles of iron oxide has shown promising results due to their small particle size and large surface area [9, 10]. Oxide mixtures, which employ magnetite nanoparticles, have allowed to remove arsenic species from aqueous medium [9]. It is expected that the use of pure magnetite nanoparticles (< 20 nm), allow such removal and also, it makes possible the study of a possible mechanisms of interaction between arsenic species and the surface of the nanoparticles.

Adsorption experiments were conducted using magnetite synthesized, with an average diameter of 11 nm. The overall purpose of this study is to corroborate that magnetite with naked surface can remove arsenic spices form water. In this sense, our attempts are primarily based on the results of FT-IR spectroscopy, although other complementary techniques were also used (mainly UV-vis spectroscopy and XRF).The developed methodology could be implemented in the water treatment industries, reducing the costs of the processes and making them more environmental friendly.

2. Experimental

2.1 Chemicals and materials. The chemicals used in this study were mostly reagent with high analytical grade and used as received (Sigma-Aldrich) without further treatment. $FeCl_3.6H_2O$ (>97%), $FeCl_2.4H_2O$ (≥99 %), NaOH (>98%), $As_2O_3.xH_2O$ (≥99 %) and As2O5 (≥99.99 %). All chemical solutions were prepared with deionized water. The As (III) and As (V) stock solutions were prepared by dissolving arsenic oxides (As_2O_3 and As_2O_5) powder in de-ionized water, using 4 g/L NaOH since both oxides have enhanced solubility in NaOH solution.

2.2 Synthesis of magnetite nanoparticles. Fe_3O_4 was synthesized from co-precipitation method which involves the simultaneous precipitation of ions Fe^{2+} and Fe^{3+} in a molar ratio of 1:2 respectively. The precipitation was performed by addition of base (NaOH) at room temperature. The reaction is shown in the following equation:

$$2FeCl_3 + FeCl_2 + 8NaOH \rightarrow Fe_3O_4\,(s) + 4H_2O + 8NaCl$$

To the reaction was used $FeCl_2.4H_2O$ (0.01 mol) and $FeCl_3.6H_2O$ (0.02 mol) dissolved in 100 mL of distilled H_2O. While stirring mixture vigorously was passed a stream of N2 at a constant flow to prevent oxidative atmosphere. NaOH (0.09 M) aqueous solution was added drop wise, as precipitant agent. After 3 hours of reaction, a black precipitate was observed, which was separated magnetically from the supernatant. Black solid was washed three times with nitric acid (HNO_3) and dried under vacuum for 24 h at 80 °C.

2.3 Characterization.

Fourier transformed infrared spectroscopy (FT-IR) was employed to analyse the structural characteristic of particles. IR spectra were recorded in transmission mode after making pellets with KBr powder with a with a Perkin Elmer Spectrum One FT-IR spectrometer in transmission mode with 16 scans at a resolution of 2 cm^{-1} in the range of 400–4000 cm^{-1}.

X-ray diffraction (XRD) powder pattern of Fe_3O_4 nanoparticles was recorded with a Bruker D8 Advance Diffractometer and CuK_α radiation (1.54183 Å). The average crystallite size was estimated from the integral intensity of the X-ray diffraction peak maximum using the Scherer equation (1):

$$D_c = \frac{k * \lambda}{\beta * \cos\Theta} \qquad (1)$$

where k is the shape factor, λ is the X-ray wavelength (0.15418 nm), β is the full width at half-maximum expresses in unit of 2, and Θ is the Bragg angle (^{0}).

The optical properties of these samples were studied using UV-visible spectroscopy technique, the UV-vis spectra were recorded on a BDA 950 UV/Vis/NIR spectrophotometer, and the software computed the Kubelka-Munk function, the range of a wavelength of work was 800 – 200 nm.

The elemental composition of samples were developed by X-ray Fluorescence (XRF) and employing Bruker AXS brand equipment S2Ranger model. In this study aluminium filter was employed.

2.4 Adsorption of As (III) and As (V) by naked Fe_3O_4 nanoparticles. (Batch adsorption experiment).

As (III) and As (V) stock solutions were prepared to concentrations of 500 ppm and 300 ppm respectively, at pH=7. Adsorption studies were performed to pH=4 and pH=7. The pH values were adjusted with acid (2M HCl) and basic (1 M NaOH). Then 15 ml of arsenic aqueous solution was added to 15 mg of nanoparticles under stirring and at room temperature, 25 °C. After a specific time elapsed (12 h), the solid was separated from the liquid by magnetic field action or centrifugation. Nanoparticles are washed three times with deionized water and ethanol, and then it were dried at room temperature for further studies.

3. Results and Discussion.

3.1 Structural characterization of Fe_3O_4.

FT-IR spectroscopy was used to characterize the as-prepared magnetite nanoparticles. **Figure 1** shows the spectra of magnetite. The spectrum of pure magnetite (**Figure 1**) presents a wide band at 3432 cm^{-1} and a small in 1622 cm^{-1}, which can be attributed to vibrations $\nu_{s(OH)}$ and dubbing in the plane $\delta_{(H\text{-}O\text{-}H)}$, respectively, by water molecules adsorbed onto the surface [12, 13]. Beside, one peak is observed at 570 cm^{-1}, which can be assigned to the stretching mode of Fe–O in Fe_3O_4 [12]. The bands at 1384 and 1230 cm^{-1} correspond to stretching vibrations ($\nu_{(NO3-)}$ and $\nu_{(ON)}$) due to NO_3^- ions adsorbed on the surface of the nanoparticles, because the final wash is performed with nitric acid [13].

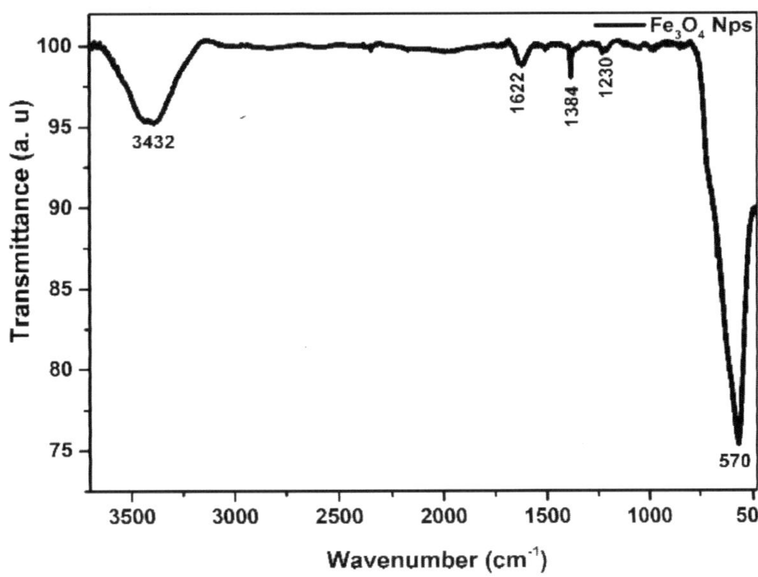

Figure 1. FT-IR spectrum of magnetite nanoparticles.

Diffraction pattern of the dried powder of magnetite nanoparticles are shown in **Figure 2**. It can be seen that, the sites and intensity of the diffraction peaks are consistent with the standard patterns for Fe_3O_4 (ICSD No. 08-4611). Cubic single phases for Fe_3O_4 powders have been obtained, which is well consistent with the FT-IR results. Moreover, the shape of the peaks in the diffractogram and the good definition it can say that the sample is crystalline. In addition, according to Debye-Scherrer equation the average particle diameter of magnetite was calculated to be 11.3 nm.

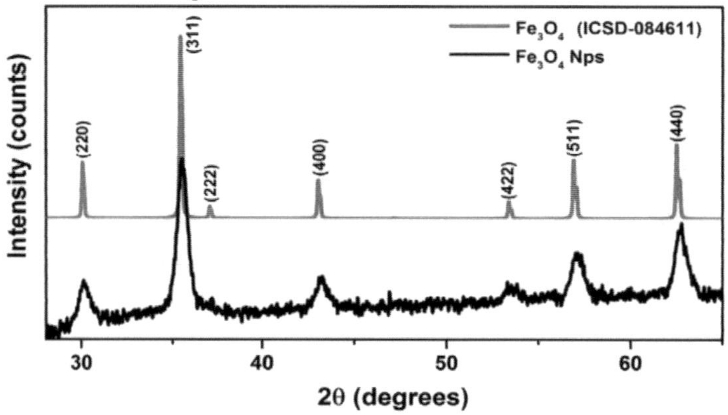

Figure 2. XRD pattern for magnetite nanoparticles.

3.1 Optic properties (UV-vis).

Figure 3 shows the measurements of absorbance in function of wavelength for magnetite nanoparticles in UV-vis zone. It is possible to observe an adsorption band at 300 nm, which is characteristic of this material, corresponding to transitions of crystalline field; in this cause due to transitions from t2g to eg orbitals (in octahedral positions) and from eg to eg orbitals (for tetrahedral positions) in Fe^{2+} y Fe^{3+} cations [12].

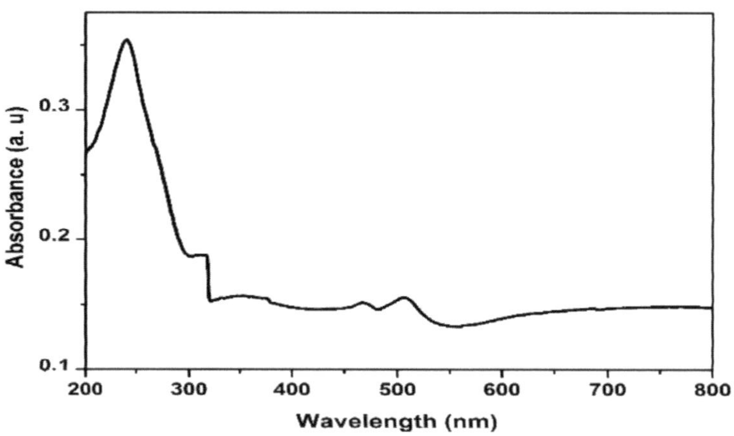

Figure 3. Absorbance in function of wavelength of pure magnetite nanoparticles.

The results of calculations for optic energy of band gap are shown in **Figure 4**, for direct and indirect transitions were, 1.5 and 0.8 eV, respectively. Both band gap values, for powder are high for this material when are compared with the value for bulk [14, 15]. This widening of the band gap is a quantum confinement effect that has been observed in many semiconductor materials having delocalized electron states near the Fermi level [15]. These results for the band gap energy of both direct and indirect transitions for magnetite nanoparticles, can permit classify this sample as a semiconductor, since it has a band gap energy between 0-3 eV [16].

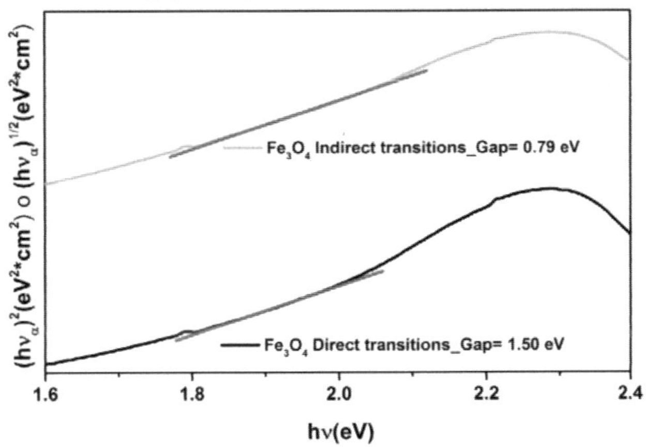

Figure 4. Determination of optic energy of band gap, for direct (E_{g1}^{opt}) and indirect (E_{g2}^{opt}) transitions.

3.2 X-Ray Fluorescence.

By XRF analysis is possible evidence the arsenic presence onto nanoparticles surface. **Table 1** show the semiquantitative estimation of arsenic after adsorption experiments, for both pH in study. From the analysis it was evidenced that in the sample there is a significant percent of arsenic, which corroborates the presence of this element in all materials. In this case, the adsorption of As (III) is favored at pH = 7, probably because in these conditions neutral species of As (III) (H_3AsO_3) are formed, while the surface of the nanoparticles is negatively charged, because at pH above the isoelectric point materials exhibit this behavior, being the case of magnetite as this pH is 6.5 [17]. Moreover, in the case of As (V), the adsorption is higher at pH = 4; in these conditions the surface of the nanoparticles is positively charged and prevail neutral species of As (V) ($H_2AsO_4^-$), which will be attracted to the oxide surface more easily, because by possessing different charges favors attraction between the species and the area. In both cases the content of Arsenic is close to 9 %.

Table 1. Semiquantitative estimation of Arsenic onto magnetite nanoparticles.

Nanopart*icles*	***Weight* (%)**
	As3d
Fe_3O_4	
Fe_3O_4 + As(III) pH= 4	6,0
Fe_3O_4 + As(III) pH= 7	9,9
Fe_3O_4 + As(V) pH= 4	7,8
Fe_3O_4 + As(V) pH= 7	5,2

3.3 Study of optic properties for magnetite nanoparticles with Arsenic onto surface.

Table 2 shows the results of band gap for direct and indirect transitions in magnetite sample at different conditions of study. As expected, no significant changes are observed in the band gap values after adsorption of arsenic in the nanoparticles, because a simple adsorption of arsenic cannot produce changes in the optical energy of the material. Therefore, it is confirmed that adsorption of arsenic species on the nanoparticles is accomplished by forming new surface species.

Tabla 2. Band gap values for direct and indirect transition in magnetite nanoparticles with and without arsenic onto surface.

Materials	Band Gap (Direct transitions) (eV)	Band Gap (Indirect transitions) (eV)
Magnetite	1.5 ± 0.1	0.8 ± 0.1
Magnetite + As(III) pH= 4	1.4 ± 0.1	0.7 ± 0.1
Magnetite + As(III) pH= 7	1.3 ± 0.1	0.6 ± 0.1
Magnetite + As(V) pH=4	1.3 ± 0.1	0.7 ± 0.1
Magnetite + As(V) pH= 7	1.2 ± 0.1	0.6 ± 0.1

Conclusion

Magnetite nanoparticles with a spherical morphology were obtained, with a diameter of 11 nm, and it was confirmed that this nanoparticles can remove arsenic species from aqueous medium. Beside, was evidenced that arsenic onto nanoparticles surface cannot changes the values of optic energy (band gap).

References

[1] Shih M C 2005 *Desalination* **172** 85-97
[2] Choong S Y et al 2007 *Desalination* **217** 139-166
[3] Lièvremont D et al 2003 *Chemosphere* **51** 419-428
[4] Tuulaikhuu B A et al 2015 *Aquatic Toxicology* **166** 72-82
[5] Shen S et al 2013 *Chemical reviews* **113** 7769-7792
[6] Bissen M and Frimmel F H 2003 *Acta hydrochimica et hydrobiology* **31** 97-107
[7] Kanel S R et al 2005 *Environmental science & technology* 39 1291-1298
[8] Feng L et al 2012 *Journal of hazardous materials* **217** 439-446
[9] Rahaman M Basu A and et al 2008 *Bioresource technology* **99** 2815-2823
[10] Shen S et al 2013 *Chemical reviews* **113** 7769-7792
[11] Cornell R M and Schewertmann U 2003 *Wiley-VC* **Second Edition** 1-659
[12] Cai W et al 2007 *Journal of Colloid and Interface Science* 305 366-370
[13] Araújo A C et al 2010 *Synthetic Metals* **160** 685–690
[14] Demortiere A et al 2011 *Nanoscale* **3** 225
[15] Diasty F et al 2009 *Solid State and Materials Science* **13** 28-34
[16] Ghandoor H et al 2012 Int. J. Electrochem. Sci. **7** 5734-5745
[17] Welch A H et al 2000 *Ground Water* **38** 589-604

Acknowledgments

We appreciate the Conacyt and Pifi economic support. We also want to thanks to CNyN, México for characterization techniques.

Evolving aerodynamic airfoils for wind turbines through a genetic algorithm

J J Hernández[1], E Gómez[2], J I Grageda[1], C Couder[1], A Solís[1], C L Hanotel[3] and JI Ledesma[2]

[1] Centro de Desarrollo Aeroespacial, Instituto Politécnico Nacional. Belisario Domínguez 22, Centro, Ciudad de México, 06610, México.
[2] Facultad de Ciencias, Universidad Nacional Autónoma de México. Circuito Exterior, Ciudad Universitaria, Coyoacán, Ciudad de México, 04510, México.
[3] Instituto de Ciencias Nucleares, Universidad Nacional Autónoma de México. Apdo. Postal 70-543, 04510, Ciudad de México, México.

E-mail: ccouder@ipn.mx

Abstract. Nowadays, genetic algorithms stand out for airfoil optimisation, due to the virtues of mutation and crossing-over techniques. In this work we propose a genetic algorithm with arithmetic crossover rules. The optimisation criteria are taken to be the maximisation of both aerodynamic efficiency and lift coefficient, while minimising drag coefficient. Such algorithm shows greatly improvements in computational costs, as well as a high performance by obtaining optimised airfoils for Mexico City's specific wind conditions from generic wind turbines designed for higher Reynolds numbers, in few iterations.

1. Introduction

The efficiency of a wind turbine depends heavily on the geometry of the airfoils used to design the blades. Historically, AO is a process that has been mainly performed in a trial and error basis in wind tunnels [1]. Actually, as many other areas of science, aerodynamics has been benefited from the vertiginous advancements in computing power. These advantages have taken two different directions:

- In CFD, to simulate more realistic flows, or flows not achievable in wind tunnel facilities.
- In the area of AO to increase their desirable physical features and response.

In the subject of CFD, there has been an enormous amount of research [2, 3, 4, 5] that has led to several open source, proprietary as well as commercial software to simulate almost any flow, for instance, flows in a wide range of Reynolds numbers.
The AO is part of what constitutes actually the field of CA, a branch of the aifoil design process [6, 7, 8]. Despite the existence of several computational techniques to optimise airfoil, most of them heavily relay on the analytical theory of aerodynamics; also, some of them are designed specifically to optimise the performance of airfoils used as wings for airplanes, helicopters, etc. The main difference between an airfoil used as a wing and those used in wind turbines is that they appear inverted in the blade. This is due to the fact that the lift force they provide in a wing, when inverted, provides the torque necessary to rotate the turbine. Some of the traditional methods to optimise airfoils are the ones that use differential gradients of functions representing the parameters to optimise; this way, one is able to modify iteratively the parameters or variables of the design being performed. This allows to map systematically the space of solutions, but they are slow to converge and are not reliable when the optimization problem is complex due to a multi-modal behaviour, non-continuities or non-

Content from this work may be used under the terms of the Creative Commons Attribution 3.0 licence. Any further distribution of this work must maintain attribution to the author(s) and the title of the work, journal citation and DOI.
Published under licence by IOP Publishing Ltd

differentiabilities in the physical behaviour of the parameters, etc. Due to such slowness, some efforts have focused in reducing execution times [9]; however, is not a common target in their development. Modern algorithms that overpass these difficulties are evolutive algorithms [9, 10, 11, 12, 13, 14, 15].

In this work we propose a novel genetic algorithm that relies on the arithmetic of lineal combinations of the theory of algebra. This algorithm is used to obtain airfoils with new geometries and with superior aerodynamic features by taking the crossover of previous airfoils already optimised for their usage in wind turbines. This set of parent-airfoils will produce son-airfoils; the latter are analysed in a flow which reproduces the mean conditions of air flows in Mexico City. The son-airfoils overpassing the efficiency criteria are taken as the new optimal airfoils.

This paper is organised as follows: in section 2 we briefly describe the algorithm we propose to obtain new airfoils through crossover techniques. In section 3 we present the most relevant results we obtained along with their corresponding aerodynamic analyses. In the last section, we present a discussion of the results as well as the main achievements in computational and human times in AO.

Listing 1. Pseudo-code for the proposed genetic algorithm

- Define a set of linearly independent one-dimensional functions.
- Generate a new bigger set of functions by performing all the possible compositions of the initial set.
- Take two parent airfoils previously known to be optimal (for wind turbines).
- Apply the crossover operator (see eq. (1)) to both parents with each of the functions previously defined, so to obtain son-airfoils.
- Do until N generations final fitness criteria met:
  * Apply fitness criteria to each airfoil in current generation.
  * Get the two fitness individuals and set them as parents.
  * Apply crossover operator to both parents and replace current generation with new airfoils.

2. Genetic algorithm

The optimization method is based on a genetic algorithm, which starts with a selected initial population, to then proceed on applying cross-over and/or mutation operators. These operators generate a new population (generation) over which, again, fittest individuals are taken. The approach taken varies in the standard genetic algorithm in that an initial set of non-random individuals are taken.

Airfoils are defined by their x, y coordinates and their chord's length c, so we propose the crossing-over operator to be a linear combination of a pair of individuals as

$$f_i y_{p1} + f_j y_{p2} = y_s \tag{1}$$

where y_{p1}, y_{p2} and y_s are the y coordinate of both of the parent-airfoils and of the resulting son-airfoil respectively, $\{f_k\}_{k=1}^{n}$ is a set of n continuous and differentiable functions such that $f_k:[0,1]\rightarrow[0,1]$, and with the constriction $f_k + g_k = 1$ so to avoid extremely thick airfoils. In listing 1 we show the pseudo-code of this algorithm.

1.1. Airfoil input for the algorithm

There exist a huge amount of airfoils, each specially designed for specific purposes, which can be found on databases and in the literature [17, 18, 19, 20, 21]. We took the set of parent-airfoils for wind turbines from the S-series provided in the public database of the NREL, designed to exhibit a maximum C_L while being relatively insensitive to roughness effects. We particularly chose an initial parent-airfoil set of 2 individuals, S809 and S804 airfoils (which can be appreciated in figs. 1 and 2), whose aerodynamic features are shown in table 1 and 2.

Table 1. Nomenclature and units.

Symbols		Abbreviations	
C_L	Lift coefficient (adimensional)	CFD	Computational Fluid Dynamics
C_D	Drag coefficient (adimensional)	CA	Computational Aerodynamics
C_P	Pressure coefficient (adimensional)	AO	Airfoil Optimization
C_M	Moment coefficient (adimensional)		
C_L/C_D	Aerodynamic efficiency parameter (adimensional)		

Table 2. Top aerodynamic parameters for the selected parent airfoil set (see figs. 1 and 2).

Airfoil	C_L	C_D	C_M	C_P	C_L/C_D
S809	1.0271 (α = 14.7)	0.0166 (α = 7.8)	-0.0019 (α = 14.5)	-6.445 (α = 15.6)	58.920 (α = 7.8)
S814	1.3743 (α = 10.2)	0.0202 (α = 6.9)	-0.0842 (α = 13.4)	-6.081 (α = 18.3)	64.964 (α = 9.6)

To increase the crossing-over possibilities, we actually define an initial set of functions with the above features, so $\{f_k\}_{k=1}^{n}$ is the set of all the possible compositions between them that preserve such properties. Our initial set of functions is $\left\{\sqrt{x}, e^x, \ln x, \operatorname{sech} x, \tanh x, \sin x, \cos x, \arctan x, J_4, J_5\right\}$, where J_k is the Bessel function of k order.

Table 3. Mexico City mean wind parameters

Parameter	Mean speed (m/s)	Density (kg/m^3)	Temperature (C)	Kinematic viscosity (Pa s)
Value	14.816	0.98	20	$1.845\text{x}10^{-5}$

3. Aerodynamics of optimised airfoils

For the purpose of this work, we simulated in XFOIL software [16] an air flow corresponding to the mean conditions in Mexico City, which are summarised in table 3.

We obtain the performance features of the parent-airfoils set as well as the best sons up to the tenth generation. The fitness criteria is set as to overcome the average of both the aerodynamic efficiency C_L/C_D and the lift coefficient C_L of parents, while not overcoming the average of C_D of parents. In figures 1 and 2, we show the coordinates of the two best son-airfoils obtained, called son1 and son2.

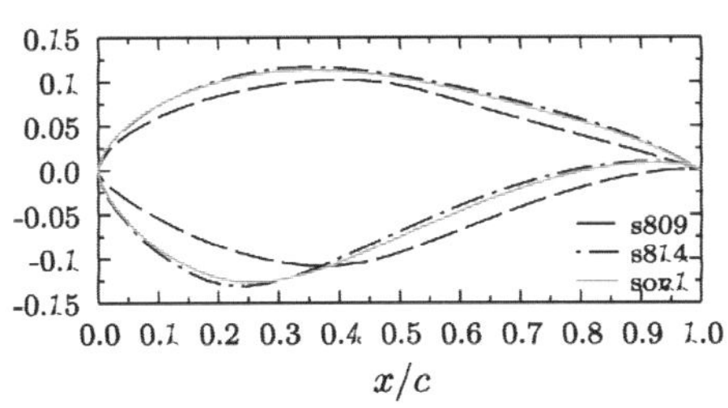

Figure 1. Shape of son1 optimised airfoil with respect to the original airfoils.

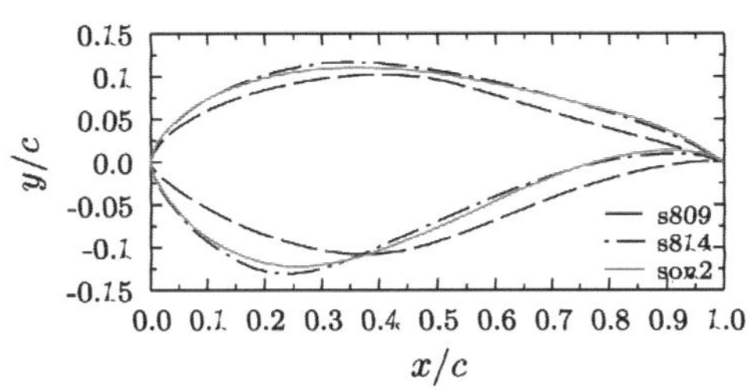

Figure 2. Shape of son2 optimised airfoil with respect to the original airfoils.

In figures 3, 4, 5 and 6 we show the plots of the aerodynamic efficiency parameter, the lift coefficient, the drag coefficient and the pressure coefficient against the attack angle, from where it is possible to observe the improvements we had in such parameters up to the tenth generation of airfoils.

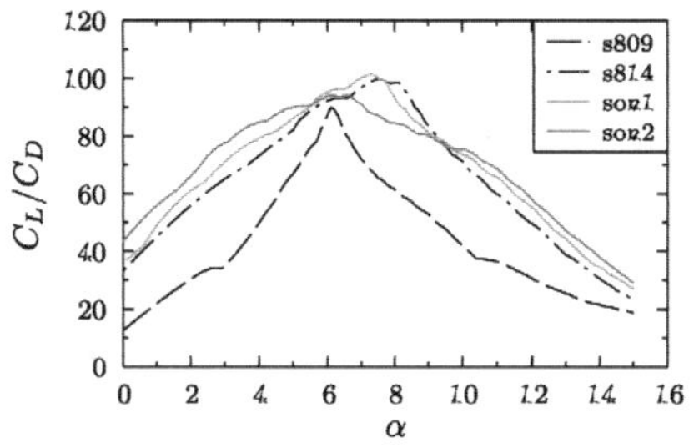

Figure 3. Efficiency parameter of the 2 optimised airfoils with respect to that of their parents.

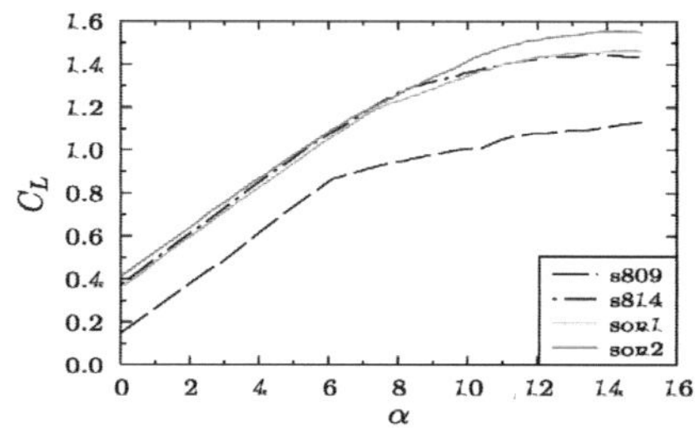

Figure 4. Lift coefficient of the 2 optimised airfoils with respect to that of their parents.

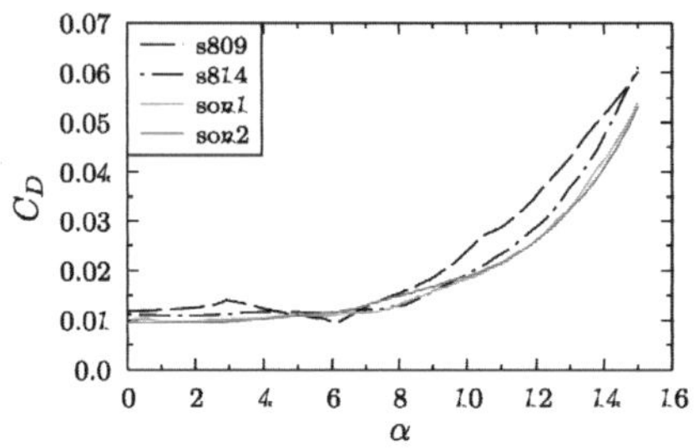

Figure 5. Drag coefficient of the 2 optimised airfoils with respect to that of their parents.

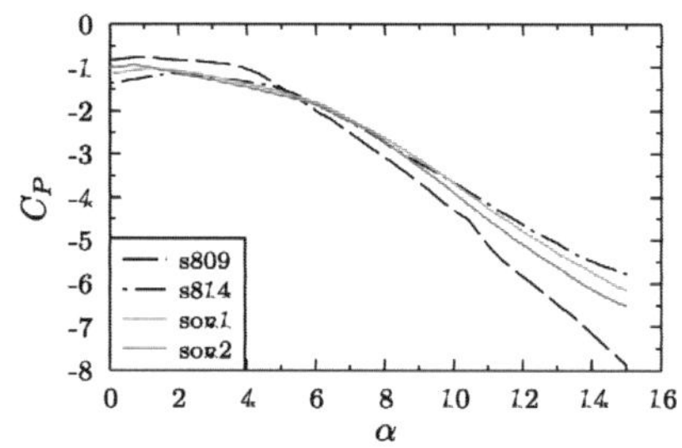

Figure 6. Pressure coefficient of the 2 optimised airfoils with respect to that of their parents.

4. Conclusions

From the two specially-designed wind turbine initial airfoils (optimal to be driven by a 100,000 Re flow), we executed 10 generations of son-airfoils in order to observe improvements in the aerodynamics of optimised airfoils. With the proposed genetic crossover rule, as early as in the tenth generation we found son-airfoils which perform better in the low angle of attack range, as can be observed from figures 3 to 6; despite son1 have a max aerodynamic efficiency greater than both parents, in the range $0 \leq \alpha \leq 6$ the performance of son2 is better than both son1 and the original parents. At this stage, son1 and son2 are the best sons of tenth generation, which already possess superior aerodynamic features for their applications on wind turbines at low Reynolds numbers as those of Mexico City. To obtain optimised airfoils with just 10 iterations is something remarkable, as other genetic algorithms report improvements only with many more iterations [22, 23], saving computing time just by taking suitably the set of parent-airfoils. It is also remarkable to be able to drive an AO process with mono-CPU programming technique. Further research on the parallel implementation of this algorithm must be conducted in order to increase the sets, both of parent-airfoils as well as of functions to augment the crossing-over possibilities so to explore a wider space of solutions to the AO problem.

Acknowledgments
JJHG and CCC acknowledge EDI grant by SIP-IPN. This research was partially supported by projects SIP 20160105 and SIP 20160576, IPN, as well as by "Caracterización del diseño, construcción y prueba de equipos aerogeneradores" CDA project.

References

[1] Jameson A and Ou K 2011 *Progress in Aerospace Sciences*, **47**:5 308
[2] Wendt JF, Anderson Jr JD, Degroote J, Degrez G, Dick E, Grundmann R and Vierendeels J 2009 *Computational Fluid Dynamics An Introduction* (Springer-Verlag Berlin Heidelberg)
[3] Tu J, Yeoh GH and Ph CL 2008 *Computational Fluid Dynamics: A Practical Approach* (Butterworth-Heinemann)
[4] Woo H 2010 *Computational Fluid Dynamics* InTech
[5] Blazek J 2015 *Computational Fluid Dynamics: Principles and Applications* (Elsevier Science)
[6] Selig MS and McGranahan BD 2004 *Journal of solar energy engineering* **126**:4 986
[7] Cambridge NJ 2004 *Aerodynamic shape optimization of complete aircraft configurations using unstructured grids*
[8] Jameson A, Kim S, Shankaran S and Leoviriyakit K 2003 *Aerodynamic shape optimization: Exploring the limits of design* (Proceedings of KSAS)
[9] Leskinen J and Périaux J 2013 *Computers & Fluids* **80**:10 190
[10] Eiben A and Smith J 2015 *Introduction to Evolutionary Computing* (Natural Computing Series, Springer Berlin Heidelberg)
[11] Kostas KV, Ginnis AI, Politis CG and Kaklis PD 2016 "Shape-optimization of 2D hydrofoils using an Isogeometric BEM solver", *Computer-Aided Design* In Press
[12] Peigin S and Epstein B 2004 *Computers & Fluids* **33**:9 1175
[13] Vučina D, Marinić-Kragić I and Milas Z 2016 *Renewable Energy* **87**:2 849
[14] Di Angelo L and Di Stefano P 2014 *Advances in Engineering Software* **67** 164
[15] Epstein B, Peigin S and Tsach S 2006 *Aerospace Science and Technology* **10**:2 100
[16] Drela M 1989 *Xfoil: An analysis and design system for low Reynolds number airfoils* (Low Reynolds number aerodynamics, Springer)
[17] Selig M, Guglielmo J, Broeren A and Giguere P *Summary of Low-Speed Airfoil Data, Volume 1* (SoarTech Publications)
[18] Selig M, Guglielmo J, Broeren A and Giguere P *Summary of Low-Speed Airfoil Data, Volume 2* (SoarTech Publications)
[19] Selig M, Guglielmo J, Broeren A and Giguere P *Summary of Low-Speed Airfoil Data, Volume 3* (SoarTech Publications)
[20] Selig MS and McGranahan BD 2004 *Wind tunnel aerodynamic tests of six airfoils for use on small wind turbines* (National Renewable Energy Laboratory)
[21] Williamson GA, McGranahan BD, Broughton BA, Deters RW, Brandt JD and Selig MS *Summary of Low-Speed Airfoil Data, Volume 5* (SoarTech Publications)
[22] Arias A 2007 *Diseño de perfiles aerodinámicos para aplicaciones a bajos números de reynolds mediante el uso de algoritmos evolutivos* (CIBIM 8, Octavo Congreso Iberoamericano de ingeniería Mecánica)
[23] Arias A, Coello C A, and Mezura A 2011 *Computational Optimization, Methods and Algorithms* Springer

IOP Conf. Series: Journal of Physics: Conf. Series **792** (2017) 012080 doi:10.1088/1742-6596/792/1/012080

Gamma radiation-induced synthesis and characterization of Polyvinylpyrrolidone nanogels

A A Ges [1], H Viltres [1, 2], R Borja [3], M Rapado [4] and Y Aguilera [1]

[1]Instituto Superior de Tecnologías y Ciencias Aplicadas, Departamento de Radioquímica, La Habana, Cuba.
[2]Centro de Investigación en Ciencia Aplicada y Tecnología Avanzada-Unidad Legaria, Instituto Politécnico Nacional, Ciudad de México, Distrito Federal, México.
[3]Centro de Nanociencias y Micro-Nanotecnologías, Instituto Politécnico Nacional, Ciudad de México, Distrito Federal, México.
[4]Centro de Aplicaciones Tecnológicas y Desarrollo Nuclear, La Habana, Cuba.

E-mail: ages@instec.cu, herlysvc.231289@gmail.com

Abstract. Due to the importance of bioactive peptides, proteins and drug for pharmaceutical purpose, there is a growing interest for suitable delivery systems, able to increase their bioavailability and to target them to the desired location. Some of the most studied delivery systems involve encapsulation or entrapment of drugs into biocompatible polymeric devices. A multitude of techniques have been described for the synthesis of nanomaterials from polymers, however, the use of ionizing radiation (γ, e-), to obtain nano- and microgels polymer is characterized by the possibility of obtaining products with a high degree of purity. Although, in the world, electronic radiation is used for this purpose, gamma radiation has not been utilized for these purposes. In this paper is developed the formulation the formulation of Polyvinylpyrrolidone (PVP) nanogels synthesized by gamma radiation techniques, for their evaluation as potential system of drug delivery. Experiments were performed in absence of oxygen using aqueous solutions of PVP (0.05% -1%). Crosslinking reactions were carried out at 25 ° C in a gamma irradiation chamber with a ^{60}Co source (MPX-γ 30). The Viscosimetry, Light Scattering, X-Ray Diffraction and Transmission Electron Microscopy (TEM), were used as characterization techniques.

1. Introduction

In the recent years, an important factor stimulating research on radiation-induced reactions of polymers in aqueous solution is the application of this technique for synthesizing new materials or modifying the properties of polymers [1], a significant advantage of the use of radiation is that the processes can be considered "solvent free" respect to conventional synthesis procedures thereby reducing or avoiding the presence of additives that could be potentially harmful. The use of radiation has involved, as might be expected, the field of nanotechnologies in the production of new materials and structures that have both fundamental interest and potential applications in areas such as biomedical sciences, electronics, optics, and material sciences [2, 3]. In this paper is develop the synthesis of Polyvynilpyrrolidone (PVP) nanogels by ionizing radiation techniques (gamma radiation), for their evaluation as potential system of drug delivery.

Content from this work may be used under the terms of the Creative Commons Attribution 3.0 licence. Any further distribution of this work must maintain attribution to the author(s) and the title of the work, journal citation and DOI.
Published under licence by IOP Publishing Ltd

Polymeric micro and nanogels are probably the most prominent examples of waterborne products obtained by radiation technology. They have a broad field of actual and potential applications ranging from filler materials in coating industry to modern biomaterials [3, 4]. Nanogels are promising novel pharmaceutical carriers for small biologically active agents and biomacromolecules. The advantages of these systems include simplicity formulation with the drugs, high loading capacity, and stability of the resulting formulation in dispersion. These systems allow immobilization of biologically active compounds of diverse structure including charge drugs, low molecular mass hydrophobes, and biopolymers. Furthermore, nanogels can be chemically modified to incorporate various ligands for targeted drug delivery. The in vivo and vitro studies suggest that nanogels can be used for efficient delivery of biopharmaceuticals in cell as well for increasing drug delivery across cellular barriers [5-7].

A multitude of techniques have been described for the synthesis of polymeric micro and nanogels, however, the use of ionizing radiation (γ, e-), to obtain nano- and polymer microgels, is characterized by the possibility of obtaining products with a high degree of purity [5]. The main factors that contribute to the success of these technologies are: water as a human-and environment-friendly solvent, lack of any potentially harmful chemicals (monomers, initiators, crosslinking agents, etc.), simple production schemes, parallel synthesis and sterilization of the products in the sealed, final packages, valuable medical properties of the products, broad range of applications, and the possibility of manufacturing "smart" gels reacting to temperature, pH, light, etc [8-10].

Particulars, the polyvynilpirrolidone nanogels can be obtained by gamma radiation, based not on polymerization, but on intramolecular crosslinking of polymers chains, in aqueous solutions. The nanogel characterization was performance by Viscosimetry, Light Scattering, X-Ray Diffraction and Transmission Electron Microscopy (TEM). The results showed that in dependence on the polymer concentration and the dose rate two different crosslinking reactions can be obtained. Irradiation experiments at room temperature in diluted solution with further increasing of the radiation dose lead to the formation of PVP nanogels due to an intramolecular crosslinking reaction. By using both light scattering and TEM the PVP nanogels were measured, these showed a size distribution of 50.10 nm. They have lots of advantages over conventional systems since they enhance the delivery, extend the bioactivity of the drug by protecting them from environmental effects in biological media, show minimal side effects, demonstrate high performance characteristics, and are more economical since minimum amount of expensive drugs are used.

2. Experimental

2.1 Chemicals and materials. The chemicals used in this study were mostly reagent with high analytical grade and used as received (Sigma-Aldrich) without further treatment. Polyvinylpyrrolidone-PVP, type K-90 (> 99.99). All chemical solutions were prepared with water purified by distillation and in order to remove any dust particles and/or polymer aggregates, solutions were filtered subsequently through filters of 0.45 and 0.22 μm pore size. Gamma irradiation were carried out with a panorama ^{60}Co source at 15 kGy dose.

2.2 Synthesis of PVP nanogels. The synthesis was carried out with a panorama ^{60}Co source of Hungarian manufacturing and a homogeneity 1.3 at room temperature. The irradiation process was monitored by Fricke dosimetry and film dosimeters using Perpex. Irradiation times were determined by calculation CaliPMMA platform, developed at the Laboratory of High Dose Dosimetry Centre for Technological Applications and Nuclear Development (CEADEN) [11]. Previously the samples were collected in glass volumetric -100 and 50mL- and saturated for 10 minutes under argon (**Figure 1**).

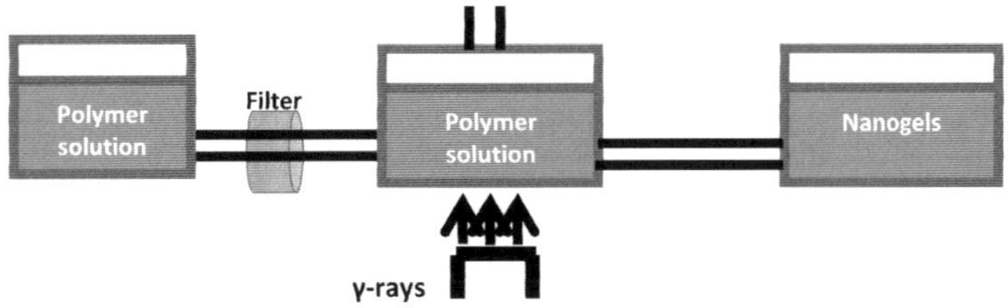

Figure 1. Scheme of experimental installation.

2.3 Characterization.

- The viscosimetric study of polymeric solutions was conducted both before and after being irradiated, according to the method described in ISO 1628-1: 1998 (E) [10] and the Cuban Norm NC 30-13 [11], using a capillary viscometer type Ubbelhode, equipped with a thermostatic bath controlled by a water recirculating at 25.0 ± 0.01 ° C and a timer. From the results obtained, the viscosimetric parameters shown in **Table 1** are calculated.

Table 1. Viscosimetric parameters.

Common name	Symbol and mathematical definition
Relative viscosity	$\eta_r = \eta/\eta_0 = t/t_0$
Specific viscosity	$\eta_{sp} = \eta_r - 1 = (\eta - \eta_0)/\eta_0 \cong (t - t_0)/t_0$
Reduced viscosity	$\eta_{red} = \eta_{sp}/C$
Inherent viscosity	$\eta_{inh} = (\ln \eta_r)/C$
Intrinsic viscosity	$[\eta] = \left(\eta_{sp}/C\right)_{C=0} = \left[(\ln \eta_{red})/C\right]_{C=0}$

-The average and distribution size of particle of irradiated solutions was measured by dynamic light scattering (DLS, by its acronym), using a particle analyser, DelsaTMNano.

-X-ray diffraction (XRD) powder patterns of sample was recorded with a Bruker D8 Advance Diffractometer and CuKα radiation (1.54183 Å).

- The morphology and particle size distribution of the produced powders were investigated by a high-resolution electron microscopy (JEOL JEM 2010), operated with an accelerating voltage of 200kV, a resolution of 1 Angstrom and an Orius camera SC200D 2K x 2K, 15 square per second incorporated. Samples for TEM were prepared by letting a drop of the samples dispersed in ethanol evaporate on top of a carbon-coated copper grid of 200 meshes.

3. Results and Discussion.

3.1 Viscosimetric study.

When comparing the values of the reduced viscosity of the solutions to 0.1% PVP unirradiated and irradiated to 10 kGy, it can be seen a considerable decrease of this magnitude, indicating the possible formation of Nanogels. Because, as nanogels particles are more compact than the spiral form of its linear polymer, the viscosity of the nanogels is smaller than linear macromolecules at the same concentration.

Table 2. Viscosimetric parameters determination before and after irradiation for PVP (0.01 %) solution.

Concentration of PVP (g/mL)	***0,001 (Before irradiation)***	***0,001 (After irradiation)***
Relative viscosity	1,11	1,02
Specific viscosity	0,11	0,02
Reduce viscosity	114,21	23,35
Inherent viscosity	108,14	23,09

4. Analysis of light scattering.

- Hydrodynamic Radius.

The hydrodynamic radius of the PVP prior to irradiating is 25.26 nm and after irradiation of PVP (0.1 %) solution to 15 kGy this parameter low, at 18.49 nm. This diminution is an indicator of the process of coil shrinkage resulting from intramolecular crosslinking and nanogels are formed [12].

-Analysis of average molecular weight Mw.

The value of the average molecular weight for the solution of PVP -before and after irradiar- were obtained by DLS using as input parameters $\alpha = 0.55$ and $\beta = 0.000565$ [11]. The value of molecular weight for PVP (K-90) unirradiated reported is of 1200000 Da [13], the solution after irradiating had a molecular weight of 3,48E + 06 Da. After irradiating of solution under study, a small increase in molecular weight is obtained, despite the decrease in reduced viscosity. Therefore, is logical to think that there is a few contribution of intermolecular recombination in spite of the used conditions. This effect has been studied in nanogel formation for some authors [14, 15].

4.1X-Ray Diffraction.

Figure 2 shows the diffractogram obtained for the solid sample of PVP with 0.1% concentration and irradiated at 15 kGy. No specific signal is observed by diffraction, which shows its amorphous nature, because it is a polymer, although a slight tendency to nano- crystallinity appears. Said result shows that both the irradiation and the polymer concentration used for obtaining these nanogels do not affect as such during initial material.

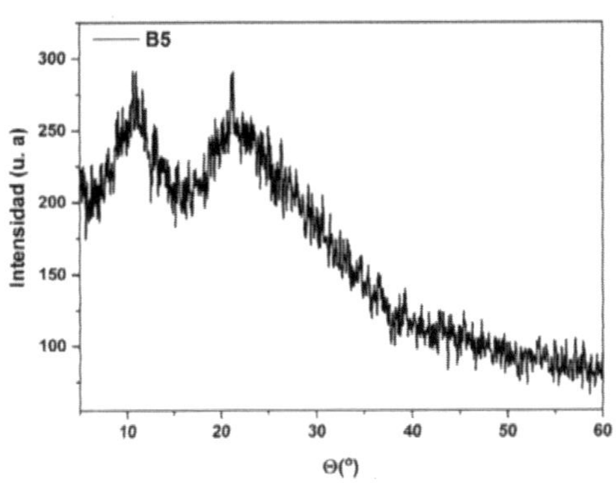

Figura 2. XRD pattern of powders obtained for solution of PVP (0.1 %) and doses (15 kGy).

4.2 Transmission Electron Microscopy (TEM).

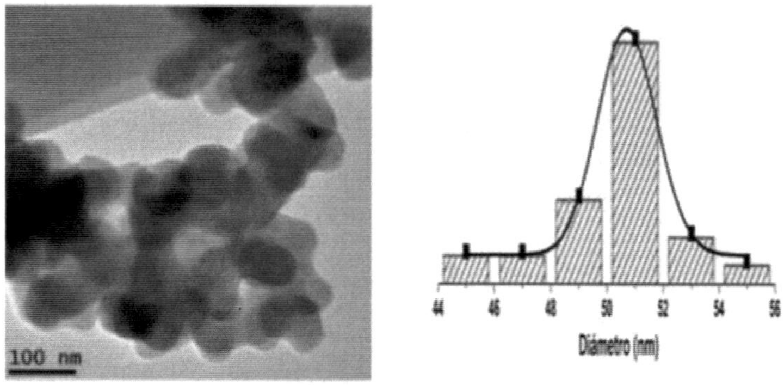

Figura 3. PVP (0.1 %) solution and y absorbed dose 15 kGy. Histogram of size particle distribution.

The TEM image of solution of PVP (0.1 %) concentration, and an irradiation dose of 15 kGy is shown in Figure 3. This image suggests that these conditions of concentration and dose, allows Nanogels formation mainly in elliptical shape, which are slightly agglomerated. The histogram in Figure 3 was obtained, based on a count of 100 particles from several images, to determine the particle size distribution, which was adjusted to a lognormal distribution. The resulting average diameter was 50.10 nm, the sample has a narrow size distribution given by a dispersion value of 1.04. This evidences a considerable uniform particle size.

Conclusion
PVP Nanogel particles were synthesized using gamma radiation, based not on polymerization, but on intramolecular crosslinking of polymers chains, in aqueous solutions. Although the concentration used in nanogel formation is below its overlap concentration, there is some contribution of intermolecular recombination in the obtained product, it is evidenced in the increase of molecular weight. The synthesized nanogels are practically amorphous because they tend at agglomerated in solid state. Their morphology is mainly elliptical forming nanoparticles in solution, for studied concentration at 15kGy. Since their properties as biomaterial the nanogel could be used for pharmaceutical or environmental purpose.

References
[1] Galego N and Peniche C 1990. *ENPES*. Cuba
[2] Galego N, Peniche C, Martínez R and et al 1987 *Científico-Técnica*. Cuba
[3] Odian G 2004. *Hoboken New Jersey: John Wiley & Sons, Inc.* 4.Ed
[4] Teraoka I 2002. *New York: John Wiley & Sons, Inc*
[5] Rosiak J M, Ulanski P and et al 2005 *Eurasian ChemTech*. 7 225-50
[6] Govind Soni and et al 2016. *Saudi Pharmaceutical Journal*. 24 133-139
[7] Alexander V, Kabanov and et al 2009. *Angew Chem Int Ed Engl*. 40 5418-5429
[8] Mozafari M R 2007 *AA Dordrecht The Netherlands: Springer*
[9] Plasencia M 2005 *Registro CENDA #1853* 1-2
[10] Norma ISO s f 1998 *ISO 1628-1*
[11] Norma Cubana s f *NC 30-13* 82.
[12] Rosiak L M 2005 *IAEA-TECDOC-1438* 141-57
[13] Gogotsi Y 2006 *Nanomaterilas Hanbook*. CRC Press Taylor & Francis Group
[14] Ulanski P and Rosiak M 1999 *Nucl. Instrum. Methods*. B 151 356-60
[15] Ulanski P, Janik I and Rosiak JM 1998 *Radiat. Phys. Chem.* 52 289-94

Acknowledgments
We want to thanks to CICATA-Legaria and CNyN in Mexico for characterization techniques.

IOP Conf. Series: Journal of Physics: Conf. Series **792** (2017) 012081 doi:10.1088/1742-6596/792/1/012081

Dynamic Modeling and Simulation of a Rotational Inverted Pendulum

J L Duart, B Montero, P A Ospina and E González.

E-mail: juan.duarte03@ustabuca.edu.co, brayan.montero@ustabuca.edu.co, paolo.ospina@ustabuca.edu.co, edwin.gonzalez@ustabuca.edu.co

Abstract. This paper presents an alternative way to the dynamic modeling of a rotational inverted pendulum using the classic mechanics known as Euler-Lagrange allows to find motion equations that describe our model. It also has a design of the basic model of the system in SolidWorks software, which based on the material and dimensions of the model provides some physical variables necessary for modeling. In order to verify the theoretical results, It was made a contrast between the solutions obtained by simulation SimMechanics-Matlab and the system of equations Euler-Lagrange, solved through ODE23tb method included in Matlab bookstores for solving equations systems of the type and order obtained. This article comprises a pendulum trajectory analysis by a phase space diagram that allows the identification of stable and unstable regions of the system.

1. Introduction

Nowadays, the underactuated mechanical systems [1] are generating interest among researchers of modern control theories. This interest is that these systems have similar problems to those found in industrial applications, such as external shocks and/or non-linear behavior in some conditions of operation. The rotational inverted pendulum [2] is a clear example of a mechanical underactuated system; this is a mechanism of two degrees of freedom (DOF) and two rotational joints. It consists of three main elements: a motor, a rotational arm and a pendulum [3]. The motor shaft is connected to one end of the rotational arm making this fully rotate on a horizontal plane; the other end of the arm has connected the pendulum that freely rotates 360 degrees in a vertical plane. Despite being a purely academic level, this system is helpful to study, apply and analyze different modeling strategies.

Some industrial applications [4] that present the previously mentioned drawbacks and behavior, are in fields such as robotics, robots balance, biped robots [5], robotic arms; aerospace, positioning rocket; telecommunications, satellite positioning; transportation, Segway [6,7], stability of ships and submarines, iBot, Self-balancing unicycle; construction, bridge cranes; and field monitoring, drones.

To carry out this work, initially a literature review of similar projects [8,9,10] was conducted in order to determine the appropriate dimensions for the prototype design in SolidWorks CAD software. Then it was performed continuously dynamic modeling of the system and using the Lagrangian mechanics as a method, two equations of motion were obtained, due to each one of the two links comprising, the inverted pendulum has a degree of freedom (DOF). To validate the equations obtained, multiple simulations were made in the Matlab software in order to observe their behavior. Thus it was possible to verify in a graphic mode, the waveform of the position and angular velocity of the arm and the pendulum.

Content from this work may be used under the terms of the Creative Commons Attribution 3.0 licence. Any further distribution of this work must maintain attribution to the author(s) and the title of the work, journal citation and DOI.
Published under licence by IOP Publishing Ltd

A fundamental part in the simulation stage, was to bring the CAD model made in SolidWorks to SimMechanics [11] Matlab environment. At the same time, the equations obtained using the Lagrangian mechanics were resolved with the ODE23tb method (Ordinary Differential Equations) belonging to the Matlab software. The use of block diagrams in Simulink allowed a representation of the system. This was intended in order to develop an overlap of results of both simulations and establish similarities and/or differences in the waveform of the variables of interest.

Finally, the graph of the effective potential of the system allowed tracing paths and obtain the phase space diagram [12] in which they can study the critical points that correspond to the maximum and minimum of the potential.

In section II the design of the basic model is presented with their respective physical variables. In section III It provides a brief definition of systems underactuated. Section IV presents the advantage of making use of Euler-Lagrange formalism. In Section V it shows the dynamic modeling of proposed rotational inverted pendulum. In section VI activities are defined for perform proposed simulations as a method of validating results. In section VII the results obtained are shown with a respective analysis of them. Finally, section VIII contains the relevant conclusions of the article.

2. Prototype Proposed

This section presents the basic design created in SolidWorks CAD software with the following features:

Table 1. Variable System

Physical characteristic	**Symbology**
Pendulum mass	m_1
Arm length	L_0
Pendulum length	L_1
Location of the center of mass of the pendulum	l_1
Moment of inertia of the arm	I_0
Moment of inertia of the pendulum	I_1
angular position of the arm	θ_0
angular position of the pendulum	θ_1

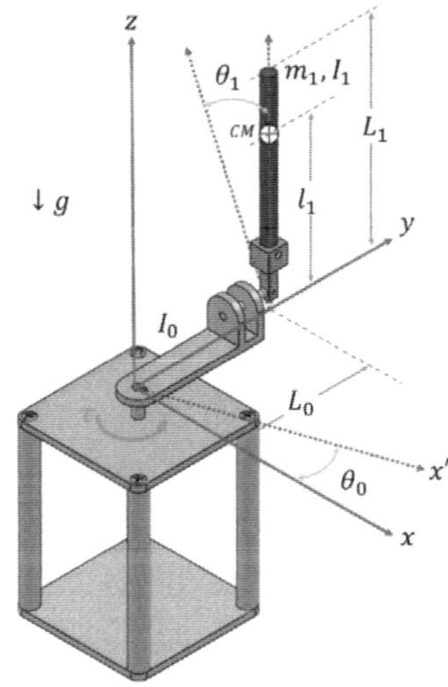

Figure 1. Rotational inverted pendulum

3. Underactuated Systems

Direct and indirect actions are two concepts that are acquired in the study of mechanisms. The first consists of movement of elements by action of an actuator, while the second consists of the action of motion transmitted by another interconnected element. Such movements are known as degrees of freedom, so that mechanical systems or mechanisms can be classified depending on the number of degrees of freedom and the number of actuators. The fully actuated mechanical systems are those having the same number of degrees of freedom and actuators. Underactuated mechanical systems are those with fewer actuators than degrees of freedom [13]. It is important to highlight the advantages of underactuated systems, since if they do not have advantages over fully actuated mechanical systems, it will not make sense its development. The main advantages present in underactuated systems are: energy saving and control efforts. However, these systems are intended to perform the same functions of fully actuated systems without their disadvantages.

4. Euler-Lagrange

The dynamic equations of any mechanical system can be obtained from the known classical mechanics (Newton), the drawback of this formalism is the use of the variables in vector form, complicating considerably the analysis when increasing the joints or there are rotations present in the system. In these cases, it is favorable to employ the Lagrange equations, which have formalism of scale, facilitating the analysis for any mechanical system.

In order to use Lagrange equations, it is necessary to follow four steps:

1) Calculation of kinetic energy.
2) Calculation of the potential energy.
3) Calculation of the Lagrangian.
4) Solve the equations.

Where the kinetic energy can be both rotationally and translational, this form of energy may be a function of both the position and the speed $K(q(t), \dot{q}(t))$.

IOP Conf. Series: Journal of Physics: Conf. Series **792** (2017) 012081 doi:10.1088/1742-6596/792/1/012081

The potential energy is due to conservative forces as the forces exerted by springs and gravity, this energy is in terms of the position $U(q(t))$.

The Lagrangian is defined as:

$$L = K - U \tag{1}$$

So the Lagrangian in general terms is defined as follows:

$$L(q(t), \dot{q}(t)) = K(q(t), \dot{q}(t)) - U(q(t)) \tag{2}$$

Finally the Euler-Lagrange equations for a n degrees of freedom system of is defined as follows:

$$\frac{d}{dt}\left(\frac{\partial L(q,\dot{q})}{\partial \dot{q}_i}\right) - \frac{\partial L(q,\dot{q})}{\partial q_i} = \tau_i \tag{3}$$

Where $i = 1, \ldots, n$, τ_i are forces or externally exercised pairs (actuators) at each joint, besides nonconservative forces such as friction, resistance to movement of an object within a fluid and generally those that depend on time or speed. It will be obtained an equal number of dynamic equations and degrees of freedom.

5. Modeling of Rotational Inverted Pendulum

Lagrangian system modeling was performed as shown in Figure 1. It is necessary to make an energy analysis. Therefore, initially the kinetic energy of each link is analyzed, so it can be identified which kinetic energies (rotational and translational) were present in each link.

5.1. *Kinetic Energy*

The kinetic energy of the system consist of a translational and rotational component for the pendulum and rotational component for the arm,

$$K = \frac{1}{2}mv^2 + \frac{1}{2}I\omega^2 \tag{4}$$

Where m is the mass of the body, v the linear velocity, I the moment of inertia, ω the angular velocity and K the kinetic energy. In this case there are two bodies, the arm and the pendulum. The kinetic energy of the arm is:

$$K_0 = \frac{1}{2}I_0\dot{\theta}_0{}^2 \tag{5}$$

The kinetic energy of the pendulum is:

$$K_1 = \frac{1}{2}m_1 v_1{}^2 + \frac{1}{2}I_1\dot{\theta}_1{}^2 \tag{6}$$

The total energy of the system is:

$$K_T = \frac{1}{2}I_0\dot{\theta}_0{}^2 + \frac{1}{2}m_1 v_1{}^2 + \frac{1}{2}I_1\dot{\theta}_1{}^2 \tag{7}$$

5.2. *Potential Energy*

This system only store gravitational potential energy in the pendulum.

$$U = mgh \tag{8}$$

The arm has in its nature a rotational movement in a horizontal plane, therefore it does not have height change in its center of mass, providing equal 0 component in equation (8) resulting in a potential energy zero .

The potential energy of the pendulum is,

$$U_1 = m_1 g l_1 (\cos\theta_1 - 1) \tag{9}$$

Where g represents the value of gravity. The total potential energy of the system is (9),

$$U_T = m_1 g l_1 (\cos\theta_1 - 1) \tag{10}$$

5.3. *Position of the pendulum*

Because the pendulum is a rigid body, the required position is the its center of mass,

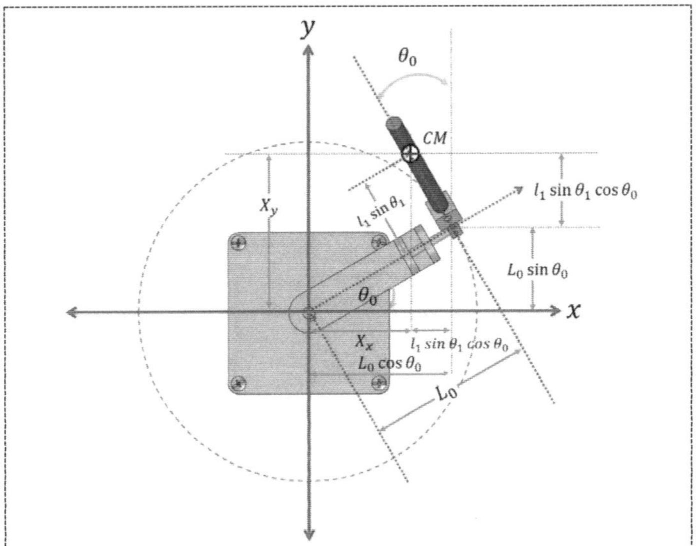

Figure 2. Projection arm and pendulum in the xy plane

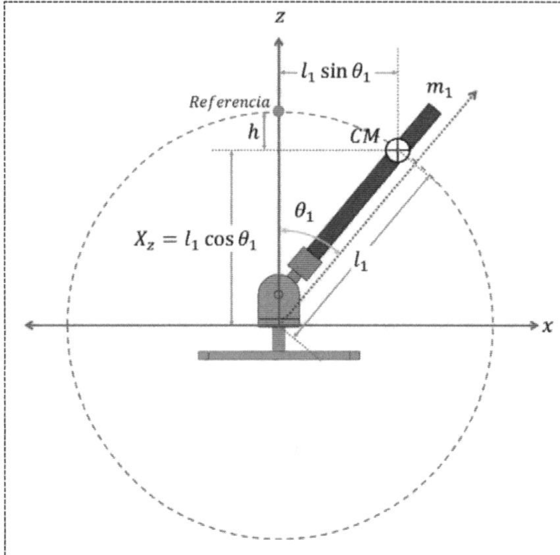

Figure 3. Projection pendulum in the xz plane

Then, it shows components the position of the center of mass.

$$X_x = L_0 \cos\theta_0 - l_1 \sin\theta_1 \sin\theta_0, \tag{11}$$

$$X_y = L_0 \sin\theta_0 + l_1 \sin\theta_1 \cos\theta_0, \tag{12}$$

$$X_z = l_1 \cos\theta_1, \tag{13}$$

The position of the center of mass,

$$X_{CM} = [X_x \quad X_y \quad X_z]^T \tag{14}$$

5.4. *Linear Speed of the center of mass*

The speed is defined as the derivative of the position with respect to time.

$$\frac{d}{dt}(X_{CM}) = V_{CM} \tag{15}$$

$$\dot{X}_{CM} = V_{CM} = [\dot{X}_x \quad \dot{X}_y \quad \dot{X}_z]^T \tag{16}$$

The velocity components are obtained by differentiating each component position (11), (12) and (13) respectively,

$$\dot{X}_x = -\dot{\theta}_0 L_0 \sin\theta_0 - l_1(\dot{\theta}_0 \sin\theta_1 \cos\theta_0 + \dot{\theta}_1 \cos\theta_1 \sin\theta_0) \tag{17}$$

$$\dot{X}_y = \dot{\theta}_0 L_0 \cos\theta_0 + l_1(\dot{\theta}_1 \cos\theta_1 \cos\theta_0 - \dot{\theta}_0 \sin\theta_1 \sin\theta_0) \tag{18}$$

$$\dot{X}_z = -\dot{\theta}_1 l_1 \sin\theta_1 \tag{19}$$

The linear velocity of the center of mass can be expressed as follows vector form,

$$V_{CM}{}^2 = [\dot{X}_x \quad \dot{X}_y \quad \dot{X}_z] \begin{bmatrix} \dot{X}_x \\ \dot{X}_y \\ \dot{X}_z \end{bmatrix} = \dot{X}_x{}^2 + \dot{X}_y{}^2 + \dot{X}_z{}^2 \tag{20}$$

They are calculated separately each of the components to the square of the velocity of the center of mass.

$$\dot{X}_x{}^2 = \dot{\theta}_0{}^2 L_0{}^2 \sin^2\theta_0 - 2(\dot{\theta}_0 L_0 \sin\theta_0)\left(l_1(\dot{\theta}_0 \sin\theta_1 \cos\theta_0 + \dot{\theta}_1 \cos\theta_1 \sin\theta_0)\right) + l_1{}^2(\dot{\theta}_0 \sin\theta_1 \cos\theta_0 + \dot{\theta}_1 \cos\theta_1 \sin\theta_0)^2 \tag{21}$$

$$\dot{X}_y{}^2 = \dot{\theta}_0{}^2 L_0{}^2 \cos^2\theta_0 + 2(\dot{\theta}_0 L_0 \cos\theta_0)\left(l_1(\dot{\theta}_1 \cos\theta_1 \cos\theta_0 - \dot{\theta}_0 \sin\theta_1 \sin\theta_0)\right) + l_1{}^2(\dot{\theta}_1 \cos\theta_1 \cos\theta_0 - \dot{\theta}_0 \sin\theta_1 \sin\theta_0)^2 \tag{22}$$

$$\dot{X}_z{}^2 = \dot{\theta}_1{}^2 l_1{}^2 \sin^2\theta_1 \tag{23}$$

Finally the following expression is obtained for the linear velocity.

$$V_{CM}{}^2 = L_0{}^2\dot{\theta}_0{}^2 + l_1{}^2(\dot{\theta}_1{}^2 + \dot{\theta}_0{}^2 \sin^2\theta_1) + 2L_0 l_1 m_1 \dot{\theta}_0 \dot{\theta}_1 \cos\theta_1 \tag{24}$$

5.5. Energy System

The total energy of the system is,

$$K_T = K_0 + K_1 \tag{25}$$

Substituting (5) and (6) in (25) is obtained,

$$K_T = \frac{1}{2} I_0 \dot{\theta}_0{}^2 + \frac{1}{2} I_1 \dot{\theta}_1{}^2 + \frac{1}{2} m_1 V_{CM}{}^2 \tag{26}$$

Total kinetic energy is obtained by substituting equation (24) in (26)

$$K_T = \frac{1}{2} I_0 \dot{\theta}_0{}^2 + \frac{1}{2} I_1 \dot{\theta}_1{}^2 + \frac{1}{2} m_1 \{L_0{}^2\dot{\theta}_0{}^2 + l_1{}^2(\dot{\theta}_1{}^2 + \dot{\theta}_0{}^2 \sin^2\theta_1) + 2L_0 l_1 m_1 \dot{\theta}_0 \dot{\theta}_1 \cos\theta_1\} \tag{27}$$

The potential energy of the system is shown in equation (10)

5.6. Euler-Lagrange Equations

The Lagrangian of the system is,

$$L = K_T - U_T \tag{28}$$

Substituting (10) and (27) (28) the Lagrangian of the system is obtained,

$$= \frac{1}{2} I_0 \dot{\theta}_0{}^2 + \frac{1}{2} I_1 \dot{\theta}_1{}^2 + \frac{1}{2}\left(L_0{}^2 m_1 \dot{\theta}_0{}^2\right) + \frac{1}{2}\left(l_1{}^2 m_1 \dot{\theta}_1{}^2\right) + \frac{1}{2}\left(l_1{}^2 m_1 \dot{\theta}_0{}^2 \sin^2\theta_1\right) \\ + L_0 l_1 m_1 \dot{\theta}_0 \dot{\theta}_1 \cos\theta_1 + m_1 g l_1 (1 - \cos\theta_1) \tag{29}$$

As there are two degrees of freedom (DOF) the Euler-Lagrange equations have the following form,

$$\frac{d}{dt}\left(\frac{\partial L}{\partial \dot{\theta}_0}\right) - \frac{\partial L}{\partial \theta_0} = \tau \tag{30}$$

$$\frac{d}{dt}\left(\frac{\partial L}{\partial \dot{\theta}_1}\right) - \frac{\partial L}{\partial \theta_1} = 0 \tag{31}$$

Where τ is the torque of the motor. Solving (30) and (31) we obtain the equations of motion which are given by,

$$I_0\ddot{\theta}_0 + L_0{}^2 m_1\ddot{\theta}_0 + \left\{l_1{}^2 m_1\left(\ddot{\theta}_0 \sin^2\theta_1 + 2\dot{\theta}_0\dot{\theta}_1 \sin\theta_1 \cos\theta_1\right)\right\} + \left\{L_0 l_1 m_1\left(\ddot{\theta}_1 \cos\theta_1 - \dot{\theta}_1{}^2 \sin\theta_1\right)\right\} = \tau \tag{32}$$

$$I_1\ddot{\theta}_1 + l_1{}^2 m_1\ddot{\theta}_1 + L_0 l_1 m_1\ddot{\theta}_0 \cos\theta_1 - l_1{}^2 m_1\dot{\theta}_0{}^2 \sin\theta_1 \cos\theta_1 - m_1 g l_1 \sin\theta_1 = 0 \tag{33}$$

Where (32) is the equation of motion of the arm (33) of the pendulum.

6. Simulation

Although the Euler-Lagrange formalism ensures a high degree of approximation of the mathematical models, is essential do comparisons to validate these results. For verification of modeling the following steps are followed:

1) Represent the system equations in the state space.
2) Define an experiment with initial conditions, natural interactions and external forces.
3) Import the CAD model of SolidWorks in SimMechanics-Matlab.
4) Add the necessary blocks to obtain the desired graphic model and applying external forces.
5) Simulating experiment.
6) Export the results of SimMechanics to Workspace Matlab.
7) Implement a block diagram Simulink-Matlab to solve the equations.
8) Export the solutions to the equations to Workspace Matlab.
9) Graphing and overlay solutions.

As can be seen in the steps above, the simulation of the model was divided into two stages: first, simulate the CAD model initially designed and the second, implement the equations obtained.

7. Results

Then, it presents each of the steps mentioned in the previous section.

1) Starting from [14],

$$M(q)\ddot{q} + C(q,\dot{q})\dot{q} + G(q) = \tau \tag{34}$$

So the equation (34) is the dynamic equation for mechanical systems of n degrees of freedom (DOF). Where M is the matrix of inertia of the system, C is the centrifugal and Coriolis matrix, G the vector of gravity and τ external forces.

Taking the equations of motion (32) and (33) and replacing in (34) is obtained representation in state space system,

$$\begin{bmatrix} I_o + m_1 L_0{}^2 + l_1{}^2 m_1 \sin^2\theta_1 & L_0 l_1 m_1 \cos\theta_1 \\ L_0 l_1 m_1 \cos\theta_1 & I_1 + m_1 l_1{}^2 \end{bmatrix}\begin{bmatrix}\ddot{\theta}_0 \\ \ddot{\theta}_1\end{bmatrix} + \begin{bmatrix} 2 l_1{}^2 m_1 \sin\theta_1 \cos\theta_1 \dot{\theta}_1 & -L_0 l_1 m_1 \sin\theta_1 \dot{\theta}_1 \\ -l_1{}^2 m_1 \sin\theta_1 \cos\theta_1 \dot{\theta}_0 & 0 \end{bmatrix}\begin{bmatrix}\dot{\theta}_0 \\ \dot{\theta}_1\end{bmatrix} + \begin{bmatrix} 0 \\ -g l_1 m_1 \sin\theta_1 \end{bmatrix} = \begin{bmatrix}\tau \\ 0\end{bmatrix} \tag{35}$$

The matrix $M(q)$ is important for the dynamic modeling and for the design of controllers. This matrix has a great relationship with the kinetic energy, also the inertia matrix is a symmetric, positive and square matrix of $n \times n$, whose elements depend only on the generalized coordinates.

Centrifugal and Coriolis matrix $C(q, \dot{q})$ it is important in the study of stability in control systems, mechanical systems, among others. This matrix is square of $n \times n$ and has dependence in its elements of the generalized coordinates and velocities.

The gravity vector $G(q)$ it is present in mechanical systems without counterweights or springs, in turn is in systems with displacement off the horizontal plane. This vector is of $n \times 1$ and has only reliance on joint positions.

2) It defined that the system would have the initial conditions shown in Figure 5, in addition to being subject to effects of gravity and to a torque step 0.2 seconds in the end of the arm that connects to the motor shaft. Finally, a simulation interval 5 seconds was established.

3) Was imported the CAD model in SimMechanics with the following code line:

mech_import('CADModel_Pendulum.xml')

The initial position of the system is Figure 5 a).

4) It was necessary add a few block the diagram SimMechanics, since the CAD model is only under the effect of gravity and not have some kind of movement, it blocks provide the step of torque arm included to start rotating, besides adding blocks to the sensing of angular displacement and velocities in an interval of 5 second of test.

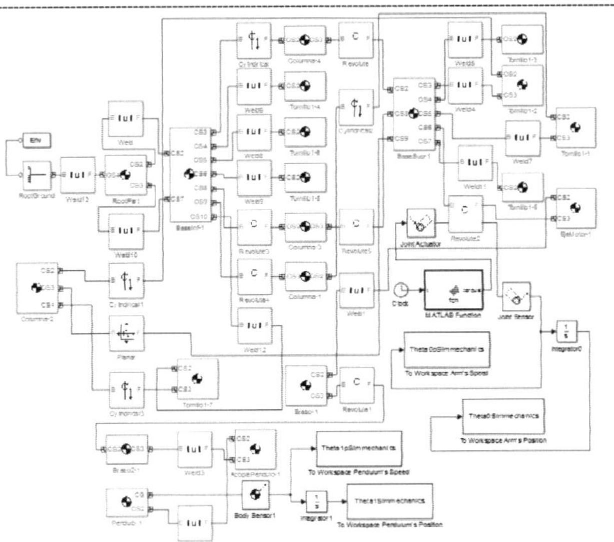

Figure 4. Final block diagram in SimMechanics

5) Some moments of the position of the system were taken in the time interval defined.

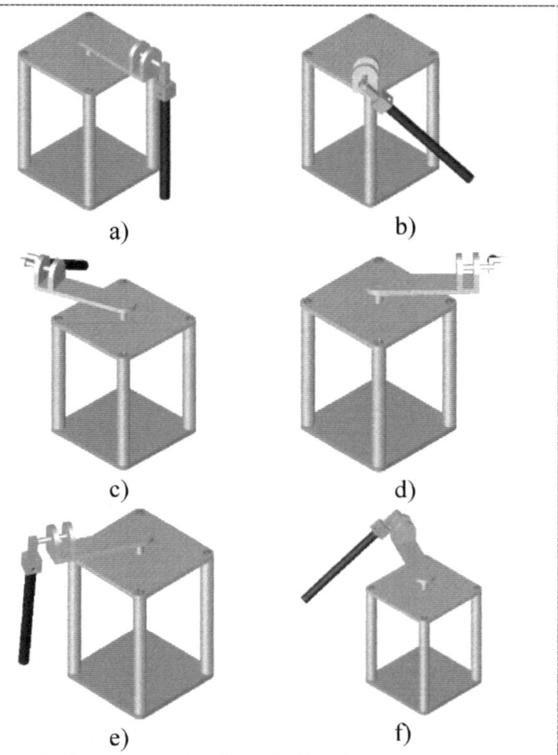

Figure 5. Experiment in SimMechanics
a) t = 0 s, b) t = 0.212227 s, c) t = 0.798791 s, d) t = 1.22854 s, e) t = 3.33321 s, f) t = 5 s

6) To export system solutions to Workspace, is need to add the blocks with the name To Workspace located in Simulink library, which will create a cell for each solution, with the respective data and time in which such data is obtained.

7) The implementation of the block diagram was performed to numerically solve the equations of the system, thus obtaining values $\theta_0, \theta_1, \dot{\theta}_0, \dot{\theta}_1$ along the defined time interval. The diagram is implemented as follows:

IOP Conf. Series: Journal of Physics: Conf. Series **792** (2017) 012081 doi:10.1088/1742-6596/792/1/012081

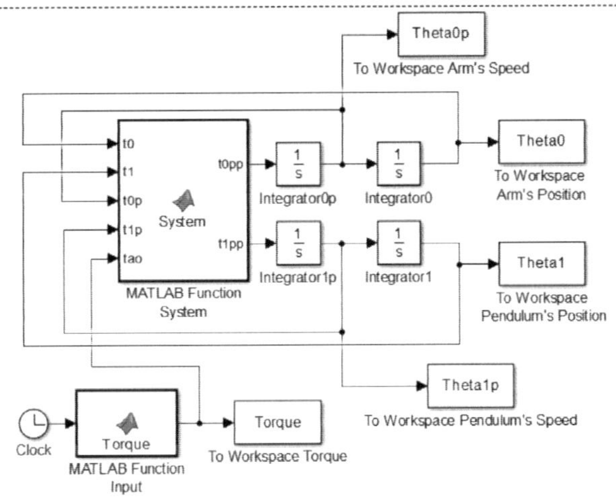

Figure 6. Block diagram Euler-Lagrange equations

Wherein the first block has the input applied to the system, which in this case is a torque step 0.2 second, which has the value 0.5 Nm which is the maximum torque provided by the engine, the block is called System contains the model equations in the state space, with 5 inputs $(\theta_0, \theta_1, \dot{\theta}_0, \dot{\theta}_1, \tau)$ and 2 outputs $(\ddot{\theta}_0, \ddot{\theta}_1)$. Internally, the block uses the ODE23tb method for solving equations.

Numerical data necessary to solve the equations of motion Euler-Lagrange are as follows:

Table 2. Numerical Variable System

Physical characteristic	**Symbology**
Pendulum mass	$0.2866\ kg$
Arm length	$0.201\ m$
Pendulum length	$0.30997\ m$
Location of the center of mass of the pendulum	$0.154985\ m$
Gravity	$9.81\ m/s^2$
Moment of inertia of the arm	$0.0052\ kg.m^2$
Moment of inertia of the pendulum	$0.0023\ kg.m^2$

Was made an approximation in the calculation of the moments of inertia, both links are taken as constant circular bars and invariant mass,

$$I_0 = \frac{1}{3} m_0 L_0^{\ 2} \tag{36}$$

$$I_1 = \frac{1}{12} m_1 L_1^{\ 2} \tag{37}$$

Where equation (36) is the moment of inertia of the arm measured from the end connected to the motor shaft to the opposite end. While equation (37) is the pendulum moment of inertia measured pendulum from the center of mass.

8) The way to export the data to Workspace of block diagram above is performed with the same aggregate block in step *6)*.

9) Taking the exported data in points *6)* and *8)* the following graphs were made.

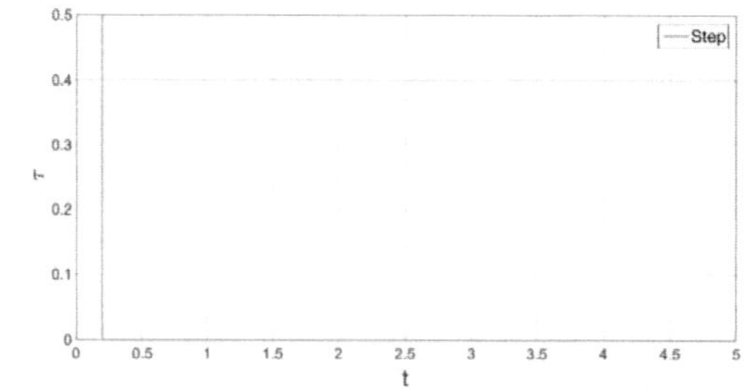

Figure 7. Step torque

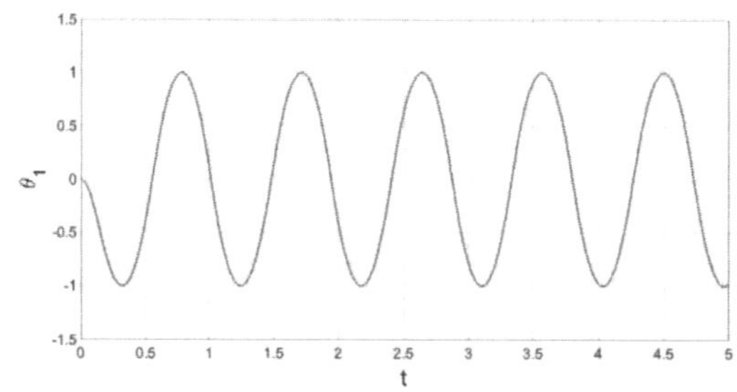

Figure 8. Angular position of the pendulum with CAD model

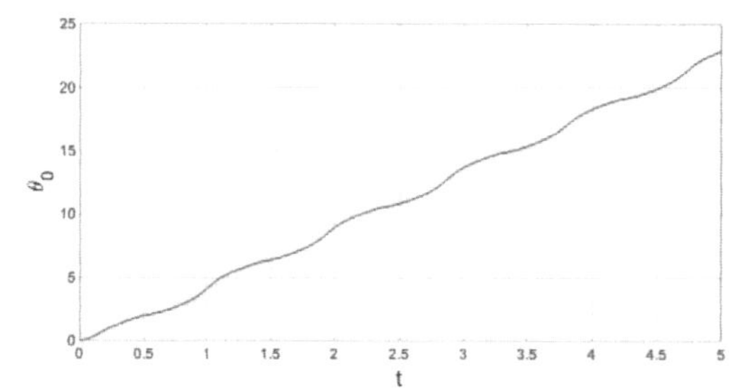

Figure 9. Angular position of the arm with CAD model

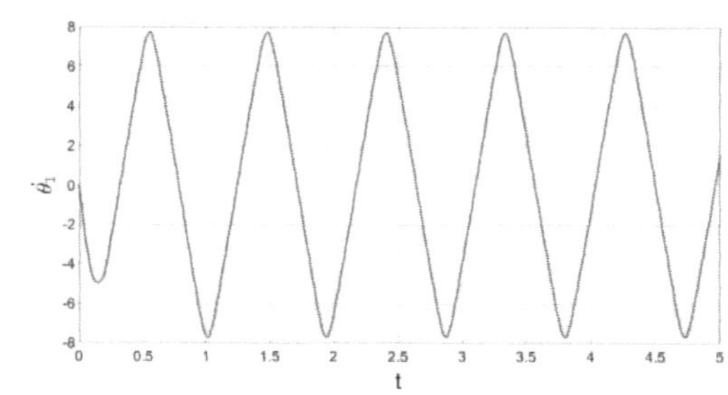

Figure 10. Angular velocity of the pendulum with CAD model

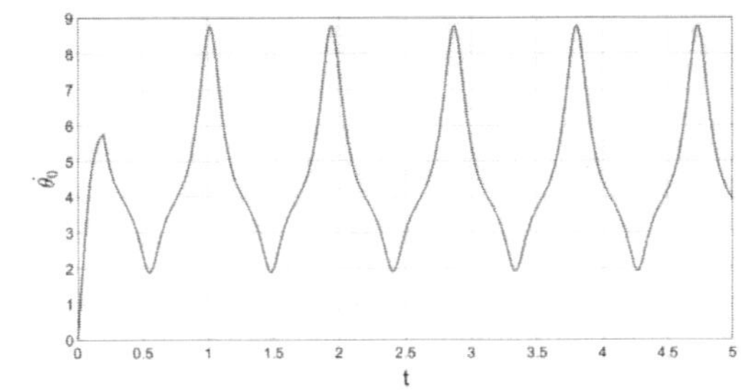

Figure 11. Angular velocity of the arm with CAD model

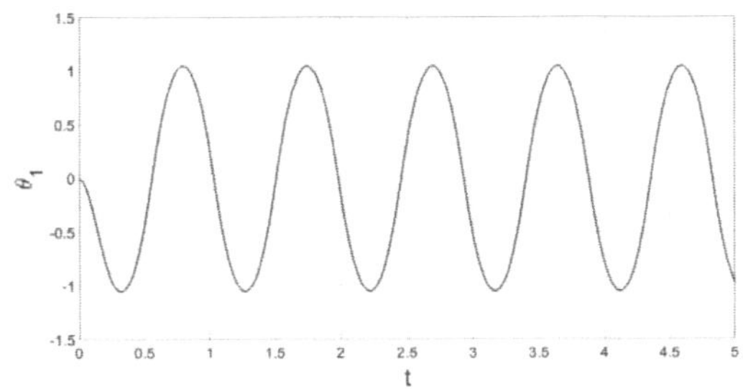

Figure 12. Angular position of the pendulum with Euler-Lagrange equations

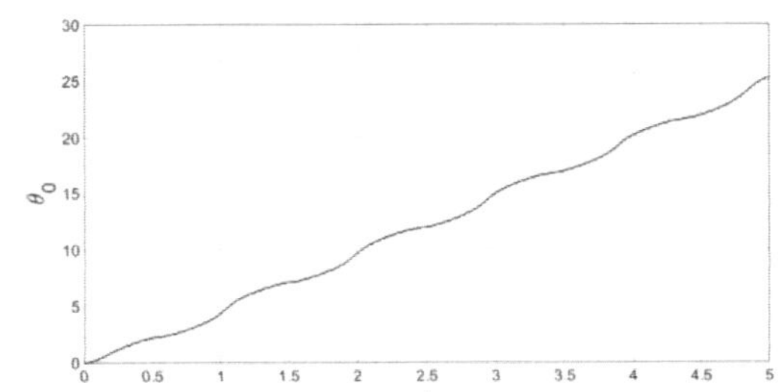

Figure 13. Angular position of the arm with Euler-Lagrange equations

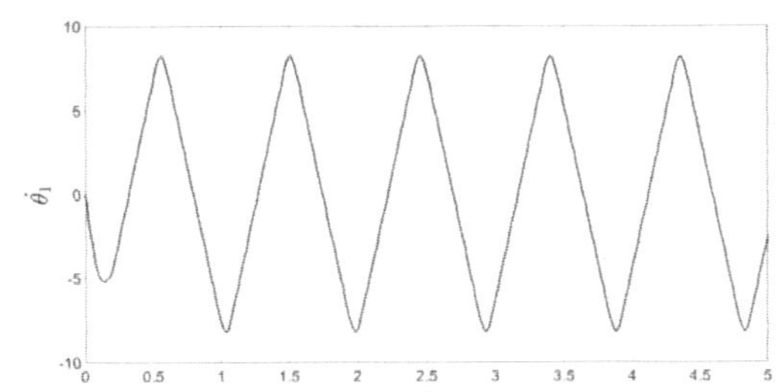

Figure 14. Angular velocity of the pendulum with Euler-Lagrange equations

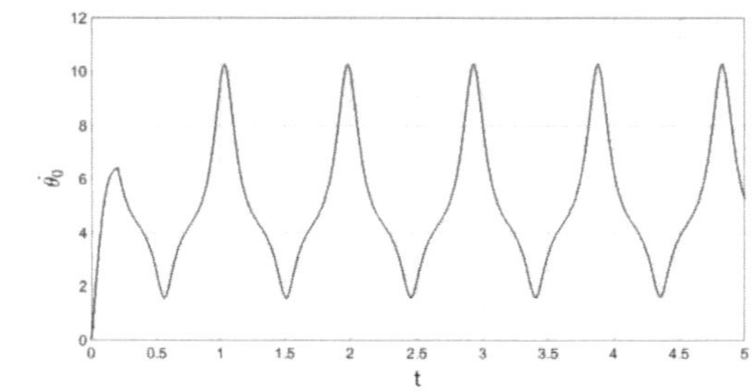

Figure 15. Angular velocity of the arm with Euler-Lagrange equations

It is noted that the system behavior is the same in both simulations. A superposition of results was performed by way of display the error between the CAD model simulated and the Euler-Lagrange equation, with the following results.

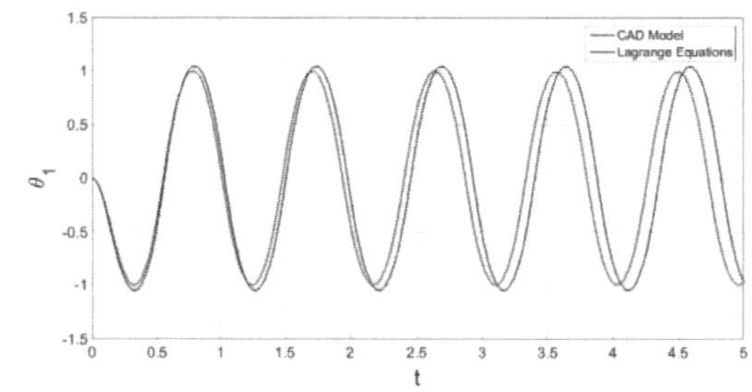

Figure 16. Comparison results angular position of the pendulum

Figure 16 allows viewing behavior overlapping displacement of the center of mass in the pendulum. The red line represents the CAD model while the blue is the result of θ_1 graph the solution over time.

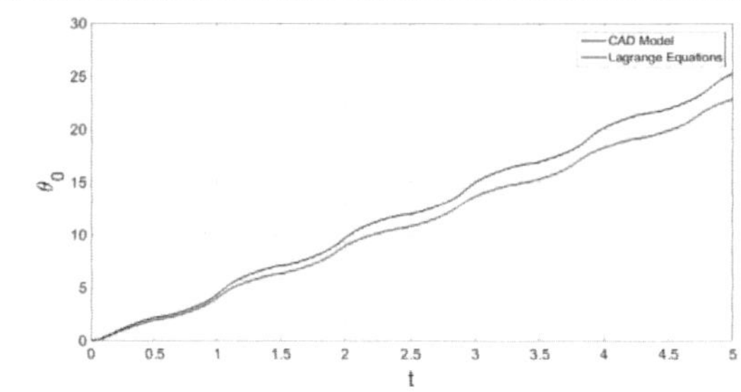

Figure 17. Comparison results angular position of the arm

Figure 17 is overlaying the results to simulate and plot θ_0 function of time, where the red line is the result of simulating in SimMechanics the CAD model and the blue line is the solution of the Euler-Lagrange equation.

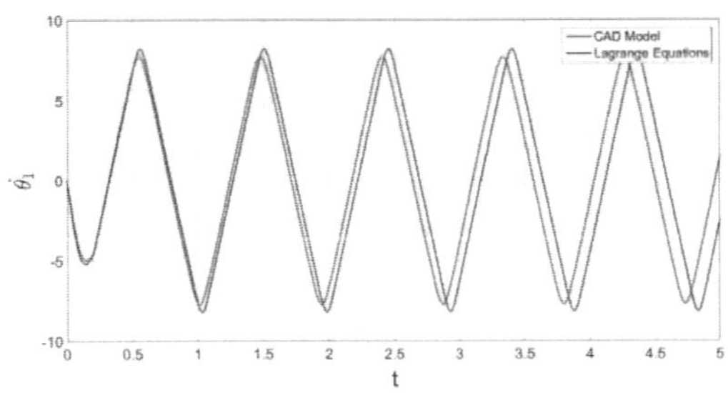

Figure 18. Comparison results angular velocity of the pendulum

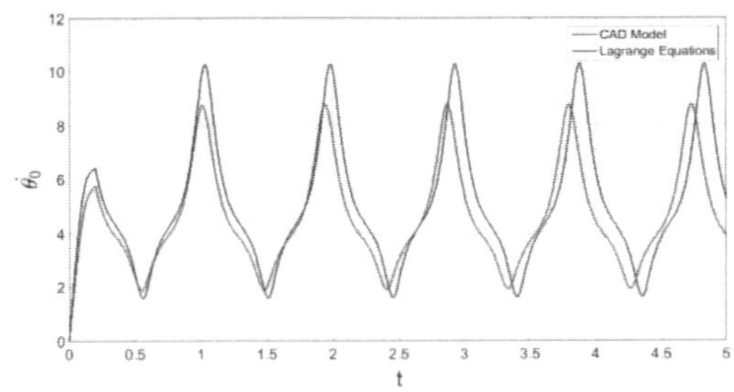

Figure 19. Comparison results angular velocity of the arm

Similarly as with Figure 16 and 17 the Figure 18 and 19 illustrate the behavior of the angular velocity with respect to time for the pendulum and the arm respectively.

It is important to note that the graphs do not have a zero error due to the approximations that were made to the calculation of the moments of inertia of each link, the difference is that the SolidWorks software performs a more accurate calculation of the moments of inertia, because the software considers the geometry of each link.

Then, the pendulum trajectory analysis shown, which part of the potential energy in the center of mass as shown in equation (9),

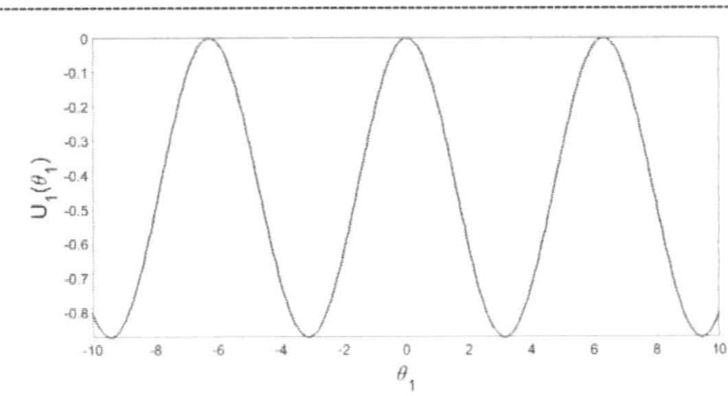

Figure 20. Potential energy

When all paths of the pendulum are collected the phase diagram, which also shows which are the critical points of the system, in this case we know that the pendulum has two critical points, a stable point at which the pendulum is in its obtained hanging position and unstable point in this case refers to a metastable point is satisfied when the pendulum is reversed. Graphically you can get the critical points in the system, we can first obtain which is the equation for calculating the minimum by Figure

20, it can be seen that the minimum occurs when the wave goes from a negative slope to a positive slope, therefore, we are obtained minimum $\pm\pi i$ where i must be odd. Similarly one can find the maximum, these occur when moving from a positive slope to a negative, therefore the maximum are present in $\pm\pi i$ where i must be equal to 0 or couple.

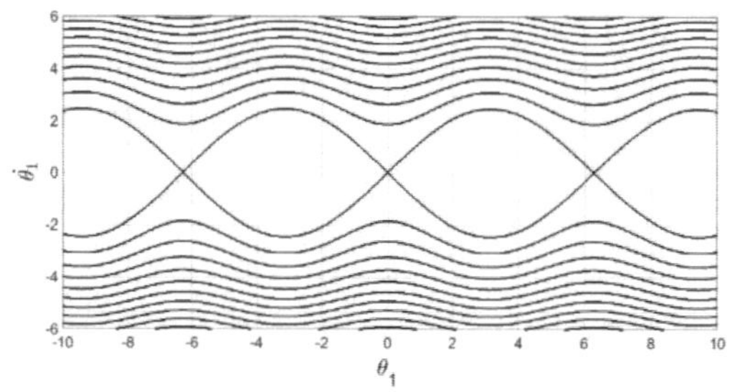

Figure 21. Pendulum phase space

Physically each point on the graph above represents a possible system status, explicitly state in which the pendulum would.

As mentioned above the pendulum has two equilibrium points, however, in the phase space endless these points are. This is because there is shown the position like a straight or R space when the real space is a circle that is not topologically equals R. Therefore, if space representing geometric positioning system is not topologically equivalent to R, for subsequent phase space will not be in R^2, since this is the Cartesian product between the geometric space and space representing the speed is generally R. Topologically speaking the pendulum has a geometric space $S1$ which is equivalent to a circle and a space velocity R, therefore, the phase space $S1 \times R$ equivalent to a cylinder and not to R^2. The correct graphic process should be done on a cylinder where its circumference is the geometric representation of the pendulum, this is equivalent to the set of possible positions of the center of mass, while the position along its axis represents the velocity of the center mass, this cylinder is topologically correct space phase. Touring the circumference of the cylinder we can see that there is only one minimum and one maximum diametrically opposed. In Figure 20 you could take energy levels were the result of drawing lines parallel to the axis of the θ_1, straight position, well, now those levels may be taken with circles surrounding the cylinder.

8. Conclusions

The Euler-Lagrange formalism allows dynamic modeling of rotational inverted pendulum in a simple way, thanks to the classical mechanics allows us to work on a climb.

This paper has presented step by step dynamic modeling of the proposed system, in addition to the respective simulation of the solution of the equations and further validation of the results using a simulation CAD model designed in SolidWorks exported in SimMechanics Matlab.

This work differs from others through the use of SimMechanics extension for simulation CAD model, this tool is easy to use and reliable enough, this is an advantage for simulations of 3D models since it is not necessary to create platforms for simulation. It also allows working in different ways on the same software, which in this case was Matlab, which allowed us to work with the CAD model and the equations of motion in the same environment (Simulink).

Something extra that is presented in this article is the analysis as to the trajectories of the pendulum, the phase space is topologically correct shown, this being a good contribution to future dynamic, kinematic and physical studies of the system.

References

[1] Gómez F 2002 *"Control de sistemas no lineales basado en la estructura hamiltoniana"* tesis doctoral, Universidad de Sevilla España

[2] Acosta J 2010 *"Furuta's pendulum: A conservative nonlinear model for theory and practice"* Mathematical Problems in Engineering

[3] Cruz M,Silva R, Merlo C, Villarreal M, Muñoz D and Hernandez V 2014 *"Modeling and Construction of a Furuta Pendulum Prototype"* International Conference on Mechatronics, Electronics and Automotive Engineering

[4] Blajer W and Kolodziejczyk K 2008 *Modeling of underactuated mechanical systems in partly specified motion* Journal of Theoretical and applied mechanics

[5] Kajita S, Morisawa M, Miura K, Nakaoka S, Harada K, Kaneko K, et al 2010 *"Biped walking stabilization based on linear inverted pendulum tracking"* In: Proceedings of the IEEE/RSJ 2010 International Conference on Intelligent Robots and Systems, Taipei, pp. 4489-449

[6] Younis W and Abdelati M. 2009. Design and implementation of an experimental segway model, AIP Conference Proceedings, 1107, p. 350-354

[7] VIguria A, Cano, Fiacchini M, Prieto A, Vela B, Rubio F, Aracil J, and Canudas de Wit C *"PPCAR (Personal Pendulum Car): Vehículo Basado en Péndulo Invertido"*. Departament Ingeniería de Sistemas y Automática, Universidad de Sevilla, España. Laboratoire d'Automatique de Grenoble (CNRS-LAG), Francia.

[8] Quanser Innovate Educate, "The Rotatory Control Lab a Modular Single Source Solution You Can Control", www.quanser.com

[9] Toro R 2009 *"Diseño y Control de un Péndulo Furuta para su Utilización en las Aulas de Clase de la Universidad EAFIT"*, Proyecto de grado para su utilización en las aulas de clase de la Universidad EAFIT. Universidad EAFIT, escuelas de ingenierías, departamento de ingeniería mecánica. Medellin, pp. 88

[10] Huang Ch, Cheng F and Ju Y 2010 *"Robust Control of a Furuta Pendulum"* SICE Annual Conference 2010, The Grand Hotel, Taipei, Taiwan

[11] Fedák V, Ďurovský F and Üveges F 2014 *"Analysis of Robotic System Motion in SimMechanics and MATLAB GUI Environment"*, Department of Electrical Engineering and Mechatronics, FEEaI, Technical University of Kosice, Slovakia

[12] Marion J 1998 *"Dinámica clásica de las partículas y sistemas"*. University of Maryland College Park. Academic Press, New York y London. Editorial Reverté, S. A, pp 643

[13] Conejo C *"Control Robusto H de Sensibilidad Mezclada Aplicado a Sistemas Lineales Invariantes en el Tiempo Subactuados"* tesis al grado de maestro en ciencias de la ingeniería eléctrica con orientación en control automático, Universidad Autónoma de Nuevo León, Facultad de Ingeniería Mecánica y Eléctrica. División de Estudios de Posgrado pp. 112

[14] Kelly R and Santibáñez V 2003 *"Control de Movimiento de Robots Manipuladores"*, PEARSON EDUCATION, S.A., Madrid, pp. 59-113

Thermodynamic Optimality criteria for biological systems in linear irreversible thermodynamics

J C Chimal[1], N Sánchez[2], PR Ramírez[2]
[1]Centro de Investigación en Computación, Instituto Politécnico Nacional, Av. Juan de Dios Bátiz s/n U.P. Zacatenco, 07738, Ciudad de México, México.
[2]Escuela Superior de Física y Matemáticas, Instituto Politécnico Nacional, Edif. 9 UP Zacatenco, 07738, Ciudad de México, México.

E-mail: jchimale@ipn.mx

Abstract. In this paper the methodology of the so-called Linear Irreversible Thermodynamics (LIT) is applied; although traditionally used locally to study general systems in non-equilibrium states in which it is consider both internal and external contributions to the entropy increments in order to analyze the efficiency of two coupled processes with generalized fluxes J_1, J_2 and their corresponding forces X_1, X_2. We extend the former analysis to takes into account two different operating regimes namely: Omega Function and Efficient Power criterion, respectively. Results show analogies in the optimal performance between and we can say that there exist a criteria of optimization which can be used specially for biological systems where a good design of the biological parameters made by nature at maximum efficient power conditions lead to more efficient engines than those at the maximum power conditions or ecological conditions.

1. Introduction

It is well known that the universal validity of the Carnot efficiency $\eta_C = 1 - T_2/T_1 \equiv 1 - \tau$ for any reversible heat engine operating between reservoirs at temperatures T_1 and $T_2 (T_1 > T_2)$ has little practical relevance since it applies to zero-power output heat devices. On the contrary, real heat engines work at nonzero power and evolve along irreversible paths coming from finite-time and finite-size unavoidable constraints. In 1975, Curzon and Ahlborn [1] introduced a Carnot-like thermal engine in which there was no thermal equilibrium between the working fluid and the thermal reservoirs at the isothermal branches of the cycle. Curzon and Ahlborn (CA) demonstrated that such an engine delivers nonzero power, a positive entropy production and has more realistic efficiency than that the Carnot engine. The efficiency of a CA engine in the performance of maximum power is given by $\eta_{CA} = 1 - \sqrt{\frac{T_2}{T_1}}$. We can thus conclude that the CA engine is a better approximation to real engines than the Carnot engine.

This seminal paper led to the establishment of a new brand of irreversible thermodynamics, known as finite-time thermodynamics (FTT), which has been developed during last decades [2–11]. One of central

Content from this work may be used under the terms of the Creative Commons Attribution 3.0 licence. Any further distribution of this work must maintain attribution to the author(s) and the title of the work, journal citation and DOI.
Published under licence by IOP Publishing Ltd

questions addressed in FTT is to identify optimal procedures for different objective functions. Originally Curzon and Ahlborn proposed to optimize the power output but some other different functions had been proposed. Besides, most of energy-transfer devices discussed by the FTT were heat engines although, some optimal performance analyses were also conducted where biochemical reaction models were analyzed [12–14].

For instance, Angulo-Brown et al. [15] proposed that the two ways of producing ATP follow different criteria of thermodynamic optimization. The anaerobic glycolisis is carried out by a chemical reaction maximizing power output, while respiration undergoes a process whose optimized thermodynamic function is that representing the best compromise between high power output and low entropy production [16]. Previously, Stucki *et. al.* [17] analyzed, with the aid of Caplan and Essig's definitions of power output and efficiency [12], some optimum working regimes different from that of minimum entropy production studied before by Prigogine [18].

In the present work we extend the former analysis to takes into account two different operating regimes namely: Omega Function and Efficient Power criterion, respectively. For the first case; the Omega Function, this feature was proposed based on a unified optimization criterion [19] which represents the best compromise between the benefits and losses of energy to any device and has the advantage that it does not explicitly require the entropy expression to be applied, or any other environmental parameter, such as the case where the Exergy or the Ecological criteria are used as objective functions [20]. The second maximization of this function provides a compromise between power and efficiency, where the designed parameters at maximum efficient power conditions lead to more efficient engines than those at the maximum power conditions [21]. Finally, the advantages and disadvantages of the mentioned regimes are discussed from a thermodynamical point of view.

2. Linear irreversible thermodynamics and optimal regimes

Let us consider, following Prigogine [18] and Yourgrau *et al.* [22], a system in a steady state consisting of two coupled processes with generalized fluxes J_1, J_2 and their corresponding forces X_1, X_2. As is well known from the theories of near-equilibrium irreversible thermodynamics, the system's entropy production rate is given by,

$$\sigma = J_1X_1 + J_2X_2 \text{ ....(1)}$$

Without loss of generality, we can choose $J_1X_1 > 0$ to be the driven process, and $J_2X_2 < 0$ to be the driver process [12]. Obviously$J_1X_1 + J_2X_2 > 0$, in accordance with the second law of thermodynamics. To have a steady state, we shall consider X_2 constant [12, 22]. Although steady states can also be obtained by setting J_2 or J_2X_2 constant [12], X_2 constant seems to be the case in many biological systems [12,18]. Caplan and Essig [12] introduced definitions of power output P and efficiency η for energy converters working in steady states at constant pressure and temperature, in the following manner:

$$P = -TJ_1X_1 \text{............(2)}$$

and

$$\eta = -\frac{J_1X_1}{J_2X_2} \text{..............(3)}$$

With the above definitions, optimum regimes like those which maximize the Omega Function and Efficient Power criterion, and in general optimize proper functions of these variables, can be studied. It is important to notice that although we take the optimum performance regimes usually studied by Finite Time Thermodynamics (FTT), the approaches of this theory and Linear Irreversible Thermodynamics (LIT) (the one here employed), are quite different. FTT studies thermal engines working in cycles and considers them as internally reversible taking into account only external contributions to the system's entropy increments. On the other hand, LIT locally studies general systems in non-equilibrium states and considers both internal and external contributions to the entropy increments. In the present work we consider only internal irreversibilities since they are responsible for the whole entropy increments of the universe. Although in general both theories led to similar results as regards the thermodynamically characteristics of the performance regimes studied.

For linear energy converters the relations between fluxes and forces, or phenomenological relations, are as follows [12, 23-26]:

$$J_i = \sum_{j=1}^{2} L_{ij} X_j \qquad i = 1,2 \text{ ....(4)}$$

with constants L_{ij} called the phenomenological coefficients. It is also assumed the validity of Onsager's reciprocity theorem which establishes the symmetry of the matrix of phenomenological coefficients, *i.e.* $L_{ij} = L_{ji}$.

From equation (4) and Onsager's reciprocity theorem, equations (1)-(3) become;

$$\sigma = L_{22} X_2^2 [(1 - q^2) + q^2(1 - x)^2] \text{ ....(5)}$$

$$P = T L_{22} X_2^2 q^2 x(1 - x) \text{ ....(6)}$$

$$\eta = -\frac{q^2 x(1-x)}{1-q^2 x} \text{ ....(7)}$$

with $q = {L_{12}^2}/{L_{11}L_{22}}$ a coefficient that measures the degree of coupling between the driver and the diven process [12], $x = {X_1}/{X_1^0}$ and $X_1^0 = -\left({L_{12}}/{L_{11}}\right) X_2$, the driven force at which $J_1 = 0$.

Now the behavior for the Omega function is presented, as we mention later thę Omega function was proposed based on a unified optimization criterion [19] which represents the best compromise between the benefits and losses of energy to any device and has the advantage that it does not explicitly require the entropy expression to be applied, it is given by,

$$\Omega = \frac{2\eta - 1}{\eta} \dot{W}_{out} \text{ ....(8)}$$

Substituting equations (6) and (7) inti equation (8) yields,

$$\Omega = -T L_{22} X_2^2 \left[(1 - q^2 x)\left(-1 + \frac{2q^2(1-x)x}{1-q^2 x}\right)\right] \text{ ....(9)}$$

For the analysis x is taken as the independent variable, and in order to obtain the maximum Omega function what must be solved is the following equation $\partial\Omega/\partial x = 0$. Then from equation (9) the condition for the Ω function to be maximum is $x = 0.75$. Now, substituting the maximum value of $x = 0.75$ into equation (7) and performing a series expansion of this efficiency in terms of q value around 0, we obtain, for instance, the expression for efficiency at maximum Omega Function given by,

$$\eta_\Omega = 0.1875\, q^2 + 0.140625\, q^4 + 0.105469\, q^6 + 0.0791016\, q^8 + 0(x^{10}) \quad .....(10)$$

In the other hand, the behavior of the under maximum efficient power regime will also be discussed. This approach was recently proposed by Yilmaz [21] in the context of heat thermal engines, and is defined as the product of the power output and efficiency, which states:

$$P_E = \eta\, \dot{W}_{out} \quad ....(11)$$

Substituting equations (6) and (7) in equation (11) yields,

$$P_E = -TL_{22}X_2^2 \left[\frac{q^4(1-x)^2x^2}{1-q^2x}\right] \quad ....(12)$$

For the analysis x is taken as the independent variable, and in order to obtain the maximum efficient power what must be solved is the following equation $\partial P_E/\partial x = 0$. Then, from equation (12) the condition for the P_E function to be maximum is,

$$x = \frac{4+q^2-\sqrt{16-16q^2+q^4}}{6q^2} \quad ....(13)$$

Now, substituting the maximum value of x into equation (7) and performing a series expansion of this efficiency in terms of q value around 0, we obtain, for instance, the expression for efficiency at maximum P_E function given by,

$$\eta_{P_E} = q^2/4 + q^4/8 + 19\, q^6/256 + 25\, q^8/512 + 0(x^{10}) \quad .....(14)$$

Fig. 1 presents optimized efficiency under the two criteria as a function of q, also maximum efficiency is shown. The figure shows that values of efficiency operating under maximum Omega criterion are greater with respect to the Maximum Efficient Power, until $q = 0.9$, where the behavior changes and form this value until $q = 1.0$ the values of efficiency operating under maximum Omega criterion are lesser with respect to the Maximum Efficient Power.

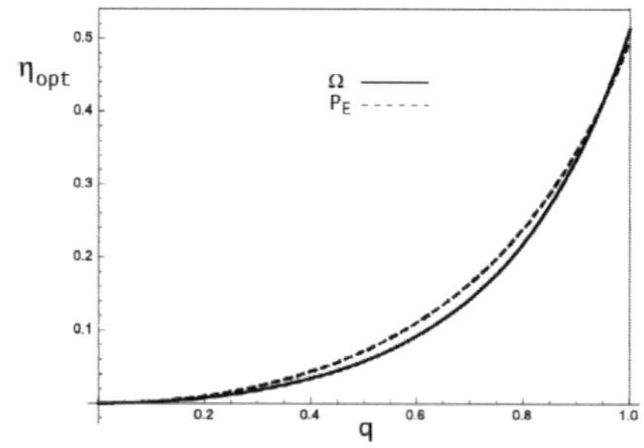

Figure 1. Efficiency under the two thermodynamic optimization criteria as a function of q, the Maximum Efficient Power represented by the dotted line and Omega Function the continuous line.

Figure 2 presents optimized efficiency under the four criteria as a function of q, the figure shows that values of efficiency operating under maximum Omega criterion (continuous), Maximum Power [14] (blue dashed), Maximum Ecological [14] (also continuous line overprint with Omega), and Function Maximum Efficient Power (dashed).

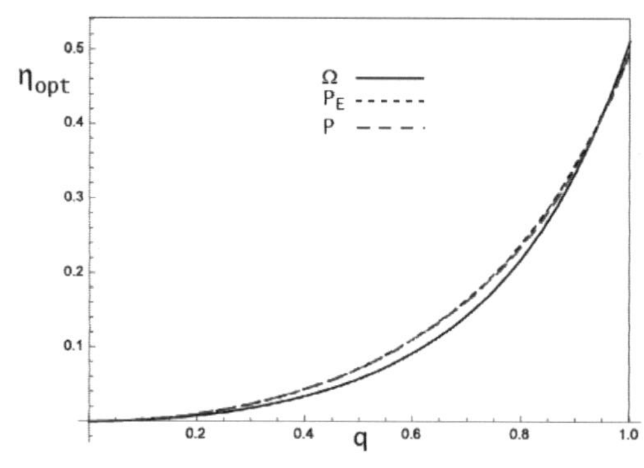

Figure 2. Efficiency under four thermodynamic optimization criteria as a function of q, namely, at Maximum Efficient Power, in the Omega Function, Ecological Function and in the Maximum Power function.

3. Concluding Remarks

In this paper the methodology of the so-called LIT is applied; although traditionally used locally to study general systems in non-equilibrium states in which it is consider both internal and external contributions to the entropy increments in order to analyze the efficiency of two coupled processes with generalized fluxes J_1, J_2 and their corresponding forces X_1, X_2. We extend the former analysis to takes into account two different operating regimes namely: Omega Function and Efficient Power criterion, respectively.

The behavior of the efficiency as a function of q is shown for the Omega and the Efficient Power functions, respectively in Figure 1. From the above analysis, it can be shown that there exists a characteristic value of x that maximizes these functions. Besides, when we obtain the efficiency in terms of the phenomenological coefficient q for both objective functions (equations 14 and 10), it seems a convex function for which the efficiency is greater when q grows.

When these two functions are compared with those obtained by Santillán et al. [14] depicted in Figure 2 (Efficiency at Maximum Power and Efficiency and Maximum Ecological Function), it is shown that for q close to one, the efficiency operating under the Eficient Power function is greater than the other three (Omega function, Maximum Power and Ecological function). Also, it is worth to notice that both the Omega function and the Ecological function have the same behavior, which is a characteristic of the isothermally of the system.

Finally, from the above we can say that there exist a criteria of optimization which can be used specially for biological systems where a good design of the biological parameters made by nature at maximum efficient power conditions lead to more efficient engines than those at the maximum power conditions or ecological conditions (where the main considerations are energetic), this suggest some differences between biological systems based on its own design.

4. References

[1] Curzon F and Ahlborn B 1975 Efficiency of a Carnot engine at maximum power output (*Am. J. Phys* vol 43) pp 22–24
[2] De Vos A. 1992, *Endoreversible Thermodynamics of Solar Energy Conversion* (Oxford University Press: New York, NY, USA)
[3] Sieniutycs S and Salamon P 1990 *Finite-Time Thermodynamics and Thermoeconomics* (New York: Taylor and Francis)
[4] Wu C, Chen L and Chen J 1999 *Recent Advances in Finite-Time Thermodynamics* (New York: Nova Science Publishers)
[5] Bejan A 1995 Entropy Generation Minimization (Florida: CRC Press)
[6] Wu F, Wu C, Guo F, Li Q and Chen L 2003 Optimization of a thermoacoustic engine with a complex heat transfer exponent. *Entropy* **5** 444–451
[7] Chen L, Zheng T, Sun F and Wu C 2004 Optimal cooling load and COP relationship of a four-heat-reservoir endoreversible absorption refrigeration cycle *Entropy* **6** 316–326
[8] Zheng T, Chen L, Sun F and Wu C 2003 Effect of heat leak and finite thermal capacity on the optimal configuration of a two-heat-reservoir heat engine for another linear heat transfer law *Entropy* **5** 519–530
[9] Ladino D 2007 On optimization of a non-endorreversible Curzon-Ahlborn cycle *Entropy* **9** 186–197
[10] Ladino D 2005 Van der Waals gas as working substance in a Curzon and Ahlborn-Novikov engine *Entropy* **7** 108–121
[11] Feidt M 2009 Optimal thermodynamics new upperbounds *Entropy* **11** 529–547
[12] Caplan S R and Essig A 1999 *Bionergetics and Linear Nonequilibrium Thermodynamics the Steady State*, iUniverse
[13] Angulo F, Santillán M and Calleja E 1995 Thermodynamic optimality in some biochemical reactions *Il Nuovo Cimento* **17**D 85
[14] Santillán M, Arias L and Angulo F 1997 Some optimization criteria for biological systems in linear irreversible thermodynamics *Il Nuovo Cimento* **19**D 99
[15] Angulo F, Santillán M and Calleja E 1995 Thermodynamic optimality in some biochemical reactions, *Il Nuovo Cimento* **17**D 85

[16] Angulo F 1991 An ecological optimization criterion for finite-time heat engines *J. Appl. Phys.* **69** 7465
[17] Stucki J W 1980 *Eur. J. Biochem.* **109** 269
[18] Prigogine I 1961 *Thermodynamic of Irreversible Processes* (New York: John Wiley and Sons)
[19] Calvo A, Medina A, Roco J M M, White J A and Velasco S 2001 Unified optimization criterion for energy converters, *Phys.Rev.E* **63** 037102
[20] Bejan A 1995 *Entropy Generation Minimization:The Method of Thermodynamic Optimization of Finite-Size Systems and Finite-Time Processes* (Florida: CRC Press)
[21] Yilmaz T 2006 A new performance criterion for heat engines: efficient power *J. Energy Inst.* **79** 38
[22] Yourgrau W, Van Der Merwe A and Raw G 1982 *Treatise on Irreversible and Statistical Thermophysics* (New York: Dover)
[23] Lebon G, Jou D and Casas J 2008 *Understanding Non-Equilibrium Thermodynamics* Springer
[24] Katchalsky A and Curran P F 1965 *Non-Equilibrium in Biophysics* (Cambridge: Harvard University Press)
[25] Gyarmati I 1970 *Non-Equilibrium Thermodynamics* Springer (New York: Berlin Heidelberg)
[26] Aguieev I P 2001 *Termodinámica Fuera de Equilibrio* (Moscú: URSS)

Acknowledgments
Authors thank CONACYT, EDI-IPN and COFAA-IPN for supporting this work. N. Sánchez-Salas is grateful for financial support from MINECO (Spain) under Grant. ENE2013-40644-R.

IOP Conf. Series: Journal of Physics: Conf. Series **792** (2017) 012083 doi:10.1088/1742-6596/792/1/012083

Digital tool for detecting diabetic retinopathy in retinography image using gabor transform

Y Morales, R Nuñez, J Suarez and C Torres
Laboratorio de óptica e informática, Universidad Popular del Cesar
Valledupar, Cesar-Colombia

E-mail: morales.yaileth@unicesar.edu.co

Abstract. Diabetic retinopathy is a chronic disease and is the leading cause of blindness in the population. The fundamental problem is that diabetic retinopathy is usually asymptomatic in its early stage and, in advanced stages, it becomes incurable, hence the importance of early detection. To detect diabetic retinopathy, the ophthalmologist examines the fundus by ophthalmoscopy, after sends the patient to get a Retinography. Sometimes, these retinography are not of good quality. This paper show the implementation of a digital tool that facilitates to ophthalmologist provide better patient diagnosis suffering from diabetic retinopathy, informing them that type of retinopathy has and to what degree of severity is find . This tool develops an algorithm in Matlab based on Gabor transform and in the application of digital filters to provide better and higher quality of retinography. The performance of algorithm has been compared with conventional methods obtaining resulting filtered images with better contrast and higher.

1. Introduction

Diabetic retinopathy is a complication ophthalmological of the diabetes. This is caused by deterioration of blood vessels that nourish the retina. This weakening of blood vessels may cause dilation of the vascular wall, leading to extravasation of fluid such as blood plasma and as a final result the appearance of bleeding [1].

Imaging techniques are based on the development of systems capable of detecting different physical signals emitted by the body or organ object study, and that are imperceptible to humans. Once detected, the own system the signals into data which, when processed in turn, allow the formation of an image. Against this background, it has been doing for the past 20 years, a large research to optimize existing technologies, and develop new medical imaging techniques [2].

The progress of medical imaging techniques is very diverse, from areas where research is still at a very conceptual level, to fields where there are well developed prototypes that are used in medical research, but not yet part of the clinical routine for various reasons, or techniques used with success in clinical practice, but has not extended its use for other applications [3]

This paper provides a technique related to computational analysis of the ophthalmoscopic images (retinography), with the aim of making improvements to those images by applying Gabor transform and

Content from this work may be used under the terms of the Creative Commons Attribution 3.0 licence. Any further distribution of this work must maintain attribution to the author(s) and the title of the work, journal citation and DOI.
Published under licence by IOP Publishing Ltd

digital filters that allow you to have higher quality and better viewing. What would enable ophthalmologists perform an analysis more detailed about the possible conditions that affect their patients, especially in our case study, diagnose "diabetic retinopathy"

The Gabor function has been used extensively in texture segmentation [4] and defect detection [5], as it can be tuned to a specific frequency and orientation, and achieve both localization in the time domain and frequency domain; two- dimensional (2-D) Gabor filters are multi-channel filters with a multi-resolution decomposition process, which is similar to wavelet analysis. Time-frequency analysis, such as the Gabor transform, plays an important role in many signal processing applications. In harmonic analysis the algorithms of Fourier transformation are often applied [6,7]. However, this method gives significantly better results for stationary systems where frequency composition does not change during the investigation.

A fair amount of research papers have been published in literature for Gabor filter-based image processing [8,9]. Besides face recognition, Gabor filters are successfully used in many other images processing and analysis domains, such as: image smoothing, image coding, texture analysis, shape analysis, edge detection, and fingerprint and iris recognition. In this paper show the implementation of a digital tool that facilitates to ophthalmologist provide better patient diagnosis suffering from diabetic retinopathy, informing them that type of retinopathy has and to what degree of severity is find. This tool develops an algorithm in Matlab based on Gabor transform and in the application of digital filters to provide better and higher quality of image retinography

2. Methodology

2.1 *Design*

The images are acquired by a retinograph that takes a picture of the fundus. The retinograph we use for testing is the TRC50DX, consisting of a digital camera of high resolution 12.1 and 16 megapixels to color, which it allows to observe the details of the fundus, storing and transferring to a computer . Here these images are converted to grayscale and segmented. 'Figure 1'

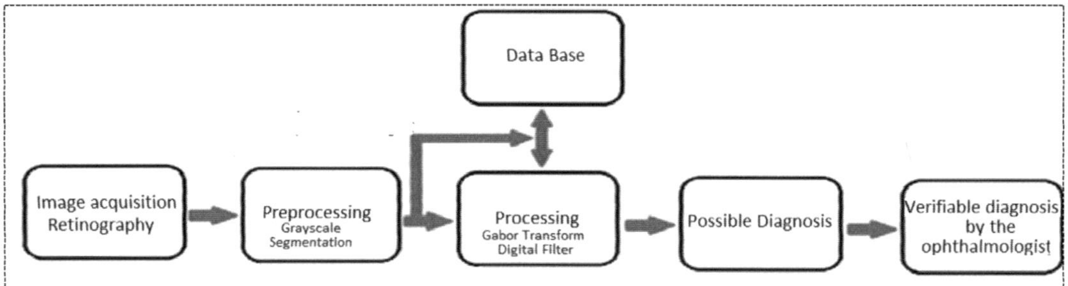

Figure 1. Block Diagram of the system design

The proposed algorithm for the detection of frequency hopping signals, perform processing of the signal segment in four stages: In the first stage, the image and the Gabor transform are defined; in the second stage the Fourier transform is computed for the image and Gabor surface; in the next stage, the convolution of the Fourier transforms is developed, and finally in the next stage the Gabor transform of the image is allowed and the skeletonization of image is finally obtain [10]. 'Figure 2'

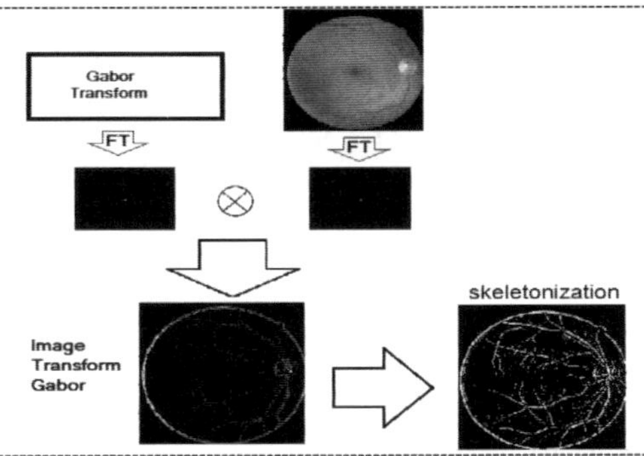

Figure 2. Block Diagram of the Image Processing

The database images were taken from Kuopio University Hospital and the eye clinic of the Doctor Flavio Piñeres, who is ophthalmologist that helped in the classification of the same

2.2 Implementation

2.2.1 *Gabor Transform*

Gabor transform was first defined by Gabor [10-13] and later extended to 2-D. In the spatial domain, the Gabor function is a Gaussian curve with aspect ratio modulated by a complex sinusoid; the Gabor filter was originally introduced by Dennis Gabor. The one-dimensional Gabor filter is defined as the multiplication of a cosine/sine (even/odd) wave with a Gaussian window as follows:

$$g_e(x) = \frac{1}{\sqrt{2\pi\sigma}} e^{-\frac{x^2}{2\sigma^2}} cos(2\pi\omega_0 x) \qquad (1)$$

$$g_0(x) = \frac{1}{\sqrt{2\pi\sigma}} e^{-\frac{x^2}{2\sigma^2}} cos(2\pi\omega_0 x) \qquad (2)$$

Where ω_0 defines the centre frequency (i.e., the frequency in which the filter yields the greatest response) and the spread of the Gaussian window. Daugman extended the Gabor filter to two dimensions as follows:

$$g_e(x,y) = \frac{1}{2\pi\sigma_x\sigma_y} e^{-\left(\frac{x^2}{\sigma_x}+\frac{y^2}{\sigma_y}\right)} cos\left(2\pi\omega_{x_0}x + 2\pi\omega_{y_0}y\right) \quad (3)$$

$$g_0(x,y) = \frac{1}{2\pi\sigma_x\sigma_y} e^{-\left(\frac{x^2}{\sigma_x}+\frac{y^2}{\sigma_y}\right)} sin\left(2\pi\omega_{x_0}x + 2\pi\omega_{y_0}y\right) \quad (4)$$

Where $\omega_{x_0}, \omega_{y_0}$ defines the centre frequency and σ_x, σ_y the (potentially asymmetric) spread of the Gaussian.

The Gabor filter represents a band-pass linear filter whose impulse response is defined by a harmonic function multiplied by a Gaussian function. Thus, a bidimensional Gabor filter constitutes a complex sinusoidal plane of particular frequency and orientation modulated by a Gaussian envelope [14]. It achieves an optimal resolution in both spatial and frequency domains. 'Figure 3'

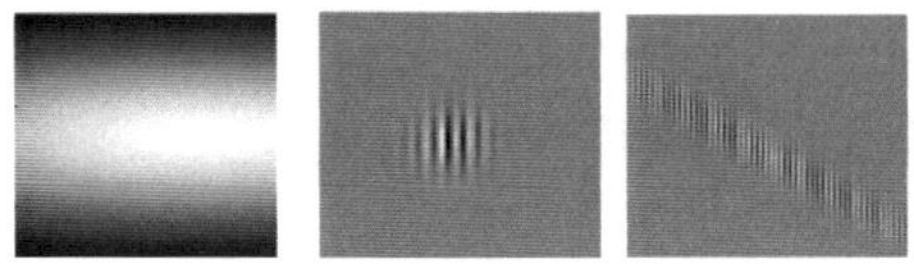

Figure 3. Gabor filter surfaces

3. Results

From the image processing point of view, many strategies and algorithms have been developed to deal with the extraction of the retinal vessel tree. An automatic segmentation of the vasculature could save workload of the ophthalmologists and may assist in characterizing the detected lesions and to identify false positives. The filters used in the test images are filters that are predetermined in matlab such as Laplacian, Prewitt, Sobel, Sharpen, among others, which finally did not produce the expected results. 'Figure 4'

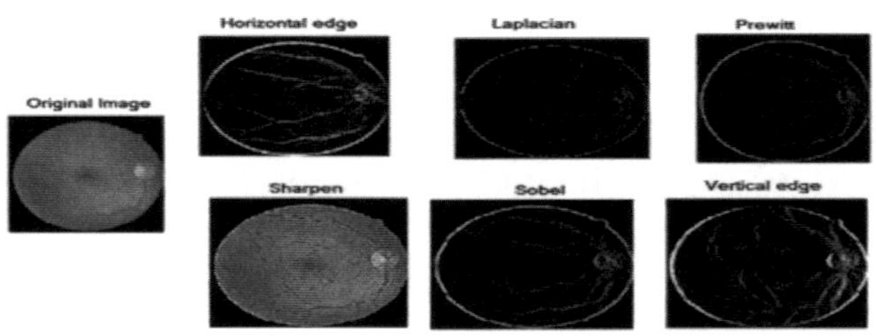

Figure 4. Retinography analyzed with different filters

For this reason, they continued testing with other filters that serve as a basis for designing a new one that would allow us to get a better picture. Among these is the canny filter and filter gabor. The Gabor filter is one efficient method for automatic segmentation of retinal blood vessels [15]. 'Figure 6'

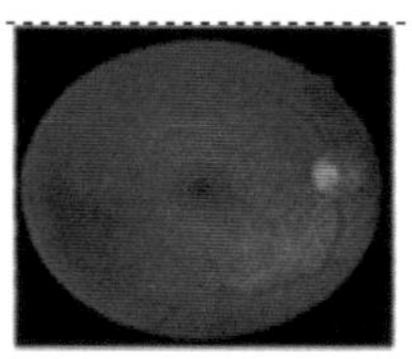

Figure 5. Original Image

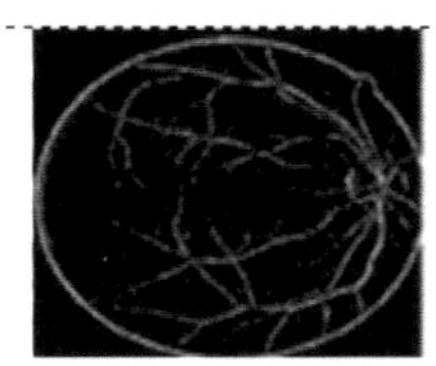

Figure 6. Gabor Filter

One of the drawbacks that occurs during the analysis of the Retinography is the optical disk, for this reason we use transforms Hung to differentiate and exclude it from our analysis. 'Figure 8'

IOP Conf. Series: Journal of Physics: Conf. Series **792** (2017) 012083 doi:10.1088/1742-6596/792/1/012083

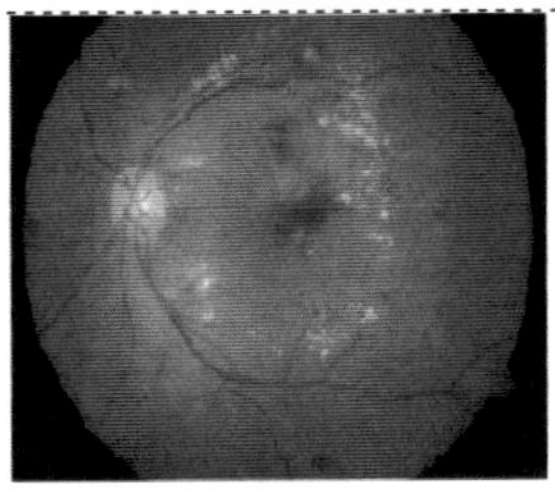

Figure 7. Image of a diseased eye

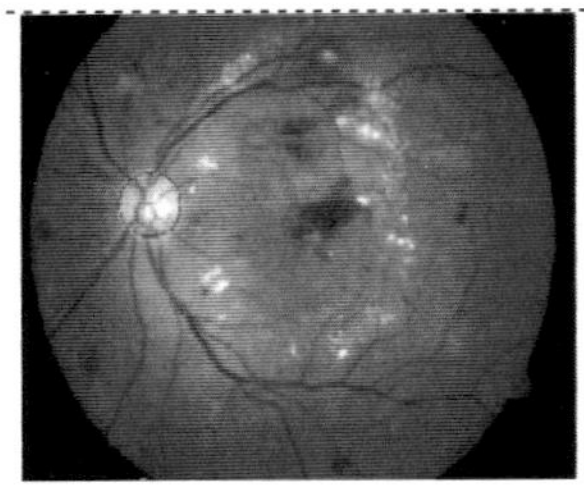

Figure 8. Detection optical disk

It has designed a graphical interface called "Computer Aided System for Early Detection of Diabetic Retinopathy" which it has two formularies, one gives it to the ophthalmologist the option to manually analyse the image; and the other it is of the software, which it is able to analyse and give their own diagnosis, This must be validated by the health professional

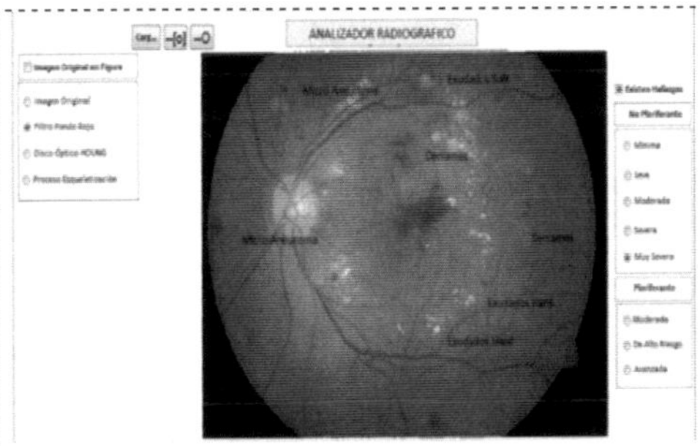

Figure 9. Computer Aided System for Early Detection of Diabetic Retinopathy. Manual Part

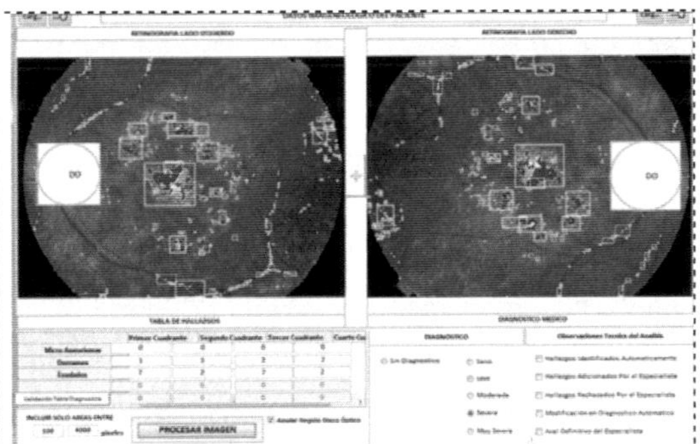

Figure 10. Computer Aided System for Early Detection of Diabetic Retinopathy. Assisted Part

In these figures are showing the final analysis made by the software. In they are showing the detection of the regions and the possible anomalies in the different quadrants. This digital tool can to perform its own diagnostic; in this case the patient has severe diabetic retinopathy. This result was validated by the ophthalmologist.

4. Conclusion

It has developed a digital tool that detects in a Retinography if a patient has diabetic retinopathy and to what degree is this. The Gabor filter is one efficient method for automatic segmentation of retinal blood vessels. This filter is implemented in Matlab. It is a powerful tool when processing digital images and this application is a proof, since it was used to make a software that helps the ophthalmologist to diagnose diabetic retinopathy. In this application you only need one retinography image and then be processed in Matlab, where it has carried out different routines to adjust the image so that it can display the constituent parts of the eye and those found to be of interest when giving an assessment. This tool does not replace the ophthalmologist, but on the contrary was developed as an aid to diagnosis of diabetic retinopathy.

Acknowledgments
The authors are grateful to the popular Cesar University for their support in this investigation and the ophthalmologist Flavio Piñeres for their support and validation in the results.

References

[1] Matthew D and Davis M D 1992 *Diabetic retinopathy* (Wisconsin: Diabetes Care **15**) 1844-1869
[2] Narvaez M, Toribio C and Benlloch J M 2009 *Diagnóstico por Imagen* (España: Servifull)
[3] Anglulo J and Iñigo R 2003 *Imágenes Digitales y Aplicaciones de la visión artificial* ed Paraninfo
[4] Hoover A, Kouznetsova V and Goldbaum M 2000 *Locating blood Vessels in retinal images by piecewise threshold probing of a matched filter response* (IEEE Trans. Med Imag **19**) 203-2 10
[5] Jiang X and Mojon D 2003 *A Local daptive thresholding by verification-based multithreshold probing with application to vessel detectionin retinal images* (IEEE Trans. Pattern Anal. Machine Intell) **25** 131-1 37
[6] Klonoff D C and Schwartz D M 2000 *An economic analysis of interventions for diabetes* (Diabetes Care **2**) 390-404
[7] Spencer T, Olson J A, McHardy K C, Sharp P F and Forrester J V 1996 *An image-processing strategy for the segmentation and quantification of microaneurysms in fluorescein angiograms of the ocular fundus* (Comput. Biomed. Res. **29** no **4**) 284-302
[8] Zana F and Klein J C 1999 *A multimodal registration algorithm of eyefundus images using Hough transform detection and Vessels* (IEEE Trans. Med. Imag. **18**) 419-428
[9] Chutatape O, Zheng L and Krishman S 1998 *Retinal detectionand blood vessel tracking by matched Gaussian and Kalman filters* (Proc.IEEE Int Biol Eng Conf Soc **20**) 3144-3149.
[10]Lasso W, Morales Y and Torres C 2014 *Image segmentation blood vessel of retinal using conventional filters Gabor transform and skeletonization XIX Symposium on Image Signal Processing and Artificial Vision* (IEEE) 1-4
[11] Tolias Y A and Panas S M 1998 *A fuzzy vessel tracking algorithm based on fuzzy images forretinal clustering* (IEEE Trans. Med. Imag. **17**) 263-273
[12]Can A, Shen H, Turner J N, Tanenbaum H L, and Roysam B 1999 *Rapid automated tracing and feature extraction from retinal fundusimages using direct exploratory algorithms* (IEEE Trans. Inform.Technol. Biomed **3**) 125-1 38
[13]Gagnon L, Lalonde M, Beaulieu M and Boucher M C 2001 *Procedure todetect anatomical structures in optical fundus images* (Proc. SPIE Med.Imaging: ImageProcessing, **4322**) 1218 -1225
[14]. Chaudhuri S, Chatterjee S, Katz N, Nelson M and Goldbaum M 1989 *Detection of Blood Vessels in retinal images using two-dimensionalmatched filters* (IEEE Trans. Med. Imag, **8**) 263-269
[15]. Kaur J and Sinha H P *2012 Automated detection of retinal blood vessels in diabetic retinopathy using Gabor filter* (International Journal of Computer Science and Network Security **4**) pp 12-14 109-116.

Design of an electronic system for acquisition of the tunneling current signal of a Scanning Tunneling Microscope STM

J F Ávila, H Aya and E M Gualteros
Facultad de Ingeniería
Universidad Distrital Francisco José de Caldas
Bogotá, Colombia

E-mail: haya@udistrital.edu.co

Abstract. An electronic system was designed to acquire tunneling current of a Scanning Tunneling Microscope. A tunneling current originates from the interaction between a conductive tip and a conductive or semiconductor sample. Lower currents than 100 nA were generated and from them, voltages were obtained in the order of 1 volt. Using this mechanism, it is possible to construct images of the atoms positions on the surface of the sample, also other properties of the materials can be studied.

1. Introduction.

The scanning tunneling microscope STM is a device for imaging surfaces at scale atomic level [1]. The scanning tunneling microscope was developed in 1981 by researchers at IBM in Switzerland and winners of the Nobel Prize in physics Gerd Binnig and Heinrich Rohrer [2]. A scanning tunneling microscope is a useful tool for research in different fields such as nanoelectronics, nanomaterials, biotechnology, and nanorobotics, among others [3]. The specific results of the STM implementation and research include the following achievements: analysis of transistor surface oxide [4], identification through images of quantum dots in diodes [5], Atomic reduction of voltage bias on a controller [6], characterization of interface of MOS devices [7] and modulated hole with semiconductor [8].

2. Theory.

Tunnel effect refers to the phenomenon of quantum mechanics in which an electron has a probability of passing through a potential barrier with a height greater than the energy of the electron [9]. In the scanning tunneling microscope (STM), tunneling current, as well as the potential barrier height, also depends on the ratio of the electronic filling levels on each side of the barrier, the filling levels are referenced Fermi level, material work function ϕ, and polarization voltage V.
The value of tunneling current depends on the distance between the sample and the tip of the microscope, it can be expressed as

$$I_t = k\, e\, V\, exp\,(-2\frac{\sqrt{2m\phi}}{\hbar}d)\,, \tag{1}$$

where V is the applied voltage, $\hbar = \frac{h}{2\pi}$, m is the electron mass, e is the electron charge, ϕ is the effective potential barrier, d is the effective wide barrier and k is a constant of proportionality. A Scanning

Content from this work may be used under the terms of the Creative Commons Attribution 3.0 licence. Any further distribution of this work must maintain attribution to the author(s) and the title of the work, journal citation and DOI.
Published under licence by IOP Publishing Ltd

Tunneling Microscope STM is composed of various mechanical and electronic devices. This instrument operates in two different modes: constant-height mode and constant-current mode. In constant-current mode, a feedback controller adjusts the height of the probe to keep the current constant [10]. In constant-height mode, the tip travels in a horizontal plane above sample while the current changes depending on the local topography of the sample [11].

3. Implementation.

The next elements were used for creating a current of the order of 10-100 nA, followed by the conditioning of this current to voltages of the order of 1 volt: current source, transimpedance amplifier TIA which converts the current signal to a voltage, logarithmic amplifier which linearizes the response of transducers with exponential response, conditioning amplifier which adjust gain and analog-digital which converter an analog signal to digital values. Figure 1 shows the elements used in this project and his connections. In the left part of the figure are observed the lumped elements that generate and transform the virtual current into a voltage signal, whereas in the right part are showed embedded systems which convert the analogous voltage signal to digital. Digital signal allows integrating to a software. In table 1, commercial references of the devices used in the different stages are given.

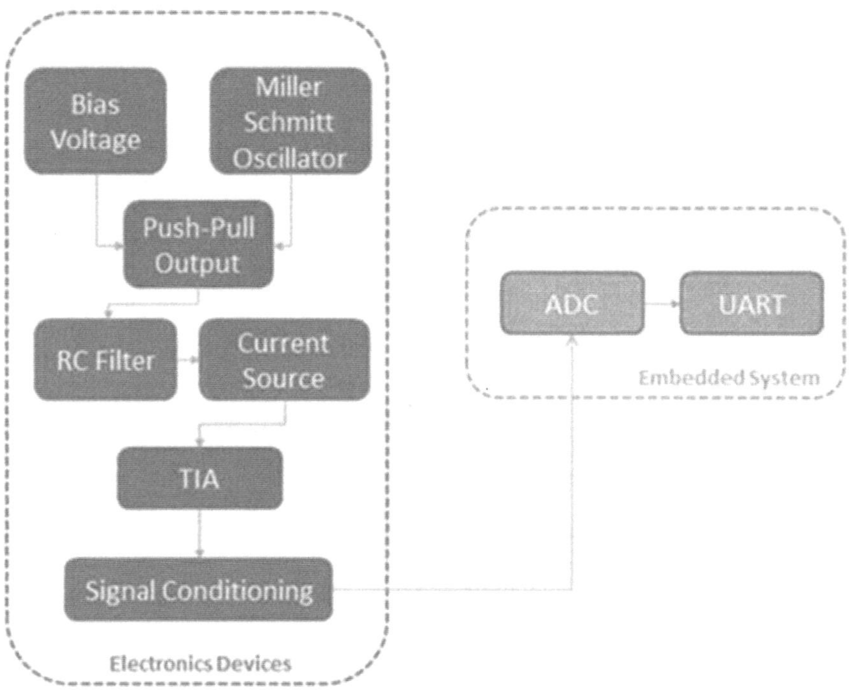

Figure 1. Electronic acquisition stage of the current tunnel signal and treatment [15].

Those operational amplifiers are characterized by a high precision, very low bias current, low noise, low offset voltage (on the order of the μV) and high open-loop gain (on the order of the nA and pA).

Table 1. Precision operational amplifiers

Stage	Reference
Bias voltage	OPA129U
Miller-Schmitt oscillator	LF353
Howland current source	OPA2277U
Transimpedance amplifier TIA	LF353 or OPA129U
Voltage follower	OPA132
Inverter with DC level adjustment	LT1012S8

4. Results

Figure 2 shows the results obtained experimentally of direct current values, lower than 100 nA, that were transduced to voltages between 0 and 2 volts approximately with different two transimpedance amplifiers TIA: LF353 (blue dots) and OPA129U (red dots). A clear linear relation is observed between the current and the output voltage. Figure 3 shows the frequency response of the system with transimpedance amplifier TIA OPA129U, from which can be concluded that the system has a high-bandwidth.

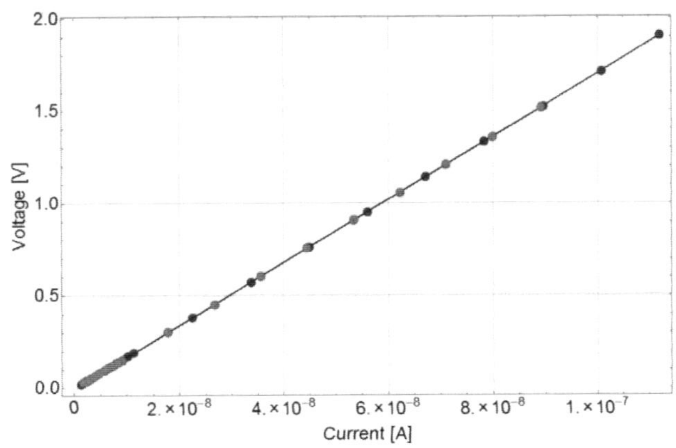

Figure 2. DC experimental analysis with transimpedance amplifier TIA LF353 (blue dots) and OPA129U (red dots).

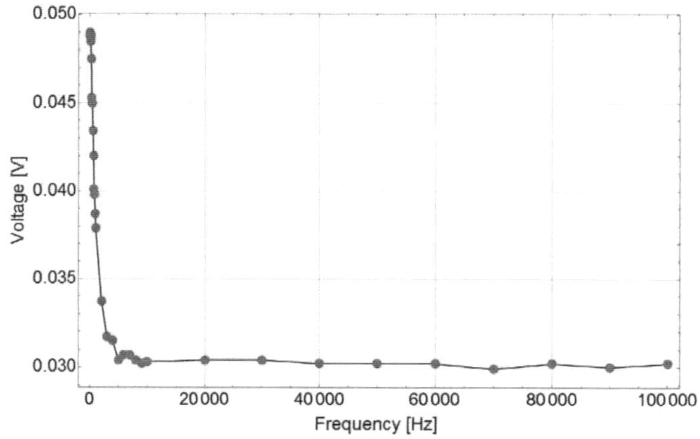

Figure 3. Frequency analysis with transimpedance amplifier TIA OPA129U.

Table 2. Mean values obtained for each stage

Stage	Mean Value
Bias voltage	10 mV
Miller-Schmitt oscillator	VOL=-13.5 V, VOH=14 V
Push-Pull output	60.812 mV
RC filter	10 mV
Howland current source	1 nA
Transimpedance amplifier TIA	20 mV
Inverter with DC level adjustment	20 mV8

5. Conclusions

Once it is designed an electronic system to transduce the tunneling current to suitable voltages, it is possible to continue the design of other elements of the microscope, such as the mechanical control device of the interaction of the tip with the sample and the software for generating images from digital signals.

6. Acknowledgements

This project is partially founded by Engineering Faculty of the Universidad Distrital Francisco José de Caldas de Bogotá, inside the Proyecto Curricular de Ingeniería Electrónica.

7. References

[1] González C A, Ruiz J P, Bonilla R S and Avila A G 2013 *Argentine School of Micro-Nanoelectronics.*
[2] Hayashi T, Tachiki M and Itozaki H 2007 *IEEE Transactions on applied superconductivity* **17** 2
[3] Martirosyan K S, Litvinov D and Lyshevski S E 2012 *12th IEEE International Conference on Nanotechnology (IEEENANO).*
[4] Shewchun and Clarke R A 1972 *The Surface Oxide Transistor (SOT).*
[5] Marchman H M and Wetsel G C 1991 *Tunneling microscope images of identifiable quantum dot diodes.*
[6] Flaxer E 1995 *Automatic bias reduction controller for a scanning tunneling microscope.*
[7] Komeda T, Gwo S and Tokumoto H 1995 *MOS interface characterization by Spanish STM.*
[8] Groenveld R H M, Prins M W J and Kempen H 1994 *Modulated photodetection semiconductor with tips in scanning tunneling microscope.*
[9] Serway R A and Jewett J W 2004 *Physics for Scientists and Engineers.*
[10] Blackford B L, Dahn D C and Jericho M H *2010 Review of Scientific Instruments* **58** 1343
[11] Mándia G and Palotá K 2014 *Applied Surface Science* 304

Environmental monitoring in interventional radiology

S Del Sol[1] **, R Garcia** [1]**, D Sánchez**[1]**, G Ramirez** [2]**, E U Chavarin** [2] **and T Rivera** [1]
[1]Centro de Investigación en Ciencia Aplicada y Tecnología Avanzada - Legaria del IPN, 11500, Ciudad de México, México.
[2]Hospital Juárez de México, 07760, Ciudad de México, México.

E-mail: susi2489@hotmail.com

Abstract. The procedures in Interventional Radiology involve long times of exposure and high number of radiographic images that bring higher radiation doses to patients, staff and environmental than those received in conventional Radiology. Currently for monitoring the dose, the thermoluminescent dosimetry use is recommended. The aim of this work was to carry out the monitoring of the environmental scattered radiation inside the IR room using two types of thermoluminescent dosimeters, TLD-100 (reference dosimeter), CaSO4:Dy (synthesized in our laboratory). The results indicate that the TLD-100 is not effective for the environmental monitoring of low-energy Rx rooms. The CaSO4:Dy presented good behaviour over the 6 months of study. The results will be specific to each room so it is recommended such studies as part of the program of quality control of each Rx room.

1. Introduction

The Interventional Radiology (IR) has had a boom in recent years, because of the multiple benefits in the treatment of different diseases with minimum risk; however, we must emphasize the fact that IR procedures provide higher doses of radiation in the diagnosis by images, which requires us to maintain a proper management of the doses given to patients, occupationally exposed workers (OEW) and work environments to reduce the risk of overexposure [1].

Dosimetry is responsible for estimating the dose of radiation received as part of the medical exposure for which relies on different types of instruments, among them the so-called thermoluminescent dosimeters (TLDs) [2, 3].

The TLDs have the property to store part of the energy they absorb when exposed to ionizing radiation. Subsequently, when they are heated they emit this energy in the form of light and, in an ideal system, this is directly proportional to the dose absorbed by the material (thermally stimulated luminescence) [3].

The main available commercial TLDs are based on oxides, fluorides and sulfate doped with rare earths. Currently, the most commonly used thermoluminescent dosimeter is the TLD-100 [4] although this material does not present a very favorable properties for all applications when working with low-energy x-ray [5]. Other materials as the CaSO4:Dy is 50 times more sensitive than the TLD-100 although are not used routinely for the monitoring of doses [6].

The Juarez Hospital in Mexico is one of the hospitals where more RI interventions are performed per year in the Mexico City, so in this work it was conducted the environmental monitoring of its IR room

Content from this work may be used under the terms of the Creative Commons Attribution 3.0 licence. Any further distribution of this work must maintain attribution to the author(s) and the title of the work, journal citation and DOI.
Published under licence by IOP Publishing Ltd

IOP Conf. Series: Journal of Physics: Conf. Series **792** (2017) 012085 doi:10.1088/1742-6596/792/1/012085

using the CaSO4:Dy synthesized in our laboratory and the TLD-100 at the same time, this last as reference dosimeter for a period of 6 months.

2. Materials and methods

For this study a total of 18 LiF: Mg, Ti chips, commercially known TLD-100 and 18 $CaSO_4$:Dy discs synthesized in our laboratory (CICATA-Legaria, IPN) were used. The materials were characterized previously dosimetrically for low-energy X-rays [5, 6].

Dosimeters received a standard annealing treatment before exposure to radiation. Depending on the type of material, thermal annealing schemes were: 300°C for 30 minutes to $CaSO_4$:Dy and 400°C for 1 hour followed by 100°C for 2 hours to TLD-100. The method of slow cooling inside the muffle was used to reach room temperature for all cases.

Thermoluminescent readings were made using a Harshaw 3500 TL analyzer. The heating rate of the TL analyzer was kept at 10°C/s for both materials. The TL emission was integrated from room temperature up to 350°C.

In order to measure the contribution of the radiation dose scattered in the X-ray room, dosimeters were placed in 9 different positions (A1, A2, A3,...,A9), in order to monitor and identify areas of greatest risk of permanency for workers and the dose distribution within the X-ray room, see the figure 1.

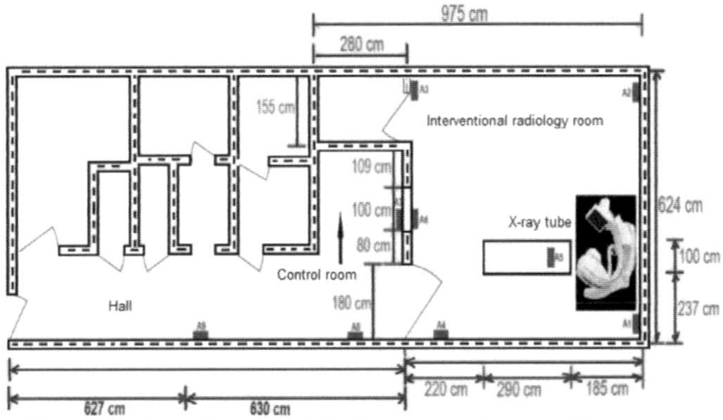

Figure 1. Distribution of dosimeters within the X-ray room.

For positioning in the room the dosimeters were included in a chassis with identical for all geometries and with a 5 mm PMMA filtration, each of these contained 2 TLD-100 chips and two discs of CaSO4:Dy.

The dosimeters were placed at a height of 150cm above the floor, and were subjected to normal conditions of work within the room for the period of 1 month. TL readings performed 24 hours after they were removed. Finally, the monitoring of the environment was conducted over 6 months.

3. Results and discussion

The dose values determined at different locations in the room of IR for the environmental monitoring of 6 months are shown in table 1.

As you can see from the first five months of monitoring, the TLD-100 showed abnormal behavior, detecting values very near the background radiation in different positions within the room except around the X-ray tube. Therefore, the TLD-100 was not already used in the last month.

Table 1. Medium doses collected by the TLD-100 (D_1) and (D_2) CaSO4-Dy in different positions within the room of IR (mGy).

Pos.	1st month		2nd month		3th month		4th month		5th month		6th month
	D_1	D_2	D_1	D_2	D_1	D_2	D_1	D_2	D_1	D_2	D_2
A1	<0.1	0.3	<0.1	0.2	0.5	0.6	0.3	0.9	0.8	1.3	1.5
A2	<0.1	0.2	<0.1	0.2	0.2	0.3	<0.1	0.1	0.3	0.8	0.9
A3	<0.1	0.2	<0.1	0.2	0.2	0.3	<0.1	0.5	<0.1	0.4	0.4
A4	<0.1	0.4	0.9	0.2	0.9	1.1	1.1	1.7	0.6	1.3	1.4
A5	1.1	5.2	1.1	1.3	5.8	12.5	12.5	18.4	11.7	18.5	21.8
A6	<0.1	0.6	0.3	0.2	1.8	0.7	0.8	0.4	0.9	0.6	0.7
A7	<0.1	0.3	<0.1	<0.1	0.1	0.1	<0.1	0.1	<0.1	0.1	0.1
A8	<0.1	0.2	<0.1	<0.1	0.3	0.4	0.2	0.5	0.2	0.6	0.5
A9	<0.1	0.01	<0.1	<0.1	<0.1	<0.1	<0.1	<0.1	<0.1	0.1	0.1

In the case of the doses reported by the $CaSO_4$:Dy thermoluminescent dosimeter, an approximate distribution of the dose of radiation in the x-ray room is displayed as shown in figure 2. As you can see, the area of greatest risk of permanence of the OEW, due to higher doses recorded over the six months of this study was around the fluoroscope and to the right side of it.

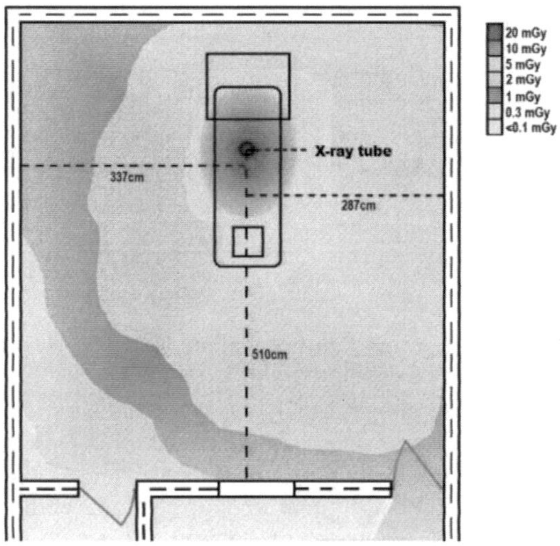

Figure 1. The distribution of the dose of radiation scattered in the room of IR of the Juarez Hospital in Mexico.

These findings were expected due to the configuration of this X-ray room in particular, that is, we think that the distance factor is one of the causes of the distribution of higher doses of the right of the equipment.

From measurements made by the A6 and A7 dosimeters we obtained that the window shield attenuates approximately 5 times scattered radiation that arrives at the control room. On the other hand, the values of the dosimeters in positions A4 and A8, which are on each side of the entrance door to the room of x-ray tube, allows us to ensure that the door is not armored, since detected doses that are attenuated due only to the law of the inverse of the square, i.e., outside the door the dose is 0. 54mGy and inside the door is 1.4mGy.

Both halls (dosimeter A9) and the inside of the control room, behind the window with leaded shield (dosimeter A7), the amount of radiation are negligible or is that it corresponds to background

radiation, so it is a safe area, although it is recommended in any way the least number of people in these areas.
The chest X-ray is diagnostic tool most widely used in the population in general. The dose on the surface of entry for one of these chest X-rays is between 0.2 mGy and 0.4 mGy [7, 8, 9]. In this way, a person who stays a full month in a position very close to the X-ray tube will receive the equivalent doses between 25 and 80 chest X-rays. Behind the window which is armored, the dose is equivalent to less than one chest X-ray. This comparison is only to point out the magnitude of the risk of permanence of the OEW in these areas.

4. Conclusions

This work is a study of the measurement of the scattered environmental radiation dose in the room of IR of the Juarez Hospital in Mexico. For the determination of the dose, two types of thermoluminescent dosimeters were used: the TLD-100 which was confirmed as a bad dosimeter to measure environmental scattered radiation and the CaSO4:Dy, which confirmed its effectiveness for this type of measurements of radiation since it is a material more sensitive to radiation.
We can conclude that the dosimeter thermoluminescent most suitable for the determination of environmental dose are the CaSO4:Dy. This type of study should be done in any room that has an X-ray tube in the country in order to know their risk areas as well as to give recommendations to the OEW in these areas.

5. References

[1] Picano E, Andreassi M G and Rehani M 2013 Radiation protection Percutaneous interventional cardiovascular medicine *The PCR-EAPCI textbook Europa Edition.*
[2] Rivera T 2012 *Applied Radiation and Isotopes* ***71*** 30
[3] McKeever S W S, Moscovitch M, Townsend P D 1995 *Thermoluminescence dosimetry materials: properties and uses* (United Kingdom: Nuclear Technology Publishing)
[4] Cameron J R, Zimmerman D, Kenney G, Buch R, Bland R and Grant R 1964 *Health Physics* **10**(1) 25
[5] Fernández S D S, Garcia R., Sanchez D, Ramírez G, Gaona E, De León M A and Rivera T 2016 *Applied Radiation and Isotopes* **107** 340
[6] Fernández S D S, García R, Mendoza J G, Sánchez D, Rodríguez G R, Gaona E and Montalvo T R 2016 *Applied Radiation and Isotopes* **111** 50
[7] European Commission 1999 Guidance on diagnostic reference levels (DRLs) for medical exposures *Radiation Protection* **109**
[8] World Health Organization 1996 International basic safety standards for protecting against ionizing radiation and for the safety of radiation sources
[9] Le J 1999 Guidelines on patient dose to promote the optimisation of protection for diagnostic medical exposures: documents of the NRPB **10**(1) 1999 *Radiation protection news and notes* **46** 17
[10] Gray J E, Archer B R, Butler P F, Hobbs B B, Mettler Jr F A, Pizzutiello Jr R J and Yaffe M J 2005 Reference values for diagnostic radiology: application and impact. *Radiology* **235**(2) 354

Acknowledgments

This work was partially supported by CONACyT-Mexico through the SNI program. We appreciate partial support from the grants COFAA, EDI, BEIFI and projects SIP-20160512 and SIP-20160522. numbering.

IOP Conf. Series: Journal of Physics: Conf. Series **792** (2017) 012086 doi:10.1088/1742-6596/792/1/012086

Numerical analysis of the concentric ring number for electric field sensing with a Split-Ring Resonator metamaterial

A Mendez[1], F Vazquez[1], S Solis[1] and A O Rodriguez[2]
[1]Facultad de Ciencias, Universidad Nacional Autónoma de México, Cd. Mx., 04519, México.
[2]Departamento de Ingeniería Eléctrica, Universidad Autónoma Metropolitana Iztapalapa, Cd. Mx., 09340, México.
e-mail: jfuv@ciencias.unam.mx

Abstract. Wireless communications have experienced a tremendous development during the last decade. The interaction of the electromagnetic waves as a function of the frequency and intensity with physical objects remains largely unknown. This poses the challenge to study the physical mechanisms involved to actually quantify the impact of the interaction with other objects caused by various electromagnetic waves-emitting sources. Here, we proposed the design of a plane SRR (Split-Ring Resonator) with metamaterial circular rings for S-Band frequency range. To guide the construction of the SRR prototype, simulations of the bandwidth were conducted using the Finite Integral technique. A logarithmic function describes the dependence of the simulated resonant frequency. The maximum simulated bandwidth of the SRR can be reached with an even number of concentric rings.

1.Introduction

Metamaterials with electrical negative indexes have had a great impact in the field of optics and electromagnetism [1]. These artificially metal-dielectric composites with periodic structures whose maximum dimension is less than the wavelength of interest have had a major impact in communications systems. In particular, the development of miniaturized antennas to improve the directivity, radiation pattern and efficiency has gained a great deal of interest due to the growing demand of more efficient communications systems. Multiband antennas can be designed with different structures such as the Split Ring Resonators (SRR) [2-5]. The experimental development of this type of antennas has been not fully investigated. In this paper, a metamaterial effect was obtained using a pair of concentric circles having a slot, one ring geometrically opposite to the other [6-7], and we studied the dependence of the number of concentric ring pairs for a flat SRR, the bandwidth and resonance frequency. The proposed design could be employed for electric field sensing in communications working in the S-Band (2-4GHz).

2.Methods and Results

A flat SRR metamaterial with circular rings was designed. This geometry minimizes the leakage currents caused by discontinuities at right angles. Figure 1 shows the schematics of an elementary structure. Table 1 summarizes the dimensions of the antenna prototype. The Finite integration technique (FIT) was used to simulate the S11-parameters with the Software Tool CST (CST Microwave Studio,

Content from this work may be used under the terms of the Creative Commons Attribution 3.0 licence. Any further distribution of this work must maintain attribution to the author(s) and the title of the work, journal citation and DOI.
Published under licence by IOP Publishing Ltd

Darmstadt, Germany). This software is able to determine the reflection coefficients based on a time domain solver, which obtains the frequency response due to a temporary stimulus. Table 1 data was also used for all electromagnetic simulations. Figure 2 shows the simulated S11-parameter for the resonance frequency as a function of the number of concentric ring pairs.

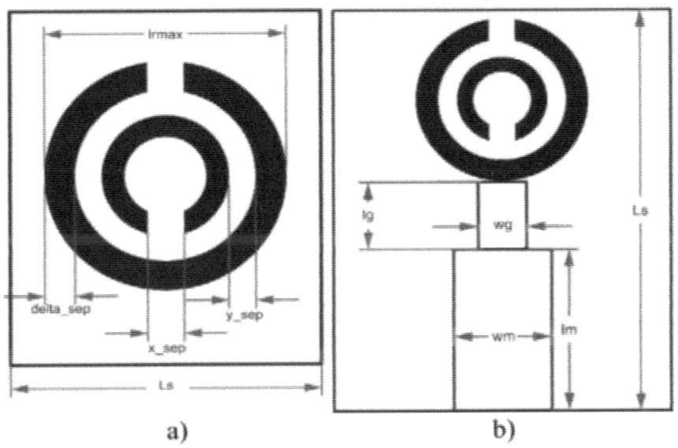

Figure 1. Design of the flat metamaterial SRR with circular rings. a) primary structure, b) current feeding microstrip.

Table 1. Physical characteristics of a flat metamaterial SRR used for the numerical analysis of finite integration in the time domain (FITD), and a FR4 substrate with an electrical permittivity ε_r=4.3.

Name	*[mm]*	Description
delta_sep	0.3	Separation of the track
h	2	Height of the substrate
lg	1.0	Length of the wave-guide
lm	5.0	Length of the Microstrip-line
lrmax	17.0	Length of the maximum radius
Ls	40.0	Length of the substrate
Mt	0.035	Thickness of the microstrip-line
wg	0.5	Width of the wave guide
wm	1.8	Width of the microstrip-line
x_sep	2.0	Separation of the split ring
y_sep	0.1	Separation between each pair of rings

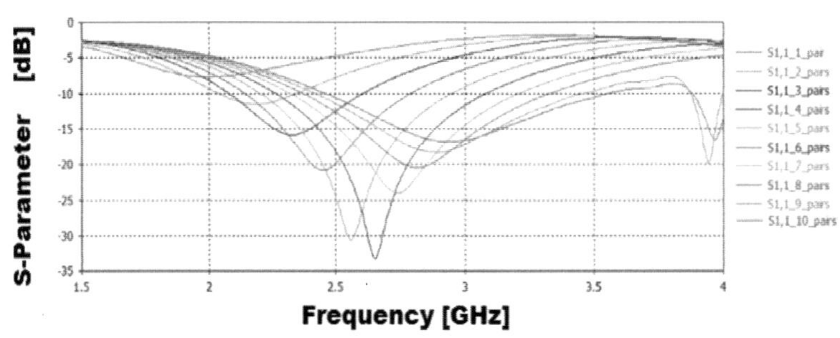

Figure 2. Attenuation plot of the resonance frequency as a function of the number of pairs of concentric rings.

From the Figure 2 data, it was obtained the S11 parameter, the bandwidth (BW) at 3 dB and summarized in Table 2. The dependence of the number of concentric ring pairs on the resonance frequency was plotted in Figure 3. Additionally, a plot of number of pairs of concentric rings vs. bandwidth is shown in Figure 4.

Table 2. Bandwidth calculation data for the number of concentric ring pairs tested.

Rings	F central [GHz]	S11 [dB]	F1 [GHz]	F2 [GHz]	BW [GHz]
1	1.9975	-8	1.6708	2.4647	0.7939
2	2.1875	-12	1.984	2.4166	0.4326
3	2.33	-16	2.1853	2.4834	0.2981
4	2.4525	-21	2.3843	2.5222	0.1379
5	2.56	-31	2.5375	2.5827	0.0452
6	2.6525	-33	2.6245	2.6824	0.0579
7	2.74	-24	2.6529	2.8289	0.176
8	2.815	-20	2.6569	2.9801	0.3232
9	2.89	-18	2.6707	3.1193	0.4486
10	2.93	-17	2.6922	3.2106	0.5184

It is noted that as the number of pairs of concentric rings (*Npac*) increases toward the inside of the maximum ring, the resonance frequency (*Fr*) increases up to a certain limit, presenting a logarithmic behavior that can be modeled by the following equation:

$$Fr = (0.4201)\ln(Npac) + 1.921 \quad Npac \in \{1,\ldots,10\} \tag{1}$$

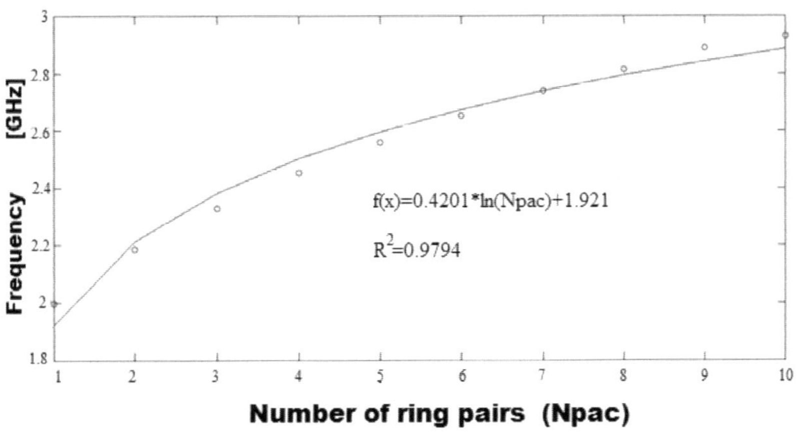

Figure 3. Relationship between the **Npac** with the**Fr**.

On the other hand, by increasing the *Npac*, the BW reaches its minimum value (ring pair 5), and after this value the BW increases again. Figure 4 shows a parabolic behavior for the simulated BW when the *Npac* is varied.

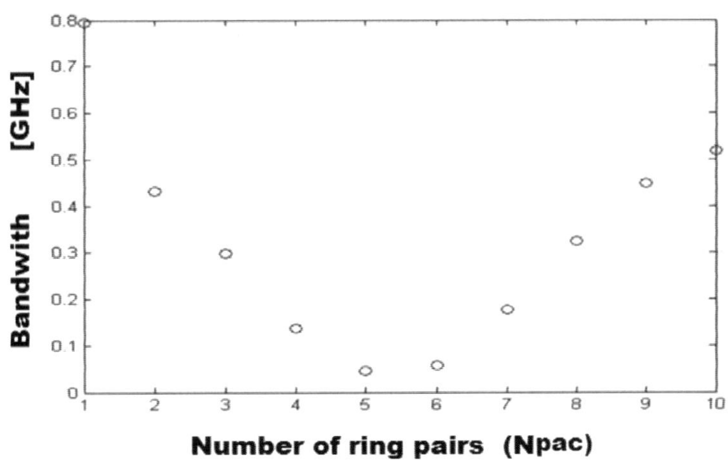

Figure 4. Relationship between the *Npac* and the simulated BW.

A symmetrical pattern can be appreciated indicating that the same BW can be achieved for different *Npac*. This also implies that a greater *Npac* is necessary to obtain a high BW, in the frequency of interest for electric field sensing.

3.Conclusions

The simulation results showed that the relation between the number of concentric rings and the resonance frequency of the SRR metamaterial with circular rings can be fitted to a logarithmic function. This allows us to compute the concentric ring number that must be added to a primary structure for a specific resonance frequency in the microwave band from 2 to 4 GHz, which is actually used in many of the devices found in home, industry and healthcare centers. The present sensor could be used for monitoring the electric field emitted by the devices and for improving the reported safe values for electric field exposure. One of the principal problems for Electric Field Sensor at high frequencies is the limited bandwidth; almost all have to be a resonant or multi-resonant structure with a restricted range of frequencies. The simulation shows that the dimensions of the antenna do not restrict its operation in the microwave band, so this antenna can be fully integrated with circuits that operate at high frequencies. These simulated data offer the means to be able to compute the bandwidth for a specific working frequency to perform the electric field sensing at certain points in the space. It is still remain to perform an experimental comparison with an actual SRR metamaterial.

4. Acknowledgments

The present work was developed with the support of program UNAM-DGAPA-PAPIME PE105516.

5. References

[1] Smith D, Pendry J and Wiltshire M 2004 *Science* **305** 788
[2] Pendry J B, Holden A J, Robbins D J and Stewart W J 1999 *IEEE Trans. Microw. Theory Tech.* **47** 2075
[3] Smith D, Padilla W, Vier D, Nemat S and Schultz S 2000 *Phys. Rev. Lett* **84** 4184
[4] Baena J, Marques R, Medina F and Martel J 2004 Artificial magnetic metamaterial design by using spiral resonators *Phys Rev* B **69** 014402
[5] Baena J, Bonache J, Martin F, Marques R, Falcone F, Lopetegi T, Laso M, Garcia J, Gil I, Portillo M and Sorolla M 2005 *IEEE Trans. Microw. Theory Tech.* **53** 1451
[6] Aznar F, García-García J, Gil M, Bonache J and Martín F *2008 Microw. Opt. Technol. Lett* **50** 1263
[7] Duran S, Naqui J, Paredes F, Bonache J and Martin F 2012 Electrically Small Resonators for Planar Metamaterial, Microwave Circuit and Antenna Design: A Comparative Analysis. *Appl. Sci.* **2** 375

IOP Conf. Series: Journal of Physics: Conf. Series **792** (2017) 012087 doi:10.1088/1742-6596/792/1/012087

Three-dimensional cellular automata as a model of a seismic fault

G Gálvez and A Muñoz
Department of Basic Sciences, UPIBI Instituto Politécnico Nacional, Av. Acueducto S/N, Barrio la Laguna Ticomán, C. P. 07340, Ciudad de México, Mexico

E-mail: ggalvezcoyt@gmail, amunozdiosdado@gmail

Abstract. The Earth's crust is broken into a series of plates, whose borders are the seismic fault lines and it is where most of the earthquakes occur. This plating system can in principle be described by a set of nonlinear coupled equations describing the motion of the plates, its stresses, strains and other characteristics. Such a system of equations is very difficult to solve, and nonlinear parts leads to a chaotic behavior, which is not predictable. In 1989, Bak and Tang presented an earthquake model based on the sand pile cellular automata. The model though simple, provides similar results to those observed in actual earthquakes. In this work the cellular automata in three dimensions is proposed as a best model to approximate a seismic fault. It is noted that the three-dimensional model reproduces similar properties to those observed in real seismicity, especially, the Gutenberg-Richter law.

Introduction

According to the theory of plate tectonics earthquakes occur in the contact surfaces between the plates, which are called seismic faults, although some of them occur between the plates and they cannot be explained in an easy way. The rocks move and break along either side of the fault, movement along the seismic fault causes seismic waves. Seismographs recorded seismic waves through seismograms.
The magnitude of an earthquake can be determined in principle from the amplitude of the seismic waves recorded on seismograms. The magnitude reflects the amount of energy released by the earthquake fault and is directly related to the caused damage.
Gutenberg and Richter developed a scale to measure the magnitude of an earthquake, which reflects the existence of invariant quantities of scale, also known as fractals.
A compilation that include at least the following details of an earthquake: date and exact time of the occurrence, geographic location, magnitude of earthquakes and the caused damage, constitutes what is called a seismic catalogue. The information is published by seismographic networks. Usually seismic catalogues are incomplete and only include earthquakes from a significant magnitude.
Although the theory of plate tectonics explains the seismic production, it says nothing about the possible magnitude of the earthquake and much less on its location and occurrence.
The proper description of a seismic fault by equations is very complicated to solve. Some seismic fault models more simple have been proposed, these simple models have some advantages, among them are relatively easy to implement, can generate seismic catalogs full synthetics, furthermore, despite its simplicity they properly reproduce some statistics properties of earthquakes, and one of these models is the two-dimensional sand pile model that is represented by a cellular automaton. This particular

Content from this work may be used under the terms of the Creative Commons Attribution 3.0 licence. Any further distribution of this work must maintain attribution to the author(s) and the title of the work, journal citation and DOI.
Published under licence by IOP Publishing Ltd

IOP Conf. Series: Journal of Physics: Conf. Series **792** (2017) 012087 doi:10.1088/1742-6596/792/1/012087

model has been very successful in describing the properties observed actual seismicity. A natural step for description of a seismic fault is extending the sandpile model to three dimensions.
In this paper, the 3-D sand pile model is used; again, it is obtained as a result synthetic seismic catalogs the analysis of these catalogs shows properties similar to those observed in the behavior of a actual seismic fault. Some parameters of 3-D sand pile model sand pile may suitably be selected for obtain a behavior more similar to that observed in actual seismicity.
The main objective of this paper is to show that the generalization of automata sandpile two to three dimensions is adequate, Besides, synthetic catalogs generated by the model can then be used to investigate more properties of the seismic faults.

Gutenberg-Richter Law
Gutenberg and Richter [1] developed the scale of seismic measurement, using the data occurred in southern California USA, between 1934 and 1943, they noted that the number of earthquakes of local magnitude between 3.5 and 7.0 decreased exponentially as the magnitude increased, based on this observation they suggested an empirical relationship in 1944 for the logarithm of the frequency density vs the local magnitude, per unit time, $\log_{10} \dot{N}(M)$, which may be expressed as:

$$\log_{10} \dot{N}(M) = a - bM \quad (1)$$

The parameters a and b vary according to factors such as sampling time and the regional level of seismicity of the region, b has characteristic values for different regions of the Earth, with values in the range of $0.75 < b < 1.5$ [2]. Subsequent studies showed that the law is fulfilled for all regions of the world. The parameter b describes the relative size distribution of the events. That is, a small slope implies that it has a greater number of large earthquakes, whereas if the slope is greater, there are fewer, so the parameter b can be related to the risk or danger of the seismic region. In Figure 1 the plot of the logarithm of the frequency against the earthquake magnitude for real data from the area shown Guerrero Mexico. The graph shows that there are two distinct regions, can establish a Guteberg-Richter law for $4 \le M \le 6.5$ and other in the region $7.5 \le M$ [3].

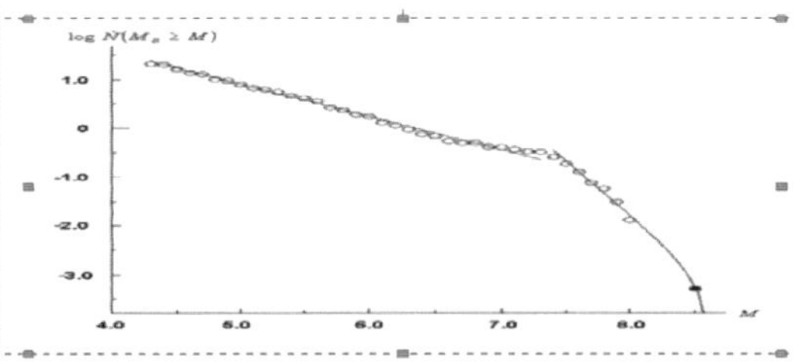

Figure 1. Logarithm of the frequency against the magnitude. is observed two different behaviors of the Gutenberg-Richter law ,one for .$4 \le M \le 6.5$ and other for $7.5 \le M$

Another important feature of the seismicity is the graph of the cumulative number of earthquakes against time for cutting magnitude Mc in Figure 2, is presented the graph for number of earthquake vs time for Mc ≥ 4.3 in the same of Guerrero México region [3]. The graph 2has a staircase shape, which a big scale follows a linear trend, as it is shown in the same figure 2, it is further noted that abrupt changes in the staircase correspond to earthquakes, some of appreciable magnitude.

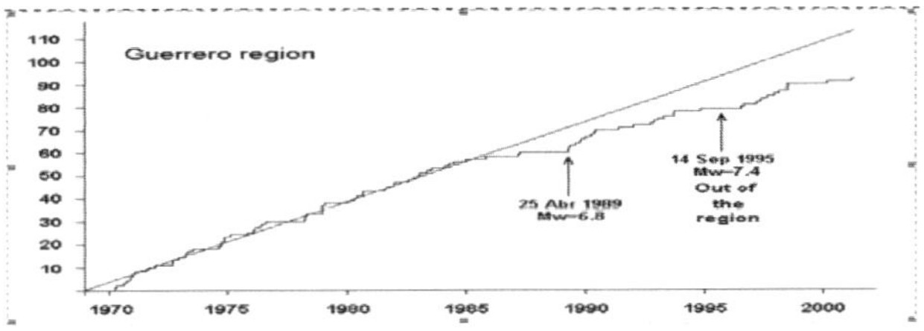

Figure 2. Earthquake number against time Ms ≥ 4.3, Guerrero region. The earthquakes depth is less or equal to 60 km, and they are located between 16.1 - 17.8°N, and 99.3 – 101.1°W, since January, 1, 1969 to June, 30, 2001. The number of events is 92 and the average annual frequency is 3.5 earthquakes per year.

Self-organized criticality

In the study of the dynamics of non-equilibrium systems (phase transitions) and heterogeneous systems has been made popular the concept of complexity, that is, systems with a large number of interacting parts, that change energy, matter or information with their environment are naturally complex systems, they self-organize its internal structure with new and surprising macroscopic properties, for example, has been observed the existence of fractal structures in space and time (which is called 1/f noise) [4-7].

The concept of self-organized criticality (SOC) was proposed by Bak, Tang and Wiesenfeld (BTW) in 1987 [8, 9] and provides a connection between nonlinear dynamics and the appearance of spatial auto similarity and $1/f$ noise in a natural and robust way.

Power laws

Mathematically a power law means that a number N can be expressed by a power of another quantity s,

$$N(s) = K\, s^{-\tau} \qquad (2)$$

For example, s can be a characteristic length and $N(s)$ the number of objects needed to cover a region with squares of length s, this fact would indicate the fractal dimension of the object, that is, fractals are characterized by power laws. If in the above expression is applied the logarithm is obtained that:

$$\log N(s) = \log(K) - \tau \log s \qquad (3)$$

That is, by plotting log $N(s)$ vs. log(s) a straight line is obtained, its slope is -τ. The scale invariance can be seen simply from the fact that the straight line looks the same at any scale.

Experimentally Utsu and Seki in 1954 found the following relationship:

$$\log_{10} A = M - 4 \qquad (4)$$

Where A is the area of replicas of large earthquakes and M is the magnitude of an earthquake. Using the relation 4 can show that the Gutenberg-Richter law is equivalent to a power law, that is:

$$\log_{10} N(A) = \log K_1 - b \log_{10} A \qquad (5)$$

In this case $\log K_1 = a - 4b$.

The explanation of the features observed in complex systems can be expressed mathematically as the explanation of the power laws and particularly the value of the exponents. In short, self-organized criticality is a hypothesis to cover a multitude of complex phenomena observed in nature. It is a theory of internal interactions of large systems. Specifically, it states that large interactive systems could self-organize in a critical state (governed by a power law). Once in this state small perturbations produce chain reactions that may affect any number of elements within the system [1].

Sand pile model

The model "sand pile" It was proposed by Per Bak [4],is normally used to represent or generate complexity and self-organization [1], and consist in adding randomly a grain of sand every time, inside a box, the first grains accumulates in separate places and then, some small mounds are formed. We continue the process; the slope of the mounds is growing. In the future, the slope locally reaches a critical value such that the sum of a grain results in an "avalanche". The avalanche filled empty box areas with sand. With the addition of even one grain of sand the box could be filled and when continuing adding grains of sand, they will leave the system. When the aggregate grains account equals the amount lost (on average) then, according to the theory, the pile of sand grains is self-organized in a critical condition. Of course the notion of nature as a heap of sand seems overly simple, but it is a good model to represent complex systems.

The two-dimensional sand pile is described below using cellular automata [1]. The space is divided by a homogeneous mesh network and each square is associated with a point (x, y), to each cell is assigned a number of grains of sand by the function $Z(x, y)$. The addition of a grain of sand to the square of the grid is at random $Z(x, y) \rightarrow Z(x, y) + 1$, this process is repeated until the number of grains on a site reaches a critical value, it is normally considered $Zc = 4$ grains. When the critical condition is achieved in one of the grid points the following rules apply:

$$\begin{aligned} Z(x,y) &\rightarrow Z(x,y)-4 \\ Z(x\pm1,y) &\rightarrow Z(x\pm1,y)+1 \\ Z(x,y\pm1) &\rightarrow Z(x,y\pm1)+1 \end{aligned} \qquad (6)$$

The process is repeated each time on sites that reach the critical state until stability is reached. The number of sites s that were activated before reaching stability is the size of the avalanche and the number of steps t_d to reach the stable position is equal to the duration of the event, plus some particles may be lost by falling outside the borders.

The process is illustrated in Figure 3, in which it is shown an automaton of size 5x5, the sequence of figures should be from left to right from top to bottom. The first drawing shows an automaton in a critical state, then, in the second, it is added a particle in the center cell (the cell increases its value 3 to 4) and thus the site reaches an unstable condition and the process of redistribution of particles to neighboring sites begins, the first distribution is shown in the third figure, the process continues until it reaches a critically stable state, it is shown in the following drawings. The avalanche size in this case is $s = 9$ because a site was relaxed twice, and the duration of the event is $t_d = 7$.

The frequency distribution of magnitude leads to a power law, $N(s) \cong s^{-\tau}$ and distribution of duration of event also results in a power law, $N(t_d) \cong t_d^{-\sigma}$.

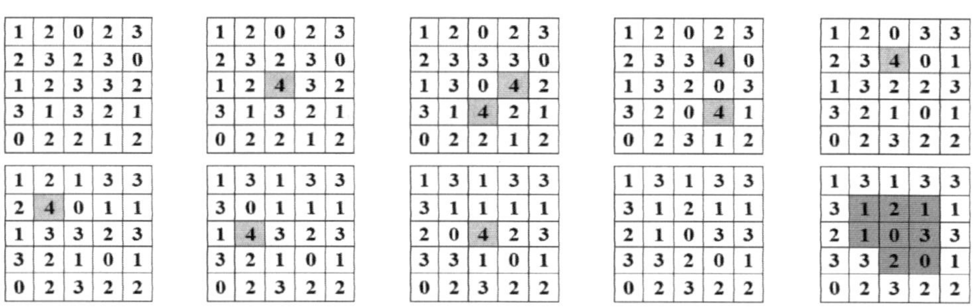

Figure 3. Illustration of an avalanche in the sand pile model 5x5. A grain falling at the center of the grid, the site changes its value from 3 to 4. The the cell becomes unstable starting an avalanche. The continuous distribution process until it reaches a stable condition. The last picture shows the eight sites exceeded the critical value. As can be seen the avalanche has a size $s = 9$ (a site relaxed twice) and a duration $t_d = 7$

A natural step in the development of the models is to generalize the previous model to the three-dimensional (3D) sand pile, in this case we have to modify some parameters, for example, you have to modify the critical value that becomes unstable the cell.
The three-dimensional cell is represented by a cube, each site in the cube has 6 direct neighbors and it is natural to choose now the critical value as 6 particles or grains. It is appropriate in the context of a seismic fault to change the idea of molehill or particle by the concept of stress to which it is subjected a cell and the logarithm of size of the avalanche as the earthquake magnitude, that is, $M=\log(s)$. In this context, the duration of an earthquake is directly related to the t_d parameter.
As in the two dimensional case s can be defined in three dimensions as the number of cells before reaching relaxed stability, this is the magnitude of the avalanche. Moreover, the number of steps to reach t_d stability is the duration of the event. Generalized rules for 3D sand pile are given below:

$$\begin{gathered} Z(x,y,z)\rightarrow Z(x,y,z)-6, \\ Z(x\pm1,y,z)\rightarrow Z(x\pm1,y,z)+1 \\ Z(x,y\pm1,z)\rightarrow Z(x,y\pm1,z)+1 \end{gathered} \tag{7}$$

Results
The cellular automaton 3D sand pile model was programmed complying with the generalized rules. The cellular automaton was test for many sizes, Figure 4 shows the graph of the number of events $N(s)$ with the size s in log-log scale to a size 15x15x15 and for 30 million of events, as shown, the relationship is a typical power law (on the left), it should be noted that this power law is equivalent to the Gutenberg-Richter law using M = log (s). The fit was made only by using the linear region of the graph to obtain a slope of the line b = 1.33, the value obtained is within the range reported for the parameter b of the Gutenberg-Richter law (0.75 <b <1.5).

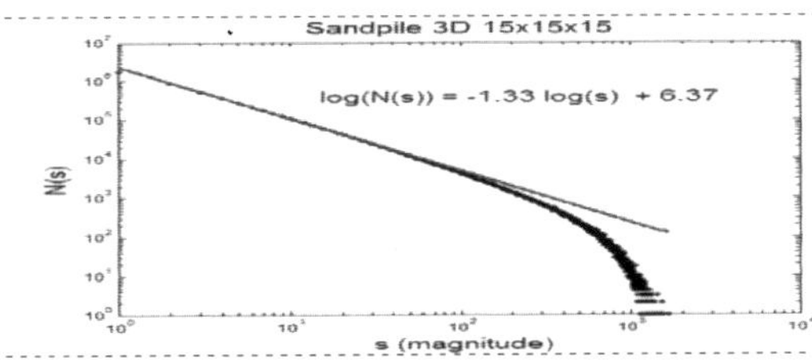

Figure 4. Graphical log-log scale of the magnitude s and the number of events N(s) for the sand pile cellular automaton in three dimensions with 15x15x15 size. By making the appropriate least squares adjustment in the linear region the slope is obtained. The slope is 1.33, value within the region of validity of the parameter b of the Gutenberg-Richter law.

The analysis of t_d (physically related to the duration of an earthquake), which also follows a power law is shown in Figure 5, done for the 3D sand pile model, size 15x15x15. The linear part for the time duration t_d leads to a slope b_t= 1.29, which as expected is slightly less in absolute value to the slope of the power law for the magnitude b = 1.33. It is expected that because the time duration t_d is less than the magnitude M (in the computer sense), the straight slope in the linear part of the log-log plot of time duration t_d is less in absolute value than that obtained in the graph of the magnitude M. Tests with different sizes of the 3-D sand pile model were performed; the results are very similar among them.
One question we want to know is, which is the relation of the straight line slope and el number Np of events?, to answer this question a model of 15x15x15 pile of sand was used, and Np took the set of values 10000, 100000, 1000000 and 10000000. Figure 6 shows the results; it can be seen the graphs for each of the values of Np. For Np=10000, the data are widely dispersed, however, the slope is -1.26. Increasing the value of Np, the data dispersion decreases and the slope improves.

With Np = 10000000, a slope of -1.33 is obtained. It is further noted that the graph of Figure 1 where actual data are analyzed is similar to the graph of Figure 6.

In real seismicity it was found that adding the magnitude of earthquakes with respect to time a graphic resembling a staircase is obtained. Using different sizes of sand pile model, we obtained graphics type staircases. It can be seen from Figure 5 that the slope of the line that can be associated to each staircase increases with model size.

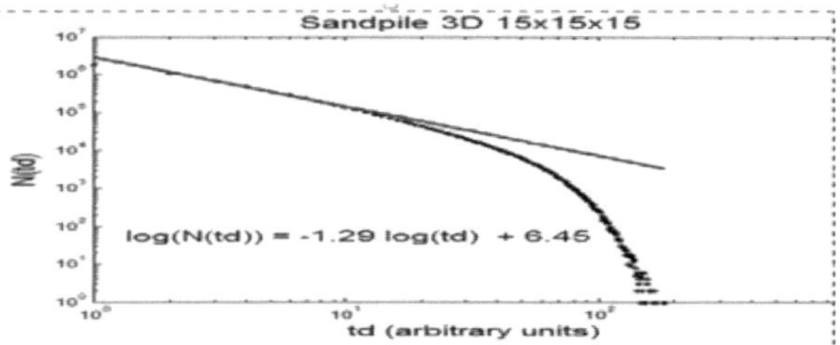

Figure 5. Graphic in log-log scale for the duration t_d of events and the number of events $N(t_d)$ for the 3D sand pile model 15x15x15. The slope is 1.29.

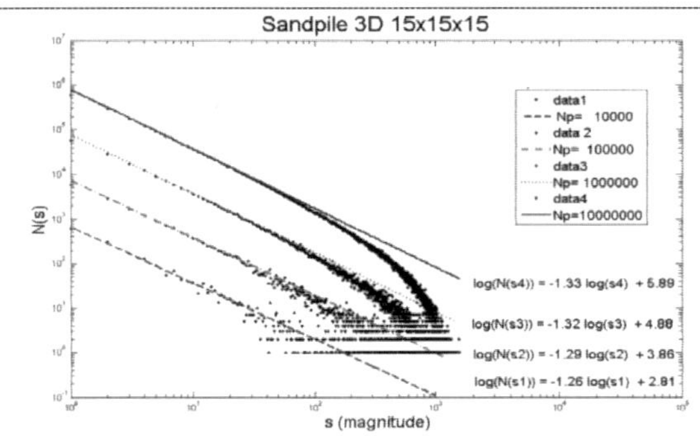

Figure 6. In this figure the dependence of the slope of the power law with respect to Np is observed.

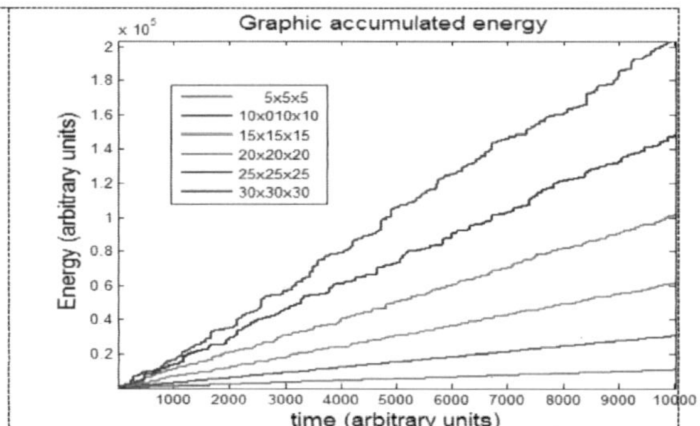

Figure 7 Staircase graphic for the 3D sand pile model 15x15x15.

7. Conclusions

The model sand pile in two dimensions can expand smoothly to three dimensions suitably modifying the distribution rules. The consideration that the number of grains of sand or particles is directly related to stress of stone makes the 3D cellular automata a suitable model to describe a real seismic fault. The power laws for the magnitude have slopes that are consistent with the expected values for the parameter b of the Gutenberg-Richter law, also the duration of avalanches t_d complies with power law, which is related to fractal properties observed in complex systems.

Several properties of seismic faults are obtained with the 3D sand pile model, as the staircase graphs for the cumulative magnitude. The 3D sand pile model produces synthetic seismic catalogs with properties very similar to the actual seismicity catalogs. Setting some parameters could help to adjust b the value of the Gutenberg-Richter law by more appropriate values. Synthetic seismic catalogs can be used to investigate properties of the actual seismicity [11-14].

Acknowledgments

We thank COFAA and EDI from the Instituto Politécnico Nacional for supporting this work.

References

[1] Gutenberg B and Richter C F 1944 *Bull. Seismol. Soc. Am.* **34** 185
[2] Olami Z, Feder H J S and Christensen K 1992 *Phys, Rev. Lett.* **68** 12441247
[3] Muñoz, A, Rudolf A H and Angulo F 2012 Rev. Mex. Fís. S **58** (1) 96–103
[4] Bak P 1996 *How nature works* (Springer-Verlag New York. USA)
[5] Mandelbrot B 1983 *The fractal geometry of nature* (Freeman, New York)
[6] Mandelbrot B and Van J W 1968 *SIAM Rev* **10** 422
[7] Mandelbrot B and Wallis J R 1969 *Water Resour. Res.* **5** 321
[8] Bak P, Tang C and Weisenfeld K 1987 *Phys. Rev. Lett.* **59**, 381
[9] Bak P, Tang C and Weisenfeld K 1988 *Phys. Rev. A* **38** 364
[10] Utsu T and Seki A 1954Journal of Seismological Society of Japan v. 7 p. 233-240.
[11] Bizzarri A 201 Earth-Science Reviews **115** 304–318
[13] Jiménez A 2013 Acta Geophysica, **61** (6) 1325-1350
[14] Asadi A 2014 Iranian Journal of Science & Technology **38** 35-41

A simple model for determining the atmospheric thermal conductivity

M A Ramírez-Moreno[1] **and F Angulo-Brown**[1]

[1] Departamento de Física, Escuela Superior de Física y Matemáticas-IPN, Zacatenco, CDMX 07738, México

E-mail: antonio_ramirezm@hotmail.com, angulo@esfm.ipn.imx

Abstract. This paper presents a simple method based on Finite-Time Thermodynamics (FTT) to determine the thermal conductivity of atmospheric air. The method considers an atmospheric heat engine in which the air moves convectively on the border day-night driven by the Sun's energy. The numerical value obtained for the thermal conductivity reasonably accords with the reported value by Van Ness by the number of Rayleigh.

1. Introduction

A finite-time thermodynamics has been developed four decades ago, from the pioneering work of Curzon and Ahlborn [1]. As it is well known, Carnot's theorem states that for thermal cycles operating between two heat reservoirs at absolute temperatures T_1 and T_2 ($T_1 > T_2$), the cycle of maximum efficiency that can work between they temperatures is the Carnot cycle, consisting of two isotherms and two adiabat that take place reversibly. This fact implies that to the Carnot cycle has zero power output, since a reversible cycle consisting of quasistatic process require an infinite time to complete. The Curzon and Ahlborn (CA) engine (see figure 1) consists of two parts, a reversible one (endoreversible) which is an internal Carnot engine operating between T_{1w} and T_{2w}, and another part formed by the two couplings between T_{1w} and T_1 and between T_{2w} and T_2 respectively, which constitute the irreversible part of the model. In addition, the heat transfer along the mentioned couplings is given by a linear heat transfer law of the Newtonian type, although other heat transfer laws can be used to model heat fluxes, so as Dulong-Petit law and Stefan-Boltzmann law among others. The endoreversible CA engine was a first step towards irreversible engine models by incorporating irreversible elements as thermal resistances at the couplings between the internal reversible cycle and the external heat reservoirs. The CA engine under maximum power conditions leads to the famous CA–efficiency given by $\eta_{CA} = 1 - \sqrt{\tau}$, being $\tau = T_2/T_1$. The present paper is organized as follows: In Section 2, we define the atmospheric winds; in Section 3 we maximize the power output; in Section 4 we analyze the tidal winds as heat engine driven by the Sun; finally in Section 5 we present some concluding remarks.

2. Atmospheric winds

The existence of two heat reservoirs at different temperatures makes possible the production of work. In fact, the planet's atmosphere is a working fluid illuminated by the Sun, thus the atmospheric gases tend to increase its temperature and they expand. Therefore, the illuminated gas by the Sun rises to the upper atmosphere. When moving to less enlightened places it is cooled, and the more density tends to

Content from this work may be used under the terms of the Creative Commons Attribution 3.0 licence. Any further distribution of this work must maintain attribution to the author(s) and the title of the work, journal citation and DOI.
Published under licence by IOP Publishing Ltd

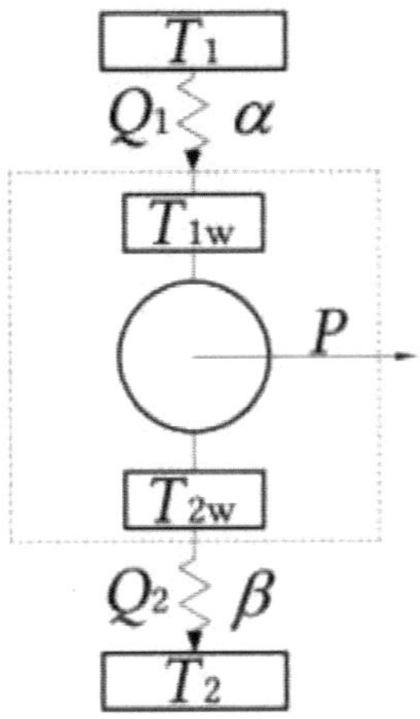

Figure 1. Curzon-Ahlborn endoreversible engine model.

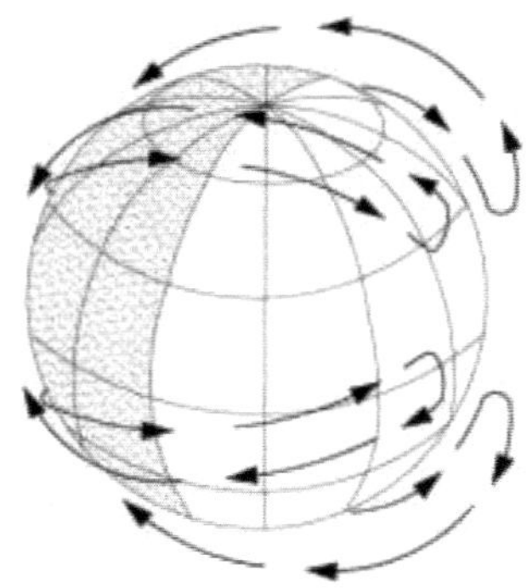

Figure 2. Model of tidal winds or thermal tides [2].

sink back to the planet's surface. That is, the movement of gases is promoted by the differences in longitudinal temperature (geographical coordinates): the day lengths (floodlit) are warmer than the night lengths (unlit). Such macroscopic cycles are one of the sources of the winds. These kinds of winds are called tidal winds or thermal tides, because air travel is much like the movement of tidal water from the sea [2], see figure 2.

Taking the model of CA as a basis for modelling tidal winds, it follows that the temperature of the hot reservoir is the temperature of the Sun ($T_1 = 5778K$) and the temperature of the cold reservoir is the temperature of the surface of the Earth at night ($T_2 = 287.15K$). On the other hand, thermal conductance α (see figure 1), is proportional to the Stefan-Boltzmann constant, $\sigma = 5.67 \times 10^{-8} Wm^{-2}K^{-4}$, that is

$$\alpha = f\sigma, \tag{1}$$

being f the dilution factor, defined by $f = R_s^2/r^2$ with R_s the Sun's radius and $r = \sqrt{bc}$, with b the half major axis and c the half minor axis of the elliptical orbit of a planet around the Sun [3]. Similarly, the thermal conductance β (see figure 1) is attenuated by the absorption coefficient a [4], thus

$$\beta = ak, \tag{2}$$

where k is the atmospheric thermal conductance, $[k] = Wm^{-2}K^{-1}$.

2.1. Power output

Figure 1 shows the CA model, which consists of two heat reservoirs (T_1 and T_2) connected by two irreversible components (α and β associated to thermal conductances) to two isotherms of the working fluid (T_{1w} and T_{2w}) with an internal reversible Carnot cycle. Thus, the endoreversibility hypothesis [3] is

$$\frac{Q_1}{T_{1w}} = \frac{Q_2}{T_{2w}}. \tag{3}$$

being Q_1 and Q_2 heat fluxes (heat transfer per unit time). From first law of thermodynamics, we have

$$Q_1 = P - Q_2, \tag{4}$$

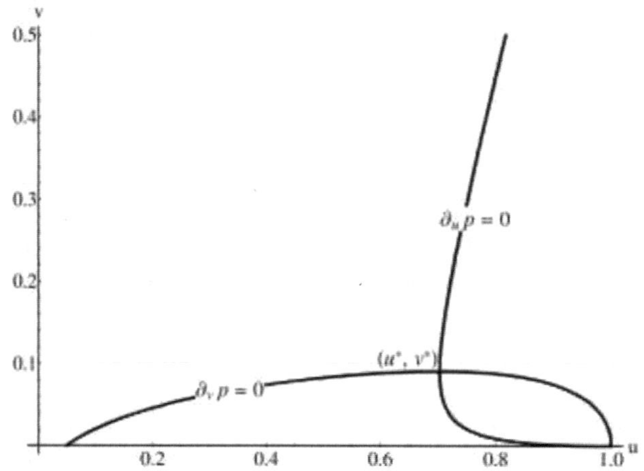

Figure 3. Overlap of implicit equations (10) and (11).

with P the power output. We propose a law of heat transfer for each of the couplings between the working fluid and heat reservoirs, of the type

$$Q_1 = \alpha(T_1^4 - T_{1w}^4), \quad Q_2 = \beta(T_{2w} - T_2). \tag{5}$$

Q_1 corresponds to a radiative transfer and Q_2 to a convective transfer in the day-night boundary [2]. Thus, the power output is given by [5,6]

$$P = \frac{\alpha\beta y[T_1-T_2-x-y][T_1^4-(T_1-x)^4]}{\alpha[T_1^4-(T_1-x)^4][T_2+y]+\beta y(T_1-x)}, \tag{6}$$

where $x = T_1 - T_{1w}$, $y = T_{2w} - T_2$ [1]. We define the dimensionless variables, $N = \alpha T_1^3/\beta$, $u = 1 - x/T_1$, $v = y/T_1$ and $\tau = T_2/T_1$. Under these parameters we rewrite equation (6) to obtain

$$P = \alpha T_1^4 \frac{v(1-u^4)(u-v-\tau)}{N(1-u^4)(v+\tau)+uv}. \tag{7}$$

3. Maximization of the normalized power output

Given the above expression we proceed to maximize the power output by using the variables u and v, taking τ and N as system parameters. To simplify the calculations, we normalize the power output in the following manner, $P/\alpha T_1^4$. Then equation (7) becomes,

$$p(u,v,\tau,N) = \frac{v(1-u^4)(u-v-\tau)}{N(1+u^4)(v+\tau u)+uv}. \tag{8}$$

This expression is analogous to the conversion efficiency of solar energy ω, defined by De Vos [3]. To maximize $p(u,v,\tau,N)$given by equation (8) we calculate

$$\frac{\partial p(u,v)}{\partial u} = 0, \quad \frac{\partial p(u,v)}{\partial v} = 0. \tag{9}$$

The following pair of transcendental equations is obtained

$$[v+\tau][u^4(3v-2N)+(1+u^8)N+v)] - 4u^5 v = 0, \tag{10}$$

$$N[1-u^4][(v+\tau)^2 - u\tau] + uv^2 = 0. \tag{11}$$

Figure 3 shows the intersection point (u_0, v_0), of the implicit equations (10) and (11), for N and τ constants. For a different value of N, it can be seen that within this set up is critical point (u_0, v_0) [7].

3.1. Tidal winds as heat engine driven by the Sun

Taking the temperature of the Sun ($5788K$) and the temperature of the surface of the Earth at night ($287.15K$), the parameter $\tau = T_2/T_1 \approx 1/20$, the coefficient of thermal conductances N, determine the appropriate configuration of the heat engine, such that the power output or the efficiency of solar energy conversion, equation (8), are maximal. Thus, for a value of $N = 0.48$ and $\tau = 1/20$. The maximum values of u and v are 0.7044 and 0.1148 respectively. By using equation (8), $p_{max} = 17.3\%$, for the efficiency of solar energy conversion. De Vos reported a value for the conversion efficiency of solar energy, of the tidal winds, of the 12.79%, [3] and Barranco-Jiménez report a value of 17.33% [9], therefore the value of N it is appropriate for the configuration of the tidal winds, such that the power output is maximum. On the other hand, from the definition of N, it follows that

$$N = \frac{\alpha T_1^3}{\beta} = \frac{f\sigma T_1^3}{ak}, \tag{13}$$

where we have used the equations (1) and (2). Then, it is obtained that the atmospheric thermal conductance is given by

$$k = \frac{f}{a}\frac{\sigma T_1^3}{N}. \tag{14}$$

The dilution factor for the Earth is given by $f = 2.16 \times 10^{-5}$ [3]. On the other hand, we know that the absorption coefficient of the atmosphere is $a = 14\%$ [4], then we finally obtain

$$k = 3.5156 W m^{-2} K^{-1}. \tag{15}$$

Which is a value reasonably close to that reported by Van Ness [8], which has a value of $3.3185 W m^{-2} K^{-1}$, for the thermal conductance of air at ambient temperature.

4. Concluding remarks

In this paper we have calculated the approximate value of the thermal conductance of air through a very simple model of heat engine operating at finite time. As is well known, although the atmosphere has a temperature distribution with a different for each one of the atmospheric layers (troposphere, the tropopause, stratosphere, etc.), the behavior of the winds can be simulated, in first approximation, by an atmospheric heat engine operating between two extreme temperatures, with it has is troposphere value [2,9,10,11]. Among these models is that of De Vos and Flater [2] which focuses on the horizontal winds at night-day border. Based on this model we have proposed a simple method for calculating the thermal conductance of air, that gives a very close to that reported by Van Ness [8] by using the Rayleigh number value; a dimensionless number used in calculations of heat transfer by natural convection in fluid mechanics. As can be seen in the present work based on the general methods of thermodynamics, they can be used to calculate some thermal properties of materials (air in our case).

5. References

[1] Curzon F and Ahlborn B 1975 *Am. J. Phys.* **43** 22

[2] De Vos, A and Flater, G 1991 *Am. J. Phys.* **59(8)** 751.

[3] De Vos A 1992 *Endorreversible Thermodynamics of Solar Energy Conversion* (Oxford:Oxford University Press)

[4] Sjoerd J P 1999 *Science* **284** 1971

[5] Arias L A and Angulo F 1994 *Rev. Mex. Fis.* **40** 829

[6] Arias L A and Angulo F 1997 *J. Appl. Phys.* **81 (7)** 2973

[7] Ramírez M, González S and Angulo F 2016 *Physica A* **444** 914

[8] Van H C 1969 *Understanding Thermodynamics* (New York: McGraw-Hill).

[9] Barranco M A and Angulo F 1996 *J. Appl. Phys.* **80(9)** 4872

[10] Gordon J M and Zarmi, Y 1989 *Am. J. Phys.* **57(11)** 995

[11] Angulo F and Páez R 1993 *J. Appl. Phys.* **74** 2216

IOP Conf. Series: Journal of Physics: Conf. Series **792** (2017) 012089 doi:10.1088/1742-6596/792/1/012089

Photothermal Techniques Used to Evaluate Quality in Dairy Products.

E. López-Romero and J. A. Balderas-López
Instituto Politécnico Nacional-UPIBI, Av. Acueducto S/N, Col. Barrio la Laguna Ticomán, C. P. 07340, México City, México.

E-mail: abrahambalderas@hotmail.com, eolopezromero@gmail.com

Abstract. Photothermal systems were used to quantify thermal and optical properties of commercial and natural dairy products. Thermal diffusivity and light absorption coefficient were analyzed. It was found that water content easily alters thermal properties in samples of milk. In addition, all samples showed strong light absorptions at 405 nm, 980 nm and 488 nm, evidencing presence of proteins, fat and vitamins (riboflavin), respectively. Therefore, it was shown that thermo-physical properties measured in this work could be used as complementary parameters for quality evaluation of dairy products.

1. Introduction

Milk is a complex colloidal suspension, it is mainly composed by fat, casein micelles, proteins, lactose, and some vitamins [1]. It is widely promoted as a main staple food for pediatric population and it is an important component of other marketed products, such as cheese, yoghurts, creams, etc., However, dairy industry faces two main problems: on one side, it is subjected to a variety of adulteration in all different processes of chain distribution, from production to final consumption; on the other side, it degrades its quality in short periods of time due to the presence of microorganisms. For these reasons, dairy industry applies a set of biochemical and biophysical tests to evaluate milk coming from stables [2,3]. Each component in milk is characterized for having specific thermal and optical properties [4-6]. Thermal diffusivity of water, for instance, is relatively larger than other components of milk; therefore, a minimal addition of water can modify greatly thermal properties of milk. Other components, such as proteins and lactose, absorb light at specific wave-lengths, since microbial activity can modify the natural proportion of these components, optical properties can also be used to study adulteration of milk. Photothermal techniques are used in this work to measure thermal diffusivity and optical absorption coefficient (at 405 nm, 488 nm and 980 nm) of dairy products. Results show the feasibility of using these properties as a tool to assess products quality in dairy industry.

2. Experimental set-ups

Two photopyroelectric techniques, for thermal diffusivity [7,8] and optical absorption coefficient measurements [9], were used. Details of these different photothermal methodologies are reported elsewhere for which just a brief description of these is provided in this work.

Content from this work may be used under the terms of the Creative Commons Attribution 3.0 licence. Any further distribution of this work must maintain attribution to the author(s) and the title of the work, journal citation and DOI.
Published under licence by IOP Publishing Ltd

Related with thermal diffusivity measurements it has been shown that [7,8], under appropriate experimental conditions, photopyroelectric signal, δP, behaves as: $\delta P = Cexp(-\sigma L)$. In this equation L is the sample´s thickness, C and $\sigma = (1+i)\sqrt{\pi f/\alpha}$, are complex expressions, where f and α are modulation frequency and sample´s thermal diffusivity, respectively. This thermal property can be obtained from linear fittings on phase and amplitude of δP, by means of the fitted parameter $M = \sqrt{\pi f/\alpha}$, taking the photopyroelectric signal as function of the sample´s thickness. For optical absorption coefficient measurements, on the other hand, it has been shown that [9], under appropriate experimental conditions, photopyroelectric signal, δP, behaves as: $\delta P = Dexp(-\beta L)$, where D is a complex expression, L is sample´s thickness and β is the sample´s optical absorption coefficient. As for thermal diffusivity measurements, the optical absorption coefficient can be measured by fitting the amplitude of δP as function of the sample´s thickness. In this experiment, pyroelctric phase signal remains constant, consequently, it can be used as an experimental criterion for the analysis.

2.1 Photopyroelectric setup for thermal diffusivity measurements

Figure 1 shows a cross section of the photopyroelectric setup for thermal diffusivity measurements [7,8], where a diode laser at 875 nm (Thorlabs, model L785P090) was used as a light source.

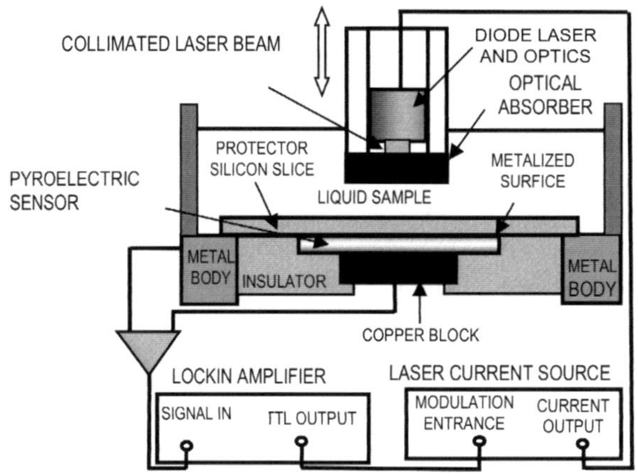

Fig. I. Cross section of the photopyroelectric setup for thermal difussivity measurements of liquids.

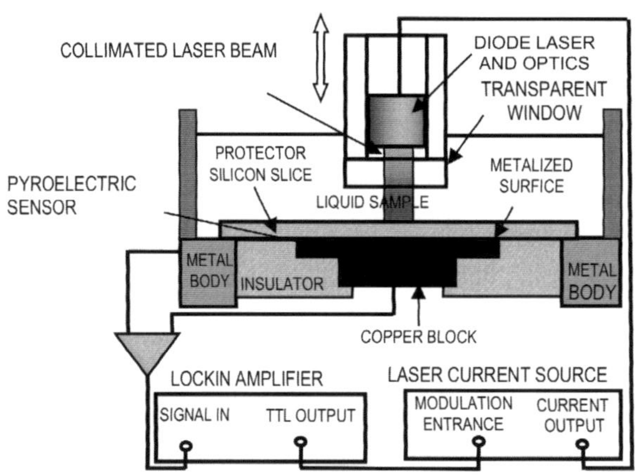

Fig. 2. Cross section of the photopyroelectric setup for optical absorption coefficient measurements of liquids.

2.2 Photopyroelectric setup for optical absorption coefficient measurements

Figure 2 shows a cross section of the photopyroelectric setup for optical absorption coefficient measurements [9]. Three diode lasers at different wave-lengths, 405 nm (laser pointer), 488 (Thorlabs, model L488P60) and 980 nm (Thorlabs, model L980P030), were used to determine optical measurements.

Intensity lasers modulation was achieved, in both cases, by means of a current source (Thorlabs, model LDC205C) controlled by the TTL output (transistor-transistor logic) of the lockin amplifier (Research Scientific Instruments, model SRS830) using a PZT element as a pyroelectric sensor. Pyroelectric signal was first pre-amplified (Stanford Research Systems, model SR550) before to be sent to the lockin amplifier.

3. Materials and Methods

Different variety of milks (Table I), bought in local markets and stables in Mexico City, were used in this work. All measurements were made at a fixed modulation frequency of 1 Hz, liquid sample´s thickness was changed in steps of 50 microns by taking 15 experimental data for thermal diffusivity measurements and 30 experimental data for optical absorption coefficient measurements.

Table I. Optical and Thermal Properties of different type of Milks as measured by means of the photothermal techniques presented in this work.

Milk type	Optical Absorption Coefficient β (cm^{-1})			Thermal Diffusivity α (cm^2/s)	
	488 nm	**405 nm**	**980 nm**	**Phase**	**Amplitude**
Raw Milk	7.03	7.06	5.07	0.00139	0.00140
"Alpura Clásica"	6.08	7.56	5.89	0.00135	0.00135
"Alpura Deslactosada"	5.58	7.87	5.21	0.00138	0.00137
"Alpura Light"	6.14	6.84	3.92	0.00141	0.00141

4. Results

Figures 3 and 4 show phase (3a and 4a) and amplitude (3b and 4b) profiles, as function of sample´s thickness, for "Alpura Clásica" and raw milks, respectively. The experimental set up shown in Fig. 1 was used to measure thermal diffusivity in samples.

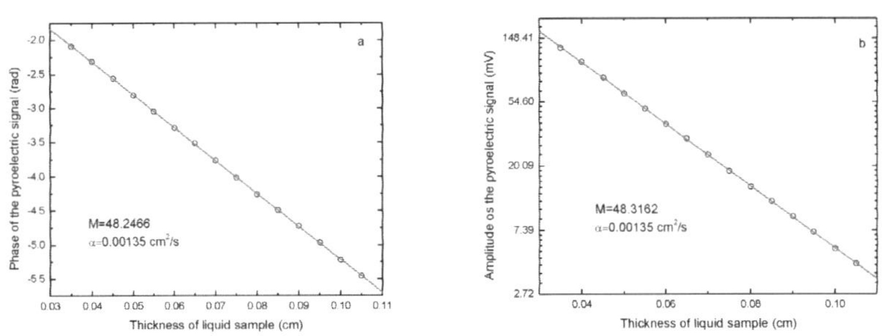

Fig. 3. Phase (a) and amplitude (b) of the photopyroelectrical signal for "Alpura Clásica" milk sample, as a function of liquid´s sample thickness. Continues lines represent linear fits as for to obtain thermal diffusivities.

Thermal diffusivities of milk samples studied in this work are summarized in Table I, columns 5 (for phase) and 6 (for amplitude). The close agreement between thermal diffusivity values from these two pyroelectric signals is evident, situation that proof the photopyroelectric technique consistency. It is also evident that thermal diffusivity values for all milk samples are very similar to each other. Thus, values of this thermal property can´t be considered as a parameter to differentiate kinds of milks.

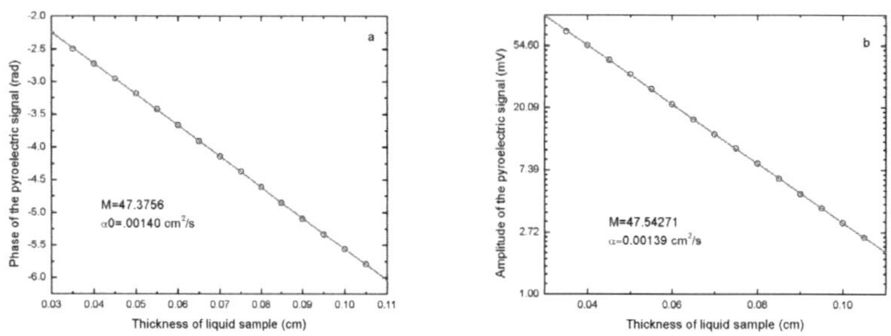

Fig. 4. Phase (a) and amplitude (b) of the photopyroelectrical signal for raw milk sample, as a function of liquid´s sample thickness. Continues lines represent linear fits as for to obtain thermal diffusivities.

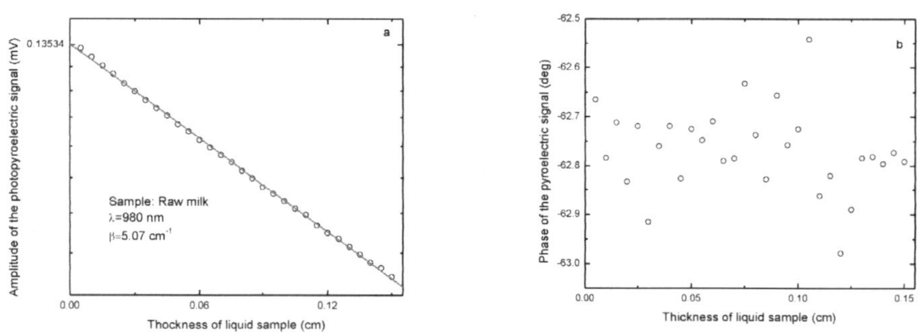

Fig. 5. Photopyroelectric signal for amplitude (a) and phase (b), as a function of sample´s thickness for raw milk sample. Continues line represents the linear fit as to obtain the sample´s optical absorption coefficient at 980 nm.

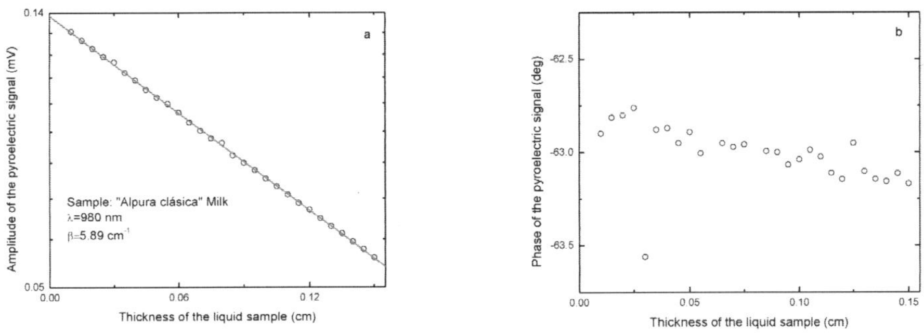

Fig. 6. Photopyroelectric signal for amplitude (a) and phase (b), as a function of sample´s thickness for "alpura clásica" milk sample. Continues line represents the linear fit as to obtain the sample´s optical absorption coefficient at 980 nm.

Figures 5 and 6 show, on the other side, photopyroelectric signal profiles, as function of sample´s thickness, for raw milk and "Alpura Clásica" milk at 980nm, by using the experimental setup shown in Fig. 2, for optical absorption measurements.

Constants phases in these two figures (5b and 5b) certify that the mathematical model for optical absorption coefficient measurements applies for these substances. The set of optical properties measured for all milk samples at 488 nm, 405 nm and 980 nm are summarized in Table I, columns 2, 3 and 4, respectively.

Important variations on optical parameter here presented are obvious for each milk sample. The optical absorption coefficient value depends on both, wavelength and kind of milk. Important differences are particularly observed for "Alpura Clásica" and "Alpura Light" milks at 980 nm (column 4). This fact can be clearly attached to the different fat content of these two kinds of commercial milks.

Another important difference in optical properties happens between raw and "Alpura Deslactosada" milk, at 488 nm. This fact could be due to their different vitamins concentration, especially for riboflavin content which has an absorbing band around 470 nm. Since this optical parameter is larger for raw milk, this difference can be explained as some reduction of this vitamin due to any removing lactose industrial process.

These conclusions shall be verified by means of some other conventional tests by quantifying the fat and lactose concentrations for these kinds of milks.

Thermal diffusivity measurements as a function of water content were also carried out for "Alpura Descremada" milk sample. Thermal diffusivity values, amplitude and phase, are summarized in Table II, columns 3 and 4, respectively. The results show that this thermal property increase with water content for which this thermal property could be used for the assessment of milk adulteration with water. Since adulteration is a common practice in dairy industry, this is a very important outcome.

Table II. Thermal diffusivity measurements for "alpura descremada" milk, as a function of water content, by means of the photothermal technique presented in this work.

Volumetric Milk fraction	Thermal Diffusivity (cm^2/s)	Thermal Diffusivity (cm^2/s)
0.0	0.00147	0.00147
0.1	0.00142	0.00142
0.2	0.00141	0.00141
0.3	0.00141	0.00141
0.4	0.00140	0.00140
0.5	0.00140	0.00139
0.6	0.00139	0.00139
0.7	0.00139	0.00139
0.8	0.00138	0.00138
0.9	0.00138	0.00138
1	0.00138	0.00137

5. Conclusions

Convenience of photothermal techniques in thermal and optical properties of dairy products has been proved. Potential of these thermophysical properties as parameters of food products quality assessment has been shown. It is important to notice that no other conventional spectroscopy technique conducts direct measurements of optical absorption coefficients as the photopyroelectric technique reported in this work does. High light dispersion is an important limitation to conventional spectroscopy but it is not in photothermal techniques. In the last case, signal generation is caused mainly by the direct light absorption.

This condition particularly applies for milk, therefore, analyzing its optical properties with conventional methods is a harder practice. Other more convenient wave-length (in the UV and infrared regions) could also be used for optical absorption coefficient measurements in milk, this could allow to establish quality criteria for these important products.

6. References

[1] Baldi A and Pinotti L 2008 *Lipophilic Microconstituents of Milk. In Bioactive Components of Milk* 125

[2] Bhandari V 2003 Analysis of milk and milk products physical methods *Dairy Sciences Enciclopedy* 93–101

[3] Biggs D 1964 Infra-red analysis of milk for fat, protein, lactose and solids-not-fat *Process Milk Industry*

[4] Fernández F and Montes F 1972 Influence of temperature and composition on some physical properties of milk and milk concentrates III *Thermal conductivity. Milchwissenschaft.* **27** 772–776

[5] Fernández F and Montes F 1977 Thermal conductivity of creams *Journal Dairy Research* **44** 103–109

[6] More G R and Prasad S 1988 Thermal conductivity of concentrated whole milk *Journal Food Process Enginnering* **10** 105–112

IOP Conf. Series: Journal of Physics: Conf. Series **792** (2017) 012089 doi:10.1088/1742-6596/792/1/012089

[7] Balderas J A, Mandelis A and García J 2000 Thermal-wave resonator cavity design and measurements of the thermal diffusivity of liquids *Review of Scientific Instruments* **71** 2933-2937
[8] Balderas J A et al 2015 Generalized photopyroelectric setup for thermal diffusivity measurements of liquids *Int. J. of Thermophys*
[9] Balderas J A 2012 Generalized 1D photopyroelectric technique for thermal and optical characetrization of liquids *Measurement Science Technology* **23** 10

IOP Conf. Series: Journal of Physics: Conf. Series **792** (2017) 012090 doi:10.1088/1742-6596/792/1/012090

Calibrating a point diffraction interferometer using aspherical wavefront reference.

D. Castán-Ricaño, Fermín S. Granados-Agustín, E. Percino-Zacarías, P. Tecuatl-Tecuatl, A. Cornejo-Rodríguez.
Instituto Nacional de Astrofísica, Óptica y Electrónica, Apdo. Postal 216, C.P. 72000, Puebla, Pue., México.
E-mail: diana.ncr@gmail.com, fermin@inaoep.mx, epercino@inaoep.mx, ptecuatl@inaoep.mx, acornejo@inaoep.mx.

Abstract. This work presents a method to calibrate the reference wavefront generated by an IDP. This wavefront is considered for quantitative evaluation of a plano-convex aspheric surface. Preliminary experimental results are presented.

1. Introducción.

The point-diffraction interferometer (PDI) divided the light in two components to produce a test and reference beams, after passing through the system under test. The PDI generates a spherical reference wave by diffraction at a point discontinuity placed in the path of a beam [1, 2]. The configuration illustrated in Fig.1.

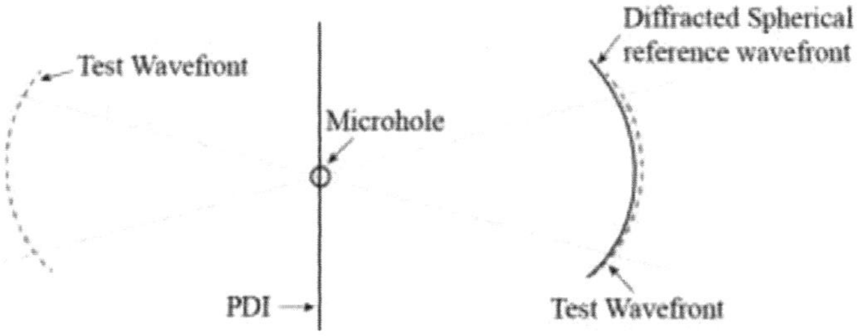

Fig. 1. The configuration of PDI.

The PDI operation consists in impinging a converging wave on a semi-transparent plate. The plate has a micro hole, which functions as a point source which generates spherical waves. After the plate an interference unlocated pattern is generated by the sum of the transmitted wave and the spherical reference wave. The PDI has several advantages such as not requiring a reference surface, like other interferometers, the PDI can measure phase variations from variations of the interference pattern and one of the most important characteristics is a that it is a common paths interferometer, therefore is insensitivity to mechanical vibrations.

Content from this work may be used under the terms of the Creative Commons Attribution 3.0 licence. Any further distribution of this work must maintain attribution to the author(s) and the title of the work, journal citation and DOI.
Published under licence by IOP Publishing Ltd

2. PDI manufacturing.

The construction of the PDI begins with cleaning coverglass, between two of them a drop of mercury is placed and pressed to form tiny droplets. With the aid of a stereomicroscope (Fig. 2), the right sizes are chosen and others are removed with compressed air, in this case the orifice size is a 12μm (Fig. 3).

Fig. 2. Stereomicroscope used in the construction process of the PDI.

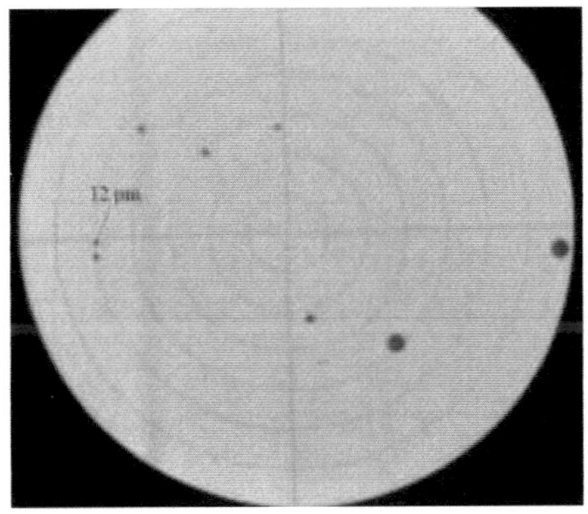

Fig. 3. Mercury chosen seen through a Stereomicroscope.

Then an aluminum coating is made by evaporation with certain percentage of a defined thickness, the evaporator with glass slides is illustrated in Fig. 4. The thickness obtained with evaporation are shown in Table 1. Finally, the sphere of mercury is separated from glass coverslips. Mercury hole seen through a stereomicroscope is show in Fig. 5.

Fig. 4. The evaporator with slides.

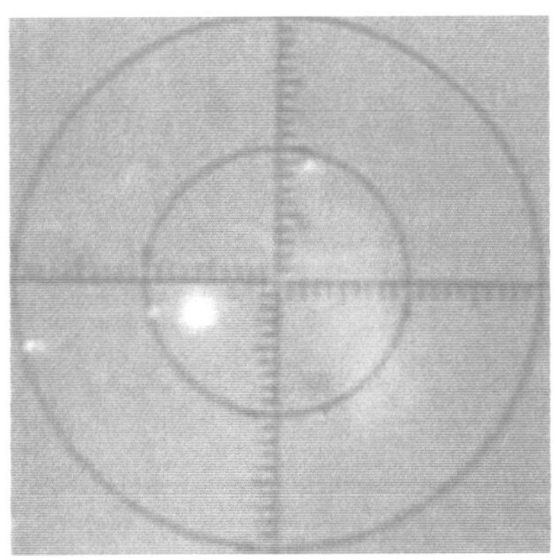

Fig. 5. The hole of 12μm seen through a stereomicroscope.

Table 1. Data coating thickness used to PDI.

	Disseminator (*Torr*)	Cold Cathode (*Torr*)	VariaC (*C*)	Rate (*Å/s*)	Tickness (*kÅ*)	Time (*Min*)
1	$1.1x10^{-2}$	$4.5x10^{-5}$	4.5	0.0	0.000	0:00
2	$1.7x10^{-2}$	$2.1x10^{-4}$	4.5	1.0	0.118	2:00
3	$1.2x10^{-2}$	$1.6x10^{-4}$	4.75	1.0	0.231	4:00
4	$1.1x10^{-2}$	$1.5x10^{-4}$	6.0	0.4	0.310	7:00

3. Calibration of the PDI.

Before taking the test with an aspherical lens, the PDI was calibrated with a spherical wavefront coming from a reference spherical lens of a commercial interferometer Zygo®, the characteristics of the lens are shown in Table 2.

Table 2. Parameters of the espherical lens [λ/20] of Zygo®.

Parameter	**Value**
R[mm]	298.03
D[mm]	101.6
f[mm]	338.08

The PDI was placed at the focus of the lens, as shown in Fig. 6, and the observed interferogram is shown in Fig. 7.

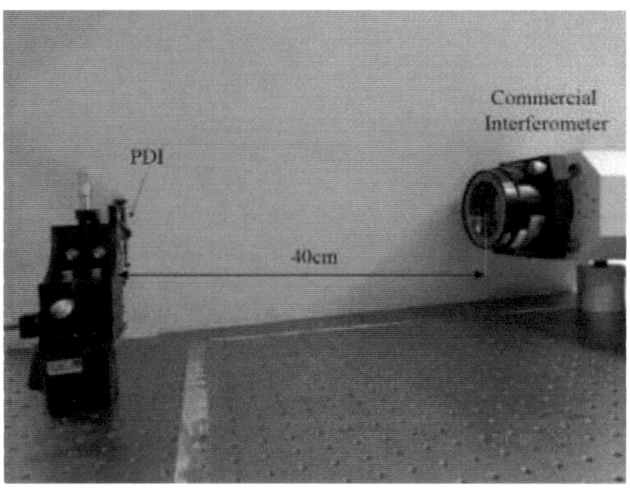

Fig. 6. The set up to calibration of the PDI.

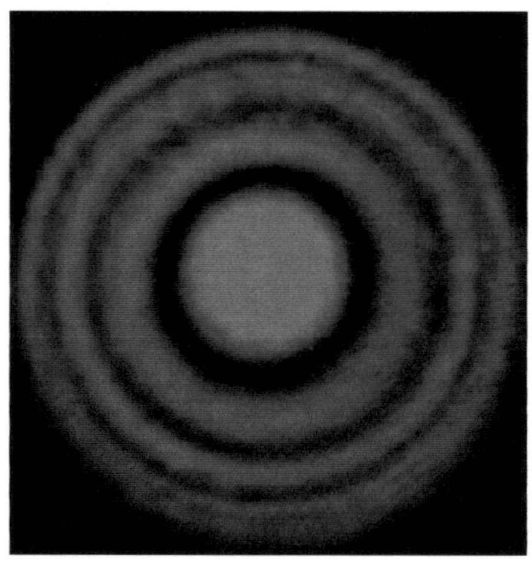

Fig. 7. Experimental interferogram obtained of PDI with 12μm hole, placed at the focus.

4. Experimental Development.

The experimental arrangement, shown in Fig. 8, for testing an aspherical lens (the parameters are shown in Table 3) was placed at a distance of 52.3cm from the spherical lens. The PDI is placed at a distance of 111cm from the aspheric lens. A camera was used to obtain interferograms produced by the PDI was placed at a distance of 4 cm from the PDI.

Table 3. Parameters of the aspherical lens A100-200LPX-U-S of Asphericon®, A_j are the three coefficients.

Parameter	Value
R[mm]	102.24
D[mm]	100
K	-1
N	1.5168
A_4	$4.9646003\text{x}10^{-008}$
A_6	$7.4017872\text{x}10^{-013}$
A_8	$9.4141703\text{x}10^{-018}$

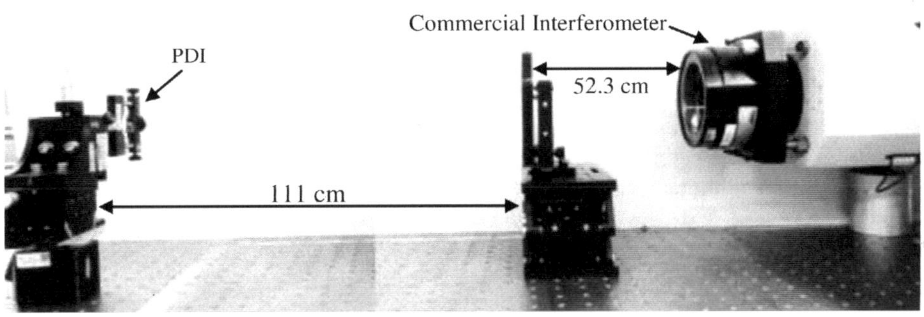

Fig. 8. The set up with the aspherical lens under test and the PDI.

5. Preliminary results.

The interferogram obtained with the PDI, with a hole of 12μm is shown in Fig. 9, where seven visible fringes and a good contrast between them. A qualitative evaluation is shown in Fig. 10 and Fig. 11, were made using the commercial program Apex®. In Fig. 12, we can see the contrast between the strips for two wavelengths along the interferogram and qualitative evaluation data by the analysis program Apex® shown in Table 4.

Table 4. Data by an analysis program Apex®.

Data	**Value**
λ[μm]	0.6328
Irregularity[μm]	6.97
RMS[μm]	1.88
Peak-to-Valley[μm]	6.99
Points	3297

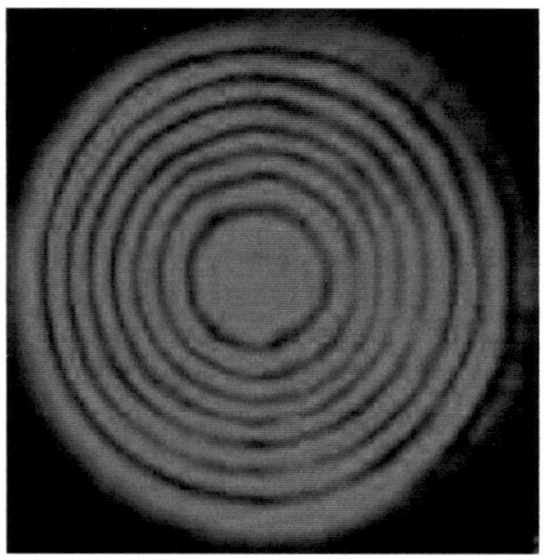

Fig. 9. Experimental interferogram obtained of PDI with 12μm hole, placed at the focus.

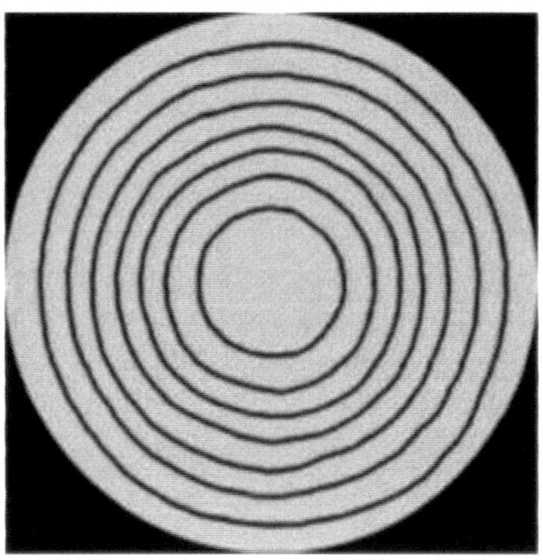

Fig. 10. Processed contourn of the interferogram obtained with PDI.

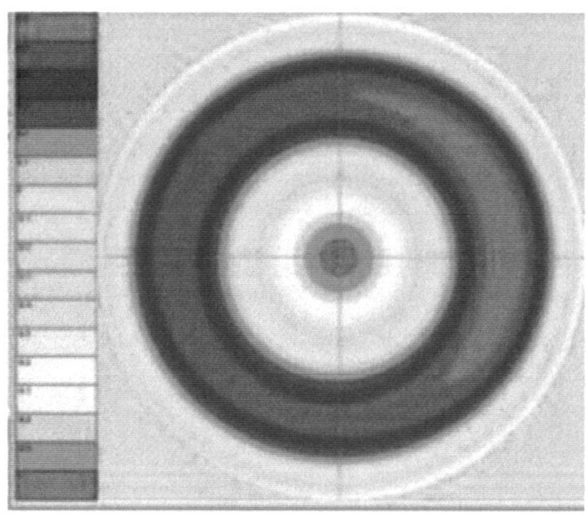

Fig. 11. Processed curves of the interferograma obtained with PDI, where red is lowest and blue is highest.

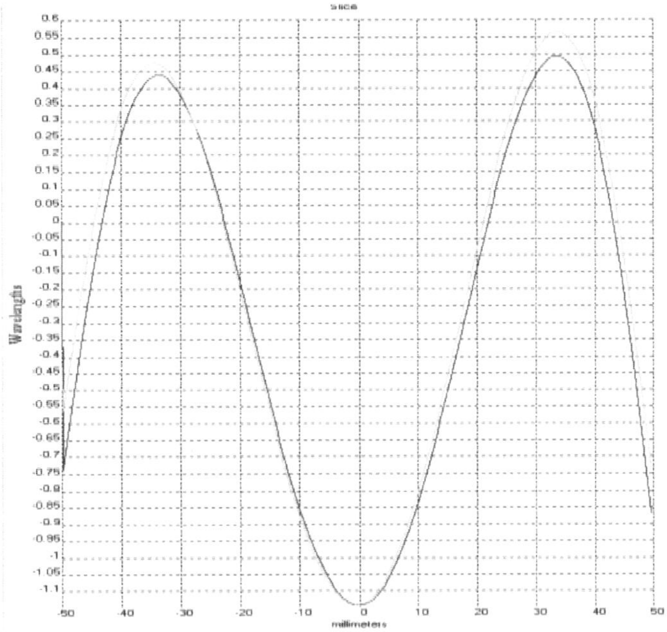

Fig. 12. Sweep through the interferograma obtained with PDI of 12μm.

6. Conclusions.

Since in the PDI is used a micropinhole to produce the reference wavefront, it is not necessary to use optical reference surface; some intrinsic errors can be eliminated. On the other hand, the results show that although IDPs are not perfect, they are suitable for evaluating aspherical surfaces. It is noteworthy that the process for evaluating the aspheric surface will be carried out as future work.

7. References.

[1] Smartt R N 1979 *Proc. of SPIE* 0192
[2] Linnik W P *A* 1933 *Simple Interferometer for the Investigation of Optical Systems* vol 5 (URSS: Proceedings of the Academy of Science) p 210
[3] Harris W,Speer R J and Stanley V 1980 *Proc. of SPIE* 0235
[4] García P 2000 *Alineación de un telescopio Newtoniano mediante un interferómetro de Difracción por Punto* (INAOE)
[5] Rueda E 2000 *Interferómetro de Difracción por Punto (IDP) para el análisis de fluidos* (INAOE)

IOP Conf. Series: Journal of Physics: Conf. Series **792** (2017) 012091 doi:10.1088/1742-6596/792/1/012091

Tunnel currents produced by defects in p-n junctions of GaAs grown on vapor phase

V R Barrales Guadarrama, E M Rodríguez Rodriguez, R Barrales Guadarrama and N Reyes Ayala
Universidad Autónoma Metropolitana – Azcapotzalco, Av. Sn. Pablo 180, Col. Reynosa-Tamaulipas, México D. F., Tel: 53 18 90 42

E-mail: vrbg@correo.azc.uam.mx

Summary. With the purpose of assessing if the epitaxy on vapor phase technique "Close Space Vapor Deposition (CSVT)" is capable of produce thin films with adequate properties in order to manufacture p-n junctions, a study of invert and direct current was developed, in a temperature range of 94K to 293K, to junctions p-n of GaAs grown through the technique CSVT. It is shown that the dominant current, within the range 10^{-7} to 10^{-2} A, is consistent with a currents model of the type of internal emission form field, which shows these currents are due to the presence of localized states in the band gap.

1. Introduction

In the last three decades, the development of manufacturing p-n junctions technology has been accelerated, linking energy saving, efficiency, and sustainability. An important part of this development has been focused on the improvement of the characteristics and physical properties of typical materials such as GaAs, which is used in the production of electronic devices. Some examples are: almost null density defects, control of the profiles of doping, increase in carriers mobility and their current density with minor current dissipation. It has allowed the production of innovative materials as InGaP and CdTe, and, at the same time, original architectural designs with nanometric dimensions. Giving place to a great quantity of devices with high efficiencies ~40.8%[1], they have been used in different applications, such as electricity supply, electronic of potency, satellites and other as communication and medical equipment, such is the case of solar cells manufactured based on compounds III-V on monolithic multi-junction tandem arrangements based on InGaP/(In)GaAs/Ge, where the most important is the design and the manufacturing of the tunnel diode structure, due to its series connection. The design of the tunnel diode structure and the current adjustment, between the upper part (InGaP) and the inferior (GaAs) of the cell, must be optimized in order to achieve a high efficiency and conversion stability[2]. Other significant example is the interest of developing applications CMOS[3] with technology high-K gate stack that provides low thermal consumption, and high carriers mobility. An achievement of this technology is obtaining high performance p+/n junctions[4], as it gets a high quality in dielectric for the gates[5,6] and low resistant contacts[7].

Content from this work may be used under the terms of the Creative Commons Attribution 3.0 licence. Any further distribution of this work must maintain attribution to the author(s) and the title of the work, journal citation and DOI.
Published under licence by IOP Publishing Ltd

Therefore, to accomplish a high performance of devices or applications is preponderant that the crystal growing systems have the capacity of obtaining materials with adequate physical properties as crystallinity, carriers concentration, mobility, electrical resistance; and allowing the possible production of different materials, without risk of contamination, to manufacture heterostructures.

This article describes the type of currents that occurs in p-n junctions of GaAs obtained through epitaxy technique in vapor phase CSVT[5,6,7]. The current-voltage measurements showed that in direct polarization the dominant current varies as $\exp(qv/nkT)$, where n = 3 at room temperature, and n > 3 for lower temperatures, indicating the presence of a currents tunnel. It is shown that the variation of current density in the pre-avalanche range is qualitatively consistent with a model of tunnel effect, through local states in the gap.

The diffusion current is not present for any range of polarization at room temperature, even setting the junction in conditions where usually the diffusion current is increased faster than the recombination; this is, increasing the temperature for a given polarization. The concentration ranges in the studied junctions were 1×10^{13}, 1×10^{16} cm^{-3}, and 3×10^{17} cm^{-3} on the layer "n" type, and 1×10^{19} cm^{-3} on the layer "p" type. A correspondence analysis between experimental and theoretical data indicates that exist defects on the junction that cause a local enlargement of the electric field and in consequence a current tunneling effect through the junction. The energies associated to local defects in the wide range of the band gap, besides the impurity accumulation disseminated within and around these defects, are probably caused by high temperature when the sample is exposed during growing (850°C <T<900°C) and generated by the metallurgical interphase.

2. Experimental procedure

The p-n Junctions were manufactured using water vapor as carrier gas, growing GaAs type "n", doped epitaxially with Te, over a GaAs (100) substrate type "p" with a Zn concentration of 1×10^{19} cm^{-3}, obtained through the Czochralski . The layer thickness type "n" had an interval between 14 y 42 μm, and the substrate thickness always was approximately 350 μm. The metallization for the contacts were evaporated. The materials and masses are presented in the Table 1.

Table 1. Materials and compositions used for ohmic contact

Material	Alloy for Ohmic Contact
GaAs type "p"	Au-Zn (92.4 % Au-7.6 % Zn). Zn(14 mg). Au(170 mg).
GaAs type "n"	Au-Ge-Ni (83 % Au-11.4 % Ge-5.4 % Ni). Au(152 mg). Ge(21 mg). Ni(10 mg).

To achieve an ohmic contact resistance, an annealing process was performed at 450 °C with "formingas", which was flowing to 5 lt/min. The diodes were cleaved in a square shape of ~78 × 10^{-4} cm^2 and the upper contact was given a star shaped, in order to promote a uniform polarization upon the p-n junction. Table 2 shows the experimental conditions of two series of representative p-n junctions, grown through the CSVT technique.

Table 2. Experimental conditions of two series of representative p-n junctions

Junction p-n	SOURCE [cm^{-3}]	SUBSTRATE [cm^{-3}]	**T_F [°C]	**T_S [°C]	**V_r [mV]	**ε [μm]	**r_c [μm/min]
*U10	High Purity (~10^{15})	P(Zn) 1×10^{19}	850	780	0.80	~60	2
*U20	High Purity (~10^{15})	P(Zn) 1×10^{19}	850	780	0.80	~20	2

*The nomenclature U# summarizes the predominant characteristics in DELs series obtained during that growing.
**T_F ≡ temperature of source, T_S ≡ temperature of substrate, V_r ≡ voltage of spray, ε ≡ thickness of the deposited film, r_c ≡ growing rate.

3. Results and discussion

3.1. Current-Voltage Reverse Characteristic

Figures 2a and 2b show the characteristic behavior of current-voltage to different temperatures to p-n junctions of the samples in Table 2. It is observed a monotonous growing of the current regarding the voltage, besides, the voltage for constant reverse current decreases with an increment of temperature.

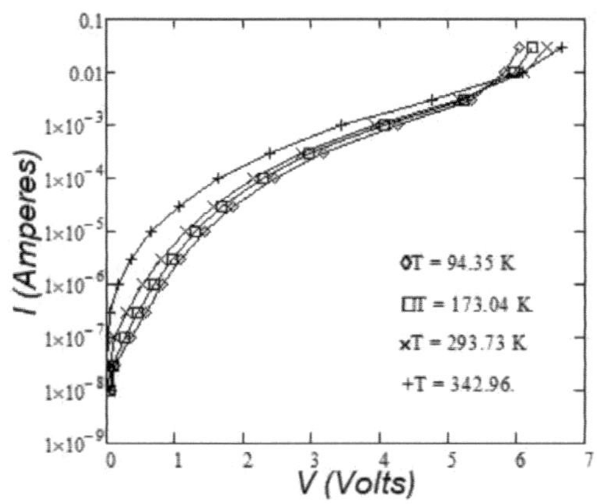

Figure 2a. Reverse current-voltage characteristic for DEL60 to different temperatures.

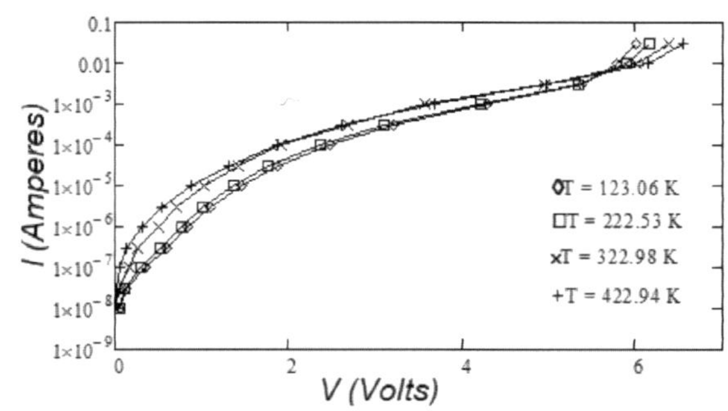

Figure 2b. Reverse current-voltage characteristic for DEL60 to different temperatures.

This behavior is shown on Figure 3, and suggests that the current in the indicated range is due to emission for internal field or Zener effect.

The emission by internal field is observed in junctions highly doped in Si [8,9], Ge [10], InSb [11] y GaAs [12] where the electric field is relatively high, even if it is a low inverse voltage. It is possible that the effect of a high electric field, the electron move from the valence band, p type, towards the conduction band, n type, of the junction. It is a consequence of tunnel effect.

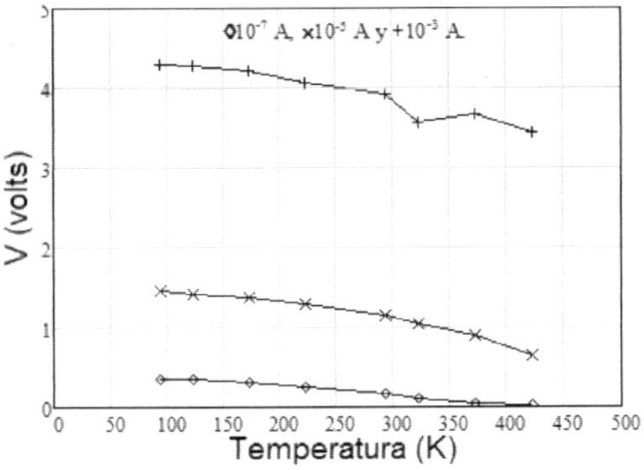

Figure 3. Voltage decreasing with temperature increment for three constant reverse currents.

The expression that allows to calculate the tunneling current (Zener), based on the tunneling probability for transitions through localized levels in the "gap", is giving by

$$I = A \cdot V_a \cdot E \cdot exp\left\{-\left(\tfrac{4}{3}\right)\cdot\left[(2m_T)^{1/2}/q\hbar\right]\left(E_b^{3/2}/E\right)\right\} \quad (1)$$

Where A is a constant for a given temperature, V_a applied voltage, E is the electric field in the junction, m_T is the tunneling effective mass, q is the electric charge, $\hbar$ is the Planck constant divided by 2π, and E_b is the energy barrier trough which is possible that the electron suffers a transition by tunnel effect.

In inverse polarization conditions, the minimum height of the energy barrier is equals to the "gap" energy of $E_g = qV_g$. The calculus of the tunneling probability is considered a uniform electric field. The selected value for the electric field is the maximum value existing in the junction, for an abrupt junction is given by:

$$E = 2(V_{bi} - V_a)/W \quad (2)$$

where V_{bi} is the inter-constructed voltage and W is the wide-band gap zone, given for:

$$W = W_l (V_{bi} - V_a)^{1/2} \quad (3)$$

being W_l a constant. From here, the Zener tunneling constant is given then for:

$$I = A \cdot V_a \cdot (V_{bi} - V_a)^{1/2} \cdot \exp\left\{-\alpha\left[V_g^{3/2}/(V_{bi} - V_a)^{1/2}\right]\right\} \quad (4)$$

and

$$\alpha = \left(2\sqrt{2}/3\right)\cdot\left(q^{1/2}/\hbar\right)\cdot m_T^{1/2} \cdot W_1 \quad (5)$$

Equation (4) predicts that given a graphic of $log[I/V_a(V_{bi} - V_a)^{1/2}]$ vs $[1/(V_{bi} - V_a)^{1/2}]$, then it may be obtained a straight line with a slope equals to $-\alpha V_g^{3/2}$. Such graphic is shown in Figure 4, which describes the characteristic I vs. V in inverse polarization to 94.35 K and 293.73 K. The experimental data are well represented through a straight line with a deviation on the range of the high currents, probably for the beginning of the breaking point due to the avalanche. The slope of data at 293.73 K is equals to -17.557 $V^{1/2}$, which corresponds to 9.53 V^{-1} for α. The slope at 94.35 K increases to -24.098 $V^{1/2}$ corresponding for $\alpha = 13.08$ V^{-1}.

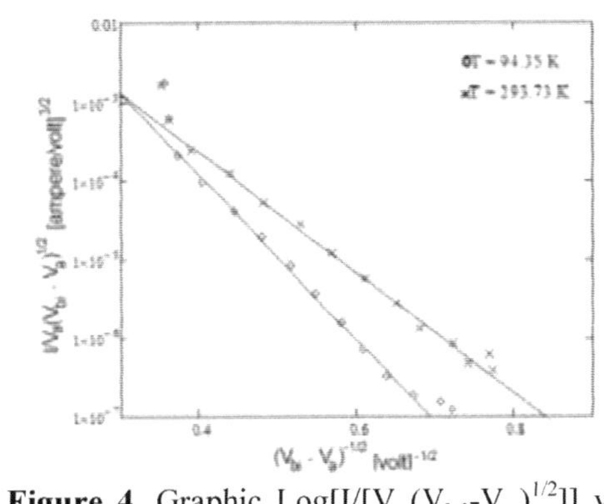

Figure 4. Graphic $Log[I/[V_a(V_{bi}-V_a)^{1/2}]]$ vs. $(V_{bi}-V_a)^{-1/2}$ on reverse polarization.

A theoretical value of α may be calculated from equation (5). Thus, assuming that the tunneling mass is equals to 0.063 m_0, and using the value of the wide-band gap of the dump area of the junction, it is obtained from the capacitance measurements. The value calculated of α is equals to 60.56 V^{-1} and it is an independent variable of temperature.

An explanation about the absence of a good approximation between the experimental value and the theoretical prediction for α will be discussed later.

It is important to point out that all diodes grown through CSVT showed a very similar behavior to the previous one described, this means, the current is qualitative consistent with the model of the Zenner effect on the range 10^{-5} A/cm^2 a 10^{-1} A/cm^2.

3.2. Current-Voltage forward Characteristic

The current-voltage characteristic to different temperatures also was determined for the same p-n junctions discussed in the previous section, Figure 5.

Figure 6 shows the I-V direct characteristic in an interval of voltage ~0.2 a 0.8 V. The experimental data may be approximated to a straight line, indicating that the relation current-voltage is determined as:

$$I = I_D + I_R + I_T \qquad (6)$$

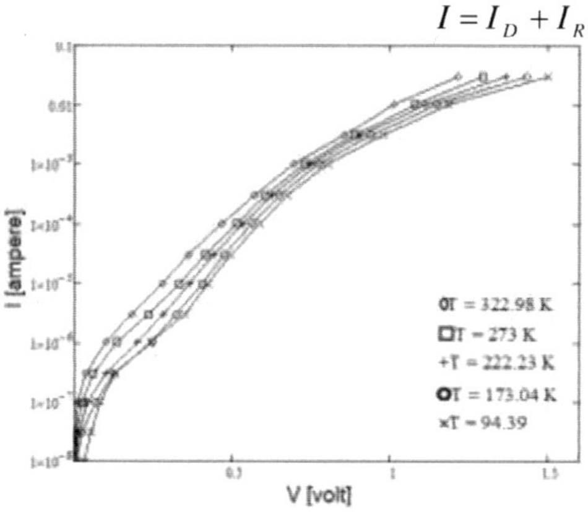

Figure 5. Characteristic I-V direct to different temperatures.

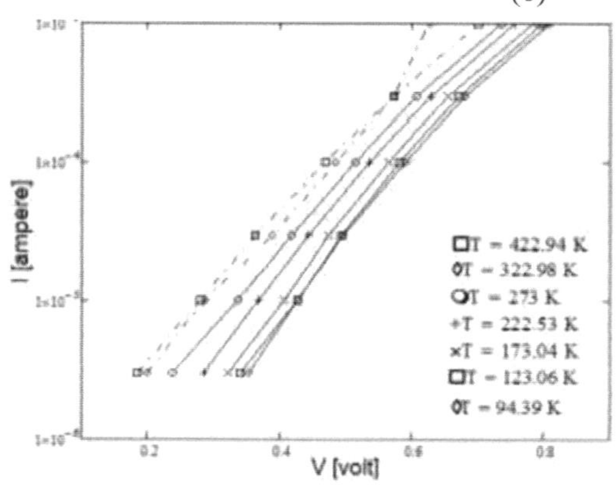

Figure 6. Characteristic I-V direct.

where:

- $I_D = I_{0D} \cdot exp\left(^{qV}/_{nkT}\right)$ represents the current due to an injection thermally activated upon the junction barrier, $n = 1$ denotes the diffusion current[12], and I_{0D} is given by $I_{0D} = \left(\frac{qD_p p_{n0}}{L_p} + \frac{qD_n n_{p0}}{L_n}\right)$, being D_p, D_n, L_p, L_n the coefficients and length diffusion on the "n" side and the electrons on the "p" side of the junction, respectively.p_{n0} and n_{p0} represent the concentration of holes and electrons in equilibrium on the side "n" and "p" of the junction, respectively;
- $I_R = I_{0R} \cdot exp\left(^{qV}/_{nkT}\right)$, with $n = 2$ for the recombination on the region of spatial charge in those deep level traps [11], and I_{0R} is given by $I_{0R} = \frac{qW}{2}\sigma v_{th} N_t n_i$, being q the electron charge, W wide-band gap, σ transversal section of the defect capture, N_T the defect concentration, and n_i the intrinsic carriers concentration of the semiconductor and
- $I_T = I_{0T} \cdot exp(\alpha V)$ represents the current due to the tunnel effect.

The curves shown on Figure 6 only predominates a region of exponential behavior, which represent n = 2.95 for T = 322.98 K and n = 9.29 for T = 94.39 K. Large values of n and a relative steady state of the slope respect to the temperature show that the mechanism of current carrier, in this region, is not dominated by the thermic effect, but a tunnel effect mechanism. The deviation in the curves of Figure 6, above 0.8 V respect to the followed behavior on the range 0.2 a 0.8 V, are due to the resistant series effects of diode.

The exponential variation of DC current with the voltage, with independent slopes of temperature has been reported for p-n junctions of diamond[15], SiGe heterojunction bipolar transistor[16], GaAs[17], and GaN[18]. Chynoweth *et a l*[19] proposes a model for this exponential current, which is based on a tunneling mechanism involving energy level localized on the band gap. The basic process involves an electron tunneling from the conduction band on "n" side to inner gap state, until it takes a state in or nearby the valence band on the p side. The exact nature of the tunneling process is no determined, in other words, it is not specified if the tunneling is band to band by real localized states, as Chynoweth *et al*[19]postulates, or, if the tunneling is band to impurity states near the valence band via intermediate virtual states, as is postulated by Butenko[20]. However, the elemental assumption is done on both cases and the expression derived from the model is experimentally verified, in its qualitative aspect, on the previous analysis.

The basic assumption done on the tunneling simple model, is that, the tunnel current is dominated by the variation of the tunneling probability, then the current can be described as:

$$I = I_T \cdot \exp\left\{-\left(\tfrac{4}{3}\right)\left[(2m_T)^{1/2} / q\hbar\right]\left(E_b^{3/2} / E\right)\right\} \tag{7}$$

where $I_T = AV_a E$.

As, in the analysis case, the reverse characteristic, the electric field that is taken into account is the maximum developed in the junction, which is given in the equation (2). For the forward polarization of the barrier is given for:

$$E_b = q \cdot (V_{bi} - V_a) \tag{8}$$

Therefore

$$I = I_T \cdot exp[\alpha(V_a - V_{bi})] \tag{9}$$

Where α is the same for the reverse characteristic and it is given for equation (5).

Qualitatively, equation (9) describes the experimental results, inasmuch indicates an exponential dependence of the forward current against the voltage, with an equal pendent to α that is relatively independent of temperature. The experimental α value is obtained from the current-voltage forward characteristic at 300 K is 6.64 V^{-1}, whose carrier concentration in the epitaxial region (layer type "n") is 2.9×10^{17} cm^{-3}.

The characteristics in the forward case were also analyzed for the rest of diodes with different carrier concentration, therefore, different junction width. For diodes with carrier concentration of 1×10^{16} cm^{-3}, the value of α at 300 K, was 10.04 V^{-1}, it corresponds the theoretical value of 326.2 V^{-1}. For diodes with concentration 1×10^{17} cm^{-3} the experimental value of α was 7.21 V^{-1}.

4. Conclusions

The study shows that the current in reverse polarization is due to Zener tunneling. The dominant current in forward polarization, for the diodes obtained, is qualitatively consistent with a simple tunneling model involving states on the band "gap". The lack of a good quantitative concordance between the experiment and the theory is probably a consequence of imperfections in the junction that causes that the intensity of the electric field increases locally. The fact that the theoretical prediction for the parameter α has been higher that the obtained from the analysis for currents, due to an internal field emission, indicating these last ones are the expected for wider minor junction than the determined correspondents from capacitance data. Inasmuch the narrow junctions imply intense electric fields, the results suggest that can exist localized regions or defects concentration distributed along the junction that tend to increase the electric field, similar to microplasma sites (areas where the current flows through imperfect regions of p-n junction) that give as a result low voltage breaking[19,21,22]. Different studies for several materials, including GaAs, have shown a correlation between microplasma and dislocations sites[19,21,22]. The defects present on junctions or its surroundings can be dislocations, vacancies or stacking faults generated by the overheating that the wafer is exposed during growing, another possibility is that the defect concentration contained in the substrate (~900 EPD/cm^2) were perturbing the p-n junction, since this coincides with the metallurgical junction or it is very close. It exists the possibility that during growing the contained Zn in the substrate, it spreads to the epitaxial layer and the contained Te in the epitaxial layer spreads to the substrate or this has lost As. No matter the case, they will form defects that will result in the intensification of the electric field in the p-n junction. For instance, it is possible to introduce dislocations (high superficial concentration conditions and temperatures > 750 ^{0}C) during the diffusion process[20], due to the net adjustment between the widespread region and the crystal non-diffused[22]. Dislocations are being observed diffusion inducted in p-n junctions of GaAs formed through Zn diffusion[23], besides the precipitated contained upon these dislocations[18]. Residual impurities have been found on CSVT films (through SIMS) of Si, Cr, Fe, Mg and S of the interface[24], with concentrations of 5×10^{14} cm^{-3} to 10^{15} cm^{-3} for Mg and Fe, of 10^{17} cm^{-3} for Cr and S and higher to 10^{17} cm^{-3} for Si. Besides the impurities accumulation in the interface are increased when growing temperature increases, which indicates that probably exist impurities diffusion towards the epitaxial layer[25] and the substrate, then is probable that the metallurgic junction, that is strongly "disturbed", influences over the p-n junction behavior, since the effect of the defects (dislocations, vacancies, micro-defects or stacking fault) upon the electrical properties of the p-n junction take a greater importance when exists an impurity accumulation around these, so as can introduce a certain density of recombination centers. Goetzberger and Shockley[26] show that the reverse characteristics of current-voltage on its p-n junctions were caused by precipitates in the dump area, and it is assumed that the current was due to Zener tunneling caused by points of a intense electric field.

5. References

[1] Geisz, J F et al 40.8% efficient inverted triple-junction solar cell with two independently metamorphic junctions *Appl. Phys. Lett.* **93**(123505) 1–3
[2] Repmann, T et al 2003 *Proceedings of the 3rd World Conference on Photovoltaic Energy Conversion* (Japan) p 1843–1846 11-18
[3] J. Oh et al 2009 *Proc. Symp. VLSI Technology* p 238–239
[4] Piyush B et al 2014 *High Performance 400 °C p+/n Ge Junctions Using Cryogenic Boron Implantation* Vol 35 (IEEE Electron Device Letters)
[5] P.H. Robinson 1963 *RCA Rev.* **24** 574
[6] Chavez F et al 1983 *J.Appl. Phys.* **54** 6646
[7] Pérez G F et al 2014 *Synthesis of In–In2O3 microstructures by close-spaced vapor transport (CSVT) and their transformation to In2O3 nanobelts at low temperature* Vol 107 (Vacuum) p 236–241
[8] R. Zhang et al 2012 *Proc. IEEE IEDM* p 371–374
[9] Bhatt P et al 2013 Germanium oxynitride gate interlayer dielectric formed on Ge(100) using decoupled plasma nitridation *Appl. Phys. Lett.* **103** 1-5
[10] Manik P et al 2012 Fermi-level unpinning and low resistivity in contacts to n-type Ge with a thin ZnO interfacial layer *Appl. Phys. Lett* **101** 1-5
[11] . Chynoweth A G 1960 *Phys. Rev.* **118** 425
[12]Singh M 1968 *Solid-State Electron* p 11 99 117
[13] Chynoweth A G and Logan R A 1960 *Phys. Rev.* **118** 1470
[15] Garino Y et al 2012 *Diamond and Related Materials* Vol 21 p 33–36
[16] Milovanovi´c V and Van R 2010 *IEEE Transactions on Electron Devices* Vol 57
[17] Park K W et al 2015 *Semiconductor Science and Technology* Vol 30
[18] Krishnamoorthy S, Sung P and Rajan S 2011 *Applied Physics Letters* **99** 233504
[19] Chynoweth A G, Feldmann W L and Logan R A 1961 *Phys. Rev.* **121** 684

IOP Conf. Series: Journal of Physics: Conf. Series **792** (2017) 012091 doi:10.1088/1742-6596/792/1/012091

[20] Butenko A V et al 2008 *Journal of Applied Physics* **103** 024506
[21] Lesniak M and Holt D B 1987 *Journal of Materials Science* **22** 3547–3555
[22] Koktavy P et al 2007 *Wseas Transactions on Electronics* Vol 4
[23] Akopyan A A et al 2009 *Semiconductors* Vol 43p 368–373.
[24] Black J F and. Jungbluth E D 1967 *Electrochem. Soc* 114-188
[25] Bourgoin J C et al 1985 *Photoluminiscence and Electrical Properties of Epitaxial GaAs Grown by Closed Space Vapor Transport*
[26] Goetzberger A and Shokley W 1960 *J. Appl. Phys* **31** 1821

IOP Conf. Series: Journal of Physics: Conf. Series **792** (2017) 012092 doi:10.1088/1742-6596/792/1/012092

Multiple velocity encoding in the phase of an MRI signal

E E Benitez-Read[1]

1 Centro de Resonancia Magnética, Facultad de Ingeniería, Universidad Autónoma de Chihuahua, Circuito No. 1 Nuevo Campus Universitario, Chihuahua, Chih., Mexico C.P. 31125

E-mail: edbenite@uach.mx

Abstract. The measurement of fluid velocity by encoding it in the phase of a magnetic resonance imaging (MRI) signal could allow the discrimination of the stationary spins signals from those of moving spins. This results in a wide variety of applications i.e. in medicine, in order to obtain more than angiograms, blood velocity images of veins, arteries and other vessels without having static tissue perturbing the signal of fluid in motion. The work presented in this paper is a theoretical analysis of some novel methods for multiple fluid velocity encoding in the phase of an MRI signal. These methods are based on a tripolar gradient (TPG) and can be an alternative to the conventional methods based on a bipolar gradient (BPG) and could be more suitable for multiple velocity encoding in the phase of an MRI signal.

1. Introduction

The velocity of moving spins is one of the most relevant characteristics of flow when using nuclear magnetic resonance (NMR) methods. It can be estimated from the magnitude and/or phase of the MRI signal. It is well known that measurement of velocity done by phase methods is more sensitive than those done by magnitude methods.

The presence of surrounding stationary structures, such as soft tissue and bones, around a vessel introduces a conspicuity problem [1] in the image of the vessel and has been one of the major problems in flow measurements by NMR.

When static and moving spins are present in the same voxel, the phase of the MRI signal, which is a function of the velocity of flowing spins, is distorted by the presence of static spins. Some novel methods for multiple velocity phase encoding of the MRI signal, based on a tripolar gradient (TPG) as alternative to the conventional bipolar gradient (BPG), are presented and analyzed in this paper.

2. Background

Phase encoding methods [2, 3], measure the attenuation of the echo amplitudes obtained from sequences of the form $90°_x$ -T- ($180°_y$ –T-)n. This type of sequence separates the effects of coherent flow and diffusion. The phase shifts at echo time increases attenuating the amplitudes of the echoes in function of the velocity. A similar method [4] uses the amplitude of the first echo of a CPMG sequence with a gradient in the direction of flow. The gradient is stepped every time that the sequence is applied. The velocity distribution of laminar and turbulent flow can be obtained from the inverse Fourier transform of the gradient dependent signal.

Two dimensional flow imaging based on these principles has been done with the flow encoded in the plane of the image [5] or perpendicular to it [6].

A summary of previous work that had been done before to suppress signals from stationary spins falls essentially in two modalities: Subtraction and Auto-cancellation and can be found in [7].

An analysis of actual methods for velocity encoding in the phase of MRI signals can be found in [8] as well as previous work on multiple velocity encoding, using a tripolar gradient , done by the author , can be found in [9].

Content from this work may be used under the terms of the Creative Commons Attribution 3.0 licence. Any further distribution of this work must maintain attribution to the author(s) and the title of the work, journal citation and DOI.
Published under licence by IOP Publishing Ltd

3. Multiple Velocity Encoding Methods

Three novel methods of multiple encoding of velocity based on a tripolar TPG gradient echo sequences shown in Figures 1b, 1c and 1d are analysed in this section and compared to the conventional velocity encoding method based on the BPG gradient echo sequence shown in Figure 1a. Each sequence will be analysed with velocity encoding only in the readout direction in order to show the constraints in this direction. Similar considerations can be taken if the velocity encoding is in the other two orthogonal directions.

Although the four sequences of Figure 1 are in terms of rectangular pulses, in actual experiments, ramps could be used to get trapezoid and sometimes half sine pulses. In all the following gradient waveforms, the amplitude of a basic gradient waveform will be given as the maximum amplitude G multiplied by a proportionality constant k. This is with the purpose of making more adaptable the pulse sequences for different gradient strengths.

3.1 Method A (Conventional BPG Sequence)

This method is based on sequence A (Fig.1a), which encodes velocity by varying the amplitude k_vG of the BPG formed by the first two lobes of $G_{re}(A)$. The last three lobes of $G_{re}(A)$ form a tripolar gradient (TPG_A) that is designed to have a null zeroeth moment given by

$$m_0 = (k_1T_{g1} - k_2T_{g2} + k_3T_{g3})G = 0 \tag{1}$$

and a null first moment given by

$$m_1 = \left[\frac{k_1}{2}\left(1+\frac{k_1}{k_2}\right)T_{g1}^2 + k_1T_aT_{g1} - k_3T_{g3}\left(T_a + \frac{T_{g3}}{2} + \frac{k_3T_{g3}}{2k_2}\right)\right] G = 0$$

(2)

In a whole body imager, the values of k_3 and T_{g3} of the readout gradient are usually set as a function of the field of view and the sampling period. For a given sampling period Δt that determines the value of Tg3, and for a given value of the field of view (L_r), in the readout direction, the value of the readout gradient amplitude is obtained by

$$k_3G = \frac{N_r}{\gamma_1 T_{gr} L_r} \tag{3}$$

where γ_l the gyromagnetic ratio is given in Hz/Gauss, G is the maximum value of the readout gradient amplitude in Gauss/cm, L_r is the readout field of view in cm, N_r is the number of samples in the readout direction, T_{gr} is the width of the readout gradient which is taken to be twice as long as T_{g3}.

Assuming for convenience , that $k_l = k_2 = K$ and setting K together with T_a to convenience, the two parameters of TPG_A left to be determined are T_{gl} and T_{g2}. Combining equations (1) and (2) and the assumptions previously mentioned, T_{gl} and T_{g2} are given by

$$T_{g1} = -\frac{T_a}{2} + \frac{1}{2}\left[T_a^2 + 4\frac{k_3T_{g3}}{K}\left(\frac{k_3T_{g3}}{2K} + \frac{T_{g3}}{2} + T_a\right)\right]^{1/2} \tag{4}$$

and

$$T_{g2} = \frac{KT_{g1}+k_3T_{g3}}{K} \tag{5}$$

Once TPG_A is "balanced" by making its zeroeth and first moments equal to zero, the multiple encoding of velocity will depend only on the amplitude of BPG_A.

At the balanced position, k_v is equal to zero. This preserves the zeroeth and the first moments of the whole Gre(A) waveform equal to zero. The zeroeth moment of BPG_A is going to remain equal to zero as mentioned previously. The first moment of BPG, and of the whole $G_{re}(A)$ waveform is then given by

$$m_1 = (T_v + T_b)T_vGk_v \tag{6}$$

A series of m_l values can then be generated by varying k_v in equation (6) in positive and negative increments in order to generate the velocity image.

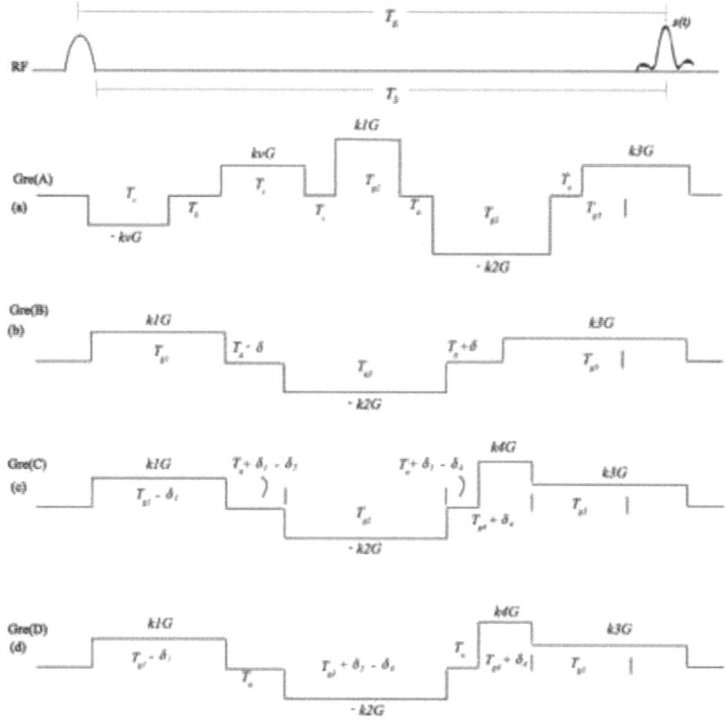

Figure 1: Gradient-echo pulse sequences with multiple encoding of velocity in the readout direction. Only the RF and the Readout gradient are shown.

3.2 Method B (Novel TPG Sequence B)

This method is based on sequence B (Fig.1b), which encodes velocity by shifting the center lobe of the tripolar gradient (TPG) without changing the width of any of the three gradient lobes. On its balanced position, TPG is designed to have a null zeroeth and first moments as given by equations (1) and (2). The considerations for k_3, T_{g3}, k_1 and k_2 are similar to those in sequence A. The parameter k_3 is obtained using equation (3) and the time parameters T_{g1}, and T_{g2}, are obtained by using equations (4) and (5). The first moment of TPG is then given by

$$m_1 = KT_{g2}G\delta \tag{7}$$

where δ is the time shift to the center lobe. A series of m_1 values can then be generated by positive and negative increments of δ in order to generate the velocity image as mentioned before.

3.3 Method C (Novel ETPG Sequence C).

This method is based on sequence C (Fig. 1c), which encodes velocity by shifting the center lobe of the extended tripolar gradient (ETPG) and adjusting the interlobe times and the widths of the side lobes.

On its balanced position, ETPG is designed to have a null zeroeth and first moments as given by

$$m_0 = (k_1T_{g1} - k_2T_{g2} + k_3T_{g3} + k_4T_{g4})G = 0 \tag{8}$$

$$m_1 = \left[\frac{k_1}{2}\left(1 + \frac{k_1}{k_2}\right)T_{g1}^2 - M - k_3T_{g3}\left(T_{g4} + \frac{T_{g3}}{2}\right) - \frac{k_4T_{g4}^2}{2}\right]G = 0 \tag{9}$$

$$M = (k_3T_{g3} + k_4T_{g4})\left(T_a + \frac{k_3T_{g3}}{2k_2} + \frac{k_4T_{g4}}{2k_2}\right) \tag{10}$$

Considerations similar to the two previous sequences apply to k_3, T_{g3}, k_1 and k_2. Combining equations (8) and (9) and the assumptions previously mentioned, T_{g1} and T_{g2}, are given by

$$T_{g1} = -\frac{T_a}{2} + \frac{1}{2}\left[T_a^2 + Q + \frac{4k_3T_{g3}}{K}\left(T_{g4} + \frac{T_{g3}}{2}\right) + \frac{2k_4T_{g4}^2}{K}\right]^{1/2} \tag{11}$$

$$T_{g2} = \frac{KT_{g1}+k_3T_{g3}+k_4T_{g4}}{K} \tag{12}$$

$$Q = \frac{4(k_3T_{g3}+k_4T_4)}{K}\left(T_a + \frac{k_3T_{g3}+k_4T_{g4}}{2K}\right) \tag{13}$$

A null zeroeth moment at a shifted position of ETPG is then given by

$$\delta_4 = \frac{k_1}{k_4}\delta_1 = \frac{K}{k_4}\delta_1 \tag{14}$$

The first moment at a shifted position of ETPG is then given by

$$m_1 = \left[\frac{K}{2}\left(1 - \frac{K}{k_4}\right)\delta_1^2 + K(T_{g2} + 2T_a)\delta_1 + KT_{g2}\delta_2\right]G \tag{15}$$

and δ_2 could also be made dependable on δ_1 or as convenient.

As before, a series of m_1 values can then be generated by positive and negative increments of the independent time shift (i.e. δ_1).

3.4 Method D (Novel ETPG2 sequence D)

This method is based on sequence D (Fig. 1d), which is even more versatile and encodes velocity by shifting the center lobe of the extended tripolar gradient (ETPG) and adjusting the width of all the three lobes of the ETPG while keeping constant the interpulse time T_a.

At its balanced position (no shifts), m_0 and m_1 are zero, as given by equations (5) and (6).

Considerations similar to the previous three sequences apply to k_3, T_{g3}, k_1 and k_2. The gradient pulse widths T_{g1} and T_{g2} are given by equations (11) and (12) respectively.

After this ETPG has been balanced, the first moment is given by

$$m_1 = \left[\frac{k_1}{2k_2}(k_1 + k_2)T_{g1}^2 - M_D - \frac{k_3T_{g3}k_4T_{g4}}{k_2} - k_3T_{g3}T_{g4}\right]G = 0 \tag{16}$$

where

$$M_D = k_3T_{g3}\left(Ta + \frac{T_{g3}}{2} + \frac{k_3T_{g3}}{2k_2}\right) + k_4T_{g4}(T_a + \frac{T_{g4}}{2} + \frac{k_4T_{g4}}{2k_2}) \tag{17}$$

A null zeroeth moment at a shifted position of ETPG results in the ratio

$$\delta_4 = (\frac{k_1+k_2}{k_4+k_2})\delta_1 = \left(\frac{2K}{k_4+K}\right)\delta_1 \tag{18}$$

The first moment at a shifted position of ETPG is given by

$$m_1 = K(1 - \frac{K}{k_4+K})\delta_1^2G + 2K\left[T_{g2} + \left(\frac{k_4T_a}{k_4+K}\right) + KT_a\right]\delta_1G \tag{19}$$

and as in the previous cases, a series of m_1 values can be generated by positive and negative increments of an independent time shift (i.e. δ_1).

The previous four velocity encoding gradient-echo sequences could also be implemented as second spin-echo sequences.

4. Discussion.

The purpose of this section is to compare the performance of the four gradient-echo velocity encoding sequences (Fig. 1) with respect to the maximum first moment m_1 of the velocity encoding gradient. The corresponding spin-echo velocity encoding sequences yield similar results.

The basis of the comparison is to obtain the maximum m_1 for a given value of the duration T_S of the velocity encoding gradient of each pulse sequence and consequently to obtain the minimum "sampling velocity", V_S, for each sequence.

The duration T_S of the velocity encoding gradient is shown in figure 1 and is the same for all the sequences. In order to obtain the maximum m_1 from each sequence for a minimum T_S, some amplitude and time parameters will be conveniently defined.

The parameters, k_3 and T_{g3} are determined by the field of view and the sampling period. The gradient amplitude factors k_1 and k_2 are set equal to one for all the sequences. The amplitude factor k_4 is also set to one for sequences C and D.

Sequence A will have its interpulse times T_b, T_c, and T_a equal to zero. The duration of its velocity encoding gradient is then given by

$$T_s = 2T_v + T_{g1} + T_{g2} + T_{g3} \tag{20}$$

and its first moment is given by

$$m_{1A} = GT_v^2 k_v \tag{21}$$

Since the extreme values that k_v can take are minus and plus one, the range of m_{1A} is given by

$$Rm_{1A} = 2GT_v^2 \tag{22}$$

Sequence B needs T_a to produce its m_1, then the duration of its velocity encoding gradient is given by

$$m_{1B} = GT_{g2}\delta \tag{23}$$

The extreme values of δ are $-T_a$ and $+T_a$ which result in a range of m_{1B} given by

$$Rm_{1B} = 2GT_{g2}T_a \tag{24}$$

Sequence C can have its T_a equal to zero and still produce its m_1. The duration of its velocity encoding gradient is given by

$$T_s = T_{g1} + T_{g2} + T_{g3} + T_{g4} \tag{25}$$

and its first moment is given by

$$m_{1C} = 2GT_{g2}\delta_1 \tag{26}$$

The extreme values of δ_1 are T_{g1} and $-T_{g4}$ which result in a range of m_{1C} given by

$$R_{m1C} = 2GT_{g2}(T_{g1} + T_{g4}) \tag{27}$$

Sequence D can also have its T_a equal to zero and produce its m_1. Under the assumptions made, the expressions for T_S, m_{1D} and Rm_{1D} are given by equations (25), (26) and (27) respectively.

The range of m_1, Rm_1, for each of the four sequences of Figure 1 were calculated for seven values of T_S and plots of m_1 vs T_S are shown in figures 2 and 3.

The plots of figure 2 were done for L_r = 24cm, T_{g3} = 2 ms and consequently k_3 = 0.626. Sequences C and D gave the highest values of Rm_1 followed by sequence B and the lowest values of Rm_1, were given by sequence A. The range of values of T_S were taken towards the shorter end in order to achieve shorter echo times.

IOP Conf. Series: Journal of Physics: Conf. Series **792** (2017) 012092 doi:10.1088/1742-6596/792/1/012092

The set of values of T_a and T_{g4} in sequences B and C/D respectively had to be adjusted so that they would give the same set of values of T_S as sequence A. Sequences A and C/D have a parabolic dependence of Rm_l with respect to T_v and T_{g4} respectively as given by equations (22) and (27). Sequence B has a linear dependence of Rm_l on T_a as given by equation (24).

The values of maximum m_l ranges (Rm_l) of each sequence for a given T_S translates into values of velocity resolution V_S as given by

$$V_S = \frac{1}{2\gamma_1 Rm_1} \tag{28}$$

where γ_l is the gyromagnetic ratio in Hz/Gauss. Values of velocity resolution vs T_S were obtained for the same ranges of T_S and the same values of L_r and T_{g3} as the ones obtained respectively (fig, 3).

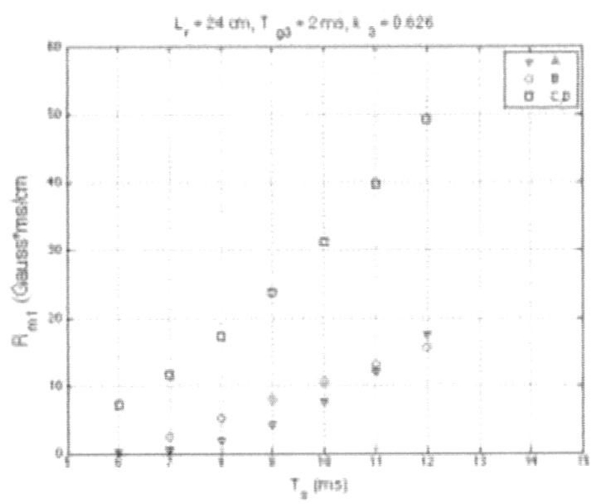

Figure 2. Plot of Rm_l vs T_s for all sequences of figure 1. L_r = 24 cm, T_{g3}=2ms ,k_3=0.626.

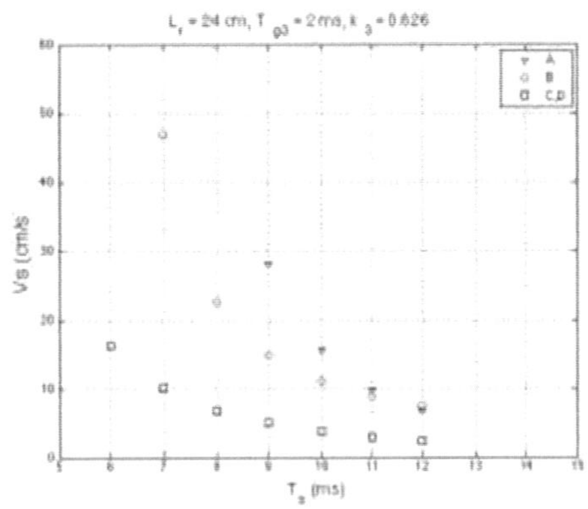

Figure 3. Plot of V_s vs T_s for all sequences of figure 1. L_r = 24 cm, T_{g3}=2ms,k_3=0.626.

5.Conclusions.

For low values of the duration of Ts of the velocity encoding gradient, which results in low values of echo time, the multiple velocity encoding methods C and D yielded higher maximum first moments (m1) than method B and this one in turn produced higher m1 than method A. This translated into methods C and D yielding lower (better) velocity resolution than method A.

Another advantage may be the reduced inconsistency in the amplitude values of the velocity encoding TPG gradients compared to the conventional BPG's. In general , the three methods B, C and D , based on their respective sequences using TPG's have been shown to be more suitable for multiple velocity encoding in the phase of an MRI signal than method A based on a conventional BPG.

Acknowledgements

The author wishes to thank M O Mata-Castillo and M X Benitez-Jones for their significant help on the typing of the manuscript and the making of the figures for this paper, which otherwise would not had been possible due to my present retinopathy.

IOP Conf. Series: Journal of Physics: Conf. Series **792** (2017) 012092 doi:10.1088/1742-6596/792/1/012092

6. References

[1] Mistretta C A 1974 *The use of a general description of the radiological transmission image for categorizing image enhancement procedures* Opt. Eng. **13** 134

[2] Garroway A N 1974 Velocity measurements in flowing fluids by NMR *J Phys D: Appl Phys* **7** L159-L163

[3] Packer K J 1969 The study of slow coherent molecular motion by pulsed magnetic resonance *Mol Phys* **17** 355-368.

[4] Fukuda K, Imani Y and Iliari A 1978 *Studies of water in biological systems by a pulsed NMR method. 1. Flow profile in a cylindrical tube* Biotelemetry and Patient Monitoring **5** 223-228

[5] Weeden V J et al 1985 MR velocity imaging by phase display *J. Comput. Assist. Tomogr*. **9** 530-536

[6] Wherli F W et al 1985 MR imaging of venous and arterial flow by a selective saturation recovery spin echo (SSRSE) method *J. Comp. Assist. Tomogr*. **9** 537-545

[7] Benitez E 1988 Suppression of static spins in NMR flow measurements Tech. Rep. BERD Albuquerque *Lovelace Med. Found*. 88-01

[8] Benítez E 2006 *IV Cong. Inter. Ing. F*is. (Mexico)

[9] Benitez E et al 1990 method for multiple velocity encoding imaging *9th Meet. SMRM, W. in P. (*New York*)* p 1291

Volcanic materials superconductivity in desert areas of the states of Sonora and Baja California

Aldo Holguín
Profesor de la Universidad Estatal de Sonora

E-mail: aldoefrainh@yahoo.com.mx

Abstract. Research was conducted to find materials in their natural state at room temperature and exhibit the effects of superconductivity in the volcanic region of deserts Altar in Sonora and Baja California Norte. 100 were collected at random samples of materials from different parts of the region and underwent tests to determine their electromagnetic parameters of electrical resistance, magnetism, temperature and conductivity. Only it has been found that the effects of superconductivity in them is only present at very low temperatures corroborating what has been done in other investigations, however no indication that there is a material or combination of materials that can produce the effects of superconductivity other temperatures so it is suggested to continue the search for such materials and / or develop a technique at room temperature to allow mimic the behavior of atoms when superconductivity occurs at.

1. Introduction

A few years ago, even before the Industrial Revolution it was very difficult to talk about energy savings or wasteful consumption of it, because there were almost no different machinery we see today. The number of cars was just a small part of what is now, and fuel consumption cannot get used to compare with today. The energy was increased and with it pollution, is why natural processes such as the greenhouse effect and acid rain, hazardous agents today are involved in very harmful consequences such as global warming. So one of the priorities in energy policy in our country and in the rest of the world is to achieve the highest degree of efficiency in energy consumption.

Currently a polluted world increasingly large amounts of toxic waste, becomes indispensable search through a thorough reflection and analysis new alternatives, strategies and techniques whose help in this scenario geared to immediate action on the sustainable use of energy, dare generation and application of knowledge, which demands prompt the benefit of the environment. To do this project, concentrates the possibility of generation and application of reliable and valid knowledge through research practice, then then the opportunity to identify the different alternatives of superconductivity and magnetism, using the tools to generate new sightings, which accompanied by new techniques, help to fill gaps in terms of sustainability exist today, and consequently, reduce the pollution from the production of electricity.

Currently the use of electricity is essential for much of our activities; thanks to this type of energy we have a better quality of life, just by pressing buttons get light, heat, cold, picture or sound. Its use is

Content from this work may be used under the terms of the Creative Commons Attribution 3.0 licence. Any further distribution of this work must maintain attribution to the author(s) and the title of the work, journal citation and DOI.
Published under licence by IOP Publishing Ltd

essential and we hardly stop to think about its importance and benefits to use it more efficiently. Saving electricity is a fundamental for the development of energy resources element; saving equivalent to reducing the consumption of fuels in electricity generation also avoiding the emission of polluting gases into the atmosphere.
Our country has a lot of energy sources. In Mexico, most of the generation of electricity is done through oil, coal and natural gas, significantly impacting the environment by relying on non-renewable resources such as fossil fuels. To use a lot of greenhouse gases, which are emitted into the atmosphere, causing global warming, whose effects are manifesting and are devastating.
Electrical energy is of great importance in the development of society, their use enables the production automation that increases productivity and improves in every way the living conditions of man. You need to save electricity, because this oil saving and currencies that can be invested in other branches of the economy, education, research and culture is saved. That is why measures for their savings are taken as the thermoelectric are our main source of electrical power, increasing electricity demand must increase the generation capacity of power plants, is why it is necessary to develop new technologies generation and one of these ways is using superconductivity [1].

Superconductivity is the property of some materials to lowering its temperature below its so-called critical temperature, abruptly stop resisting the passage of an electric current, thus forming a superconductor and to approximate a magnetic field presents a perfect diamagnetic phenomenon. By consistently have a considerable increase in electricity and likewise better control over the movements without mechanical friction [2].

2. Some History

If we take history that use the resistivity as a thermometric magnitude was what allowed it to be included in the various measures of various substances in a wide temperature range. In fact Dewar and Fleming since 1893 had indicated the possibility that the resistivity be annulled at absolute zero, although the measures taken subsequently on liquid hydrogen in which observed that the decline in this became smaller as down the temperature did suspect that it could remain finite at absolute zero. Interest Kammerlingh in measuring the electrical conductivity at low temperature was also stimulated by a suggestion Lord Kelvin (1902) that the resistivity should decrease with temperature, but below a certain value it should start increasing due to condensation of electrons [3]. This idea of electrons forming a susceptible gas be condensed, also could be suggested by theories Riecke (1898) and Drude (1900), who analyzed the phenomenon of electrical conductivity treating electrons as a gas obeys the equation state of ideal gases, thus being able to explain the old empirical Wiedemann Franz law which states that at the same temperature, the ratio of the thermal conductivity and electrical conductivity is the same for all metals. The first measures of resistance in the new temperature range carried for platinum showed that this apparently tends to a constant value against the expected behavior or approach zero by extrapolating the behavior at higher temperature or increase until it becomes infinite as Kelvin's suggestion Kammerlingh Onnes, who attributed the residual resistance to impurities in the metal and decided to repeat the experiments with two of the substances that can be put in the form of a thread with more high purity, gold and mercury. For gold, the results were similar as for platinum, but a striking change mercury was observed near 4.2 K, decreasing resistance abruptly from a finite value to zero. Repetition of the experiment allowed us to confirm that the phenomenon observed was similar to the behavior of platinum with a gradual decrease from the boiling point of hydrogen to helium boiling while around 4.2 K resistance quickly disappeared. These experiments were repeated in 1913 coining the publication of the first term superconductivity. Kammerlingh tried to explain the new phenomenon observed by using the theory of quanta of Planck, and subsequent discussions took place on the phenomenon in the Solvay conferences 1911 and 1912. Investigations of Leiden are interrupted until 1921 that the work is restarted this subject. However, Kammerlingh and elderly died in 1926 without the superconductivity phenomenon was explained, and his disciples, mainly Wilhem Keesom

(1876-1956) continued to work on the subject systematizing the study of the disappearance of superconductivity with the application of an external magnetic field and the thermodynamic treatment of the same as a phase transition [4]. However the next quantum leap in the interpretation of this phenomenon produced it in 1933 Fritz W. Meissner (1882-1974) when provided experimental evidence to the effect that the magnetic field vanishes inside the superconductor and therefore the transition a superconductor is a transition to a state of perfect diamagnetism (a is a diamagnetic substance for which an external applied field decreases inside) [5].

3. Materials and methods

100 samples of volcanic materials from different parts of the region were collected and subjected to tests to determine their electromagnetic parameters of electrical resistance, magnetism, temperature and conductivity. Because the objective of research to find superconductive materials at room temperature only were recorded for each of the samples the two main effects of superconductivity are the zero resistance to the passage of electric current and the diamagnetic effect it is to repel magnetic fields. This was done at room temperature and liquid nitrogen temperature, considering the sample as a superconductor when simultaneously present the two effects; electrical resistance that tends to zero and repels the magnetic fields [6].

4. Results

It was found that the effects of superconductivity in materials subjected to laboratory tests is only present at very low temperatures corroborating what has been done in other investigations

Table 1. Material at room temperature.

Sample	Electrical resistance tends to zero	Diamagnetic		Superconductivity
From M1 to M100	No	No		No

Table 2. Material to liquid nitrogen temperature.

Sample	Electrical resistance tends to zero	Diamagnetic	Superconductivity
From M1 to M33	No	No	No
M34	Yes	Yes	Yes
From M35 to M68	No	No	No
M69	Yes	Yes	Yes
From M70 to M100	No	No	No

5. Discussion

In April 1986 the discovery of a new superconducting materials which were ceramic and presenting a temperature greater than any of the existing materials announced transition. As of this writing the critical temperature of superconducting transition highest reported is around 135K, well above the boiling point of liquid nitrogen, a refrigerant very economical price and easy to obtain. There are also promising signs that may perhaps achieve transition temperatures above 200K.

The discovery of this new type of superconductor was made by J. C. Bednorz and K. A. Müller in a research laboratory of the company IBM in Zurich, Switzerland. First, after over 12 years was possible to find a substance having a transition temperature above 23.3 Kelvin. In his research they read a scientific article that is key in their work. It was because French scientists C. Michel, L. Er-Rakho and B. Raveau, and he presented a new material whose characteristics of being a new metal oxide mixed valence copper made him ideal candidate to present superconductivity, according to the working hypothesis Bednorz and Müller. The composition of this material is BaLa4Cu5 O13 · 4. Bednorz and Müller began to explore its properties by varying the concentration of Ba. In the spring of 1986 they published his article announcing superconductivity at a temperature of 35 Kelvin in this class of compounds. In these, the arrangement of the ions typically corresponds to a geometry known as perouvskita and that is very common among so-called ferroelectric materials [7].

The rapid progress that has been achieved to find such materials with temperatures ever higher superconducting transition has been truly amazing. Very few scientific advances, if there have been any, have generated almost frantic flow of such scientific activity worldwide and at the same time, an immediate and very great among the general public interest. What the vast majority thought as impossible and is now something real and palpable: having superconductivity at higher temperatures than liquid nitrogen. The work of Bednorz and Müller earned them the Nobel Prize in Physics 1987.

Almost immediately after the announcement of the discovery of Bednorz and Müller, many groups of scientists in the world were launched to try to get higher transition temperatures. One of the most successful groups was that of Dr. Paul Chu of the University of Houston, one of the first to realize the importance of the discovery of Bednorz and Müller, who devoted himself to the investigation of these materials. They soon found that the critical temperature could be increased to 57 Kelvin applying pressure to the material. Both the magnitude of change in critical temperature, such that with increased pressure applied were abnormal when compared with the known superconductors before these new materials. With this in mind, Chu and his colleagues began looking for ways to simulate an "internal pressure" in these materials replacing the lanthanum (La) with similar ions, such as yttrium (Y). In late February 1987, Chu announced it had found a compound that had a higher transition temperature superconductor state of 90 Kelvin. The composition of this material is given by YBa2Cu3Ox. Almost simultaneously obtaining a material of similar composition and similar properties in China was announced. In a few days, with compositions that were variants reported by Chu and his collaborators, a dozen groups around the world reported on the production of ceramic superconductors with transition temperatures above 90 Kelvin, which have been prepared in the National Autonomous University of Mexico; how to synthesise is very simple and can be done with the technology that is available to the countries of the Third World.

It is very clear that having critical temperature superconductors above liquid nitrogen is a reality in our country and in many other third world nations. It also begins to be very clear that with them the world will not be the same. It is very likely that, once again, the physical change our way of living as happened with the advent of the electric motor, transistor, etc. [8].

It is worth noting that perouvskitas copper and oxygen, the new superconducting materials had been studied, especially Raveau, Michel et al. Much of his work laid the foundation for achieving rapid progress immediately after the discovery of Bednorz and Müller. The initial interest in these materials settled in the high mobility of oxygen at elevated temperatures, which alters its electrical behavior, so that had been suggested as one of its possible applications, the oxygen sensor. Many studies have made it clear now that the superconducting properties of the compound of yttrium (Y), barium (Ba) and copper (Cu) (very widely known as 1-2-3, its composition: YBa2Cu3Ox) depend critically on the amount and oxygen in the system, which in turn depends on the details of the process for obtaining it.

6. Conclusión

The effects of superconductivity until now only present in some materials and very low temperatures, however no indication that there is a material or combination of materials that can produce the effects of superconductivity at other temperatures so it is suggested to continue with the search for such materials and / or develop a technique at room temperature to allow mimic the behavior of atoms when superconductivity occurs at.

7. References

[1] Paul C 2009 A cook's tale Canfield Nature Physics Commentary Vol 5
[2] Magaña L 2012 *Los superconductores* 4ª ed. (México: Conacyt)
[3] Serway R and Beichner R 2004 *Física para ciencias e ingeniería vol 2* ed Mc Graw Hill (México)
[4] Gazquez J et al 2016 *Emerging dilute ferromagnetism in high Tc superconductors driven by point defect clusters Advanced Science* DOI: 10.1002/advs.201500295
[5] Superconductividad. Diamagnetismo perfecto [on line]. [Consulted: June 25, 2016] . Available in: http://deliquios.blogspot.com/2008/07/superconductividad-diamagnetismo.html
[6] Nothing Up My Sleeve: The Meissner Effect [en línea]. La Florida: National High Magnetic Field Laboratory, 2014. [on line]. [Consulted: June 25, 2016] . Available in: http://www.magnet.fsu.edu/education/tutorials/magnetacademy/superconductivity101/page7.html/
[7] Gómez C, Valencia J and Jesús S. Síntesis de óxidos tipo perovskita mediante polimerización con ácido cítrico y propiónico [pdf]. [on line]. [Consulted: June 25, 2016] . Available in: http://www.redalyc.org/pdf/3090/309026681008.pdf
[8] Comisión Nacional para el Uso Eficiente de la Energía, CONUEE 2011. Guía para elaborar un diagnóstico energético en instalaciones. [on line]. [Consulted: June 25, 2016] . Available in: http://www.conae.gob.mx/work/sites/CONAE/resources/LocalContent/7406/1/R_GUIA3_Diagnostico_Instalacion.pdf.

Characterization of a outer detector (outriggers) for HAWC

T Capistrán[1], I Torres[1], E Moreno[2], and for the HAWC Collaborator[4]

[1]Instituto Nacional de Astrofísica, Óptica y Electrónica, Luis Enrique Erro 1, Tonantzintla, Puebla 72840, México.
[2] Facultad de Ciencias Físico Matemáticas, Benemérita Universidad Autónoma de Puebla, Ciudad Universitaria, Colonia San Manuel Puebla, México.
[3]For a complete author list, see www.hawc-observatory.org/collaboration/icrc2015.php

E-mail: tcapistran@inaoep.mx, ibrahim@inaoep.mx and emoreno@fcfm.buab.mx

Abstract. The High Altitude Water Cherenkov (HAWC) Observatory is a ground-based air-shower array deployed on the hillside of the Sierra Negra Volcano in the state of Puebla, Mexico. HAWC comprises of 300 water Cherenkov detectors (WCDs), each WCD is equipped with four photomultiplier tubes (PMTs) to detect secondary particles of the air-showers that are produced by the interaction of a primary particle (gamma-rays and cosmic rays) with the atmosphere. HAWC is able to reconstruct gamma-ray showers in the 100 GeV - 100 TeV energy range, but suffers from a lack of sensitivity when the particle showers core develop outside the WCD array. A proposed upgrade to fix this issue is to build an external array of smaller water detectors, called outriggers. A Outrigger comprises a PMT on the bottom of the tank. In this work the instrumentation and characterizing an Outrigger is presented, in order to know the behavior of this detector with as function of threshold voltage.

1.Introduction

The high altitude water Cherenkov (HAWC) is a gamma-ray observatory located at 4,100 meters above sea level on the hillside of the Sierra Negra volcano in the state of Puebla, Mexico. HAWC is an array of 300 water Cherenkov detector (WCDs) that began full operation in March of 2015 [2]. Each WCD consists of a metal cylindrical structure of 7.3 in diameter, and 4.5 in height, a dark bag (called bladder), high-purity water, and 4 photomultiplier tubes (PMTs) on the bottom. When a primary particle arrives in the Earth's atmosphere, they interact with the atoms of the air, and this interaction produces secondary particles that depend on the kind of primary particle (for example if a gamma ray is the primary particle, these secondary particles are made up of electrons). These particles penetrate the WCDs, and emit Cherenkov light that can be detected by the PMTs [3].

One event is considered when a primary particle arrives at the atmosphere, then it has developed an air-shower, finally it is detected by HAWC. Sandoval in [1] said "The PMTs activated during an event depends mainly upon the primary particle's energy. If the latter has a high energy, the most of the PMTs are activated. However, the shower may be poorly reconstructed if the shower core falls outside the array". Therefore an array of smaller detectors, called outriggers, is currently being deployed around HAWC (see the Figure 1). These outrigger will be comprised of a cylindrical polyethylene

Content from this work may be used under the terms of the Creative Commons Attribution 3.0 licence. Any further distribution of this work must maintain attribution to the author(s) and the title of the work, journal citation and DOI.
Published under licence by IOP Publishing Ltd

tank with approximate 2 m of diameter, filled with purified water and equipped with a single PMT at the bottom (see Figure 2). These outrigger will be able to detect the secondary particles which land outside of HAWC, which will improve the core reconstruction and improve 3-4 times the sensitivity to gamma-rays above 10 TeV [1]. In this work, we present a test for characterizing one of outrigger, and understand the behavior of the rate and charge distribution of the new detector as a function of threshold voltage.

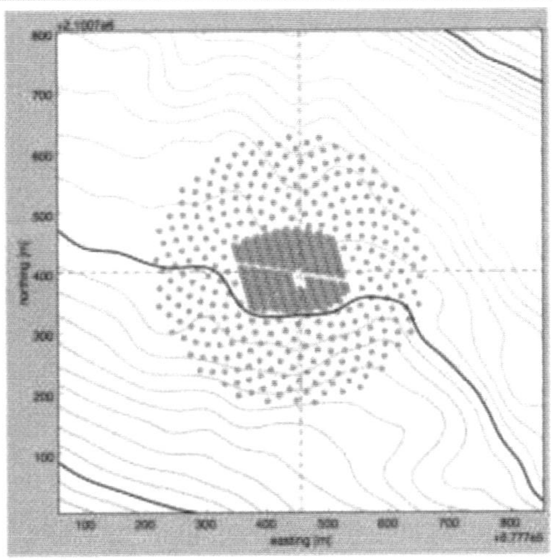

Figure 1. Contour lines of the terrain at the HAWC site, with the HAWC array and the layout of a 300 tank outrigger array in the geometry of a sunflower spiral [1] .

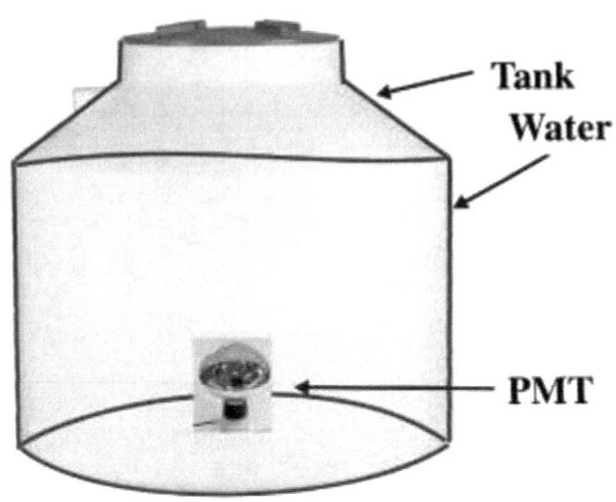

Figure 2. An outside detector, called outrigger, is comprised of one tank, purified water, and PMT on the bottom.

2.Experimental setup

The experimental setup is shown in Figure 3 in the form of block diagram illustrating the flow the information from PMT to the data acquisition module. CAEN electronics were used for collecting these informations basen on VME and an amplifier basen on NIM. First the PMT is powered by a CAEN V6533 High-voltage VME module, at an optimal tension of 1617 V. The signal cable of the PMT is isolated from the HV supply with the help of a decoupler, and fed through a 10x amplifier. The resulting output signal is then connected to an N625 Fan-In/Fan-Out Unit, because this unit output generates the sum of the input signals (in this case only one) times the gain that is 1. Therefore, there are four independent output signals: one is plugged in as input of a V1751 digitizer, and the other is connected to a discriminator (V812) for distinguishing between the low signal and high signal: a high signal is defined as the signal that is bigger than a threshold voltage (Vth), and the discriminator sends a pulse when a high signal is at the input. A high signal pulse will be sent at two locations: a V820 Scaler and a V1751 digitizer. The Scaler counts the number of pulses in a time windows of one second. Data is only acquired by the digitizer when the system detects a high-signal pulse.
Upon Data acquisition, two measurements are saved in a data file.

- The high pulses is shown in the Figure 5.
- The rate that is number of high pulses that were passed during a time window of one second.

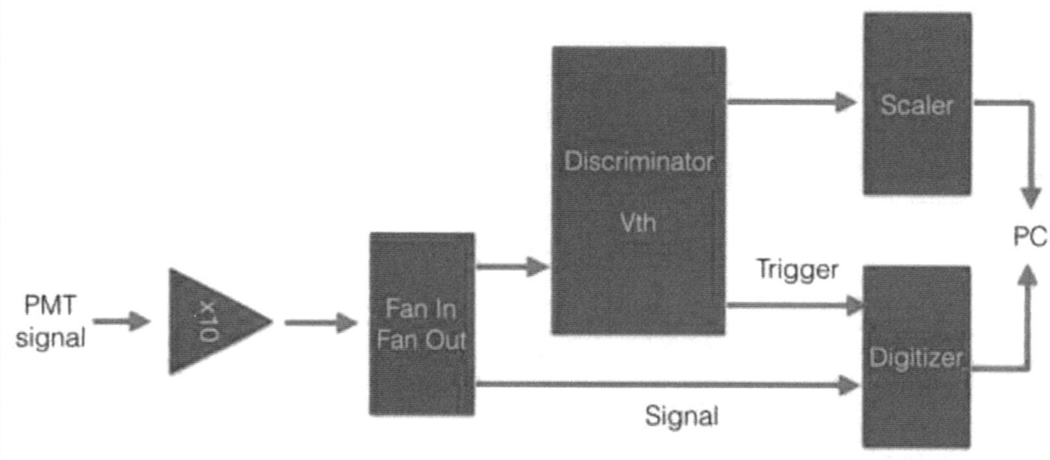

Figure 3. Schematic diagram of the CAEN electronic that were used for collecting the data.

Different values of threshold voltage were used from -5 mV to -100 mV with a step of 5 mV. Each Vth, the next informations were saved:

- In the case of Scalar: approximate 300 rate were measured, that is about 5 minutes of data. These measurements of rate were put into a histogram and fit a gaussian function (see the Figure 4), and exported the mean and sigma for an analysis.
- In the case of Digitizer: 50,000 pulses were saved with a time window of 800 ns (see in the Figure 5). Each pulse was integrated to obtain its charge.

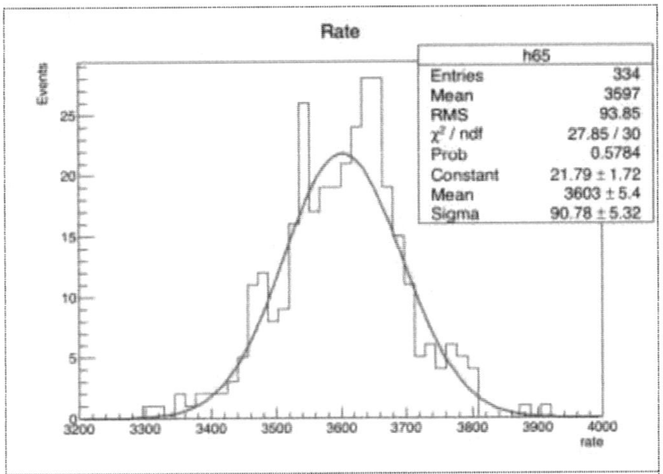

Figure 4. This is a histogram of the rate. 334 measurements were taken with a Vth of -65 mV. The red line is a gaussian fit.

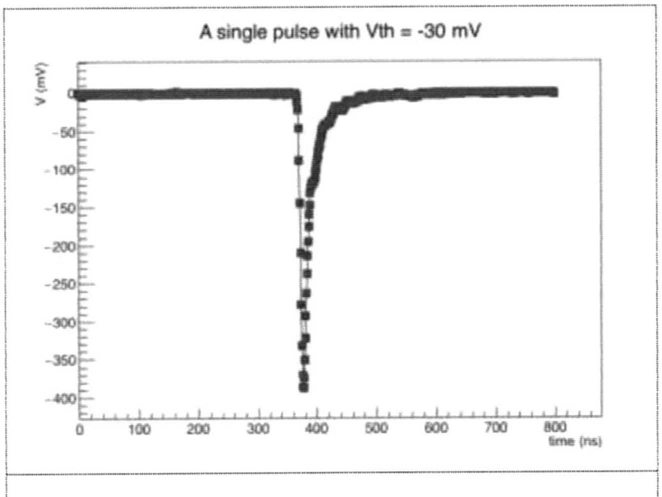

Figure 5. Is shown a high signal with a time window of 800 ns.

3.Results

In the Figure 6 is shown the rate Vs Vth. In this graph, the response of single photoelectron (PE) could be found, it is around 20 mV, but more measurements with step smaller will have to be made. On other hand when a Vth is bigger, the rate is lower, due to there are less signal with big energy that was detected by the PMT.

Figure 7 shows the histogram of the integrated pulse charge (Q) as a function of threshold voltage. When Vth is -30 mV (red histogram), there is a small peak between 0.05 and 0.1 nC, there are a possibility that will be vertical muons, but more analysis will need to do.

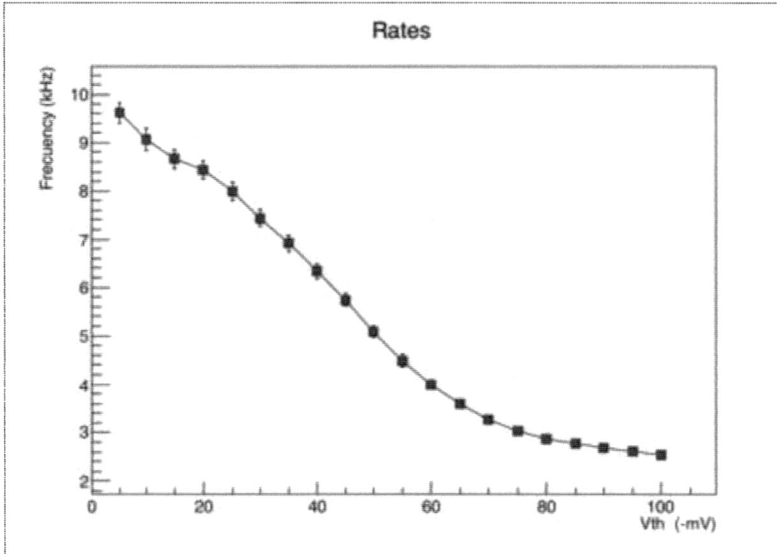

Figure 6. Bias curve of an outrigger tank. The plot shows rate vs Vth. When the voltage is low, that is, close to cero, the rate is increased. Because there are more signal with low energy were detected by the PMT.

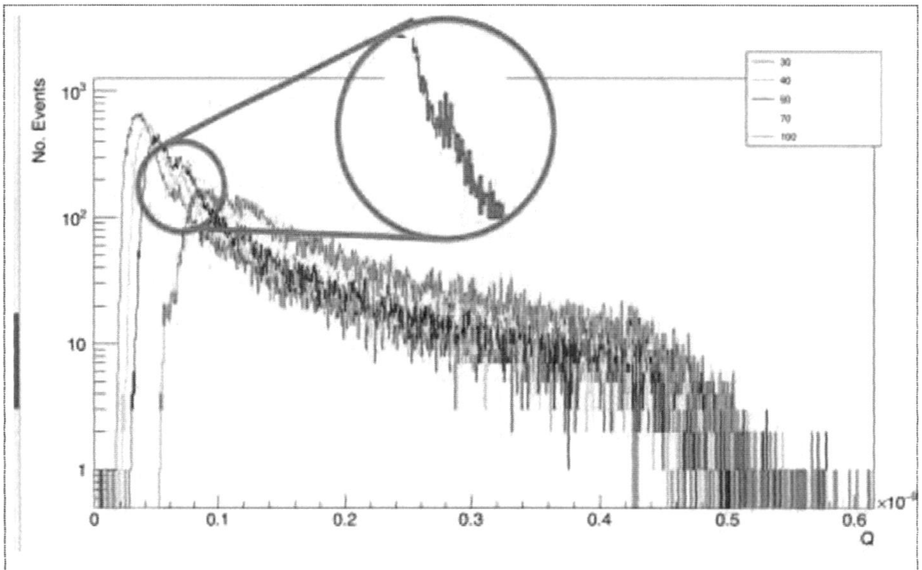

Figure 7. There are some histograms with different Vth. If the threshold voltage is increased, the small signal does not pass the trigger and is not saved therefore the numbers of events is smaller. A zoom of the -30 mV histogram is made because it is possible that are vertical muons but it is necessary make more analysis.

4.Conclusion

In this work, we started analyzing the response of an outrigger detector to various trigger threshold voltage, which is the first step in understanding the behavior of the new instrument. In addition to the measurements presented, a thorough inspection of the tank seals and light shielding needs to be made, in order to avoid flawed detector which would systematically trigger at a very high rate.

Measurements of the bias curves for one outrigger shows that the events rate increases when the Vth is low because of there are less particles with high energy. Finally, the integrated charge distribution hint at a possible contamination of the signal by muons, although more analysis needs to be done to confirm this hypothesis.

Acknowledgments

We acknowledge the support from: the US National Science Foundation (NSF); the US Department of Energy Office of High-Energy Physics; the Laboratory Directed Research and Development (LDRD) program of Los Alamos National Laboratory; Consejo Nacional de Ciencia y Tecnología (Mexico (grants 239762,260378, 55155, 105666, 122331, 132197, 167281, 167733); Red de Física de Altas Energías, Mexico; DGAPA-UNAM (grants IG100414-3, IN108713, IN121309, IN115409, IN111315); VIEP-BUAP (grant 161-EXC-2011); the University of Wisconsin Alumni Research Foundation; the Institute of Geophysics, Planetary Physics, and Signatures at Los Alamos National Laboratory; the Luc Binette Foundation UNAM Postdoctoral Fellowship program.

References

1. Sandoval for the HAWC Collaboration 2016 *Proc. 34th International Cosmic Ray Conference* (The Hague, Netherlands) 977
2. Abeysekara A U et al 2013 *Sensitivity of the high altitude water Cherenkov detector to sources of multi-TeV gamma rays* Astroparticle Physics **50** 26–32
3. Gaisser, T 1991 *Cosmic Rays and Particle Physics* (Cambridge University Press) 279

Thermal and optical analysis of selective absorber coatings based on soot for applications in solar cookers

H Servín[1,3], M Peña [1], H Sobral[2] and M González[3]

[1]Universidad Michoacana de San Nicolás de Hidalgo FCFM-UMSNH, Morelia, Michoacán-México.
[2]Universidad Nacional Autónoma de México CCADET-UNAM. Ciudad de México.
[3]Universidad Intercultural Indígena de Michoacán LDS-UIIM, Pátzcuaro, Michoacán-México.

E-mail: merlysc@gmail.com

Abstract. The thermal and optical properties of selective absorber coatings of a solar cooker have been investigated. Coatings have been prepared using soot from pine resin, wood stove and sugarcane, previously separated by size. Results show that the cooking power and the overall efficiency of these pots are higher than others painted with black primer. Besides, by using an integrating sphere, the diffuse reflectance of absorbers has been obtained. Lower values of the reflectance have been measured for the pots covered with soot, showing a high correlation with the results achieved from the thermal tests, considering the measurement errors.

1. Introduction

Solar cookers are devices that use solar radiation to cook food and are an alternative to fires fueled by wood in rural areas. The most common types of solar cookers are box cookers, panel cookers and parabolic cookers. Parabolic solar cookers use a reflector to focus the sunlight directly onto a cook pot. Parabolic geometry allows reach much higher temperatures than the other types of solar cookers. Taking into account the portability of the system, the compound parabolic collector (CPC), is a suitable alternative [1]. Designs based on CPC collectors have been implemented in rural communities regarding the simplicity of the stove design so that it can be built by the user [2].

One of the most important parameter in a solar cooker is the selective solar absorber coating that allows converting the sunlight into thermal power. For efficient photothermal conversion solar absorber surfaces must have high solar absorptance for wavelengths shorter than 3 μm and a low thermal emittance at the operational temperature [3]. Materials used in solar cookers are usually expensive and difficult to obtain for rural communities. As a result, the identification and characterization of low-cost materials allows that low-income families could have access to solar cookers. Inexpensive designs as the stove *Jorhejpatarnskua* [2] and the rural solar cooker (RSC) [4] have been reported to have thermal parameters better than to commercial cookers. These designs are constructed with a mirror aluminum foil and paint for solar applications. However, the manufacturing cost could be reduced if local materials are used. Untreated materials from rural areas such as soot from biomass combustion have been proposed as a selective absorber coating for solar cookers. In a previous work [4], the standard cooking power using different absorber coatings including soot, have been compared. It has been found that copper oxide has the best

Content from this work may be used under the terms of the Creative Commons Attribution 3.0 licence. Any further distribution of this work must maintain attribution to the author(s) and the title of the work, journal citation and DOI.
Published under licence by IOP Publishing Ltd

thermal parameters. Nevertheless, this coating is toxic and is harder to get than soot. Furthermore, Thermogravimetric analysis of pine soot resin prove that low material decomposition occurs up to 600°C; thus, showing the high resistance of soot coatings to sun heating [5].

Solar cookers performance should be evaluated according to the international standard procedure which was proposed at the Third World Conference on Solar Cooking [6]. The methodology includes rigorous procedures to determine thermal parameters such as standard cooking power (P_s) and the overall efficiency (η). P_s has to be measured by exposing the pot or container to sunlight with suitable climatic conditions for several hours.

In this work the standard cooking power of the stove *Jorhejpatarnskua* is obtained for different pots coated with absorbers based on soot from pine resin, wood stove and sugarcane. The effect on P_s as a function of the soot particle size is also evaluated. Furthermore, diffuse reflectance technique was successfully employed as an alternative to characterize the thermal performance of the coatings.

2. Metodology

2.1. Diffuse Reflectance Spectroscopy

The Reflectance is defined as the ratio of reflected irradiance, I_r, to incident irradiance, I_i, and could be used to analyze opaque samples. There are two reflectance components for a surface: *specular* and *diffuse*. Reflection light of smooth surface (such as a mirror) follows the law of reflection and the specular reflectance is the principal component. On the other hand, diffuse reflectance takes place when the surface is microscopically rough and the light is reflected and scattered in many different directions. The roughness degree of the surface will define which component will have a greater contribution to the overall reflectance [7].

Reflectance measurements are typically performed with an integrating sphere and a diffuse reflectance standard. The cavity of the sphere is covered with a diffuse white reflective coating to obtain a uniform scattering or diffusing effect with a few small ports. Multiple scattering reflections from the sample and the cavity will be equally distributed to all points inside the sphere and the scattered light is collected by a spectrophotometer or a photomultiplier. Reflectance is usually measured relative to a reference standard settled at the sample port, which is taken to be 100%.

Thus, the sample reflectance $R(\lambda)$ can be calculated by

$$R(\lambda) = \frac{I_c(\lambda) - I_b(\lambda)}{I_w(\lambda) - I_b(\lambda)}, \tag{1}$$

where λ is the wavelength, $I_c(\lambda)$ and $I_W(\lambda)$ are the detected intensity corresponding to the coating and reference standard, respectively. $I_b(\lambda)$ is the background intensity obtained with the lamp off.

2.2. Thermal Parameters

Thermal parameters are useful tools to compare different solar cookers designs. The most important parameters to evaluate the performance of solar cookers are the standard cooking power P_s and overall efficiency η. According to the American Society of Agricultural and Biological Engineers standard (ASAE S580) [8], P_s could be obtained from:

$$P_s = P_c \frac{700\ \mathrm{W/m^2}}{I}, \tag{2}$$

where P_c is the cooking power and I is the insolation falling on cooker aperture in W/m^2. The P_c can be calculated through:

$$P_c = mC_p \frac{dT}{dt},$$

where m is the mass of water, C_p is the specific heat of water and T is the water temperature at a given time t.

The other relevant thermal parameter is the *overall efficiency*, η. This parameter takes into account the amount of solar heat reaching the cooker. This can be defined by [9]:

$$\eta = \frac{mC_p(T_{wf} - T_{wi})}{A \int I dt}, \quad (3)$$

where T_{wf} and T_{wi} is water final and initial temperatures and A is the collector area.

3. Experimental Section

3.1. Materials Preparation

Soot from pine resin, wood stove and sugarcane have been investigated to employ as a solar radiation absorber. Soot is composed of particles of different size, which is then separated by a vibratory sieve shaker (Reisch AS 200). Samples have been separated according to the particle size: less than 32 μm and between 150 to 250 μm. Afterward, the coatings were prepared with a mix of 50 wt% of soot and 50 wt% of black primer paint. The paint was applied to a pressure aluminum pot by means of an air spray gun. Additionally, to compare the performance of soot coatings, two cooking pots were painted with commercial black paint and black primer paint.

3. 2. Measurement of Diffuse Reflectance

Figure 1 shows a schematic of the used device to measure the reflectance of the coated samples. The system consists of a Deuterium Halogen lamp (DH2000-DUV, from Ocean Optics), emitting in 190-2500 nm range; a 50 mm diameter integrating sphere (ISP-50-8-REFL, from Mikropack); and a miniature spectrometer (USB4000-XR1-ES- Ocean Optics), which spans the 200-1025 nm wavelength range with an optical resolution of 1.5 nm. Two 600 μm fused-silica optical fiber bundles were used to direct the light from the lamp to the integrating sphere and the spectrometer. Reflectance was obtained by using the Eq. (1) using a Polytetrafluoroethylene (PTFE) sample as the reference standard for the whole investigated wavelength range. Measurements of the cooking pots were performed at 30 different sample positions and the obtained values were averaged.

3.3. Measurement of thermal parameters

Measurements of thermal parameters were performed by using the procedure proposed by Funk [8]. The experimental arrangement is shown in figure 2. The solar cooker employed a pressure cooker pot coated with the prepared paint with soot and a black paint. Initially, the pot was filled with 3.5 *l* of water. The test was performed by exposing the solar cooker to the sun at 10:00 solar time. Two thermocouples type K have been used to register the temperature inside and outside of pot. The incident radiation was measured with a pyranometer (LP02-LI-19). All experimental parameters, such as water temperature within the pot, ambient temperature, wind speed and solar flux radiation were measured every 5 min until the water reaches 90°C.

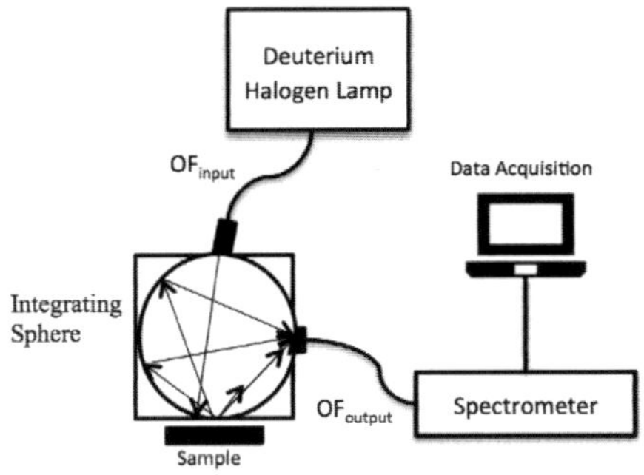

Figure 1. Experimental setup employed to measure the reflectance of coated samples.

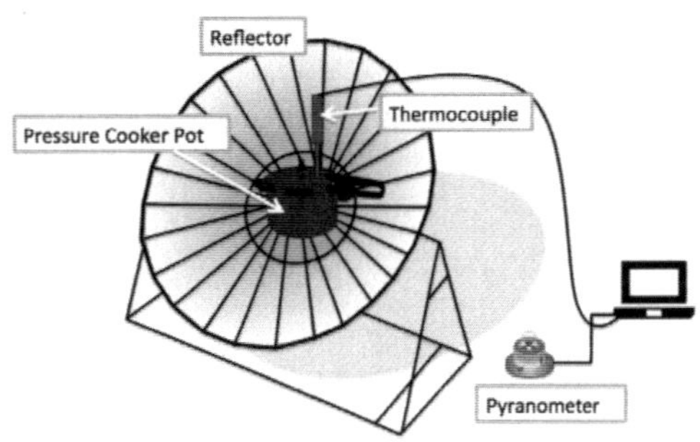

Figure 2. Solar Coker implemented to measure the thermal parameters.

4. Results

4.1. Diffuse Reflectance

Figure 3 depicts the dependence of the reflectance for particle sizes in the range 150-250 μm and less than 32 μm, for coatings of soot from sugarcane and pine resin. The estimated experimental error is about 13% which was obtained from the relative standard deviation from 30 different sample regions. Result shows that reflectance is greater for larger particle sizes. Also, reflectance values are higher for coating of soot obtained from sugarcane than the produced from pine resin. In Figure 4 the dependence of the reflectance for the different investigated coatings is presented. It was observed that coatings of black paint and black primer paint have higher reflectance values in relation to the other tested coatings, with values of 6% and 4.4% on average, respectively. The lower reflectance value (3%) was obtained for coating of soot from pine resin.

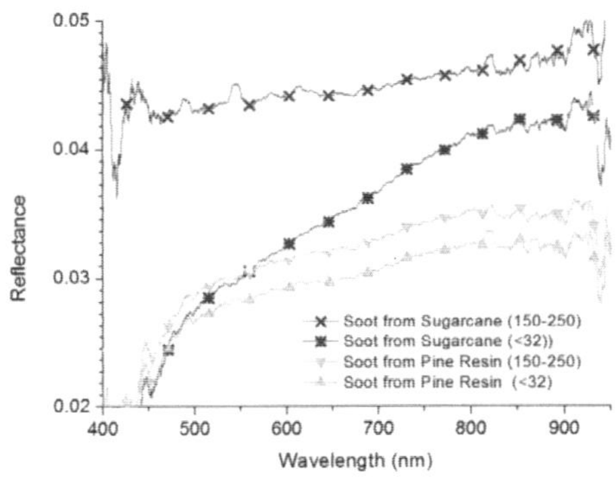

Figure 3. Reflectance for coatings with different particle size (μm) of soot from sugarcane and pine resin.

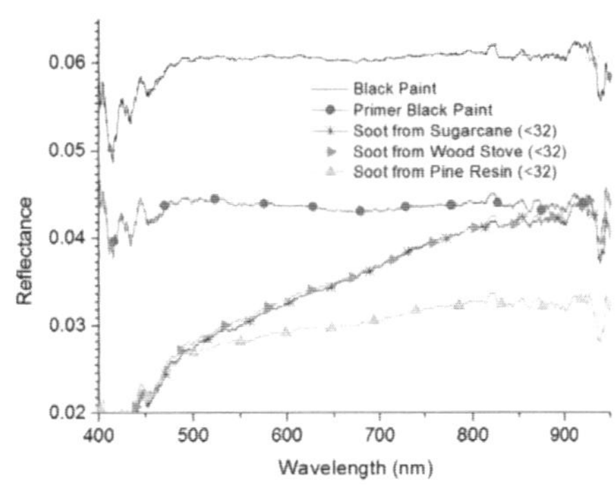

Figure 4. Reflectance for different types of coatings.

4.2. Thermal Parameters

The performance of coatings was analyzed by its thermal parameters. The design of solar cooking outlined in Figure 2 was used for thermal test. Three solar cookers were used to simultaneous test the different investigated coatings. The standard cooking power and overall efficiency were determined with equations (2) and (3), respectively. Here, the obtained experimental error has

IOP Conf. Series: Journal of Physics: Conf. Series **792** (2017) 012095 doi:10.1088/1742-6596/792/1/012095

determined to be ~10%. Results of P_s and η show that coatings of soot with particle size less to 32 μm have better thermal performance than coatings of particle size of 150-250 μm.
Table 1 shows a summary of thermal parameters and reflectance for coatings of soot and primer paint.

Table 1. Summary of thermal and optical parameters.

Coatings Parameters	Sugarcane	Wood Stove	Pine Resin	Sugarcane	Wood Stove	Pine Resin	Primer
Soot particle size (μm)	<32	<32	<32	150-250	150-250	150-250	--------
P_s (Watts)	79	75	74	63	69	66	59
η (%)	27.6	24.8	25.6	22.9	23.4	23.5	20.5
Reflectance (%)	3.6	3.5	2.9	4.4	-----	3.2	6.1

5. Conclusions

The thermal efficiency and the cooking power of different selective solar absorber coatings based on soot have been investigated. The coatings with soot show better thermal parameters than the tested black paints. Besides, diffuse reflectance has also been used to characterize the samples and the preliminary results show a high correlation between the optical properties and the thermal parameters, considering the experimental errors. Soot from pine resin has the lower reflectance which is translated into higher absorptance and a higher cooking power. This method has the advantage that could be implemented in the Laboratory regardless of the ambient weather conditions.
Although the efficiency of coatings with smaller particle size is higher, this difference is about 15%; thus, unprocessed materials could be used as absorber coatings without sacrificing its thermal efficiency.

Acknowledgments
This work was supported by the National Council of Science and Technology of Mexico (CONACyT 221506-F and 24919), the National Autonomous University of Mexico (DGAPA-UNAM: IN105515), the Universidad Michoacana de San Nicolás de Hidalgo (CIC-UMSNH) and the Universidad Intercultural Indígena de Michoacán.

6. References

[1] Panwar L, Kaushik C and Kothari S 2012 *State of the art of solar cooking: An overview* Renewable and Sustainable Energy Reviews, **16**(6) 3776-85
[2] Servín H, and González M *2014 Development of the Solar Cooker Jorhejpatarnskua: Thermal Standard Analysis of Solar Cooker with Several Absorber Pots Energy Procedia*, **57**, 1573-82
[3] Kennedy C E 2002 Rewiew of Mid to High Temperature Solar Selective Absorber Materials *Technical Report NREL/TP-520-31267*
[4] López Sosa L B L, González-Avilés M, González Pérez D y Solis Gutiérrez Y 2014 *Rural Solar Cookers, an Alternative to Reduce the Timber Resource Extraction through the Use of Renewable Energy Sources: Technology Transfer and Monitoring Project Energy Procedia* **57** 1593-02
[5] Duffie A and Beckman A 1980 *Solar engineering of thermal processes* Wiley (Wisconsin-Madison)
[6] Correa F, González M, Servín H, Márquez F, Rutiaga G, Lemus A, and Alonso V 2014 *Development of a selective low cost absorbing surface based on soot for solar thermal applications Energy Procedia* **57** 1565-72
[7] Torrent J and Barrón V 2008 *Diffuse reflectance spectroscopy: Methods of soil analysis* Soil Society of America **5** 367-85

[8] Funk A 2000 *Evaluating the international standard procedure for testing solar cookers and reporting performance Solar Energy* **68** 1-7
[9] Kundapur A and Sudhir C V 2009 *Proposal for new world standard for testing solar cookers Journal of engineering science and technology* **4**(3) 272-81

IOP Conf. Series: Journal of Physics: Conf. Series **792** (2017) 012096 doi:10.1088/1742-6596/792/1/012096

Characteristics of the oil transport network in the South of Mexico

R Juárez[1], I Y Fernández[1] and L Guzmán[1,2]

[1] Laboratorio de Sistemas Complejos, Instituto Politécnico Nacional, Unidad Interdisciplinaria en Ingeniería y Tecnologías Avanzadas, Av. IPN No. 2580, L. Ticomán, México, D.F., 07340, México

E-mail: lguzmanv@ipn.mx

Abstract. We present a study of some organizational properties of the oil transport network of the Mexican oil company (PEMEX) in a region of the State of Tabasco. Particularly, the generalized centrality and the distribution of connectivities are calculated in order to evaluate some aspects of the structure of the network. We find that the connectivities (k) are characterized by a degree distribution which follows a power-law function of the form, $P(k) \sim k^{-\lambda}$, with $\lambda =$ 2.6. Moreover, our procedure permits to evalute the importance of lines (ducts) and nodes, which can be wells, production headers, separation batteries and petrochemical complexes.

1. Introduction

In recent years there has been a growing interest in the study of systems that can be modelled by means of networks or interconnected systems [1–3]. Particularly, the transport systems have attracted the attention of the investigators from different areas of science, as they analyze systems to explore from the information flow and transportation of products to the movement of people. The growth of many of these systems depends on external factors, reason why the emergent configurations reflect irregular aspects. For example, it is recognized that the growth population in cities together with its spatial expansion generate the sprouting of supply routes, which are adapted to the existing ones; the same happens in systems where the transport of substances or products is necessary. In this sense, the oil transportation takes a great importance for the operation, distribution and commercialization. Moreover, the use of some optimization criteria during the design can not only impact reducing investment but also improving the efficiency of the system. There are few studies in different parts of the world related with the evaluation of the topological properties of pipeline networks for the characterization of the transport of petroleum, which represents a strategic information for any oil company. To our knowledge, the pipeline transport network of the Mexican oil company (Pemex) has not been analyzed by means of network methodologies, where some topologic and organizational properties can be extracted, providing information of the transport [4,5].

In the present study the oil pipeline network is modeled like a complex network, where the nodes represent wells, production header, separation batteries, petrochemical complexes or terminals and the connections are equivalent to lines of ducts connecting the nodes. Our results show that the

[2] To whom any correspondence should be addressed

Content from this work may be used under the terms of the Creative Commons Attribution 3.0 licence. Any further distribution of this work must maintain attribution to the author(s) and the title of the work, journal citation and DOI.
Published under licence by IOP Publishing Ltd

connectivities in the pipeline network are characterized by a degree distribution which follows a power-law function, denoting a scale-free character in the connectivities. We also found that the generalized centrality provides a criterion to identify the importance of both nodes or links, in terms of dynamical situations like the transport of oil in a pipeline network. The paper is organized as follows. In Section 2, we present a description of the pipeline network and the main results are also described. Finally, some final remarks are given in the Conclusions.

2. Results

2.1. Degree distribution of the oil pipeline network

Usually a network is conceived as a set of nodes which are connected by means of links. The degree of a node is defined as the number of links entering or exiting a node, when the connections are not directed. The degree distribution $P(k)$, where k represents the degree, allows to describe the connectivity of the network, that is, if the relative frequencies are plotted against the degree, one obtains a distribution that characterizes the network, therefore, the structure of $P(k)$ offers information about how connections [6,8] are distributed. It is known that random networks have a characteristic scale in the connectivity, with a degree distribution of the Poisson's type, where the average degree is representative for the connectivity of the network. Many other networks observed in the real world do not have a single scale in the connectivities, i.e., they are scale-free networks, in which many nodes with few connections coexist with a small number of nodes with high connectivity (hubs).

Recent studies of the transport properties in complex networks have shown that scale-free networks have better conditions for transport compared to random networks, this is mainly related to the presence of hubs in scale-free configurations [9,10]. A simplified analysis of transport in a network, can be conceived as flow subject to the direction of each link and a potential difference between nodes. With this framework, the problem of finding the features that improve or worsen transportation, satisfying capacity constraints on the links and flow conservation at intermediate nodes [11,12], is an important topic and can be used to model and characterize some real transportation networks.

Here, we study the emergent configuration of an oil pipeline network. First, we recall that this type of network is a directed network because the oil is collected at the wellhead and from the wellhead it is pumped through a pipe to other intermediate sites, and finally, is discharged at its final destination, which in most of the cases is a refinery. We collected data of the pipeline network for the oil transportation from the wells to the refinery. This infrastructure is owned and operated by PEMEX, the Mexican oil company, and it is located in the South of Mexico. Specifically, our analysis is based on the Bellota-Jujo (B-J) system located in the area close to Comalcalco City in the Southern state of Tabasco [5]. The oil fields are located in different municipalities. This transport infrastructure is comprised by 165 wells, 26 production headers, 13 separation batteries and 3 petrochemical complexes, which correspond to nodes connected by pipelines (see figure 1 and 2).

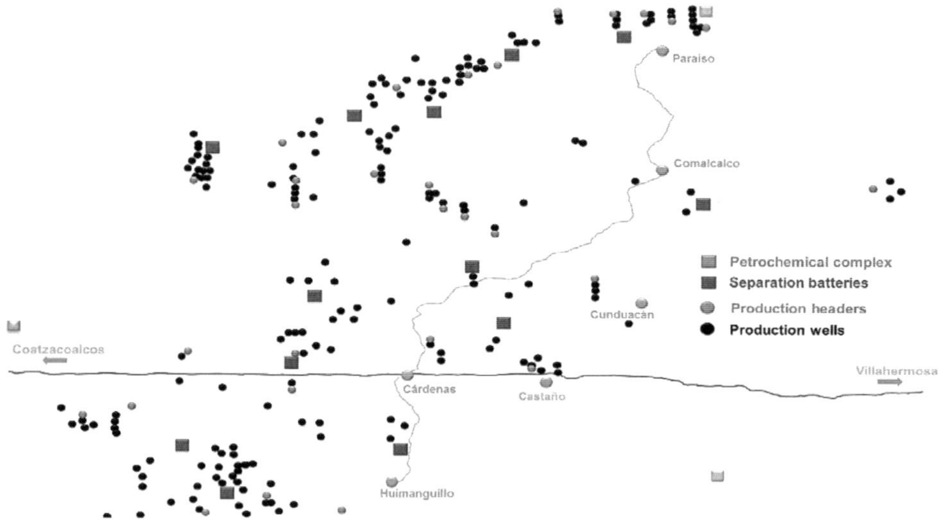

Figure 1. Oil production system in the State of Tabasco, Mexico. We show the location of the nodes: wells (black squares), production headers (blue circles), separation batteries (red squares) and petrochemical complexes (green squares).

Figure 2. Representation of the oil transportation network. Here the nodes represent wells (yellow), production headers (red), separation batteries (green) and petrochemical complexes (blue).

The corresponding network is constructed and depicted in figure 2. Here, we observe that a few nodes (production headers colored in red) tend to have more connections. This visualization suggests that there are a variety of ranges for the connectivities, and especially if one considers the incoming or outgoing links. In order to characterize the connectivities, we first consider the incoming degrees and calculate the degree distribution $P(k)$, with k the in-degree. The results presented in figure 3 show a clear indication that the network topology for the system, is a scale-free configuration. More precisely, the incoming degree distribution can be described as a power law function $P(k) \sim k^{-\lambda}$, with $\lambda = 2.6 \pm 0.03$ (see figure 3). It is worth to mention that the distribution for the outgoing degrees is very simple, because all the nodes, except petrochemical complexes, have out-degree equal to one.

2.2. The generalized centrality

In order to explore the topological properties of the transport, beyond the description offered by the distribution of connectivities, we evaluate the behavior of the transport in terms of the variables of design and the capacities installed in the pipes that compose the network. The B-J oil transport network expands on hundreds of kilometers with different measures in diameter that go from the 3 inches until the 36 inches at the final node. In order to evaluate the pipeline capacity for transportation, in figure 4 we show the behavior of the flow (in barrels per day) in terms of the diameter of pipes entering or leaving a node. A great dispersion of the data is observed for points corresponding to wells, however, an increasing tendency can be identified for the other nodes (production headers, separation batteries and petrochemical complex). The best fit to the data in a log-log scale, excluding wells, is $\gamma = 1.95 \pm 0.13$, which is general agreement with values reported for real pipeline systems [16].

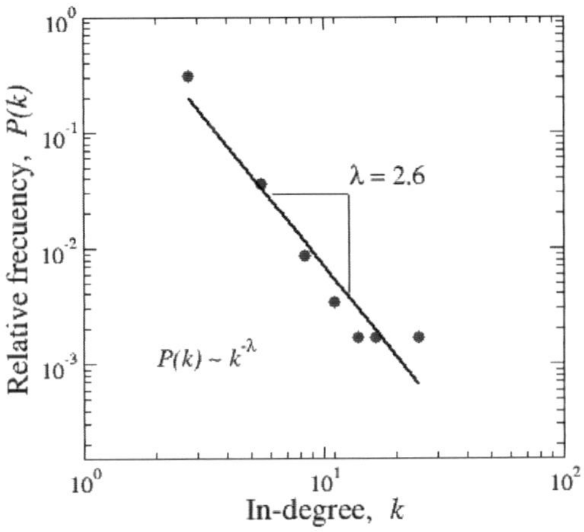

Figure 3. Log-log plot of the degree distribution vs. in-degree for the B-J oil transportation network. The solid line corresponds to the best fit with exponent $\lambda = 2.6$.

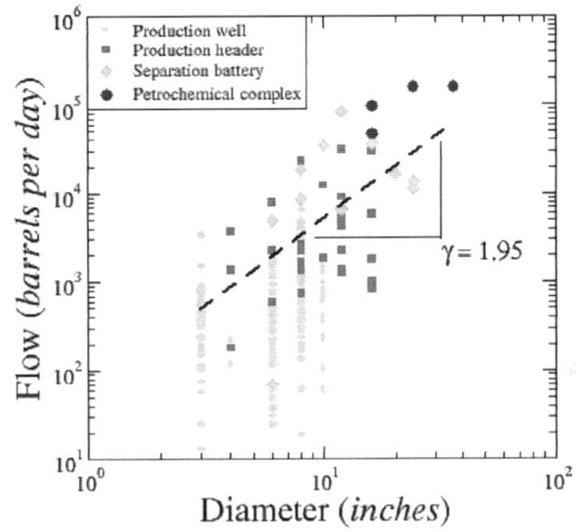

Figure 4. Plot of flow vs. diameter for the B-J system in a log-log scale. We show separately the cases of pipelines entering or existing different nodes. As expected, a growing trend is observed for the flow when the diameter is increased, suggesting that there is a power-law relationship, the straight line is a linear regression of the data, excluding wells. The value of the exponent is $\gamma = 1.95$.

Next, we evaluate the generalized centrality proposed in reference [16], which describes the topological importance of a specific connection within a transport network with appropriate information of the flows between sources and sinks. Particularly, for the oil transport network, the generalized centrality gives information about the importance of pipelines, which connect nodes (wells, production header, separation batteries or petrochemical complex). Formally, the generalized centrality of the connection e_{ij} is given by the following expression [16]:

$$G_{ij}^{f} = \sum_{e_{ST} \in E} \sum_{\substack{s \in V_S \\ t \in V_T}} \frac{f_{ST}}{|V_S||V_T|} \frac{\sigma_{st}(e_{ij})}{\sigma_{st}}$$

where E is the set of links in the network; V_S is the set of source nodes predecessors; V_T is the set of sink nodes; σ_{st} is the number of shortest-paths from node origin s to target node t; $\sigma(e_{ij})$ is the number of shortest-paths from s to t, passing through the connection e_{ij}, and f_{ST} is the existing flow in the connection $e(i,j)$.

For simplicity, we consider a normalization of the generalized centrality given by:

$$G_{ij}^{f} = \frac{G_{ij}^{f}}{maxG_{ij}^{f}}$$

where $maxG_{ij}^{f}$ represents the maximum value of the set of connections.

The values of G_{ij} are calculated for all the links in the B-J system. We find that the distribution of the generalized centrality values can be described by a power-law function, $H(G) \sim G^{\beta}$, with exponent $\beta = 2.5 \pm 0.05$ (see figure 5a), indicating that there is no preferential scale for the centrality and a few connections are responsible of high values for this quantity.

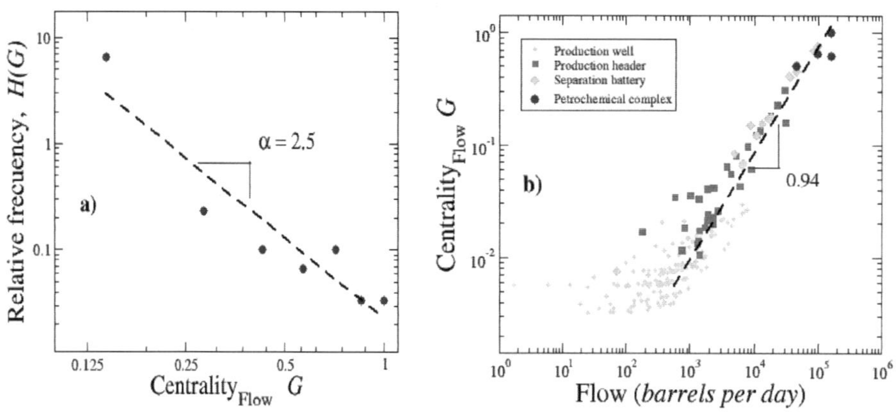

Figure 5. (a) Probability distribution of the generalized centrality in a log-log scale. The power-law fit for the data gives an exponent $\alpha = 2.5$. (b) Log-log plot of the generalized centrality vs. flow for the B-J system. The straight line is a regression of the data (excluding wells) with slope close to 1. We notice that for low values of the flow, there is more dispersion for the points corresponding to wells.

Moreover, when we evaluate the generalized centrality in terms of the flow entering or leaving a node, we find that there is an almost linear relationship at large values of both quantities (see figure 5b). Specifically, for those nodes connected by pipelines with high flow values (production headers, batteries and petrochemical complex), the corresponding centralities are proportional, whereas more dispersion is observed for wells. Finally, we construct a representation of the centrality values, i.e., we plotted the real B-J oil pipeline system and related the color of each pipeline proportional to its generalized centrality according to the colorbar (see figure 6). We remark that the highest value is associated to the link El Misterio-Cactus, which is responsible of handling a major load.

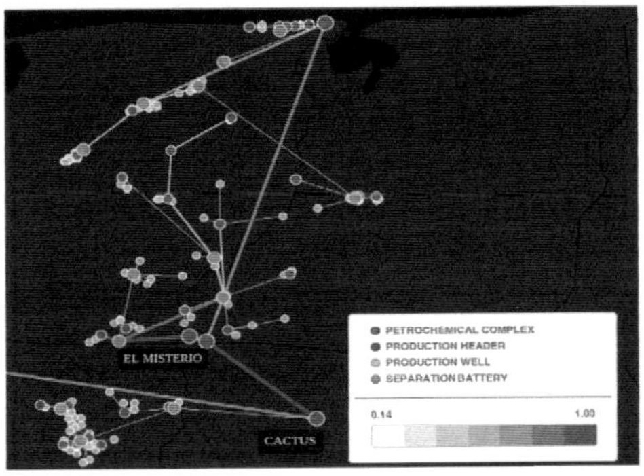

Figure 6. Visualization of the B-J oil-transport network. The color of the links represents the value of the generalized centrality according with the colorbar. We observe that the link connecting El Misterio-Cactus has the highest centrality value, i.e., is the most important connection.

3. Conclusions

We have studied the oil transport network from the B-J system. Our results shown that the degree distribution is characterized by a scale-free behavior, where a few nodes have high number of connections, which correspond to the production headers since they are connected to several wells. This finding is not unexpected, since different studies have reported the emergence of scale-free topologies in many man-made systems [1,13,16]. In our case, the growth of the oil transport network is strongly influenced by spatial distribution of wells and also economic constraints. By means of the generalized centrality, we have estimated the importance of links in terms of the flow it can handle. Interestingly, the distribution of centrality values also follows a power-law function, which indicates that there is no a characteristic scale for the centrality. Finally, we remark that this type of analysis is of great importance in order to understand the role of the topology in the pipeline transport system, with implications over the design and evaluation of the system's efficiency.

Acknowledgments
This work was partially supported by COFAA-IPN, EDI-IPN and SNI-Conacyt.

References

[1] Albert R and Barabási L 2002 *Rev. Mod. Phys.* **74** 47
[2] Dorogovtsev S and Mendes J 2003 *Evolution of Networks: From Biological Networks to the Internet and WWW* (Oxford, Oxford University Press)
[3] Newman E 2003 *SIAM Rev.* **45** 167–256
[4] Hernández B 2010 *Administración de la Integridad en Sistemas de Transporte de Hidrocarburos* (IPN México, D. F. IPN-ESIA)
[5] Petróleos Mexicanos, PEP 2015 *Activo Integral Bellota-Jujo*, Proyectos: Bellota-Chinchorro, Golpe-Puerto Ceiba y Jujo-Tecoaminoacán, Av. Prolongación Juárez, Ranchería Sur, sin número, C.P.:86360, Comalcalco, Tabasco, México
[6] Erdős P and Rényi A 1960 *Publ. Math. Inst. Hungar. Acad. Sci* **5** 17-61
[7] Watts J and Strogatz H 1998 *Nature* **393** 6684 440-442
[8] Barabási L and Albert R 1999 *Science* **8** 509-512
[9] López E, Buldyrev S, Havlin S and Stanley E 2005 *Phys. Rev. Lett.* **94** 24 248701
[10] Juárez R, Obregón B, Hernández R, Reyes I and Guzmán L 2014 *Ch. Sol. & Frac.* **69** 100-106
[11] Ahuja R K, Magnanti T L and Orlin J B 1994 *J. Oper. Res. Soc.* **45** 1340
[12] Rockafellar R T 1984 *Network flows and monotropic optimization* (John Wiley & Sons, Inc., Halsted Press)
[13] Andrade Jr J S, Bezerra D M, Ribeiro Filho J and Moreira, A A 2006 *Physica A* **360** 2 637-643
[14] Liu S, Liao Z, Feng Y and Rong G 2010 Topological Properties of Refinery System: A Complex Network Approach. 8th IEEE International Conference on Control and Automation
[15] Jiang Q, Zhou X, Xu B and Yuan K 2007 *AIChE J.* **53** 2 423-428

[16] Carvalho R, Buzna, L, Bono F, Gutiérrez E, Just W and Arrowsmith D 2009 *Phys. Rev. E* **80** 1 016106

Thermoeconomical analysis of an endoreversible chemical engine model with a diffusive transport law of particles

M A Barranco[2], A Ocampo[2] J. C. Pacheco[2] and F Angulo[2]

[1]Departamento de Formación Básica, Escuela Superior de Cómputo del Instituto Politécnico Nacional, Av. Juan de Dios Bátiz s/n. UP Zacatenco, CP 07738, D. F., México. *mbarrancoj@ipn.mx,*
[2]Departamento de Física, Escuela Superior de Física y Matemáticas del Instituto Politécnico Nacional, Edif. 9, 2° Piso, UP Zacatenco CP 07738, D.F., México.

E-mail: *aocampog@esfm.ipn.mx, jcpacheco@esfm.ipn.mx, angulo@esfm.ipn.mx*

Abstract. In 1995, De Vos introduced a thermoeconomical analysis for heat engines working in finite time. This analysis was made by the maximization of a benefit function defined as the ratio of the power output and the total costs which are associated to the investment and the fuel consumption. The De Vos methodology has been applied to several heat engine models working at different regimes of performance. In the present work we apply this methodology to a simple model of a chemical engine which works between two reservoirs of constant chemical potential. In this work we take into account a diffusive transport law of particles. From the optimization of the benefit chemical functions defined as the ratio of some characteristic function (power output, efficient power and ecological function) and the total cost involved in the performance of the chemical engine, we obtain the optimal chemical efficiencies under different regimes of performance in an analogous way to the thermo-economic method proposed by De Vos.

1. Introduction

In 1995 [1], De Vos introduced a thermoeconomical analysis of an irreversible heat engine model, the so-called Novikov plant model (see Figure 1) [2], in terms of the maximization of an objective function defined by the ratio of the power output and the running costs of the plant exploitation. De Vos [1] assumed that the running costs of the plant consist of two parts: a capital cost that is proportional to the investment and, therefore, to the size of the plant and a fuel cost that is proportional to the fuel consumption. De Vos [1] showed how the optimal efficiency, η_{opt} (in the economical sense) smoothly increases from the maximum-power point (Curzon-Ahlborn efficiency, η_{CA} [3]) corresponding to energy sources where the investment is the preponderant cost up to the Carnot value (Carnot efficiency, η_C), that is, for energy sources where the fuel is the predominant cost when the heat fluxes in the Novikov model are given by a linear Newtonian heat transfer law; that is, $\eta_{MP} < \eta_{opt} < \eta_C$, where the subscripts MP, opt and C mean maximum power, optimal efficiency and Carnot respectively. De Vos [1] found for the Novikov's model that η_{opt} is given by,

Content from this work may be used under the terms of the Creative Commons Attribution 3.0 licence. Any further distribution of this work must maintain attribution to the author(s) and the title of the work, journal citation and DOI.
Published under licence by IOP Publishing Ltd

$$\eta_{opt}(\tau, f) = 1 - \frac{f}{2}\tau - \frac{\sqrt{4(1-f)\tau + f^2\tau^2}}{2}, \tag{1}$$

where f is the fractional fuel cost (see Table 1 of the Ref. [1]), which is defined as the ratio of the cost of the fuel consumption and the running costs of the power plant; $\tau = T_L/T_H$ with T_H and T_L the temperatures of the hot and cold thermal reservoirs respectively (see Fig. 1). Later, Barranco-Jiménez and Angulo-Brown [4, 5] also studied the Novikov engine model following the thermoeconomical approach used by De Vos, but by means of the so-called ecological optimization criterion. The ecological optimization criterion consists in maximizing the well known ecological function [6, 7] defined as $E = W - T_L\sigma$, where W is the plant's power output and σ is the total entropy production of the endoreversible power plant model. Barranco-Jiménez and Angulo-Brown [4, 5] also obtained an optimal efficiency η_{opt}^E between η_{MP} and η_C that drastically reduces the entropy production of the engine. Besides the optimal efficiency satisfy the following inequality $\eta_{MP} < \eta_{opt} < \eta_{opt}^E < \eta_C$ [8].

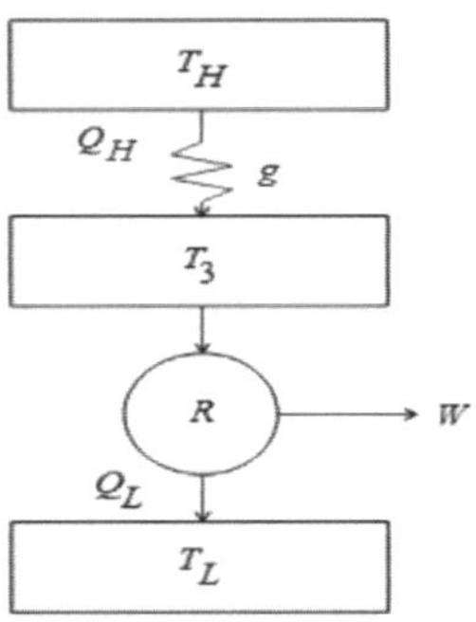

Figure 1. Novikov's model for a thermal power plant.

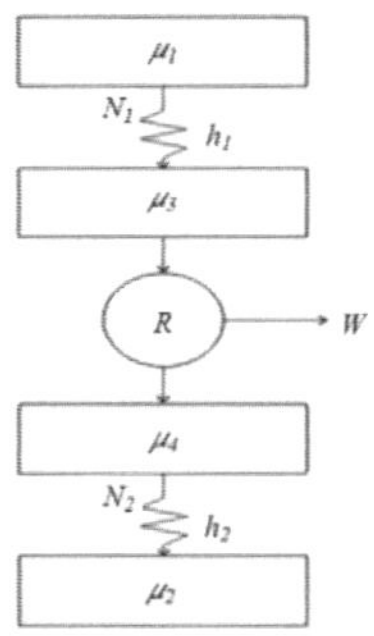

Figure 2. Endoreversible chemical engine.

The concept of heat engine working between two thermal reservoirs has been extended to chemical engines working between two reservoirs of constant chemical potential μ [9]. This model of chemical engine has been studied under the methodology of finite-time thermodynamics (FTT) [9-12]. Some authors [13] have considered that these chemical engine models can have practical value for many devices such as chemically reactive devices, mass exchangers, photochemical cells and solid-state devices. Recently, Barranco-Jiménez et al. [14] also studied the thermoeconomics of an endoreversible chemical engine model by applying the De Vos methodolody. In that study, they also take into account a linear mass transfer law that comes from the linear irreversible thermodynamics [9, 13]. Besides, in their study they consider three different regimes of performance: maximum power output [3], maximum efficient power [15-17] and maximum ecological function [6]. In the present work we also study the thermoeconomics of an endoreversible chemical engine but taking into account a diffusive transport law of particles. From the optimization of the benefit chemical functions defined as the ratio of a characteristic function (power output, efficient power and ecological function) and the total cost involved in the performance of the chemical engine, we obtain the optimal chemical efficiencies under different regimes of performance by using the De Vos methodology.

2. Thermoeconomical optimization of an endoreversible chemical engine.

The schematic diagram of an isothermal chemical engine model is shown in Figure 2. The two mass reservoirs associated to the chemical engine have chemical potentials μ_1 and μ_2. The configuration of the chemical engine is analogous to that of thermal engines with heat fluxes. The chemical engine receives the mass flux N_1 at chemical potential μ_3 from the high chemical potential μ_1 in time t_1 and rejects the mass flux N_2 at chemical potential μ_4 to the low chemical potential μ_2 in time t_2. In his

1991 paper [9] De Vos studied a chemical reactor by means of a Fick-type particle diffusive law. Following Fig. 2 the fluxes N_1 y N_2 are given by,

$$N_1 = h_1(n_1 - n_3), \tag{2}$$
$$N_2 = h_2(n_4 - n_2), \tag{3}$$

where h_1 y h_2 are particle conductances also called coefficients of diffusivity, and n_i (i=1, 2, 3, 4) are the particle concentrations per unit volume, and they are related with the chemical potentials by means of,

$$\mu_i = \mu_0 + kT \ln \frac{n_i}{n_0}, \tag{4}$$

where μ_0 denotes the chemical potential at normalized concentration n_0, being k the Boltzmann constant. From mass conservation we have,

$$N_1 = N_2 = N. \tag{5}$$

Applying the firts law of thermodynamics to the model of Figure 2 we obtain,

$$\mu_4 N + W - \mu_3 N = 0. \tag{6}$$

By combining Eqs. (4) and (6) we get,

$$\xi = \mu_3 - \mu_4 = \xi_0 + kT \ln\left(\frac{n_3}{n_4}\right), \tag{7}$$

being ξ the work production per mole of fuel, which is given by $\xi = W/N$, and from Eq. (4) immediately we obtain

$$\xi_0 = \mu_{03} - \mu_{04} = \mu_{01} - \mu_{02}, \tag{8}$$

by using Eqs. (2), (3) and (5) we get,

$$h_1(n_1 - n_3) = h_2(n_4 - n_2), \tag{9}$$

by solving Eqs. (7) and (9) for the two unknown concentrations n_3 and n_4 we obtain [9],

$$n_3 = \frac{h_1 n_1 + h_2 n_2}{h_1 + h_2 exp\left(-\frac{\xi - \xi_0}{kT}\right)}, \tag{10}$$
$$n_4 = \frac{h_1 n_1 + h_2 n_2}{h_2 + h_1 exp\left(\frac{\xi - \xi_0}{kT}\right)}. \tag{11}$$

By substituting Eq. (9) into Eq. (2) we have,

$$N = \frac{n_1 - n_2 exp\left(\frac{\xi - \xi_0}{kT}\right)}{\frac{1}{h_1} + \frac{1}{h_2} exp\left(\frac{\xi - \xi_0}{kT}\right)}. \tag{12}$$

From Eq. (12) and by using the definition of ξ we obtain,

$$W = \xi \frac{n_1 - n_2 exp\left(\frac{\xi - \xi_0}{kT}\right)}{\frac{1}{h_1} + \frac{1}{h_2} exp\left(\frac{\xi - \xi_0}{kT}\right)}. \tag{13}$$

On the other hand, when we apply the second law of thermodynamics to Fig. 2, the rate of entropy production of the chemical engine is given by [9],

$$\sigma = (\xi_C - \xi)\frac{N}{T}, \tag{14}$$

where ξ_c is the "Carnot" chemical efficiency and by using Eq. (4) it results [9],

$$\xi_c = \mu_1 - \mu_2 = \xi_0 + kTln\left(\frac{n_1}{n_2}\right), \tag{15}$$

by substituting Eq. (12) into Eq. (14) we arrive to,

$$\sigma = \left[\frac{\xi_0}{T} + k\ln\left(\frac{n_1}{n_2}\right) - \frac{\xi}{T}\right]\frac{n_1 - n_2 exp\left(\frac{\xi-\xi_0}{kT}\right)}{\frac{1}{h_1}+\frac{1}{h_2}exp\left(\frac{\xi-\xi_0}{kT}\right)}. \tag{16}$$

For our thermoeconomical optimization, we apply the De Vos methodology to the chemical engine model by maximizing three benefit chemical functions defined by the ratio of the corresponding characteristic function (Power Output (MP), [1], Efficient Power (EP), [15, 16] and Ecological Function (ME) [6]) and the total costs involved in the performance of the chemical engine, that is, $q_{MP} = \frac{W}{C}$, $q_{EP} = \frac{\xi W}{C}$ and $q_{ME} = \frac{W - T\sigma}{C}$ [14]. Besides in the maximization of the benefit chemical functions, the total costs C are taken as $C = C_{inv} + C_{fuel} = aN_{max} + bN$, where the proportionality constants a and b have units of $\$ \cdot time$ ($:currency) and N_{max} is the maximum mass transfer without supplying chemical work. To illustrate the mathematical form for the benefit chemical functions we only show the case of q_{MP},

$$q_{MP} = \frac{\xi n_1 h_1 \frac{1-exp\left(\frac{\xi-\xi_C}{kT}\right)}{1+nh exp\left(\frac{\xi-\xi_C}{kT}\right)}}{\xi_c n_1 h_1 \frac{1-exp\left(\frac{-\xi_C}{kT}\right)}{1+nh exp\left(\frac{-\xi_C}{kT}\right)} + \beta n_1 h_1 \frac{1-exp\left(\frac{\xi-\xi_C}{kT}\right)}{1+nh exp\left(\frac{\xi-\xi_C}{kT}\right)}}. \tag{17}$$

We see that the curve of q_{MP} presents a maximum point ξ^* between 0 and ξ_C. The maximization condition $dq_{MP}/d\xi = 0$ results in a transcendental equation. The same occur for q_{EP} and q_{ME} (see Fig. 3). Figure 3 shows the behavior of the three profit chemical functions and we can observe that there exists an optimal chemical efficiency value which depends on the parameter $\beta = b/a$ and the optimum value of the chemical efficiency ξ^*. We can also observe that the maximum of benefits tends to the reversible value $\xi = \xi_C$ as the parameter β tends to infinity, such as it occurs for thermal engines [1-4].

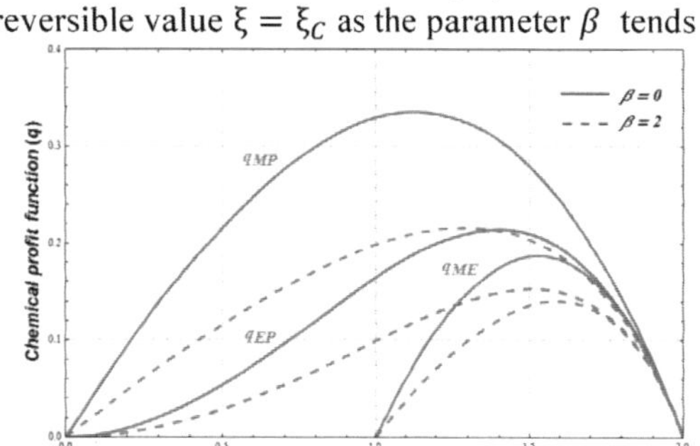

Figure 3. Comparison between the three chemical benefit functions versus chemical efficiency for two values of β.

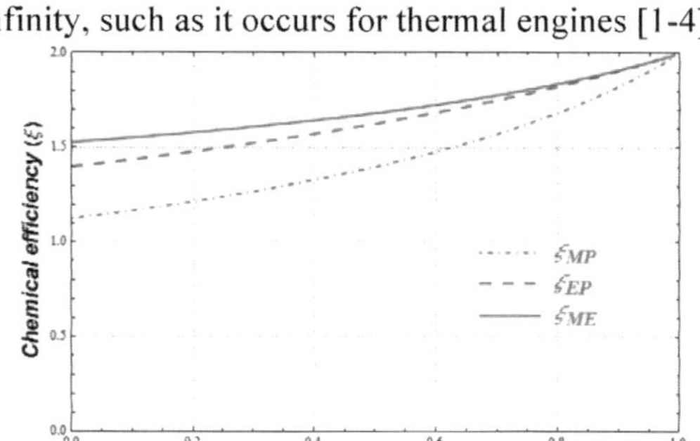

Figure 4. Optimal efficiencies for ecological regime, efficient power and maximum power output conditions.

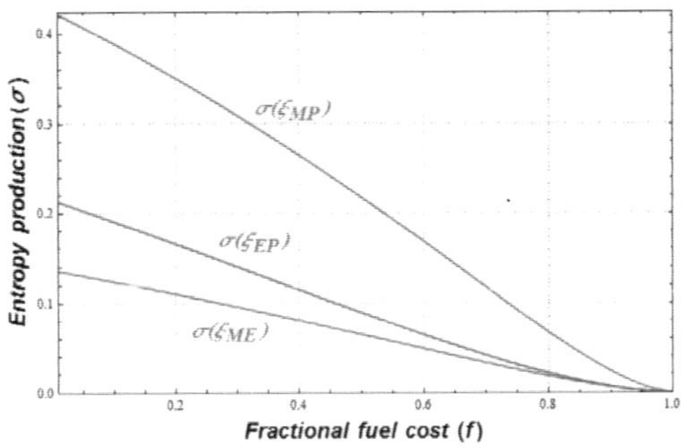

Figure 5. Entropy production under the three regimes of performance vs. fractional fuel cost

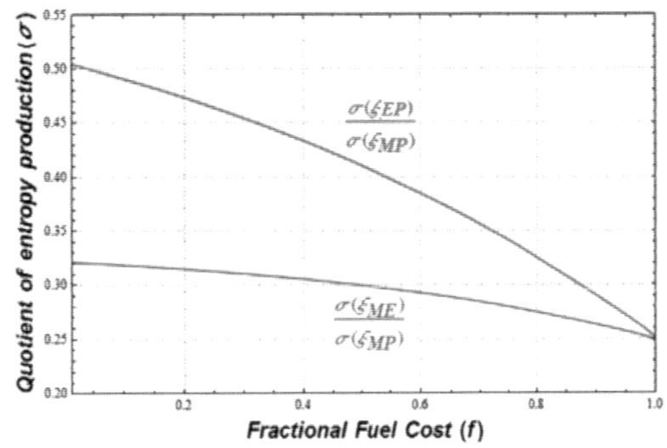

Figure 6. Ratio of entropy production vs. fractional fuel cost.

The optimal chemical efficiencies are obtained in terms of the dimensionless parameter β but now with a chemical meaning. Also in these cases we can define a fractional cost defined as the ratio between the reactants cost and the total costs of the plant, that is,

$$f = \frac{bN}{aN_{max}+bN} = \frac{\beta N}{N_{max}+\beta N}. \tag{18}$$

In Figure 4, we show the optimal efficiencies under the three regimes of performance. As it is shown in these figures the three optimal efficiencies smoothly increase from the maximum efficient point (in each regime), corresponding to energy sources where the investment is the preponderant cost up to the "Carnot" chemical efficiency (ξ_c), that is, for energy sources where the reactants are the predominant cost [1], and we can also observe that for all the values of the fractional cost, the optimal efficiencies satisfy the following inequality,

$$\frac{\xi_C}{2} < \xi_{MP}^* < \xi_{EP}^* < \xi_{ME}^* < \xi_C. \tag{19}$$

Finally, we can calculate the total entropy production and the power output under the three regimes of performance. In Fig. 5 we see how the inequality $\sigma_{MP}^* > \sigma_{EP}^* > \sigma_{ME}^*$ is hold [6, 7, 14]. In Fig. 6, we can see how the ME-regime produces less entropy than the other two regimes.

3. Concluding Remarks

In the present work, we applied the De Vos methodology to an endoreversible chemical engine model, i. e., an engine operating between two reservoirs of constant chemical potential. In our chemical engine model we take into account a diffusive transport law of particles. From the maximization of chemical profit-functions defined in terms of some model characteristic functions (power output, efficient power and ecological function) and the costs related to investment and reactants consumption, we obtain the optimum efficiencies for this model under different performance regimes. Besides, we analyze both, the effect on the reduction of power output and the reduction in the total entropy production under the three considered regimes of performance. The extension of thermoeconomic studies from heat engines to chemical engines can be useful to analyze some aspects of biochemical systems, such as photosynthetic processes.

Acknowledgments

This work was supported in part by CONACYT, COFAA and EDI-IPN-México.

[1] De Vos A 1995 *Endoreversible thermoeconomics. Energy. Convers. Magnagement.*;36:(1)1-5
[2] Novikov I I 1957 *The efficiency of atomic power stations* (a review) *Atomimaya Energiya* **3** 409 (English translation 1958. *Nuclear Energy* 7 125)

[3] Curzon F. and Ahlborn B 1975 *Efficiency of a Carnot engine at maximum power output*. Am. J. Phys.;43:22-24
[4] Barranco M and Angulo F 2007 *"Thermoeconomic optimization of a Novikov power plant model under maximum ecological conditions"* J. of the Energy Inst. 80:2(1-4):96-104
[5] Barranco M 2009 "Finite-time thermoeconomic optimization of a non endoreversible heat engine", Rev. Mex. Fis. 55 (3) 211-220
[6] Angulo F 1991 *An ecological optimization criterion for finite-time heat engines*. J. Appl. Phys. 69:7465-7469
[7] Arias L and Angulo F 2001 Reply to "Comment on A general property of endoreversible thermal engines" [J. Appl. Phys. 89, 1518]
[8] M. A. Barranco-Jiménez, N. Sánchez-Salas, F. Angulo-Brown, *"On the optimum operation conditions of an endoreversible heat engine with different heat transfer laws in the thermal couplings"*, Rev. Mex. Fis. 54 (4), 284-292
[9] De Vos A, Endoreversible 1992 *Thermodynamics of a solar energy conversion*. Oxford University: Oxford University Press
[10] Chen L, Sun F and Wu C 1998 *Performance of chemical engines with a mass leak*, J. Phys. D: Appl. Phys. 31, 1596-1600
[11] Chen L, Sun F, Wu C and Yu J 1997 *Performance characteristic of isothermal chemical engines, Energy*. Convers. Mgmt, 38 (18), 1841-1846
[12] Xia S, Chen L and Sun F 2009 *Optimal configuration of a finite mass reservoir isothermal chemical engine for maximum work output with linear mass transfer law*. Rev. Mex. Fis. 55 (5), 399-408
[13] Xia S, Chen L and Sun F 2010 *Optimal performance of a generalized irreversible chemical engine with diffusive mass transfer law, Mathematical and Computer Modelling*, 51, 127-136
[14] Barranco M. A., Ocampo A and Angulo F 2016 *Thermoeconomical optimization for an endoreversible chemical engine model*. Proceedings of ECOS2014, Portoroz Eslovenia, June 19-23
[15] Pacheco J, Angulo F and Barranco M 20015 *"Thermoeconomical analysis of a non-endoreversible Novikov power plant model under different regimes of performance"*, J. Phys.: Conf. Ser. 582 012050
[16] Yilmaz T 2006 *A new performance criterion for heat engine: efficient power*. J. Energy Inst. 79(1):38-41
[17] Arias L, Barranco M and Angulo F 2009 *"Comparison between two ecological-type modes of performance for a simple energy converter"*, Journal of the Energy Institute, 82 (3) 193-197

Radiation monitoring in interventional cardiology: a requirement

T. Rivera[1*] and E. S. Uruchurtu[2]
Instituto Politécnico Nacional, CICATA-Legaria, 11500 México, MX
Hospital Juarez de Mexico, Unidad de Hemodinamia, 07760 México, MX

E-mail: riveramt@hotmail.com

Abstract. The increasing of procedures using fluoroscopy in interventional cardiology procedures may increase medical and patients to levels of radiation that manifest in unintended outcomes. Such outcomes may include skin injury and cancer. The cardiologists and other staff members in interventional cardiology are usually working close to the area under examination and they receive the dose primarily from scattered radiation from the patient. Mexico does not have a formal policy for monitoring and recording the radiation dose delivered in hemodynamic establishments. Deterministic risk management can be improved by monitoring the radiation delivered from X-ray devices. The objective of this paper is to provide cardiologist, techniques, nurses, and all medical staff an information on DR levels, about X-ray risks and a simple a reliable method to control cumulative dose.

Introduction

The story began in 1895, when a German physicist discovered a new kind of rays. This is almost 120 years. The medical use of X-ray beam, while offering great benefit to patients, also contributes significantly to radiation exposure of individuals and populations. According to the results published by the United Nations Scientific Committee on the Effects of Atomic Radiation in 2012, interventional procedures contribute only 10% to the frequency of radiation use on the medical field whereas their contribution to collective dose is 19%. The cardiologists and other staff members in interventional cardiology are usually working close to the area under examination and they receive the dose primarily from scattered radiation from the patient. Mexico does not have a formal policy for monitoring and recording the radiation dose delivered in an interventional cardiology laboratory. The objective of this paper is to provide cardiologist, techniques, nurses, and all medical staff the new findings and brings them up a simple a reliable method to control cumulative dose for radiation safety.

Interventional Cardiology

Fluoroscopy is the method that provides real-time X ray imaging that is especially useful for guiding a variety of diagnostic and interventional procedures. Since the early 20th century, fluoroscopy has been integral to the practice of diagnostic radiology [1]. For the most part, fluoroscopic procedures were primarily diagnostic and involved relatively small risks to patients and personnel. However, over the

* Corresponding Author, T. Rivera, riveramt@hotmail.com.

Content from this work may be used under the terms of the Creative Commons Attribution 3.0 licence. Any further distribution of this work must maintain attribution to the author(s) and the title of the work, journal citation and DOI.
Published under licence by IOP Publishing Ltd

past 10–15 years, fluoroscopic procedure mixes have included an increasing fraction that are primarily therapeutic [2]. Thus, radiation exposure is a significant concern for interventional cardiologists and patients due to the increasing workloads and the complexity of procedures over the last decade [3]. Interventional radiology and interventional cardiology (IC) contributes a significant proportion of the collective dose of the population from medical exposures. According to the results published by the United Nations Scientific Committee on the Effects of Atomic Radiation, interventional procedures contribute only 10% to the frequency of radiation use on the medical field whereas their contribution to collective dose is 19% [4]. When complex procedures are performed or procedures are repeated for the same patient, high radiation dose levels can occur because procedures often require long fluoroscopy time and a large number of images.

Radiation effects and humans

Much has been said about the dangers of radiation, and particularly about the role played by diagnostic X-ray radiology, which contributes 99% of man-made radiation exposure to humans [5]. The adverse risks of radiation exposure may be described in terms of stochastic and deterministic effects. The higher dose delivered in IC observes its severity only deterministic effects. It severity is proportional to the dose and the dose exposure is usually described in terms of fluoroscopic time, cumulative air kerma, and dose-area product. In 2011, the ICRP alerted the radiological community to the epidemiological evidence of higher incidence of cataracts in interventional cardiologists [6]. The degree of exposure has been correlated with a linear increase in the development of both benign and malignant thyroid neoplasms, structural and functional changes as a result of radiation exposure have been reported in the thyroid gland [7, 8].

Recognizing radiation injury and effects

The Center for Device Regulation, Radiation Health and Research (CDRRHR)-Food and Drug Administration (FDA) of the Department of Health (DOH) [9] reports the potential effects of prolonged exposure to radiation during interventional fluoroscopic procedures described in Table 1.

TABLE1. Potential effects of X-rays during interventional fluoroscopic procedures

Effect	Single-dose Threshold (Gy)	Onset
SKIN		
Early transient erythema	2	2-24 Hours
Main Erythema	6	~10 d (-1.5wk)
Temporary hair loss	3	~3 wk
Permanent hair loss	7	~3 wk
Dry desquamation	14	~4 wk
Moist desquamation	18	~4 wk
Secondary ulceration	24	>6 wk
Late erythema	15	~8 – 10 wk
Ischemic dermal necrosis	18	>10 wk
Dermal atrophy (1st phase)	10	>14 wk
Dermal atrophy (2nd phase)	10	>1 y
Induration (Invasive Fibrosis)	10	>1 y
Telangiectasia	10	>1 y
Late dermal necrosis	>12?	
Skin cancer	not known	>5 y
EYE		
Lens opacity (detectable)	>1-2	>5y
Lens/cataract (debilitating)	>5	>5y

Radiation dosimetry and principles

Monitoring of the radiation exposure induced by interventional cardiology procedures has become more important due to the rapidly increasing number of X-ray devices in use. Therefore, the dose delivered is not standardized and, in some cases, could reach levels high enough to produce deterministic effects to the patient. Surface determination of the dose is often measured by detector, but, often it is not possible to measure the dose at an organ or tumour directly due the obvious obstacles for introducing dosimeters into the patient. In this case the dose can be calculated taking into account the appropriate models of the anatomy of the patient anatomy as well as the radiation field parameters from measurements on the skin of the patient [10]. In an IC laboratory can be used different dosimetric methods in order to radiation monitoring. In this paper are described some of them. Historically, the first issue to measure radiation dose IC laboratory was radiographic film. X-ray film consists of a base of thin plastic sheet with a radiation sensitive emulsion coated uniformly on one or both sides of the base. Radiographic film finds use in an immense variety of diagnostic, having the advantages of high reliability, versatility and ability to provide a permanent information record. The radiographic film has a working range from 0.01 Gy up to 2 Gy which is rather narrow range that can be exceeded in the highest dose fluoroscopic procedures. Disadvantage, the radiographic film is difficult to measure the absolute of X-ray dose. Traditionally, the ionization chamber (ICH) has been the reference dosimeter for interventional cardiology and can be used for quality assurance and calibration of other dosimeters. The sensible volume of ionization chamber could be 0.6 or 0.18 cc. The range of measurable radiation dose is from 1uGy up to 16kGy. However, inconvenient of IC in clinical practice is it size. The ICH is connected via a cable with an electrometer to make the measure should be used that ensure appropriate levels of accuracy and long term stability. Semiconductor detectors Dosimeters are other detectors proposed for diagnostic measurements [11]. They produce large signals from modest amounts of radiation, they are rigid and do not require pressure correction, which makes them suitable for some clinical applications. The simplest of semiconducting devices is the diode, which based on a p–n junction between the p-type and n-type parts of a semiconductor. As ionizing radiation strikes the semiconductor, electron hole pairs are induced. The diode is more sensitive and has small size compared to ionization chamber. A disadvantage of diode as dosimeter in interventional cardiology is also uses cable. The metal-oxide semiconductor field effect transistor (MOSFET) dosimeter for its an excellent spatial resolution is suggested in diagnostic radiology. The use of MOSFET dosimeter in clinical practice also needs correction factors for energy dependence and field size, this is not tissue equivalent. The MOSFET dosimeter is useful in the range from 1.5 mGy up to 2Gy. Disadvantage of MOSFETD is visible in radiographs and it has temperature dependence [12, 13]. The diamond dosimeter (DD) is other detector can be used to measure equivalent dose in interventional cardiology procedures with a good agreement with IC. DD is suitable for in vivo dosimetry due to its small size, resistance to the radiation damage, and tissue equivalent. However, the use of the diamond for low energy X-ray needs a correction factor another disadvantage is expensive [14]. The most commonly used radiochromic film is the Gaf-Crhomic^{TM} film. Basically, the radiochromic films are rectangular sheets that change color when exposed to a sufficient absorbed dose. Because color indicated cumulative dose, dosimetry is usually qualitative to some extent, which challenges the accuracy of its measurement. Disadvantages, radiochromic film is insensitive to low doses, below 3 Gy, which is an important range for accurate tracking of diagnostic radiation dose [15. A practicable used type of dosimeter in the past and present is the thermoluminescent dosimeter (TLD) [16, 17]. The thermoluminescent dosimeters (TLDs) are based on the physical property of certain crystals to absorbe energy in metastable states that are a result of defects in the material lattice. These trapped electrons or holes remain at the metastable state until they are heated to recombine and emit the energy in the form of light. The quantity of light emitted is proportional to the energy deposited in to the crystal during irradiation. Dosimeters with thermoluminescence and semiconductor detectors are considered here the small size of TL dosimeters (TLDs) allows their application for conducting measurements on patients. TLDs are available in various forms (e.g. powder, chips, rods, ribbons) and made of various materials [18]. Dosimeters most commonly used in medical applications are based on lithium fluoride doped with magnesium and titanium (LiF:Mg,Ti) because of its tissue equivalence (Z_{eff}=8) but other high sensitivity materials such as LiF:Mg,Cu,P; Li2B4O7:Mn; CaSO4:Dy and CaF2:Mn have also been used. It can be used to measure the dose ranging from 10 uGy to 10 kGy [19, 20]. The main disadvantage of these devices has been their energy dependence of response which differs considerably from that of ionization chambers. More recently, optically stimulated luminescent dosimeters (OSLDs) have been adopted for this role. Both types of dosimeters operate by recording an energy reading upon irradiation and emitting an output of the recorded dose upon appropriate stimulation. TLDs require a heating stimulus to produce readouts of their recorded doses while OSLDs require an optical stimulus, enabling more rapid, accurate, and reproducible measurements [21, 22, 23]. Radiation dosimetry as well as dose distribution for interventional dosimetry purposes has steadily evolved over the last few decades with the introduction of various new detectors. Concerning quality assurance, The Society of International Radiology Standards of Practice Committee guidelines for establishing quality improvement program in interventional radiology [24]. The quality assurance (QC) of diagnostic X ray beams in IC is fully characterized by their spectrum as well as X-ray beam parameters: tube voltage, the first and the second HVLs (HVL1 and HVL2), total filtration and their combinations are used for this purpose. A quality assurance program (QAP) for interventional cardiology includes all of the aspects of radiological protection (RP) of patients and staff in addition to the usual clinical aspects. The Working Group on Interventional Cardiology of ISEMIR [25] has produced a set of recommendations for occupational radiological protection and concluded after a wide international survey that the dose received by cardiologists during percutaneous coronary interventions, electrophysiology procedures, and other interventional cardiology procedures can differ by more than an order of magnitude for the same type of procedure and for similar patient doses. Simple methods for reducing or minimizing the occupational radiation dose are included in the recommendations.

Dose reduction

Radiation safety is the concern of all health care providers who perform procedures associated with radiation imaging, whether for diagnostic purposes or therapeutic procedures. Appropriately, there has been increasing public and societal interest in limiting patient radiation. During the last few years, several scientific and professional societies have produced guidelines on radiation safety (including patient dosimetry) for interventional radiology [26]. Some of these guidelines have been adopted simultaneously by American and European interventional radiology societies [24, 25] and others have been produced by groups of experts and later endorsed by professional societies [26, 27, 1]. In order to help in the optimization of the interventional practices. Diagnostic reference levels (DRLs) are still a challenge for interventional radiology. The ICRP proposed their application in interventional radiology in 2009 [5], but it still is a long way to their effective application. The National Council on Radiation Protection and Measurements in the USA has published a document on this issue [6] and the ICRP launched a working party in 2012 to also give more specific advice on the use of DRLs in interventional procedures and new imaging techniques. In a workshop organized by the IAEA in 2010 for pediatric interventional cardiology [24] involving 11 Latin American countries, it was reported that only 64 % of the cardiologists used their personal dosimeters regularly. In Mexico not have a formal policy for monitoring and recording the radiation dose delivered in an interventional cardiology laboratory. To manage the radiation dose, it must be measured. All physicians and staff involved with interventional fluoroscopy must be properly trained on the basic principles of radiation physics and safety. A qualified medical physicist must be involved with the physician in equipment selection and staff education. The best quality image with the most effective radiation dose provides the best patient care.

Future

Concerning the future of interventional cardiology is picts in good chance if form an interventional team group: Cardiologists, medical physics, technician, nurses and industry together commit to clinical care; and work with many other associates with continued cooperative efforts and research to optimization dose. A data base on a real project is necessary to set up and develop acceptable methods for future surveys of population exposure from medical X-rays. Electro technical Commission (IEC), the digital imaging and communications in medicine (DICOM) for radiation dose reporting in radiology and the profiles from form the RIS systems around the country may then be input to databases. Radiation dose data collected by the real-time dosimeter during each intervention is the new generation dosimeter. One of the important initiatives under radiological protection is the SmartCard project, this program was proposed by IAEA [28-33]. The role of dosimetry in diagnostic radiology has to be consistent with the Bonn Call-for-Action proposed during last meeting in 2012. Nowadays, electronic dosimeters are also suggested for personal dosimetry. Education and training programs will be consist in two levels. The first provides recommendations for continuing education and training after qualification and when new techniques are implemented. The second level, specific educational objectives to be included in specific diagnostic area. The program will contain specific topics for radiation dosimetry and radiation protection, recognizing the importance of reducing patient doses while maintaining the desired level of guarantee quality in medical exposures (good therapeutic treatments and images of sufficient quality for diagnosis). I suggest in all ICL an environmental program dosimetry, an estrictic clinical dosimetry and physical dosimetry to maintain radiation doses As Low As Reasonably Achievable (ALARA). The professional works in different establishment of practical radiological sources in the case of ICL to radiation, take account of the relevant recommendations of the international organizations. Patient dosimetry could will be the main topic in the next generations. PD will be the main objective for radiation protection programs. Radiation dosimetry in diagnostic radiology will provide recommendations taking into account diagnostic reference levels (DRL) and it requires acceptable quality dose (AQD) [34-37, 3]. For personal radiation monitoring, I suggest the use of two personal dosimeters for occupational dosimetry interventional cardiology laboratories: one worn on the trunk of the body inside the apron and the other worn outside the apron at the level of the collar or the left shoulder. The dosimeter under the apron provides an estimate of the dose to the organs of the shielded region. The dosimeter worn outside the apron supplies an estimate of the dose to the organs of the head and neck, including the thyroid and lens of the eyes (if unshielded), but greatly overestimates the doses to organs of the trunk.

Conclusion

We are seeing an impressive increase in the use of interventional procedures for more complex procedures and they are being used by more medical specialties owing to the undisputed clinical benefits. However, practitioners, scientists, manufacturers, and regulators have an obligation to promote the best level of radiation safety for patients and staff. For dose optimization a Monte Carlo simulation can be used to estimate the absorbed dose conversion coefficients. Dose determinations should be performed using a calibrated instrument and a dosimetry code of practice such as the one recommended by the IAEA. The assessment of patient dose is an essential aspect of the justification and the optimization process in X-ray diagnostic radiology. Medical physics support is needed for accurate dose determination and for a better understanding of how patient dose is affected in clinical practice by examination protocols and exposure parameters. A training program is an essential part of the dose reduction for the interventional cardiology laboratory. A team collaborative effort involving physicians, staff, medical or health physicists, quality assurance personnel, and hospital administration. Interventional cardiologists are an essential part of this process and need to ensure the best possible outcomes for ourselves and for our patients.

References.

[1] Levine G N, Bates E R, Blankenship J C, et al., 2011 *J Am Coll Cardiol.* **58** e44–122
[2] Miller D L 2008 *Health Phys* **95** 638-44
[3] Ryan T J 2002 *Circulation.* **106** 752–756
[4] ICRP 118. *International Commission on Radiological Protection* 2012 Ann ICRP **41**(1–2) 1–322.
[5] ICRP 117. *International Commission on Radiological Protection* 2010 Ann ICRP **40**(6) 1–102.
[6] ICRP 113, *International Commission on Radiological Protection* 2009 ANN ICRP 39 5
[7] UNSCEAR, 2013. *United Nations Scientific Committee on the Effects of Atomic Radiation.* 2013 Report to the General Assembly Scientific. Vol 1, New York
[8] Balter S, Hopewell JW, Miller DL, et al. 2010 *Radiology* **254** 326-41
[9] WHO 1982. *World Health Organization* (WHO). 1982 Geneva
[10] NCRPM 168. *National Council on Radiation Protection and Measurements* 2010 NCRP Report No. 168.
[11] Ron E, Brenner A. 2010 *Radiat Res* **174**:877-88.
[12] Jacob S, Boveda S, Bar O, et al. 2013 *Int J Cardiol* **167** 1843-7.
[13] Chougule A A 1993 *Bull. Radiation Protection* **16** (2) 135-138
[14] FDA 1994. Food and Drug Administration 1994. Accessed May 20, 2016
[15] STUK Radiation and Nuclear Safety Authority. Accessed May 20, 2016
[16] ACR 2008. *American College of Radiology 2008* Reston, Va: American College of Radiology 1143-1149.
[17] IAEA 2010 *International Atomic Energy Agency* 2010 IAEA-TECDOC-1641 Report, Vienna
[18] AAPM 1991 American Association of Physics in Medicine 1991 Med. Phys. 19 231–241
[19] Rivera T. 2012 *Appl Radiat Isot* **71** (S) 30–34
[20] Salazar G and Abujudeh H 2012 *American Journal of Roentgenology* **199** W441-W443
[21] ISEMIR *International Atomic Energy Agency*: Accessed May 20, 2016
[22] Meyer P, Regal R, Jung M, et al. 2001 *Med. Phys.* **28** 2002-6.
[23] Bower M W, Hintenlang D E 1998 *Heath. Phys.* **75** 197-204
[24] Hugtenburg RP, Johnston K, Chalmers GJ, Beddoe AH. 2001 *Phys. Med. Biol.* **46** 2489-501
[25] Bordier C, Klausz R, Desponds L. 2015 *Radiat Prot Dosim* **163** (3) 306-18.
[26] Del Sol S, García R, Sánchez D, Ramírez G, Gaona E, de León M A and Rivera T 2016 *Appl. Radiat. Isot.* **107** 340-345
[27] Del Sol S, García R, Guzmán J, Sánchez D, Ramírez G, Gaona E and Rivera T 2016 *Appl. Radiat. Isot.* **111** 50-55
[28] Rivera T 2016 *Appl. Radiat. Isot.* (In press). *http: //dx.doi.org/10.1016/j.apradiso.2016.03.008i*
[29] Elona I.A., and Morales A.A. Jr. 2015 *World Congress on Medical Physics and Biomedical Engineering, June 7-12, 2015, Toronto, Canada*, (D.A. Jaffray ed.)
[30] Jursinic P A 2007 *Med. Phys.* **34** 4594-4604
[31] Zoetelief, J. et al., 2003. *Procc. Int. Symp. Vienna*: Standards and Codes of Practice in Medical radiation Dosimetry Vienna
[32] Reeves R., Mahmud 2016 *J Am Coll Cardiol Intv.* **9**(3):301-302.
[33] Rehani M.M 2015 *AQD. Br J. Radiol.* **88** 20140344
[34] IAEA 2016 *International Atomic Energy Agency:.* Accessed 22 May 2016
[35] Rehani M.M., Frush D.P. 2011 *Radiat. Prot. Dosim.*, **147** (1–2) 314–316
[36] Renani 2012. Rehani, M.M., Frush, D.P., Berris, T., Einstein, A.J. 2012. IAEA survey, **81** 10 968–976
[37] IAEA 2014. *International Atomic Energy Agency. International.* IAEA Safety Series 115, Vienna (2014)

IOP Conf. Series: Journal of Physics: Conf. Series **792** (2017) 012099 doi:10.1088/1742-6596/792/1/012099

Diffusion in monodisperse ferrofluids

R Peredo-Ortíz[1], M Hernández-Contreras[1] and R Hernández Gómez[2]

[1] Departamento de Física, Centro de Investigación y Estudios Avanzados del Instituto Politécnico Nacional, Apartado Postal 14-740, México Distrito Federal, México.
[2] Departamento de Computación, Centro de Investigación y Estudios Avanzados del Instituto Politécnico Nacional, Apartado Postal 14-740, México Distrito Federal, México.

E-mail: marther@fis.cinvestav.mx

Abstract. The diffusion coefficients characterizing the translational and rotational Brownian motion of a particle in a concentrated suspension were determined in the long time diffusive regime for a ferrofluid suspension with the use of a Langevin equation approach. These dynamical properties depend on the equilibrium micro-structural information of the suspension and take into account the effect of direct anisotropic inter-particle's interactions on self-diffusion. The comparison of this theory with Brownian dynamic simulations results is made in terms of colloid density and dipole interaction strength.

1. Introduction

Nowdays, the effect of the direct anisotropic interactions among magnetic nanoparticles and external magnetic field gradients, on their collective diffusion is currently investigated with forced Rayleigh scattering techniques [1] and dynamic light scattering [2] in ferrofluid suspensions of up to 10% volume fractions. Yet, the non-direct long range hydrodynamic interaction (HI) in the system is experimentally measured by means of x-ray correlation spectroscopy [3] and with neutron spin-echo experiments [4,5,6]. Whereas, the description of the collective diffusion has been accomplished with gradient diffusion approach on concentrated suspensions [7,8], and through the mean field kinetic Smoluchowski equation for low and moderate volume fractions [9]. However, there is a clear absence of experimental studies of the rotational and translational self- and tracer diffusion coefficients in mono- and poli-disperse ferrofluids which remains to be performed in the already well characterized experimental systems. The above cited theories have been useful to describe the translational self-diffusion of Brownian dynamic simulations were dissipative HI was ignored for model systems of homogeneous moderately concentrated suspensions [9]. Thus, due to the lack of HI there are no cross correlation of distinct particle's velocities and therefore the collective gradient diffusion coefficient coincides with the self-diffusion expression [10,9]. The simulation studies of reference [9] confirmed good agreement with

[1] To whom any correspondence should be addressed.

Content from this work may be used under the terms of the Creative Commons Attribution 3.0 licence. Any further distribution of this work must maintain attribution to the author(s) and the title of the work, journal citation and DOI.
Published under licence by IOP Publishing Ltd

gradient diffusion theory. In the present paper we show that a Langevin equation description of the translational and rotational Brownian dynamics (BD) of a colloidal particle in concentrated ferrofluid suspensions without HI provides, in the long time diffusive regime the translational self-diffusion coefficient which yields good quantitative prediction as compared to BD results. We provide expressions both for the self- translational and rotational coefficients which depend in the equilibrium microstructural information of the fluid through the anisotropic pair correlation function. In ferrofluids the rotational diffusion of the particles can be measured with transient magnetic birefringence relaxation [11] and small-angle neutron-scattering [6].

2. Langevin equation of particle velocities and of their local concentration

The colloidal suspension contained in a volume V consist of a carrier fluid plus the monodisperse system of N particles of equal mass M and moment of inertia I, where each particle can be either of spherical or rod-like shape having an embedded optical anisotropy or a permanent dipolar moment. The orientation of a particle's main axis of symmetry is defined with respect to tracer's body fixed frame. Thus, another particle's orientation will be denoted in this frame by the polar angles of its orientation $\Omega_i = (\theta_i, \phi_i)$ for $i = 1, \ldots, N$. Then the total potential energy of the system is $U = \sum_{i,j}^{N} \psi_{ij}/2$ where it is assumed that ψ_{ij} is the direct pairwise interaction potential between particles. Notice that we will not consider in this paper the contribution of the non-additive HI among particles. The equation of motion governing the particle translational $\boldsymbol{v}(t)$ and angular $\boldsymbol{\omega}(t)$ velocity of its rotation can be written as

$$M \cdot \frac{d\mathbf{V}(t)}{dt} = -\varsigma^0 \cdot \mathbf{V}(t) + \mathbf{f}(t) + \int d\mathbf{r} d\Omega[\text{"} \psi(\mathbf{r},\Omega)]\delta n(\mathbf{r},\Omega,t) \quad (1)$$

where the dynamical fluctuation in concentration about its equilibrium value is $\delta n(\mathbf{r},\Omega,t) = n(\mathbf{r},\Omega,t) - n^{eq}(\mathbf{r},\Omega)$ since the equilibrium concentration n^{eq} does not contribute to the net force and torque component. The third term in the Langevin equation above represents the direct forces and torques on the diffusing particle by the host colloidal particles around it. We defined the generalized velocity $\mathbf{V}(t) = (\mathbf{V}, \omega)$, and mass $\mathbf{M}_{ij} = M\delta_{ij}$ $(i, j = 1,2,3)$, $\delta_{ij} I_{i-3}$ with I_{i-3} the principal moments of inertia of the particle ($i, j = 4,5,6$). The free particle's friction coefficient is diagonal with elements $\varsigma^0{}_{11} = \varsigma^0{}_{22} = \varsigma^0{}_{\perp}$, $\varsigma^0{}_{33} = \varsigma^0$, $\varsigma^0{}_{44} = \varsigma^0{}_{55} = \varsigma^0{}_{R}$, and $\varsigma^0{}_{66} = 0$, which are external inputs to the theory. The general random term $f(t)$ represents the force and torque imparted on the particle by its collision with the solvent molecules on its surface. The Langevin equation above is coupled to the fluctuation in concentration whose evolution equation can be written in its most general form as

$$\frac{\partial \delta n(\mathbf{r},\Omega,t)}{\partial t} = [\nabla n^{eq}(\mathbf{r},\Omega)] \cdot \mathbf{V}(t) - \int_0^t dt' d\mathbf{r}' d\Omega' d\mathbf{r}'' d\Omega'' L(\mathbf{r},\Omega;\mathbf{r}',\Omega',t-t') \times$$

$$\sigma^{-1}(\mathbf{r}',\Omega';\mathbf{r}'',\Omega'')\delta n(\mathbf{r}'',\Omega'';\mathbf{r}',\Omega',t') + h(\mathbf{r},\Omega,t), \quad (2)$$

where σ^{-1} is the inverse of the static correlation function $\sigma(\mathbf{r},\Omega;\mathbf{r}',\Omega') := \langle \delta n(\mathbf{r},\Omega;0)\delta n(\mathbf{r}',\Omega';0)\rangle$ [12] which will be denoted simply as σ and similarly for the other quantities $\delta n(t), \chi(t), n^{eq}, \psi, L(t), h(t)$. Thus, it fulfills the identity

$$\int d\mathbf{r}''\Omega''\sigma(\mathbf{r},\Omega;\mathbf{r}'',\Omega'')\sigma^{-1}(\mathbf{r}'',\Omega'';\mathbf{r}',\Omega') = \delta(\mathbf{r}-\mathbf{r}')\delta(\Omega-\Omega'). \tag{3}$$

In equation (2) L is an Onsager coefficient that satisfies the fluctuation-dissipation relation with the stationary random term of zero mean $\langle h(\mathbf{r},\Omega;t)h(\mathbf{r}',\Omega';t')\rangle = L(\mathbf{r},\Omega;\mathbf{r}',\Omega';t-t')$.

3. Effective friction function

The formal solution of equation (2) is known and given in references [13,14] as

$$\delta n(t) = \chi(t)\ \ \delta n(0) + \int_0^t dt'\chi(t')\ \ [\nabla n^{eq}]\cdot V(t') + \int_0^t dt'\chi(t\ \ t') - h(t'), \tag{4}$$

where $= \int d\mathbf{r}d\Omega$, and the propagator χ satisfies

$$\frac{\partial\chi(t)}{\partial t} = -\int_0^t dt'L(t-t')\ \ \sigma^{-1}\ \ \chi(t'), \tag{5}$$

with initial condition $\chi(t=0) = \delta(\mathbf{r}-\mathbf{r}')\delta(\Omega-\Omega')$. Substituting equation (4) in (1), we obtain the effective Langevin equation of the particle

$$M\cdot\frac{d\mathbf{V}(t)}{dt} = -\zeta^0\cdot\mathbf{V}(t) + \mathbf{f}(t) - \int_0^t dt'\Delta\zeta(t-t')\cdot\mathbf{V}(t)\ \ \mathbf{F}(t). \tag{6}$$

Where the dynamical friction function can be expressed in different equivalent forms with the use of the Wertheim-Lovett equation $\beta\nabla\psi = -\sigma^{-1}\ \ \nabla n^{eq}$ [15],

$$\begin{aligned}\Delta\varsigma(t) &= \frac{1}{\Omega^2}\beta[\nabla\psi]\ \ \chi(t)\ \ \sigma\ \ [\nabla\psi]^\dagger \\ &= \frac{1}{\Omega^2}\beta[\nabla\psi]\ \ C(t)\ \ [\nabla\psi]^\dagger \\ &= \frac{k_BT}{\Omega^2}[\nabla n^{eq}]\ \ \sigma^{-1}\ \ \chi(t)\ \ [\nabla n^{eq}]^\dagger.\end{aligned} \tag{7}$$

Since the observable $\Delta\varsigma(t)$ is related to the measured mean squared displacement, it's average over the solid $\Omega = 4\pi$ is performed. Here $\beta = 1/k_BT$ with k_B the Boltzmann constant, T the temperature, and $F(t)$ satisfying the fluctuation-dissipation relationship $\langle F(t)F^\dagger(0)\rangle = k_BT\Delta\varsigma(t)$. The general expressions for the friction function in (7) can also be used to study tracer diffusion on rod-shaped particles suspensions [16]. Such as fd viruses on which the experimental techniques of birefringence, and forced Rayleigh scattering measure the diffusion properties. On the other hand, the propagator $C(t) = \chi(t)\ \ \sigma := \langle\delta n(t)\delta n(0)\rangle$ governs the collective relaxation of the particle's configuration variables ($\mathbf{r},\Omega$) due to thermal fluctuations, and it has initial condition $C(0) = \sigma := \langle\delta n(0)\delta n(0)\rangle$. The equation of $C(t)$ is obtained from (5) yielding

$$\frac{\partial C(\mathbf{r},\Omega,\Omega',\Omega_r;t)}{\partial t} = -\rho[D^{*0}\nabla^2 + D_R^{*0}\nabla_\Omega^2]\int dt'd\mathbf{r}'d\Omega''\sigma^{-1}(\mathbf{r}-\mathbf{r}',\Omega,\Omega'',\Omega_{\mathbf{r}-\mathbf{r}'})\times C(\mathbf{r}''-\mathbf{r}',\Omega'',\Omega',\Omega_{\mathbf{r}''-\mathbf{r}'};t-t'). \tag{8}$$

4. Monodisperse ferrofluid

For a hard ferrofluid suspension made of spherical particles with permanent magnetic moment of magnitude, μ, we obtain a hard sphere (HS) diameter d with a Lennard-Jones short range (sr) repulsive interaction. Thus, the total pair direct interaction is $\psi_{12} = u_{sr} + \phi_{dd}$

$$u_{sr} = 4\varepsilon_0\left(\left(\frac{d}{r}\right)^{12} - \left(\frac{d}{r}\right)^{6}\right), r \quad 2.5d \tag{9}$$

$1/\beta\varepsilon_0 = 1.35$, Plus the dipolar interaction $\phi_{dd} = -\mu_0\mu^2 D(\Omega_1\Omega_2,\Omega_r)/4\pi r^3$. μ_0 is the magnetic permeability of vacuum. $D(\Omega_1,\Omega_2,\Omega_r) := 3(\mathbf{r}_{12}\cdot\mathbf{u}_1)(\mathbf{r}_{12}\cdot\mathbf{u}_2) - (\mathbf{u}_1\cdot\mathbf{u}_2)$, Where $\mathbf{r} = \mathbf{r}/r$ is a unitary vector along vector $\mathbf{r}$ and with orientation Ω_r. $\mathbf{u} = \mathbf{u}(\theta,\phi)$ is a unitary vector of dipole orientation. At equilibrium, the bulk profile concentration of particles around the tracer $n^{eq} = \rho g(\mathbf{r},\Omega_1,\Omega_2)$ is proportional to the pair correlation function which for a dipolar fluid is provided by its first three terms in a series expansion of spherical harmonics. Here $\rho = N/V$ is the number density of particles and $g(\mathbf{r},\Omega_1,\Omega_2)$ has the fixed space rotational invariant expansion of Wertheim [17,18]

$$g(\mathbf{r},\Omega_1,\Omega_2) = g(r) + h_\Delta(r)\Delta + h_D(r)D, \tag{10}$$

with $\Delta := \mathbf{u}_1 \cdot \mathbf{u}_2$. There is an equivalent rotational invariant expansion of (10) given in [19]

$$f(\mathbf{r},\Omega,\Omega') = (4\pi)^{3/2}\sum_{mnl}\frac{f^{mnl}(r)}{\sqrt{(2l+1)}}\sum_{\mu\nu\lambda}\begin{pmatrix} m & n & l \\ \mu & \nu & \lambda \end{pmatrix} Y_{m\mu}(\Omega)Y_{n\nu}(\Omega')Y_{l\lambda}(\Omega_r), \tag{11}$$

With the identification $g^{000} = g(r), h^{110} = -h_\Delta(r)/\sqrt{3}, h^{112} = h_D(r)\sqrt{10/3}$ [20]. We determined these micro-structural properties using the reaction field method as boundary condition [21, 22] in the Brownian dynamic simulation [16]. The vector positions were decomposed into its parallel (||) and perpendicular(⊥) projections along the particle's axis of symmetry (in this case $\mathbf{u}_i$), $\mathbf{r} = \mathbf{r}^{\parallel} + \mathbf{r}^{\perp}$ where for particle i is $\mathbf{r}_i^{\parallel} = [\mathbf{u}_i \cdot \mathbf{r}_i]\mathbf{u}_i$, therefore [16]

$$\mathbf{r}_i^{\parallel}(t+\Delta t) = \mathbf{r}_i^{\parallel}(t) + \frac{\Delta t}{\zeta^0}\mathbf{F}_i^{\parallel}(t) + \Delta r^{\parallel}\mathbf{u}_i(t)$$

$$\mathbf{r}_i^{\perp}(t+\Delta t) = \mathbf{r}_i^{\perp}(t) + \frac{\Delta t}{\zeta^0}\mathbf{F}_i^{\perp}(t) + \Delta r_1^{\perp}\mathbf{e}_{i1}(t) + \Delta r_2^{\perp}\mathbf{e}_{i2}(t). \tag{12}$$

Here $\mathbf{F}_i = -\nabla U$, $r^{\parallel}$, $r_1^{\perp}$, $r_2^{\perp}$ are three uncorrelated Gaussian random numbers of variance $2\Delta t/\zeta^0$ and $\mathbf{e}_{i1}(t)$, $\mathbf{e}_{i2}(t)$ two orthogonal unit vectors which are perpendicular to $\mathbf{u}_i(t)$,

$$\mathbf{u}_i(t+\Delta t)=\mathbf{u}_i(t)+\frac{\Delta t}{\zeta_R^0}\mathbf{T}_i(t)\quad \mathbf{u}_i(t)+x_1\mathbf{e}_{i1}(t)+x_2\mathbf{e}_{i2}(t). \tag{13}$$

$\mathbf{T}_i=-L_{\mathbf{u}}U$. The angular operator $L_{\mathbf{u}}=\mathbf{u}\times\partial/\partial\mathbf{u}$. x_1, x_2 are two uncorrelated Gaussian random numbers of variance $2\Delta t/\zeta_R^0$. We used for $\zeta^0=3\pi\eta d$, $\zeta_R^0=\pi\eta d^3$ and $N=125$ particles. To equilibrate the system 400000 time steps of size $\Delta t^*:=\Delta t/t_0=5\times10^{-4}$, $t_0=3\pi\eta d^3/k_BT$ are performed. After this, 3 million more new particle's configurations were generated and used 30000 configurations for statistics to get the micro-structural functions of figure 1(a), and dynamical properties such as the long time and time-dependent translational and rotational self-diffusion. Using the definitions [23]

$$g(r)=\frac{\langle\sum_{i,j(i\neq j)}\delta(r-|\mathbf{r}_{ij}|)\rangle}{N4\pi\rho^* r^2},$$

$$h(r)=3\frac{\langle\sum_{i,j(i\neq j)}\delta(r-|\mathbf{r}_{ij}|)\mathbf{u}_i\cdot\mathbf{u}_j\rangle}{N4\pi\rho^* r^2},$$

$$h_D(r)=\frac{3}{2}\frac{\langle\sum_{i,j(i\neq j)}\delta(r-|\mathbf{r}_{ij}|)[3\mathbf{u}_i\cdot\mathbf{r}_{ij}\mathbf{u}_j\cdot\mathbf{r}_{ij}-\mathbf{u}_i\cdot\mathbf{u}_j]\rangle}{N4\pi\rho^* r^2}. \tag{14}$$

We determined typical correlation functions as displayed in figure 1(a) for a dimensionless number density $\rho^*=nd^3=.02$ and reduced dipolar strength $\mu^{*2}=\mu^2\mu_0/4\pi k_BTd^3=2.75$. The term $g(r)$ is calculated as the usual isotropic pair correlation function [21]. By taking the Fourier-Bessel $f^{mnl}(k)=4\pi i^l\int_0^\infty r^2 j_l(kr)f^{mnl}(r)dr$ [22] with j_l being the spherical Bessel function, and Laplace transform $C_{,\alpha}^{mnl}(k,w)=\int_0^\infty dte^{iwt}C_{,\alpha}^{mnl}(k,t)$ of (8) we find

$$[-iwC_{,\alpha}^{mnl}(k,w)-C_{,\alpha}^{mnl}(k,w=0)]=-\rho4\pi(2l+1)[D^{*0}k^2+D_R^{*0}m(m+1)]\times\sum_{n_1}(-1)^\alpha C_{,\alpha}^{n_1n}(k,w)[\sigma^{-1}]_{,\alpha}^{mn_1}, \tag{15}$$

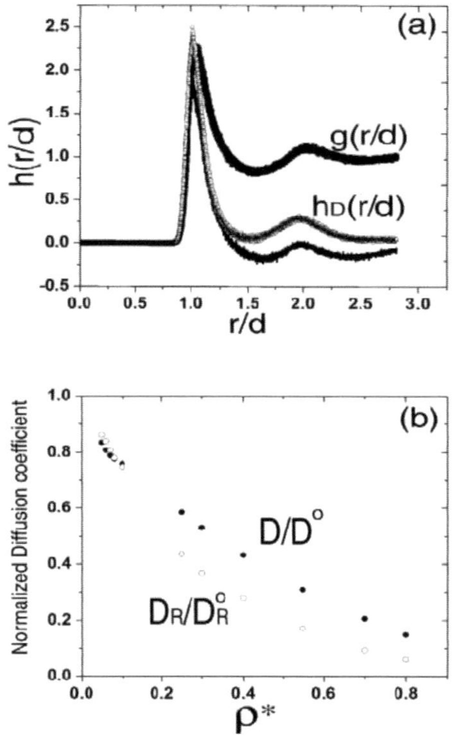

Figure 1. Figure (a) gives $g(r)$ (black dots •) $h_D(r)$ (open circles °) and bottom curve for $h_\Delta(r)$ (continuous line) from BD simulations of equations (12-13) for the model ferrofluid of soft Lennard-Jones short range repulsion and dipolar interaction with $\rho^* = .02$ and $\mu^{*2} = 2.75$. Notice that all functions are different from zero for r/d â 1, which fixes the lower limit in the radial dependence of equations (20-23). Figure (b) depicts the translational D/D^0 (symbol •) and rotational D_R/D_R^0 (symbol ° self-diffusion coefficients calculated from theory (20-23) for $\mu^{*2} = 2.75$ and as a function of reduced density ρ^*. No HI's are included in the calculation on the properties of (b).

Where $i = \sqrt{(-1)}$ and $C^{mnl}_{,\alpha}(k,w) = \sum_{l=|m-n|}^{m+n} \begin{pmatrix} m & n & l \\ \alpha & -\alpha & 0 \end{pmatrix} C^{mnl}(k,w)$ [22]. Using the above values of $mnl = 000, 110, 112$, then for dipolar liquids $\alpha = 0, \pm 1$ [19], and using the approximation $D^{*0} = 2D^0$, $D_R^{*0} = 2D_R^0$ [14], we get for equation (15)

$$C^{11}_{,\alpha}(k,w) = \frac{[\sigma(k)]^{11}_{,\alpha}}{-iw + \rho 4\pi(-1)^{\alpha}(D^{*0}k^2 \;\; 2D_R^{*0})[\sigma^{-1}(k)]^{11}_{,\alpha}}. \tag{16}$$

Moreover, the inverse relation holds

$$C^{mnl}(k,w) = (2l+1) \sum_{\alpha=-inf(m,n)}^{inf(m,n)} \begin{pmatrix} m & n & l \\ \alpha & -\alpha & 0 \end{pmatrix} C^{mnl}_{,\alpha}(k,w). \tag{17}$$

Thus

$$C^{000}(k,w) = C^{00}_{,0}(k,w)$$
$$C^{110}(k,w) = \frac{1}{\sqrt{(3)}}[2C^{11}_{,1}(k,w) - C^{11}_{,0}(k,w)]$$

$$C^{112}(k,w) = \frac{10}{\sqrt{(30)}}[C^{11}_{,1}(k,w) + C^{11}_{,0}(k,w)]. \tag{18}$$

The structure factor $S = \sigma / \rho$ and total correlation function h are related by

$$\sigma^{mn}_{,\alpha}(k) = \rho S^{mn}_{,\alpha}(k) = \rho[1 + (\ 1)^{\alpha} h^{mn}_{,\alpha}(k)],$$

$$S^{00}_{,0}(k) = 1 + \frac{\rho^*}{d^3} h^{000}(k),$$
$$S^{11}_{,0}(k) = 1 - \frac{\rho^*}{d^3}\left(\frac{1}{\sqrt{(3)}} h^{110}(k) - \frac{2}{\sqrt{(30)}} h^{112}\right),$$
$$S^{11}_{,\pm 1}(k) = 1 - \frac{\rho^*}{d^3}\left(\frac{1}{\sqrt{(3)}} h^{110}(k) \quad \frac{1}{\sqrt{(30)}} h^{112}\right), \tag{19}$$

with $h^{000} = g^{000} - 1$. Therefore from the Laplace transform of the second identity of (7) we attain in the long time limit $w = 0$ [14]

$$\Delta\zeta_{\gamma}(w=0) = \Delta\zeta^{HS} + \quad \zeta_{\gamma}, \tag{20}$$

for $\gamma = \perp,$, where [24]

$$\Delta\zeta^{HS} = \frac{\zeta^0}{36\pi\phi}\int_0^{\infty} dx x^2 \frac{[S^{00}_{,0}(x) - 1]^2}{1 + S^{00}_{,0}(x)}. \tag{21}$$

Notice that in the above equation (23) the first term is the hard sphere contribution [24] that results from the third form of $\Delta\zeta$ in equation (7) which depends on the gradient of particle concentration ∇n^{eq}. This first contribution was used here instead of the Lennard-Jones short range contribution to the friction that would is attained from the second identity of (7) that depends on ∇u_{sr}. The second part $\Delta\zeta_{\gamma}$ results from the last identity of (7). Here $x := k\sigma$. Whereas the contribution to the translational friction, in the direction perpendicular to the main axis of symmetry of the particle, due to direct interactions is [14]

$$\Delta\zeta_{\perp} = \frac{96}{5}\frac{\zeta^0\phi(\mu^{*2})^2}{\Omega^2}\int_0^{\infty} dx \frac{j_2(x)^2}{\left(x^2 + 2\frac{D_R^{*0}}{D^{*0}}d^2\right)} \times [6(S^{11}_{,1}(x))^2 \quad 9(S^{11}_{,0}(x))^2]. \tag{22}$$

On the other hand, the component of the friction parallel to the main body axis is $\Delta\zeta \ = \frac{4}{3}\Delta\zeta_{\perp}$.

Moreover, for the rotational friction we get

$$\Delta\zeta_R = \frac{96}{105}\frac{\zeta_R^0\phi(\mu^{*2})^2}{\Omega}\int_0^{\infty} dx \frac{j_1(x)^2}{\left(x^2 + 2\frac{D_R^{*0}}{D^{*0}}d^2\right)} \times [67(S^{11}_{,1}(x))^2 \quad 38(S^{11}_{,0}(x))^2]. \tag{23}$$

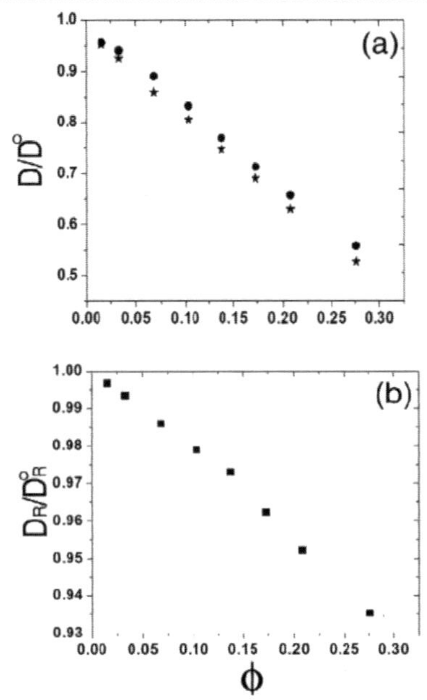

Figure 2. Figure (a) is the translational self-diffusion coefficient of the ferrofluid as a function of particle's volume fraction ϕ. Comparison is made between theory prediction of equations (20-22) (symbol •) and BD simulations taken from reference [9] (symbol A) for dipole magnitude $\mu^{*2} = 0.58$. Figure (b) describes the rotational diffusion coefficient that results from equation (23). There is no HI among particles included.

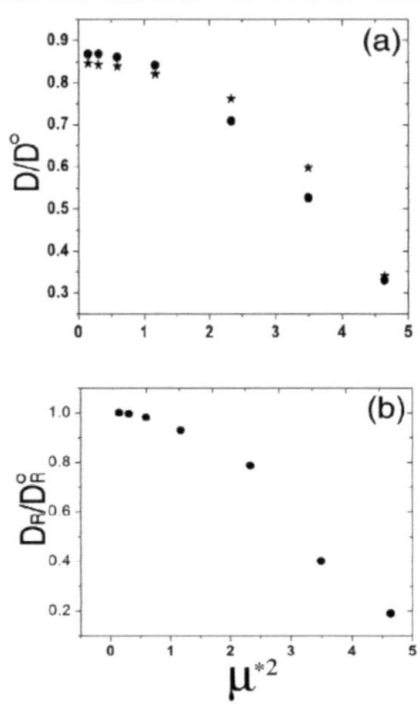

Figure 3. Figure (a) is the calculated normalized translational diffusion coefficient from equations (20-23) (symbol •), and BD results are from reference [9] (symbol A), as a function of dipole moment μ^{*2} for fixed volume fraction $\phi = 0.05/0.58$. Corresponding rotational diffusion from equation (23) is depicted in (b).

Thus, the components of the self-diffusion properties are $D_\gamma / D_\gamma^0 = (1 + \Delta\zeta_\gamma(\omega \to 0)/\zeta_\gamma^0)^{-1}$, $\gamma = \parallel, \perp, R, \zeta^0 = \zeta_\perp^0 = \zeta^0$, and the average total translational and rotational diffusion coefficient are, respectively, $D = (2D_\perp + D_\parallel)/3$, D_R / D_R^0 and $D^0 = k_B T / \zeta^0$. Figure 1(b) is the plot of the average translational property D / D^0 (black dots), and rotational self-diffusion D_R / D_R^0 (open circles) obtained from the theory, equations (20-23), as a function of ρ^* for $\mu^{*2} = 2.75$. In general, both translational and rotational diffusion of the particle diminish by an increase in ferrofluid density. In figure 2(a) we compared theory (20-22) for D / D^0 (symbol •) versus the Brownian dynamics results of reference [9] denoted with A, as function of volume fraction and fixed dipolar strength $\mu^{*2} = 0.58$. Notice that in reference [9] the reduced dipolar strength is denoted as $\lambda = \mu_0 \mu^2 / 4\pi k_B T d_c^3$ with $d_c = 1.2d$, and its value is related to ours $\mu^{*2} = \lambda \times 0.58$. Thus, they used the value $\lambda = 1$. The fraction volume definition used in reference [9] is $\rho d_c^3\pi/6$ which is related to our expression for this quantity as $\phi=(\rho d_c^3\pi/6)/0.58$. Figure 2(b) is the theoretical prediction of (23) for the rotational diffusion. Figure 3(a) provides the comparison of theory (20-22) (symbol •) and Brownian dynamic simulation (symbol A) of reference [9] for translational diffusion at volume fraction $\phi=0.05/0.58$ where the value of 0.05 was used by authors of reference [9]. We plotted this property as function of dipolar strength $\mu^{*2}=\lambda\times 0.58$. Whereas figure 3(b) is the corresponding theory, results for D_R/D_R^0. We consider now a ferrofluid under the influence of a constant external magnetic field $\mathbf{H}$. Therefore, the energy of the system is. Figure 4 depicts the time-dependent self-diffusion coefficients results from Brownian dynamic simulation only at volume fraction $\phi=0.05/0.58$ versus reduced time t/t_0 and three magnitudes of particle's dipole moment μ^{*2}, and under applied magnetic field in dimensionless units of $H^*=(\mu_0\mu H/k_BT)=4$, where $H=|\mathbf{H}|$. The field orientation is along the Z axis. For translational diffusion, we determined the coefficient as $D(t)=(1/N)\sum_{i=1}^{N}\langle[\mathbf{r}_i(t)-\mathbf{r}_i(0)]^2\rangle/6t$, where the component $D(t)_\perp=D(t)_{xx},D(t)_{yy}$ [25] refers to diffusion parallel to axis X and Y but perpendicular to the diffusion along the main particle symmetry axis which is along Z [9], direction about which we defined the component $D=D_{zz}$. For the rotational motion, we used $D^r(t)=-\ln[\langle\mathbf{u}_{i=1}(t)\cdot\mathbf{u}_{i=1}(0)\rangle]/2t$ [16]. Figure 4(a) shows that at the lowest dipole strength $\mu^{*2}=0.14$, and $H^*=0$ the short time diffusion limit presents a plateau in the diffusion coefficient (upper continuous line) where it remains constant and then drops continuously due to direct inter-particle's interactions that restrain its movement. For $H^*=4$ the short time section of the diffusion, curve lowers (dotted curve) and then converges for longer times to the zero field case above. For a slightly larger dipole, strength of $\mu^{*2}=0.59$ either with field (dot-dash line) or not (dash line) both curves almost coincide in general. For the highest dipolar case of $\mu^{*2}=4.64$, in contrast to the lowest dipole moment case before, the field now increases, in general, the average translational diffusion $D(t)/D^0$ at all times (second curve from bottom up continuous line curve) as compared to the case of the zero field homogeneous suspension (bottom curve thick black line). Figure 4(b) depicts the rotational diffusion without magnetic field at the three dipole moments discussed before. In this plot and for the two curves with corresponding $\mu^{*2}=0.14,0.59$ values the diffusion coefficient $D_R(t)$ is given correctly only in the small time interval $0\le t/t_0\le 20$ where the mean square displacement is well defined whereas for larger times is incorrectly provided by the remaining part of those two curves. At the highest dipole moment (bottom line) the rotational diffusion lowers as

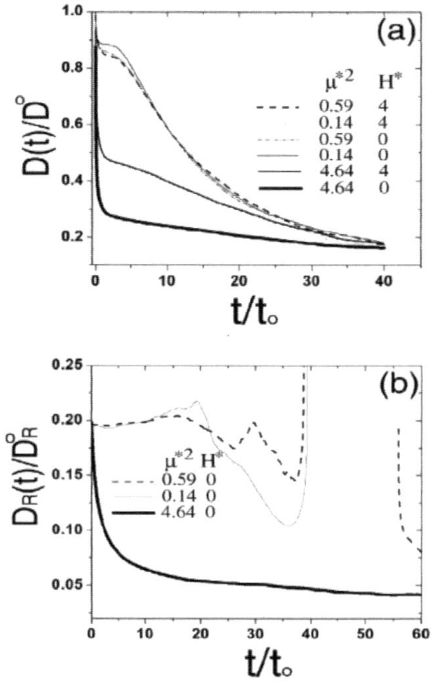

Figure 4. Figure (a) is the BD simulation results from equations (12-13) for the time-dependent translational self-diffusion coefficient versus reduced time t/t_0 at volume fraction $\phi = 0.05/0.58$ for five values of reduced dipole moment μ^{*2} with and without magnetic field H^*. Figure (b) yields the corresponding time-dependent rotational diffusion of the particle.

A function of time due to the accumulative effect of inter- particle's interactions that lowers the rotational movement of the particle. It should be notice that similar time-dependent translational diffusion properties have been measured in a super-paramagnetic colloid confined to a surface [26], and through BD simulations in suspensions of hard spherocylinders [25].

5. Conclusions

A Langevin equation theory allowed us the determination of the friction contribution due to direct inter-particle's interactions that experience a colloid particle during its diffusion in a concentrated mono-disperse suspension. In the long time diffusive limit the resulting expressions of the particle self-diffusion coefficient of its translation and rotational movement shows a good comparison with Brownian dynamic results of reference [9] as a function of colloid concentration and dipole strength of interaction. The expression we derived for the friction function apply to colloidal systems made of rod-shaped particles with anisotropic pairwise inter- particle's interactions. Its generalization for colloidal mixtures is straightforward. Future work incorporates the effect of external gradient magnetic field and the dissipative HI. These dynamical properties are feasible to be measured experimentally [1].

Acknowledgments

The authors acknowledge to the General Coordination of Information and Communications Technologies (CGSTIC) at CINVESTAV for providing HPC resources on the Hybrid Supercomputer " Xiuhcoatl" that have contributed to the research results reported within this paper.

References

[1] Bacri C, Cebers A, Bourdon A, Demouchy G, Heegaard B M, Kashevsky B and Perzynski R 1995 *Phys. Rev.* E **52** 3936
[2] Mertelj A, Cmok L and Copic M 2009 *Phys. Rev. E* **79** 041402
[3] Wagner J, Fischer B, Autenrieth T and Hempelmann R 2006 *J. Phys.: Condens. Matter* **18** S2697
[4] Gazeau F, Boué F, Dubois E and Perzynski R 2003 *J. Phys.: Condens. Matter* **15** S1305
[5] Gazeau F, Dubois E, Bacri J C, Boué F, Cebers A and Perzynski R 2002 *Phys. Rev.* E **65** 031403
[6] Mériguet G, Dubois E, Jardat M, Bourdon A, Demouchy G, Dupois V, Farago B, Perzynski R and Turq P 2006 *J. Phys. Condens. Matter* **18** S2685
[7] Morozov K I 1996 *Phys. Rev.* E **53** 3841
[8] Pshenichnikov A F, Elfimova E A and Ivanov A O 2011 *J. Chem. Phys.* **134** 184508
[9] Ilg P and Kroger M 2005 *Phys. Rev.* E **72** 031504
[10] Gomer R 1990 *Rep. Prog. Phys.* **53** 917
[11] Bacri J C, Perzynski R, Salin D and Servais J 1987 *J. Phys. (Paris)* **48** 1385
[12] Hansen J P and McDonald I R 1980 *The Theory of Simple Liquids* (London: Academic Press)
[13] Hernández-Contreras M, Medina-Noyola M and Vizcarra-Rendón A 1996 *Physica* A **234** 271
[14] Hernández-Contreras M and Ruíz-Estrada H 2003 *Phys. Rev.* E **68** 031202
[15] Gubbins K E 1980 *Chem. Phys. Lett.* **76** 329
[16] Löwen H 1994 *Phys. Rev.* E **50** 1232
[17] Wertheim M S 1971 *J. Chem. Phys.* **55** 4291
[18] Levesque D, Patey G N and Weis J J 1977 *Mol. Phys.* **34** 1077
[19] Blum L 1972 *J. Chem. Phys.* **75** 1862
[20] Gray C G, Gubbins K E and Joslin C G 2011 *Theory of Molecular Fluids* (USA: Oxford University Press)
[21] Allen M P and Tildesley D J 1989 *Computer Simulation of Liquids* (Clarendon Press)
[22] Fries P H and Patey G N 1985 *J. Chem. Phys.* **82** 429
[23] Weis J J and Levesque D 1993 *Phys. Rev.* E **48** 3728
[24] Nägele G, Medina-Noyola M, Klein R and Arauz-Lara J L 1988 *Physica* A **149** 123
[25] Kirchhoff T, Löwen H and Klein R 1996 *Phys. Rev.* E **53** 5011
[26] Kollmann M, Hund R, Rinn B, Nägele G, Zahn K, Krönig H, Maret G, Klein R and Dhont J K G 2002 *Europhys. Lett.* **58** 919

Effective potentials of Hofmeister salts in aqueous solution

M A Marquina-Carmona, M Hernández-Contreras[1]

[1] Departamento de Física, Centro de Investigación y Estudios Avanzados del Instituto Politécnico Nacional, Apartado Postal 14-740, México Distrito Federal, México.

E-mail: marther@fis.cinvestav.mx

Abstract. The effective pair interaction potentials of monovalent halide anions and alkali cations in bulk solution of a molecular solvent model were determined using atomistic simulations. These properties result from the microstructure of the ionic liquid which is the main ingredient in the hypernetted chain approach of liquid theory. We used an optimized set of Lennard-Jones parameters that are thermodynamically reliable for bulk phases of the Hofmeister salts.

1. Introduction

Experimental light scattering techniques allows the study of the microstructure of colloidal suspensions [1]. These properties are feasible to be determined either for synthetic particles of polystyrene or biological macromolecules in aqueous solvents containing electrolytes. The modeling of such systems to reproduce the observed microstructure is a major task to their basic comprehension and for their industrial processing. With the assumption of monoatomic bare particles pair interactions, it is possible to get agreement with the observed structural factor and of the dynamics of the colloids through computer simulations [2]. In the simulations, the particles can be considered to interact through Lennard-Jones potential that takes into account the repulsion at contact between them, and with a Van der Waals short distance attraction. Also, it is included their long-range electrostatic interactions [3]. However, the simulation of complex multicomponent systems of a vast number of particles is a challenging task. References [4,5] shows that a successful statistical mechanical interpretation of the experimental structural static properties using only the colloid's configurational variables is satisfactory if the no relevant ions and solvent variables are coarse grained from the statistical description. With this method, we considered Hofmeister electrolytes and determined effective pair interaction potentials of ions in bulk solution for different concentrations in a SPC/E molecular solvent model. Such potentials are a key ingredient in complementary theories to study the stability and interaction of surfaces of colloids, osmotic and activity coefficients of Hofmeister salts [6,7].

Content from this work may be used under the terms of the Creative Commons Attribution 3.0 licence. Any further distribution of this work must maintain attribution to the author(s) and the title of the work, journal citation and DOI.
Published under licence by IOP Publishing Ltd

Table 1. Gromacs forcefield OPLSAA/L was modified in reference [9] to have ion's Lennard-Jones parameters optimized for water model SPC/E. These parameters for the hydrogen of water are zero. They are constructed using the geometric rule such that the interaction of an ion with an oxygen molecule of water is given by $\sigma_{io} = \sqrt{\sigma_i \sigma_O}$, $\varepsilon_{io} = \sqrt{\varepsilon_i \varepsilon_O}$ [9].

Ion	$\sigma_i[10^{-7}cm]$	$\varepsilon_i[10^3 J/mol]$	$\sigma_{iO}[10^{-7}cm]$	$\varepsilon_{io}[10^3 J/mol]$
F^-	0.344	0.46	0.33	0.55
Cl^-	0.452	0.42	0.378	0.52
Br^-	0.505	0.21	0.40	0.37
I^-	0.571	0.16	0.425	0.32
Li^+	0.163	1.54	0.227	1.0
Na^+	0.222	1.54	0.265	1.0
K^+	0.278	1.54	0.297	1.0
C_s^+	0.330	1.54	0.323	1.0
	Model of water molecule		SPC/E	
O of water	0.3169	0.65		
H of water	0.0	0.0		

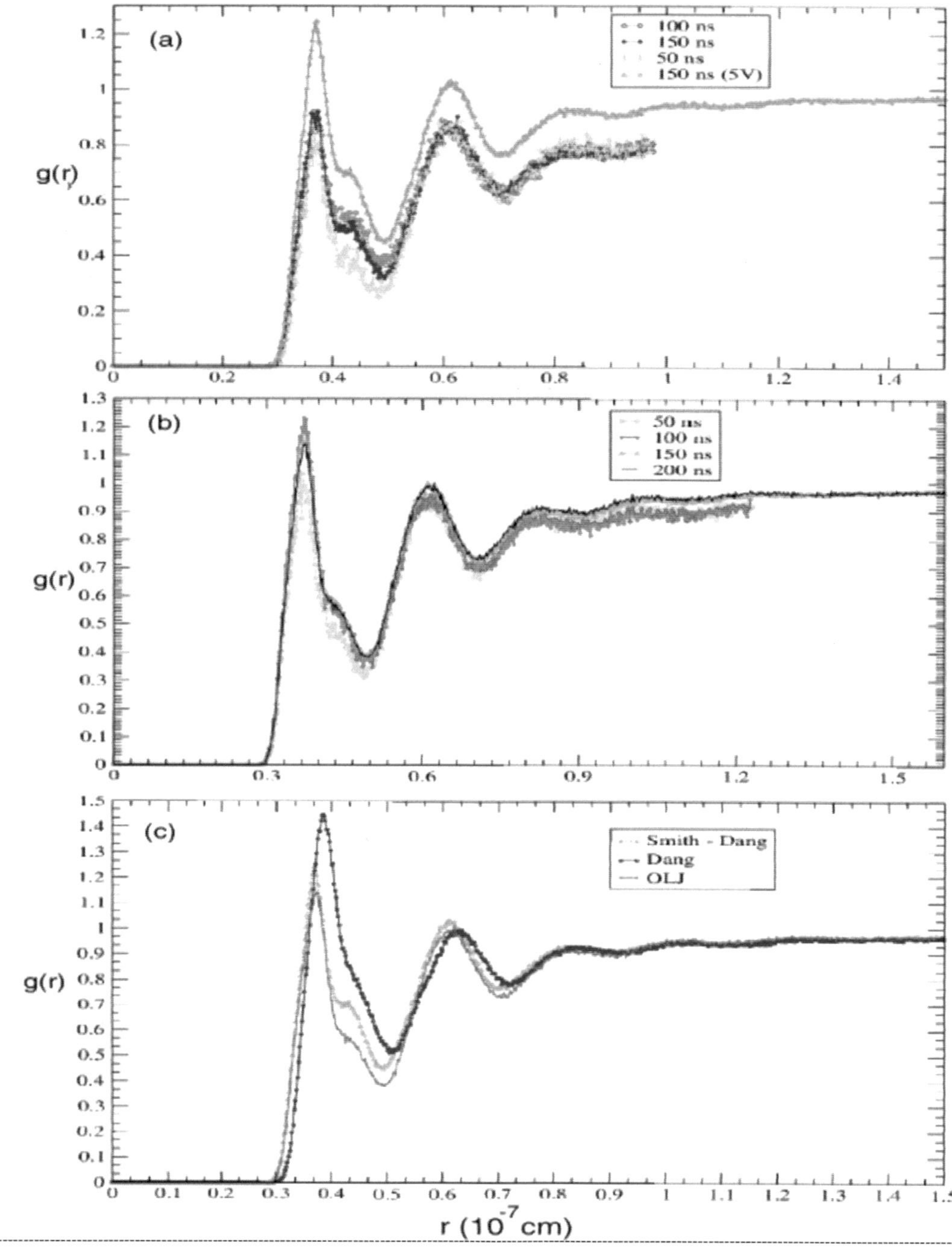

Figure 1. $g(r)$ of pair of Na-Na ions for NaCl at 0.87M and cell volume $V = 7.7154$ nm^3 for the three lower curves from bottom up in (a). Moreover, volume $5V$ yellow curve in (a) at different simulation times. In figure (b) the two curves from bottom up (green and purple) were calculated with

$2V = 15.4308$ nm³. The yellow and black curves correspond to $5V$. In figure (c) we compared the more realistic $g(r)$ obtained with the optimized Lennard-Jones parameters (OLJ) versus Smith and Dang, and Dang parameters for same cell volume of 3.3789 nm³. We used 1242 water molecules of SPC/E model and 20 pairs of NaCl.

2. Effects of size of the simulation box and Lennard-Jones (LJ) parameters on microstructure

We observed that the first maximum of the pair correlation function of ions $g(r)$ is sensitive to the LJ parameters used in the atomic force field, the system size considered and equilibration time, which it should not be the case for fixed molar concentration, pressure, and temperature conditions. In order to get corrected equilibrium thermodynamic properties independently of the biases produced by all these factors we first determined with Gromacs [8] the distribution functions for NaCl electrolyte solution using optimized LJ parameters [9] and compared with published results by other authors for the same system at 0.87 M concentration. This calculation corresponds to 248 water molecules within SPC/E (Extended Simple Point Charge Model. In all calculations we used only this solvent model [10]) and 4 salt pairs, with a cubic simulation cell size $L = 1.976$ nm and volume $V = 7.7154$ nm³. We use a cut of Coulomb and Van der Waals interactions of 0.9 nm and fixed pressure of 1 bar and a temperature of 300K. Lyubartsev et al. used the same values [4], and it is one of our reference for comparison. These authors took 0.2 ns of equilibration time for the pressure and temperature and 1 ns for further data statistical processing. Their Lennard-Jones parameters used were the same as those of Smith and Dang for ions [11] with $\sigma_{ij} = (\sigma_i + \sigma_j)/2$ energy parameter $\varepsilon_{Na} = 0.544kJ/M$, ion diameter for Na $\sigma_{Na} = 2.35$ Å, $\varepsilon_{Cl} = 0.419kJ/M$, ion diameter for Cl $\sigma_{Cl} = 4.4$ Å. Whereas in reference [12] the authors considered overall 50 ns of simulation with the first 5 ns of equilibration. Moreover, Kalcher and Dzubiella [13] performed among 150-200 ns of simulation with 5 ns of equilibration for the same system. In all cases the obtained correlation functions first maximum were different. For instance, our radial distribution of Na-Na has the first maximum at $r = 3.7$ Å with a high of 0.8 whereas Lyubartsev plot yields lower $g(r)$ at $r = 3.2$ Å with a larger amplitude of 2 [4]. However, our result is more consistent with Horinek *et al.* [9] calculations at 4.1 M concentration who used the same Lennard-Jones parameter values as ours of table 1. The interatomic potentials implemented in all the above mentioned molecular dynamics simulations are [3]

$$U(r_{ij}) = \frac{q_i q_j}{r_{ij}} + 4\varepsilon_{ij}\left[\left(\frac{\sigma_{ij}}{r_{ij}}\right)^{12} - \left(\frac{\sigma_{ij}}{r_{ij}}\right)^{6}\right], \varepsilon_{ij} \quad \sqrt{\varepsilon_i \varepsilon_j}. \qquad (1)$$

With q_i being the electric charge of ion of species i of diameter σ_i separated a distance r_{ij} from atom j. The first term of equation (1) is the long-range Coulomb pair interaction and the second Lennard-Jones term represents the repulsion at contact plus the van der Waals attraction of strength ε_{ij} at short distances of separation of particles. Figure 1 (a) depicts our simulations of this system for runs of 50 (green symbol, symbol + for 50 ns), 100 (purple color, symbol o) and 150 ns (blue color, symbol ⇌). We observed that as longer the simulation time is, the noisier the resulting correlation functions are, and they do not approach the value of one as it should be. On the other hand, we got just 0.6415 bar of equilibrium pressure and below of our 1 bar reference pressure fixed from the start of our run. We used the exponential Berendsen barostat. Because of the increase in five times of the original system volume (V), and with a simulation of 150 ns long (orange color plot, symbol ſ) a 1 bar of equilibration pressure we considered this to be the correctly determined $g(r)$ independent of system size. We checked that volumes as less as twice $2V = 15.4308\,\text{nm}^3$ from the original one leads to the same pair correlation function, see figure 1. We can also observe that the high of this correlation function is larger than the ones reported by other authors at the same electrolyte conditions but with different LJ parameters [4,12], see figure 1 (b). Therefore, we expect the thermodynamic properties to differ on the published data.

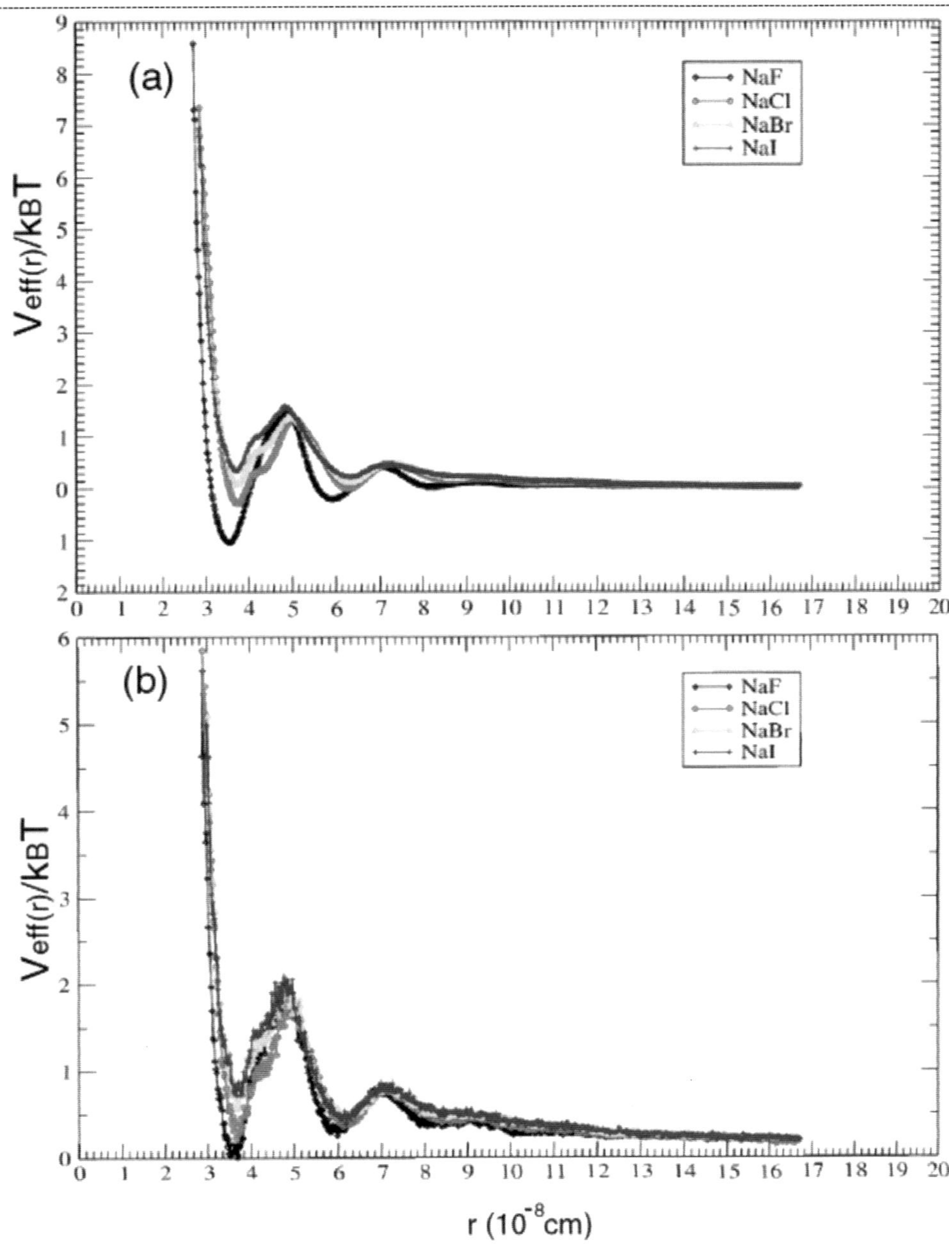

Figure 2. Effective potential of pair of cations Na-Na in presence of different anions at (a) 0.87M, (b) 0.2175M salt concentration.

2.1 Optimized Lennard-Jones parameters

Typically the parameters of the Lennard-Jones potential ε_{ij} and σ_{ij} are adjusted by fitting to the experimental thermodynamic property with molecular dynamic simulation. For instance, Smith and Dang reconstructed the potential of ion water in NaCl such that the binding enthalpy in vapor phase of small ion-water clusters coincides with the experimental value [11]. However, their results depend also on the water model employed. Therefore, the description of just a single ion may end up with very different parameter values depending on the solvent model. Horinek *et al.* [9] determined a set of optimized LJ parameters (OLJ, table 1) that reproduce both the ion solvation energy and entropy such that they minimize the error of fitting such experimental thermodynamic properties with simulations by using the SPC/E water model at low and moderate electrolyte concentrations. Because static properties such as the pair correlation function yields also effective pair ion potentials at finite salt concentrations, in addition to the osmotic coefficient and free energies that can be experimentally measured, in figure 1 (c) we provide the radial distribution function for NaCL. We calculated the correlation function with three set of LJ parameters.

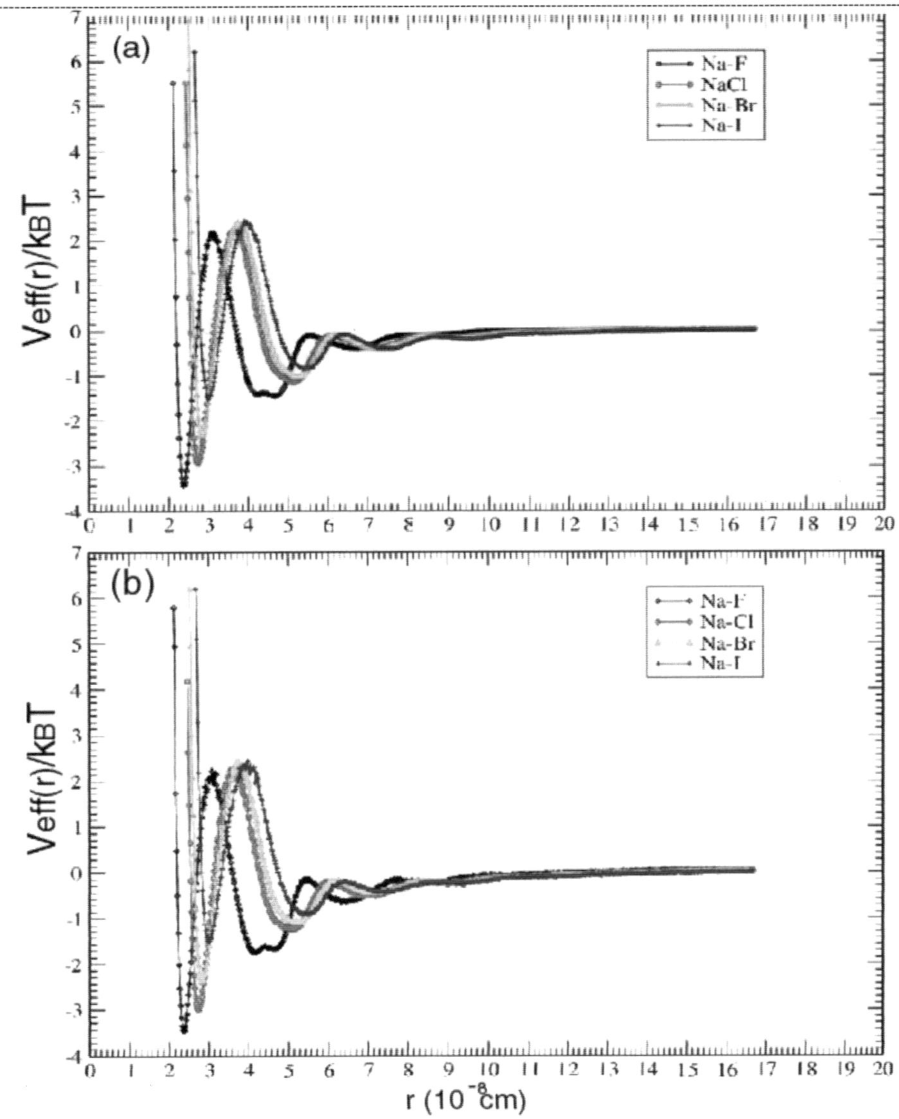

Figure 3. Effective potential of pair of cation and anion, Na-anion, in presence of different anions at (a) 0.87M, (b) 0.2175M salt concentration.

We observe that the optimized LJ parameters [9] yield the lowest high for the first maximum of $g(r)$ for pairs Na-Na as compared to calculations using Smith and Dang [11], and Dang [14] parameters. However, our pair correlation function is greater than previous calculations given in reference [4] with Smith-Dang LJ parameters and for shorter simulation times than ours (250 ns).

IOP Conf. Series: Journal of Physics: Conf. Series **792** (2017) 012100 doi:10.1088/1742-6596/792/1/012100

3.Effective pair potential of ions of Hofmeister salts

We used Gromacs version 5.0.4 together with OPLS-AA/L force field [15,16] and LJ parameters of table 1 with SPC/E molecular solvent to calculate the pair correlation functions of electrolyte solutions NaBr, NaI, NaCl, NaF at three salt concentrations 0.217, 0.52 and 0.87M. Our simulation cell size was $L = 3.38$ nm, and we applied periodic boundary conditions. Temperature equilibration was reached in the NVT ensemble after 10 ns with the V-rscale thermostat that fixed the temperature to 300K. This ensemble was then coupled to NPT ensemble to equilibrate the system to 1 bar pressure in the Berendsen barostat. We reached this equilibrium pressure after 200 ns of an overall 1femtosecond run. Electrostatic interactions were calculated with the Ewald sum. A ratio cut of Coulomb and Van der Waals interactions of 1.0 nm was required. With the obtained, $g(r)$ the effective potential of ions interaction was derived from the effective direct correlation function $c_{ij}^{eff}(r)$ that satisfy the Ornstein-Zernike equation [7]

$$h_{ij}(r) = c_{ij}^{eff}(r) + \sum_{m} \rho_m \int d\mathbf{r}' h_{im}(|\mathbf{r} - \mathbf{r}'|) c_{mj}^{eff}(r), \tag{2}$$

$h_{im} = g_{im} - 1$, and ρ_m are the solute densities. The effective ion potentials are defined as

$$V_{ij}^{eff}(r) = -k_B T \ln\left(g_{ij}(r)\right) + h_{ij}(r) \quad c_{ij}^{eff}(r), \tag{3}$$

Moreover, k_B is the Boltzmann constant. The effective potential is the average interaction of two ions in the presence of molecular solvent and the other ions [7]. We determined from equation (3) the effective potential $V_{ij}^{eff}(r) = -k_B T \ln\left(g_{ij}(r)\right) + h_{ij}(r) \quad c_{ij}^{eff}(r)$, We notice from figure 2 (a) that the first minimum of the effective interactions of cation pairs Na-Na at 0.87M concentration follow the trends of Hofmeister order NaF<NaCl<NaBR<NaI. This inequality is valid for the interaction of Na-anion pairs (figure 3). The pair interaction Na-Na is strongly affected by the other anions and also by variation in concentration, see figure 2 (b) for 0.2175M. There occurs a shift to larger absolute values of the minimum of the interaction of Na-Na pairs as the concentration increases. This behavior is not, however, reproduced by the anion-anion potential (figure 4). Also, we observed that the first minimum of the potential of anion pairs of type F-F, Cl-Cl, Br-Br, I-I does not follow a Hofmeister series.

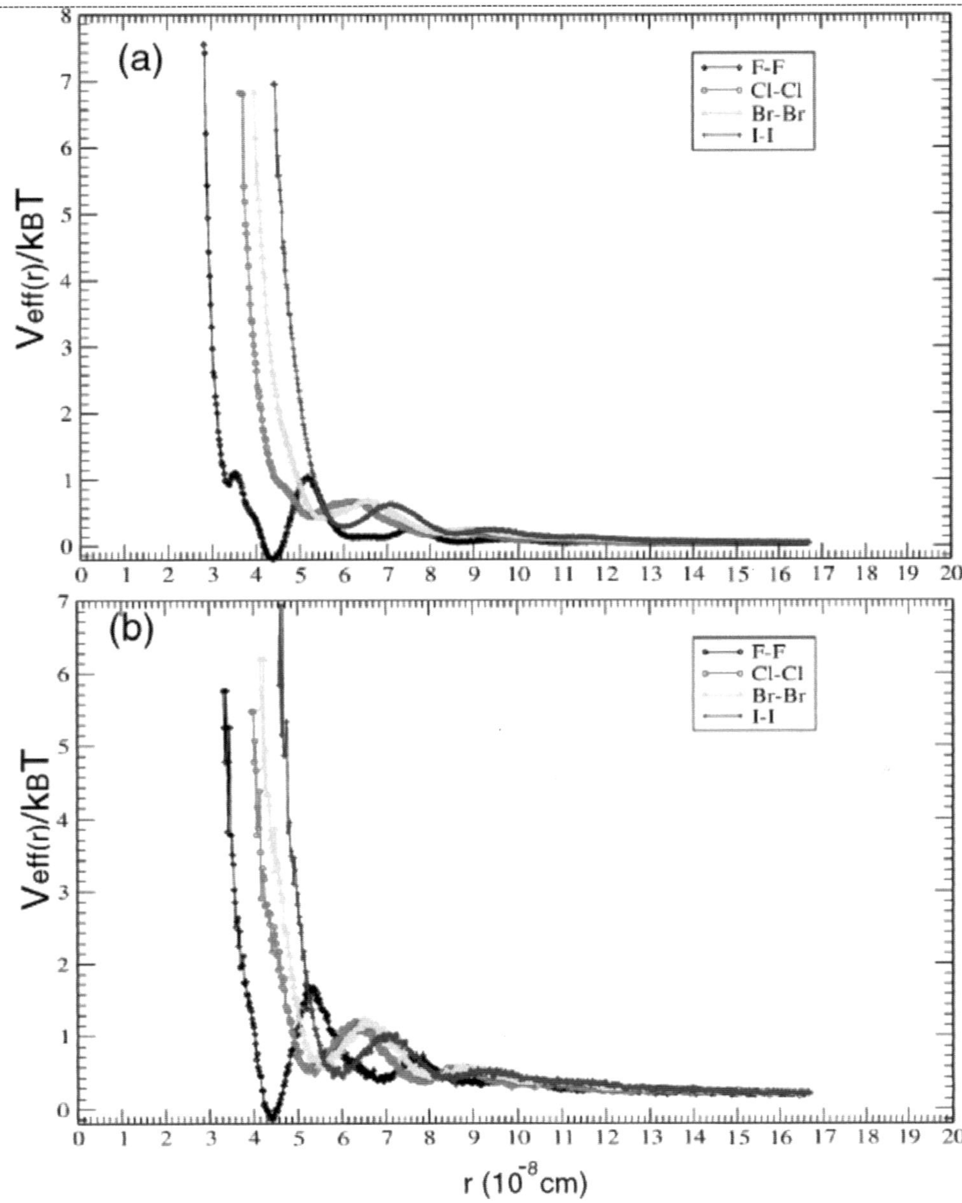

Figure 4. Effective potential of pair of anions in presence of different anions at (a) 0.87M, (b) 0.2175M salt concentration.

Acknowledgments

The authors acknowledge to the General Coordination of Information and Communications Technologies (CGSTIC) at CINVESTAV for providing HPC resources on the Hybrid Supercomputer " Xiuhcoatl" that have contributed to the research results reported within this paper.

References

[1] Pecora R 1985 *Dynamic Light Scattering* (New York, Plenum Press)
[2] Allen M P and Tildesley D J 1989 *Computer Simulation of Liquids* (Oxford, University Press)
[3] Barthel J M G, Krienke H and Kunz W 2002 *Physical Chemistry of Electrolyte Solutions* (Springer, Steinkopf)
[4] Lyubartsev A P and Laaksonen A 1995 *Phys. Rev.* E **52** 3730
[5] Andelman S 1976 *Chem. Phys. Lett.* *3*8 567
[6] Lyubartsev A P and Laaksonen A 1997 *Phys. Rev.* E **55** 5689
[7] Lyubartsev A P and Marcelja S 2002 *Phys. Rev.* E **65** 041202
[8] http://www.gromacs.org
[9] Horinek D, Mamatkulov S I and Netz R 2009 *J. Chem. Phys.* **130** 124507
[10] Berendsen H J C, Grigera J R and Straatsma T P 1987 *J. Phys. Chem.* **91** 6269
[11] Smith D E and Dang L X 1994 *J. Chem. Phys.* **100** 3757
[12] Mirzoev A and Lyubartsev A P 2009 *Phys. Chem. Chem. Phys.* **13** 5722
[13] Kalcher I and Dzubiella J 2009 *J. Chem. Phys.* **130** 134507
[14] Dang C 1995 *J. Am. Chem. Soc.* **117** 6954
[15] Jorgensen W L, Maxwell D S and Tirado-Rives J 1996 *J. Am. Chem. Soc.* **118** 11225
[16] Kamisnki G A, Friesner R A, Tirado-Rives J and Jorgensen W L 2001 *J. Phys. Chem.* B **105** 6474

Institute of Physics
Dirac House, Temple Back
Bristol BS1 6BE UK

ISSN: 1742-6588
ISBN 978-1-5108-3753-9

CURRAN ASSOCIATES INC.
proceedings.com